Seventh Edition

DIGITAL
FUNDAMENTALS

Prentice Hall International, Inc.

Editor: Scott Sambucci
Developmental Editor: Katie E. Bradford
Production Editor: Rex Davidson
Design Coordinator: Karrie Converse-Jones
Text Designer: John Edeen
Cover Photo: © Photobank
Cover Designer: Rod Harris
Production Manager: Patricia A. Tonneman
Marketing Manager: Ben Leonard
Illustrations: Jane Lopez

This book was set in Times Roman by The Clarinda Company and was printed and bound by R. R. Donnelley & Sons. The cover was printed by Phoenix Color Corp.

Printed in the United States of America
10   9   8   7   6   5   4   3

ISBN: 0-13-085268-6

Prentice-Hall International (UK) Limited, *London*
Prentice-Hall of Australia Pty. Limited, *Sydney*
Prentice-Hall of Canada, Inc., *Toronto*
Prentice-Hall Hispanoamericana, S. A., *Mexico*
Prentice-Hall of India Private Limited, *New Delhi*
Prentice-Hall of Japan, Inc., *Tokyo*
Prentice-Hall (Singapore) Pte. Ltd., *Singapore*
Editora Prentice-Hall do Brasil, Ltda., *Rio de Janeiro*
Prentice-Hall, *Upper Saddle River, NJ*

PREFACE

This seventh edition of *Digital Fundamentals* provides a comprehensive coverage in a clear, concise, well-illustrated format. As in previous editions, this one contains a balanced treatment of basic concepts, up-to-date technology, practical applications, and troubleshooting. In this edition, many topics have been strengthened or enhanced and numerous improvements are reflected throughout the book. Also, some completely new features have been added, including Electronics Workbench (EWB) exercises, Computer Notes, and Hands-On Tips. This edition has been carefully reviewed and checked to ensure up-to-date accurate coverage.

Most likely, you will find more topics than can be covered in one term. This extensive range of topics provides flexibility to accommodate a variety of technology programs. For example, some of the mathematical, design-oriented, or system application topics may not be appropriate for certain programs. Other programs may not cover PLDs or may not require an introduction to microprocessors. Many programs cover the circuit details of integrated circuits while others do not deal with component-level topics at all. These topics can be omitted or lightly covered without affecting other coverage. A background in transistor circuits is not a prerequisite for this text.

New to This Edition

In addition to general updating and improvements in text and graphics throughout the book, the seventh edition includes the following new features.

Electronics Workbench (EWB) Tutorials   There is a Website tutorial associated with many chapters in the text that can be downloaded for student use. These tutorials introduce aspects and elements of EWB as needed on a chapter by chapter basis and also contain examples and exercises. These tutorials may be found at *http://www.prenhall.com/floyd.*

System Application EWB Simulation   For many System Applications in the text, the circuit board logic is simulated in EWB and is available on the CD-ROM accompanying the textbook. To observe the circuit in operation for a given System Application, the reader simply opens the specified file, connects any necessary EWB instruments, and runs the simulation. The CD-ROM contains free Electronics Workbench Demonstration software. The full version of EWB software is also on the CD-ROM, with access available for a fee.

EWB Troubleshooting Problems   At the end of most chapters, new troubleshooting problems reference circuits on the CD-ROM. Generally, these circuits have hidden faults, and students must use troubleshooting techniques to identify the faults. In some cases, the circuits do not have faults and are working properly, and the student must determine this also. Results can be found in the Instructor's Resource Manual (IRM).

Computer Notes   This feature provides interesting and instructional information about computer technology as it relates to text coverage. These optional Computer Notes, identified by a special logo and design treatment, are found throughout the book, and most chapters have one or more.

Hands-On Tips Called "HOTip" for short, this feature provides useful and practical information interspersed throughout the book. HOTips are identified by a special logo and design treatment. They generally relate to the text coverage but can be skipped over without affecting the instructor's presentation and students' understanding of the material.

Other new or improved features include

- Key term list at the end of each chapter
- General revisions and improvements throughout
- Revised calculator tutorials using the TI-85 calculator
- Revised coverage of integrated circuit logic gates in Chapter 3
- Revised coverage of memories and storage devices in Chapter 12
- Revised coverage of standard buses in Chapter 13
- Revised coverage of microprocessors and computers in Chapter 14

Additional Features

- Full-color format
- Two chapters on PLDs (Chapters 7 and 11)
- Unique "floating chapter" concept with Chapter 15 provides optional coverage of IC technology at any point.
- Overview and objectives at the opening of each chapter
- Introduction and objectives at the beginning of each section within a chapter
- Review questions at the end of each section in a chapter
- Related Problem in each worked example
- Digital System Applications sections at the end of most chapters
- Chapter summaries
- Multiple-choice self-test at the end of each chapter
- Extensive sectionalized chapter problem sets
- Comprehensive glossary at the end of the book

Improved Ancillary Package

- Two lab manuals: *Experiments in Digital Fundamentals,* Fifth Edition, by David Buchla, and *Digital Experiments Emphasizing Troubleshooting,* Fifth Edition, by Jerry Cox
- EWB exercise CD-ROM accompanying the lab manual by David Buchla
- Instructor's Resource Manual with solutions to end-of-chapter problems and System Applications, worksheet masters, introduction to CUPL, and electronics standards
- Test Item File (hard copy)
- Prentice Hall Custom Test (electronic testbank copy)
- PowerPoint slides
- Bergwall Video

Illustration of the Chapter Features

Chapter Opener Each chapter begins with a two-page spread, as shown in Figure P–1. The left page includes a list of the sections in the chapter, a list of chapter objectives, and a website reference for the EWB tutorial. The right page has an overview of the chapter, a list of specific devices introduced in the chapter (each new device is indicated by an IC logo at the point where it is introduced), and a brief description of the Digital System Application with related art.

List of chapter objectives.

Chapter overview and a
list of devices introduced
in this chapter.

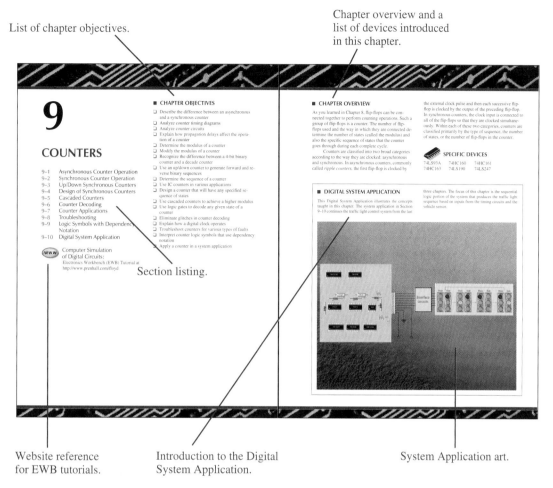

Section listing.

Website reference
for EWB tutorials.

Introduction to the Digital
System Application.

System Application art.

FIGURE P–1
Chapter opener.

Section Opener   Each of the sections in a chapter begins with a brief introduction that includes a general overview and section objectives. An example is shown in Figure P–2.

Section Review   Each section ends with a review consisting of questions or exercises that emphasize the main concepts presented in the section. This is shown in Figure P–2. Answers to the Section Reviews are at the end of the chapter.

Worked Examples and Related Problems   Numerous worked examples help illustrate and clarify basic concepts or specific procedures. Each example concludes with a Related Problem that reinforces or expands on the example. Some Related Problems require a repetition of the example using different parameters or conditions. Others focus on a more limited part of the example or encourage further thought. A typical worked example and Related Problem are shown in Figure P–3. Answers to Related Problems are at the end of the chapter.

Troubleshooting Section   Many chapters include a troubleshooting section that relates to the topics covered in the chapter and that emphasizes troubleshooting techniques and the use of test instruments. A portion of a typical troubleshooting section is illustrated in Figure P–4 on page vii.

Digital System Application   The last section of most chapters presents a practical application of the concepts and devices covered in the chapter. Each of these sections presents a real-world system in which analysis, troubleshooting, and design elements are

FIGURE P–2
Section opener and section review.

Introductory paragraph and a list of performance-based section objectives begin each section.

Computer Notes are found throughout the text (not part of section opener).

Review exercises end each section.

FIGURE P–3
An example and related problem.

Each example is contained within a ruled box.

Each example contains a problem related to the example.

IC icon indicates introduction to new devices.

FIGURE P–4
Representative pages from a portion of a typical Troubleshooting section.

implemented in a series of activities called *System Assignments.* Some System Applications are limited to a single chapter, and others extend over two or more chapters. Specific Digital System Applications and the chapters in which they appear are as follows:

■ Tablet counting and control system: Chapters 1, 2, 3, and 4
■ Digital motor control system: Chapter 5
■ Traffic light control system: Chapters 6, 7, 8, 9, and 11
■ Security entry system: Chapters 10 and 12
■ Satellite antenna positioning system: Chapter 13

Although they are not intended or designed for use as laboratory projects (except the laboratory of the mind), many of the System Applications utilize realistic representations of printed circuit boards and instruments to provide experience in relating schematics to actual boards, identifying IC packages, and obtaining data from certain instrument readings and displays. Because their omission will not affect any other material in the text, System Applications may be treated as optional. Figure P–5 shows a portion of a Digital System Application section.

Chapter End Matter The following study aids end each chapter:

■ Chapter Summary
■ Key Terms
■ Self-Test
■ Problem set that includes some or all of the following categories: Basic problems, Troubleshooting problems, System Application problems, Design problems, and EWB Troubleshooting problems.
■ Answers to Section Reviews
■ Answers to Related Problems for Examples
■ Answers to Self-Test

"HOTips" are found throughout the text (not always part of System Application).

Opener with a list of objectives.

System Assignment provides a series of analysis, design, or troubleshooting activities. Many applications include a printed circuit board and instrumentation.

Reference to EWB simulation on CD-ROM.

An overall introduction to the system application is provided before the System Assignments.

FIGURE P–5
Representative pages from a typical Digital System Application section.

Content and Organization

The fifteen chapters in *Digital Fundamentals* cover a comprehensive range of topics, beginning with basic digital concepts and progressing through number systems, logic gates, Boolean algebra, combinational logic, introduction to PLDs, sequential logic, implementation of sequential logic with PLDs, memories and storage, interfacing, an introduction to microprocessors and computers, and integrated circuit technologies. Further discussion of a few specific areas may be helpful.

Integrated Circuit Technologies Chapter 15 is unique in that it is intended to be used as a "floating chapter" that can be covered in whole or in part at any point in the text, or it can be completely omitted without affecting any other topics. Notices are placed throughout the text to suggest points where topics in Chapter 15 may be introduced, although it can be introduced at any time after Chapter 3. Its placement as the last chapter in the text is intended to facilitate this flexible usage, and a tab edge design provides quick and easy reference. Also, if Chapter 15 is omitted, this book can be covered without requiring a student to have a background in transistor circuits. Chapter 3 provides sufficient coverage of digital ICs for those without a transistor circuits background. For those wishing to go further into the details of integrated circuit technology, Chapter 15 provides that coverage.

Programmable Logic Devices   Chapters 7 and 11 provide an introduction to the important topic of PLDs; however, this coverage can be treated as optional. Either one or both of these chapters can be omitted without affecting other topics. Chapter 7 follows the coverage of combinational logic and provides an introduction to PLDs and PLD programming as applied to combinational logic only. Chapter 11 follows the coverage of sequential logic and continues the PLD coverage from Chapter 7 with an introduction to using PLDs to implement sequential logic. A number of PLD programming languages are available; ABEL is the one used to illustrate PLD programming in this text. A brief tutorial on CUPL, another popular PLD programming language, is available in the Instructor's Resource Manual (IRM), for those who are interested. Some instructors may prefer to delay coverage of Chapter 7 until after Chapter 10 so that both PLD chapters are covered consecutively.

Microprocessors   Chapter 14 provides a brief introduction to microprocessors and computers. Both hardware and software aspects are discussed. The 8086/8088 microprocessor family is used as a basic model for introducing concepts that are valid even in the latest microprocessors. The chapter covers the features of all Intel devices up through the Pentiums and introduces the Motorola microprocessor family. This chapter, of course, is not intended to provide a complete coverage of the topic as it would take one or more entire textbooks to do so. It can be used, however, as an introduction to microprocessors for students who later will take a full course devoted to the subject.

End Matter   At the end of the book are appendices that contain several representative IC data sheets, a table of code conversions, and a table of powers-of-two. The appendices are followed by the Answers to Selected Odd-Numbered Problems (solutions to all end-of-chapter problems are in the IRM). A comprehensive glossary includes definitions of the key terms for each chapter in addition to other definitions. The comprehensive glossary is followed by the index.

Suggestions for Teaching with *Digital Fundamentals*

Generally, time limitations or course emphasis dictates the topics that can be covered in a term. Also, it is not uncommon to alter the sequence of certain topics as they appear in the text. The following suggestions for selective coverage, light coverage, or omission do not imply that a given topic is less important than others, but in the context of a specific program, the topic *may* not require the emphasis that the more fundamental topics do. Also, these suggestions do not necessarily reflect all possibilities for sequence alteration, selective or light coverage, or omission; in any particular program, there may be other areas that can be considered.

Suggestions for altering the sequence of chapters:

1. If you wish to cover logic gates earlier in the course, Chapter 1 can be lightly covered and Chapter 3 on logic gates can be covered before Chapter 2 on number systems, operations, and codes.
2. Coverage of Chapter 7 on PLDs can be postponed until after Chapter 10. This way, both PLD chapters (7 and 11) can be covered consecutively, if that works best in your course.
3. Chapter 15 can be covered at just about any point in the text. For example, it can be covered partially or completely after Chapter 3.

Chapters that may be considered for selective coverage:

1. Chapter 1, Introductory Digital Concepts
2. Chapter 2, Number Systems, Operations and Codes
3. Chapter 4, Boolean Algebra and Logic Simplification

4. Chapter 13, Interfacing

5. Chapter 15, Integrated Circuit Technologies

Chapters that may be considered for omission without affecting other coverage:

1. Chapter 7, Introduction to Programmable Logic Devices (Because of the growing importance of this topic, serious consideration should be given before omitting either Chapter 7 or Chapter 11.)

2. Chapter 11, Sequential Logic Applications of PLDs

3. Chapter 13, Interfacing

4. Chapter 14, Introduction to Microprocessors and Computers

5. Chapter 15, Integrated Circuit Technologies

Depending on your program, there may be additional topics that can be treated lightly or omitted. For example, the Digital System Applications sections can be omitted without affecting any other topics, or they may be assigned for extra credit or as special projects.

Acknowledgments

Many very capable people have been involved with this seventh edition of *Digital Fundamentals.* It has been completely reviewed and checked for both content and accuracy. Those at Prentice Hall who have been instrumental in moving this project through its many phases include Rex Davidson, Katie Bradford, and Scott Sambucci. They deserve much credit. My appreciation, once again to Lois Porter who has done another wonderful job of editing the manuscript. Also, thanks to Jane Lopez for the great job on the graphics art used in the text. Gary Snyder has again been my accuracy checker on this project, and I commend him for his outstanding work. In addition, Gary created the EWB circuit files for the Electronics Workbench features that are new to this edition. Last, but certainly not least, thanks to my colleague Dave Buchla for his contribution to the revision of Chapter 14, his consultations on many topics, and his thorough review of the manuscript.

As always, we depend on expert input from many users and nonusers of *Digital Fundamentals.* My thanks to the following reviewers who made many valuable suggestions and provided much constructive criticism.

William H. Brownlowe, Montgomery County Community College

Hikmat Chedid, Lorain County Community College

John P. Corgan, Luzerne County Community College

Gene Craft, Lima Technical College

Lance Crimm, Southern Polytechnic State University

Paul A. Dilsner, Seminole Community College

Floyd Martin, Santa Ana College

Keith Quigley, Midlands Technical College

Nelson A. Riedel, Central Ohio Technical College

Tom Floyd

TO DEBBIE AND CYNDI

CONTENTS

1

INTRODUCTORY DIGITAL CONCEPTS

 Computer Simulation of Digital Circuits:
Electronics Workbench (EWB) Tutorial at http://www.prenhall.com/floyd

■ CHAPTER OBJECTIVES

❏ Explain the basic differences between digital and analog quantities

❏ Show how voltage levels are used to represent digital quantities

❏ Describe various parameters of a pulse waveform such as rise time, fall time, pulse width, frequency, period, and duty cycle

❏ Explain the basic logic operations of NOT, AND, and OR

❏ Describe the basic functions of the comparator, adder, code converter, encoder, decoder, multiplexer, demultiplexer, counter, and register

❏ Identify digital integrated circuits according to their complexity and the type of circuit packaging

❏ Identify pin numbers on integrated circuit packages

❏ Recognize digital instruments and understand how they are used in troubleshooting digital circuits and systems

❏ Show how a complete digital system is formed from the basic functions in a practical application

■ CHAPTER OVERVIEW

The concept of a digital computer can be traced back to Charles Babbage, who developed a crude mechanical computation device in the 1830s. The first functioning digital computer was built in 1944 at Harvard University, but it was electromechanical, not electronic. Modern digital electronics began in 1946 with an electronic digital computer called ENIAC, which was implemented with vacuum-tube circuits. Even though it took up an entire room, ENIAC didn't have the computing power that your hand-held calculator does.

The term *digital* is derived from the way computers perform operations, by counting digits. For many years, applications of digital electronics were confined to computer systems. Today, digital technology is applied in a wide range of areas. Such applications as television, communications systems, radar, navigation and guidance systems, military systems, medical instrumentation, industrial process control, and consumer electronics use digital techniques. Digital technology has progressed from vacuum-tube circuits to discrete transistors to complex integrated circuits, some of which contain millions of transistors.

■ DIGITAL SYSTEM APPLICATION

The last section in most chapters of this textbook uses a system application to bring together many of the principal topics covered in the chapter. Each system is designed to fit the particular chapter to illustrate how the theory and devices can be used. Throughout the book, five different systems are introduced, some covering two or more chapters.

All of the systems are simplified to make them manageable in the context of the chapter material. Although they are based on actual system requirements, they are de-

signed to accommodate the topical coverage of the chapter and are not intended to necessarily represent the most efficient or ultimate approach in a given application.

Section 1–7 introduces the first system, which is an industrial control system for counting and controlling items for packaging on a conveyor line. It is designed to incorporate all of the logic functions that are introduced in this chapter so that you can see how they are used and how they work together to achieve an overall objective.

1–1 ■ DIGITAL AND ANALOG QUANTITIES

Electronic circuits can be divided into two broad categories, digital and analog. Digital electronics involves quantities with discrete values, and analog electronics involves quantities with continuous values. Although you will be studying digital fundamentals in this book, you should also know about analog because many applications require both.

After completing this section, you should be able to

■ Define *analog*   ■ Define *digital*   ■ Explain the difference between digital and analog quantities   ■ State the advantages of digital over analog   ■ Give examples of how digital and analog quantities are used in electronics

An **analog*** quantity is one having continuous values. A **digital** quantity is one having a discrete set of values. Most things that can be measured quantitatively appear in nature in analog form. For example, the air temperature changes over a continuous range of values. During a given day, the temperature does not go from, say, 70° to 71° instantaneously; it takes on all the infinite values in between. If you graphed the temperature on a typical summer day, you would have a smooth, continuous curve similar to the curve in Figure 1–1. Other examples of analog quantities are time, pressure, distance, and sound.

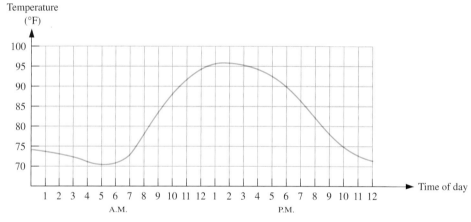

FIGURE 1–1
Graph of an analog quantity (temperature versus time).

Rather than graphing the temperature on a continuous basis, suppose you just take a temperature reading every hour. Now you have sampled values representing the temperature at discrete points in time (every hour) over a 24-hour period, as indicated in Figure 1–2. You have effectively converted an analog quantity to a form that can now be digitized by representing each sampled value by a digital code. It is important to realize that Figure 1–2 itself is not the digital representation of the analog quantity.

The Digital Advantage   Digital has certain advantages over analog in electronics applications. For one thing, digital data can be processed and transmitted more efficiently and reliably than analog data. Also, digital data has a great advantage when storage is necessary. For example, music when converted to digital form can be stored more compactly and reproduced with greater accuracy and clarity than is possible when it is in analog form.

---

*Boldface terms in the text are defined in the end-of-book glossary. Certain terms have been selected to also be included in a Key Term list at the end of each chapter.

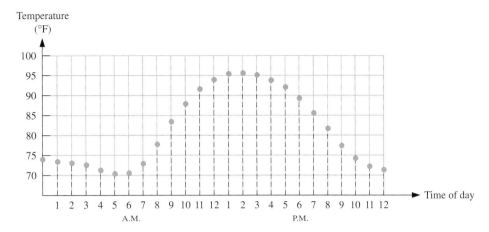

FIGURE 1–2
Sampled-value representation (quantization) of the analog quantity in Figure 1–1. Each value represented by a dot can be digitized by representing it as a digital code consisting of a series of 1s and 0s.

Noise (unwanted voltage fluctuations) does not affect digital data nearly as much as it does analog signals.

An Analog Electronic System

A public address system, used to amplify sound so that it can be heard by a large audience, is one example of an application of analog electronics. The basic diagram in Figure 1–3 illustrates that sound waves, which are analog in nature, are picked up by a microphone and converted to a small analog voltage called the audio signal. This voltage varies continuously as the volume and frequency of the sound changes and is applied to the input of a linear amplifier. The output of the amplifier, which is an increased reproduction of input voltage, goes to the speaker(s). The speaker changes the amplified audio signal back to sound waves having a much greater volume than the original sound waves picked up by the microphone.

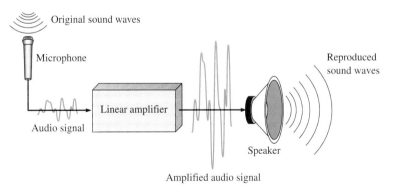

FIGURE 1–3
A basic audio public address system.

A System Using Digital and Analog Methods

The compact disk (CD) player is an example of a system in which both digital and analog circuits are used. The simplified diagram in Figure 1–4 illustrates the basic principle. Music in digital form is stored on the compact disk. A laser diode optical system picks up the

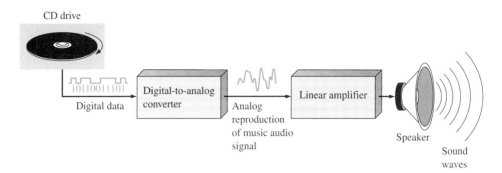

FIGURE 1–4
Basic principle of a CD player. Only one channel is shown.

digital data from the rotating disk and transfers it to the **digital-to-analog converter (DAC).** The DAC changes the digital data into an analog signal that is an electrical reproduction of the original music. This signal is amplified and sent to the speaker for you to enjoy. When the music was originally recorded on the CD, a process, essentially the reverse of the one described here, using an **analog-to-digital converter (ADC)** was used.

**SECTION 1–1
REVIEW**

Answers to section reviews are found at the end of the chapter.

1. Define *analog*.
2. Define *digital*.
3. Explain the difference between a digital quantity and an analog quantity.
4. Give an example of a system that is analog and one that is a combination of both digital and analog. Name a system that is entirely digital.

1–2 ■ BINARY DIGITS, LOGIC LEVELS, AND DIGITAL WAVEFORMS

Digital electronics involves circuits and systems in which there are only two possible states. These states are represented by two different voltage levels: A HIGH and a LOW. The two states can also be represented by current levels, open and closed switches, or lamps turned on and off. In digital systems, combinations of the two states, called codes, are used to represent numbers, symbols, alphabetic characters, and other types of information. The two-state number system is called binary, and its two digits are 0 and 1. A binary digit is called a bit.

After completing this section, you should be able to

■ Define *binary* ■ Define *bit* ■ Name the bits in a binary system ■ Explain how voltage levels are used to represent bits ■ Explain how voltage levels are interpreted by a digital circuit ■ Describe the general characteristics of a pulse ■ Determine the amplitude, rise time, fall time, and width of a pulse ■ Identify and describe the characteristics of a digital waveform ■ Determine the amplitude, period, frequency, and duty cycle of a digital waveform ■ Explain what a timing diagram is and state its purpose ■ Explain serial and parallel data transfer and state the advantage and disadvantage of each

Binary Digits

The two digits in the **binary** system, 1 and 0, are called **bits,** which is a contraction of the words *binary digit.* In digital circuits, two different voltage levels are used to represent the two bits. A 1 is represented by the higher voltage, which we will refer to as a HIGH and a

0 is represented by the lower voltage level, which we will refer to as a LOW. This is called **positive logic** and will be used throughout the book.

$$\textbf{HIGH} = \textbf{1} \quad \text{and} \quad \textbf{LOW} = \textbf{0}$$

A less common system in which a 1 is represented by a LOW and a 0 is represented by a HIGH is called *negative logic.*

Groups of bits (combinations of 1s and 0s), called *codes,* are used to represent numbers, letters, symbols, instructions, and anything else required in a given application.

Logic Levels

The voltages used to represent a 1 and a 0 are called *logic* levels. Ideally, one voltage level represents a HIGH and another voltage level represents a LOW. In a practical digital circuit, however, a HIGH can be any voltage between a specified minimum value and a specified maximum value. Likewise, a LOW can be any voltage between a specified minimum and a specified maximum. There can be no overlap between the accepted HIGH levels and the accepted LOW levels.

Figure 1–5 illustrates the general range of LOWs and HIGHs for a digital circuit. The variable $V_{H(max)}$ represents the maximum HIGH voltage value, and $V_{H(min)}$ represents the minimum HIGH voltage value. The maximum LOW voltage value is represented by $V_{L(max)}$, and the minimum LOW voltage value is represented by $V_{L(min)}$. The voltage values between $V_{L(max)}$ and $V_{H(min)}$ are unacceptable for proper operation. A voltage in the unacceptable range can appear as either a HIGH or a LOW to a given circuit. Therefore, these unacceptable values are never used. For example, the HIGH values for a certain type of digital circuit called TTL may range from 2 V to 5 V and the LOW values may range from 0 V to 0.8 V. So, for example, if a voltage of 3.5 V is applied, the circuit will accept it as a HIGH or binary 1. If a voltage of 0.5 V is applied, the circuit will accept it as a LOW or binary 0. For this type of circuit, voltages between 0.8 V and 2 V are unacceptable and are never used.

FIGURE 1–5
Logic level ranges of voltage for a digital circuit.

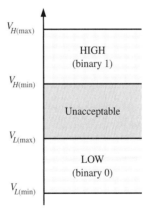

Digital Waveforms

Digital waveforms consist of voltage levels that are changing back and forth between the HIGH and LOW levels or states. Figure 1–6(a) shows that a single positive-going **pulse** is generated when the voltage (or current) goes from its normally LOW level to its HIGH level and then back to its LOW level. The negative-going pulse in Figure 1–6(b) is generated when the voltage goes from its normally HIGH level to its LOW level and back to its HIGH level. A digital waveform is made up of a series of pulses.

The Pulse As indicated in Figure 1–6, the pulse has two edges: a **leading edge** that occurs first at time t_0 and a **trailing edge** that occurs last at time t_1. For a positive-going pulse, the leading edge is a rising edge, and the trailing edge is a falling edge. The pulses

FIGURE 1–6
Ideal pulses.

in Figure 1–6 are ideal because the rising and falling edges are assumed to change in zero time (instantaneously). In practice, these transitions never occur instantaneously, although for most digital work you can assume ideal pulses.

Figure 1–7 shows a nonideal pulse. The time required for the pulse to go from its LOW level to its HIGH level is called the **rise time** (t_r), and the time required for the transition from the HIGH level to the LOW level is called the **fall time** (t_f). In practice, it is common to measure rise time from 10% of the pulse **amplitude** (height from baseline) to 90% of the pulse amplitude and to measure the fall time from 90% to 10% of the pulse amplitude, as indicated in Figure 1–7. The bottom 10% and the top 10% of the pulse are not included in the rise and fall times because of the nonlinearities in the waveform in these areas. The **pulse width** (t_W) is a measure of the duration of the pulse and is often defined as the time interval between the 50% points on the rising and falling edges, as indicated in Figure 1–7.

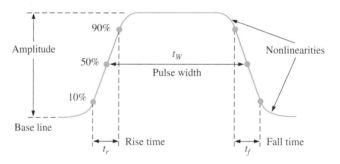

FIGURE 1–7
Nonideal pulse characteristics.

Waveform Characteristics   Most waveforms encountered in digital systems are composed of series of pulses, sometimes called *pulse trains,* and can be classified as either periodic or nonperiodic. A **periodic** pulse waveform is one that repeats itself at a fixed interval, called a **period** (T). The frequency (f) is the rate at which it repeats itself and is measured in hertz (Hz). A nonperiodic pulse waveform, of course, does not repeat itself at fixed intervals and may be composed of pulses of randomly differing pulse widths and/or randomly differing time intervals between the pulses. An example of each type is shown in Figure 1–8.

The **frequency** (f) of a pulse (digital) waveform is the reciprocal of the period. The relationship between frequency and period is expressed as follows:

$$f = \frac{1}{T} \qquad\qquad \textbf{(1–1)}$$

$$T = \frac{1}{f} \qquad\qquad \textbf{(1–2)}$$

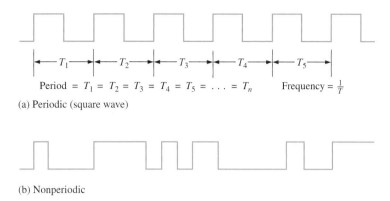

Period $= T_1 = T_2 = T_3 = T_4 = T_5 = \ldots = T_n$ Frequency $= \frac{1}{T}$

(a) Periodic (square wave)

(b) Nonperiodic

FIGURE 1–8
Examples of digital waveforms.

An important characteristic of a periodic digital waveform is its duty cycle. The **duty cycle** is defined as the ratio of the pulse width (t_W) to the period (T) and can be expressed as a percentage.

$$\text{Duty cycle} = \left(\frac{t_W}{T}\right)100\% \qquad\qquad (1\text{–}3)$$

EXAMPLE 1–1

A portion of a periodic digital waveform is shown in Figure 1–9. The measurements are in milliseconds. Determine the following:
(a) period   **(b)** frequency   **(c)** duty cycle

FIGURE 1–9

Solution
(a) The period is measured from the edge of one pulse to the corresponding edge of the next pulse. In this case T is measured from leading edge to leading edge, as indicated. T equals **10 ms.**

(b) $f = \dfrac{1}{T} = \dfrac{1}{10 \text{ ms}} = \textbf{100 Hz}$

(c) Duty cycle $= \left(\dfrac{t_W}{T}\right)100\% = \left(\dfrac{1 \text{ ms}}{10 \text{ ms}}\right)100\% = \textbf{10\%}$

Related Problem*   A periodic digital waveform has a pulse width of 25 μs and a period of 150 μs. Determine the frequency and the duty cycle.

---

*Answers to all related problems are found at the end of the chapter.

COMPUTER NOTE
The speed at which a computer can operate depends on the type of microprocessor used in the system. The speed specification, for example 300 MHz, of a computer is the maximum clock frequency at which the microprocessor can run.

A Digital Waveform Carries Binary Information

Binary information that is handled by digital systems appears as waveforms that represent sequences of bits. When the waveform is HIGH, a binary 1 is present; when the waveform is LOW, a binary 0 is present. Each bit in a sequence occupies a defined time interval called a **bit time.**

The Clock   In digital systems, all waveforms are synchronized with a basic timing waveform called the **clock.** The clock is a periodic waveform in which each interval between pulses (the period) equals the time for one bit.

   An example of a clock waveform is shown in Figure 1–10. Notice that, in this case, each change in level of waveform A occurs at the leading edge of the clock waveform. In other cases, level changes occur at the trailing edge of the clock. During each bit time of the clock, waveform A is either HIGH or LOW. These HIGHs and LOWs represent a sequence of bits as indicated. A group of several bits can be used as a piece of binary information, such as a number or a letter. The clock waveform itself does not carry information.

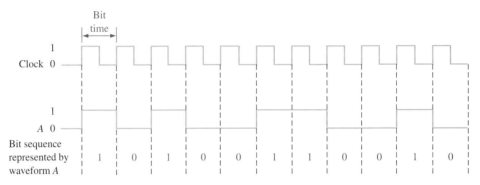

FIGURE 1–10
Example of a clock waveform synchronized with a waveform representation of a sequence of bits.

Timing Diagrams

A **timing diagram** is a graph of digital waveforms showing the actual time relationship of two or more waveforms and how each waveform changes in relation to the others. Figure 1–10 is an example of a simple timing diagram that shows how the clock waveform and waveform A are related on a time base.

   By looking at a timing diagram, you can determine the states (HIGH or LOW) of all the waveforms at any specified point in time and the exact time that a waveform changes state relative to the other waveforms. Figure 1–11 is an example of a timing diagram made up of four waveforms. From this timing diagram you can see, for example, that all three waveforms (A, B, and C) are HIGH only during bit time 7 and they all change back LOW at the end of bit time 7.

FIGURE 1–11
Example of a timing diagram.

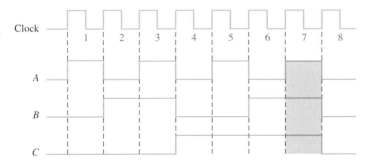

Data Transfer

Data refers to groups of bits that convey some type of information. Binary data, which are represented by digital waveforms, must be transferred from one circuit to another within a digital system or from one system to another in order to accomplish a given purpose. For example, numbers stored in binary form in the memory of a computer must be transferred to the central processing unit in order to be added. The sum of the addition must then be transferred to a monitor for display and/or transferred back to the memory. In computer systems as in Figure 1–12, binary data are transferred in two ways: serial and parallel.

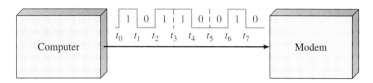

(a) Serial transfer of binary data from computer to modem. Interval t_0 to t_1 is first.

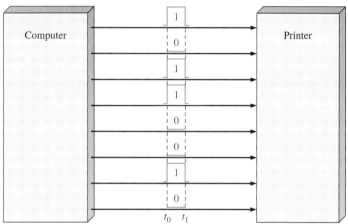

(b) Parallel transfer of binary data from computer to printer. t_0 is the beginning time.

FIGURE 1–12
Illustration of serial and parallel transfer of binary data. Only the data lines are shown.

When bits are transferred in **serial** form from one point to another, they are sent one bit at a time along a single conductor, as illustrated in Figure 1–12(a) for the case of a computer-to-modem transfer. During the time interval from t_0 to t_1, the first bit is transferred. During the time interval from t_1 to t_2, the second bit is transferred, and so on. To transfer eight bits in series, it takes eight time intervals.

When bits are transferred in **parallel** form, all the bits in a group are sent out on separate lines at the same time. There is one line for each bit, as shown in Figure 1–12(b) for the example of eight bits being transferred from a computer to a printer. To transfer eight bits in parallel, it takes one time interval compared to eight time intervals for the serial transfer.

To summarize, the advantage of serial transfer of binary data is that a minimum of only one line is required. In parallel transfer, a number of lines equal to the number of bits to be transferred at one time is required. The disadvantage of serial transfer is that it takes longer to transfer a given number of bits than with parallel transfer. For example, if one bit can be transferred in 1 μs, then it takes 8 μs to serially transfer eight bits but only 1 μs to parallel transfer eight bits. The disadvantage of parallel transfer is that it takes more lines.

EXAMPLE 1–2

(a) Determine the total time required to serially transfer the eight bits contained in waveform A of Figure 1–13, and indicate the sequence of bits. The left-most bit is the first to be transferred. The 100 kHz clock is used as reference.

(b) What is the total time to transfer the same eight bits in parallel?

FIGURE 1–13

Solution

(a) Since the frequency of the clock is 100 kHz, the period is

$$T = \frac{1}{f} = \frac{1}{100 \text{ kHz}} = 10 \ \mu s$$

It takes $10 \ \mu s$ to transfer each bit in the waveform. The total transfer time for 8 bits is

$$8 \times 10 \ \mu s = \mathbf{80 \ \mu s}$$

To determine the sequence of bits, examine the waveform in Figure 1–13 during each bit time. If waveform A is HIGH during the bit time, a 1 is transferred. If waveform A is LOW during the bit time, a 0 is transferred. The bit sequence is illustrated in Figure 1–14. The left-most bit is the first to be transferred.

FIGURE 1–14

(b) A parallel transfer would take **10 μs** for all eight bits.

Related Problem If binary data are transferred at the rate of 10 million bits per second (10 Mbits/s), how long will it take to parallel transfer 16 bits on 16 lines? How long will it take to serially transfer 16 bits?

SECTION 1–2 REVIEW

1. Define *binary*.
2. What does *bit* mean?
3. What are the bits in a binary system?
4. How are the rise time and fall time of a pulse measured?
5. Knowing the period of a waveform, how do you find the frequency?
6. Explain what a clock waveform is.
7. What is the purpose of a timing diagram?
8. What is the main advantage of parallel transfer over serial transfer of binary data?

1-3 ■ BASIC LOGIC OPERATIONS

In its basic form, logic is the realm of human reasoning that tells you a certain propo-sition (declarative statement) is true if certain conditions are true. Propositions can be classified as true or false. Many situations and processes that you encounter in your daily life can be expressed in the form of propositional, or logic, functions. Since such functions are true/false or yes/no statements, digital circuits with their two-state char-acteristics are applicable.

After completing this section, you should be able to

■ List three basic logic operations ■ Define the NOT operation ■ Define the AND operation ■ Define the OR operation

Several propositions, when combined, form propositional, or logic, functions. For exam-ple, the propositional statement "The light is on" will be true if "The bulb is not burned out" is true and if "The switch is on" is true. Therefore, this logical statement can be made: *The light is on only if the bulb is not burned out and the switch is on.* In this example the first statement is true only if the last two statements are true. The first statement ("The light is on") is then the basic proposition, and the other two statements are the conditions on which the proposition depends.

In the 1850s, the Irish logician and mathematician George Boole developed a math-ematical system for formulating logic statements with symbols so that problems can be written and solved in a manner similar to ordinary algebra. Boolean algebra, as it is known today, is applied in the design and analysis of digital systems and will be covered in detail in Chapter 4.

The term **logic** is applied to digital circuits used to implement logic functions. Sev-eral kinds of digital logic **circuits** are the basic elements that form the building blocks for such complex digital systems as the computer. We will now look at these elements and dis-cuss their functions in a very general way. Later chapters will cover these circuits in detail.

Three basic logic operations are indicated by standard distinctive shape symbols in Figure 1–15. Other standard symbols for these logic operations will be introduced in Chap-ter 3. The lines connected to each symbol are the **inputs** and **outputs.** The inputs are on the left of each symbol and the output is on the right. A circuit that performs a specified logic operation (AND, OR) is called a logic **gate.** AND and OR gates can have any number of inputs, as indicated by the dashes in the figure.

NOT   AND   OR

FIGURE 1–15
The basic logic operations and symbols.

In logic operations, the true/false conditions mentioned earlier are represented by a HIGH (true) and a LOW (false). Each of the three basic logic operations produces a unique response to a given set of conditions.

NOT

The **NOT** operation changes one logic level to the opposite logic level, as indicated in Figure 1–16. When the input is HIGH (1), the output is LOW (0). When the input is LOW, the output is HIGH. In either case, the output is *not* the same as the input. The NOT operation is implemented by a logic circuit known as an **inverter.**

FIGURE 1–16
The NOT operation.   HIGH (1) ———▷o—— LOW (0)   LOW (0) ———▷o—— HIGH (1)

AND

The **AND** operation produces a HIGH output only if all the inputs are HIGH, as indicated in Figure 1–17 for the case of two inputs. When one input is HIGH *and* the other input is HIGH, the output is HIGH. When any or all inputs are LOW, the output is LOW. The AND operation is implemented by a logic circuit known as an *AND gate.*

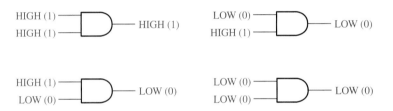

FIGURE 1–17
The AND operation.

OR

The **OR** operation produces a HIGH output when any of the inputs is HIGH, as indicated in Figure 1–18 for the case of two inputs. When one input is HIGH *or* the other input is HIGH *or* both inputs are HIGH, the output is HIGH. When both inputs are LOW, the output is LOW. The OR operation is implemented by a logic circuit known as an *OR gate.*

FIGURE 1–18
The OR operation.

HIGH (1) ——┐
            ⟩— HIGH (1)   LOW (0) ——┐
HIGH (1) ——┘   ⟩— HIGH (1)
                                 HIGH (1) ——┘

HIGH (1) ——┐
            ⟩— HIGH (1)   LOW (0) ——┐
LOW (0) ——┘   ⟩— LOW (0)
                                 LOW (0) ——┘

**SECTION 1–3
REVIEW**

1. When does the NOT operation produce a HIGH output?
2. When does the AND operation produce a HIGH output?
3. When does the OR operation produce a HIGH output?
4. What is an inverter?
5. What is a logic gate?

1–4 ■ BASIC LOGIC FUNCTIONS

The three basic logic elements AND, OR, and NOT can be combined to form more complex logic circuits that perform many useful operations and that are used to build complete digital systems. Some of the common logic functions are comparison, arithmetic, code conversion, encoding, decoding, data selection, storage, and counting. This section provides a general overview of these important functions so that you can begin to see how they form the building blocks of digital systems such as computers. Each of the basic logic functions will be covered in detail in later chapters.

After completing this section, you should be able to

■ Identify eight basic types of logic functions ■ Describe a basic magnitude comparator ■ List the four arithmetic functions ■ Describe a basic adder ■ Describe a basic encoder ■ Describe a basic decoder ■ Define multiplexing and demultiplexing ■ State how data storage is accomplished ■ Describe the function of a basic counter

The Comparison Function

Magnitude comparison is performed by a logic circuit called a **comparator,** covered in Chapter 6. A comparator compares two quantities and indicates whether or not they are equal. For example, suppose you have two numbers and wish to know if they are equal or not equal and, if not equal, which is greater. The comparison function is represented in Figure 1–19. One number in binary form is applied to input *A*, and the other number in binary form is applied to input *B*. The outputs indicate the relationship of the two numbers by producing a HIGH level on the proper output line. Suppose that a binary representation of the number 2 is applied to input *A* and a binary representation of the number 5 is applied to input *B*. (We will discuss the binary representation of numbers and symbols in the next chapter.) A HIGH level will appear on the $A < B$ (*A* is less than *B*) output, indicating the relationship between the two numbers (2 is less than 5). The wide arrows represent a group of parallel lines on which the bits are transferred.

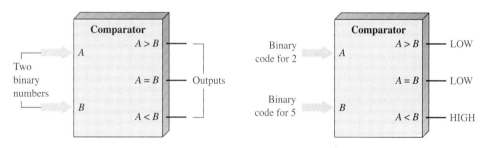

(a) Basic magnitude comparator

(b) Example: *A* less than *B* (2 < 5) as indicated by HIGH output ($A < B$)

FIGURE 1–19
The comparison function.

The Arithmetic Functions

Addition   Addition is performed by a logic circuit called an **adder,** covered in Chapter 6. An adder adds two binary numbers (on inputs *A* and *B* with a carry input C_{in}) and generates a sum (Σ) and a carry output (C_{out}), as shown in Figure 1–20(a). Figure 1–20(b) illustrates the addition of 3 and 9. You know that the sum is 12; the adder indicates this result by producing 2 on the sum output and 1 on the carry output. Assume that the carry input in this example is 0.

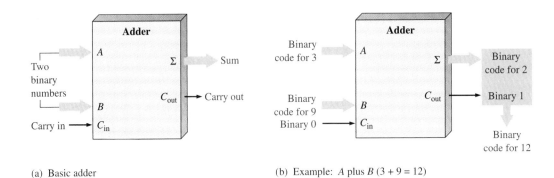

(a) Basic adder   (b) Example: A plus B $(3 + 9 = 12)$

FIGURE 1–20
The addition function.

COMPUTER NOTE
In a microprocessor, the arithmetic logic unit (ALU) performs the operations of add, subtract, multiply, and divide as well as the logic operations on digital data as directed by a series of instructions. A typical ALU is constructed of many thousands of logic gates.

Subtraction   Subtraction is also performed by a digital circuit. A **subtracter** requires three inputs: the two numbers that are to be subtracted and a borrow input. The two outputs are the difference and the borrow output. When, for instance, 5 is subtracted from 8 with no borrow input, the difference is 3 with no borrow output. You will see in Chapter 2 how subtraction can actually be performed by an adder because subtraction is simply a special case of addition.

Multiplication   Multiplication is performed by a digital circuit called a *multiplier*. Numbers are always multiplied two at a time, so two inputs are required. The output of the multiplier is the product. Because multiplication is simply a series of additions with shifts in the positions of the partial products, it can be performed by using an adder in conjunction with other circuits.

Division   Division can be performed with a series of subtractions, comparisons, and shifts, and thus it can also be done using an adder in conjunction with other circuits. Two inputs to the divider are required, and the outputs generated are the quotient and the remainder.

The Code Conversion Function

A **code** is a set of bits arranged in a unique pattern and used to represent specified information. A code converter changes one form of coded information into another coded form. Examples are conversion between binary and other codes such as the binary coded decimal (BCD) and the Gray code. Various types of codes are covered in Chapter 2, and code converters are covered in Chapter 6.

The Encoding Function

The encoding function is performed by a logic circuit called an **encoder,** covered in Chapter 6. The encoder converts information, such as a decimal number or an alphabetic character, into some coded form. For example, one certain type of encoder converts each of the decimal digits, 0 through 9, to a binary code. A HIGH level on the input corresponding to a specific decimal digit produces the proper binary code on the output lines.

Figure 1–21 is a simple illustration of an encoder used to convert (encode) a calculator keystroke into a binary code that can be processed by the calculator circuits.

The Decoding Function

The decoding function is performed by a logic circuit called a **decoder,** covered in Chapter 6. The decoder converts coded information, such as a binary number, into a noncoded form, such as a decimal form. For example, one particular type of decoder converts a 4-bit binary code into the appropriate decimal digit.

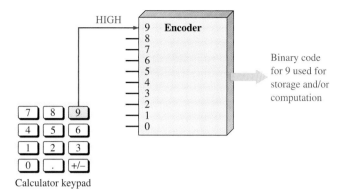

FIGURE 1–21
An encoder used to encode a calculator keystroke into a binary code for storage or for calculation.

Figure 1–22 is a simple illustration of one type of decoder that is used to activate a 7-segment display. Each of the seven segments of the display is connected to an output line from the decoder. When a particular binary code appears on the decoder inputs, the appropriate output lines are activated and light the proper segments to display the decimal digit corresponding to the binary code.

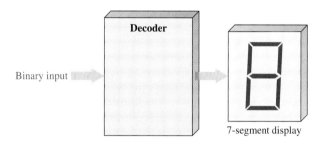

FIGURE 1–22
A decoder used to convert a special binary code into a 7-segment decimal readout.

The Data Selection Function

Two types of circuits that select data are the multiplexer and the demultiplexer. The **multiplexer,** or mux for short, is a logic circuit that switches digital data from several input lines onto a single output line in a specified time sequence. Functionally, a multiplexer can be represented by an electronic switch operation that sequentially connects each of the input lines to the output line. The **demultiplexer** (demux) is a logic circuit that switches digital data from one input line to several output lines in a specified time sequence. Essentially, the demux is a mux in reverse.

Multiplexing and demultiplexing are used when data from several sources are to be transmitted over one line to a distant location and redistributed to several destinations. Figure 1–23 illustrates this type of application where digital data from three computers are sent out along a single line to three other computers at another location.

In Figure 1–23, binary data from computer A are connected to the output line during time interval Δt_1 and transmitted to the demultiplexer that connects them to computer D. Then, during interval Δt_2, the multiplexer switches to the input from computer B and the demultiplexer switches the output to computer E. During interval Δt_3, the multiplexer switches to the input from computer C and the demultiplexer switches the output to computer F.

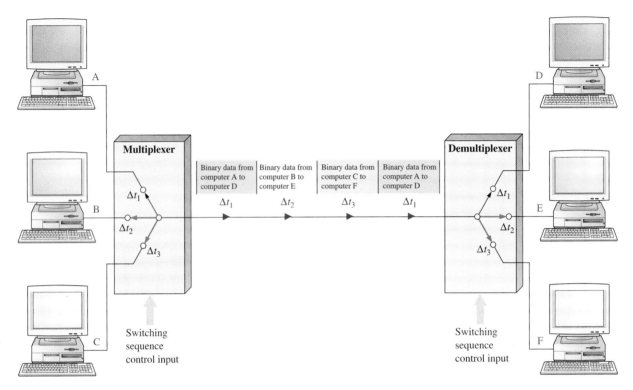

FIGURE 1–23
Illustration of a basic multiplexing/demultiplexing application.

To summarize, during the first time interval, computer A sends data to computer D. During the second time interval, computer B sends data to computer E. During the third time interval, computer C sends data to computer F. After this, the first two computers again communicate and the sequence repeats. Because the time is divided up among several sets of systems where each has its turn to send and receive data, this process is called *time division multiplexing* (TDM).

The Storage Function

COMPUTER NOTE
The internal computer memories, RAM and ROM, as well as the smaller caches are semiconductor memories. The registers in a microprocessor are constructed of semiconductor flip-flops. Magnetic disk memories are used in the internal hard drive, the floppy drive, and for the CD-ROM. Magnetic tape memories are often used for data backup.

Storage is a function that is required in most digital systems, and its purpose is to retain binary data for a period of time. Some storage devices are used for short-term storage and some for long-term storage. A storage device can "memorize" a bit or a group of bits and retain the information as long as necessary. Common types of storage devices are flip-flops, registers, semiconductor memories, magnetic disks, magnetic tape, and optical disks (CDs).

Flip-flops   The **flip-flop** is a bistable (two stable states) logic circuit that can store only one bit at a time, either a 1 or a 0. The output of a flip-flop indicates which bit it is storing. A HIGH output indicates that a 1 is stored and a LOW output indicates that a 0 is stored. Flip-flops are implemented with logic gates and are covered in Chapter 8.

Registers   A **register** is formed by combining several flip-flops so that groups of bits can be stored. For example, an 8-bit register is constructed from eight flip-flops. In addition to storing bits, registers can be used to shift the bits from one position to another within the register or out of the register to another circuit; therefore, these devices are known as *shift registers*. Shift registers are covered in Chapter 10.

The two basic types of shift registers are serial and parallel. The bits are stored in a serial shift register one at a time, as illustrated in Figure 1–24. A good analogy to this is loading passengers onto a bus single file through the door. They also exit the bus single file.

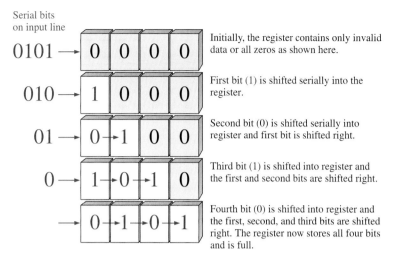

FIGURE 1–24
Example of the operation of a 4-bit serial shift register. Each block represents one storage "cell" or flip-flop.

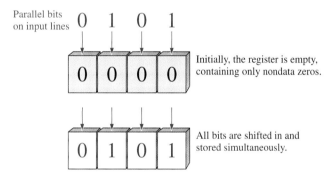

FIGURE 1–25
Example of the operation of a 4-bit parallel shift register.

The bits are stored in a parallel register simultaneously from parallel lines, as shown in Figure 1–25. For this case, a good analogy is loading passengers on a roller coaster where they enter all of the cars in parallel.

Semiconductor Memories Semiconductor memories are devices typically used for storing large numbers of bits. In one type of memory, called the *read-only m*emory or ROM, the binary data are permanently or semipermanently stored and cannot be readily changed. In the *random-access m*emory or RAM, the binary data are temporarily stored and can be easily changed. Memories are covered in Chapter 12.

Magnetic Memories Magnetic disk memories are used for mass storage of binary data. Examples are the so-called floppy disks used in computers and the computer's internal hard disk. Magneto-optical disks use laser beams to store and retrieve data. Magnetic tape is still used in memory applications and for backing up data from other storage devices.

The Counting Function

The counting function is very important in digital systems. There are many types of digital **counters,** but their basic purpose is to count events represented by changing levels or pulses or to generate a particular code sequence. To count, the counter must "remember" the present number so that it can go to the next proper number in sequence. Therefore,

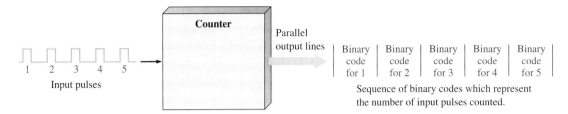

FIGURE 1–26
Illustration of basic counter operation.

storage capability is an important characteristic of all counters, and flip-flops are generally used to implement them. Figure 1–26 illustrates the basic idea of counter operation. Counters are covered in Chapter 9.

<table>
<tr><td>

**SECTION 1–4
REVIEW**

</td><td>

1. What does a comparator do?
2. What are the four basic arithmetic operations?
3. Describe encoding and give an example.
4. Describe decoding and give an example.
5. Explain the basic purpose of multiplexing and demultiplexing.
6. Name four types of storage devices.
7. What does a counter do?

</td></tr>
</table>

1–5 ■ DIGITAL INTEGRATED CIRCUITS

All the logic elements and functions that have been discussed—and many more— are available in integrated circuit (IC) form. Modern digital systems use ICs almost exclusively in their designs because of their small size, high reliability, low cost, and low power consumption. It is important to be able to recognize the IC packages and to know how the pin connections are numbered, as well as to be familiar with the way in which circuit complexities and circuit technologies determine the various IC classifications.

After completing this section, you should be able to

■ Recognize the difference between through-hole devices and surface-mount devices ■ Identify dual-in-line packages (DIP) ■ Identify small-outline integrated circuit packages (SOIC) ■ Identify plastic leaded chip carrier packages (PLCC) ■ Identify leadless ceramic chip carrier packages (LCCC) ■ Identify flat packs (FP) ■ Determine pin numbers on various types of IC packages ■ Explain the circuit complexity classifications of integrated circuits

A monolithic **integrated circuit (IC)** is an electronic circuit that is constructed entirely on a single small chip of silicon. All the components that make up the circuit—transistors, diodes, resistors, and capacitors—are an integral part of that single chip.

Figure 1–27 shows a cutaway view of one type of IC package, with the circuit chip shown within the package. Points on the chip are connected to the package pins to allow input and output connections to the outside world.

FIGURE 1–27
Cutaway view of one type of IC package showing the chip mounted inside, with connections to input and output pins.

IC Packages

Integrated circuit (IC) packages are classified according to the way they are mounted on printed circuit (PC) boards as either through-hole mounted or surface mounted. The through-hole type packages have pins (leads) that are inserted through holes in the PC board and can be soldered to conductors on the opposite side. The most common type of through-hole package is the dual-in-line package **(DIP)** shown in Figure 1–28(a).

(a) Dual-in-line package (DIP)

(b) Small-outline IC (SOIC)

FIGURE 1–28
Examples of through-hole and surface-mounted devices. The DIP is larger than the SOIC with the same number of leads. This particular DIP is approximately 0.785 in. long, and the SOIC is approximately 0.385 in. long.

Another type of IC package uses surface-mount technology **(SMT)**. Surface mounting is a newer, space-saving alternative to through-hole mounting. The holes through the PC board are unnecessary for SMT. The pins of surface-mounted packages are soldered directly to conductors on one side of the board, leaving the other side free for additional circuits. Also, for a circuit with the same number of pins, a surface-mounted package is much smaller than a dual-in-line package because the pins are placed closer together. An example of a surface-mounted package is the small-outline integrated circuit (SOIC) shown in Figure 1–28(b).

Four common types of SMT packages are the **SOIC** (small-outline IC), the **PLCC** (plastic leaded chip carrier), the **LCCC** (leadless ceramic chip carrier), and the **flat pack** (FP). These types of packages are available in various sizes depending on the number of leads (more leads are required for more complex circuits). Examples of each type are shown in Figure 1–29. As you can see, the leads of the SOIC are formed into a "gull-wing" shape. The leads of the PLCC are turned under the package in a J-type shape. Instead of leads, the LCCC has metal contacts molded into its ceramic body. The leads of the flat pack extend straight out from the body. Other variations of SMT packages include **SSOP** (shrink small-outline package), **TSSOP** (thin shrink small-outline package), and **TVSOP** (thin very small-outline package).

(a) SOIC with "gull-wing" leads

(b) PLCC with J-type leads

(c) LCCC with no leads (contacts are part of case)

(d) Flat pack (FP) with straight leads

FIGURE 1–29
Examples of SMT package configurations.

Pin Numbering

All IC packages have a standard format for numbering the pins (leads). The dual-in-line packages (DIPs), the small-outline IC packages (SOICs), and the flat packs have the numbering arrangement illustrated in Figure 1–30(a) for a 16-pin package. Looking at the top of the package, pin 1 is indicated by an identifier that can be either a small dot, a notch, or a beveled edge. The dot is always next to pin 1. Also, with the notch oriented upward, pin 1 is always the top left pin, as indicated. Starting with pin 1, the pin numbers increase as you go down, then across and up. The highest pin number is always to the right of the notch or opposite the dot.

FIGURE 1–30
Pin numbering for two standard types of IC packages. Top views are shown.

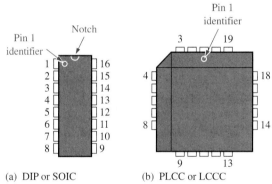

(a) DIP or SOIC

(b) PLCC or LCCC

The PLCC and LCCC packages have leads arranged on all four sides. Pin 1 is indicated by a dot or other index mark and is located at the center of one set of leads. The pin numbers increase going counterclockwise as viewed from the top of the package. The highest pin number is always to the right of pin 1. Figure 1–30(b) illustrates this format for a 20-pin PLCC package.

Integrated Circuit Complexity Classifications

Integrated circuits are classified according to their complexity. They are listed here from the least complex to the most complex. The complexity figures stated here for SSI, MSI, LSI, VLSI, and ULSI are generally accepted, but definitions may vary from one source to another.

- **Small-scale integration (SSI)** describes integrated circuits that have up to twelve equivalent gate circuits on a single chip, and they include basic gates and flip-flops.

■ **Medium-scale integration (MSI)** describes integrated circuits that have from 12 to 99 equivalent gates on a chip. They include logic functions such as encoders, decoders, counters, registers, multiplexers, arithmetic circuits, small memories, and others.

■ **Large-scale integration (LSI)** is a classification of ICs with complexities of 100 to 9999 equivalent gates per chip, including memories.

■ **Very large-scale integration (VLSI)** describes integrated circuits with complexities of 10,000 to 99,999 equivalent gates per chip.

■ **Ultra large-scale integration (ULSI)** describes very large memories, larger **micro-processors,** and larger single-chip computers. Complexities of 100,000 equivalent gates and greater are classified as ULSI.

Integrated Circuit Technologies

The types of transistors with which all integrated circuits are implemented are either bipolar junction transistors or MOSFETs (metal-oxide semiconductor field-effect transistors). Two types of digital circuit technology that use bipolar junction transistors are TTL (transistor-transistor logic) and ECL (emitter-coupled logic). Of these two, TTL is more widely used. The major circuit technologies that use MOSFETs are CMOS (complementary MOS) and NMOS (*n*-channel MOS). Microprocessors use MOS technology.

All gates and other functions can be implemented with either type of circuit technology. SSI and MSI circuits are generally available in both TTL and CMOS. LSI, VLSI, and ULSI are generally implemented with CMOS or NMOS because it requires less area on a chip and consumes less power. There is more on these integrated technologies in Chapter 3. In addition, Chapter 15 acts as a "floating" chapter that provides a complete circuit-level coverage that can be introduced at various points throughout the book.

Because of their particular structure, MOS devices are very sensitive to static charge and can be damaged by electrostatic discharge (ESD) if not handled properly. The following precautions should be taken when working with MOS devices:

❑ MOS devices should be shipped and stored in conductive foam.
❑ All instruments and metal benches used in testing should be connected to earth ground.
❑ The handler's wrist should be connected to earth ground with a length of wire and high-value series resistor.
❑ Do not remove a MOS device (or any device for that matter) from a circuit while the dc power is on.
❑ Do not connect ac or signal voltages to a MOS device while the dc power supply is off.

**SECTION 1–5
REVIEW**

1. What is an integrated circuit?
2. Define the terms DIP, SMT, SOIC, SSI, MSI, LSI, VLSI and ULSI.
3. Generally, in what classification does an IC with the following number of equivalent gates fall?
   (a) 75   **(b)** 500   **(c)** 10   **(d)** 15,000   **(e)** 200,000

1–6 ■ TESTING AND TROUBLESHOOTING INSTRUMENTS

Troubleshooting is the technique of systematically isolating, identifying, and correcting a fault in a circuit or system. A variety of instruments are available for use in digital troubleshooting and testing. Some typical equipment is presented in this section.

After completing this section, you should be able to

■ Recognize common oscilloscope controls ■ Distinguish between an analog and digital oscilloscope ■ Determine the amplitude, period, frequency, and duty cycle of a digital waveform with an oscilloscope. ■ Discuss what a logic analyzer does ■ List common logic analyzer formats ■ Explain how the logic probe, pulser, and current probe are used ■ State the purposes of the dc power supply, function generator, and digital multimeter

The Oscilloscope

The oscilloscope (scope for short) is one of the most widely used instruments for general testing and **troubleshooting.** It is basically a graph-displaying device that traces a graph of an electrical signal on its screen. In most applications, the graphs show how signals change over time. The vertical axis of the screen represents voltage and the horizontal axis represents time. In digital applications, pulse waveforms can be displayed and parameters such as amplitude, period, frequency, rise and fall times, and duty cycle can be measured. Also, two pulse waveforms (with 2-channel oscilloscopes) can be displayed simultaneously so that their time relationship can be determined.

There are two basic types of oscilloscope, analog and digital. Both types can be used to view pulse waveforms found in digital circuits. An analog scope works by applying the signal voltage being measured to an electron beam that is sweeping across the screen. The measured voltage deflects the beam up and down proportional to the amount of voltage, immediately tracing the waveform on the screen. A digital scope samples the signal voltage and uses an analog-to-digital converter (ADC) to convert the voltage being measured into digital information. The digital information is then used to reconstruct the waveform on the screen. Typical bench type and handheld oscilloscopes are shown in Figure 1–31.

Although there are still some analog scopes being manufactured and many older ones that are still in use, the digital scope is widely used and is becoming the instrument of choice to replace older analog instruments.

Analog Oscilloscope

Figure 1–32 shows a generic front panel view of a typical analog oscilloscope. There are two VOLTS/DIV controls, one for each of the two channels. These controls determine the attenuation or amplification of the signal, and each of the settings on the dial indicate the number of volts (V) or millivolts (mV) represented by each vertical division on the screen. Typically, there are two brackets on the VOLTS/DIV dial, one for a times 10 (×10) voltage probe and one for a times 1 (×1) voltage probe. The most commonly used probe is a ×10, which attenuates the input signal by ten to reduce loading effects on the circuit. When using a ×10 probe, you must read the VOLTS/DIV setting in the ×10 bracket.

The time base or SEC/DIV control sets the number of seconds (s), milliseconds (ms), or microseconds (μs) represented by each horizontal division as indicated by the dial settings. You can measure the period of a pulse waveform on the horizontal scale and then calculate the frequency.

(a) Digital oscilloscope

(b) Analog oscilloscope

(c) Handheld oscilloscope

FIGURE 1–31
Typical oscilloscopes. (Parts (a) and (c) copyright 1998 Tektronix, Inc. All rights reserved. Reproduced by permission. Part (b) courtesy of B&K Precision Corp.)

FIGURE 1–32
A typical dual-channel analog oscilloscope.

EXAMPLE 1–3

Based on the control dial settings, determine the amplitude and the period of the pulse waveform on the screen of an analog oscilloscope as shown in Figure 1–33. Also, calculate the frequency. Assume measurements are made with a ×10 voltage probe.

FIGURE 1–33

Solution The VOLTS/DIV setting is 1 V. The pulses are three divisions high. Since each division represents 1 V, the pulse amplitude is

$$\text{Amplitude} = (3 \text{ divisions})(1 \text{ volt/division}) = \mathbf{3\ V}$$

The SEC/DIV setting is 10 μs. A full cycle of the waveform (from beginning of one pulse to the beginning of the next) covers four divisions; therefore, the period is

$$\text{Period} = (4 \text{ divisions})(10\ \mu\text{s/division}) = \mathbf{40\ \mu s}$$

The frequency is calculated as

$$f = \frac{1}{T} = \frac{1}{40\ \mu s} = \mathbf{25\ kHz}$$

Related Problem For a VOLTS/DIV setting of 4 V and a SEC/DIV setting of 2 ms, determine the amplitude and period of the pulse shown on the screen in Figure 1–33. Assume a ×10 probe is used.

Other Basic Controls The following descriptions refer to the analog scope in Figure 1–32. However, many of these are also common to digital scopes but with some functions determined by a menu-driven format.

Power switch The power switch turns the power on and off. A light indicates when the power is on.

Intensity The intensity control varies the brightness of the trace on the screen. Caution should be used so that the intensity is not left too high for an extended period of time, especially when the beam forms a motionless dot on the screen.

Focus This control focuses the beam so that it converges to a tiny point at the screen. An out-of-focus condition results in a fuzzy trace.

Horizontal and vertical positions These controls are used to move a trace horizontal across the screen and up or down for easier viewing or measurement.

AC-GND-DC switch This switch allows the input signal to be ac coupled, dc coupled, or grounded. The ac coupling eliminates any dc component on the input signal. The dc coupling permits dc values to also be displayed. The ground position allows a 0 V reference to be established on the screen. *For digital signals, dc coupling should be used.*

Signal inputs The signal voltages to be displayed are connected into the channel 1 (CH1) and/or channel 2 (CH2). As mentioned previously, these input connections are done with an attenuator probe that minimizes the loading effect of the scope on the circuit being measured. The common type of oscilloscope voltage probe attenuates (reduces) the signal voltage by a factor of ten and is called a times 10 ($\times$10) probe. Times 1 ($\times$1) probes are also occasionally used when measuring very small voltages. A typical voltage probe is shown in Figure 1–34.

FIGURE 1–34
An oscilloscope voltage probe.

Mode switches These switches provide for displaying either or both channel input signals, inverting channel 2 signal (INVERT), adding two waveforms (ADD), and selecting between alternate (ALT) and chopped (CHOP) mode of sweep. *As a rule, ALT is selected when the frequency is above 100 Hz; otherwise, CHOP is used.*

Trigger controls These controls allow the scope to be triggered from various selected sources. Triggering causes the trace to begin its sweep across the screen. Triggering can occur from an internal signal, an external signal, or from the line voltage. Typical trigger modes are *normal, auto, single-sweep,* and *TV.* In the auto mode, sweep occurs in the absence of an adequate trigger signal. In the normal mode, a trigger signal must be present for the sweep to occur. The TV mode provides triggering on the TV field or TV line signals. The *slope* switch sets the trigger to occur on either the positive-going or the negative-going slope of the trigger waveform. The *level* control selects the voltage level on the trigger signal at which the triggering occurs. Basically, the trigger controls provide for synchronization of the horizontal sweep waveform in the scope and the input signals. As a result, the display of the input signal is stable on the screen, rather than appearing to drift.

Digital Oscilloscope

A generic front-panel view of a typical digital oscilloscope is shown in Figure 1–35. Notice that the Vertical and the Horizontal sections each consist of a position control, a menu button, and a scale control.

In the *vertical* section, the controls apply to either channel as determined by channel selection buttons. The position control lets you move the waveform up or down on the screen. The menu button provides for the selection of several items that appear on the screen, such as coupling (ac, dc, or ground), waveform inversion, and other parameters. The scale control adjusts the number of volts represented by each vertical division, but in this type of scope the volts/division setting for a channel is displayed on the screen rather than having to read it from the dial. Probes for digital scopes have a special element that detects whether the probe is a $\times$1 or a $\times$10 and automatically adjusts the readouts.

In the *horizontal* section, the controls apply to both channels. The position control lets you move the waveform horizontally on the screen. The menu button provides for the

FIGURE 1–35
A typical dual-channel digital oscilloscope.

selection of several items that appear on the screen, such as main or delayed sweep and other parameters. The scale control adjusts the time represented by each horizontal division, but in this type of scope the sec/division is displayed on the screen rather than having to read it from the dial.

In the *trigger* section, the level control determines where the trigger point occurs. The menu button provides for selection of items that appear on the screen, such as normal mode, auto mode, video mode, trigger source, trigger slope, and other parameters.

Many digital scopes have a diskette drive that allows you to store waveforms for later viewing, analysis, or for importing to desktop publishing or other programs on a computer.

Most measurements you make with an oscilloscope require a ×10 attenuator probe which reduces the signal voltage and circuit loading by a factor of 10. Before making any measurement, you should make sure the probe is compensated because an improperly compensated probe can distort the waveforms that you see on the scope screen. On most probes, there is a screwdriver adjustment or other means for compensation and there is normally a calibrated square wave output available on the front panel of the oscilloscope. Scope waveforms are shown here for three probe conditions: properly compensated, undercompensated, and overcompensated.

Properly compensated

Undercompensated

Overcompensated

FIGURE 1–36
Typical logic analyzers. (Copyright 1998 Tektronix, Inc. All rights reserved. Reproduced by permission.)

The Logic Analyzer

Typical logic analyzers are shown in Figure 1–36. This type of instrument can detect and display digital data in several formats.

Oscilloscope Format The logic analyzer can be used to display single or dual waveforms on the screen so that characteristics of individual pulses or waveform parameters can be measured.

Timing Diagram Format The logic analyzer can display typically up to sixteen waveforms in proper time relationship so that you can analyze sets of waveforms and determine how they change in time with respect to each other.

State Table Format The logic analyzer can display binary data in tabular form. For example, various memory locations in a microprocessor-based system can be examined to determine the contents. The data can be displayed in a variety of number systems and codes such as binary, hexadecimal, octal, binary coded decimal (BCD), and ASCII. These number systems and codes are the topics of the next chapter.

The Logic Probe, Pulser, and Current Probe

The logic probe is a convenient, hand-held tool that provides a means of troubleshooting a digital circuit by sensing various conditions at a point in a circuit, as illustrated in Figure 1–37. The probe can detect high-level voltage, low-level voltage, single pulses, repetitive pulses, and opens on a PC board. The probe lamp indicates the condition that exists at a certain point, as indicated in the figure.

The logic pulser produces a repetitive pulse waveform that can be applied to any point in a circuit. You can apply pulses at one point in a circuit with the pulser and check another point for resulting pulses with a logic probe. Also, the pulser can be used in conjunction with the current probe as indicated in Figure 1–38. The current probe senses when there is pulsating current in a line and is particularly useful for locating shorts on a PC board.

The DC Power Supply

The dc power supply is an indispensable instrument on any test bench. The power supply converts ac power from the standard wall outlet into regulated dc voltage. All digital circuits require dc voltage to operate. For example, TTL circuits and many CMOS circuits

FIGURE 1–37
Illustration of how a logic probe is used to detect various voltage conditions at a given point in a circuit.

FIGURE 1–38
Illustration of how a logic pulser and a current probe can be used to pulse a given point and check for resulting current in another part of the circuit.

Programmable power supplies

Dual power supply

Hand-held DMM

Function generator

DMM

FIGURE 1–39
Typical test instruments. (Copyright 1998 Tektronix, Inc. All rights reserved. Reproduced by permission.)

require approximately +5 V. The power supply is used when a new circuit is breadboarded or when a PC board is pulled from a system for testing and is no longer operating from the internal system power supply. Typical test bench dc power supplies are shown in Figure 1–39.

The Function Generator

The function generator is a versatile signal source that provides pulse waveforms, as well as sine wave and triangular waveforms. Many function generators have logic-compatible outputs to provide proper level waveforms as inputs to digital circuits in order to check the operation. One type of function generator is shown in Figure 1–39.

The Digital Multimeter

No test bench is complete without a digital multimeter (DMM). This instrument is used for measuring dc and ac voltage, dc and ac current, and resistance. Figure 1–39 shows typical test bench and handheld DMMs.

**SECTION 1–6
REVIEW**

1. How many major horizontal divisions are there on an oscilloscope screen?
2. How many major vertical divisions are there on an oscilloscope screen?
3. What type of instrument can display up to sixteen waveforms on its screen?
4. What is indicated when the lamp on a logic probe is flashing repetitively?

1–7 ■ DIGITAL SYSTEM APPLICATION

In this section, an interesting but simplified system application of the logic elements and functions that were discussed in previous sections is presented. It is important that an electronic technician or technologist understand how various digital functions can operate together as a total system to perform a specified task. It is also important to begin to think in terms of system-level operation because, in practice, a large part of your work will involve systems rather than individual functions. Of course, to understand systems, you must first understand the basic elements and functions that make up a system.

This section introduces you to the system concept. The example here will show you how logic functions can work together to perform a higher-level task and will get you started thinking at the system level. The specific system used here to illustrate the system concept serves as an instructional model and is not necessarily the approach that would be used in practice, although it could be. In modern industrial applications like the one discussed here, instruments known as programmable controllers are often used.

After completing this section, you should be able to

■ Begin to think in terms of system-level operation   ■ Explain the purpose of each logic function in the example system   ■ Discuss how the various functions work together to accomplish a desired task

The System

Let's imagine that a pharmaceutical company uses the system shown in the simplified block diagram of Figure 1–40 for automatically counting and bottling tablets. The tablets are fed into a large funnel-like hopper. The narrow neck of the funnel allows only one tablet at a time to fall into a bottle on the conveyor belt below.

The digital system controls the number of tablets going into each bottle and displays a continually updated total near the conveyor line as well as at a remote location in another part of the plant. As you can see, this system utilizes all the basic logic functions that were introduced in Section 1–4, and its only purpose is to show you how these functions may be combined to achieve a desired result.

The general operation is as follows. An optical sensor at the bottom of the funnel neck detects each tablet that passes and produces an electrical pulse. This pulse goes to the counter and advances it by one count; thus, at any time during the filling of a bottle, the counter contains the binary representation of the number of tablets in the bottle. The binary count is transferred from the counter on parallel lines to the B input of the comparator (comp). A preset binary number equal to the number of tablets that are to go into each bottle is placed on the A input of the comparator. The preset number comes from the keypad and the associated circuits, which include the encoder, register A, and code converter A. When the desired number of tablets is entered on the keypad, it is encoded and then stored by parallel register A until a change in the quantity of tablets per bottle is required.

Suppose, for example, that each bottle is to hold fifty tablets. When the number in the counter reaches fifty, the $A = B$ output of the comparator goes HIGH, indicating that the bottle is full.

The HIGH output of the comparator immediately closes the valve in the neck of the funnel to stop the flow of tablets, and at the same time it activates the conveyor to move the next bottle into place under the funnel. When the next bottle is positioned properly under the neck of the funnel, the conveyor control circuit produces a pulse that resets the counter to zero. The $A = B$ output of the comparator goes back LOW, opening the funnel valve to restart the flow of tablets.

FIGURE 1–40
Simplified basic block diagram for a tablet-counting and bottling control system.

In the display portion of the system, the number in the counter is transferred in parallel to the *A* input of the adder. The *B* input of the adder comes from parallel register B that holds the total number of tablets bottled, up through the last bottle filled. For example, if ten bottles have been filled and each bottle holds fifty tablets, register B contains the binary representation for 500. Then, when the next bottle has been filled, the binary number for 50 appears on the *A* input of the adder, and the binary number for 500 is on the *B* input. The adder produces a new sum of 550, which is stored in register B, replacing the previous sum of 500.

The binary number in register B is transferred in parallel to the code converter and decoder, which changes it from binary form to decimal form for display on a readout near the conveyor line. The binary number in the register is also transferred to a multiplexer (mux) so that it can be converted from parallel to series form and transmitted along a single line to a remote location some distance away. It is more economical to run a single line than to run several parallel lines when significant distances are involved, and speed of data transmission is not a factor in this application. At the remote location, the serial data are demultiplexed and sent to register C. From there the data are then decoded for display on the remote readout.

Keep in mind that this system is an instructional model and does not necessarily represent the ultimate or most efficient way to implement this hypothetical process. Although there are certainly other approaches, this particular approach has been used in order to illustrate all of the logic functions that were introduced in Section 1–4 and that will be covered in detail in future chapters. It shows you one application of the various functional devices at the system level and how they can be connected to accomplish a specific objective. You will see this system again in the next chapter.

SECTION 1–7 REVIEW

1. Explain the purpose of the comparator (comp) in the system in Figure 1–40.
2. What actions take place when the $A = B$ output of the comparator goes HIGH?
3. What is the content of each register at any given time?

■ SUMMARY

- An analog quantity has continuous values.
- A digital quantity has a discrete set of values.
- A binary digit is called a bit.
- A pulse is characterized by rise time, fall time, pulse width, and amplitude.
- The frequency of a periodic waveform is the reciprocal of the period. The formulas relating frequency and period are

$$f = \frac{1}{T} \quad \text{and} \quad T = \frac{1}{f}$$

- The duty cycle of a pulse waveform is the ratio of the pulse width to the period, expressed by the following formula as a percentage:

$$\text{Duty cycle} = \left(\frac{t_W}{T}\right)100\%$$

- A timing diagram is an arrangement of two or more waveforms showing their relationship with respect to time.
- Three basic logic operations are NOT, AND, and OR. The standard symbols for these are given in Figure 1–41.

FIGURE 1–41

NOT   AND   OR

- The basic logic functions are comparison, arithmetic, code conversion, decoding, encoding, data selection, storage, and counting.
- The two broad physical categories of IC packages are through-hole mounted and surface mounted.
- The categories of ICs in terms of circuit complexity are SSI (small-scale integration), MSI (medium-scale integration), LSI, VLSI, and ULSI (large-scale, very large-scale, and ultra large-scale integration).
- Common instruments used in testing and troubleshooting digital circuits are the oscilloscope, logic analyzer, logic probe, pulser, current probe, dc power supply, function generator, and digital multimeter.

■ KEY TERMS

These terms are also in the end-of-book glossary.

Adder   A logic circuit used to add two binary numbers.

Amplitude   In a pulse waveform, the height or maximum value of the pulse as measured from its low level.

Analog   Being continuous or having continuous values.

AND   A basic logic operation in which a true (HIGH) output occurs only if all the input conditions are true (HIGH).

Binary   Having two values or states; describes a number system that has a base of two and utilizes 1 and 0 as its digits.

Bit   A binary digit, which can be either a 1 or a 0.

Bit time   The interval of time occupied by a single bit in a sequence of bits; the period of the clock.

Clock   The basic timing signal in a digital system.

Data   Information in numeric, alphabetic, or other form.

Digital   Related to digits or discrete quantities; having a set of discrete values.

DIP   Dual-in-line package. A type of IC package whose leads must pass through holes to the other side of a PC board.

Frequency   The number of pulses in one second for a periodic digital waveform. The unit is hertz (Hz).

Gate   A logic circuit that performs a specified logic operation such as AND or OR.

Input   The signal or line going into a circuit.

Integrated circuit (IC)   A type of circuit in which all of the components are integrated on a single chip of semiconductive material of extremely small size.

Inverter   A NOT circuit; a circuit that changes a HIGH to a LOW or vice versa.

Logic   In digital electronics, the decision-making capability of gate circuits, in which a HIGH represents a true statement and a LOW represents a false one.

NOT   A basic logic operation that performs inversions.

OR   A basic logic operation in which a true (HIGH) output occurs when one or more of the input conditions are true (HIGH).

Output   The signal or line coming out of a circuit.

Parallel   In digital systems, data occurring simultaneously on several lines; the transfer or processing of several bits simultaneously.

Period   The time required for a periodic waveform to repeat itself.

Pulse   A sudden change from one level to another, followed after a time, called the pulse width, by a sudden change back to the original level.

Serial   Having one element following another, as in a serial transfer of bits; occurring in sequence rather than simultaneously.

SMT   Surface-mount technology. A category of integrated circuit package having leads that are connected on the surface of the PC board.

Troubleshooting   The technique or process of systematically identifying, isolating, and correcting a fault in a circuit or system.

■ SELF-TEST

Answers are found at the end of the chapter.

1. A quantity having continuous values is
   (a) a digital quantity   (b) an analog quantity
   (c) a binary quantity   (d) a natural quantity

2. The term *bit* means

(a) a small amount of data   (b) a 1 or a 0

(c) binary digit   (d) both answers (b) and (c)

3. The time interval on the leading edge of a pulse between 10% and 90% of the amplitude is the

(a) rise time   (b) fall time   (c) pulse width   (d) period

4. A pulse in a certain waveform occurs every 10 ms. The frequency is

(a) 1 kHz   (b) 1 Hz   (c) 100 Hz   (d) 10 Hz

5. In a certain digital waveform, the period is twice the pulse width. The duty cycle is

(a) 100%   (b) 200%   (c) 50%

6. An inverter

(a) performs the NOT operation   (b) changes a HIGH to a LOW

(c) changes a LOW to a HIGH   (d) does all of the above

7. The output of an AND gate is HIGH when

(a) any input is HIGH   (b) all inputs are HIGH

(c) no inputs are HIGH   (d) both answers (a) and (b)

8. The output of an OR gate is HIGH when

(a) any input is HIGH   (b) all inputs are HIGH

(c) no inputs are HIGH   (d) both answers (a) and (b)

9. The device used to convert a binary number to a 7-segment display format is

(a) multiplexer   (b) encoder   (c) decoder   (d) register

10. An example of a data storage device is

(a) the logic gate   (b) the flip-flop   (c) the comparator

(d) the register   (e) both answers (b) and (d)

11. An IC package containing four AND gates is an example of

(a) MSI   (b) SMT   (c) SOIC   (d) SSI

12. An LSI device has a circuit complexity of

(a) 12 to 99 equivalent gates   (b) 100 to 9999 equivalent gates

(c) 2000 to 5000 equivalent gates   (d) 10,000 to 99,999 equivalent gates

---

■ PROBLEMS

Answers to most odd-numbered problems are found at the end of the book.

SECTION 1–1   Digital and Analog Quantities

1. Name two advantages of digital data as compared to analog data.

2. Name an analog quantity other than temperature and sound.

SECTION 1–2   Binary Digits, Logic Levels, and Digital Waveforms

3. Define the sequence of bits (1s and 0s) represented by each of the following sequences of levels:

(a) HIGH, HIGH, LOW, HIGH, LOW, LOW, LOW, HIGH

(b) LOW, LOW, LOW, HIGH, LOW, HIGH, LOW, HIGH, LOW

4. List the sequence of levels (HIGH and LOW) that represent each of the following bit sequences:

(a) 1 0 1 1 1 0 1   (b) 1 1 1 0 1 0 0 1

5. For the pulse shown in Figure 1–42, graphically determine the following:

(a) rise time   (b) fall time   (c) pulse width   (d) amplitude

6. Determine the period of the digital waveform in Figure 1–43.

7. What is the frequency of the waveform in Figure 1–43?

8. Is the pulse waveform in Figure 1–43 periodic or nonperiodic?

9. Determine the duty cycle of the waveform in Figure 1–43.

10. Determine the bit sequence represented by the waveform in Figure 1–44. A bit time is 1 μs in this case.

11. What is the total serial transfer time for the eight bits in Figure 1–44? What is the total parallel transfer time?

FIGURE 1–42

FIGURE 1–43

FIGURE 1–44

SECTION 1–3   Basic Logic Operations

12. A logic circuit requires HIGHs on all its inputs to make the output HIGH. What type of logic circuit is it?

13. A basic 2-input logic circuit has a HIGH on one input and a LOW on the other input, and the output is LOW. Identify the circuit.

14. A basic 2-input logic circuit has a HIGH on one input and a LOW on the other input, and the output is HIGH. What type of logic circuit is it?

SECTION 1–4   Basic Logic Functions

15. Name the function of each block in Figure 1–45 based on your observation of the inputs and outputs.

16. A pulse waveform with a frequency of 10 kHz is applied to the input of a counter. During 100 ms, how many pulses are counted?

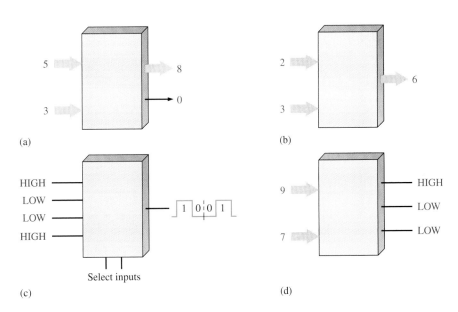

FIGURE 1–45

17. Consider a register that can store eight bits. Assume that it has been reset so that it contains zeros in all positions. If you transfer four alternating bits (0101) serially into the register, beginning with a 1 and shifting to the right, what will the total content of the register be as soon as the fourth bit is stored?

SECTION 1–5   Digital Integrated Circuits

18. An integrated circuit chip has a complexity of 200 equivalent gates. How is it classified?

19. Explain the main difference between the DIP and SMT packages.

20. Label the pin numbers on the packages in Figure 1–46. Top views are shown.

FIGURE 1–46

(a)   (b)

SECTION 1–6   Testing and Troubleshooting Instruments

21. A pulse is displayed on the screen of an oscilloscope, and you measure the base line as 1 V and the top of the pulse as 8 V. What is the amplitude?

22. A logic probe is applied to a contact point on an IC that is operating in a system. The lamp on the probe flashes repeatedly. What does this indicate?

SECTION 1–7   Digital System Application

23. Define the term *system*.

24. In the system depicted in Figure 1–40, why are the multiplexer and demultiplexer necessary?

25. What action can be taken to change the number of tablets per bottle in the system of Figure 1–40?

■ **ANSWERS TO SECTION REVIEWS**

SECTION 1–1

1. Analog means continuous.   **2.** Digital means discrete.

3. A digital quantity has a discrete set of values and an analog quantity has continuous values.

4. A PA system is analog. A CD player is analog and digital. A computer is all digital.

SECTION 1–2

1. Binary means having two states or values.   **2.** A bit is a binary digit.

3. The bits are 1 and 0.

4. Rise time: from 10% to 90% of amplitude. Fall time: from 90% to 10% of amplitude.

5. Frequency is the reciprocal of the period.

6. A clock waveform is a basic timing waveform from which other waveforms are derived.

7. A timing diagram shows time relationships of waveforms.

8. Parallel transfer is faster than serial transfer.

SECTION 1–3

1. When the input is LOW   **2.** When all inputs are HIGH

3. When any or all inputs are HIGH   **4.** An inverter is a NOT circuit.

5. A logic gate is a circuit that performs a logic operation (AND, OR).

SECTION 1–4

1. A comparator compares the magnitudes of two input numbers.
2. Add, subtract, multiply, and divide
3. Encoding is changing a familiar form such as decimal to a coded form such as binary.
4. Decoding is changing a code to a familiar form such as binary to decimal.
5. Multiplexing puts data from many sources onto one line. Demultiplexing takes data from one line and distributes it to many destinations.
6. Flip-flops, registers, semiconductor memories, magnetic disks
7. A counter counts events with a sequence of binary states.

SECTION 1–5

1. An IC is an electronic circuit with all components integrated on a single silicon chip.
2. DIP—dual-in-line package; SMT—surface-mount technology; SOIC—small-outline integrated circuit; SSI—small-scale integration; MSI—medium-scale integration; LSI—large-scale integration; VLSI—very large-scale integration; ULSI—ultra large-scale integration
3. **(a)** MSI   **(b)** LSI   **(c)** SSI   **(d)** VLSI   **(e)** ULSI

SECTION 1–6

1. Ten   2. Eight   3. Logic analyzer
4. A repetitive pulse waveform is indicated.

SECTION 1–7

1. The comparator determines when the tablet count reaches the preset number of tablets per bottle.
2. The dispenser valve is closed, the next bottle is moved into place by the conveyor, and the new sum is stored in register B.
3. Register A stores the preset number of tablets per bottle. Register B stores the total number of tablets bottled.

■ **ANSWERS TO RELATED PROBLEMS FOR EXAMPLES**

1–1 $f = 6.67$ kHz; Duty cycle $= 16.7\%$
1–2   Parallel transfer: 100 ns; Serial transfer: 1.6 μs
1–3   Amplitude $= 12$ V; $T = 8$ ms

■ **ANSWERS TO SELF-TEST**

1. (b)   2. (d)   3. (a)   4. (c)   5. (c)   6. (d)   7. (b)   8. (d)
9. (c)   10. (e)   11. (d)   12. (d)

2

NUMBER SYSTEMS, OPERATIONS, AND CODES

 Computer Simulation of Digital Circuits: Electronics Workbench (EWB) Tutorial at: http://www.prenhall.com/floyd

■ CHAPTER OBJECTIVES

❏ Review the decimal number system
❏ Count in the binary number system
❏ Convert from decimal to binary and from binary to decimal
❏ Apply arithmetic operations to binary numbers
❏ Determine the 1's and 2's complements of a binary number
❏ Express signed binary numbers in sign-magnitude, 1's complement, 2's complement, and floating-point format
❏ Carry out arithmetic operations with signed binary numbers
❏ Convert between the binary and hexadecimal number systems
❏ Add numbers in hexadecimal form
❏ Convert between the binary and octal number systems
❏ Express decimal numbers in binary coded decimal (BCD) form
❏ Add BCD numbers
❏ Convert between the binary system and the Gray code
❏ Interpret the American Standard Code for Information Interchange (ASCII)
❏ Use binary numbers and codes in a system application

■ CHAPTER OVERVIEW

The binary number system and digital codes are fundamental to computers and to digital electronics in general. In this chapter, the binary number system and its relationship to other number systems such as decimal, hexadecimal, and octal is the principal focus. Arithmetic operations with binary numbers are covered to provide a basis for understanding how computers and many other types of digital systems work. Also, digital codes such as binary coded decimal (BCD), the Gray code, and the ASCII are covered. The parity method for detecting errors in codes is introduced. The tutorials on the use of the calculator in certain operations are based on the TI-85 and TI-86 graphics calculators. The procedures shown may vary on other types.

■ DIGITAL SYSTEM APPLICATION

This Digital System Application helps to illustrate the concepts taught in this chapter. To demonstrate how binary numbers and codes are used in an application, the tablet-counting and control system that was introduced in Chapter 1 is used as an example. In the digital system application (Section 2–12) you will examine the binary numbers and codes at various points in the system and see how the system sequences through a counting cycle. You will analyze the system to determine the binary states for certain specified conditions.

2–1 ■ DECIMAL NUMBERS

You are familiar with the decimal number system because you use decimal numbers every day. Although decimal numbers are commonplace, their weighted structure is often not understood. In this section, the structure of decimal numbers is reviewed. This review will help you more easily understand the structure of the binary number system, which is important in computers and digital electronics.

After completing this section, you should be able to

■ Explain why the decimal number system is a weighted system   ■ Define *powers of ten*   ■ Determine the weight of each digit in a decimal number

In the **decimal** number system each of the ten digits, 0 through 9, represents a certain quantity. As you know, the ten symbols (**digits**) do not limit you to expressing only ten different quantities because you use the various digits in appropriate positions within a number to indicate the magnitude of the quantity. You can express quantities up through nine before running out of digits; if you wish to express a quantity greater than nine, you use two or more digits, and the position of each digit within the number tells you the magnitude it represents. If, for example, you wish to express the quantity twenty-three, you use (by their respective positions in the number) the digit 2 to represent the quantity twenty and the digit 3 to represent the quantity three, as illustrated below.

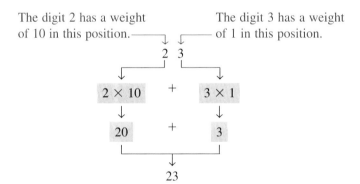

The position of each digit in a decimal number indicates the magnitude of the quantity represented and can be assigned a **weight.** The weights for whole numbers are positive powers of ten that increase from right to left, beginning with $10^0 = 1$.

$$\ldots 10^5 \ 10^4 \ 10^3 \ 10^2 \ 10^1 \ 10^0$$

For fractional numbers, the weights are negative powers of ten that decrease from left to right beginning with 10^{-1}.

$$10^2 \ 10^1 \ 10^0.10^{-1} \ 10^{-2} \ 10^{-3} \ldots$$
$$\llcorner \text{Decimal point}$$

The value of a decimal number is the sum of the digits after each digit has been multiplied by its weight, as Examples 2–1 and 2–2 illustrate.

EXAMPLE 2–1

Express the decimal number 47 as a sum of the values of each digit.

Solution   The digit 4 has a weight of 10, which is 10^1, as indicated by its position. The digit 7 has a weight of 1, which is 10^0, as indicated by its position.

$$47 = (4 \times 10^1) + (7 \times 10^0)$$
$$= (4 \times 10) + (7 \times 1) = \mathbf{40 + 7}$$

Related Problem   Determine the value of each digit in 939.

EXAMPLE 2–2

Express the decimal number 568.23 as a sum of the values of each digit.

Solution   The whole number digit 5 has a weight of 100, which is 10^2, the digit 6 has a weight of 10, which is 10^1, the digit 8 has a weight of 1, which is 10^0, the fractional digit 2 has a weight of 0.1, which is 10^{-1}, and the fractional digit 3 has a weight of 0.01, which is 10^{-2}.

$$568.23 = (5 \times 10^2) + (6 \times 10^1) + (8 \times 10^0) + (2 \times 10^{-1}) + (3 \times 10^{-2})$$
$$= (5 \times 100) + (6 \times 10)\ + (8 \times 1)\ \ + (2 \times 0.1)\ \ + (3 \times 0.01)$$
$$=\ \ \ \ \mathbf{500}\ \ \ \ +\ \ \mathbf{60}\ \ \ \ +\ \ \ \mathbf{8}\ \ \ \ +\ \ \ \mathbf{0.2}\ \ \ \ +\ \ \ \mathbf{0.03}$$

Related Problem   Determine the value of each digit in 67.924.

Calculator Tutorial

Powers of ten can be found on the calculator* using the [10ˣ] key.

Example   Find the value of 10^3.

Step 1.   Press [2nd] then [10ˣ].

Step 2.   Press [3] then [ENTER].

Display

10 ^ 3
1000 ⟵——Result

SECTION 2–1 REVIEW

1. What weight does the digit 7 have in each of the following numbers?
   (a) 1370   (b) 6725   (c) 7051   (d) 58.72
2. Express each of the following decimal numbers as a sum of the products obtained by multiplying each digit by its appropriate weight:
   (a) 51   (b) 137   (c) 1492   (d) 106.58

2–2 ■ BINARY NUMBERS

The binary number system is simply another way to represent quantities. The binary system is less complicated than the decimal system because it has only two digits. It may seem more difficult at first because it is unfamiliar to you. The decimal system with its ten digits is a base-ten system; the binary system with its two digits is a base-two

*The TI-85 is used in these tutorials.

system. The two binary digits (bits) are 1 and 0. The position of a 1 or 0 in a binary number indicates its weight, or value within the number, just as the position of a decimal digit determines the value of that digit. The weights in a binary number are based on powers of two.

After completing this section, you should be able to

▪ Count in binary ▪ Determine the largest decimal number that can be represented by a given number of bits ▪ Convert a binary number to a decimal number

Counting in Binary

To learn to count in the **binary** system, first look at how you count in the decimal system. You start at zero and count up to nine before you run out of digits. You then start another digit position (to the left) and continue counting 10 through 99. At this point you have exhausted all two-digit combinations, so a third digit position is needed to count from 100 through 999.

A comparable situation occurs when you count in binary, except that you have only two digits, called **bits.** Begin counting: 0, 1. At this point you have used both digits, so include another digit position and continue: 10, 11. You have now exhausted all combinations of two digits, so a third position is required. With three digit positions you can continue to count: 100, 101, 110, and 111. Now you need a fourth digit position to continue, and so on. A binary count of zero through fifteen is shown in Table 2–1. Notice the patterns with which the 1s and 0s alternate in each column.

TABLE 2–1

Decimal Number	Binary Number			
0	0	0	0	0
1	0	0	0	1
2	0	0	1	0
3	0	0	1	1
4	0	1	0	0
5	0	1	0	1
6	0	1	1	0
7	0	1	1	1
8	1	0	0	0
9	1	0	0	1
10	1	0	1	0
11	1	0	1	1
12	1	1	0	0
13	1	1	0	1
14	1	1	1	0
15	1	1	1	1

As you have seen in Table 2–1, four bits are required to count from zero to 15. In general, with n bits you can count up to a number equal to $2^n - 1$.

$$\text{Largest decimal number} = 2^n - 1$$

For example, with five bits ($n = 5$) you can count from zero to thirty-one.

$$2^5 - 1 = 32 - 1 = 31$$

With six bits ($n = 6$) you can count from zero to sixty-three.

$$2^6 - 1 = 64 - 1 = 63$$

A table of powers of two is given in Appendix B.

Calculator Tutorial

Powers of two can be found on the calculator using the [∧] key.

Example   Find the value of 2^5.

Step 1.   Press [2] then [∧].

Step 2.   Press [5] then [ENTER].

Display

2 ∧ 5

32

Application Example

Learning to count in binary will help you to basically understand how digital circuits can be used to count events. This can be anything from counting items on an assembly line to counting operations in a computer. Let's take a simple example of counting tennis balls going into a box from a conveyor belt. Assume that nine balls are to go into each box.

The counter in Figure 2–1 counts the pulses from a sensor that detects the passing of a ball and produces a sequence of logic levels (digital waveforms) on each of its four parallel outputs. Each set of logic levels represents a 4-bit binary number (HIGH = 1 and LOW = 0), as indicated. As the decoder receives these waveforms, it decodes each set of four bits and converts it to the corresponding decimal number in the 7-segment display. When the counter gets to the binary state of 1001, it has counted nine tennis balls, the display shows decimal 9, a new box is moved under the conveyor, the counter goes back to its zero state (0000), and the process starts over. (The number 9 was used only in the interest of single-digit simplicity.)

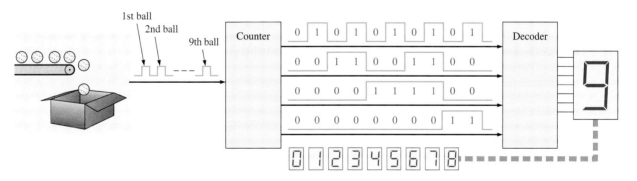

FIGURE 2–1
Simple illustration of a binary counting application.

The Weighting Structure of Binary Numbers

A binary number is a weighted number. The right-most bit is the **least significant bit (LSB)** in a binary whole number and has a weight of $2^0 = 1$. The weights increase from right to left by a power of two for each bit. The left-most bit is the **most significant bit (MSB);** its weight depends on the size of the binary number.

Fractional numbers can also be represented in binary by placing bits to the right of the binary point, just as fractional decimal digits are placed to the right of the decimal point. The

COMPUTER NOTE

Computers use binary numbers to select memory locations. Each location is assigned a unique number called an *address*. The Pentium microprocessor, for example, has 32 address lines which can select 2^{32} (4,294,967,296) unique locations.

left-most bit is the MSB in a binary fractional number and has a weight of $2^{-1} = 0.5$. The fractional weights decrease from left to right by a negative power of two for each bit.

The weight structure of a binary number is

$$2^{n-1} \ldots 2^3 \, 2^2 \, 2^1 \, 2^0 . \, 2^{-1} \, 2^{-2} \ldots 2^{-n}$$
$$\llcorner \text{ Binary point}$$

where n is the number of bits from the binary point. Thus, all the bits to the left of the binary point have weights that are positive powers of two, as previously discussed for whole numbers. All bits to the right of the binary point have weights that are negative powers of two, or fractional weights.

The powers of two and their equivalent decimal weights for an 8-bit binary whole number and a 6-bit binary fractional number are shown in Table 2–2. Notice that the weight doubles for each positive power of two and that the weight is halved for each negative power of two. You can easily extend the table by doubling the weight of the most significant positive power of two and halving the weight of the least significant negative power of two; for example, $2^9 = 512$ and $2^{-7} = 0.0078125$.

TABLE 2–2
Binary weights.

Positive Powers of Two (whole numbers)									Negative Powers of Two (fractional number)					
2^8	2^7	2^6	2^5	2^4	2^3	2^2	2^1	2^0	2^{-1}	2^{-2}	2^{-3}	2^{-4}	2^{-5}	2^{-6}
256	128	64	32	16	8	4	2	1	1/2	1/4	1/8	1/16	1/32	1/64
									0.5	0.25	0.125	0.0625	0.03125	0.015625

Binary-to-Decimal Conversion

The decimal value of any binary number can be found by adding the weights of all bits that are 1 and discarding the weights of all bits that are 0. The following two examples will illustrate this.

EXAMPLE 2–3

Convert the binary whole number 1101101 to decimal.

Solution Determine the weight of each bit that is a 1, and then find the sum of the weights to get the decimal number.

$$\text{Weight: } 2^6 \, 2^5 \quad 2^4 \, 2^3 \, 2^2 \, 2^1 \, 2^0$$
$$\text{Binary number: } 1 \quad 1 \quad 0 \quad 1 \quad 1 \quad 0 \quad 1$$
$$1101101 = 2^6 + 2^5 + 2^3 + 2^2 + 2^0$$
$$= 64 + 32 + 8 + 4 + 1 = \mathbf{109}$$

Related Problem Convert the binary number 10010001 to decimal.

EXAMPLE 2–4

Convert the fractional binary number 0.1011 to decimal.

Solution Determine the weight of each bit that is a 1, and then sum the weights to get the decimal fraction.

$$\text{Weight: } \quad\quad 2^{-1} \, 2^{-2} \, 2^{-3} \, 2^{-4}$$
$$\text{Binary number: } 0 \, . \, 1 \quad 0 \quad 1 \quad 1$$

$$0.1011 = 2^{-1} + 2^{-3} + 2^{-4}$$
$$= 0.5 + 0.125 + 0.0625 = \mathbf{0.6875}$$

Related Problem Convert the binary number 10.111 to decimal.

1. What is the largest decimal number that can be represented in binary with eight bits?
2. Determine the weight of the 1 in the binary number 10000.
3. Convert the binary number 10111101.011 to decimal.

2–3 ■ DECIMAL-TO-BINARY CONVERSION

In Section 2–2 you learned how to convert a binary number to the equivalent decimal number. Now you will learn two ways of converting from a decimal number to a binary number.

After completing this section, you should be able to

■ Convert a decimal number to binary using the sum-of-weights method ■ Convert a decimal whole number to binary using the repeated division-by-2 method ■ Convert a decimal fraction to binary using the repeated multiplication-by-2 method

Sum-of-Weights Method

One way to find the binary number that is equivalent to a given decimal number is to determine the set of binary weights whose sum is equal to the decimal number. An easy way to remember binary weights is that the lowest weight is 1, which is 2^0, and that by doubling any weight, you get the next higher weight; thus, a list of seven binary weights would be 64, 32, 16, 8, 4, 2, 1 as you learned in the last section. The decimal number 9, for example, can be expressed as the sum of binary weights as follows:

$$9 = 8 + 1 \quad \text{or} \quad 9 = 2^3 + 2^0$$

Placing 1s in the appropriate weight positions, 2^3 and 2^0, and 0s in the 2^2 and 2^1 positions determines the binary number for decimal 9.

$$2^3 \quad 2^2 \quad 2^1 \quad 2^0$$
$$1 \quad \ 0 \quad \ 0 \quad \ 1 \qquad \text{Binary number for nine}$$

EXAMPLE 2–5

Convert the following decimal numbers to binary:
(a) 12   **(b)** 25   **(c)** 58   **(d)** 82

Solution
(a) $12 = 8 + 4 = 2^3 + 2^2$ ⟶ **1100**
(b) $25 = 16 + 8 + 1 = 2^4 + 2^3 + 2^0$ ⟶ **11001**
(c) $58 = 32 + 16 + 8 + 2 = 2^5 + 2^4 + 2^3 + 2^1$ ⟶ **111010**
(d) $82 = 64 + 16 + 2 = 2^6 + 2^4 + 2^1$ ⟶ **1010010**

Related Problem Convert the decimal number 125 to binary.

Repeated Division-by-2 Method

A systematic method of converting whole numbers from decimal to binary is the *repeated division-by-2* process. For example, to convert the decimal number 12 to binary, begin by dividing 12 by 2. Then divide each resulting quotient by 2 until there is a 0 whole-number quotient. The **remainders** generated by each division form the binary number. The first remainder to be produced is the least significant bit (LSB) in the binary number, and the last remainder to be produced is the most significant bit (MSB). This procedure is shown in the following steps for converting the decimal number 12 to binary.

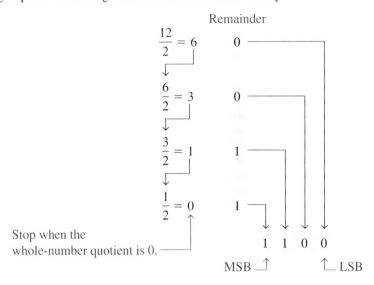

EXAMPLE 2–6   Convert the following decimal numbers to binary: **(a)** 19   **(b)** 45

Solution

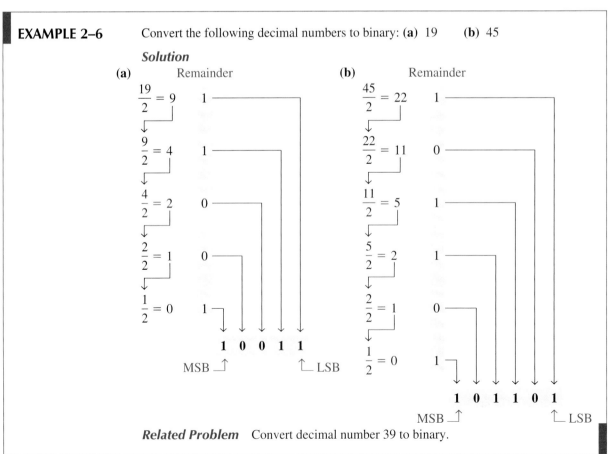

Related Problem   Convert decimal number 39 to binary.

Calculator Tutorial	A decimal number can be converted to a binary number by first using the base conversion key sequence 2nd BASE F3.

Example Convert decimal 57 to binary.

Step 1. Press 5 7.

Step 2. Press F1.

Step 3. Press ENTER.

Converting Decimal Fractions to Binary

Examples 2–5 and 2–6 demonstrated whole-number conversions. Now let's look at fractional conversions. An easy way to remember fractional binary weights is that the most significant weight is 0.5, which is 2^{-1}, and that by halving any weight, you get the next lower weight; thus a list of four fractional binary weights would be 0.5, 0.25, 0.125, 0.0625.

Sum-of-Weights The sum-of-weights method can be applied to fractional decimal numbers, as shown in the following example:

$$0.625 = 0.5 + 0.125 = 2^{-1} + 2^{-3} = 0.101$$

There is a 1 in the 2^{-1} position, a 0 in the 2^{-2} position, and a 1 in the 2^{-3} position.

Repeated Multiplication by 2 As you have seen, decimal whole numbers can be converted to binary by repeated division by 2. Decimal fractions can be converted to binary by repeated multiplication by 2. For example, to convert the decimal fraction 0.3125 to binary, begin by multiplying 0.3125 by 2 and then multiplying each resulting fractional part of the product by 2 until the fractional product is zero or until the desired number of decimal places is reached. The carried digits, or **carries,** generated by the multiplications produce the binary number. The first carry produced is the MSB, and the last carry is the LSB. This procedure is illustrated as follows:

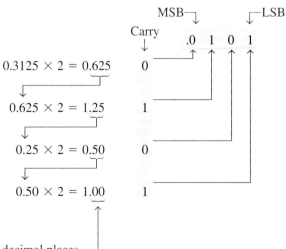

Continue to the desired number of decimal places —| or stop when the fractional part is all zeros.

1. Convert each decimal number to binary by using the sum-of-weights method:
   (a) 23   (b) 57   (c) 45.5
2. Convert each decimal number to binary by using the repeated division-by-2 method (repeated multiplication-by-2 for fractions):
   (a) 14   (b) 21   (c) 0.375

2–4 ■ BINARY ARITHMETIC

Binary arithmetic is essential in all digital computers and in many other types of digital systems. To understand digital systems, you must know the basics of binary addition, subtraction, multiplication, and division. This section provides an introduction that will be expanded in later sections.

After completing this section, you should be able to

■ Add binary numbers   ■ Subtract binary numbers   ■ Multiply binary numbers
■ Divide binary numbers

Binary Addition

The four basic rules for adding binary digits (bits) are as follows:

$$
\begin{aligned}
0 + 0 &= 0 \quad &\text{Sum of 0 with a carry of 0}\\
0 + 1 &= 1 \quad &\text{Sum of 1 with a carry of 0}\\
1 + 0 &= 1 \quad &\text{Sum of 1 with a carry of 0}\\
1 + 1 &= 10 \quad &\text{Sum of 0 with a carry of 1}
\end{aligned}
$$

Notice that the first three rules result in a single bit and in the fourth rule the addition of two 1s yields a binary two (10). When binary numbers are added, the last condition creates a sum of 0 in a given column and a carry of 1 over to the next column to the left, as illustrated in the following addition of 11 + 1:

In the right column, $1 + 1 = 0$ with a carry of 1 to the next column to the left. In the middle column, $1 + 1 + 0 = 0$ with a carry of 1 to the next column to the left. In the left column, $1 + 0 + 0 = 1$.

When there is a carry of 1, you have a situation in which three bits are being added (a bit in each of the two numbers and a carry bit). This situation is illustrated as follows:

Carry bits

$$
\begin{aligned}
1 + 0 + 0 &= 01 \quad &\text{Sum of 1 with a carry of 0}\\
1 + 1 + 0 &= 10 \quad &\text{Sum of 0 with a carry of 1}\\
1 + 0 + 1 &= 10 \quad &\text{Sum of 0 with a carry of 1}\\
1 + 1 + 1 &= 11 \quad &\text{Sum of 1 with a carry of 1}
\end{aligned}
$$

Example 2–7 illustrates binary addition.

EXAMPLE 2–7

Add the following binary numbers:
(a) 11 + 11   (b) 100 + 10   (c) 111 + 11   (d) 110 + 100

Solution   The equivalent decimal addition is also shown for reference.

(a)
$$\begin{array}{r} 11 \\ +11 \\ \hline \mathbf{110} \end{array} \qquad \begin{array}{r} 3 \\ +3 \\ \hline 6 \end{array}$$

(b)
$$\begin{array}{r} 100 \\ +10 \\ \hline \mathbf{110} \end{array} \qquad \begin{array}{r} 4 \\ +2 \\ \hline 6 \end{array}$$

(c)
$$\begin{array}{r} 111 \\ +11 \\ \hline \mathbf{1010} \end{array} \qquad \begin{array}{r} 7 \\ +3 \\ \hline 10 \end{array}$$

(d)
$$\begin{array}{r} 110 \\ +100 \\ \hline \mathbf{1010} \end{array} \qquad \begin{array}{r} 6 \\ +4 \\ \hline 10 \end{array}$$

Related Problem   Add 1111 and 1100.

Binary Subtraction

The four basic rules for subtracting bits are as follows:

$$0 - 0 = 0$$
$$1 - 1 = 0$$
$$1 - 0 = 1$$
$$10 - 1 = 1 \qquad 0 - 1 \text{ with a borrow of } 1$$

When subtracting numbers, you sometimes have to borrow from the next column to the left. A borrow is required in binary only when you try to subtract a 1 from a 0. In this case, when a 1 is borrowed from the next column to the left, a 10 is created in the column being subtracted, and the last of the four basic rules just listed must be applied. Examples 2–8 and 2–9 illustrate binary subtraction, with the equivalent decimal subtraction also shown.

EXAMPLE 2–8

Perform the following binary subtractions:
(a) 11 − 01   (b) 11 − 10

Solution

(a)
$$\begin{array}{r} 11 \\ -01 \\ \hline \mathbf{10} \end{array} \qquad \begin{array}{r} 3 \\ -1 \\ \hline 2 \end{array}$$

(b)
$$\begin{array}{r} 11 \\ -10 \\ \hline \mathbf{01} \end{array} \qquad \begin{array}{r} 3 \\ -2 \\ \hline 1 \end{array}$$

No borrows were required in this example. The binary number 01 is the same as 1.

Related Problem   Subtract 100 from 111.

EXAMPLE 2–9

Subtract 011 from 101.

Solution

$$\begin{array}{r} 101 \\ -011 \\ \hline \mathbf{010} \end{array} \qquad \begin{array}{r} 5 \\ -3 \\ \hline 2 \end{array}$$

Let's examine exactly what was done to subtract the two binary numbers since a borrow is required. Begin with the right column.

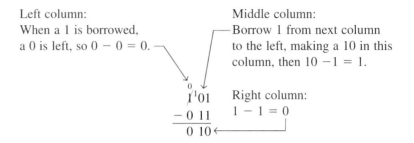

Left column:
When a 1 is borrowed,
a 0 is left, so $0 - 0 = 0$.

Middle column:
Borrow 1 from next column to the left, making a 10 in this column, then $10 - 1 = 1$.

$$\begin{array}{r} \overset{0}{1}{}^{1}01 \\ -\ 0\ 11 \\ \hline 0\ 10 \end{array}$$

Right column:
$1 - 1 = 0$

Related Problem Subtract 101 from 110.

Binary Multiplication

The four basic rules for multiplying bits are as follows:

$$0 \times 0 = 0$$
$$0 \times 1 = 0$$
$$1 \times 0 = 0$$
$$1 \times 1 = 1$$

Multiplication is performed with binary numbers in the same manner as with decimal numbers. It involves forming partial products, shifting each successive partial product left one place, and then adding all the partial products. Example 2–10 will illustrate the procedure, with the equivalent decimal multiplication shown for reference.

EXAMPLE 2–10

Perform the following binary multiplications:
(a) 11×11 **(b)** 101×111

Solution

(a)

$$\begin{array}{r} 11 \\ \times\ 11 \\ \hline 11 \\ +\ 11 \\ \hline 1001 \end{array} \qquad \begin{array}{r} 3 \\ \times\ 3 \\ \hline 9 \end{array}$$

Partial products

(b)

$$\begin{array}{r} 111 \\ \times\ 101 \\ \hline 111 \\ 000 \\ +\ 111 \\ \hline 100011 \end{array} \qquad \begin{array}{r} 7 \\ \times\ 5 \\ \hline 35 \end{array}$$

Partial products

Related Problem Multiply 1101×1010.

Binary Division

Division in binary follows the same procedure as division in decimal, as Example 2–11 illustrates. The equivalent decimal divisions are also given.

EXAMPLE 2–11   Perform the following binary divisions:

(a) $110 \div 11$ (b) $110 \div 10$

Solution

(a) $\begin{array}{r} 10 \\ 11\overline{)110} \\ \underline{11} \\ 000 \end{array}$ $\begin{array}{r} 2 \\ 3\overline{)6} \\ \underline{6} \\ 0 \end{array}$ (b) $\begin{array}{r} 11 \\ 10\overline{)110} \\ \underline{10} \\ 10 \\ \underline{10} \\ 00 \end{array}$ $\begin{array}{r} 3 \\ 2\overline{)6} \\ \underline{6} \\ 0 \end{array}$

Related Problem   Divide 1100 by 100.

SECTION 2–4 REVIEW

1. Perform the following binary additions:
   (a) $1101 + 1010$ (b) $10111 + 01101$
2. Perform the following binary subtractions:
   (a) $1101 - 0100$ (b) $1001 - 0111$
3. Perform the indicated binary operations:
   (a) 110×111 (b) $1100 \div 011$

2–5 ■ 1'S AND 2'S COMPLEMENTS OF BINARY NUMBERS

The 1's complement and the 2's complement of a binary number are important because they permit the representation of negative numbers. The method of 2's complement arithmetic is commonly used in computers to handle negative numbers.

After completing this section, you should be able to

■ Convert a binary number to its 1's complement   ■ Convert a binary number to its 2's complement using either of two methods

Finding the 1's Complement of a Binary Number

The 1's **complement** of a binary number is found by changing all 1s to 0s and all 0s to 1s, as illustrated below:

$$\begin{array}{cccccccc} 1 & 0 & 1 & 1 & 0 & 0 & 1 & 0 \\ \downarrow & \downarrow & \downarrow & \downarrow & \downarrow & \downarrow & \downarrow & \downarrow \\ 0 & 1 & 0 & 0 & 1 & 1 & 0 & 1 \end{array} \quad \begin{array}{l} \text{Binary number} \\ \\ \text{1's complement} \end{array}$$

Application Example

The simplest way to obtain the 1's complement of a binary number with a digital circuit is to use parallel inverters (NOT circuits), as shown in Figure 2–2 for an 8-bit binary number.

FIGURE 2–2

Example of inverters used to obtain the 1's complement of a binary number.

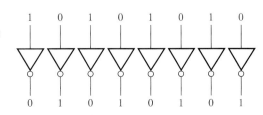

Finding the 2's Complement of a Binary Number

The 2's complement of a binary number is found by adding 1 to the LSB of the 1's complement.

$$2\text{'s complement} = (1\text{'s complement}) + 1$$

Example 2–12 shows how to find the 2's complement.

EXAMPLE 2–12

Find the 2's complement of 10110010.

Solution

10110010	Binary number
01001101	1's complement
+ 1	Add 1
01001110	2's complement

Related Problem   Determine the 2's complement of 11001011.

An alternative method of finding the 2's complement of a binary number is as follows:

1. Start at the right with the LSB and write the bits as they are up to and including the first 1.
2. Take the 1's complements of the remaining bits.

Example 2–13 illustrates these steps.

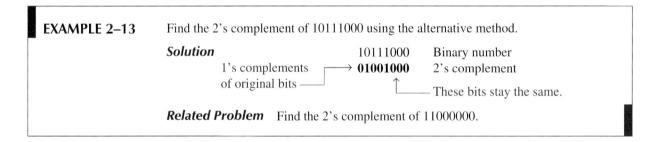

EXAMPLE 2–13

Find the 2's complement of 10111000 using the alternative method.

Solution

                              10111000   Binary number

1's complements ⟶ **01001000**   2's complement
of original bits ─────┘   ↑
                                   └──── These bits stay the same.

Related Problem   Find the 2's complement of 11000000.

Application Example

The 2's complement of a negative binary number can be realized using inverters and an adder, as indicated in Figure 2–3. This illustrates how an 8-bit number can be converted to its 2's complement by first inverting each bit (taking the 1's complement) and then adding 1 to the 1's complement with an adder circuit.

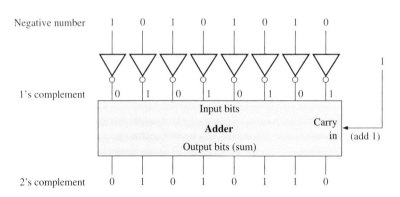

FIGURE 2–3
Example of obtaining the 2's complement of a negative binary number.

To convert from a 1's or 2's complement back to the true (uncomplemented) binary form, use the same two procedures described previously. To go from the 1's complement back to true binary, reverse all the bits. To go from the 2's complement form back to true binary, take the 1's complement of the 2's complement number and add 1 to the least significant bit.

SECTION 2–5 REVIEW

1. Determine the 1's complement of each binary number:
   (a) 00011010   (b) 11110111   (c) 10001101
2. Determine the 2's complement of each binary number:
   (a) 00010110   (b) 11111100   (c) 10010001

2–6 ■ SIGNED NUMBERS

Digital systems, such as the computer, must be able to handle both positive and negative numbers. A signed binary number consists of both sign and magnitude information. The sign indicates whether a number is positive or negative and the magnitude is the value of the number. There are three ways in which signed integer (whole) numbers can be represented in binary form: sign-magnitude, 1's complement, and 2's complement. Noninteger and very large or small numbers can be expressed in floating-point format.

After completing this section, you should be able to

■ Express positive and negative numbers in sign-magnitude   ■ Express positive and negative numbers in 1's complement   ■ Express positive and negative numbers in 2's complement   ■ Determine the decimal value of signed binary numbers   ■ Express a binary number in floating-point format

The Sign Bit

The left-most bit in a signed binary number is the **sign bit**, which tells you whether the number is positive or negative.

A 0 is for positive, and a 1 is for negative.

Sign-Magnitude System

When a signed binary number is represented in sign-magnitude, the left-most bit is the sign bit and the remaining bits are the magnitude bits. The magnitude bits are in true (uncomplemented) binary for both positive and negative numbers. For example, the decimal number $+25$ is expressed as an 8-bit signed binary number using the sign-magnitude system as

$$00011001$$

Sign bit ↑   ↑ Magnitude bits

The decimal number -25 is expressed as

$$10011001$$

Notice that the only difference between $+25$ and -25 is the sign bit because the magnitude bits are in true binary for both positive and negative numbers.

> **In the sign-magnitude system, a negative number has the same magnitude bits as the corresponding positive number but the sign bit is a 1 rather than a zero.**

COMPUTER NOTE
Computers use the 2's complement for negative integer numbers in all arithmetic operations. The reason is that subtraction of a number is the same as adding the 2's complement of the number. Computers form the 2's complement by inverting the bits and adding one using special instructions that produce the same result as the adder in the last application example.

1's Complement System

Positive numbers in the 1's complement system are represented the same way as the positive sign-magnitude numbers. Negative numbers, however, are the 1's complements of the corresponding positive numbers. For example, using eight bits, the decimal number -25 is expressed as the 1's complement of $+25$ (00011001) as

$$11100110$$

In the 1's complement system, a negative number is the 1's complement of the corresponding positive number.

2's Complement System

Positive numbers in the 2's complement system are represented the same way as in the sign-magnitude and 1's complement systems. Negative numbers are the 2's complements of the corresponding positive numbers. Again, using eight bits, let's take decimal number -25 and express it as the 2's complement of $+25$ (00011001).

$$11100111$$

In the 2's complement system, a negative number is the 2's complement of the corresponding positive number.

EXAMPLE 2–14

Express the decimal number -39 as an 8-bit number in the sign-magnitude, 1's complement, and 2's complement systems.

Solution   First, write the 8-bit number for $+39$.

$$00100111$$

In the *sign-magnitude system*, -39 is produced by changing the sign bit to a 1 and leaving the magnitude bits as they are. The number is

10100111

In the *1's complement system*, -39 is produced by taking the 1's complement of $+39$ (00100111).

11011000

In the *2's complement system*, -39 is produced by taking the 2's complement of $+39$ (00100111) as follows:

$$
\begin{array}{ll}
11011000 & \text{1's complement} \\
\underline{+\qquad 1} & \\
\mathbf{11011001} & \text{2's complement}
\end{array}
$$

Related Problem   Express $+19$ and -19 in sign-magnitude, 1's complement, and 2's complement.

The Decimal Value of Signed Numbers

Sign-magnitude   Decimal values of positive and negative numbers in the sign-magnitude system are determined by summing the weights in all the magnitude bit positions where there are 1s and ignoring those positions where there are zeros. The sign is determined by examination of the sign bit. This is illustrated by Example 2–15.

EXAMPLE 2–15

Determine the decimal value of this signed binary number expressed in sign-magnitude: 10010101.

Solution   The seven magnitude bits and their powers-of-two weights are as follows:

$$2^6 \ 2^5 \ 2^4 \ 2^3 \ 2^2 \ 2^1 \ 2^0$$
$$0 \ \ 0 \ \ 1 \ \ 0 \ \ 1 \ \ 0 \ \ 1$$

Summing the weights where there are 1s,

$$16 + 4 + 1 = 21$$

The sign bit is 1; therefore, the decimal number is **−21.**

Related Problem   Determine the decimal value of the sign-magnitude number 01110111.

1's Complement   Decimal values of positive numbers in the 1's complement system are determined by summing the weights in all bit positions where there are 1s and ignoring those positions where there are zeros. Decimal values of negative numbers are determined by assigning a negative value to the weight of the sign bit, summing all the weights where there are 1s, and adding 1 to the result. This is illustrated by Example 2–16.

EXAMPLE 2–16

Determine the decimal values of the signed binary numbers expressed in 1's complement: **(a)** 00010111   **(b)** 11101000.

Solution
(a) The bits and their powers-of-two weights for the positive number are as follows:

$$-2^7 \ 2^6 \ 2^5 \ 2^4 \ 2^3 \ 2^2 \ 2^1 \ 2^0$$
$$0 \ \ \ 0 \ \ 0 \ \ 1 \ \ 0 \ \ 1 \ \ 1 \ \ 1$$

Summing the weights where there are 1s,

$$16 + 4 + 2 + 1 = \textbf{+23}$$

(b) The bits and their powers-of-two weights for the negative number are as follows: Notice that the negative sign bit has a weight of -2^7 or -128.

$$-2^7 \ 2^6 \ 2^5 \ 2^4 \ 2^3 \ 2^2 \ 2^1 \ 2^0$$
$$1 \ \ \ 1 \ \ 1 \ \ 0 \ \ 1 \ \ 0 \ \ 0 \ \ 0$$

Summing the weights where there are 1s,

$$-128 + 64 + 32 + 8 = -24$$

Adding 1 to the result, the final decimal number is

$$-24 + 1 = \textbf{−23}$$

Related Problem   Determine the decimal value of the 1's complement number 11101011.

2's Complement   Decimal values of positive and negative numbers in the 2's complement system are determined by summing the weights in all bit positions where there are 1s and ignoring those positions where there are zeros. The weight of the sign bit in a negative number is given a negative value. This is illustrated by Example 2–17.

EXAMPLE 2–17   Determine the decimal values of the signed binary numbers expressed in 2's complement: **(a)** 01010110   **(b)** 10101010.

Solution

(a) The bits and their powers-of-two weights for the positive number are as follows:

$$-2^7 \quad 2^6 \quad 2^5 \quad 2^4 \quad 2^3 \quad 2^2 \quad 2^1 \quad 2^0$$
$$0 \quad\ 1 \quad\ 0 \quad\ 1 \quad\ 0 \quad\ 1 \quad\ 1 \quad\ 0$$

Summing the weights where there are 1's,

$$64 + 16 + 4 + 2 = \mathbf{+86}$$

(b) The bits and their powers-of-two weights for the negative number are as follows. Notice that the negative sign bit has a weight of $-2^7 = -128$.

$$-2^7 \quad 2^6 \quad 2^5 \quad 2^4 \quad 2^3 \quad 2^2 \quad 2^1 \quad 2^0$$
$$1 \quad\ 0 \quad\ 1 \quad\ 0 \quad\ 1 \quad\ 0 \quad\ 1 \quad\ 0$$

Summing the weights where there are 1's,

$$-128 + 32 + 8 + 2 = \mathbf{-86}$$

Related Problem   Determine the decimal value of the 2's complement number 11010111.

From these examples, you can see why the 2's complement system is preferred for representing signed integer numbers: To convert to decimal, it simply requires a summation of weights regardless of whether the number is positive or negative. The sign-magnitude system requires two steps—sum the weights of the magnitude bits and examine the sign bit to determine if the number is positive or negative. The 1's complement system requires adding 1 to the summation of weights for negative numbers but not for positive numbers. Also, the 1's complement system is generally not used because two representations of zero (00000000 or 11111111) are possible.

Range of Signed Integer Numbers That Can Be Represented

We have used 8-bit numbers for illustration because the 8-bit grouping is common in most computers and has been given the special name **byte.** Using eight bits, 256 different numbers can be represented. Combining two bytes to get sixteen bits, 65,536 different numbers can be represented. Combining four bytes to get 32 bits, 4.295×10^9 different numbers can be represented, and so on. The formula for finding the number of different combinations of n bits is

$$\text{Total combinations} = 2^n$$

For 2's complement signed numbers, the range of values for n-bit numbers is

$$-(2^{n-1}) \quad \text{to} \quad +(2^{n-1} - 1)$$

where in each case there is one sign bit and $n - 1$ magnitude bits. For example, with four bits you can represent numbers in 2's complement ranging from $-(2^3) = -8$ to $2^3 - 1 = +7$. Similarly, with eight bits, you can go from -128 to $+127$, with sixteen bits you can go from $-32,768$ to $+32,767$, and so on.

Floating-Point Numbers

To represent very large **integer** (whole) numbers, many bits are required. There is also a problem when numbers with both integer and fractional parts, such as 23.5618, need to be represented. The floating-point number system, based on scientific notation, is capable of representing very large and very small numbers without an increase in the number of bits and also for representing numbers that have both integer and fractional components.

A **floating-point number** (also known as a *real number*) consists of two parts plus a sign. The **mantissa** is the part of a floating-point number that represents the magnitude of the number. The **exponent** is the part of a floating-point number that represents the number of places that the decimal point (or binary point) is to be moved.

A decimal example will be helpful in understanding the basic concept of floating-point numbers. Let's consider a decimal number which, in integer form, is 241,506,800. The mantissa is .2415068 and the exponent is 9. When the integer is expressed as a floating-point number, it is normalized by moving the decimal point to the left of all the digits so that the mantissa is a fractional number and the exponent is the power of ten. The floating-point number is written as

$$0.2415068 \times 10^9$$

For binary floating-point numbers, the format is defined by ANSI/IEEE Standard 754-1985 in three forms: *single-precision, double-precision,* and *extended-precision.* These all have the same basic formats except for the number of bits. Single-precision floating-point numbers have 32 bits, double-precision numbers have 64 bits, and extended-precision numbers have 80 bits. We will restrict our discussion to the single-precision floating-point format.

Single-Precision Floating-Point Binary Numbers In the standard format for a single-precision binary number, the sign bit (S) is the left-most bit, the exponent (E) includes the next eight bits, and the mantissa or fractional part (F) includes the remaining 23 bits, as shown next.

In the mantissa or fractional part, the binary point is understood to be to the left of the 23 bits. Effectively, there are 24 bits in the mantissa because in any binary number the left-most (most significant) bit is always a 1. Therefore, this 1 is understood to be there although it does not occupy an actual bit position.

The eight bits in the exponent represent a *biased exponent,* which is obtained by adding 127 to the actual exponent. The purpose of the bias is to allow very large or very small numbers without requiring a separate sign bit for the exponents. The biased exponent allows a range of actual exponent values from -126 to $+128$.

To illustrate how a binary number is expressed in floating-point format, let's use 1011010010001 as an example. First, it can be expressed as 1 plus a fractional binary number by moving the binary point 12 places to the left and then multiplying by the appropriate power of two.

$$1011010010001 = 1.011010010001 \times 2^{12}$$

Assuming that this is a positive number, the sign bit (S) is 0. The exponent, 12, is expressed as a biased exponent by adding it to 127 ($12 + 127 = 139$). The biased exponent (E) is expressed as the binary number 10001011. The mantissa is the fractional part (F) of the binary number, .011010010001. Because there is always a 1 to the left of the binary point in the power-of-two expression, it is not included in the mantissa. The complete floating-point number is

S	E	F
0	10001011	01101001000100000000000

Next, let's see how to evaluate a binary number that is already in floating-point format. The general approach to determining the value of a floating-point number is expressed by the following formula:

$$\text{Number} = (-1)^S(1 + F)(2^{E-127})$$

To illustrate, let's consider the following floating-point binary number:

1	10010001	10001110001000000000000

The sign bit is 1. The biased exponent is $10010001 = 145$. Applying the formula, we get

$$\text{Number} = (-1)^1(1.10001110001)(2^{145-127})$$
$$= (-1)(1.10001110001)(2^{18}) = -1100011100010000000$$

This floating-point binary number is equivalent to $-407,688$ in decimal. Since the exponent can be any number between -126 and $+128$, extremely large and small numbers can be expressed. A 32-bit floating-point number can replace a binary integer number having 129 bits. Because the exponent determines the position of the binary point, numbers containing both integer and fractional parts can be represented.

There are two exceptions to the format for floating-point numbers: The number 0.0 is represented by all 0s, and infinity is represented by all 1s in the exponent and all 0s in the mantissa.

EXAMPLE 2–18

Convert the decimal number 3.248×10^4 to a single-precision floating-point binary number.

Solution
Convert the decimal number to binary.

$$3.248 \times 10^4 = 32480 = 111111011100000_2 = 1.11111011100000 \times 2^{14}$$

The MSB will not occupy a bit position because it is always a 1. Therefore, the mantissa is the fractional 23-bit binary number 11111011100000000000000 and the biased exponent is

$$14 + 127 = 141 = 10001101_2$$

The complete floating-point number is

0	10001101	11111011100000000000000

Related Problem Determine the binary value of the following floating-point binary number:

0 10011000 10000100010100110000000

**SECTION 2–6
REVIEW**

1. Express the decimal number +9 as an 8-bit binary number in the sign-magnitude system.
2. Express the decimal number −33 as an 8-bit binary number in the 1's complement system.
3. Express the decimal number −46 as an 8-bit binary number in the 2's complement system.
4. List the two parts of a signed, floating-point number.

2–7 ■ ARITHMETIC OPERATIONS WITH SIGNED NUMBERS

In the last section, you learned how signed numbers are represented in three different systems. In this section, you will learn how signed numbers are added, subtracted, multiplied, and divided. Because the 2's complement system for representing signed numbers is the most widely used in computers and microprocessor-based systems, the coverage in this section is limited to 2's complement arithmetic. The processes covered can be extended to the other systems if necessary.

After completing this section, you should be able to

- Add signed binary numbers ■ Explain how computers add strings of numbers
- Define *overflow* ■ Subtract signed binary numbers ■ Multiply signed binary numbers using the direct addition method ■ Multiply signed binary numbers using the partial products method ■ Divide signed binary numbers

COMPUTER NOTE
In addition to the CPU (central processing unit), computers use *coprocessors* to perform complicated mathematical calculations using floating-point numbers. The purpose is to increase performance by freeing up the CPU for other tasks. The mathematical coprocessor is also known as the floating-point unit (FPU).

Addition

The two numbers in an addition are the **addend** and the **augend.** The result is the **sum.** There are four cases that can occur when two signed binary numbers are added.

1. Both numbers positive
2. Positive number with magnitude larger than negative number
3. Negative number with magnitude larger than positive number
4. Both numbers negative

Let's take one case at a time using 8-bit signed numbers as examples. The equivalent decimal numbers are shown for reference.

Both numbers positive:

$$
\begin{array}{rr}
00000111 & 7 \\
+\ 00000100 & +\ 4 \\
\hline
00001011 & 11
\end{array}
$$

The sum is positive and is therefore in true (uncomplemented) binary.

Positive number with magnitude larger than negative number:

$$
\begin{array}{rr}
00001111 & 15 \\
+\ 11111010 & +\ -6 \\
\hline
\text{Discard carry} \longrightarrow\ 1\ \ 00001001 & 9
\end{array}
$$

The final carry bit is discarded. The sum is positive and therefore in true (uncomplemented) binary.

Negative number with magnitude larger than positive number:

$$
\begin{array}{rr}
00010000 & 16 \\
+\ 11101000 & +\ -24 \\
\hline
11111000 & -8
\end{array}
$$

The sum is negative and therefore in 2's complement form.

Both numbers negative:

$$
\begin{array}{rr}
11111011 & -5 \\
+\ 11110111 & +\ -9 \\
\hline
\text{Discard carry} \longrightarrow 1\ \ 11110010 & -14
\end{array}
$$

The final carry bit is discarded. The sum is negative and therefore in 2's complement form.

In a computer, the negative numbers are stored in 2's complement form so, as you can see, the addition process is very simple: *Add the two numbers and discard any final carry bit.*

Overflow Condition   When two numbers are added and the number of bits required to represent the sum exceeds the number of bits in the two numbers, an **overflow** results as indicated by an incorrect sign bit. An overflow can occur only when both numbers are positive or both numbers are negative. The following 8-bit example will illustrate this condition.

$$
\begin{array}{rr}
01111101 & 125 \\
+\ 00111010 & +\ 58 \\
\hline
10110111 & 183
\end{array}
$$

Sign incorrect

Magnitude incorrect

In this example the sum of 183 requires eight magnitude bits. Since there are seven magnitude bits in the numbers (one bit is the sign), there is a carry into the sign bit which produces the overflow indication.

Numbers Are Added Two at a Time   Now let's look at the addition of a string of numbers, added two at a time. This can be accomplished by adding the first two numbers, then adding the third number to the sum of the first two, then adding the fourth number to this result, and so on. This is how computers add strings of numbers. The addition of numbers taken two at a time is illustrated in Example 2–19.

EXAMPLE 2–19   Add the signed numbers: 01000100, 00011011, 00001110, and 00010010.

Solution   The equivalent decimal additions are given for reference.

$$
\begin{array}{rll}
68 & 01000100 & \\
+\ 27 & +\ 00011011 & \text{Add 1st two numbers} \\
\hline
95 & 01011111 & \text{1st sum} \\
+\ 14 & +\ 00001110 & \text{Add 3rd number} \\
\hline
109 & 01101101 & \text{2nd sum} \\
+\ 18 & +\ 00010010 & \text{Add 4th number} \\
\hline
127 & \mathbf{01111111} & \text{Final sum}
\end{array}
$$

Related Problem   Add 00110011, 10111111, and 01100011. These are signed numbers.

Subtraction

Subtraction is a special case of addition. For example, subtracting +6 (the **subtrahend**) from +9 (the **minuend**) is equivalent to adding −6 to +9. Basically, *the subtraction operation changes the sign of the subtrahend and adds it to the minuend.* The result of a subtraction is called the **difference**.

The sign of a positive or negative binary number is changed by taking its 2's complement.

For example, when you take the 2's complement of the positive number 00000100 (+4), you get 11111100, which is −4 as the following sum-of-weights evaluation shows:

$$-128 + 64 + 32 + 16 + 8 + 4 = -4$$

As another example, when you take the 2's complement of the negative number 11101101 (−19), you get 00010011, which is +19 as the following sum-of-weights evaluation shows:

$$16 + 2 + 1 = 19$$

Since subtraction is simply an addition with the sign of the subtrahend changed, the process is stated as follows:

> **To subtract two signed numbers, take the 2's complement of the subtrahend and add. Discard any final carry bit.**

Example 2–20 illustrates the subtraction process.

EXAMPLE 2–20

Perform each of the following subtractions of the signed numbers:
(a) 00001000 − 00000011   **(b)** 00001100 − 11110111
(c) 11100111 − 00010011   **(d)** 10001000 − 11100010

Solution   Like in other examples, the equivalent decimal subtractions are given for reference.
(a) In this case, 8 − 3 = 8 + (−3) = 5.

00001000	Minuend (+8)
+ 11111101	2's complement of subtrahend (−3)
Discard carry ⟶ **1 00000101**	Difference (+5)

(b) In this case, 12 − (−9) = 12 + 9 = 21.

00001100	Minuend (+12)
+ 00001001	2's complement of subtrahend (+9)
00010101	Difference (+21)

(c) In this case, −25 − (+19) = −25 + (−19) = −44.

11100111	Minuend (−25)
+ 11101101	2's complement of subtrahend (−19)
Discard carry ⟶ **1 11010100**	Difference (−44)

(d) In this case, −120 − (−30) = −120 + 30 = −90.

10001000	Minuend (−120)
+ 00011110	2's complement of subtrahend (+30)
10100110	Difference (−90)

Related Problem   Subtract 01000111 from 01011000.

Multiplication

The numbers in a multiplication are the **multiplicand,** the **multiplier,** and the **product.** These are illustrated in the following decimal multiplication:

8	Multiplicand
× 3	Multiplier
24	Product

The multiplication operation in most computers is accomplished using addition. As you have already seen, subtraction is done with an adder; now let's see how multiplication is done.

Direct addition and *partial products* are two basic methods for performing multiplication using addition. In the direct addition method, you add the multiplicand a number of times equal to the multiplier. In the previous decimal example (3 × 8), three multiplicands are added: 8 + 8 + 8 = 24. The disadvantage of this approach is that it becomes very lengthy if the multiplier is a large number. For example, to multiply 75 × 350, you must add 350 to itself 75 times. Incidentally, this is why the term *times* is used to mean multiply.

The partial products method is perhaps the more common one because it reflects the way you multiply long hand. The multiplicand is multiplied by each multiplier digit beginning with the least significant digit. The result of the multiplication of the multiplicand by a multiplier digit is called a *partial product*. Each successive partial product is moved (shifted) one place to the left and when all the partial products have been produced, they are added to get the final product. Here is a decimal example.

$$
\begin{array}{rl}
239 & \text{Multiplicand} \\
\times\ 123 & \text{Multiplier} \\
\hline
717 & \text{1st partial product (3 × 239)} \\
478 & \text{2nd partial product (2 × 239)} \\
+\ 239 & \text{3rd partial product (1 × 239)} \\
\hline
29{,}397 & \text{Final product}
\end{array}
$$

The sign of the product of a multiplication depends on the signs of the multiplicand and the multiplier according to the following two rules:

- **If the signs are the same, the product is positive.**
- **If the signs are different, the product is negative.**

When two binary numbers are multiplied, both numbers must be in true (uncomplemented) form. The direct addition method is illustrated in Example 2–21 adding two binary numbers at a time.

EXAMPLE 2–21

Multiply the signed binary numbers: 01001101 (multiplicand) and 00000100 (multiplier) using the direct addition method.

Solution   Since both numbers are positive, they are in true form, and the product will be positive. The decimal value of the multiplier is 4, so the multiplicand is added to itself four times as follows:

$$
\begin{array}{rl}
01001101 & \text{1st time} \\
+\ 01001101 & \text{2nd time} \\
\hline
10011010 & \text{Partial sum} \\
+\ 01001101 & \text{3rd time} \\
\hline
11100111 & \text{Partial sum} \\
+\ 01001101 & \text{4th time} \\
\hline
\mathbf{100110100} & \text{Product}
\end{array}
$$

Since the sign bit of the multiplicand is 0, it has no effect on the outcome. All of the bits in the product are magnitude bits.

Related Problem   Multiply 01100001 by 00000110.

Now let's look at the partial products method of binary multiplication. The basic steps in the process are as follows:

Step 1. Determine if the signs of the multiplicand and multiplier are the same or different. This determines what the sign of the product will be.

Step 2. Change any negative number to true (uncomplemented) form. Because most computers store negative numbers in 2's complement, a 2's complement operation is required to get the negative number into true form.

Step 3. Starting with the least significant multiplier bit, generate the partial products. When the multiplier bit is 1, the partial product is the same as the multiplicand. When the multiplier bit is 0, the partial product is zero. Shift each successive partial product one bit to the left.

Step 4. Add each successive partial product to the sum of the previous partial products to get the final product.

Step 5. If the sign bit that was determined in step 1 is negative, take the 2's complement of the product. If positive, leave the product in true form. Attach the sign bit to the product.

Example 2–22 illustrates these steps.

EXAMPLE 2–22

Multiply the signed binary number: 01010011 (multiplicand) and 11000101 (multiplier).

Solution

Step 1. The sign bit of the multiplicand is 0 and the sign bit of the multiplier is 1. The sign bit of the product will be 1 (negative).

Step 2. Take the 2's complement of the multiplier to put it in true form.

$$11000101 \longrightarrow 00111011$$

Steps 3 and 4. The multiplication proceeds as follows. Notice that only the magnitude bits are used in these steps.

```
       1010011      Multiplicand
     × 0111011      Multiplier
       1010011      1st partial product
   +   1010011      2nd partial product
      11111001      Sum of 1st and 2nd
   +  0000000       3rd partial product
     011111001      Sum
   + 1010011        4th partial product
    1110010001      Sum
   + 1010011        5th partial product
   100011000001     Sum
   + 1010011        6th partial product
   1001100100001    Sum
   + 0000000        7th partial product
   1001100100001    Final product
```

Step 5. Since the sign of the product is a 1 as determined in step 1, take the 2's complement of the product.

$$1001100100001 \longrightarrow 0110011011111$$

Attach the sign bit.

$$\longrightarrow \mathbf{1\ 0110011011111}$$

Related Problem Verify the multiplication is correct by converting to decimal numbers and performing the multiplication.

Division

The numbers in a division are the **dividend,** the **divisor,** and the **quotient.** These are illustrated in the following standard division format.

$$\frac{dividend}{divisor} = quotient$$

The division operation in computers is accomplished using subtraction. Since subtraction is done with an adder, division can also be accomplished with an adder.

The result of a division is called the *quotient;* the quotient is the number of times that the divisor will go into the dividend. This means that the divisor can be subtracted from the dividend a number of times equal to the quotient, as illustrated by dividing 21 by 7.

21	Dividend
− 7	1st subtraction of divisor
14	1st partial remainder
− 7	2nd subtraction of divisor
7	2nd partial remainder
− 7	3rd subtraction of divisor
0	Zero remainder

In this simple example, the divisor was subtracted from the dividend three times before a remainder of zero was obtained. Therefore, the quotient is 3.

The sign of the quotient depends on the signs of the dividend and the divisor according to the following two rules:

- **If the signs are the same, the quotient is positive.**
- **If the signs are different, the quotient is negative.**

When two binary numbers are divided, both numbers must be in true (uncomplemented) form. The basic steps in a division process are as follows:

Step 1. Determine if the signs of the dividend and divisor are the same or different. This determines what the sign of the quotient will be. Initialize the quotient to zero.

Step 2. Subtract the divisor from the dividend using 2's complement addition to get the first partial remainder and add 1 to the quotient. If this partial remainder is positive, go to step 3. If the partial remainder is zero or negative, the division is complete.

Step 3. Subtract the divisor from the partial remainder and add 1 to the quotient. If the result is positive, repeat for the next partial remainder. If the result is zero or negative, the division is complete.

Continue to subtract the divisor from the dividend and the partial remainders until there is a zero or a negative result. Count the number of times that the divisor is subtracted and you have the quotient. Example 2–23 illustrates these steps using 8-bit signed binary numbers.

EXAMPLE 2–23

Divide 01100100 by 00011001.

Solution
Step 1. The signs of both numbers are positive, so the quotient will be positive. The quotient is initially zero: 00000000.

Step 2. Subtract the divisor from the dividend using 2's complement addition (remember that final carries are discarded).

01100100	Dividend
+ 11100111	2's complement of divisor
01001011	Positive 1st partial remainder

Add 1 to quotient: 00000000 + 00000001 = 00000001.

Step 3. Subtract the divisor from the 1st partial remainder using 2's complement addition.

$$
\begin{array}{ll}
01001011 & \text{1st partial remainder} \\
+\ 11100111 & \text{2's complement of divisor} \\
\hline
00110010 & \text{Positive 2nd partial remainder}
\end{array}
$$

Add 1 to quotient: 00000001 + 00000001 = 00000010.

Step 4. Subtract the divisor from the 2nd partial remainder using 2's complement addition.

$$
\begin{array}{ll}
00110010 & \text{2nd partial remainder} \\
+\ 11100111 & \text{2's complement of divisor} \\
\hline
00011001 & \text{Positive 3rd partial remainder}
\end{array}
$$

Add 1 to quotient: 00000010 + 00000001 = 00000011.

Step 5. Subtract the divisor from the 3rd partial remainder using 2's complement addition.

$$
\begin{array}{ll}
00011001 & \text{3rd partial remainder} \\
+\ 11100111 & \text{2's complement of divisor} \\
\hline
00000000 & \text{Zero remainder}
\end{array}
$$

Add 1 to quotient: 00000011 + 00000001 = **00000100** (final quotient). The process is complete.

Related Problem Verify that the process is correct by converting to decimal numbers and performing the division.

SECTION 2–7 REVIEW

1. List the four cases when numbers are added.
2. Add 00100001 and 10111100.
3. Subtract 00110010 from 01110111.
4. What is the sign of the product when two negative numbers are multiplied?
5. Multiply 01111111 by 00000101.
6. What is the sign of the quotient when a positive number is divided by a negative number?
7. Divide 00110000 by 00001100.

2–8 ■ HEXADECIMAL NUMBERS

The hexadecimal number system has sixteen digits and is used primarily as a compact way of displaying or writing binary numbers because it is very easy to convert between binary and hexadecimal. As you are probably aware, long binary numbers are difficult to read and write because it is easy to drop or transpose a bit. Since computers and microprocessors understand only 1s and 0s, it is necessary to use these digits when you program in "machine language." Imagine writing a sixteen bit instruction for a microprocessor system in 1s and 0s. It is much more efficient to use hexadecimal or octal; octal numbers are covered in Section 2–9. Hexadecimal is widely used in computer and microprocessor applications.

After completing this section, you should be able to

■ List the hexadecimal digits ■ Count in hexadecimal ■ Convert from binary to hexadecimal ■ Convert from hexadecimal to binary ■ Convert from hexadecimal to decimal ■ Convert from decimal to hexadecimal ■ Add hexadecimal numbers ■ Subtract hexadecimal numbers

The **hexadecimal** system has a base of sixteen; that is, it is composed of 16 digits and alphabetic **characters.** Most digital systems process binary data in groups that are multiples of four bits, making the hexadecimal number very convenient because each hexadecimal digit represents a 4-bit binary number (as listed in Table 2–3).

TABLE 2–3

Decimal	Binary	Hexadecimal
0	0000	0
1	0001	1
2	0010	2
3	0011	3
4	0100	4
5	0101	5
6	0110	6
7	0111	7
8	1000	8
9	1001	9
10	1010	A
11	1011	B
12	1100	C
13	1101	D
14	1110	E
15	1111	F

Ten **numeric** digits and six alphabetic characters make up the hexadecimal number system. The use of letters A, B, C, D, E, and F to represent numbers may seem strange at first, but keep in mind that any number system is only a set of sequential symbols. If you understand what quantities these symbols represent, then the form of the symbols themselves is less important once you get accustomed to using them. We will use the subscript 16 to designate hexadecimal numbers to avoid confusion with decimal numbers. Sometimes you may see an "h" following a hexadecimal number.

COMPUTER NOTE

With computer memories in the Gigabyte range, specifying a memory address in binary is quite cumbersome. For example it takes 32 bits to specify an address in a 4 Gbyte memory. It is much easier to express a 32-bit code using 8 hexadecimal digits.

Counting in Hexadecimal

How do you count in hexadecimal once you get to F? Simply start over with another column and continue as follows:

10, 11, 12, 13, 14, 15, 16, 17, 18, 19, 1A, 1B, 1C, 1D, 1E, 1F, 20, 21, 22, 23
24, 25, 26, 27, 28, 29, 2A, 2B, 2C, 2D, 2E, 2F, 30, 31, . . .

With two hexadecimal digits, you can count up to FF_{16}, which is decimal 255. To count beyond this, three hexadecimal digits are needed. For instance, 100_{16} is decimal 256, 101_{16} is decimal 257, and so forth. The maximum 3-digit hexadecimal number is FFF_{16}, or decimal 4095. The maximum 4-digit hexadecimal number is $FFFF_{16}$, which is decimal 65,535.

Binary-to-Hexadecimal Conversion

Converting a binary number to hexadecimal is a very straightforward procedure. Simply break the binary number into 4-bit groups, starting at the right-most bit and replace each 4-bit group with the equivalent hexadecimal symbol as illustrated in Example 2–24.

EXAMPLE 2–24

Convert the following binary numbers to hexadecimal:
(a) 1100101001010111   **(b)** 111111000101101001

Solution

(a)
    C   A   5   7 = **CA57**$_{16}$

(b)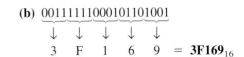
  3   F   1   6   9 = **3F169**$_{16}$

Two zeros have been added in part (b) to complete a 4-bit group at the left.

Related Problem Convert the binary number 1001111011110011100 to hexadecimal.

Hexadecimal-to-Binary Conversion

To convert from a hexadecimal number to a binary number, reverse the process and replace each hexadecimal symbol with the appropriate four bits as illustrated in Example 2–25.

EXAMPLE 2–25

Determine the binary numbers for the following hexadecimal numbers:
(a) 10A4$_{16}$ **(b)** CF8E$_{16}$ **(c)** 9742$_{16}$

Solution

(a) 1   0   A   4   **(b)** C   F   8   E   **(c)** 9   7   4   2

1000010100100   **1100111110001110**   **1001011101000010**

In part (a), the MSB is understood to have three zeros preceding it, thus forming a 4-bit group.

Related Problem Convert the hexadecimal number 6BD3 to binary.

It should be clear that it is much easier to deal with a hexadecimal number than the equivalent binary number. Since conversion is so easy, the hexadecimal system is widely used for representing binary numbers in programming, printouts, and displays.

Hexadecimal-to-Decimal Conversion

One way to find the decimal equivalent of a hexadecimal number is to first convert the hexadecimal number to binary and then convert from binary to decimal. Example 2–26 illustrates this procedure.

EXAMPLE 2–26

Convert the following hexadecimal numbers to decimal:
(a) 1C$_{16}$ **(b)** A85$_{16}$

Solution Remember, convert the hexadecimal number to binary first, then to decimal.

(a)   1   C

    $\overline{00011100}$ = $2^4 + 2^3 + 2^2 = 16 + 8 + 4 = $ **28**$_{10}$

(b)   A   8   5
         ↓   ↓   ↓

$$\overline{101010000101} = 2^{11} + 2^9 + 2^7 + 2^2 + 2^0 = 2048 + 512 + 128 + 4 + 1 = \mathbf{2693_{10}}$$

Related Problem   Convert the hexadecimal number 6BD to decimal.

Another way to convert a hexadecimal number to its decimal equivalent is to multiply the decimal value of each hexadecimal digit by its weight and then take the sum of these products. The weights of a hexadecimal number are increasing powers of 16 (from right to left). For a 4-digit hexadecimal number, the weights are

$$16^3 \quad 16^2 \quad 16^1 \quad 16^0$$
$$4096 \quad 256 \quad 16 \quad 1$$

Example 2–27 shows this conversion method.

EXAMPLE 2–27

Convert the following hexadecimal numbers to decimal:
(a) $E5_{16}$ **(b)** $B2F8_{16}$

Solution   Recall from Table 2–3 that letters A through F represent decimal numbers 10 through 15, respectively.

(a) $E5_{16} = (E \times 16) + (5 \times 1) = (14 \times 16) + (5 \times 1) = 224 + 5 = \mathbf{229_{10}}$
(b) $B2F8_{16} = (B \times 4096) + (2 \times 256) + (F \times 16) + (8 \times 1)$
$$= (11 \times 4096) + (2 \times 256) + (15 \times 16) + (8 \times 1)$$
$$= \quad 45,056 \quad + \quad 512 \quad + \quad 240 \quad + \quad 8 = \mathbf{45,816_{10}}$$

Related Problem   Convert $60A_{16}$ to decimal.

Calculator Tutorial

Powers of sixteen can be found on the calculator using the ⌃ key.

Example   Find the value of 16^3.
Step 1.   Press ⊡ ⊡.
Step 2.   Press ⌃.
Step 3.   Press ③ then ⌷ENTER⌷.

16^3

4096

Decimal-to-Hexadecimal Conversion

Repeated division of a decimal number by 16 will produce the equivalent hexadecimal number, formed by the remainders of the divisions. The first remainder produced is the least significant digit (LSD). Each successive division by 16 yields a remainder that becomes a digit in the equivalent hexadecimal number. This procedure is similar to repeated division by 2 for decimal-to-binary conversion that was covered in Section 2–3. Example 2–28 illustrates the procedure. Note that when a quotient has a fractional part, the fractional part is multiplied by the divisor to get the remainder.

EXAMPLE 2–28

Convert the decimal number 650 to hexadecimal by repeated division by 16.

Solution

Hexadecimal
remainder

$$\frac{650}{16} = 40.625 \rightarrow 0.625 \times 16 = 10 = \text{A}$$

$$\frac{40}{16} = 2.5 \longrightarrow 0.5 \times 16 = 8 = 8$$

$$\frac{2}{16} = 0.125 \longrightarrow 0.125 \times 16 = 2 = 2$$

Stop when whole number
quotient is zero.

2 8 A   Hexadecimal number

MSD ⌐   ⌐ LSD

Related Problem Convert decimal 2591 to hexadecimal.

Calculator Tutorial

A decimal number can be converted to a hexadecimal number by first using the base conversion key sequence [2nd] [BASE] [F3].

Example Convert decimal 650 to hexadecimal.

Step 1. Press [6] [5] [0].

Step 2. Press [F2].

Step 3. Press [ENTER].

650 ►Hex

28Ah

A-F	TYPE	CONV	BOOL	BIT
►Bin	►Hex	►Oct	►Dec	

Hexadecimal Addition

Addition can be done directly with hexadecimal numbers by remembering that the hexadecimal digits 0 through 9 are equivalent to decimal digits 0 through 9 and that hexadecimal digits A through F are equivalent to decimal numbers 10 through 15. When adding two hexadecimal numbers, use the following rules. (Decimal numbers are indicated by a subscript 10.)

1. In any given column of an addition problem, think of the two hexadecimal digits in terms of their decimal values. For instance, $5_{16} = 5_{10}$ and $C_{16} = 12_{10}$.
2. If the sum of these two digits is 15_{10} or less, bring down the corresponding hexadecimal digit.
3. If the sum of these two digits is greater than 15_{10}, bring down the amount of the sum that exceeds 16_{10} and carry a 1 to the next column.

EXAMPLE 2–29

Add the following hexadecimal numbers:

(a) $23_{16} + 16_{16}$ **(b)** $58_{16} + 22_{16}$ **(c)** $2B_{16} + 84_{16}$ **(d)** $DF_{16} + AC_{16}$

Solution

(a) 23_{16}
     $+\ 16_{16}$
     $\overline{\mathbf{39}_{16}}$

right column: $3_{16} + 6_{16} = 3_{10} + 6_{10} = 9_{10} = 9_{16}$

left column: $2_{16} + 1_{16} = 2_{10} + 1_{10} = 3_{10} = 3_{16}$

(b) 58_{16} right column: $8_{16} + 2_{16} = 8_{10} + 2_{10} = 10_{10} = A_{16}$

$\underline{+\ 22_{16}}$ left column: $5_{16} + 2_{16} = 5_{10} + 2_{10} = 7_{10} = 7_{16}$

$\mathbf{7A}_{16}$

(c) $2B_{16}$ right column: $B_{16} + 4_{16} = 11_{10} + 4_{10} = 15_{10} = F_{16}$

$\underline{+\ 84_{16}}$ left column: $2_{16} + 8_{16} = 2_{10} + 8_{10} = 10_{10} = A_{16}$

$\mathbf{AF}_{16}$

(d) DF_{16} right column: $F_{16} + C_{16} = 15_{10} + 12_{10} = 27_{10}$

$\underline{+\ AC_{16}}$ $27_{10} - 16_{10} = 11_{10} = B_{16}$ with a 1 carry

$\mathbf{18B}_{16}$ left column: $D_{16} + A_{16} + 1_{16} = 13_{10} + 10_{10} + 1_{10} = 24_{10}$

    $24_{10} - 16_{10} = 8_{10} = 8_{16}$ with a 1 carry

Related Problem   Add $4C_{16}$ and $3A_{16}$.

Hexadecimal Subtraction Using 2's Complement Method

Since a hexadecimal number can be used to represent a binary number, it can also be used to represent the 2's complement of a binary number. For instance, the hexadecimal representation of 11001001_2 is $C9_{16}$. The 2's complement of this binary number is 00110111, which is written in hexadecimal as 37_{16}.

As you have learned, the 2's complement allows you to subtract by adding binary numbers. You can also use this method for hexadecimal subtraction, as in Example 2–30.

EXAMPLE 2–30   Subtract the following hexadecimal numbers:

(a) $84_{16} - 2A_{16}$ **(b)** $C3_{16} - 0B_{16}$

Solution

(a) $2A_{16} = 00101010$

  2's complement of $2A_{16} = 11010110 = D6_{16}$

84_{16}

$\underline{+\ D6_{16}}$ Add

$\cancel{1}5A_{16}$ Drop carry, as in 2's complement addition

The difference is $\mathbf{5A}_{16}$.

(b) $0B_{16} = 00001011$

  2's complement of $0B_{16} = 11110101 = F5_{16}$

$C3_{16}$

$\underline{+\ F5_{16}}$ Add

$\cancel{1}B8_{16}$ Drop carry

The difference is $\mathbf{B8}_{16}$.

Related Problem   Subtract 173_{16} from BCD_{16}.

**SECTION 2–8
REVIEW**

1. Convert the following binary numbers to hexadecimal:
   (a) 10110011   (b) 110011101000
2. Convert the following hexadecimal numbers to binary:
   (a) 57_{16} (b) $3A5_{16}$ (c) $F80B_{16}$
3. Convert $9B30_{16}$ to decimal.
4. Convert the decimal number 573 to hexadecimal.
5. Add the following hexadecimal numbers directly:
   (a) $18_{16} + 34_{16}$ (b) $3F_{16} + 2A_{16}$
6. Subtract the following hexadecimal numbers:
   (a) $75_{16} - 21_{16}$ (b) $94_{16} - 5C_{16}$

2–9 ■ OCTAL NUMBERS

Like the hexadecimal system, the octal system provides a convenient way to express binary numbers and codes. However, it is used less frequently than hexadecimal in conjunction with computers and microprocessors to express binary quantities for input and output purposes.

After completing this section, you should be able to

■ Write the digits of the octal number system   ■ Convert from octal to decimal
■ Convert from decimal to octal   ■ Convert from octal to binary   ■ Convert from binary to octal

The **octal** number system is composed of eight digits, which are

$$0, 1, 2, 3, 4, 5, 6, 7$$

To count above 7, begin another column and start over:

$$10, 11, 12, 13, 14, 15, 16, 17, 20, 21$$

and so on. Counting in octal is similar to counting in decimal, except that the digits 8 and 9 are not used. To distinguish octal numbers from decimal numbers or hexadecimal numbers, we will use the subscript 8 to indicate an octal number. For instance, 15_8 in octal is equivalent to 13_{10} in decimal and D in hexadecimal. Sometimes you may see an "o" or a "Q" following an octal number.

Octal-to-Decimal Conversion

Since the octal number system has a base of eight, each successive digit position is an increasing power of eight, beginning in the right-most column with 8^0. The evaluation of an octal number in terms of its decimal equivalent is accomplished by multiplying each digit by its weight and summing the products, as illustrated here for 2374_8.

$$\text{Weight: } 8^3 \ 8^2 \ 8^1 \ 8^0$$
$$\text{Octal number: } 2 \ 3 \ 7 \ 4$$

$$
\begin{aligned}
2374_8 &= (2 \times 8^3) + (3 \times 8^2) + (7 \times 8^1) + (4 \times 8^0) \\
&= (2 \times 512) + (3 \times 64) + (7 \times 8) + (4 \times 1) \\
&= \quad 1024 \quad + \quad 192 \quad + \quad 56 \quad + \quad 4 \quad = 1276_{10}
\end{aligned}
$$

Decimal-to-Octal Conversion

A method of converting a decimal number to an octal number is the repeated division-by-8 method, which is similar to the method used in the conversion of decimal numbers to binary or to hexadecimal. To show how it works, let's convert the decimal number 359 to octal. Each successive division by 8 yields a remainder that becomes a digit in the equivalent octal number. The first remainder generated is the least significant digit (LSD).

Calculator Tutorial	A decimal number can be converted to an octal number by using the base conversion key sequence [2nd] [BASE] [F3].

Example Convert decimal 439 to octal.

Step 1. Press [4] [3] [9].

Step 2. Press [F3].

Step 3. Press [ENTER].

439 ►Oct

667o

A-F	TYPE	CONV	BOOL	BIT
►Bin	►Hex	►Oct	►Dec	

Octal-to-Binary Conversion

Because each octal digit can be represented by a 3-bit binary number, it is very easy to convert from octal to binary. Each octal digit is represented by three bits as shown in Table 2–4.

TABLE 2–4

Octal/binary conversion.

Octal Digit	0	1	2	3	4	5	6	7
Binary	000	001	010	011	100	101	110	111

To convert an octal number to a binary number, simply replace each octal digit with the appropriate three bits. This procedure is illustrated in Example 2–31.

EXAMPLE 2–31

Convert each of the following octal numbers to binary:
(a) 13_8 (b) 25_8 (c) 140_8 (d) 7526_8

Solution

(a)   (b)   (c)   (d)

Related Problem   Convert each of the binary numbers to decimal and verify that each value agrees with the decimal value of the corresponding octal number.

Binary-to-Octal Conversion

Conversion of a binary number to an octal number is the reverse of the octal-to-binary conversion. The procedure is as follows: Start with the right-most group of three bits and, moving from right to left, convert each 3-bit group to the equivalent octal digit. If there are not three bits available for the left-most group, add either one or two zeros to make a complete group. These leading zeros do not affect the value of the binary number.

EXAMPLE 2–32

Convert each of the following binary numbers to octal:
(a) 110101   (b) 101111001   (c) 100110011010   (d) 11010000100

Solution

(a) $\underline{110}\underline{101}$
    ↓   ↓
    6   5 $= \mathbf{65_8}$

(b) $\underline{101}\underline{111}\underline{001}$
    ↓   ↓   ↓
    5   7   1 $= \mathbf{571_8}$

(c) $\underline{100}\underline{110}\underline{011}\underline{010}$
    ↓   ↓   ↓   ↓
    4   6   3   2 $= \mathbf{4632_8}$

(d) $\underline{011}\underline{010}\underline{000}\underline{100}$
    ↓   ↓   ↓   ↓
    3   2   0   4 $= \mathbf{3204_8}$

Related Problem   Convert the binary number 1010101000111110010 to octal.

SECTION 2–9 REVIEW

1. Convert the following octal numbers to decimal:
    (a) 73_8 (b) 125_8
2. Convert the following decimal numbers to octal:
    (a) 98_{10} (b) 163_{10}
3. Convert the following octal numbers to binary:
    (a) 46_8 (b) 723_8 (c) 5624_8
4. Convert the following binary numbers to octal:
    (a) 110101111   (b) 1001100010   (c) 10111111001

2–10 ■ BINARY CODED DECIMAL (BCD)

Binary coded decimal (BCD) is a way to express each of the decimal digits with a binary code. Since there are only ten code groups in the BCD system, it is very easy to convert between decimal and BCD. Because we like to read and write in decimal, the BCD code provides an excellent interface to binary systems. Examples of such interfaces are keypad inputs and digital readouts.

After completing this section, you should be able to

■ Convert each decimal digit to BCD   ■ Express decimal numbers in BCD   ■ Convert from BCD to decimal   ■ Add BCD numbers

The 8421 Code

The 8421 code is a type of **binary coded decimal (BCD)** code. Binary coded decimal means that each decimal digit, 0 through 9, is represented by a binary **code** of four bits. The designation 8421 indicates the binary weights of the four bits ($2^3, 2^2, 2^1, 2^0$). The ease of conversion between 8421 code numbers and the familiar decimal numbers is the main advantage of this code. All you have to remember are the ten binary combinations that represent the ten decimal digits as shown in Table 2–5. The 8421 code is the predominant BCD code, and when we refer to BCD, we always mean the 8421 code unless otherwise stated.

TABLE 2–5
Decimal/BCD conversion.

Decimal Digit	0	1	2	3	4	5	6	7	8	9
BCD	0000	0001	0010	0011	0100	0101	0110	0111	1000	1001

Invalid Codes   You should realize that, with four bits, sixteen numbers (0000 through 1111) can be represented but that, in the 8421 code, only ten of these are used. The six code combinations that are not used—1010, 1011, 1100, 1101, 1110, and 1111—are invalid in the 8421 BCD code.

To express any decimal number in BCD, simply replace each decimal digit with the appropriate 4-bit code, as shown by Example 2–33.

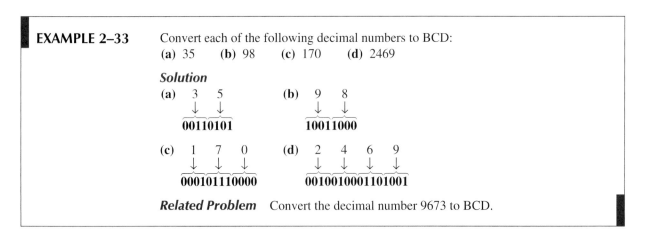

EXAMPLE 2–33   Convert each of the following decimal numbers to BCD:
(a) 35   **(b)** 98   **(c)** 170   **(d)** 2469

Solution

(a)   3   5
00110101

(b)   9   8
10011000

(c)   1   7   0
000101110000

(d)   2   4   6   9
0010010001101001

Related Problem   Convert the decimal number 9673 to BCD.

It is equally easy to determine a decimal number from a BCD number. Start at the right-most bit and break the code into groups of four bits. Then write the decimal digit represented by each 4-bit group. Example 2–34 illustrates.

EXAMPLE 2–34

Convert each of the following BCD codes to decimal:
(a) 10000110   (b) 001101010001   (c) 1001010001110000

Solution
(a) <u>1000</u><u>0110</u>   (b) <u>0011</u><u>0101</u><u>0001</u>   (c) <u>1001</u><u>0100</u><u>0111</u><u>0000</u>
  ↓   ↓   ↓   ↓   ↓   ↓   ↓   ↓   ↓
  8 **6**   **3** **5** **1**   **9** **4** **7** **0**

Related Problem   Convert the BCD code 10000010001001110110 to decimal.

BCD Addition

BCD is a numerical code and can be used in arithmetic operations. Addition is the most important operation because the other three operations (subtraction, multiplication, and division) can be accomplished by the use of addition. Here is how to add two BCD numbers:

Step 1. Add the two BCD numbers, using the rules for binary addition in Section 2–4.
Step 2. If a 4-bit sum is equal to or less than 9, it is a valid BCD number.
Step 3. If a 4-bit sum is greater than 9, or if a carry out of the 4-bit group is generated, it is an invalid result. Add 6 (0110) to the 4-bit sum in order to skip the six invalid states and return the code to 8421. If a carry results when 6 is added, simply add the carry to the next 4-bit group.

Example 2–35 illustrates BCD additions in which the sum in each 4-bit column is equal to or less than 9, and the 4-bit sums are therefore valid BCD numbers. Example 2–36 illustrates the procedure in the case of invalid sums (greater than 9 or a carry).

EXAMPLE 2–35

Add the following BCD numbers:
(a) 0011 + 0100   (b) 00100011 + 00010101
(c) 10000110 + 00010011 (d) 010001010000 + 010000010111

Solution   The decimal number addition is shown for comparison.

(a)   0011   3   (b)   0010 0011   23
     + 0100   + 4   + 0001 0101   + 15
      0111   7   **0011** **1000**   38

(c)   1000 0110   86   (d)   0100 0101 0000   450
  + 0001 0011 + 13   + 0100 0001 0111 + 417
   1001 **1001**   99   **1000** **0110** **0111**   867

Note that in each case the sum in any 4-bit column does not exceed 9, and the results are valid BCD numbers.

Related Problem   Add the BCD numbers: 1001000001000011 + 0000100100100101.

EXAMPLE 2–36

Add the following BCD numbers:
(a) 1001 + 0100   (b) 1001 + 1001
(c) 00010110 + 00010101 (d) 01100111 + 01010011

Solution   The decimal number additions are shown for comparison.

(a)

```
      1001                                          9
    + 0100                                        + 4
      1101    Invalid BCD number (>9)              13
    + 0110    Add 6
0001  0011    Valid BCD number
 ↓     ↓
 1     3
```

(b)

```
        1001                                        9
      + 1001                                      + 9
   1    0010    Invalid because of carry           18
      + 0110    Add 6
0001  1000    Valid BCD number
 ↓     ↓
 1     8
```

(c)

```
  0001   0110                                         16
+ 0001   0101                                       + 15
  0010   1011    Right group is invalid (>9),         31
                 left group is valid.
       + 0110    Add 6 to invalid code. Add
                 carry, 0001, to next group.
  0011   0001    Valid BCD number
   ↓      ↓
   3      1
```

(d)

```
  0110   0111                                         67
+ 0101   0011                                       + 53
  1011   1010    Both groups are invalid (>9)        120
+ 0110 + 0110    Add 6 to both groups
0001  0010  0000  Valid BCD number
 ↓     ↓     ↓
 1     2     0
```

Related Problem   Add the BCD numbers: 01001000 + 00110100.

SECTION 2–10 REVIEW

1. What is the binary weight of each 1 in the following BCD numbers?
   (a) 0010   **(b)** 1000   **(c)** 0001   **(d)** 0100
2. Convert the following decimal numbers to BCD:
   (a) 6   **(b)** 15   **(c)** 273   **(d)** 849
3. What decimal numbers are represented by each BCD code?
   (a) 10001001   **(b)** 001001111000   **(c)** 000101010111
4. In BCD addition, when is a 4-bit sum invalid?

2–11 ■ DIGITAL CODES AND PARITY

There are many specialized codes used in digital systems. You have just learned about the BCD code; now let's look at a few others. Some codes are strictly numeric, like BCD, and others are alphanumeric; that is, they are used to represent numbers, letters, symbols, and instructions. The codes introduced in this section are the Gray code and the ASCII code. Also, the detection of errors in codes using a parity bit is covered.

After completing this section, you should be able to

- ▪ Explain the advantage of the Gray code ▪ Convert between Gray code and binary
- ▪ Use the ASCII code ▪ Identify errors in codes based on the parity method

The Gray Code

The **Gray code** is unweighted and is not an arithmetic code; that is, there are no specific weights assigned to the bit positions. The important feature of the Gray code is that *it exhibits only a single bit change from one code number to the next.* This property is important in many applications, such as shaft position encoders, where error susceptibility increases with the number of bit changes between adjacent numbers in a sequence.

Table 2–6 is a listing of the 4-bit Gray code for decimal numbers 0 through 15. Binary numbers are shown in the table for reference. Like binary numbers, *the Gray code can have any number of bits.* Notice the single-bit change between successive Gray code numbers. For instance, in going from decimal 3 to decimal 4, the Gray code changes from 0010 to 0110, while the binary code changes from 0011 to 0100, a change of three bits. The only bit change is in the third bit from the right in the Gray code; the others remain the same.

TABLE 2–6
Four-bit Gray code.

Decimal	Binary	Gray Code	Decimal	Binary	Gray Code
0	0000	0000	8	1000	1100
1	0001	0001	9	1001	1101
2	0010	0011	10	1010	1111
3	0011	0010	11	1011	1110
4	0100	0110	12	1100	1010
5	0101	0111	13	1101	1011
6	0110	0101	14	1110	1001
7	0111	0100	15	1111	1000

Binary-to-Gray Code Conversion Conversion between binary code and Gray code is sometimes useful. The following rules explain how to convert from a binary number to a Gray code number:

1. The most significant bit (left-most) in the Gray code is the same as the corresponding MSB in the binary number.
2. Going from left to right, add each adjacent pair of binary code bits to get the next Gray code bit. Discard carries.

For example, the conversion of the binary number 10110 to Gray code is as follows:

$$1 - + \rightarrow 0 - + \rightarrow 1 - + \rightarrow 1 - + \rightarrow 0 \qquad \text{Binary}$$
$$\downarrow \qquad \downarrow \qquad \downarrow \qquad \downarrow \qquad \downarrow$$
$$1 \qquad 1 \qquad 1 \qquad 0 \qquad 1 \qquad \text{Gray}$$

The Gray code is 11101.

Gray-to-Binary Conversion To convert from Gray code to binary, use a similar method; however, there are some differences. The following rules apply:

1. The most significant bit (left-most) in the binary code is the same as the corresponding bit in the Gray code.
2. Add each binary code bit generated to the Gray code bit in the next adjacent position. Discard carries.

For example, the conversion of the Gray code number 11011 to binary is as follows:

The binary number is 10010.

EXAMPLE 2–37

(a) Convert the binary number 11000110 to Gray code.
(b) Convert the Gray code 10101111 to binary.

Solution
(a) Binary to Gray code:

(b) Gray code to binary:

Related Problem (a) Convert binary 101101 to Gray code. (b) Convert Gray code 100111 to binary.

Application Example

A simplified diagram of a 3-bit shaft position encoder mechanism is shown in Figure 2–4. Basically, there are three concentric conductive rings that are segmented into eight sectors. The more sectors there are, the more accurately the position can be represented, but we are using only eight for purposes of illustration. Each sector of each ring is fixed at either a high-level or a low-level voltage to represent 1s and 0s. A 1 is indicated by a color sector and a 0 by a white sector. As the rings rotate with the shaft, they make contact with a brush arrangement that is in a fixed position and to which output lines are connected. As the shaft rotates counterclockwise through 360°, the eight sectors move past the three brushes producing a 3-bit binary output that indicates the shaft position.

In Figure 2–4(a), the sectors are arranged in a straight binary pattern, so that the brushes go from 000 to 001 to 010 to 011, and so on. When the brushes are on color sectors, they output a 1 and when on white sectors, they output a 0. If one brush is slightly ahead of the others during the transition from one sector to the next, an erroneous output can occur. Consider what happens when the brushes are on the 111 sector and about to enter the 000 sector. If the MSB brush is slightly ahead, the position would be incorrectly indicated by a transitional 011 instead of a 111 or a 000. In this type of application, it is virtually impossible to maintain precise mechanical alignment of all the brushes; therefore, some error will always occur at many of the transitions between sectors.

The Gray code is used to eliminate the error problem which is inherent in the binary code. As shown in Figure 2–4(b), the Gray code assures that only one bit will change between adjacent sectors. This means that even though the brushes may not be in precise alignment, there will never be a transitional error. For example, let's again consider what happens when the brushes are on the 111 sector and about to move into the next sector, 101. The only two possible outputs during the transition are 111 and 101, no matter how the brushes are aligned. A similar situation occurs at the transitions between each of the other sectors.

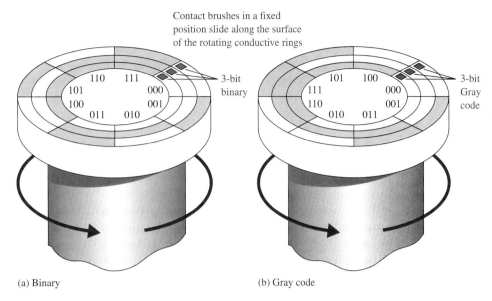

(a) Binary   (b) Gray code

FIGURE 2–4
A simplified illustration of how the Gray code solves the error problem in shaft position encoders.

Alphanumeric Codes

In order to communicate, you need not only numbers, but also letters and other symbols. In the strictest sense, **alphanumeric** codes are codes that represent numbers and alphabetic characters (letters). Most such codes, however, also represent other characters such as symbols and various instructions necessary for conveying information.

At a minimum, an alphanumeric code must represent 10 decimal digits and 26 letters of the alphabet, for a total of 36 items. This number requires six bits in each code combination because five bits are insufficient ($2^5 = 32$). There are 64 total combinations of six bits, so there are 28 unused code combinations. Obviously, in many applications, symbols other than just numbers and letters are necessary to communicate completely. You need spaces, periods, colons, semicolons, question marks, etc. You also need instructions to tell the receiving system what to do with the information. So with codes that are six bits long, you can handle decimal numbers, the alphabet, and 28 other symbols. This should give you an idea of the requirements for a basic alphanumeric code. The ASCII is the most common alphanumeric code and is covered next.

ASCII

The **American Standard Code for Information Interchange (ASCII)**—pronounced "askee"— is a universally accepted alphanumeric code used in most computers and other electronic equipment. Most computer keyboards are standardized with the ASCII. When you enter a letter, a number, or control command, the corresponding ASCII code goes into the computer.

ASCII has 128 characters and symbols represented by a 7-bit binary code. Actually, ASCII can be considered an 8-bit code with the MSB always 0. This 8-bit code is 00 through 7F in hexadecimal. The first thirty-two ASCII characters are nongraphic commands that are never printed or displayed and used only for control purposes. Examples of the control characters are "null," "line feed," "start of text," and "escape." The other characters are graphic symbols that can be printed or displayed and include the letters of the alphabet (lowercase and uppercase), the ten decimal digits, punctuation signs, and other commonly used symbols.

COMPUTER NOTE
A computer keyboard has a dedicated microprocessor that constantly scans keyboard circuits to detect when a key has been pressed and released. A unique scan code is produced by computer software representing that particular key. The scan code is then converted to an alphanumeric code (ASCII) for use by the computer.

Table 2–7 is a listing of the ASCII code showing the decimal, hexadecimal, and binary representations for each character and symbol. The left section of the table lists the names of the 32 control characters (00 through 1F hexadecimal). The graphic symbols are listed in the rest of the table (20 through 7F hexadecimal).

EXAMPLE 2–38

Determine the binary ASCII codes that are entered from the computer's keyboard when the following BASIC program statement is typed in. Also express each code in hexadecimal.

```
20 PRINT "A=";X
```

Solution   The ASCII code for each symbol is found in Table 2–7.

Symbol	Binary	Hexadecimal
2	0110010	32_{16}
0	0110000	30_{16}
Space	0100000	20_{16}
P	1010000	50_{16}
R	1010010	52_{16}
I	1001001	49_{16}
N	1001110	$4E_{16}$
T	1010100	54_{16}
Space	0100000	20_{16}
"	0100010	22_{16}
A	1000001	41_{16}
=	0111101	$3D_{16}$
"	0100010	22_{16}
;	0111011	$3B_{16}$
X	1011000	58_{16}

Related Problem   Determine the sequence of ASCII codes required for the following program statement and express them in hexadecimal:

```
80 INPUT Y
```

The ASCII Control Characters   The first thirty-two codes in the ASCII table (Table 2–7) represent the control characters. These are used to allow devices such as a computer and printer to communicate with each other when passing information and data. Table 2–8 lists the control characters and the control key function that allows them to be entered directly from an ASCII keyboard by pressing the control key (CTRL) and the corresponding symbol. A brief description of each control character is also given.

Extended ASCII Characters

In addition to the 128 standard ASCII characters, there are an additional 128 characters that were adopted by IBM for use in their PCs (personal computers). Because of the popularity of the PC, these particular extended ASCII characters are also used in applications other than PCs and have become essentially an unofficial standard.

The extended ASCII characters are represented by an 8-bit code series from hexadecimal 80 to hexadecimal FF.

TABLE 2–7
American Standard Code for Information Interchange (ASCII).

Control Characters

Name	Dec	Hex	Binary
NUL	0	00	0000000
SOH	1	01	0000001
STX	2	02	0000010
ETX	3	03	0000011
EOT	4	04	0000100
ENQ	5	05	0000101
ACK	6	06	0000110
BEL	7	07	0000111
BS	8	08	0001000
HT	9	09	0001001
LF	10	0A	0001010
VT	11	0B	0001011
FF	12	0C	0001100
CR	13	0D	0001101
SO	14	0E	0001110
SI	15	0F	0001111
DLE	16	10	0010000
DC1	17	11	0010001
DC2	18	12	0010010
DC3	19	13	0010011
DC4	20	14	0010100
NAK	21	15	0010101
SYN	22	16	0010110
ETB	23	17	0010111
CAN	24	18	0011000
EM	25	19	0011001
SUB	26	1A	0011010
ESC	27	1B	0011011
FS	28	1C	0011100
GS	29	1D	0011101
RS	30	1E	0011110
US	31	1F	0011111

Graphic Symbols

Symbol	Dec	Binary	Hex	Symbol	Dec	Binary	Hex	Symbol	Dec	Binary	Hex
space	32	0100000	20	@	64	1000000	40	`	96	1100000	60
!	33	0100001	21	A	65	1000001	41	a	97	1100001	61
"	34	0100010	22	B	66	1000010	42	b	98	1100010	62
#	35	0100011	23	C	67	1000011	43	c	99	1100011	63
$	36	0100100	24	D	68	1000100	44	d	100	1100100	64
%	37	0100101	25	E	69	1000101	45	e	101	1100101	65
&	38	0100110	26	F	70	1000110	46	f	102	1100110	66
'	39	0100111	27	G	71	1000111	47	g	103	1100111	67
(	40	0101000	28	H	72	1001000	48	h	104	1101000	68
)	41	0101001	29	I	73	1001001	49	i	105	1101001	69
*	42	0101010	2A	J	74	1001010	4A	j	106	1101010	6A
+	43	0101011	2B	K	75	1001011	4B	k	107	1101011	6B
,	44	0101100	2C	L	76	1001100	4C	l	108	1101100	6C
-	45	0101101	2D	M	77	1001101	4D	m	109	1101101	6D
.	46	0101110	2E	N	78	1001110	4E	n	110	1101110	6E
/	47	0101111	2F	O	79	1001111	4F	o	111	1101111	6F
0	48	0110000	30	P	80	1010000	50	p	112	1110000	70
1	49	0110001	31	Q	81	1010001	51	q	113	1110001	71
2	50	0110010	32	R	82	1010010	52	r	114	1110010	72
3	51	0110011	33	S	83	1010011	53	s	115	1110011	73
4	52	0110100	34	T	84	1010100	54	t	116	1110100	74
5	53	0110101	35	U	85	1010101	55	u	117	1110101	75
6	54	0110110	36	V	86	1010110	56	v	118	1110110	76
7	55	0110111	37	W	87	1010111	57	w	119	1110111	77
8	56	0111000	38	X	88	1011000	58	x	120	1111000	78
9	57	0111001	39	Y	89	1011001	59	y	121	1111001	79
:	58	0111010	3A	Z	90	1011010	5A	z	122	1111010	7A
;	59	0111011	3B	[	91	1011011	5B	{	123	1111011	7B
<	60	0111100	3C	\	92	1011100	5C	|	124	1111100	7C
=	61	0111101	3D	]	93	1011101	5D	}	125	1111101	7D
>	62	0111110	3E	^	94	1011110	5E	~	126	1111110	7E
?	63	0111111	3F	_	95	1011111	5F	Del	127	1111111	7F

TABLE 2–8
ASCII control characters.

Name	Decimal	Hex	Key	Description
NUL	0	00	CTRL @	null character
SOH	1	01	CTRL A	start of header
STX	2	02	CTRL B	start of text
ETX	3	03	CTRL C	end of text
EOT	4	04	CTRL D	end of transmission
ENQ	5	05	CTRL E	enquire
ACK	6	06	CTRL F	acknowledge
BEL	7	07	CTRL G	bell
BS	8	08	CTRL H	backspace
HT	9	09	CTRL I	horizontal tab
LF	10	0A	CTRL J	line feed
VT	11	0B	CTRL K	vertical tab
FF	12	0C	CTRL L	form feed (new page)
CR	13	0D	CTRL M	carriage return
SO	14	0E	CTRL N	shift out
SI	15	0F	CTRL O	shift in
DLE	16	10	CTRL P	data link escape
DC1	17	11	CTRL Q	device control 1
DC2	18	12	CTRL R	device control 2
DC3	19	13	CTRL S	device control 3
DC4	20	14	CTRL T	device control 4
NAK	21	15	CTRL U	negative acknowledge
SYN	22	16	CTRL V	synchronize
ETB	23	17	CTRL W	end of transmission block
CAN	24	18	CTRL X	cancel
EM	25	19	CTRL Y	end of medium
SUB	26	1A	CTRL Z	substitute
ESC	27	1B	CTRL [	escape
FS	28	1C	CTRL /	file separator
GS	29	1D	CTRL]	group separator
RS	30	1E	CTRL ^	record separator
US	31	1F	CTRL _	unit separator

The extended ASCII contains characters in the following general categories:

1. Foreign (non-English) alphabetic characters
2. Foreign currency symbols
3. Greek letters
4. Mathematical symbols
5. Drawing characters
6. Bar graphing characters
7. Shading characters

Table 2–9 is a list of the extended ASCII character set with the decimal and hexadecimal representations.

Parity Method for Error Detection

Many systems use a parity bit as a means for bit **error detection.** Any group of bits contain either an even or an odd number of 1s. A **parity** bit is attached to a group of bits to make the total number of 1s in a group always even or always odd. An even parity bit makes the total number of 1s even, and an odd parity bit makes the total odd.

A given system operates with even or odd parity, but not both. For instance, if a system operates with even parity, a check is made on each group of bits received to make sure the total number of 1s in that group is even. If there is an odd number of 1s, an error has occurred.

As an illustration of how parity bits are attached to a code, Table 2–10 lists the parity bits for each BCD number for both even and odd parity. The parity bit for each BCD number is in the *P* column.

TABLE 2–9
Extended ASCII characters.

Symbol	Dec	Hex	Symbol	Dec	Hex	Symbol	Dec	Hex	Symbol	Dec	Hex
Ç	128	80	á	160	A0	└	192	C0	α	224	E0
ü	129	81	í	161	A1	┴	193	C1	β	225	E1
é	130	82	ó	162	A2	┬	194	C2	Γ	226	E2
â	131	83	ú	163	A3	├	195	C3	π	227	E3
ä	132	84	ñ	164	A4	─	196	C4	Σ	228	E4
à	133	85	Ñ	165	A5	┼	197	C5	σ	229	E5
å	134	86	a	166	A6	╞	198	C6	μ	230	E6
ç	135	87	o	167	A7	╟	199	C7	τ	231	E7
ê	136	88	¿	168	A8	╚	200	C8	Φ	232	E8
ë	137	89	⌐	169	A9	╒	201	C9	Θ	233	E9
è	138	8A	¬	170	AA	╩	202	CA	Ω	234	EA·
ï	139	8B	½	171	AB	╦	203	CB	δ	235	EB
î	140	8C	¼	172	AC	╠	204	CC	∞	236	EC
ì	141	8D	¡	173	AD	═	205	CD	φ	237	ED
Ä	142	8E	«	174	AE	╬	206	CE	ε	238	EE
Å	143	8F	»	175	AF	╧	207	CF	∩	239	EF
É	144	90		176	B0	╨	208	D0	≡	240	F0
æ	145	91		177	B1	╤	209	D1	±	241	F1
Æ	146	92		178	B2	╥	210	D2	≥	242	F2
ô	147	93	│	179	B3	╙	211	D3	≤	243	F3
ö	148	94	┤	180	B4	╘	212	D4	∫	244	F4
ò	149	95	╡	181	B5	╒	213	D5	⌡	245	F5
û	150	96	╢	182	B6	╓	214	D6	÷	246	F6
ù	151	97	╖	183	B7	╫	215	D7	≈	247	F7
ÿ	152	98	╕	184	B8	╪	216	D8	°	248	F8
Ö	153	99	╣	185	B9	┘	217	D9	•	249	F9
Ü	154	9A	║	186	BA	┌	218	DA	·	250	FA
¢	155	9B	╗	187	BB	█	219	DB	√	251	FB
£	156	9C	╝	188	BC	▄	220	DC	η²	252	FC
¥	157	9D	╜	189	BD	▌	221	DD	²	253	FD
Pτ	158	9E	╛	190	BE	▐	222	DE	■	254	FE
ƒ	159	9F	┐	191	BF	▀	223	DF		255	FF

TABLE 2–10
The BCD code with parity bits.

Even Parity		Odd Parity	
P	BCD	P	BCD
0	0000	1	0000
1	0001	0	0001
1	0010	0	0010
0	0011	1	0011
1	0100	0	0100
0	0101	1	0101
0	0110	1	0110
1	0111	0	0111
1	1000	0	1000
0	1001	1	1001

The parity bit can be attached to the code at either the beginning or the end, depending on system design. Notice that the total number of 1s, including the parity bit, is always even for even parity and always odd for odd parity.

Detecting an Error A parity bit provides for the detection of a single bit error (or any odd number of errors, which is very unlikely) but cannot check for two errors in one group. For instance, let's assume that we wish to transmit the BCD code 0101. (Parity can be used with any number of bits; we are using four for illustration.) The total code transmitted, including the even parity bit, is

Now let's assume that an error occurs in the third bit from the left (the 1 becomes a 0) as follows:

When this code is received, the parity check circuitry determines that there is only a single 1 (odd number), when there should be an even number of 1s. Because an even number of 1s does not appear in the code when it is received, an error is indicated.

An odd parity bit also provides in a similar manner for the detection of a single error in a given group of bits.

EXAMPLE 2–39

Assign the proper even parity bit to the following code groups:

(a) 1010   (b) 111000   (c) 101101
(d) 100011100101   (e) 101101011111

Solution   Make the parity bit either 1 or 0 as necessary to make the total number of 1s even. The parity bit will be the left-most bit (color).

(a) **0**1010   (b) **1**111000   (c) **0**101101

(d) **0**100011100101   (e) **1**101101011111

Related Problem   Add an even parity bit to the 7-bit ASCII code for the letter K.

EXAMPLE 2–40

An odd parity system receives the following code groups: 10110, 11010, 110011, 110101110100, and 1100010101010. Determine which groups, if any, are in error.

Solution   Since odd parity is required, any group with an even number of 1s is incorrect. The following groups are in error: **110011** and **1100010101010.**

Related Problem   The following ASCII character is received by an odd parity system: 00110111. Is it correct?

SECTION 2–11 REVIEW

1. Convert the following binary numbers to the Gray code:
   (a) 1100   (b) 1010   (c) 11010
2. Convert the following Gray codes to binary:
   (a) 1000   (b) 1010   (c) 11101
3. What is the ASCII representation for each of the following characters? Express each as a bit pattern and in hexadecimal notation.
   (a) K   (b) r   (c) $ (d) +
4. Add an even parity bit to each ASCII code found in question 3.

2–12 ■ DIGITAL SYSTEM APPLICATION

The system that was introduced in Chapter 1 provides a simple way to illustrate how binary numbers and codes can be used in digital systems. Recall that the system controls the number of tablets for each bottle on an assembly line in a pharmaceutical plant. In this section, you will see an example of the binary numbers and codes that exist at various points in this system during a part of its sequence. You will also determine numbers and codes throughout the system under certain specified conditions.

After completing this section, you should be able to

■ Explain the purpose of each functional block in the system in terms of the numbers and codes ■ Analyze the system under specified conditions in terms of the numbers and codes at various points

System Operation

The block diagrams in Figures 2–5, 2–6, and 2–7 show the state of the system at certain points in its sequence. The binary numbers (red) and BCD codes (blue) indicate the inputs and outputs of each block and the contents of the storage devices. The decimal value for each binary or BCD code is given in black. The wide arrows represent parallel lines that connect

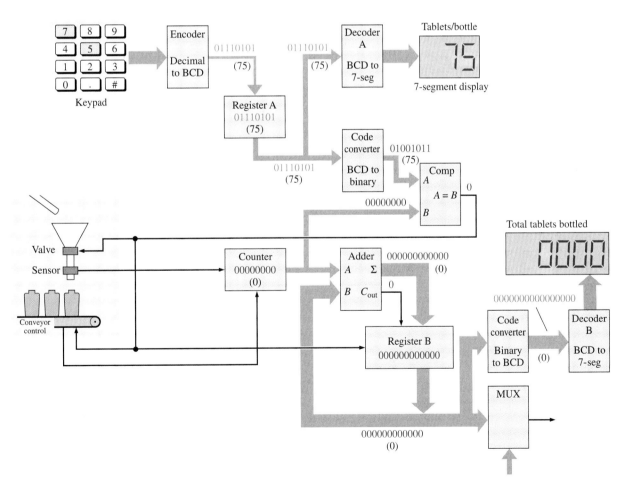

FIGURE 2–5
The system is in its initial state.

the various blocks in the system and provide paths for parallel transfer of data. For example, eight bits of a given binary number are simultaneously present on eight parallel lines. Parallel lines such as these are commonly called *buses* in digital systems terminology.

The block diagram of the system is shown in a simplified form, leaving out many details that are not essential at this time. Also, the remote unit that was shown in Chapter 1 is omitted to keep things as simple as possible.

Initializing the System At the beginning, the system must be initialized so that it starts in the proper binary states. First, the counter and the registers must be reset so the count can start from zero. Next, the number of tablets (up to 99) that will be put into each bottle is entered on the keypad, converted from decimal to BCD by the encoder, and stored in register A.

In Figure 2–5, the decimal number 75 is entered on the keypad, indicating that 75 tablets will go into each bottle. Each keystroke is encoded into BCD and sent, four bits at a time, to be stored in register A. The 8-bit BCD code is placed on the register's output lines and is applied to the decoder which converts it from BCD to a 7-segment code to light the display. The same BCD code is simultaneously applied to the code converter which converts it to binary. The resulting binary number for 75 appears at the *A* input to the comparator (comp). Now the system is initialized and ready to start counting tablets.

The System at Work Now assume that the system has counted a total of 50 bottles (3750 tablets), and it is in the process of counting tablets for the fifty-first bottle. In fact, it has counted 74 tablets as indicated by the contents of the counter in Figure 2–6. Notice that

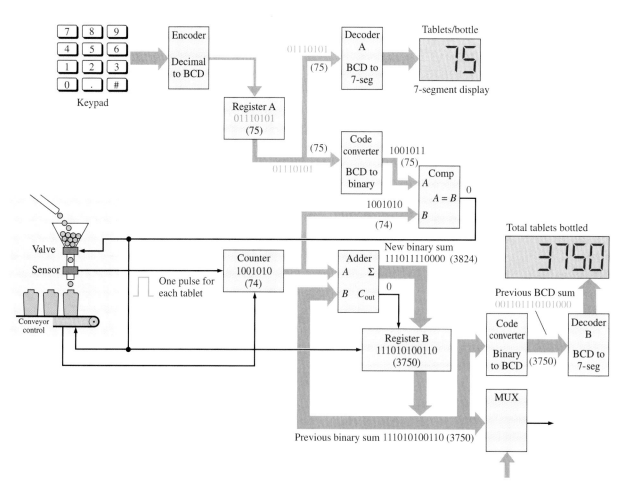

FIGURE 2–6
The system has counted 50 bottles of tablets and is working on the next bottle.

register B is holding the previous binary sum (3750) which is converted to BCD and then to 7-segment format for the display. Register B is updated each time another bottle is filled but not during the filling of a bottle.

The output of the comparator is LOW (0) because the *A* and *B* inputs are not equal at this time. Input *A* has a binary 75 and input *B* has a binary 74. The output of the adder always indicates the most current sum, in this case 3824, but this number is not stored in register B until the bottle is full.

In the next state, as shown in Figure 2–7, the counter has reached a binary count of 75. As a result, the *A* = *B* output of the comparator goes HIGH (1) and causes several things to happen simultaneously. It stops the flow of tablets, it allows the new binary sum on the output of the adder (3825) to be stored in register B, and it initiates an advance of the conveyor to get the next bottle into position. The binary number in register B, which is now the current sum, is applied to the code converter and converted to BCD and then to 7-segment format to update the display to 3825. As soon as the next bottle is in place, the counter is reset to zero and the cycle starts over.

FIGURE 2–7
The system has just counted its fifty-first bottle of tablets.

■ SYSTEM ASSIGNMENT

■ *Activity 1:   Initialization*

(a) The system is set up for 60 tablets per bottle. Specify the contents of register A and the data on input *A* of the comparator.

(b) What is the maximum binary number to which the counter will go when the system is initialized for 60 tablets per bottle?

■ *Activity 2:   Maximum Count*

(a) Determine the maximum number of full bottles that can be counted by the system if it is set for 75 tablets per bottle. The number of bits shown at each point is the maximum number of bits.

(b) Specify the binary number in register B at the time the last bottle has been filled.

(c) Specify the BCD number at the output of the binary to BCD code converter at this time.

■ *Activity 3:   Sequencing*

(a) If the system is set for 30 tablets per bottle and at a given point, 523 bottles have been counted, determine the BCD number in register A and the binary number in register B.

(b) If the counter is at the binary number 00010110, determine the binary output of the adder for the condition in part (a).

(c) How many total tablets have actually been counted in part (b)?

(d) What is the current number on the total display and what will be the next number?

■ *Activity 4:   Hexadecimal*

(a) Express each binary number in Figure 2–5 in hexadecimal.

(b) Express each binary number in Figure 2–6 in hexadecimal.

(c) Express each binary number in Figure 2–7 in hexadecimal.

(d) If the counter is at $4C_{16}$ and then it counts nine more tablets, what will be its hexadecimal count?

(e) If the display showing the total tablets in Figure 2–6 were a hexadecimal display, what would it show?

SECTION 2–12 REVIEW

1. In the system discussed in this section, when does the comparator output go to a HIGH level?

2. Under what condition does the display showing number of tablets per bottle change?

3. Under what condition does the display showing total tablets change?

■ SUMMARY

- A binary number is a weighted number in which the weight of each whole number digit is a positive power of two and the weight of each fractional digit is a negative power of two. The whole number weights increase from right to left—from least significant digit to most significant.
- A binary number can be converted to a decimal number by summing the decimal values of the weights of all the 1s in the binary number.
- A decimal whole number can be converted to binary by using the sum-of-weights or the repeated division-by-2 method.
- A decimal fraction can be converted to binary by using the sum-of-weights or the repeated multiplication-by-2 method.
- The basic rules for binary addition are as follows:

$$0 + 0 = 0$$
$$0 + 1 = 1$$
$$1 + 0 = 1$$
$$1 + 1 = 10$$

- The basic rules for binary subtraction are as follows:

$$0 - 0 = 0$$
$$1 - 1 = 0$$
$$1 - 0 = 1$$
$$10 - 1 = 1$$

- The 1's complement of a binary number is derived by changing 1s to 0s and 0s to 1s.
- The 2's complement of a binary number can be derived by adding 1 to the 1's complement.
- Binary subtraction can be accomplished with addition by using the 1's or 2's complement method.
- A positive binary number is represented by a 0 sign bit.
- A negative binary number is represented by a 1 sign bit.
- For arithmetic operations, negative binary numbers are represented in 1's complement or 2's complement form.
- In an addition operation, an overflow is possible when both numbers are positive or when both numbers are negative. An incorrect sign bit in the sum indicates the occurrence of an overflow.
- The hexadecimal number system consists of 16 digits and characters, 0 through 9 followed by A through F.
- One hexadecimal digit represents a 4-bit binary number, and its primary usefulness is in simplifying bit patterns and making them easier to read.
- A decimal number can be converted to hexadecimal by the repeated division-by-16 method.
- The octal number system consists of eight digits, 0 through 7.
- A decimal number can be converted to octal by using the repeated division-by-8 method.
- Octal-to-binary conversion is accomplished by simply replacing each octal digit with its 3-bit binary equivalent. The process is reversed for binary-to-octal conversion.
- A decimal number is converted to BCD by replacing each decimal digit with the appropriate 4-bit binary code.
- The ASCII is a 7-bit alphanumeric code that is widely used in computer systems for input and output of information.

■ KEY TERMS

These terms are also in the end-of-book glossary.

Alphanumeric   Consisting of numerals, letters, and other characters.

ASCII   American Standard Code for Information Interchange; the most widely used alphanumeric code.

BCD   Binary coded decimal; a digital code in which each of the decimal digits, 0 through 9, is represented by group of four bits.

Byte   A group of eight bits.

Carry   The digit generated when the sum of two binary digits exceeds 1.

Complement   The inverse or opposite of a number.

Decimal   Describes a number system with a base of ten.

Digit   A symbol used to express a quantity.

Exponent   The part of a floating-point number that represents the number of places that the decimal point (or binary point) is to be moved.

Floating-point number   A number representation based on scientific notation in which the number consists of an exponent and a mantissa.

Gray code   An unweighted digital code characterized by a single bit change between adjacent code numbers in a sequence.

Hexadecimal   Describes a number system with a base of 16.

Integer   A whole number.

Least significant bit (LSB)   Generally, the right-most bit in a binary whole number or code.

Mantissa   The magnitude of a floating-point number.

Most significant bit (MSB)   The left-most bit in a binary whole number or code.

Octal   Describes a number system with a base of eight.

Overflow   The condition that occurs when the number of bits in a sum exceeds the number of bits in each of the numbers added.

Parity   In relation to binary codes, the condition of evenness or oddness of the number of 1s in a code group.

Sign bit   The left-most bit of a binary number that designates whether the number is positive (0) or negative (1).

Weight   The value of a digit in a number based on its position in the number.

■ SELF-TEST

1. $2 \times 10^1 + 8 \times 10^0$ is equal to
   (a) 10   **(b)** 280   **(c)** 2.8   **(d)** 28

2. The binary number 1101 is equal to the decimal number
   (a) 13   **(b)** 49   **(c)** 11   **(d)** 3

3. The binary number 11011101 is equal to the decimal number
   (a) 121   **(b)** 221   **(c)** 441   **(d)** 256

4. The decimal number 17 is equal to the binary number
   (a) 10010   **(b)** 11000   **(c)** 10001   **(d)** 01001

5. The decimal number 175 is equal to the binary number
   (a) 11001111   **(b)** 10101110   **(c)** 10101111   **(d)** 11101111

6. The sum of 11010 + 01111 equals
   (a) 101001   **(b)** 101010   **(c)** 110101   **(d)** 101000

7. The difference of 110 − 010 equals
   (a) 001   **(b)** 010   **(c)** 101   **(d)** 100

8. The 1's complement of 10111001 is

    (a) 01000111   **(b)** 01000110   **(c)** 11000110   **(d)** 10101010

9. The 2's complement of 11001000 is

    (a) 00110111   **(b)** 00110001   **(c)** 01001000   **(d)** 00111000

10. The decimal number $+122$ is expressed in the 2's complement system as

    (a) 01111010   **(b)** 11111010   **(c)** 01000101   **(d)** 10000101

11. The decimal number -34 is expressed in the 2's complement system as

    (a) 01011110   **(b)** 10100010   **(c)** 11011110   **(d)** 01011101

12. A single-precision floating-point binary number has a total of

    (a) 8 bits   **(b)** 16 bits   **(c)** 24 bits   **(d)** 32 bits

13. In the 2's complement system, the binary number 10010011 is equal to the decimal number

    (a) -19 **(b)** $+109$ **(c)** $+91$ **(d)** -109

14. The binary number 101100111001010100001 can be written in octal as

    (a) 5471230_8 **(b)** 5471241_8 **(c)** 2634521_8 **(d)** 23162501_8

15. The binary number 1000110101000110111 can be written in hexadecimal as

    (a) $AD467_{16}$ **(b)** $8C46F_{16}$ **(c)** $8D46F_{16}$ **(d)** $AE46F_{16}$

16. The binary number for $F7A9_{16}$ is

    (a) 1111011110101001   **(b)** 1110111110101001

    (c) 1111111010110001   **(d)** 1111011010101001

17. The BCD number for decimal 473 is

    (a) 111011010   **(b)** 110001110011   **(c)** 010001110011   **(d)** 010011110011

18. Refer to Table 2–7. The word STOP in ASCII is

    (a) 1010011101010010011111010000   **(b)** 1010010100110010011101010000

    (c) 1001010110110110011101010001   **(d)** 1010011101010010011101100100

■ PROBLEMS

SECTION 2–1   Decimal Numbers

1. What is the weight of the digit 6 in each of the following decimal numbers?

    (a) 1386   **(b)** 54,692   **(c)** 671,920

2. Express each of the following decimal numbers as a power of ten:

    (a) 10   **(b)** 100   **(c)** 10,000   **(d)** 1,000,000

3. Give the value of each digit in the following decimal numbers:

    (a) 471   **(b)** 9356   **(c)** 125,000

4. How high can you count with four decimal digits?

SECTION 2–2   Binary Numbers

5. Convert the following binary numbers to decimal:

    (a) 11   **(b)** 100   **(c)** 111   **(d)** 1000

    (e) 1001   **(f)** 1100   **(g)** 1011   **(h)** 1111

6. Convert the following binary numbers to decimal:

    (a) 1110   **(b)** 1010   **(c)** 11100   **(d)** 10000

    (e) 10101   **(f)** 11101   **(g)** 10111   **(h)** 11111

7. Convert each binary number to decimal:

    (a) 110011.11   **(b)** 101010.01   **(c)** 1000001.111

    (d) 1111000.101   **(e)** 1011100.10101   **(f)** 1110001.0001

    (g) 1011010.1010   **(h)** 1111111.11111

8. What is the highest decimal number that can be represented by each of the following numbers of binary digits (bits)?

    (a) two   **(b)** three   **(c)** four   **(d)** five   **(e)** six

    (f) seven   **(g)** eight   **(h)** nine   **(i)** ten   **(j)** eleven

9. How many bits are required to represent the following decimal numbers?

(a) 17   (b) 35   (c) 49   (d) 68

(e) 81   (f) 114   (g) 132   (h) 205

10. Generate the binary sequence for each decimal sequence:

(a) 0 through 7   (b) 8 through 15   (c) 16 through 31

(d) 32 through 63   (e) 64 through 75

SECTION 2–3   Decimal-to-Binary Conversion

11. Convert each decimal number to binary by using the sum-of-weights method:

(a) 10   (b) 17   (c) 24   (d) 48

(e) 61   (f) 93   (g) 125   (h) 186

12. Convert each decimal fraction to binary using the sum-of-weights method:

(a) 0.32   (b) 0.246   (c) 0.0981

13. Convert each decimal number to binary using repeated division by 2:

(a) 15   (b) 21   (c) 28   (d) 34

(e) 40   (f) 59   (g) 65   (h) 73

14. Convert each decimal fraction to binary using repeated multiplication by 2:

(a) 0.98   (b) 0.347   (c) 0.9028

SECTION 2–4   Binary Arithmetic

15. Add the binary numbers:

(a) $11 + 01$ (b) $10 + 10$ (c) $101 + 11$

(d) $111 + 110$ (e) $1001 + 101$ (f) $1101 + 1011$

16. Use direct subtraction on the following binary numbers:

(a) $11 - 1$ (b) $101 - 100$ (c) $110 - 101$

(d) $1110 - 11$ (e) $1100 - 1001$ (f) $11010 - 10111$

17. Perform the following binary multiplications:

(a) 11×11 (b) 100×10 (c) 111×101

(d) 1001×110 (e) 1101×1101 (f) 1110×1101

18. Divide the binary numbers as indicated:

(a) $100 \div 10$ (b) $1001 \div 11$ (c) $1100 \div 100$

SECTION 2–5   1's and 2's Complements of Binary Numbers

19. Determine the 1's complement of each binary number:

(a) 101   (b) 110   (c) 1010

(d) 11010111   (e) 1110101   (f) 00001

20. Determine the 2's complement of each binary number using either method:

(a) 10   (b) 111   (c) 1001   (d) 1101

(e) 11100   (f) 10011   (g) 10110000   (h) 00111101

SECTION 2–6   Signed Numbers

21. Express each decimal number in binary as an 8-bit sign-magnitude number:

(a) $+29$ (b) -85 (c) $+100$ (d) -123

22. Express each decimal number as an 8-bit number in the 1's complement system:

(a) -34 (b) $+57$ (c) -99 (d) $+115$

23. Express each decimal number as an 8-bit number in the 2's complement system:

(a) $+12$ (b) -68 (c) $+101$ (d) -125

24. Determine the decimal value of each signed binary number in the sign-magnitude system:

(a) 10011001   (b) 01110100   (c) 10111111

25. Determine the decimal value of each signed binary number in the 1's complement system:
   (a) 10011001   (b) 01110100   (c) 10111111

26. (a) Determine the decimal value of each signed binary number in the 2's complement system:
   (a) 10011001   (b) 01110100   (c) 10111111

27. Express each of the following sign-magnitude binary numbers in single-precision floating-point format:
   (a) 0111110000101011   (b) 100110000011000

28. Determine the values of the following single-precision floating-point numbers:
   (a) 1 10000001 01001001110001000000000
   (b) 0 11001100 10000111110100100000000

SECTION 2–7   Arithmetic Operations with Signed Numbers

29. Convert each pair of decimal numbers to binary and add using the 2's complement system:
   (a) 33 and 15   (b) 56 and −27   (c) −46 and 25   (d) −110 and −84

30. Perform each addition in the 2's complement system:
   (a) 00010110 + 00110011   (b) 01110000 + 10101111

31. Perform each addition in the 2's complement system:
   (a) 10001100 + 00111001   (b) 11011001 + 11100111

32. Perform each subtraction in the 2's complement system:
   (a) 00110011 − 00010000   (b) 01100101 − 11101000

33. Multiply 01101010 by 11110001 in the 2's complement system.

34. Divide 01000100 by 00011001 in the 2's complement system.

SECTION 2–8   Hexadecimal Numbers

35. Convert each hexadecimal number to binary:
   (a) 38_{16} (b) 59_{16} (c) $A14_{16}$ (d) $5C8_{16}$
   (e) 4100_{16} (f) $FB17_{16}$ (g) $8A9D_{16}$

36. Convert each binary number to hexadecimal:
   (a) 1110   (b) 10   (c) 10111
   (d) 10100110   (e) 1111110000   (f) 100110000010

37. Convert each hexadecimal number to decimal:
   (a) 23_{16} (b) 92_{16} (c) $1A_{16}$ (d) $8D_{16}$
   (e) $F3_{16}$ (f) EB_{16} (g) $5C2_{16}$ (h) 700_{16}

38. Convert each decimal number to hexadecimal:
   (a) 8   (b) 14   (c) 33   (d) 52
   (e) 284   (f) 2890   (g) 4019   (h) 6500

39. Perform the following additions:
   (a) $37_{16} + 29_{16}$ (b) $A0_{16} + 6B_{16}$ (c) $FF_{16} + BB_{16}$

40. Perform the following subtractions:
   (a) $51_{16} − 40_{16}$ (b) $C8_{16} − 3A_{16}$ (c) $FD_{16} − 88_{16}$

SECTION 2–9   Octal Numbers

41. Convert each octal number to decimal:
   (a) 12_8 (b) 27_8 (c) 56_8 (d) 64_8 (e) 103_8
   (f) 557_8 (g) 163_8 (h) 1024_8 (i) 7765_8

42. Convert each decimal number to octal by repeated division by 8:
   (a) 15   (b) 27   (c) 46   (d) 70
   (e) 100   (f) 142   (g) 219   (h) 435

43. Convert each octal number to binary:
   (a) 13_8 (b) 57_8 (c) 101_8 (d) 321_8 (e) 540_8
   (f) 4653_8 (g) 13271_8 (h) 45600_8 (i) 100213_8

44. Convert each binary number to octal:

(a) 111   (b) 10   (c) 110111

(d) 101010   (e) 1100   (f) 1011110

(g) 101100011001   (h) 10110000011   (i) 111111101111000

SECTION 2–10   Binary Coded Decimal (BCD)

45. Convert each of the following decimal numbers to 8421 BCD:

(a) 10   (b) 13   (c) 18   (d) 21   (e) 25   (f) 36

(g) 44   (h) 57   (i) 69   (j) 98   (k) 125   (l) 156

46. Convert each of the decimal numbers in Problem 45 to straight binary, and compare the number of bits required with that required for BCD.

47. Convert the following decimal numbers to BCD:

(a) 104   (b) 128   (c) 132   (d) 150   (e) 186

(f) 210   (g) 359   (h) 547   (i) 1051

48. Convert each of the BCD numbers to decimal:

(a) 0001   (b) 0110   (c) 1001

(d) 00011000   (e) 00011001   (f) 00110010

(g) 01000101   (h) 10011000   (i) 100001110000

49. Convert each of the BCD numbers to decimal:

(a) 10000000   (b) 001000110111

(c) 001101000110   (d) 010000100001

(e) 011101010100   (f) 100000000000

(g) 100101111000   (h) 0001011010000011

(i) 1001000000011000   (j) 0110011001100111

50. Add the following BCD numbers:

(a) 0010 + 0001   (b) 0101 + 0011

(c) 0111 + 0010   (d) 1000 + 0001

(e) 00011000 + 00010001   (f) 01100100 + 00110011

(g) 01000000 + 01000111   (h) 10000101 + 00010011

51. Add the following BCD numbers:

(a) 1000 + 0110   (b) 0111 + 0101

(c) 1001 + 1000   (d) 1001 + 0111

(e) 00100101 + 00100111   (f) 01010001 + 01011000

(g) 10011000 + 10010111   (h) 010101100001 + 011100001000

52. Convert each pair of decimal numbers to BCD, and add as indicated:

(a) 4 + 3   (b) 5 + 2   (c) 6 + 4   (d) 17 + 12

(e) 28 + 23   (f) 65 + 58   (g) 113 + 101   (h) 295 + 157

SECTION 2–11   Digital Codes and Parity

53. In a certain application a 4-bit binary sequence cycles from 1111 to 0000 periodically. There are four bit changes, and because of circuit delays, these changes may not occur at the same instant. For example, if the LSB changes first, the number will appear as 1110 during the transition from 1111 to 0000 and may be misinterpreted by the system. Illustrate how the Gray code avoids this problem.

54. Convert each binary number to Gray code:

(a) 11011   (b) 1001010   (c) 1111011101110

55. Convert each Gray code to binary:

(a) 1010   (b) 00010   (c) 11000010001

56. Convert each of the following decimal numbers to ASCII. Refer to Table 2–7.

(a) 1   (b) 3   (c) 6   (d) 10   (e) 18

(f) 29   (g) 56   (h) 75   (i) 107

57. Determine each ASCII character. Refer to Table 2–7.

    (a) 0011000   **(b)** 1001010   **(c)** 0111101

    (d) 0100011   **(e)** 0111110   **(f)** 1000010

58. Decode the following ASCII coded message:

        1001000 1100101 1101100 1101100 1101111 0101110
        0100000 1001000 1101111 1110111 0100000 1100001
        1110010 1100101 0100000 1111001 1101111 1110101
        0111111

59. Write the message in Problem 58 in hexadecimal.

60. Convert the following computer program statement to ASCII:

    `30 INPUT A, B`

61. Determine which of the following even parity codes are in error:

    (a) 100110010   **(b)** 011101010   **(c)** 10111111010001010

62. Determine which of the following odd parity codes are in error:

    (a) 11110110   **(b)** 00110001   **(c)** 01010101010101010

63. Attach the proper even parity bit to each of the following bytes of data:

    (a) 10100100   **(b)** 00001001   **(c)** 11111110

SECTION 2–12   Digital System Application

64. In the tablet-counting and control system, what limits the maximum number of tablets per bottle that can be preset into the system?

65. Assume that the system is set up for 25 tablets per bottle. Determine the binary numbers at each of the following locations in the system after 15 tablets have gone into the 1000th bottle?

    (a) In register A   **(b)** Input *A* to comparator   **(c)** In the counter

    (d) In register B   **(e)** Output of adder

66. Determine the maximum number of total tablets that the system can count. Explain what changes would be required to increase this number.

■ ANSWERS TO SECTION REVIEWS

SECTION 2–1

1. (a) 1370: 10   **(b)** 6725: 100   **(c)** 7051: 1000   **(d)** 58.72: 0.1

2. (a) $51 = (5 \times 10) + (1 \times 1)$ **(b)** $137 = (1 \times 100) + (3 \times 10) + (7 \times 1)$

    (c) $1492 = (1 \times 1000) + (4 \times 100) + (9 \times 10) + (2 \times 1)$

    (d) $106.58 = (1 \times 100) + (0 \times 10) + (6 \times 1) + (5 \times 0.1) + (8 \times 0.01)$

SECTION 2–2

1. $2^8 - 1 = 255$ **2.** Weight is 16.   **3.** $10111101.011 = 189.375$

SECTION 2–3

1. (a) $23 = 10111$ **(b)** $57 = 111001$ **(c)** $45.5 = 101101.1$

2. (a) $14 = 1110$ **(b)** $21 = 10101$ **(c)** $0.375 = 0.011$

SECTION 2–4

1. (a) $1101 + 1010 = 10111$ **(b)** $10111 + 01101 = 100100$

2. (a) $1101 - 0100 = 1001$ **(b)** $1001 - 0111 = 0010$

3. (a) $110 \times 111 = 101010$ **(b)** $1100 \div 011 = 100$

SECTION 2–5

1. (a) 1's comp of 00011010 = 11100101   **(b)** 1's comp of 11110111 = 00001000
   (c) 1's comp of 10001101 = 01110010
2. (a) 2's comp of 00010110 = 11101010   **(b)** 2's comp of 11111100 = 00000100
   (c) 2's comp of 10010001 = 01101111

SECTION 2–6

1. Sign-magnitude: $+9 = 00001001$ **2.** 1's comp: $-33 = 11011110$
3. 2's comp: $-46 = 11010010$ **4.** Exponent and mantissa

SECTION 2–7

1. Cases of addition: positive number is larger, negative number is larger, both are positive, both are negative
2. $00100001 + 10111100 = 11011101$
3. $01110111 - 00110010 = 01000101$
4. Sign of product is positive.
5. $00000101 \times 01111111 = 01001111011$
6. Sign of quotient is negative.
7. $00110000 \div 00001100 = 00000100$

SECTION 2–8

1. (a) $10110011 = B3_{16}$ **(b)** $110011101000 = CE8_{16}$
2. (a) $57_{16} = 01010111$ **(b)** $3A5_{16} = 001110100101$
   (c) $F8OB_{16} = 1111100000001011$
3. $9B30_{16} = 39,728_{10}$ **4.** $573_{10} = 23D_{16}$
5. (a) $18_{16} + 34_{16} = 4C_{16}$ **(b)** $3F_{16} + 2A_{16} = 69_{16}$
6. (a) $75_{16} - 21_{16} = 54_{16}$ **(b)** $94_{16} - 5C_{16} = 38_{16}$

SECTION 2–9

1. (a) $73_8 = 59_{10}$ **(b)** $125_8 = 85_{10}$ **2.   (a)** $98_{10} = 142_8$ **(b)** $163_{10} = 243_8$
3. (a) $46_8 = 100110$ **(b)** $723_8 = 111010011$ **(c)** $5624_8 = 101110010100$
4. (a) $110101111 = 657_8$ **(b)** $1001100010 = 1142_8$ **(c)** $10111111001 = 2771_8$

SECTION 2–10

1. (a) 0010: 2   **(b)** 1000: 8   **(c)** 0001: 1   **(d)** 0100: 4
2. (a) $6_{10} = 0110$ **(b)** $15_{10} = 00010101$ **(c)** $273_{10} = 001001110011$
   (d) $849_{10} = 100001001001$
3. (a) $10001001 = 89_{10}$ **(b)** $001001111000 = 278_{10}$ **(c)** $000101010111 = 157_{10}$
4. A 4-bit sum is invalid when it is greater than 9_{10}.

SECTION 2–11

1. (a) $1100_2 = 1010$ Gray   **(b)** $1010_2 = 1111$ Gray   **(c)** $11010_2 = 10111$ Gray
2. (a) 1000 Gray = 1111_2 **(b)** 1010 Gray = 1100_2 **(c)** 11101 Gray = 10110_2
3. (a) K: $1001011 \rightarrow 4B_{16}$ **(b)** r: $1110010 \rightarrow 72_{16}$
   (c) \$: $0100100 \rightarrow 24_{16}$ **(d)** +: $0101011 \rightarrow 2B_{16}$
4. (a) *0*1001011   **(b)** *0*1110010   **(c)** *00*100100   **(d)** *00*101011

SECTION 2-12

1. The comparator output goes HIGH when the actual tablets counted equals the preset number.

2. The number of tablets/bottle display changes when a new number is entered on the keypad.

3. The total tablets display changes each time another bottle is filled.

■ **ANSWERS TO RELATED PROBLEMS FOR EXAMPLES**

2-1 9 has a value of 900, 3 has a value of 30, 9 has a value of 9.

2-2 6 has a value of 60, 7 has a value of 7, 9 has a value of 9/10 (0.9), 2 has a value of 2/100 (0.02), 4 has a value of 4/1000 (0.004).

2-3 $10010001 = 128 + 16 + 1 = 145$ **2-4** $10.111 = 2 + 0.5 + 0.25 + 0.125 = 2.875$

2-5 $125 = 64 + 32 + 16 + 8 + 4 + 1 = 1111101$ **2-6** $39 = 100111$

2-7 $1111 + 1100 = 11011$ **2-8** $111 - 100 = 011$ **2-9** $110 - 101 = 001$

2-10 $1101 \times 1010 = 10000010$ **2-11** $1100 \div 100 = 11$ **2-12** 00110101

2-13 01000000 **2-14** See Table 2-11.

TABLE 2-11

	Sign-Magnitude	1's comp	2's comp
+19	00010011	00010011	00010011
−19	10010011	11101100	11101101

2-15 $01110111 = +119_{10}$ **2-16** $11101011 = -20_{10}$ **2-17** $11010111 = -41_{10}$

2-18 $11000010001010011000000000$

2-19 01010101 **2-20** 00010001 **2-21** 1001000110

2-22 $(83)(-59) = -4897$ (10110011011111 in 2's comp)   **2-23** $100 \div 25 = 4$ (0100)

2-24 $4F79C_{16}$ **2-25** 0110101111010011_2

2-26 $6BD_{16} = 011010111101 = 2^{10} + 2^9 + 2^7 + 2^5 + 2^4 + 2^3 + 2^2 + 2^0$
$= 1024 + 512 + 128 + 32 + 16 + 8 + 4 + 1 = 1725_{10}$

2-27 $60A_{16} = (6 \times 256) + (0 \times 16) + (10 \times 1) = 1546_{10}$

2-28 $2591_{10} = A1F_{16}$ **2-29** $4C_{16} + 3A_{16} = 86_{16}$ **2-30** $BCD_{16} - 173_{16} = A5A_{16}$

2-31 **(a)** $001011_2 = 11_{10} = 13_8$ **(b)** $010101_2 = 21_{10} = 25_8$
(c) $001100000_2 = 96_{10} = 140_8$ **(d)** $111101010110_2 = 3926_{10} = 7526_8$

2-32 1250762_8 **2-33** 1001011001110011

2-34 $82,276_{10}$ **2-35** 1001100101101000 **2-36** 10000010

2-37 **(a)** 111011 (Gray)   **(b)** 111010_2

2-38 The sequence of codes for 80 INPUT Y is $38_{16}30_{16}20_{16}49_{16}4E_{16}50_{16}55_{16}54_{16}20_{16}59_{16}$

2-39 01001011 **2-40** Yes

■ **ANSWERS TO SELF-TEST**

1. (d)   **2.** (a)   **3.** (b)   **4.** (c)   **5.** (c)   **6.** (a)   **7.** (d)   **8.** (b)
9. (d)   **10.** (a)   **11.** (c)   **12.** (d)   **13.** (d)   **14.** (b)   **15.** (c)   **16.** (a)
17. (c)   **18.** (a)

3

LOGIC GATES

 Computer Simulation
of Digital Circuits:
Electronics Workbench (EWB) Tutorial at:
http://www.prenhall.com/floyd

■ CHAPTER OBJECTIVES

❑ Describe the operation of the inverter, the AND gate, and the OR gate
❑ Describe the operation of the NAND gate and the NOR gate
❑ Express the operation of NOT, AND, OR, NAND, and NOR gates with Boolean algebra
❑ Describe the operation of the exclusive-OR and exclusive-NOR gates
❑ Recognize and use both the distinctive shape logic gate symbols and the rectangular outline logic gate symbols of ANSI/IEEE Standard 91–1984
❑ Construct timing diagrams showing the proper time relationships of inputs and outputs for the various logic gates
❑ Make basic comparisons between the major IC technologies—CMOS and TTL
❑ Explain how the different series within the CMOS and TTL families differ from each other
❑ Define *propagation delay time, power dissipation, speed-power product,* and *fan-out* in relation to logic gates
❑ List specific devices that contain the various logic gates
❑ Use each logic gate in simple applications
❑ Troubleshoot logic gates for opens and shorts by using the logic pulser and probe or the oscilloscope
❑ Apply logic gates in a system application

■ CHAPTER OVERVIEW

The emphasis in this chapter is on the logical operation, application, and troubleshooting of logic gates. The relationship of input and output waveforms of a gate using timing diagrams is thoroughly covered.

Logic symbols used to represent the logic gates are in accordance with ANSI/IEEE Standard 91–1984. This standard has been adopted by private industry and the military for use in internal documentation as well as published literature.

Because integrated circuits (IC) are used in all applications, the logic function of a device is generally of greater importance to the technician or technologist than the details of the component-level circuit operation within the IC package. Therefore, detailed coverage of the devices at the component level can be treated as an optional topic. For those who need it and have the time, a thorough coverage of digital integrated circuit technologies is available in Chapter 15. Portions or all of this "floating" chapter can be used after this chapter or at any appropriate point throughout the text. For those choosing to do so, Chapter 15 may be omitted or partially omitted without affecting any of the text material.

■ SPECIFIC DEVICES (CMOS AND TTL SERIES)

74XX00	74XX02	74XX04
74XX08	74XX10	74XX11
74XX20	74XX21	74XX27
74XX30	74XX32	74XX86
74XX133		

■ DIGITAL SYSTEM APPLICATION

This Digital System Application illustrates concepts taught in this chapter. The system applications in this text are intended to present "virtual" systems that provide some perspective on how logic circuits may be used in purposeful and functional ways. This will help develop the ability to comprehend the "system concept" and put the study of individual circuits in the context of realism.

To illustrate how logic gates are the basic building blocks for more complex functions, the comparator function in the tablet-counting and control system introduced earlier is implemented and analyzed in Section 3–9. Also a certain system timing problem is examined and a modification is presented using basic logic elements. You will troubleshoot a specified system malfunction using the logic pulser and probe on the system printed circuit board.

3–1 ■ THE INVERTER

The inverter (NOT circuit) performs the operation called inversion or complementation. The inverter changes one logic level to the opposite level. In terms of bits, it changes a 1 to a 0 and a 0 to a 1.

After completing this section, you should be able to

■ Identify negation and polarity indicators ■ Identify an inverter by either its distinctive shape symbol or its rectangular outline symbol ■ Produce the truth table for an inverter ■ Describe the logical operation of an inverter

Standard logic symbols for the **inverter** are shown in Figure 3–1. Part (a) shows the *distinctive shape* symbols, and part (b) shows the *rectangular outline* symbols. In this textbook, distinctive shape symbols are generally used; however, the rectangular outline symbols are found in many industry publications, and you should become familiar with them as well. (Logic symbols are in accordance with **ANSI/IEEE** Standard 91–1984.)

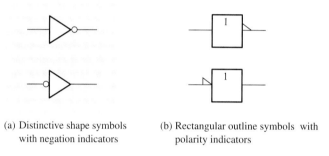

(a) Distinctive shape symbols with negation indicators

(b) Rectangular outline symbols with polarity indicators

FIGURE 3–1
Standard logic symbols for the inverter (ANSI/IEEE Std. 91–1984).

The Negation and Polarity Indicators

The negation indicator is a "bubble" (○) that indicates **inversion** or **complementation** when it appears on the input or output of a logic element, as shown in Figure 3–1(a). Generally, inputs are on the left of a logic symbol and the output is on the right. When appearing on the input, the bubble means that a 0 is the active or *asserted* input state. When appearing on the output, the bubble means that a 0 is the active or asserted output state. The absence of a bubble on the input or output means that a 1 is the active or asserted state.

The polarity or level indicator is a "triangle" (◁) that indicates inversion when it appears on the input or output of a logic element, as shown in Figure 3–1(b). When appearing on the input, it means that a LOW level is the active or asserted input state. When appearing on the output, it means that a LOW level is the active or asserted output state.

Either indicator (bubble or triangle) can be used both on distinctive shape symbols and on rectangular outlines. Figure 3–1(a) indicates the principal inverter symbols used in this text. Note that a change in the placement of the negation or polarity indicator does not imply a change in the way an inverter operates.

Inverter Truth Table

When a HIGH level is applied to an inverter input, a LOW level will appear on its output. When a LOW level is applied to its input, a HIGH will appear on its output. This operation is summarized in Table 3–1, which shows the output for each possible input in terms of levels and corresponding bits. A table such as this is called a **truth table.**

TABLE 3–1	Input	Output
Inverter truth table.	LOW (0)	HIGH (1)
	HIGH (1)	LOW (0)

Inverter Operation

Figure 3–2 shows the output of an inverter for a pulse input, where t_1 and t_2 indicate the corresponding points on the input and output pulse waveforms.

When the input is LOW, the output is HIGH; when the input is HIGH, the output is LOW, thereby producing an inverted output pulse.

FIGURE 3–2
Inverter with a pulse input.

Timing Diagrams

Recall from Chapter 1 that a **timing diagram** is basically *a graph that accurately displays the relationship of two or more waveforms with respect to each other on a time basis.* For example, the time relationship of the output pulse to the input pulse in Figure 3–2 can be shown with a simple timing diagram by aligning the two pulses so that the occurrences of the pulse edges appear in the proper time relationship. The rising edge of the input pulse and the falling edge of the output pulse occur at the same time (ideally). Similarly, the falling edge of the input pulse and the rising edge of the output pulse occur at the same time (ideally). This timing relationship is shown in Figure 3–3. Timing diagrams are especially useful for illustrating the relationship of digital waveforms with multiple pulses, as Example 3–1 illustrates.

FIGURE 3–3
Timing diagram for the case in Figure 3–2.

EXAMPLE 3–1

A waveform is applied to an inverter in Figure 3–4. Determine the output waveform corresponding to the input and show the timing diagram. According to the placement of the bubble, what is the active output state?

FIGURE 3–4

Solution The output waveform is exactly opposite to the input (inverted), as shown in Figure 3–5, which is the basic timing diagram. The active or asserted output state is **0.**

FIGURE 3–5

Related Problem If the inverter is shown with the negative indicator (bubble) on the input instead of the output, how is the timing diagram affected?

Logic Expression for the Inverter

In **Boolean algebra,** which is the mathematics of logic circuits and will be covered thoroughly in Chapter 4, a variable is designated by a letter. The **complement** of a variable is designated by a bar over the letter. A variable can take on a value of either 1 or 0. If a given variable is 1, its complement is 0 and vice versa.

The operation of an inverter (NOT circuit) can be expressed as follows: If the input variable is called A and the output variable is called X, then

$$X = \overline{A}$$

This expression states that the output is the complement of the input, so if $A = 0$, then $X = 1$, and if $A = 1$, then $X = 0$. Figure 3–6 illustrates this. The complemented variable $\overline{A}$ can be read as "A bar" or "not A."

FIGURE 3–6
The inverter complements an input variable.

Application Example

Figure 3–7 shows a circuit for producing the 1's complement of an 8-bit binary number. The bits of the binary number are applied to the inverter inputs and the 1's complement of the number appears on the outputs.

FIGURE 3–7
Example of a 1's complement circuit using inverters.

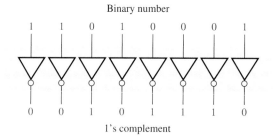

**SECTION 3–1
REVIEW**

1. When a 1 is on the input of an inverter, what is the output?
2. An active HIGH pulse (HIGH level when asserted, LOW level when not) is required on an inverter input.
   (a) Draw the appropriate logic symbol, using the distinctive shape and the negation indicator, for the inverter in this application.
   (b) Describe the output when a positive-going pulse is applied to the input of an inverter.

3–2 ■ THE AND GATE

The AND gate is one of the basic gates from which all logic functions are constructed. An AND gate can have two or more inputs and performs what is known as logical multiplication.

After completing this section, you should be able to

■ Identify an AND gate by its distinctive shape symbol or by its rectangular outline symbol ■ Describe the logical operation of an AND gate ■ Generate the truth table for an AND gate with any number of inputs ■ Produce a timing diagram for an AND gate with any specified input waveforms ■ Write the logic expression for an AND gate with any number of inputs ■ Discuss examples of AND gate applications

COMPUTER NOTE
Logic gates are the building blocks of computers. Most of the functions in a computer, with the exception of certain types of memory, are implemented with logic gates used on a very large scale. For example, a microprocessor, which is the main part of a computer, is made up of hundreds of thousands of logic gates.

The term **gate** is used to describe a circuit that performs a basic logic operation. The AND gate is composed of two or more inputs and a single output, as indicated by the standard logic symbols shown in Figure 3–8. Inputs are on the left, and the output is on the right in each symbol. Gates with two inputs are shown; however, an AND gate can have any number of inputs greater than one. Although examples of both distinctive shape symbols and rectangular outline symbols are shown, the distinctive shape symbol, shown in part (a), is used predominantly in this book.

(a) Distinctive shape

(b) Rectangular outline with the AND (&) qualifying symbol

FIGURE 3–8
Standard logic symbols for the AND gate showing two inputs (ANSI/IEEE Std. 91–1984).

Logical Operation of the AND Gate

The **AND gate** produces a HIGH output *only* when *all* of the inputs are HIGH. When any of the inputs is LOW, the output is LOW. Therefore, the basic purpose of an AND gate is to determine when certain conditions are simultaneously true, as indicated by HIGH levels on all of its inputs, and to produce a HIGH on its output to indicate that all these conditions are true. The inputs of the 2-input AND gate in Figure 3–8 are labeled *A* and *B*, and the output is labeled *X*. The gate operation can be stated as follows:

For a 2-input AND gate, output *X* is HIGH if inputs *A* and *B* are HIGH; *X* is LOW if either *A* or *B* is LOW, or if both *A* and *B* are LOW.

Figure 3–9 illustrates a 2-input AND gate with all four possibilities of input combinations and the resulting output for each.

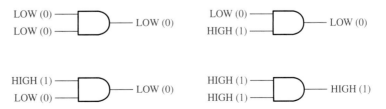

FIGURE 3–9
All possible logic levels for a 2-input AND gate.

AND Gate Truth Table

The logical operation of a gate can be expressed with a truth table that lists all input combinations with the corresponding outputs, as illustrated in Table 3–2 for a 2-input AND gate. The truth table can be expanded to any number of inputs. Although the terms HIGH and LOW tend to give a "physical" sense to input and output states, the truth table is shown with 1s and 0s, since a HIGH is equivalent to a 1 and a LOW is equivalent to a 0 in positive logic. For any AND gate, regardless of the number of inputs, the output is HIGH *only* when *all* inputs are HIGH.

TABLE 3–2
Truth table for a 2-input AND gate.

Inputs		Output
A	B	X
0	0	0
0	1	0
1	0	0
1	1	1

1 = HIGH, 0 = LOW

The total number of possible combinations of binary inputs to a gate is determined by the following formula:

$$N = 2^n \qquad\qquad (3\text{–}1)$$

where N is the number of possible input combinations and n is the number of input variables. To illustrate,

For two input variables: $N = 2^2 = 4$ combinations

For three input variables: $N = 2^3 = 8$ combinations

For four input variables: $N = 2^4 = 16$ combinations

You can determine the number of input bit combinations for gates with any number of inputs by using Equation (3–1). On your calculator, press $\boxed{2}$ $\boxed{\wedge}$ and then enter the value of n, followed by $\boxed{\text{ENTER}}$.

EXAMPLE 3–2

(a) Develop the truth table for a 3-input AND gate.

(b) Determine the total number of possible input combinations for a 4-input AND gate.

Solution

(a) There are eight possible input combinations ($2^3 = 8$) for a 3-input AND gate. The input side of the truth table (Table 3–3) shows all eight combinations of three bits. The output side is all 0s except when all three input bits are 1s.

TABLE 3–3

Inputs			Output
A	B	C	X
0	0	0	0
0	0	1	0
0	1	0	0
0	1	1	0
1	0	0	0
1	0	1	0
1	1	0	0
1	1	1	1

(b) $N = 2^4 = \mathbf{16}$. There are 16 possible combinations of input bits for a 4-input AND gate.

Related Problem   Develop the truth table for a 4-input AND gate.

Pulsed Operation

In a majority of applications, the inputs to a gate are not stationary levels but are voltage waveforms that change frequently between HIGH and LOW logic levels. Now let's look at the operation of AND gates with pulse waveform inputs, keeping in mind that an AND gate obeys the truth table operation regardless of whether its inputs are constant levels or levels that change back and forth.

Let's examine the pulsed operation of the AND gate by looking at the inputs with respect to each other in order to determine the output level at any given time. In Figure 3–10, the inputs are both HIGH (1) during the time interval t_1, making the output HIGH (1) during this interval. During time interval t_2, input A is LOW (0) and input B is HIGH (1), so the output is LOW (0). During time interval t_3, both inputs are HIGH (1) again, and therefore the output is HIGH (1). During time interval t_4, input A is HIGH (1) and input B is LOW (0), resulting in a LOW (0) output. Finally, during time interval t_5, input A is LOW (0), input B is LOW (0), and the output is therefore LOW (0). As you know, a diagram of input and output waveforms showing time relationships is called a *timing diagram.*

FIGURE 3–10
Example of pulsed AND gate operation with a timing diagram showing input and output relationships.

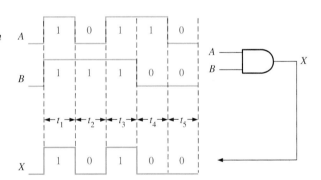

EXAMPLE 3–3   If two waveforms, A and B, are applied to the AND gate inputs as in Figure 3–11, what is the resulting output waveform?

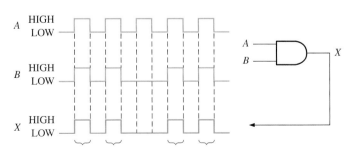

A and B are both HIGH during these four time intervals. Therefore X is HIGH.

FIGURE 3–11

Solution The output waveform *X* is HIGH only when both *A* and *B* waveforms are HIGH as shown in the timing diagram in Figure 3–11.

Related Problem Determine the output waveform and show a timing diagram if the second and fourth pulses in waveform *A* of Figure 3–11 are replaced by LOW levels.

Remember, when analyzing the pulsed operation of logic gates, it is important to pay careful attention to the time relationships of all the inputs with respect to each other and to the output.

EXAMPLE 3–4

For the two input waveforms, *A* and *B*, in Figure 3–12, show the output waveform with its proper relation to the inputs.

FIGURE 3–12

Solution The output waveform is HIGH only when both of the input waveforms are HIGH as shown in the timing diagram.

Related Problem Show the output waveform if the *B* input to the AND gate in Figure 3–12 is always HIGH.

EXAMPLE 3–5

For the 3-input AND gate in Figure 3–13, determine the output waveform in relation to the inputs.

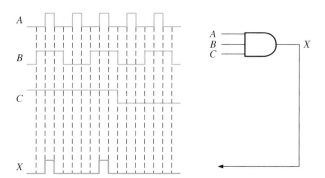

FIGURE 3–13

Solution The output waveform *X* of the 3-input AND gate is HIGH only when all three input waveforms *A*, *B*, and *C* are HIGH.

Related Problem What is the output waveform of the AND gate in Figure 3–13 if the *C* input is always HIGH?

Logic Expressions for the AND Gate

The logical AND function of two variables is represented mathematically either by placing a dot between the two variables, as $A \cdot B$, or by simply writing the adjacent letters without the dot, as AB. We will normally use the latter notation because it is easier to write.

Boolean multiplication follows the same basic rules governing binary multiplication, which were discussed in Chapter 2 and are as follows:

$$0 \cdot 0 = 0$$
$$0 \cdot 1 = 0$$
$$1 \cdot 0 = 0$$
$$1 \cdot 1 = 1$$

Boolean multiplication is the same as the AND function.

The operation of a 2-input AND gate can be expressed in equation form as follows: If one input variable is A, the other input variable is B, and the output variable is X, then the Boolean expression is

$$X = AB$$

Figure 3–14(a) shows the gate with the input and output variables indicated.

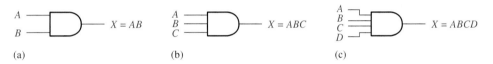

(a) (b) (c)

FIGURE 3–14
Boolean expressions for AND gates with two, three, and four inputs.

To extend the AND expression to more than two input variables, simply use a new letter for each input variable. The function of a 3-input AND gate, for example, can be expressed as $X = ABC$, where A, B, and C are the input variables. The expression for a 4-input AND gate can be $X = ABCD$, and so on. Parts (b) and (c) of Figure 3–14 show AND gates with three and four input variables, respectively.

You can evaluate an AND gate operation by using the Boolean expressions for the output. For example, each variable on the inputs can be either a 1 or a 0; so for the 2-input AND gate, make substitutions in the equation for the output, $X = AB$, as shown in Table 3–4. This evaluation shows that the output X of an AND gate is a 1 (HIGH) only when both inputs are 1s (HIGHs). A similar analysis can be made for any number of input variables.

TABLE 3–4

A	B	AB = X
0	0	$0 \cdot 0 = 0$
0	1	$0 \cdot 1 = 0$
1	0	$1 \cdot 0 = 0$
1	1	$1 \cdot 1 = 1$

Application Examples

The AND Gate as an Enable/Inhibit Device   A common application of the AND gate is to **enable** (that is, to allow) the passage of a signal (pulse waveform) from one point to another at certain times and to inhibit (prevent) the passage at other times.

A simple example of this particular use of the AND gate is shown in Figure 3–15, where the AND gate controls the passage of a signal (waveform A) to a digital counter. The

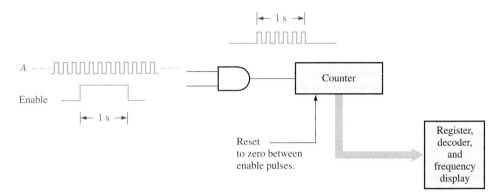

FIGURE 3–15

An AND gate performing an enable/inhibit function for a frequency counter.

purpose of this circuit is to measure the frequency of waveform *A*. The enable pulse has a width of precisely 1 s. When the enable pulse is HIGH, waveform *A* passes through the gate to the counter, and when the enable pulse is LOW, the signal is prevented from passing through (inhibited).

During the 1 second (1 s) interval of the enable pulse, a certain number of pulses in waveform *A* pass through the AND gate to the counter. The number of pulses passing through during 1 s is equal to the frequency of waveform *A*. For example, Figure 3–15 shows six pulses in one second, which is a frequency of 6 Hz. If 1000 pulses pass through the gate in the 1 s interval of the enable pulse, there are 1000 pulses/s, or a frequency of 1000 Hz.

The counter counts the number of pulses per second and produces a binary output that goes to a decoding and display circuit to produce a readout of the frequency. The enable pulse repeats at certain intervals and a new updated count is made so that if the frequency changes, the new value will be displayed. Between enable pulses, the counter is reset so that it starts at zero each time an enable pulse occurs. The current frequency count is stored in a register so that the display is unaffected by the resetting of the counter.

A Seat Belt Alarm System In Figure 3–16, an AND gate is used in a simple automobile seat belt alarm system to detect when the ignition switch is on *and* the seat belt is unbuckled. If the ignition switch is on, a HIGH is produced on input *A* of the AND gate. If the seat belt is not properly buckled, a HIGH is produced on input *B* of the AND gate. Also, when the ignition switch is turned on, a timer is started that produces a HIGH on input *C* for 30 s. If all three conditions exist—that is, if the ignition is on *and* the seat belt is unbuckled *and* the timer is running—the output of the AND gate is HIGH, and an audible alarm is energized to remind the driver.

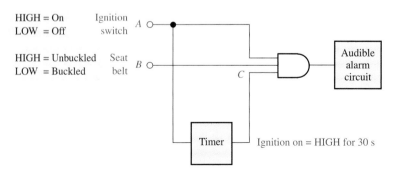

FIGURE 3–16

A simple seat belt alarm circuit using an AND gate.

SECTION 3–2 REVIEW

1. When is the output of an AND gate HIGH?
2. When is the output of an AND gate LOW?
3. Describe the truth table for a 5-input AND gate.

3–3 ■ THE OR GATE

The OR gate is another of the basic gates from which all logic functions are constructed. An OR gate can have two or more inputs and performs what is known as logical addition.

After completing this section, you should be able to

- Identify an OR gate by its distinctive shape symbol or by its rectangular outline symbol
- Describe the logical operation of an OR gate ■ Generate the truth table for an OR gate with any number of inputs ■ Produce a timing diagram for an OR gate with any specified input waveforms ■ Write the logic expression for an OR gate with any number of inputs ■ Discuss examples of OR gate applications

An **OR gate** has two or more inputs and one output, as indicated by the standard logic symbols in Figure 3–17, where OR gates with two inputs are illustrated. An OR gate can have any number of inputs greater than one. Although both distinctive shape and rectangular outline symbols are shown, the distinctive shape OR gate symbol is used in this textbook.

(a) Distinctive shape

(b) Rectangular outline with the OR (≥ 1) qualifying symbol

FIGURE 3–17
Standard logic symbols for the OR gate showing two inputs (ANSI/IEEE Std. 91–1984).

Logical Operation of the OR Gate

The OR gate produces a HIGH on the output when *any* of the inputs is HIGH. The output is LOW only when all of the inputs are LOW. Therefore, an OR gate determines when one or more of its inputs are HIGH and produces a HIGH on its output to indicate this condition. The inputs of the 2-input OR gate in Figure 3–17 are labeled A and B, and the output is labeled X. The operation of the gate can be stated as follows:

> **For a 2-input OR gate, output X is HIGH if either input A or input B is HIGH, or if both A and B are HIGH; X is LOW if both A and B are LOW.**

The HIGH level is the active or asserted output level for the OR gate. Figure 3–18 illustrates the logic operation for a 2-input OR gate for all four possible input combinations.

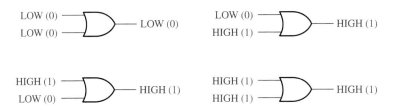

FIGURE 3–18
All possible logic levels for a 2-input OR gate.

OR Gate Truth Table

The logical operation of the 2-input OR gate is described in Table 3–5. This truth table can be expanded for any number of inputs; but regardless of the number of inputs, the output is HIGH when one or more of the inputs are HIGH.

TABLE 3–5
Truth table for a 2-input OR gate.

Inputs		Output
A	B	X
0	0	0
0	1	1
1	0	1
1	1	1

1 ≡ HIGH, 0 ≡ LOW.

Pulsed Operation

Now let's look at the operation of an OR gate with pulsed inputs, keeping in mind its logical operation. Again, the important thing in the analysis of gate operation with pulsed waveforms is the time relationship of all the waveforms involved. For example, in Figure 3–19, inputs A and B are both HIGH (1) during time interval t_1, making the output HIGH (1). During time interval t_2, input A is LOW (0), but because input B is HIGH (1), the output is HIGH (1). Both inputs are LOW (0) during time interval t_3, so there is a LOW (0) output during this time. During time interval t_4, the output is HIGH (1) because input A is HIGH (1).

FIGURE 3–19
Example of pulsed OR gate operation with a timing diagram showing input and output time relationships.

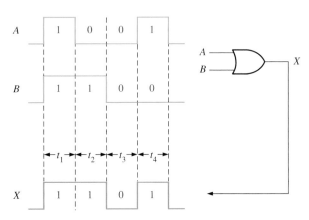

In this illustration, we have simply applied the truth table operation of the OR gate to each of the time intervals during which the levels are nonchanging. Examples 3–6 through 3–8 further illustrate OR gate operation with waveforms on the inputs.

EXAMPLE 3–6

If the two input waveforms, *A* and *B*, in Figure 3–20 are applied to the OR gate, what is the resulting output waveform?

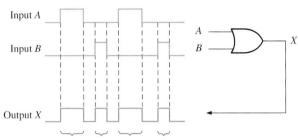

When either input or both inputs are HIGH, the output is HIGH.

FIGURE 3–20

Solution   The output waveform *X* of a 2-input OR gate is HIGH when either or both input waveforms are HIGH as shown in the timing diagram. In this case, both input waveforms are never HIGH at the same time.

Related Problem   Determine the output waveform and show the timing diagram if input *A* is changed such that it is HIGH from the beginning of the existing first pulse to the end of the existing second pulse.

EXAMPLE 3–7

For the two input waveforms, *A* and *B*, in Figure 3–21, show the output waveform with its proper relation to the inputs.

FIGURE 3–21

Solution   When either or both input waveforms are HIGH, the output is HIGH as shown by the output waveform *X* in the timing diagram.

Related Problem   Determine the output waveform and show the timing diagram if the middle pulse of input *A* is replaced by a LOW level.

EXAMPLE 3–8

For the 3-input OR gate in Figure 3–22, determine the output waveform in proper time relation to the inputs.

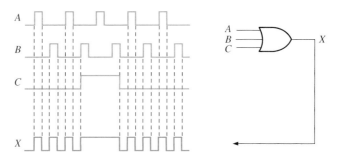

FIGURE 3–22

Solution The output is HIGH when one or more of the input waveforms are HIGH as indicated by the output waveform X in the timing diagram.

Related Problem Determine the output waveform and show the timing diagram if input C is always LOW.

Logic Expressions for the OR Gate

The logical OR function of two variables is represented mathematically by a + between the two variables, for example, $A + B$.

Addition in Boolean algebra involves variables whose values are either binary 1 or binary 0. The basic rules for **Boolean addition** are as follows:

$$0 + 0 = 0$$
$$0 + 1 = 1$$
$$1 + 0 = 1$$
$$1 + 1 = 1$$

Boolean addition is the same as the OR function.

Notice that Boolean addition differs from binary addition in the case where two 1s are added. There is no carry in Boolean addition.

The operation of a 2-input OR gate can be expressed as follows: If one input variable is A, if the other input variable is B, and if the output variable is X, then the Boolean expression is

$$X = A + B$$

Figure 3–23(a) shows the gate logic symbol with input and output variables labeled.

FIGURE 3–23
Boolean expressions for OR gates with two, three, and four inputs.

COMPUTER NOTE

Another mask operation that is used in computer programming to selectively make certain bits in a data byte equal to 1 (called setting) while not affecting any other bit is done with the OR operation. A mask is used that contains a 1 in any position where a data bit is to be set. For example, if you want to force the most significant bit in a data byte to equal 1, but leave all other bits unchanged, you can OR the data byte with the mask 10000000.

To extend the OR expression to more than two input variables, a new letter is used for each additional variable. For instance, the function of a 3-input OR gate can be expressed as $X = A + B + C$. The expression for a 4-input OR gate can be written as $X = A + B + C + D$, and so on. Parts (b) and (c) of Figure 3–23 show OR gates with three and four input variables, respectively.

OR gate operation can be evaluated by using the Boolean expressions for the output X by substituting all possible combinations of 1 and 0 values for the input variables, as shown in Table 3–6 for a 2-input OR gate. This evaluation shows that the output X of an OR gate is a 1 (HIGH) when any one or more of the inputs are 1 (HIGH). A similar analysis can be extended to OR gates with any number of input variables.

TABLE 3–6

A	B	$A + B = X$
0	0	$0 + 0 = 0$
0	1	$0 + 1 = 1$
1	0	$1 + 0 = 1$
1	1	$1 + 1 = 1$

Application Example

A simplified portion of an intrusion detection and alarm system is shown in Figure 3–24. This system could be used for one room in a home—a room with two windows and a door. The sensors are magnetic switches that produce a HIGH output when open and a LOW output when closed. As long as the windows and the door are secured, the switches are closed and all three of the OR gate inputs are LOW. When one of the windows or the door is opened, a HIGH is produced on that input to the OR gate and the gate output goes HIGH. It then activates and latches an alarm circuit to warn of the intrusion.

FIGURE 3–24
A simplified intrusion detection system using an OR gate.

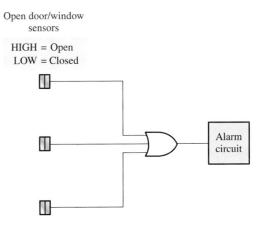

SECTION 3–3 REVIEW

1. When is the output of an OR gate HIGH?
2. When is the output of an OR gate LOW?
3. Describe the truth table for a 3-input OR gate.

3–4 ■ THE NAND GATE

The NAND gate is a popular logic element because it can be used as a universal gate; that is, NAND gates can be used in combination to perform the AND, OR, and inverter operations. The universal property of the NAND gate will be examined thoroughly in Chapter 5.

After completing this section, you should be able to

■ Identify a NAND gate by its distinctive shape symbol or by its rectangular outline symbol ■ Describe the logical operation of a NAND gate ■ Develop the truth table for a NAND gate with any number of inputs ■ Produce a timing diagram for a NAND gate with any specified input waveforms ■ Write the logic expression for a NAND gate with any number of inputs ■ Describe NAND gate operation in terms of its negative-OR equivalent ■ Discuss examples of NAND gate applications

The term *NAND* is a contraction of NOT-AND and implies an AND function with a complemented (inverted) output. The standard logic symbol for a 2-input NAND gate and its equivalency to an AND gate followed by an inverter are shown in Figure 3–25(a). A rectangular outline symbol is shown in part (b).

(a) Distinctive shape, 2-input NAND gate and its NOT/AND equivalent

(b) Rectangular outline, 2-input NAND gate with polarity indicator

FIGURE 3–25
Standard NAND gate logic symbols (ANSI/IEEE Std. 91–1984).

Logical Operation of the NAND Gate

The **NAND gate** produces a LOW output only when all the inputs are HIGH. When any of the inputs is LOW, the output will be HIGH. For the specific case of a 2-input NAND gate, as shown in Figure 3–25 with the inputs labeled A and B and the output labeled X, the operation can be stated as follows:

For a 2-input NAND gate, output X is LOW if inputs A and B are HIGH; X is HIGH if either A or B is LOW, or if both A and B are LOW.

Note that this operation is opposite that of the AND in terms of the output level. In a NAND gate, the LOW level (0) is the active or asserted output level, as indicated by the bubble on the output. Figure 3–26 illustrates the logical operation of a 2-input NAND gate for all four input combinations, and Table 3–7 is the truth table summarizing the logical operation of the 2-input NAND gate.

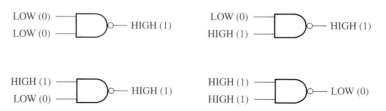

FIGURE 3–26
Logical operation of a 2-input NAND gate.

TABLE 3–7
Truth table for a 2-input NAND gate.

Inputs		Output
A	*B*	*X*
0	0	1
0	1	1
1	0	1
1	1	0

1 = HIGH, 0 = LOW.

Pulsed Operation

Now let's look at the pulsed operation of the NAND gate. Remember from the truth table that the only time a LOW output occurs is when all of the inputs are HIGH. Examples 3–9 and 3–10 illustrate pulsed operation.

EXAMPLE 3–9

If the two waveforms *A* and *B* shown in Figure 3–27 are applied to the NAND gate inputs, determine the resulting output waveform.

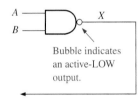

Bubble indicates an active-LOW output.

A and *B* are both HIGH during these four time intervals. Therefore *X* is LOW.

FIGURE 3–27

Solution Output waveform *X* is LOW only during the four time intervals when both input waveforms *A* and *B* are HIGH as shown in the timing diagram.

Related Problem Determine the output waveform and show the timing diagram if input waveform *B* is inverted.

EXAMPLE 3–10

Show the output waveform for the 3-input NAND gate in Figure 3–28 with its proper time relationship to the inputs.

FIGURE 3–28

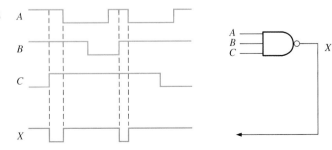

Solution   The output waveform X is LOW only when all three input waveforms are HIGH as shown in the timing diagram.

Related Problem   Determine the output waveform and show the timing diagram if input waveform A is inverted.

Negative-OR Equivalent Operation of the NAND Gate   Inherent in the NAND gate's operation is the fact that one or more LOW inputs produce a HIGH output. Table 3–7 shows that output X is HIGH (1) when any of the inputs, A and B, are LOW (0). From this viewpoint, the NAND gate can be used for an OR operation that requires one or more LOW inputs to produce a HIGH output. This mode of operation is referred to as **negative-OR.** The term *negative* in this context means that the inputs are defined to be in the active or asserted state when LOW.

> **For a 2-input NAND gate functioning as a negative-OR gate, output X is HIGH if either input A or input B is LOW, or if both A and B are LOW.**

When the NAND gate is "looking" for one or more LOWs on its inputs rather than for all HIGHs, it is effectively acting as a negative-OR gate and is represented by the standard logic symbol in Figure 3–29. Although the two symbols in Figure 3–29 represent the same physical gate, they serve to define its role or mode of operation in a particular application, as illustrated by Examples 3–11 through 3–13.

FIGURE 3–29
Standard symbols representing the two equivalent operations of the NAND gate.

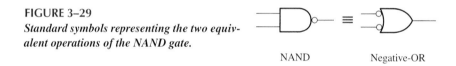

NAND   Negative-OR

EXAMPLE 3–11   A manufacturing plant uses two tanks to store a certain liquid chemical that is required in a manufacturing process. Each tank has a sensor that detects when the chemical level drops to 25% of full. The sensors produce a 5 V level when the tanks are more than one-quarter full. When the volume of chemical in a tank drops to one-quarter full, the sensor puts out a 0 V level.

It is required that a single green light-emitting diode (LED) on an indicator panel show when both tanks are more than one-quarter full. Show how a NAND gate can be used to implement this function.

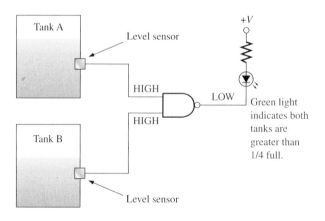

FIGURE 3–30

Solution   Figure 3–30 shows a NAND gate with its two inputs connected to the tank level sensors and its output connected to the indicator panel. The operation can be stated as follows: If tank *A and* tank *B* are above one-quarter full, the LED is on.

As long as *both* sensor outputs are HIGH (5 V), indicating that both tanks are more than one-quarter full, the NAND gate output is LOW (0 V). The green LED circuit is arranged so that a LOW voltage turns it on.

Related Problem   How can the circuit of Figure 3–30 be modified to monitor the levels in three tanks rather than two?

EXAMPLE 3–12

The supervisor of the manufacturing process described in Example 3–11 has decided that he would prefer to have a red LED display come on when at least one of the tanks falls to the quarter-full level rather than have the green LED display indicate when both are above one quarter. Show how this requirement can be implemented.

Solution   Figure 3–31 shows the NAND gate operating as a negative-OR gate to detect the occurrence of at least one LOW on its inputs. A sensor puts out a LOW voltage if the volume in its tank goes to one-quarter full or less. When this happens, the gate output goes HIGH. The red LED circuit in the panel is arranged so that a HIGH voltage turns it on. The operation can be stated as follows: If tank *A or* tank *B or* both are below one-quarter full, the LED is on.

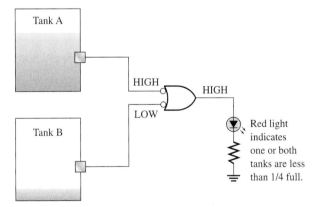

FIGURE 3–31

Notice that, in this example and in Example 3–11, the same 2-input NAND gate is used, but a different gate symbol is used in the schematic, illustrating the different way in which the NAND and negative-OR operations are used.

Related Problem   How can the circuit in Figure 3–31 be modified to monitor four tanks rather than two?

EXAMPLE 3–13   For the 4-input NAND gate in Figure 3–32, operating as a negative-OR, determine the output with respect to the inputs.

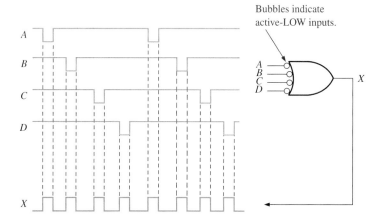

FIGURE 3–32

Solution   The output waveform X is HIGH any time an input waveform is LOW as shown in the timing diagram.

Related Problem   Determine the output waveform if input waveform A is inverted before it is applied to the gate.

Logic Expressions for the NAND Gate

The Boolean expression for the output of a 2-input NAND gate is

$$X = \overline{AB}$$

This expression says that the two input variables, A and B, are first ANDed and then complemented, as indicated by the bar over the AND expression. This is a logical description in equation form of the operation of a NAND gate with two inputs. Evaluating this expression for all possible values of the two input variables, you get the results shown in Table 3–8.

TABLE 3–8

A	B	$\overline{AB} = X$
0	0	$\overline{0 \cdot 0} = \overline{0} = 1$
0	1	$\overline{0 \cdot 1} = \overline{0} = 1$
1	0	$\overline{1 \cdot 0} = \overline{0} = 1$
1	1	$\overline{1 \cdot 1} = \overline{1} = 0$

Once an expression is determined for a given logic function, that function can be evaluated for all possible values of the variables. The evaluation tells you exactly what the output of the logic circuit is for each of the input conditions, and it therefore gives you a complete description of the circuit's logical operation. The NAND expression can be extended to more than two input variables by including additional letters to represent the other variables.

**SECTION 3–4
REVIEW**

1. When is the output of a NAND gate LOW?
2. When is the output of a NAND gate HIGH?
3. Describe the functional differences between a NAND gate and a negative-OR gate. Do they both have the same truth table?
4. Write the output expression for a NAND gate with inputs A, B, and C.

3–5 ■ THE NOR GATE

The NOR gate, like the NAND gate, is a useful logic element because it can also be used as a universal gate; that is, NOR gates can be used in combination to perform the AND, OR, and inverter operations. The universal property of the NOR gate will be examined thoroughly in Chapter 5.

After completing this section, you should be able to

■ Identify a NOR gate by its distinctive shape symbol or by its rectangular outline symbol ■ Describe the logical operation of a NOR gate ■ Develop the truth table for a NOR gate with any number of inputs ■ Produce a timing diagram for a NOR gate with any specified input waveforms ■ Write the logic expression for a NOR gate with any number of inputs ■ Describe NOR gate operation in terms of its negative-AND equivalent ■ Discuss examples of NOR gate applications

The term *NOR* is a contraction of NOT-OR and implies an OR function with an inverted (complemented) output. The standard logic symbol for a 2-input NOR gate and its equivalent OR gate followed by an inverter are shown in Figure 3–33(a). A rectangular outline symbol is shown in part (b).

(a) Distinctive shape, 2-input NOR gate and its NOT/OR equivalent

(b) Rectangular outline, 2-input NOR gate with polarity indicator

FIGURE 3–33
Standard NOR gate logic symbols (ANSI/IEEE Std. 91–1984).

Logical Operation of the NOR Gate

The **NOR gate** produces a LOW output when *any* of its inputs is HIGH. Only when all of its inputs are LOW is the output HIGH. For the specific case of a 2-input NOR gate, as shown in Figure 3–33 with the inputs labeled A and B and the output labeled X, the operation can be stated as follows:

> **For a 2-input NOR gate, output X is LOW if either input A or input B is HIGH, or if both A and B are HIGH; X is HIGH if both A and B are LOW.**

This operation results in an output level opposite that of the OR gate. In a NOR gate, the LOW output is the active or asserted output level as indicated by the bubble on the output.

Figure 3–34 illustrates the logical operation of a 2-input NOR gate for all four possible input combinations, and Table 3–9 is the truth table for the 2-input NOR gate.

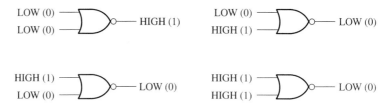

FIGURE 3–34
Logical operation of a 2-input NOR gate.

TABLE 3–9
Truth table for a 2-input NOR gate.

Inputs		Output
A	B	X
0	0	1
0	1	0
1	0	0
1	1	0

1 ≡ HIGH, 0 ≡ LOW.

Pulsed Operation

The next two examples illustrate the logic operation of the NOR gate with pulsed inputs. Again, as with the other types of gates, we will simply follow the truth table operation to determine the output waveforms in the proper time relationship to the inputs.

EXAMPLE 3–14

If the two waveforms shown in Figure 3–35 are applied to the NOR gate, what is the resulting output waveform?

FIGURE 3–35

Solution Whenever any input of a NOR gate is HIGH, the output is LOW as shown by the output waveform *X* in the timing diagram.

Related Problem Invert input *B* and determine the output waveform in relation to the inputs.

EXAMPLE 3–15

Show the output waveform for the 3-input NOR gate in Figure 3–36 with the proper time relation to the inputs.

FIGURE 3–36

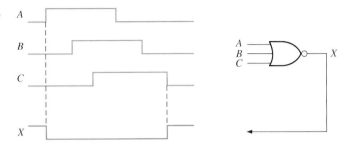

Solution The output *X* is LOW when any input is HIGH as shown by the output waveform *X* in the timing diagram.

Related Problem With the *B* and *C* inputs inverted, determine the output and show the timing diagram.

Negative-AND Equivalent Operation of the NOR Gate The NOR gate, like the NAND, has another aspect of its operation that is inherent in the way it logically functions. Table 3–9 shows that a HIGH is produced on the gate output only if all of the inputs are LOW. From this viewpoint, the NOR gate can be used for an AND operation that requires all LOW inputs to produce a HIGH output. This mode of operation is called **negative-AND.** The term *negative* in this context means that the inputs are defined to be in the active or asserted state when LOW.

> **For a 2-input NOR gate functioning as a negative-AND gate, output *X* is HIGH if both inputs *A* and *B* are LOW.**

When the NOR gate is looking for all LOWs on its inputs rather than for one or more HIGHs, it is effectively acting as a negative-AND gate and is represented by the standard symbol in Figure 3–37. It is important to remember that the two symbols in Figure 3–37 represent the same physical gate and serve only to distinguish between the two modes of its logical operation. The following three examples illustrate this.

FIGURE 3–37
Standard symbols representing the two equivalent operations of the NOR gate.

NOR   Negative-AND

EXAMPLE 3–16

A device is needed to indicate when two LOW levels occur simultaneously on its inputs and to produce a HIGH output as an indication. Specify the device.

Solution A negative-AND gate is required to produce a HIGH output when both inputs are LOW, as shown in Figure 3–38.

FIGURE 3–38   LOW ——○⊃) —— HIGH
                  LOW ——○

Related Problem A device is needed to indicate when one or two HIGH levels occur on its inputs and to produce a LOW output as an indication. Specify the device.

EXAMPLE 3–17

As part of an aircraft's functional monitoring system, a circuit is required to indicate the status of the landing gears prior to landing. A green LED display turns on if all three gears are properly extended when the "gear down" switch has been activated in preparation for landing. A red LED display turns on if any of the gears fail to extend properly prior to landing. When a landing gear is extended, its sensor produces a LOW voltage. When a landing gear is retracted, its sensor produces a HIGH voltage. Implement a circuit to meet this requirement.

Solution   Power is applied to the circuit only when the "gear down" switch is activated. Use a NOR gate for each of the two requirements as shown in Figure 3–39. One NOR gate operates as a negative-AND to detect a LOW from each of the three landing gear sensors. When all three of the gate inputs are LOW, the three landing gears are properly extended and the resulting HIGH output from the negative-AND gate turns on the green LED display. The other NOR gate operates as a NOR to detect if one or more of the landing gears remain retracted when the "gear down" switch is activated. When one or more of the landing gears remain retracted, the resulting HIGH from the sensor is detected by the NOR gate, which produces a LOW output to turn on the red LED warning display.

FIGURE 3–39

Related Problem   What type of gate should be used to detect if all three landing gears are retracted after takeoff, assuming a LOW output is required to activate an LED display?

When driving a load such as an LED with a logic gate, the manufacturer's data sheet should be consulted for maximum drive capabilities (output current). A regular IC logic gate may not be capable of handling the current required by certain loads such as some LEDs. Logic gates with a buffered output, such as an open-collector (OC) or open drain (OD) output are available in many types of IC logic gate configurations. The output current capability of typical IC logic gates is limited to the μA or relatively low mA range. For example, standard TTL can handle output currents up to 16 mA. Most LEDs require currents in the range of about 10 mA to 50 mA.

EXAMPLE 3–18

For the 4-input NOR gate operating as a negative-AND in Figure 3–40, determine the output relative to the inputs.

FIGURE 3–40

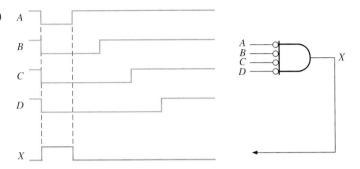

Solution Any time all of the input waveforms are LOW, the output is HIGH as shown by output waveform X in the timing diagram.

Related Problem Determine the output with input D inverted and show the timing diagram.

Logic Expressions for the NOR Gate

The Boolean expression for the output of a 2-input NOR gate can be written as

$$X = \overline{A + B}$$

This equation says that the two input variables are first ORed and then complemented, as indicated by the bar over the OR expression. Evaluating this expression, you get the results shown in Table 3–10. The NOR expression can be extended to more than two input variables by including additional letters to represent the other variables.

TABLE 3–10

A	B	$\overline{A + B} = X$
0	0	$\overline{0 + 0} = \overline{0} = 1$
0	1	$\overline{0 + 1} = \overline{1} = 0$
1	0	$\overline{1 + 0} = \overline{1} = 0$
1	1	$\overline{1 + 1} = \overline{1} = 0$

SECTION 3–5 REVIEW

1. When is the output of a NOR gate HIGH?
2. When is the output of a NOR gate LOW?
3. Describe the functional difference between a NOR gate and a negative-AND gate. Do they both have the same truth table?
4. Write the output expression for a 3-input NOR with input variables A, B, and C.

3–6 ■ THE EXCLUSIVE-OR AND EXCLUSIVE-NOR GATES

Exclusive-OR and exclusive-NOR gates are formed by a combination of other gates already discussed, as you will see in Chapter 5. However, because of their fundamental importance in many applications, these gates are often treated as basic logic elements with their own unique symbols.

After completing this section, you should be able to

▪ Identify the exclusive-OR and exclusive-NOR gates by their distinctive shape symbols or by their rectangular outline symbols ▪ Describe the logical operation of exclusive-OR and exclusive-NOR gates ▪ Show the truth tables for exclusive-OR and exclusive-NOR gates ▪ Produce a timing diagram for an exclusive-OR or exclusive-NOR gate with any specified input waveforms ▪ Discuss examples of exclusive-OR and exclusive-NOR gate applications

COMPUTER NOTE

Exclusive-OR gates connected to form an adder circuit allow a computer to perform addition, subtraction, multiplication, and division in its Arithmetic Logic Unit (ALU). An exclusive-OR gate is implemented with the basic logic circuits of AND, OR, and NOT.

The Exclusive-OR Gate

Standard symbols for the exclusive-OR (XOR for short) gate are shown in Figure 3–41. The XOR gate has only two inputs.

(a) Distinctive shape | (b) Rectangular outline with the XOR qualifying symbol (= 1)

FIGURE 3–41
Standard logic symbols for the exclusive-OR gate.

The output of an **exclusive-OR gate** is HIGH *only* when the two inputs are at opposite logic levels. This operation can be stated as follows with reference to inputs A and B and output X:

> **For an exclusive-OR gate, output X is HIGH if input A is LOW and input B is HIGH, or if input A is HIGH and input B is LOW; X is LOW if A and B are both HIGH or both LOW.**

The four possible input combinations and the resulting outputs for the XOR gate are illustrated in Figure 3–42. The HIGH level is the active or asserted output level and occurs only when the inputs are at opposite levels. The logical operation of the XOR gate is summarized in the truth table shown in Table 3–11.

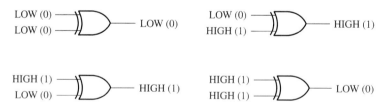

FIGURE 3–42
All possible logic levels for an exclusive-OR gate.

TABLE 3–11
Truth table for an exclusive-OR gate.

Inputs		Output
A	*B*	*X*
0	0	0
0	1	1
1	0	1
1	1	0

EXAMPLE 3–19

A certain system contains two identical circuits operating in parallel. As long as both are operating properly, the outputs of both circuits are always the same. If one of the circuits fails, the outputs will be at opposite levels at some time. Devise a way to detect that a failure has occurred in one of the circuits.

Solution The outputs of the circuits are connected to the inputs of an XOR gate as shown in Figure 3–43. A failure in either one of the circuits produces differing outputs, which cause the XOR inputs to be at opposite levels. This condition produces a HIGH on the output of the XOR gate, indicating a failure in one of the circuits.

FIGURE 3–43

Related Problem Will the exclusive-OR gate always detect simultaneous failures in both circuits of Figure 3–43? If not, under what condition?

The Exclusive-NOR Gate

Standard symbols for the **exclusive-NOR** (XNOR) **gate** are shown in Figure 3–44. Like the XOR gate, the XNOR has only two inputs. The bubble on the output of the XNOR symbol indicates that its output is opposite that of the XOR gate. When the two input logic levels are opposite, the output of the exclusive-NOR gate is LOW. The operation can be stated as follows (A and B are inputs, X is output):

> **For an exclusive-NOR gate, output X is LOW if input A is LOW and input B is HIGH, or if A is HIGH and B is LOW; X is HIGH if A and B are both HIGH or both LOW.**

FIGURE 3–44
Standard logic symbols for the exclusive-NOR gate.

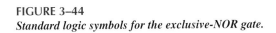

(a) Distinctive shape (b) Rectangular outline

The four possible input combinations and the resulting outputs for the XNOR gate are shown in Figure 3–45. The logical operation is summarized in Table 3–12. Notice that the output is HIGH when the same level is on both inputs.

FIGURE 3–45
All possible logic levels for an exclusive-NOR gate.

TABLE 3–12
Truth table for an exclusive-NOR gate.

Inputs		Output
A	B	X
0	0	1
0	1	0
1	0	0
1	1	1

Pulsed Operation

As we have done with the other gates, let's examine the operation of the XOR and XNOR gates under pulsed input conditions. As before, we apply the truth table operation during each distinct time interval of the pulsed inputs, as illustrated in Figure 3–46 for an XOR gate. You can see that the input waveforms A and B are at opposite levels during time intervals t_2 and t_4. Therefore, the output X is HIGH during these two times. Since both inputs are at the same level, either both HIGH or both LOW, during time intervals t_1 and t_3, the output is LOW during those times as shown in the timing diagram.

FIGURE 3–46
Example of pulsed exclusive-OR gate operation.

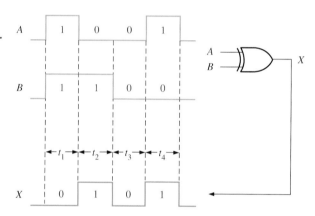

EXAMPLE 3–20   Determine the output waveforms for the XOR gate and for the XNOR gate, given the input waveforms, A and B, in Figure 3–47.

FIGURE 3–47

Solution The output waveforms are shown in Figure 3–47. Notice that the XOR output is HIGH only when both inputs are at opposite levels. Notice that the XNOR output is HIGH only when both inputs are the same.

Related Problem Determine the output waveforms if the two input waveforms, *A* and *B*, are inverted.

Application Example

The exclusive-OR gate can be used as a two-bit adder. Recall from Chapter 2 that the basic rules for binary addition are as follows: $0 + 0 = 0, 0 + 1 = 1, 1 + 0 = 1$, and $1 + 1 = 10$. An examination of the truth table for the XOR gate will show you that its output is the binary sum of the two input bits. In the case where the inputs are both 1s, the output is the sum 0, but you lose the carry of 1. In Chapter 6 you will see how XOR gates are combined to make complete adding circuits. Figure 3–48 illustrates the XOR gate used as a basic adder.

FIGURE 3–48
The XOR gate used to add two bits.

Input bits		Output (sum)
A	*B*	Σ
0	0	0
0	1	1
1	0	1
1	1	0 (without 1 carry)

SECTION 3–6 REVIEW

1. When is the output of an XOR gate HIGH?
2. When is the output of an XNOR gate HIGH?
3. How can you use an XOR gate to detect when two bits are different?

3–7 ■ INTEGRATED CIRCUIT LOGIC GATES

There are three digital integrated circuit (IC) technologies that are used to implement the basic logic gates. Two of these, CMOS and TTL, are the most widely used and the third, ECL, is used in more specialized applications. The logic operations of NOT, AND, OR, NAND, NOR, and exclusive-OR are the same regardless of the IC technology used; that is, an AND gate has the same logic function whether it is implemented with CMOS, TTL, or ECL. We will not cover the IC technologies at the circuit component level in this section. However, those wishing to study this topic further may refer to Chapter 15, Integrated Circuit Technologies, for detailed circuit information.

After completing this section, you should be able to

■ Identify the most common CMOS and TTL series ■ Compare CMOS and TTL in terms of device types and performance parameters ■ Define *propagation delay time* ■ Define *power dissipation* ■ Define *fan-out* ■ Define *speed-power product* ■ Interpret basic data sheet information

CMOS stands for Complementary Metal-Oxide Semiconductor and is implemented with a type of field-effect transistor. **TTL** stands for Transistor-Transistor Logic and is implemented with bipolar junction transistors. **ECL,** Emitter-Coupled Logic, is also a **bipolar** technology. The coverage in this section is restricted to CMOS and TTL because they are the dominant digital IC technologies in use today. Keep in mind that CMOS and TTL differ only in the type of circuit components and values of parameters and not in the basic logic operation. A CMOS AND gate has the same logic operation as a TTL AND gate. This is true for all the other basic logic operations. The difference in CMOS and TTL is in performance characteristics such as switching speed (propagation delay), power dissipation, noise immunity, and other parameters.

CMOS

There is little disagreement about which circuit technology, CMOS or TTL, is the most widely used. It appears that CMOS has become the dominant technology and may eventually replace TTL. This is most certainly true in large-scale ICs and in microprocessors that are all implemented with MOS technology. Although TTL dominated for many years mainly because it had faster switching speeds and a greater selection of device types, CMOS always had the advantage of much lower power dissipation. The switching speeds of CMOS have been greatly improved and are now competitive with TTL, while low power dissipation and other desirable factors have been retained as the technology has progressed.

CMOS Series The two categories of CMOS in terms of the dc supply voltage are the 5 V CMOS and the 3.3 V CMOS. The 3.3 V CMOS is a more recent development and is the result of an effort to reduce the power dissipation. Since power dissipation is proportional to the square of the voltage, a reduction from 5 V to 3.3 V cuts the power by 34%, with other factors remaining the same.

Within each supply voltage category, several series of CMOS logic gates are available. These series within the CMOS family differ in their performance characteristics and are designated by the prefix 74 or 54 followed by a letter or letters that indicate the series and then a number that indicates the type of logic device. The prefix 74 indicates commercial grade for general use, and the prefix 54 indicates military grade for more severe environments. We will refer only to the 74-prefixed devices in this textbook. The basic CMOS series for the 5 V category and their designations are as follows:

- 74HC and 74HCT—High-speed CMOS (the "T" indicates TTL compatibility)
- 74AC and 74ACT—Advanced CMOS
- 74AHC and 74AHCT—Advanced High-speed CMOS

The basic CMOS series for the 3.3 V category and their designations are as follows:

- 74LV—Low-voltage CMOS
- 74LVC—Low-voltage CMOS
- 74ALVC—Advanced Low-voltage CMOS

In addition to the 74 series there is a 4000 series, which is an older, low-speed CMOS technology that is still available, although in limited use. In addition to the "pure" CMOS, there is a series that combines both CMOS and TTL called BiCMOS. The basic BiCMOS series and their designations are as follows:

- 74BCT—BiCMOS
- 74ABT—Advanced BiCMOS
- 74LVT—Low-voltage BiCMOS
- 74ALB—Advanced Low-voltage BiCMOS

TTL

TTL has been and still is a popular digital IC technology. One advantage of TTL is that it is not sensitive to electrostatic discharge as CMOS is and, therefore, is more practical in most laboratory experimentation and prototyping because you do not have to worry about handling problems.

TTL Series   Like CMOS, several series of TTL logic gates are available, all which operate from a 5 V dc supply. These series within the TTL family differ in their performance characteristics and are designated by the prefix 74 or 54 followed by a letter or letters that indicate the series and a number that indicates the type of logic device within the series. A TTL IC can be distinguished from CMOS by the letters that follow the 74 or 54 prefix.

The basic TTL series and their designations are as follows:

- 74—standard TTL (no letter)
- 74S—Schottky TTL
- 74AS—Advanced Schottky TTL
- 74LS—Low-power Schottky TTL
- 74ALS—Advanced Low-power Schottky TTL
- 74F—Fast TTL

Types of IC Logic Gates

All of the basic logic operations, NOT, AND, OR, NAND, NOR, and exclusive-OR (XOR), are available in the integrated circuit technologies. In addition to these, buffered output gates are also available for driving loads that require high currents. The types of gate configurations typically available in IC packages are identified by the last two or three digits in the series designation. For example, 74LS04 is a low-power **Schottky** hex inverter package. Some of the common logic gate configurations and their standard identifier digits are as follows:

- Quad 2-input NAND—**00**
- Quad 2-input NOR—**02**
- Hex inverter—**04**
- Quad 2-input AND—**08**
- Triple 3-input NAND—**10**
- Triple 3-input AND—**11**
- Dual 4-input NAND—**20**

- Dual 2-input AND—**21**
- Triple 3-input NOR—**27**
- Single 8-input NAND—**30**
- Quad 2-input OR—**32**
- Quad 2-input XOR—**86**
- Single 13-input NAND—**133**

IC Packages   All of the 74 series CMOS are pin-compatible with the same types of devices in TTL. This means that a CMOS digital IC such as the 74AHC00 (quad 2-input NAND), which contains four 2-input NAND gates in one IC package, has the identical package pin numbers for each input and output as does the corresponding TTL device. Typical IC gate packages, the dual in-line package (DIP) for plug-in or feedthrough mounting and the small-outline integrated circuit (SOIC) package for surface mounting, are shown in Figure 3–49. In some cases, other types of packages are also available. Although not shown to scale, the SOIC package is significantly smaller than the DIP. The pin configuration diagrams for most of the logic devices listed above are shown in Figure 3–50.

Single-Gate Logic   A limited selection of CMOS gates is available in single-gate packages. With one gate to a package, this series comes in tiny 5-pin packages that are intended for use in last-minute modifications for squeezing logic into tight spots where available space is limited.

(a) 14-pin dual in-line package (DIP) for feedthrough mounting

(b) 14-pin small outline package (SOIC) for surface mounting

FIGURE 3–49

Typical dual in-line (DIP) and small-outline (SOIC) packages showing pin numbers and basic dimensions.

FIGURE 3–50

Pin configuration diagrams for some common IC gate configurations.

Logic Symbols The logic symbols for digital integrated circuits use the standard gate symbols and show the number of gates in the IC package and the associated pin numbers for each gate as well as the pin numbers for V_{CC} and ground. An example is shown in Figure 3–51 for the hex inverter and for the quad 2-input NAND gate. Both the distinctive shape and the rectangular outline formats are shown. Regardless of the logic family, all devices with the same suffix are pin-compatible; in other words, they will have the same arrangement of pin numbers. For example, the 7400, 74S00, 74LS00, 74ALS00, 74F00, 74HC00, and 74AHC00 are all pin-compatible quad 2-input NAND gate packages.

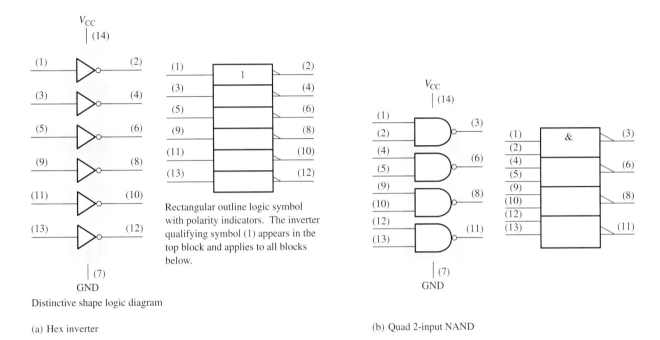

Rectangular outline logic symbol with polarity indicators. The inverter qualifying symbol (1) appears in the top block and applies to all blocks below.

Distinctive shape logic diagram

(a) Hex inverter

(b) Quad 2-input NAND

FIGURE 3–51
Logic symbols for the hex inverter (04 suffix) and quad 2-input NAND (00 suffix). The symbol applies to the same device in any CMOS or TTL series.

Performance Characteristics and Parameters

Several things define the performance of a logic circuit. These performance characteristics are the switching speed measured in terms of the propagation delay time, the power dissipation, the fan-out or drive capability, the speed-power product, the dc supply voltage, and the input/output logic levels. We will define each of these and then compare the performance of CMOS and TTL.

Propagation Delay Time This parameter is a result of the limitation on switching speed or frequency at which a logic circuit can operate. The terms *low speed* and *high speed,* applied to logic circuits refer to the propagation delay time. The shorter the propagation delay, the higher the speed of the circuit and the higher the frequency at which it can operate.

Propagation delay time, t_P, of a logic gate is the time interval between the application of an input pulse and the occurrence of the resulting output pulse. There are two different measurements of propagation delay time associated with a logic gate that apply to all the types of basic gates:

■ t_{PHL}: The time between a specified reference point on the input pulse and a corresponding reference point on the resulting output pulse, with the output changing from the HIGH level to the LOW level (HL).

■ t_{PLH}: The time between a specified reference point on the input pulse and a corresponding reference point on the resulting output pulse, with the output changing from the LOW level to the HIGH level (LH).

EXAMPLE 3–21   Show the propagation delay times of the inverter in Figure 3–52(a).

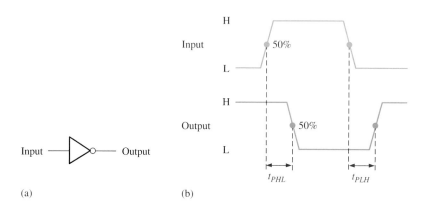

(a)   (b)

FIGURE 3–52

Solution   The propagation delay times, t_{PHL} and t_{PLH}, are indicated in part (b) of the figure. In this case, the delays are measured between the 50% points of the corresponding edges of the input and output pulses. The values of t_{PHL} and t_{PLH} are not necessarily equal but in many cases they are the same.

Related Problem   One type of logic gate has a specified maximum t_{PLH} and t_{PHL} of 10 ns. For another type of gate the value is 4 ns. Which gate can operate at the highest frequency?

For standard-series TTL gates, the typical propagation delay is 11 ns and for F-series gates it is 3.3 ns. For HCT-series CMOS, the propagation delay is 7 ns, for the AC series it is 5 ns, and for the ALVC series it is 3 ns. All specified values are dependent on certain operating conditions as stated on a data sheet. For comparison, one type of ECL has a typical propagation delay of 0.22 ns.

DC Supply Voltage (V_{CC})   The typical dc supply voltage for CMOS is either 5 V or 3.3 V, depending on the category. An advantage of CMOS is that the supply voltages can vary over a wider range than for TTL. The 5 V CMOS can tolerate supply variations from 2 V to 6 V and still operate properly although propagation delay time and power dissipation are significantly affected. The 3.3 V CMOS can operate with supply voltages from 2 V to 3.6 V. The typical dc supply voltage for TTL is 5.0 V with a minimum of 4.5 V and a maximum of 5.5 V.

Power Dissipation   The **power dissipation,** P_D, of a logic gate is the product of the dc supply voltage and the average supply current. Normally, the supply current when the gate output is LOW is greater than when the gate output is HIGH. The manufacturer's data sheet usually designates the supply current for the LOW output state as I_{CCL} and for the HIGH state as I_{CCH}. The average supply current is determined based on a 50% duty cycle (output LOW half the time and HIGH half the time), so the average power dissipation of a logic gate is

$$P_D = V_{CC}\left(\frac{I_{CCH} + I_{CCL}}{2}\right) \qquad \textbf{(3–2)}$$

CMOS series gates have very low power dissipations compared to the TTL series. However, the power dissipation of CMOS is dependent on the frequency of operation. At zero frequency the quiescent power is typically in the microwatt/gate range, and at the maximum operating frequency it can be in the low milliwatt range; therefore, power is sometimes specified at a given frequency. The HC series, for example, has a power of 2.75 μW/gate at 0 Hz (quiescent) and 600 μW/gate at 1 MHz.

Power dissipation for TTL is independent of frequency. For example, the ALS series uses 1.4 mW/gate regardless of the frequency and the F series uses 6 mW/gate.

Input and Output Logic Levels V_{IL} is the LOW level input voltage for a logic gate, and V_{IH} is the HIGH level input voltage. The 5 V CMOS accepts a maximum voltage of 1.5 V as V_{IL} and a minimum voltage of 3.5 V as V_{IH}. TTL accepts a maximum voltage of 0.8 V as V_{IL} and a minimum voltage of 2 V as V_{IH}.

V_{OL} is the LOW level output voltage and V_{OH} is the HIGH level output voltage. For 5 V CMOS, the maximum V_{OL} is 0.33 V and the minimum V_{OH} is 4.4 V. For TTL, the maximum V_{OL} is 0.4 V and the minimum V_{OH} is 2.4 V. All values depend on operating conditions as specified on the data sheet.

Speed-Power Product (SPP) This parameter (**speed-power product**) can be used as a measure of the performance of a logic circuit taking into account the propagation delay time and the power dissipation. It is especially useful for comparing the various logic gate series within the CMOS or TTL family or for comparing a CMOS gate to a TTL gate.

The SPP of a logic circuit is the product of the propagation delay time and the power dissipation and is expressed in joules (J), which is the unit of energy. The formula is

$$SPP = t_P P_D \qquad \textbf{(3–3)}$$

EXAMPLE 3–22

A certain gate has a propagation delay of 5 ns and $I_{CCH} = 1$ mA and $I_{CCL} = 2.5$ mA with a dc supply voltage of 5 V. Determine the speed-power product.

Solution

$$P_D = V_{CC}\left(\frac{I_{CCH} + I_{CCL}}{2}\right) = 5\text{ V}\left(\frac{1\text{ mA} + 2.5\text{ mA}}{2}\right) = 5\text{ V}(1.75\text{ mA}) = 8.75\text{ mW}$$

$$SPP = (5\text{ ns})(8.75\text{ mW}) = \textbf{43.75 pJ}$$

Related Problem If the propagation delay of a gate is 15 ns and its SPP is 150 pJ, what is its average power dissipation?

Fan-Out and Loading The **fan-out** of a logic gate is the maximum number of inputs of the same series in an IC family that can be connected to a gate's output and still maintain the output voltage levels within specified limits. Fan-out is a significant parameter only for TTL because of the type of circuit technology. Since very high impedances are associated with CMOS circuits, the fan-out is very high but depends on frequency because of capacitive effects.

Fan-out is specified in terms of **unit loads.** A unit load for a logic gate equals one input to a like circuit. For example, a unit load for a 7400 NAND gate equals *one* input to another logic gate in the standard 74 series (not necessarily a NAND gate). Because, as specified in the data sheet, the current to a HIGH input (I_{IH}) of a 7400 gate is 40 μA and the current from a LOW input (I_{IL}) is 1.6 mA, the number of unit loads that a 7400 gate can drive in both output states is

$$\text{Unit loads} = \frac{I_{\text{OH}}}{I_{\text{IH}}} = \frac{I_{\text{OL}}}{I_{\text{IL}}} = \frac{400 \ \mu A}{40 \ \mu A} = \frac{16 \ mA}{1.6 \ mA} = 10$$

Figure 3–53 shows LS logic gates driving a number of other gates of the same circuit technology, where the number of gates depends on the particular circuit technology. For example, as you have seen the maximum number of gate inputs (unit loads) that a standard 74 series TTL gate can drive is 10. Most of the other TTL series, such as the LS, can drive 20 unit loads.

FIGURE 3–53
The LS TTL NAND gate output fans out to a maximum of 20 LS TTL gate inputs.

Comparison of Propagation Delay and Power in CMOS and TTL

The graph in Figure 3–54 provides a relative comparison of the propagation delay times and the power dissipations for several CMOS series of logic gates. The shorter the bars, the less the propagation delay time (faster speed) and the less power dissipation. You can, for example, see that ALVC is three times faster than LV, which is the slowest CMOS series shown here. The graph also shows that AC CMOS dissipates the least power and AHC dissipates the most.

FIGURE 3–54
Relative comparison of switching speeds and power dissipations within the CMOS family. The green bars are for propagation delay time and the yellow bars are for power dissipation.

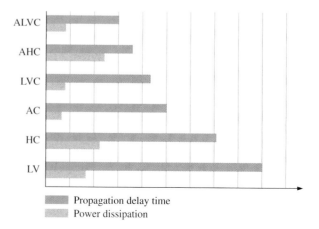

The graph in Figure 3–55 gives relative comparisons of the propagation delay times and power dissipations for several TTL series of logic gates.

The graph in Figure 3–56 is a relative comparison of the switching speeds (propagation delay times) for CMOS and TTL. As you can see, CMOS has devices that are faster than the fastest TTL. No power comparison bars are shown because CMOS values are in the μW range and TTL values are in the mW range.

FIGURE 3–55
Relative comparison of switching speeds and power dissipations within the TTL family.

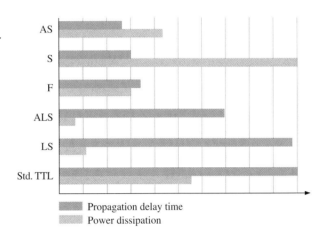

FIGURE 3–56
Relative comparison of switching speeds for CMOS and TTL.

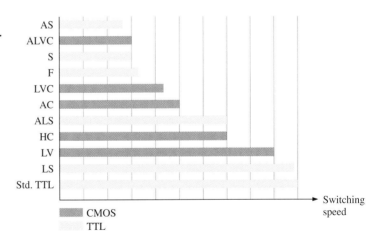

Unused gate inputs for TTL and CMOS should be connected to the appropriate logic level (HIGH or LOW). For AND/NAND, it is recommended that unused inputs be connected to V_{CC} (through a 1.0 kΩ resistor with TTL) and for OR/NOR, unused inputs should be connected to ground.

Data Sheets

A typical data sheet consists of an information page that shows, among other things, the logic diagram and packages, the recommended operating conditions, the electrical characteristics, and the switching characteristics. Partial data sheets for a 74LS00 and a 74HC00A are shown in Figures 3–57 and 3–58, respectively. The length of data sheets vary and some have much more information than others. Additional data sheets are provided in Appendix A.

QUAD 2-INPUT NAND GATE

• ESD > 3500 Volts

SN54/74LS00

QUAD 2-INPUT NAND GATE
LOW POWER SCHOTTKY

J SUFFIX
CERAMIC
CASE 632-08
14
1

N SUFFIX
PLASTIC
CASE 646-06
14
1

D SUFFIX
SOIC
CASE 751A-02
14
1

ORDERING INFORMATION
SN54LSXXJ   Ceramic
SN74LSXXN   Plastic
SN74LSXXD   SOIC

SN54/74LS00

DC CHARACTERISTICS OVER OPERATING TEMPERATURE RANGE (unless otherwise specified)

Symbol	Parameter		Min	Typ	Max	Unit	Test Conditions
V_{IH}	Input HIGH Voltage		2.0			V	Guaranteed Input HIGH Voltage for All Inputs
V_{IL}	Input LOW Voltage	54			0.7	V	Guaranteed Input LOW Voltage for All Inputs
		74			0.8		
V_{IK}	Input Clamp Diode Voltage			−0.65	−1.5	V	V_{CC} = MIN, I_{IN} = −18 mA
V_{OH}	Ouput HIGH Voltage	54	2.5	3.5		V	V_{CC} = MIN, I_{OH} = MAX, V_{IN} = V_{IH} or V_{IL} per Truth Table
		74	2.7	3.5		V	
V_{OL}	Ouput LOW Voltage	54, 74		0.25	0.4	V	I_{OL} = 4.0 mA $\quad$ V_{CC} = V_{CC} MIN, V_{IN} = V_{IL}
		74		0.35	0.5	V	I_{OL} = 8.0 mA $\quad$ or V_{IH} per Truth Table
I_{IH}	Input HIGH Current				20	μA	V_{CC} = MAX, V_{IN} = 2.7 V
					0.1	mA	V_{CC} = MAX, V_{IN} = 7.0 V
I_{IL}	Input LOW Current				−0.4	mA	V_{CC} = MAX, I_{N} = 0.4 V
I_{OS}	Short Circuit Current (Note 1)		−20		−100	mA	V_{CC} = MAX
I_{CC}	Power Supply Current Total, Output HIGH				1.6	mA	V_{CC} = MAX
	Total, Output LOW				4.4		

NOTE 1: Not more than one output should be shorted at a time, nor for more than 1 second.

AC CHARACTERISTICS (T_A = 25°C)

Symbol	Parameter	Min	Typ	Max	Unit	Test Conditions
t_{PLH}	Turn-Off Delay, Input to Output		9.0	15	ns	V_{CC} = 5.0 V
t_{PHL}	Turn-On Delay, Input to Output		10	15	ns	C_L = 15 pF

GUARANTEED OPERATING RANGES

Symbol	Parameter		Min	Typ	Max	Unit
V_{CC}	Supply Voltage	54	4.5	5.0	5.5	V
		74	4.75	5.0	5.25	
T_A	Operating Ambient Temperature Range	54	−55	25	125	°C
		74	0	25	70	
I_{OH}	Output Current — High	54, 74			−0.4	mA
I_{OL}	Output Current — Low	54			4.0	mA
		74			8.0	

FIGURE 3–57
The partial data sheet for a 74LS00.

Quad 2-Input NAND Gate High-Performance Silicon–Gate CMOS

The MC54/74HC00A is identical in pinout to the LS00. The device inputs are compatible with Standard CMOS outputs; with pullup resistors, they are compatible with LSTTL outputs.

- Output Drive Capability: 10 LSTTL Loads
- Outputs Directly Interface to CMOS, NMOS and TTL
- Operating Voltage Range: 2 to 6 V
- Low Input Current: 1 μA
- High Noise Immunity Characteristic of CMOS Devices
- In Compliance With the JEDEC Standard No. 7A Requirements
- Chip Complexity: 32 FETs or 8 Equivalent Gates

LOGIC DIAGRAM

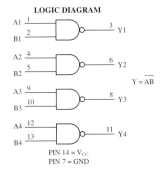

$Y = \overline{AB}$

PIN 14 = V_{CC}
PIN 7 = GND

Pinout: 14–Load Packages (Top View)

MC54/74HC00A

J SUFFIX
CERAMIC PACKAGE
CASE 632-08

N SUFFIX
PLASTIC PACKAGE
CASE 646-06

D SUFFIX
SOIC PACKAGE
CASE 751A-03

DT SUFFIX
TSSOP PACKAGE
CASE 948G-01

ORDERING INFORMATION

MC54HCXXAJ	Ceramic
MC74HCXXAN	Plastic
MC74HCXXAD	SOIC
MC74HCXXADT	TSSOP

FUNCTION TABLE

Inputs		Output
A	B	Y
L	L	H
L	H	H
H	L	H
H	H	L

MAXIMUM RATINGS*

Symbol	Parameter	Value	Unit
V_{CC}	DC Supply Voltage (Referenced to GND)	−0.5 to + 7.0	V
V_{in}	DC Input Voltage (Referenced to GND)	−0.5 to V_{CC} + 0.5	V
V_{out}	DC Output Voltage (Referenced to GND)	−0.5 to V_{CC} + 0.5	V
I_{in}	DC Input Current, per Pin	± 20	mA
I_{out}	DC Output Current, per Pin	± 25	mA
I_{CC}	DC Supply Current, V_{CC} and GND Pins	± 50	mA
P_D	Power Dissipation in Still Air, Plastic or Ceramic DIP†	750	mW
	SOIC Package†	500	
	TSSOP Package†	450	
T_{stg}	Storage Temperature	−65 to + 150	°C
T_L	Lead Temperature, 1 mm from Case for 10 Seconds		°C
	Plastic DIP, SOIC or TSSOP Package	260	
	Ceramic DIP	300	

* Maximum Ratings are those values beyond which damage to the device may occur.
 Functional operation should be restricted to the Recommended Operating Conditions.
† Derating — Plastic DIP: – 10 mW/°C from 65° to 125° C
 Ceramic DIP: – 10 mW/°C from 100° to 125° C
 SOIC Package: – 7 mW/°C from 65° to 125° C
 TSSOP Package: – 6.1 mW/°C from 65° to 125° C

RECOMMENDED OPERATING CONDITIONS

Symbol	Parameter		Min	Max	Unit
V_{CC}	DC Supply Voltage (Referenced to GND)		2.0	6.0	V
V_{in}, V_{out}	DC Input Voltage, Output Voltage (Referenced to GND)		0	V_{CC}	V
T_A	Operating Temperature, All Package Types		−55	+125	°C
t_r, t_f	Input Rise and Fall Time	V_{CC} = 2.0 V	0	1000	ns
		V_{CC} = 4.5 V	0	500	
		V_{CC} = 6.0 V	0	400	

DC CHARACTERISTICS (Voltages Referenced to GND)

MC54/74HC00A

Symbol	Parameter	Condition	V_{CC} V	Guaranteed Limit			Unit
				−55 to 25°C	≤85°C	≤125°C	
V_{IH}	Minimum High-Level Input Voltage	V_{out} = 0.1V or V_{CC} − 0.1V $|I_{out}| \leq 20\mu A$	2.0	1.50	1.50	1.50	V
			3.0	2.10	2.10	2.10	
			4.5	3.15	3.15	3.15	
			6.0	4.20	4.20	4.20	
V_{IL}	Maximum Low-Level Input Voltage	V_{out} = 0.1V or V_{CC} − 0.1V $|I_{out}| \leq 20\mu A$	2.0	0.50	0.50	0.50	V
			3.0	0.90	0.90	0.90	
			4.5	1.35	1.35	1.35	
			6.0	1.80	1.80	1.80	
V_{OH}	Minimum High-Level Output Voltage	V_{in} = V_{IH} or V_{IL} $|I_{out}| \leq 20\mu A$	2.0	1.9	1.9	1.9	V
			4.5	4.4	4.4	4.4	
			6.0	5.9	5.9	5.9	
		V_{in} = V_{IH} or V_{IL} $|I_{out}| \leq 2.4mA$	3.0	2.48	2.34	2.20	
		$|I_{out}| \leq 4.0mA$	4.5	3.98	3.84	3.70	
		$|I_{out}| \leq 5.2mA$	6.0	5.48	5.34	5.20	
V_{OL}	Maximum Low-Level Output Voltage	V_{in} = V_{IH} or V_{IL} $|I_{out}| \leq 20\mu A$	2.0	0.1	0.1	0.1	V
			4.5	0.1	0.1	0.1	
			6.0	0.1	0.1	0.1	
		V_{in} = V_{IH} or V_{IL} $|I_{out}| \leq 2.4mA$	3.0	0.26	0.33	0.40	
		$|I_{out}| \leq 4.0mA$	4.5	0.26	0.33	0.40	
		$|I_{out}| \leq 5.2mA$	6.0	0.26	0.33	0.40	
I_{in}	Maximum Input Leakage Current	V_{in} = V_{CC} or GND	6.0	±0.1	±1.0	±1.0	μA
I_{CC}	Maximum Quiescent Supply Current (per Package)	V_{in} = V_{CC} or GND I_{out} = 0μA	6.0	1.0	10	40	μA

AC CHARACTERISTICS (C_L = 50 pF, Input t_r = t_f = 6 ns)

Symbol	Parameter	V_{CC} V	Guaranteed Limit			Unit
			−55 to 25°C	≤85°C	≤125°C	
t_{PLH}, t_{PHL}	Maximum Propagation Delay, Input A or B to Output Y	2.0	75	95	110	ns
		3.0	30	40	55	
		4.5	15	19	22	
		6.0	13	16	19	
t_{TLH}, t_{THL}	Maximum Output Transition Time, Any Output	2.0	75	95	110	ns
		3.0	27	32	36	
		4.5	15	19	22	
		6.0	13	16	19	
C_{in}	Maximum Input Capacitance		10	10	10	pF

		Typical @ 25°C, V_{CC} = 5.0 V, V_{EE} = 0 V	
C_{PD}	Power Dissipation Capacitance (Per Buffer)	22	pF

FIGURE 3–58
The partial data sheet for a 74HC00A.

SECTION 3–7 REVIEW

1. List three types of IC technologies and name the two that are the most widely used.
2. Identify the following IC logic designators:
   (a) LS   (b) ALS   (c) F   (d) HC   (e) AC   (f) HCT   (g) LV
3. Identify the following devices according to logic function:
   (a) 74LS04   (b) 74HC00   (c) 74LV08   (d) 74ALS10
   (e) 7432   (f) 74ACT11   (g) 74AHC02
4. Which IC technology generally has the lowest power dissipation? The highest speed?
5. What does the term *hex inverter* mean? What does *quad 2-input NAND* mean?
6. A positive pulse is applied to an inverter input. The time from the leading edge of the input to the leading edge of the output is 10 ns. The time from the trailing edge of the input to the trailing edge of the output is 8 ns. What are the values of t_{PLH} and t_{PHL}?
7. A certain gate has a propagation delay time of 6 ns and a power dissipation of 3 mW. Determine the speed-power product?
8. Define I_{CCL} and I_{CCH}.
9. Define V_{IL} and V_{IH}.
10. Define V_{OL} and V_{OH}.

All or portions of Chapter 15, "Integrated Circuit Technologies," may be covered at this point. The coverage in Chapter 15 is not a prerequisite to any other material in this text, and its omission or partial omission will not affect any other material.

3–8 ■ TROUBLESHOOTING

Troubleshooting is the process of recognizing, isolating, and correcting a fault or failure in a circuit or system. To be an effective troubleshooter, you must understand how the circuit or system is supposed to work and be able to recognize incorrect performance. For example, to determine whether or not a certain logic gate is faulty, you must know what the output should be for given inputs. At this point it may be helpful to review Section 1–6 on pulsers, probes, and oscilloscopes.

After completing this section, you should be able to

■ Test for internally open inputs and outputs in IC gates using the logic pulser and probe
■ Recognize the effects of a shorted IC input or output   ■ Test for external faults on a PC board   ■ Troubleshoot a simple frequency counter using an oscilloscope in addition to the logic pulser and probe

Internal Failures of IC Logic Gates

Opens and shorts are the most common types of internal gate failures. These can occur on the inputs or on the output of a gate inside the IC package. *Before attempting any troubleshooting, check for proper dc supply voltage and ground.*

Effects of an Internally Open Input   An internal open is the result of an open component on the chip or a break in the tiny wire connecting the IC chip to the package pin. An open input prevents a pulser signal on that input from getting to the output of the gate, as illustrated in Figure 3–59(a) for the case of a 2-input NAND gate. An open TTL input acts effectively as a HIGH level, so pulses applied to the good input get through to the NAND gate output as shown in Figure 3–59(b).

(a) Pulsing the open input will produce no pulses on the output.

(b) Pulsing the good input will produce output pulses for TTL NAND and AND gates because an open input acts as a HIGH. It is uncertain for CMOS.

FIGURE 3–59
The effect of an open input on a NAND gate.

Conditions for Testing Gates When testing a NAND gate or an AND gate, always make sure that the inputs that are not being pulsed are HIGH to enable the gate. When checking a NOR gate or an OR gate, always make sure that the inputs that are not being pulsed are LOW. When checking an XOR or XNOR gate, the level of the nonpulsed input does not matter because the pulses on the other input will force the inputs to alternate between the same level and opposite levels.

Troubleshooting an Open Input Troubleshooting this type of failure is most easily accomplished with a logic pulser and probe, as demonstrated in Figure 3–60 for the case of a 2-input NAND gate package.

(a) Pin 13 input and pin 11 output OK. (b) Pin 12 input open.

FIGURE 3–60
Troubleshooting a NAND gate for an open input with a logic pulser and probe.

The first step in troubleshooting an IC that is suspected of being faulty is to make sure that the dc supply voltage (V_{CC}) and ground are at the appropriate pins of the IC. Next, using a logic pulser to apply continuous pulses to one of the inputs to the gate, make sure that the other inputs are HIGH (in the case of a NAND gate). In Figure 3–60(a), start by pulsing pin 13, which has been determined to be one of the inputs to the suspected gate. If pulse activity is indicated on the output (pin 11 in this case) by a flashing probe, then the pin 13 input is not open. By the way, this also proves that the output is not open. Next, pulse the other gate input (pin 12). The probe lamp is off indicating that there are no pulses on the output at pin 11 and that the output is LOW, as shown in Figure 3–60(b). Notice that the input not being pulsed must be HIGH for the case of a NAND gate or AND gate. If this were a NOR gate, the input not being pulsed would have to be LOW.

Effects of an Internally Open Output   An internally open gate output prevents a signal on any of the inputs from getting to the output. Therefore, no matter what the input conditions are, the output is unaffected. The level at the output pin of the IC will depend upon what it is externally connected to. It could be either HIGH, LOW, or floating (not fixed to any reference). In any case, a logic probe at the output pin will not be flashing.

Troubleshooting an Open Output   Figure 3–61 illustrates troubleshooting an open NOR gate output with a pulser and probe. In part (a), one of the inputs of the suspected gate (pin 11 in this case) is pulsed, and the probe on the output (pin 13) indicates no pulse activity. In part (b), the other input (pin 12) is pulsed and again there is no indication of pulses on the output. Under the condition that the input that is not being pulsed is at a LOW level, this test shows that the output is internally open.

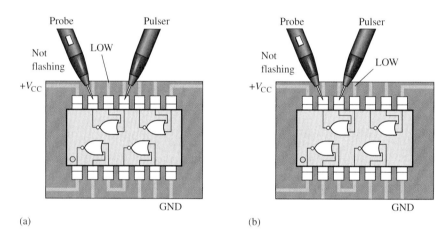

FIGURE 3–61
Troubleshooting a NOR gate for an open output with a logic pulser and probe.

Shorted Input or Output   Although not as common as an open, an internal short to the dc supply voltage, the ground, another input, or an output can occur. When an input or output is shorted to the supply voltage, it will be stuck in the HIGH state. If an input or output is shorted to the ground, it will be stuck in the LOW state (0 V). If two inputs or an input and an output are shorted together, they will always be at the same level.

We will reserve further discussion of troubleshooting shorts until Chapter 5, where the use of a current tracer to locate a short will be demonstrated in situations in which several gates are interconnected.

External Opens and Shorts

Many failures involving digital ICs are due to faults that are external to the IC package. These include bad solder connections, solder splashes, wire clippings, improperly etched printed circuit (PC) boards, and cracks or breaks in wires or printed circuit interconnections. These open or shorted conditions have the same effect on the logic gate as the internal faults, and troubleshooting is done in basically the same ways. A visual inspection of any circuit that is suspected of being faulty is the first thing a technician should do.

EXAMPLE 3–23

You are checking a 74LS10 triple 3-input NAND gate IC that is one of many ICs located on a PC board. You have checked pins 1 and 2 with your logic probe, and they are both HIGH. Now your logic pulser is placed on pin 13, and your logic probe is placed first on pin 12 and then on the connecting PC board trace as indicated in Figure 3–62. Based on the responses of the probe, what is the most likely problem?

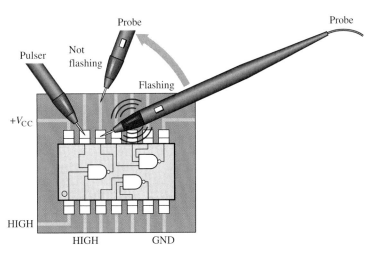

FIGURE 3–62

Solution The flashing indicator on the probe shows that there is pulse activity on the gate output at pin 12 but no activity on the PC board trace. The gate is working properly, but the signal is not getting from pin 12 of the IC to the PC board trace.

Most likely there is a bad solder connection between pin 12 of the IC and the PC board, which is creating an open. You should resolder that point and check it again.

Related Problem If the probe does not flash at either point in Figure 3–62, what fault(s) does this indicate?

In most cases, you will be troubleshooting ICs that are mounted on PC boards or prototype assemblies and interconnected with other ICs. As you progress through this book, you will learn how different types of digital ICs are used together to perform system functions. At this point, however, we are concentrating on individual IC gates. This limitation does not prevent us from looking at the system concept at a very basic and simplified level, as we have already done several times.

To continue the emphasis on systems, Examples 3–24 and 3–25 deal with troubleshooting the frequency counter that was introduced in Section 3–2.

EXAMPLE 3–24

After trying to operate the frequency counter shown in Figure 3–63, you find that it constantly reads out all 0s on its display, regardless of the input frequency. Determine the cause of this malfunction.

(a) This is how the counter should be working with a 12 Hz input signal.

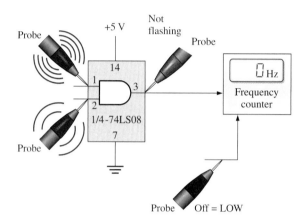

(b) The display is showing a frequency of 0 Hz and the logic probe indications are shown.

FIGURE 3–63

Solution   Here are three possible causes:

1. A constant active or asserted level on the counter reset input, which keeps the counter at zero.
2. No pulse signal on the input to the counter because of an internal open or short in the counter. This problem would keep the counter from advancing after being reset to zero.
3. No pulse signal on the input to the counter because of an open AND gate output or the absence of input signals, again keeping the counter from advancing from zero.

Figure 3–63(a) gives an example of how the frequency counter should be working with a 12 Hz pulse waveform on the input to the AND gate. Part (b) shows that the display is improperly indicating 0 Hz.

The first step is to make sure that V_{CC} and ground are connected to all the right places; assume that they are found to be okay. Next, check for pulses on both inputs to the AND gate. The probe indicates that there is pulse activity on both of these inputs. A logic probe check of the counter reset shows a LOW level which is known to be the unasserted level (assume this information was found on the data sheet for the counter) and, therefore, this is not the problem. The next probe check on pin 3 of the 74LS08 shows that there is no pulse activity on the output of the AND gate, indicating that the gate output is internally open. Replace the 74LS08 IC and check the operation again.

Related Problem   If pin 2 of the 74LS08 AND gate is open, what indication should you see on the display?

EXAMPLE 3–25

The frequency counter shown in Figure 3–64 appears to measure the frequency of input signals incorrectly. It is found that when a signal with a precisely known frequency is applied to pin 1 of the AND gate, the display indicates a higher frequency. Determine what is wrong. The readings on the screen indicate sec/div.

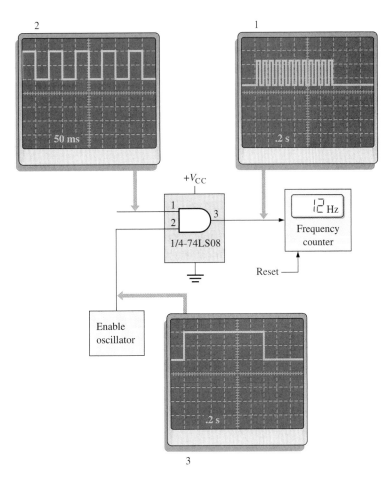

FIGURE 3–64

Solution   Recall from Section 3–2 that the input pulses were allowed to pass through the AND gate for exactly 1 s. The number of pulses counted in 1 s is equal to the frequency in hertz (cycles per second). Therefore, the 1 s interval, which is produced by the enable pulse on pin 2 of the AND gate, is very critical to an accurate frequency measurement. The enable pulses are produced internally by a precision oscillator circuit. The pulse must be exactly 1 s in width and in this case it occurs every 3 s to update the count. Just prior to each enable pulse, the counter is reset to zero so that it starts a new count each time.

Since the counter appears to be counting more pulses than it should to produce a frequency readout that is too high, the enable pulse is the primary suspect. Exact time-interval measurements must be made, so an oscilloscope is used instead of a logic probe in this situation. The logic probe indicates only the presence of pulses; it does not provide for frequency or time measurement.

An input pulse waveform of exactly 10 Hz is applied to pin 1 of the AND gate and the display incorrectly shows 12 Hz. The first scope measurement, on the output of the AND gate, shows that there are 12 pulses for each enable pulse. In the second scope measurement, the input frequency is verified to be precisely 10 Hz (period = 100 ms).

In the third scope measurement, the width of the enable pulse is found to be 1.2 s rather than 1 s.

The conclusion is that the oscillator circuit that produces the enable pulse is out of calibration for some reason and must be repaired or replaced.

Related Problem What would you suspect if the readout were indicating a frequency less than it should be?

Proper grounding is very important when setting up to take measurements or work on a circuit. Properly grounding the oscilloscope protects you from shock and grounding yourself protects your circuits from damage. Grounding the oscilloscope means to connect it to earth ground by plugging the three-prong power cord into a grounded outlet. As you are already aware, grounding yourself means using a wrist-type grounding strap, particularly when working with CMOS circuits.

Also, for accurate measurements, make sure that the ground in the circuit you are testing is the same as the scope ground. This can be done by connecting the ground lead on the scope probe to a known ground point in the circuit, such as the metal chassis or a ground point on the circuit board. You can also connect the circuit ground to the GND jack on the front panel of the scope.

SECTION 3–8 REVIEW

1. What are the most common types of failures in ICs?
2. If two different input waveforms are applied to a 2-input TTL NAND gate and the output waveform is just like one of the inputs, but inverted, what is the most likely problem?
3. Name one advantage of the oscilloscope over the logic probe.
4. Name one advantage of the logic probe over the oscilloscope.

3–9 ■ DIGITAL SYSTEM APPLICATION

The tablet-counting and control system that you studied in Chapters 1 and 2 is again the subject of the system application, and you may wish to review those sections before beginning this one. In this section, the focus is on the application of logic gates in certain system functions.

After completing this section, you should be able to

■ Implement the comparator function with the basic gates covered in this chapter ■ Explain how an AND gate and inverter can be used to solve a certain timing problem in the system ■ Use a timing diagram to analyze the comparator logic and system operation ■ Troubleshoot the implemented comparator logic using a logic probe and pulser

The Comparator Function

The **comparator** in this system detects when the binary number in the counter, which is the current number of tablets in the bottle, is equal to the preset binary number that represents the maximum number of tablets per bottle. When the equality of these two numbers occurs, the bottle has been filled with the preset number of tablets and the valve must be closed to cut off the flow of tablets into the bottle.

Although there are ICs designed specifically as comparators, which you will study later, the comparator in this application is implemented using individual gates in keeping with the topics of this chapter. Recall that the output of an exclusive-OR gate is LOW (0) when both input bits are the same, as shown in Figure 3–65. This operational feature of the exclusive-OR (XOR) can be used to advantage in the comparator application because you are looking for the occurrence of two equal binary numbers.

FIGURE 3–65
The output of an XOR gate is 0 when its in-puts are equal.

In this system application, the number of tablets entered on the keypad appears as a fixed 8-bit binary number on the output of the BCD-to-binary code converter. The binary counter also contains an 8-bit number that represents the number of tablets actually in the bottle at any given time and this number increases as each tablet is counted. Both of these 8-bit binary numbers are applied to the inputs of the comparator as shown in Figure 3–66.

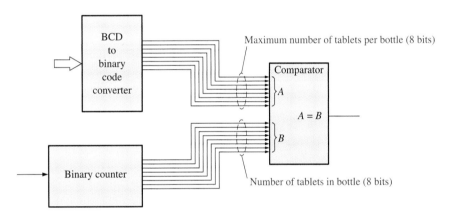

FIGURE 3–66
Block diagram of the portion of the system containing the comparator, counter, and code converter.

Implementing the Comparator

One XOR gate can accept two bits on its inputs and indicate by its output level if the two bits are equal or not equal. To compare the two 8-bit numbers, eight XOR gates are required. Each gate compares the corresponding bits of the two numbers; that is, one gate compares the LSBs of the two numbers, the next gate compares the next two bits, and so on. Finally, the eighth gate compares the MSBs of the two numbers. An 8-input negative-AND (NOR) gate is then used to detect when all the XOR outputs are LOW, indicating equal bits. The basic logic is shown in Figure 3–67(a).

An 8-input NOR gate is not available in an IC package to implement the required negative-AND; therefore, you must use an 8-input NAND gate. Since an XOR gate produces a LOW (0) when its inputs are the same (both 1 or both 0), an inverter is required on the output of each XOR gate to produce a HIGH for the NAND gate.

Figure 3–67(b) shows the actual implementation using two 74LS86 quad XOR gate packages, two 74LS04 hex inverter packages, and one 74LS30 8-input NAND gate package. This actual implementation is logically the same as the basic implementation shown in part (a). When the two 8-bit input numbers are equal, the output of the NAND gate is

FIGURE 3–67
Implementation of the comparator logic.

LOW. The NAND gate output must be inverted to provide the required HIGH to control the valve and other circuits. There is a way to implement the comparator with fewer devices, but this is reserved as a system assignment activity.

A System Problem

The HIGH output from the comparator causes the valve in the tablet "hopper" to close and shut off the flow, causes the conveyor to move the next bottle into position, and causes the adder output to be stored in register B (see Figure 2–6).

A timing problem results when the output of the comparator goes HIGH. When the last tablet passes the sensor and the counter advances to the binary count equal to the preset number of tablets, there is a possibility that another tablet has already passed through the valve and will cause the counter to advance to its next number, thus making the comparator output go back LOW.

Solution   To correct this timing problem, the inverter and AND gate shown in Figure 3–68 are used to inhibit the sensor pulse to the counter as soon as the binary numbers are equal. As long as the comparator output is LOW, the HIGH on the inverter output enables the AND gate and allows the sensor pulses to pass through the gate to the counter. When the comparator output goes HIGH, the inverter output goes LOW and prevents another sensor pulse from getting through the AND gate, thus keeping the counter from advancing if one more tablet passes the sensor.

FIGURE 3–68
Partial system diagram showing the counter inhibit logic.

Counter inhibit logic

■ SYSTEM ASSIGNMENTS

Analysis and Design

The system is assumed to be operating properly. The comparator logic and the counter inhibit logic are implemented on the portion of the system board shown in System Assignment 1. The comparator logic consists of two 74LS86 quad XOR packages, two 74LS04 hex inverter packages, and one 74LS30 8-input NAND gate package. The counter inhibit circuit uses one of the inverters left over from the 74LS04 and one of the AND gates from a 74LS08 quad AND gate package. This design will be simplified in System Assignment 1 Activity 6.

The darker traces on the system board connect corresponding points on the back side of the board via feedthroughs at each pad. The backside connections for V_{CC} and ground (GND) are not shown.

Perform the activities for System Assignment 1.

Troubleshooting

Observation A system problem has developed in which the tablets continue to flow after the bottle has been filled and the bottles do not advance on the conveyor.

Conclusion There appears to be two problems: (1) the valve is not closing and (2) the conveyor is not advancing. However, it is unlikely that both of these elements will fail at the same time, so you must look for a common cause and assume that the valve and the conveyor are not faulty. What controls both the valve and the conveyor advance? When the output of the comparator goes HIGH, it should cause the valve to close and the conveyor to move the next bottle into position. Something may be keeping the comparator output from going HIGH.

Verify the Conclusion If the output of the comparator stays LOW, the conclusion is correct. If it goes HIGH, perhaps both the valve and the conveyor have indeed failed and the conclusion is wrong. Assume that the comparator output is found to be always LOW and continue.

Priority Number One The first priority after you have determined that the fault appears to be in the system circuit board and not in the valve or the conveyor mechanism is to minimize the system down-time (down-time means lost money). First, check to make sure the tablet sensor is producing pulses to the counter and that the system power supply is working. Assuming that they are, turn off the power to the system and remove the system circuit board. Insert a spare system board that is known to be good (assume there is one always available). Turn the power back on and initialize the system. Assume that it now runs properly.

The next step is to find the problem with the faulty board and repair it so that there will be a good spare available when another problem occurs.

Possible Faults The next step is to list failures in the comparator and counter inhibit logic and failures external to the logic circuits that could keep the comparator output LOW when it should be HIGH. Refer to the schematic in System Assignment 2.

1. Binary counter not working
2. Inverter 9 output or input open or output shorted
3. Inverter 10 input open or output shorted
4. NAND gate output open
5. An input open or output shorted on one of the inverters 1 through 8
6. An output open on one of the XOR gates
7. An input on one of the XOR gates is open or shorted in such a way as to prevent a comparison from occurring
8. The sensor input to the AND gate open
9. AND gate output open

■ SYSTEM ASSIGNMENT 1: Analysis and Design

■ **Activity 1:** Check the portion of the system board against the schematic to verify that they correspond.

■ **Activity 2:** The logic analyzer screen shows a portion of the timing diagram of the waveforms from the counter to the comparator inputs and the comparator output. Determine how many tablets are going into each bottle.

■ **Activity 3:** Determine the bit on each of the inputs (a, b, c, etc.) to this portion of the system board from the BCD-to-binary code converter.

■ **Activity 4:** Identify the three lines labeled A, B, and C at the bottom of this portion of the system board using the schematic as reference.

■ **Activity 5:** Verify that the DMM readings are within data sheet specifications for each of the points on the system board that are being measured.

■ **Activity 6:** The comparator and counter inhibit logic can be simplified. Modify the circuit to reduce the number of ICs without changing the overall logic function. This redesign will require that different types of gates be substituted for the existing ones. You may wish to refer to a data book to determine the ICs available to accomplish this redesign.

The comparator PC board logic is in file SA03.EWB on your EWB CD. Use EWB to simulate a test of this circuit.

The portion of the system board containing the comparator and counter inhibit logic

■ SYSTEM ASSIGNMENT 2:   Troubleshooting

■ *Activity 1:*   A pulser and logic probe are applied to the system board as shown. The probe light is off, indicating no pulse activity. Determine whether or not this test indicates a fault. If a fault is indicated, specify the faulty device or devices. If you can't isolate the fault (assuming one exists) to a single IC, explain why you can't and state what further tests are needed.

■ *Activity 2:*   The table indicates seven *independent* sets of pulse/probe measurements on the portion of the system board that contains the comparator and counter inhibit logic. For each set of measurements, determine whether or not a fault is indicated. If a fault is indicated, specify what the fault is and how to repair it. If a set of measurements is inconclusive, explain why.

	IC#	Pin #	IC#	Pin #	Indication	IC#	Pin #	Indication	IC#	Pin #	Indication
Measurement set 1	1	1	1	3	Flashing	4	1	Flashing	4	2	Off
Measurement set 2	1	4	1	6	Flashing	4	3	Off			
Measurement set 3	2	9	2	8	Off						
Measurement set 4	4	11	4	10	Flashing	6	5	Flashing	6	8	On
Measurement set 5	5	1	5	2	Flashing	5	3	Flashing	5	4	Off
Measurement set 6	6	1	6	8	On						
Measurement set 7	3	9	3	8	Off						

Isolate the Faulty Element With the faulty system board on the workbench, power and ground are connected and a pulse is applied to the sensor input. A DMM can be used to make sure that V_{CC} and ground are getting to each IC, and a probe or scope can be used to check the counter. Assuming that power and ground are OK and that the counter appears to be working, check the comparator and counter inhibit logic circuits using a logic pulser and probe as illustrated by various measurements in System Assignment 2. Although the probability is high that only one fault exists, several possibilities are presented to provide practice in arriving at a conclusion based on a set of measurements. Keeping in mind that all possible faults are not included, perform the activities listed in System Assignment 2.

SECTION 3–9 REVIEW

1. In System Assignment 1, why do BIT4, BIT5, BIT6, and BIT7 never change?
2. In System Assignment 1, what is the maximum voltage that should be displayed by the DMM on the left? By the DMM on the right?
3. Why can't the NAND gate in the comparator logic be fully tested with a pulser and a logic probe?

■ SUMMARY

- The inverter output is the complement of the input.
- The AND gate output is HIGH only if all the inputs are HIGH.
- The OR gate output is HIGH if any of the inputs is HIGH.
- The NAND gate output is LOW only if all the inputs are HIGH.
- The NAND can be viewed as a negative-OR whose output is HIGH when any input is LOW.
- The NOR gate output is LOW if any of the inputs is HIGH.
- The NOR can be viewed as a negative-AND whose output is HIGH only if all the inputs are LOW.
- The exclusive-OR gate output is HIGH when the inputs are not the same.
- The exclusive-NOR gate output is LOW when the inputs are not the same.
- Distinctive shape symbols and truth tables for various logic gates (limited to 2 inputs) are shown in Figure 3–69.

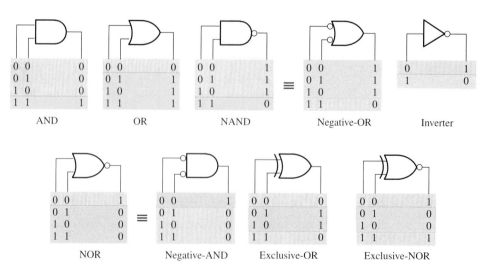

Note: Active states are shown in color.

FIGURE 3–69

- TTL and ECL are made with bipolar junction transistors.
- CMOS is made with MOS field-effect transistors.
- As a rule, ECL has the fastest switching speed, and TTL and CMOS are neck-and-neck with CMOS rapidly gaining an advantage.
- As a rule, CMOS has the lowest power consumption, TTL is next, and ECL has the highest power consumption.
- The average power dissipation of a logic gate is

$$P_D = V_{CC}\left(\frac{I_{CCH} + I_{CCL}}{2}\right) \qquad (3\text{--}2)$$

- The speed-power product of a logic gate is

$$SPP = t_P P_D \qquad (3\text{--}3)$$

■ KEY TERMS

These terms are also in the end-of-book glossary.

AND gate   A logic gate that produces a HIGH output only when all of the inputs are HIGH.

ANSI   American National Standards Institute.

Boolean addition   In Boolean algebra, the OR operation.

Boolean algebra   The mathematics of logic circuits.

Boolean multiplication   In Boolean algebra, the AND operation.

CMOS   Complementary metal-oxide semiconductor; a class of integrated logic circuit that is implemented with a type of field-effect transistor.

Complementation   Inversion. LOW is the complement of HIGH, and 0 is the complement of 1.

ECL   Emitter-coupled logic; a class of integrated logic circuit that is implemented with nonsaturating bipolar junction transistors.

Enable   To activate or put into an operational mode; an input on a logic circuit that enables its operation.

Exclusive-NOR (XNOR) gate   A logic gate that produces a LOW output only when its two inputs are at opposite levels.

Exclusive-OR (XOR) gate   A logic gate that produces a HIGH output only when its two inputs are at opposite levels.

Fan-out   The number of equivalent gate inputs of the same family series that a logic gate can drive.

IEEE   Institute of Electrical and Electronic Engineers.

Inversion   The conversion of a HIGH (1) level to a LOW (0) level or vice versa.

NAND gate   A logic gate which produces a LOW output only when all the inputs are HIGH.

Negative-AND   An equivalent NOR gate operation in which the HIGH is the active output when all inputs are LOW.

Negative-OR   An equivalent NAND gate operation in which the HIGH is the active input when one or more of the inputs are LOW.

NOR gate   A logic gate in which the output is LOW when one or more of the inputs are HIGH.

OR gate   A logic gate that produces a HIGH output when one or more inputs are HIGH.

Power dissipation   The product of the dc supply voltage and the dc supply current in an electronic circuit; the amount of power required by a circuit.

Propagation delay time The time interval between the occurrence of an input transition and the occurrence of the corresponding output transition in a logic circuit.

Speed-power product A performance parameter that is the product of the propagation delay time and the power dissipation in a logic circuit.

Truth table A table showing the inputs and corresponding output(s) of a logic circuit.

TTL Transistor-transistor logic; a class of integrated logic circuit that uses bipolar junction transistors.

Unit load A measure of fan-out. One gate input represents one unit load to the output of a gate within the same IC family.

■ **SELF-TEST**

1. When the input to an inverter is HIGH (1), the output is
   (a) HIGH or 1   (b) LOW or 1   (c) HIGH or 0   (d) LOW or 0

2. An inverter performs an operation known as
   (a) complementation   (b) assertion
   (c) inversion   (d) both answers (a) and (c)

3. The output of an AND gate with inputs A, B, and C is a 1 (HIGH) when
   (a) $A = 1, B = 1, C = 1$ (b) $A = 1, B = 0, C = 1$ (c) $A = 0, B = 0, C = 0$

4. The output of an OR gate with inputs A, B, and C is a 1 (HIGH) when
   (a) $A = 1, B = 1, C = 1$ (b) $A = 0, B = 0, C = 1$ (c) $A = 0, B = 0, C = 0$
   (d) answers (a), (b), and (c)   (e) only answers (a) and (b)

5. A pulse is applied to each input of a 2-input NAND gate. One pulse goes HIGH at $t = 0$ and goes back LOW at $t = 1$ ms. The other pulse goes HIGH at $t = 0.8$ ms and goes back LOW at $t = 3$ ms. The output pulse can be described as follows:
   (a) It goes LOW at $t = 0$ and back HIGH at $t = 3$ ms.
   (b) It goes LOW at $t = 0.8$ ms and back HIGH at $t = 3$ ms.
   (c) It goes LOW at $t = 0.8$ ms and back HIGH at $t = 1$ ms.
   (d) It goes HIGH at $t = 0.8$ ms and back LOW at $t = 1$ ms.

6. A pulse is applied to each input of a 2-input NOR gate. One pulse goes HIGH at $t = 0$ and goes back LOW at $t = 1$ ms. The other pulse goes HIGH at $t = 0.8$ ms and goes back LOW at $t = 3$ ms. The output pulse can be described as follows:
   (a) It goes LOW at $t = 0$ and back HIGH at $t = 3$ ms.
   (b) It goes LOW at $t = 0.8$ ms and back HIGH at $t = 3$ ms.
   (c) It goes LOW at $t = 0.8$ ms and back HIGH at $t = 1$ ms.
   (d) It goes HIGH at $t = 0.8$ ms and back LOW at $t = 1$ ms.

7. A pulse is applied to each input of an exclusive-OR gate. One pulse goes HIGH at $t = 0$ and goes back LOW at $t = 1$ ms. The other pulse goes HIGH at $t = 0.8$ ms and goes back LOW at $t = 3$ ms. The output pulse can be described as follows:
   (a) It goes HIGH at $t = 0$ and back LOW at $t = 3$ ms.
   (b) It goes HIGH at $t = 0$ and back LOW at $t = 0.8$ ms.
   (c) It goes HIGH at $t = 1$ ms and back LOW at $t = 3$ ms.
   (d) both answers (b) and (c)

8. A positive-going pulse is applied to an inverter. The time interval from the leading edge of the input to the leading edge of the output is 7 ns. This parameter is
   (a) speed-power product   (b) propagation delay, t_{PHL}
   (c) propagation delay, t_{PLH} (d) pulse width

9. The CMOS family with the fastest switching speed is
   (a) AC   (b) HC   (c) ACT   (d) ALVC

10. If power consumption were the major criterion in the design of a digital system, the logic family that you would probably use is
   (a) HCT   (b) ALVC   (c) LV   (d) LVC

11. To measure the period of a pulse waveform, you must use

    (a) a DMM   **(b)** a logic probe   **(c)** an oscilloscope   **(d)** a pulser

12. Once you measure the period of a pulse waveform, the frequency is found by

    (a) using another setting   **(b)** measuring the duty cycle

    (c) finding the reciprocal of the period   **(d)** using another type of instrument

---

■ PROBLEMS

SECTION 3–1   The Inverter

1. The input waveform shown in Figure 3–70 is applied to an inverter. Sketch the timing diagram of the output waveform in proper relation to the input.

2. A network of cascaded inverters is shown in Figure 3–71. If a HIGH is applied to point A, determine the logic levels at points B through F.

FIGURE 3–70 V_{IN} HIGH   LOW

FIGURE 3–71

SECTION 3–2   The AND Gate

3. Determine the output, X, for a 2-input AND gate with the input waveforms shown in Figure 3–72. Show the proper relationship of output to inputs with a timing diagram.

4. Repeat Problem 3 for the waveforms in Figure 3–73.

FIGURE 3–72

FIGURE 3–73 A

                     B

5. The input waveforms applied to a 3-input AND gate are as indicated in Figure 3–74. Show the output waveform in proper relation to the inputs with a timing diagram.

6. The input waveforms applied to a 4-input gate are as indicated in Figure 3–75. Show the output waveform in proper relation to the inputs with a timing diagram.

FIGURE 3–74 A

                     B

                     C

FIGURE 3–75

SECTION 3–3   The OR Gate

7. Determine the output for a 2-input OR gate when the input waveforms are as in Figure 3–73 and draw a timing diagram.

8. Repeat Problem 5 for a 3-input OR gate.

9. Repeat Problem 6 for a 4-input OR gate.

10. For the five input waveforms in Figure 3–76, determine the output for a 5-input AND gate and the output for a 5-input OR gate. Draw the timing diagram.

FIGURE 3–76

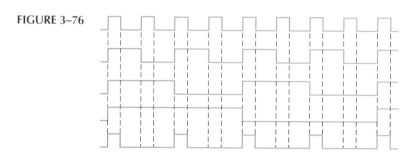

SECTION 3–4   The NAND Gate

11. For the set of input waveforms in Figure 3–77, determine the output for the gate shown and draw the timing diagram.

12. Determine the gate output for the input waveforms in Figure 3–78 and draw the timing diagram.

FIGURE 3–77

FIGURE 3–78

13. Determine the output waveform in Figure 3–79.

14. As you have learned, the two logic symbols shown in Figure 3–80 represent equivalent operations. The difference between the two is strictly from a functional viewpoint. For the NAND symbol, look for two HIGHs on the inputs to give a LOW output. For the negative-OR, look for at least one LOW on the inputs to give a HIGH on the output. Using these two functional viewpoints, show that each gate will produce the same output for the given inputs.

FIGURE 3–79

FIGURE 3–80

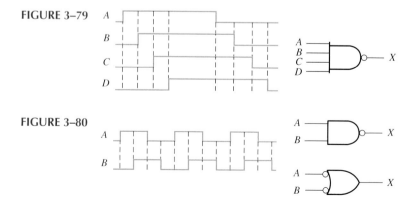

SECTION 3–5 The NOR Gate

15. Repeat Problem 11 for a 2-input NOR gate.

16. Determine the output waveform in Figure 3–81 and draw the timing diagram.

17. Repeat Problem 13 for a 4-input NOR gate.

18. The NAND and the negative-OR symbols represent equivalent operations, but they are functionally different. For the NOR symbol, look for at least one HIGH on the inputs to give a LOW on the output. For the negative-AND, look for two LOWs on the inputs to give a HIGH output. Using these two functional points of view, show that both gates in Figure 3–82 will produce the same output for the given inputs.

FIGURE 3–81

FIGURE 3–82

SECTION 3–6 The Exclusive-OR and Exclusive-NOR Gates

19. How does an exclusive-OR gate differ from an OR gate in its logical operation?

20. Repeat Problem 11 for an exclusive-OR gate.

21. Repeat Problem 11 for an exclusive-NOR gate.

22. Determine the output of an exclusive-OR gate for the inputs shown in Figure 3–73 and draw a timing diagram.

SECTION 3–7 Integrated Circuit Logic Gates

23. In the comparison of certain logic devices, it is noted that the power dissipation for one particular type increases as the frequency increases. Is the device TTL or CMOS?

24. Using the data sheets in Figures 3–57 and 3–58, determine the following:

   (a) 74LS00 power dissipation at maximum supply voltage and a 50% duty cycle

   (b) Minimum HIGH level output voltage for a 74LS00

   (c) Maximum propagation delay for a 74LS00

   (d) Maximum LOW level output voltage for a 74HC00A

   (e) Maximum propagation delay for a 74HC00A

25. Determine t_{PLH} and t_{PHL} from the oscilloscope display in Figure 3–83. The readings indicate V/div and sec/div for each channel.

FIGURE 3–83

Input

Output

2 V 5 ns   2 V 5 ns

26. Gate A has $t_{PLH} = t_{PHL} = 6$ ns. Gate B has $t_{PLH} = t_{PHL} = 10$ ns. Which gate can be operated at a higher frequency?

27. If a logic gate operates on a dc supply voltage of $+5$ V and draws an average current of 4 mA, what is its power dissipation?

28. The variable I_{CCH} represents the dc supply current from V_{CC} when all outputs of an IC are HIGH. The variable I_{CCL} represents the dc supply current when all outputs are LOW. For a 74LS00 IC, determine the typical power dissipation when all four gate outputs are HIGH. (See data sheet in Appendix A.)

SECTION 3–8   Troubleshooting

29. Examine the conditions indicated in Figure 3–84, and identify the faulty gates.

30. Determine the faulty gates in Figure 3–85 by analyzing the timing diagrams.

31. Using a logic probe and pulser, you make the observations indicated in Figure 3–86. For each observation determine the most likely gate failure.

32. The seat belt alarm circuit in Figure 3–16 has malfunctioned. You find that when the ignition switch is turned on and the seat belt is unbuckled, the alarm comes on and will not go off. What is the most likely problem? How do you troubleshoot it?

33. Every time the ignition switch is turned on in the circuit of Figure 3–16, the alarm comes on for thirty seconds, even when the seat belt is buckled. What is the most probable cause of this malfunction?

FIGURE 3–84

FIGURE 3–85

SECTION 3–9   Digital System Application

34. In relation to the system problem discussed in the section, the counter inhibit logic actually solves only one part of the timing problem. It keeps the counter from advancing an extra count which causes the comparator output to go back LOW. What is the remaining problem?

35. Assume that the tablet-counting and control system is set for 75 tablets per bottle. Draw the timing diagram that you should see on the logic analyzer for the ten sensor pulses up to and including the final count. Use the logic analyzer display in System Assignment 1 as a guide.

36. Determine the logic levels (expressed as 1s and 0s) that should appear on the inputs a through h of the system board in System Assignment 1 for each of the following tablets-per-bottle settings:

    (a) 30 tablets/bottle   **(b)** 75 tablets/bottle   **(c)** 99 tablets/bottle

FIGURE 3–86

(a)

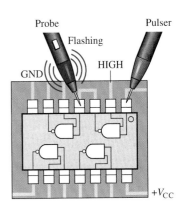

(b)

37. Assume that the tablet-counting and control system is set for 50 tablets per bottle but it is putting only 18 tablets in each bottle. List the possible system malfunctions that would cause this problem in terms of inputs to the system board in System Assignment 1.

38. With the board pulled from the system and on your workbench, suggest a procedure for testing the comparator logic.

Special Design Problems

39. Sensors are used to monitor the pressure and the temperature of a chemical solution stored in a vat. The circuitry for each sensor produces a HIGH voltage when a specified maximum value is exceeded. An alarm requiring a LOW voltage input must be activated when either the pressure or the temperature is excessive. Design a circuit for this application?

40. In a certain automated manufacturing process, electrical components are automatically inserted in a PC board. Before the insertion tool is activated, the PC board must be properly positioned, and the component to be inserted must be in the chamber. Each of these prerequisite conditions is indicated by a HIGH voltage. The insertion tool requires a LOW voltage to activate it. Design a circuit to implement this process.

41. Modify the frequency counter in Figure 3–15 to operate with an enable pulse that is active-LOW rather than HIGH during the 1 s interval.

42. Assume that the enable signal in Figure 3–15 has the waveform shown in Figure 3–87. Assume that waveform B is also available. Devise a circuit that will produce an active-HIGH reset pulse to the counter only during the time that the enable signal is LOW.

FIGURE 3–87

43. Design a circuit to fit in the colored block of Figure 3–88 that will cause the headlights of an automobile to be turned off automatically 15 s after the ignition switch is turned off, if the light switch is left on. Assume that a LOW is required to turn the lights off.

FIGURE 3–88

LOW turns off the lights.

Ignition switch HIGH = On / LOW = Off

Light switch HIGH = On / LOW = Off

Headlight control

44. Modify the logic circuit for the intrusion alarm in Figure 3–24 so that two additional rooms, each with two windows and one door, can be protected.

45. In the tablet-counting and control system, the tablet feeder and conveyor mechanisms have been changed to a different model. In the new model, the valve requires a LOW level to close and the conveyor requires a LOW to advance. Register B still requires a HIGH level to store the new total count. Modify the comparator logic to accommodate this requirement.

Electronics Workbench Troubleshooting

46. Open file PR03_46.EWB and observe the operation of the AND gate using the existing test set-up. Based on the observed inputs and output, determine the most likely fault in the gate.

47. Open file PR03_47.EWB and observe the operation of the NAND gate using the existing test set-up. Based on the observed inputs and output, determine the most likely fault in the gate.

48. Open file PR03_48.EWB and observe the operation of the NOR gate using the existing test set-up. Based on the observed inputs and output, determine the most likely fault in the gate.

49. Open file PR03_49.EWB and observe the operation of the exclusive-OR gate using the existing test set-up. Based on the observed inputs and output, determine the most likely fault in the gate.

■ **ANSWERS TO SECTION REVIEWS**

SECTION 3–1

1. When the inverter input is 1, the output is 0.

2. (a)

(b) A negative-going pulse is on the output (HIGH to LOW and back HIGH).

SECTION 3–2

1. An AND gate output is HIGH when all inputs are HIGH.

2. An AND gate output is LOW when one or more inputs are LOW.

3. Five-input AND: $X = 1$ when $ABCDE = 11111$ and $X = 0$ for all other combinations of $ABCDE$.

SECTION 3–3

1. An OR gate output is HIGH when one or more inputs are HIGH.

2. An OR gate output is LOW when all inputs are LOW.

3. Three-input OR: $X = 0$ when $ABC = 000$, and $X = 1$ for all other combinations of ABC.

SECTION 3–4

1. A NAND output is LOW when all inputs are HIGH.
2. A NAND output is HIGH when one or more inputs are LOW.
3. NAND: active-LOW output for all HIGH inputs; negative-OR: active-HIGH output for one or more LOW inputs; they have the same truth tables.
4. $X = \overline{ABC}$

SECTION 3–5

1. A NOR output is HIGH when all inputs are LOW.
2. A NOR output is LOW when one or more inputs are HIGH.
3. NOR: active-LOW output for one or more HIGH inputs; negative-AND; active-HIGH output for all LOW inputs; they have the same truth tables.
4. $X = \overline{A + B + C}$

SECTION 3–6

1. An XOR output is HIGH when the inputs are at opposite levels.
2. An XNOR output is HIGH when the inputs are at the same levels.
3. Apply the bits to the XOR inputs; when the output is HIGH, the bits are different.

SECTION 3–7

1. TTL, CMOS, and ECL; CMOS and TTL
2. **(a)** LS—Low-power Schottky   **(b)** ALS—Advanced LS
   (c) F—fast TTL   **(d)** HC—High-speed CMOS
   (e) AC—Advanced CMOS   **(f)** HCT—HC CMOS TTL compatible
   (g) LV—Low-voltage CMOS
3. **(a)** 74LS04—Hex inverter   **(b)** 74HC00—Quad 2-input NAND
   (c) 74LV08—Quad 2-input AND   **(d)** 74ALS10—Triple 3-input NAND
   (e) 7432—Quad 2-input OR   **(f)** 74ACT11—Triple 3-input AND
   (g) 74AHC02—Quad 2-input NOR
4. Lowest power—CMOS; highest speed—ECL
5. Six inverters in a package; four 2-input NAND gates in a package
6. $t_{PLH} = 10$ ns; $t_{PHL} = 8$ ns
7. 18 pJ
8. I_{CCL}—dc supply current for LOW output state; I_{CCH}—dc supply current for HIGH output state
9. V_{IL}—LOW input voltage; V_{IH}—HIGH input voltage
10 V_{OL}—LOW output voltage; V_{OH}—HIGH output voltage

SECTION 3–8

1. Opens and shorts are the most common failures.
2. An open input which effectively makes input HIGH
3. Waveforms can be observed with a scope but not with a logic probe.
4. A probe is easier to use.

SECTION 3–9

1. BIT4, BIT5, BIT6, and BIT7 never change because the count does not exceed 60: 00111100.
2. $V_{left} = 5.5$ V; $V_{right} = 0.4$ V
3. Too many inputs to drive to HIGH state simultaneously

■ **ANSWERS TO RELATED PROBLEMS FOR EXAMPLES**

3–1 The timing diagram is not affected. **3–2** See Table 3–13.

3–3 See Figure 3–89. **3–4** The output waveform is the same as input *A*.

TABLE 3–13

Inputs ABCD	Output X	Inputs ABCD	Output X
0000	0	1000	0
0001	0	1001	0
0010	0	1010	0
0011	0	1011	0
0100	0	1100	0
0101	0	1101	0
0110	0	1110	0
0111	0	1111	1

FIGURE 3–89

3–5 See Figure 3–90. **3–6** See Figure 3–91. **3–7** See Figure 3–92.

3–8 See Figure 3–93. **3–9** See Figure 3–94. **3–10** See Figure 3–95.

C = HIGH

FIGURE 3–90

FIGURE 3–91

FIGURE 3–92

C = LOW

FIGURE 3–93

FIGURE 3–94

FIGURE 3–95

FIGURE 3–96

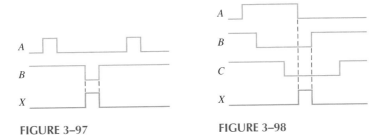

FIGURE 3–97   FIGURE 3–98

3–11 Use a 3-input NAND gate.

3–12 Use a 4-input NAND gate operating as a negative-OR gate.   **3–13** See Figure 3–96.

3–14 See Figure 3–97.   **3–15** See Figure 3–98.

3–16 Use a 2-input NOR gate.   **3–17** A 3-input NAND gate.

3–18 The output is always LOW. The timing diagram is a straight line.

3–19 The exclusive-OR gate will not detect simultaneous failures if both circuits produce the same outputs.

3–20 The outputs are unaffected.

3–21 The gate with 4 ns t_{PLH} and t_{PHL} can operate at the highest frequency.

3–22 10 mW

3–23 The gate output or pin 13 input is internally open.

3–24 The display will show an erratic readout because the counter continues until reset.

3–25 The enable pulse is too short or the counter is reset too soon.

■ **ANSWERS**
   TO SELF-TEST

1. (d)   **2.** (d)   **3.** (a)   **4.** (e)   **5.** (c)   **6.** (a)   **7.** (d)   **8.** (b)

9. (d)   **10.** (d)   **11.** (c)   **12.** (c)

4

BOOLEAN ALGEBRA AND LOGIC SIMPLIFICATION

 Computer Simulation of Digital Circuits:
Electronics Workbench (EWB) Tutorial at:
http://www.prenhall.com/floyd

■ CHAPTER OBJECTIVES

❑ Apply the basic laws and rules of Boolean algebra

❑ Apply DeMorgan's theorems to Boolean expressions

❑ Describe gate networks with Boolean expressions

❑ Evaluate Boolean expressions

❑ Simplify expressions by using the laws and rules of Boolean algebra

❑ Convert any Boolean expression into a sum-of-products (SOP) form

❑ Convert any Boolean expression into a product-of-sums (POS) form

❑ Use a Karnaugh map to simplify Boolean expressions

❑ Use a Karnaugh map to simplify truth table functions

❑ Utilize "don't care" conditions to simplify logic functions

❑ Apply Boolean algebra and the Karnaugh map method to a system application

■ CHAPTER OVERVIEW

In 1854, George Boole published a work titled *An Investigation of the Laws of Thought, on Which Are Founded the Mathematical Theories of Logic and Probabilities.* It was in this publication that a "logical algebra," known today as Boolean algebra, was formulated. Boolean algebra is a convenient and systematic way of expressing and analyzing the operation of logic circuits. Claude Shannon was the first to apply Boole's work to the analysis and design of logic circuits. In 1938, Shannon wrote a thesis at MIT titled *A Symbolic Analysis of Relay and Switching Circuits.*

This chapter covers the laws, rules, and theorems of Boolean algebra and their application to digital circuits. You will learn how to define a given circuit with a Boolean expression and then evaluate its operation. You will also learn how to simplify logic circuits using the methods of Boolean algebra and Karnaugh maps.

■ DIGITAL SYSTEM APPLICATION

This Digital System Application illustrates concepts taught in this chapter. The 7-segment display logic in the tablet-counting and control system is a good way to illustrate the application of Boolean algebra and the Karnaugh map method to obtain the simplest possible implementation in the design of logic circuits. Therefore, in Section 4–12, the focus is on the BCD-to-7-segment logic that drives the two system displays indicated below.

4–1 ■ BOOLEAN OPERATIONS AND EXPRESSIONS

Boolean algebra is the mathematics of digital systems. A basic knowledge of Boolean algebra is indispensable to the study and analysis of logic circuits. In the last chapter, Boolean operations and expressions in terms of their relationship to NOT, AND, OR, NAND, and NOR gates were introduced. This section reviews that material and provides additional definitions and information.

After completing this section, you should be able to

- Define *variable* ■ Define *literal* ■ Identify a sum term ■ Evaluate a sum term
- Identify a product term ■ Evaluate a product term ■ Explain Boolean addition
- Explain Boolean multiplication

COMPUTER NOTE

In a microprocessor, the arithmetic logic unit (ALU) performs arithmetic and Boolean logic operations on digital data as directed by program instructions. Logical operations are equivalent to the basic gate operations that you are familiar with, but deal with a minimum of 8 bits at a time. Examples of Boolean logic instructions are AND, OR, NOT, and XOR which are called *mnemonics*. An assembly language program uses the mnemonics to specify an operation. Another program called an *assembler* translates the mnemonics into a binary code that can be understood by the microprocessor.

Variable, complement, and *literal* are terms used in Boolean algebra. A **variable** is a symbol (usually an italic uppercase letter) used to represent a logical quantity. Any single variable can have a 1 or a 0 value. The **complement** is the inverse of a variable and is indicated by a bar over the variable (overbar). For example, the complement of the variable is A is $\overline{A}$. If $A = 1$, then $\overline{A} = 0$. If $A = 0$, then $\overline{A} = 1$. The complement of the variable A is read as "not A" or "A bar." Sometimes a prime symbol rather than an overbar is used to denote the complement of a variable; for example, B' indicates the complement of B. In this book, only the overbar is used. A **literal** is a variable or the complement of a variable.

Boolean Addition

Recall from Chapter 3 that **Boolean addition** is equivalent to the OR operation and the basic rules are illustrated with their relation to the OR gate as follows:

$$0 + 0 = 0 \qquad 0 + 1 = 1 \qquad 1 + 0 = 1 \qquad 1 + 1 = 1$$

In Boolean algebra, a **sum term** is a sum of literals. In logic circuits, a sum term is produced by an OR operation with no AND operations involved. Some examples of sum terms are $A + B$, $A + \overline{B}$, $A + B + \overline{C}$, and $\overline{A} + B + C + \overline{D}$.

A sum term is equal to 1 when one or more of the literals in the term are 1. A sum term is equal to 0 only if each of the literals is 0.

EXAMPLE 4–1

Determine the values of A, B, C, and D which make the sum term $A + \overline{B} + C + \overline{D}$ equal to 0.

Solution   For the sum term to be 0, each of the literals in the term must be 0. Therefore, $A = \mathbf{0}$, $B = \mathbf{1}$ so that $\overline{B} = 0$, $C = \mathbf{0}$, and $D = \mathbf{1}$ so that $\overline{D} = 0$.

$$A + \overline{B} + C + \overline{D} = 0 + \overline{1} + 0 + \overline{1} = 0 + 0 + 0 + 0 = 0$$

Related Problem   Determine the values of A and B which make the sum term $\overline{A} + B$ equal to 0.

Boolean Multiplication

Also recall from Chapter 3 that **Boolean multiplication** is equivalent to the AND operation and the basic rules are illustrated with their relation to the AND gate as follows:

$$0 \cdot 0 = 0 \quad 0 \cdot 1 = 0 \quad 1 \cdot 0 = 0 \quad 1 \cdot 1 = 1$$

In Boolean algebra, a **product term** is the product of literals. In logic circuits, a product term is produced by an AND operation with no OR operations involved. Some examples of product terms are AB, $A\overline{B}$, ABC, and $A\overline{B}C\overline{D}$.

A product term is equal to 1 only if each of the literals in the term is 1. A product term is equal to 0 when one or more of the literals are 0.

EXAMPLE 4–2

Determine the values of A, B, C, and D which make the product term $A\overline{B}C\overline{D}$ equal to 1.

Solution   For the product term to be 1, each of the literals in the term must be 1. Therefore, $A = \mathbf{1}$, $B = \mathbf{0}$ so that $\overline{B} = 1$, $C = \mathbf{1}$, and $D = \mathbf{0}$ so that $\overline{D} = 1$.

$$A\overline{B}C\overline{D} = 1 \cdot \overline{0} \cdot 1 \cdot \overline{0} = 1 \cdot 1 \cdot 1 \cdot 1 = 1$$

Related Problem   Determine the values of A and B which make the product term $\overline{A}\,\overline{B}$ equal to 1.

SECTION 4–1 REVIEW

1. If $A = 0$, what does $\overline{A}$ equal?
2. Determine the values of A, B, and C that make the sum term $\overline{A} + \overline{B} + C$ equal to 0.
3. Determine the values of A, B, and C that make the product term $A\overline{B}C$ equal to 1.

4–2 ■ LAWS AND RULES OF BOOLEAN ALGEBRA

As in other areas of mathematics, there are certain well-developed rules and laws that must be followed in order to properly apply Boolean algebra. The most important of these are presented in this section.

After completing this section, you should be able to

■ Apply the commutative laws of addition and multiplication   ■ Apply the associative laws of addition and multiplication   ■ Apply the distributive law   ■ Apply twelve basic rules of Boolean algebra

Laws of Boolean Algebra

The basic laws of Boolean algebra—the **commutative laws** for addition and multiplication, the **associative laws** for addition and multiplication, and the **distributive law**—are the same as in ordinary algebra. Each of the laws is illustrated with two or three variables, but the number of variables is not limited to this.

Commutative Laws The *commutative law of addition* for two variables is written as

$$A + B = B + A \qquad \text{(4–1)}$$

This law states that the order in which the variables are ORed makes no difference. Remember, in Boolean algebra as applied to logic circuits, addition and the OR operation are the same. Figure 4–1 illustrates the commutative law as applied to the OR gate and shows that it doesn't matter to which input each variable is applied. (The symbol $\equiv$ means "equivalent to.")

FIGURE 4–1
Application of commutative law of addition.

The *commutative law of multiplication* for two variables is

$$AB = BA \qquad \text{(4–2)}$$

This law states that the order in which the variables are ANDed makes no difference. Figure 4–2 illustrates this law as applied to the AND gate.

FIGURE 4–2
Application of commutative law of multiplication.

Associative Laws The *associative law of addition* is written as follows for three variables:

$$A + (B + C) = (A + B) + C \qquad \text{(4–3)}$$

This law states that when ORing more than two variables, the result is the same regardless of the grouping of the variables. Figure 4–3 illustrates this law as applied to 2-input OR gates.

FIGURE 4–3
Application of associative law of addition.

The *associative law of multiplication* is written as follows for three variables:

$$A(BC) = (AB)C \qquad \text{(4–4)}$$

This law states that it makes no difference in what order the variables are grouped when ANDing more than two variables. Figure 4–4 illustrates this law as applied to 2-input AND gates.

FIGURE 4–4
Application of associative law of multiplication.

Distributive Law The distributive law is written for three variables as follows:

$$A(B + C) = AB + AC \qquad (4–5)$$

This law states that ORing two or more variables and then ANDing the result with a single variable is equivalent to ANDing the single variable with each of the two or more variables and then ORing the products. The distributive law also expresses the process of *factoring* in which the common variable A is factored out of the product terms, for example, $AB + AC = A(B + C)$. Figure 4–5 illustrates the distributive law in terms of gate implementation.

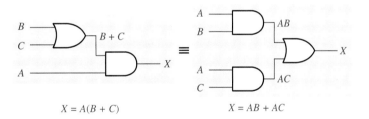

$$X = A(B + C) \qquad\qquad X = AB + AC$$

FIGURE 4–5
Application of distributive law.

Rules of Boolean Algebra

Table 4–1 lists 12 basic rules that are useful in manipulating and simplifying **Boolean expressions.** Rules 1 through 9 will be viewed in terms of their application to logic gates. Rules 10 through 12 will be derived in terms of the simpler rules and the laws previously discussed.

TABLE 4–1
Basic rules of Boolean algebra.

1. $A + 0 = A$	**7.** $A \cdot A = A$
2. $A + 1 = 1$	**8.** $A \cdot \overline{A} = 0$
3. $A \cdot 0 = 0$	**9.** $\overline{\overline{A}} = A$
4. $A \cdot 1 = A$	**10.** $A + AB = A$
5. $A + A = A$	**11.** $A + \overline{A}B = A + B$
6. $A + \overline{A} = 1$	**12.** $(A + B)(A + C) = A + BC$

A, B, or *C* can represent a single variable or a combination of variables.

1. $A + 0 = A$ A variable ORed with 0 is always equal to the variable. If the input variable A is 1, the output variable X is 1, which is equal to A. If A is 0, the output is 0, which is also equal to A. This rule is illustrated in Figure 4–6, where the lower input is fixed at 0.

FIGURE 4–6
Rule 1: $A + 0 = A$.

$$X = A + 0 = A$$

2. $A + 1 = 1$ A variable ORed with 1 is always equal to 1. A 1 on an input to an OR gate produces a 1 on the output, regardless of the value of the variable on the other input. This rule is illustrated in Figure 4–7, where the lower input is fixed at 1.

FIGURE 4–7
Rule 2: $A + 1 = 1$.

$$X = A + 1 = 1$$

3. $A \cdot 0 = 0$ A variable ANDed with 0 is always equal to 0. Any time one input to an AND gate is 0, the output is 0, regardless of the value of the variable on the other input. This rule is illustrated in Figure 4–8, where the lower input is fixed at 0.

FIGURE 4–8
Rule 3: $A \cdot 0 = 0$.

$$X = A \cdot 0 = 0$$

4. $A \cdot 1 = A$ A variable ANDed with 1 is always equal to the variable. If A is 0, the output of the AND gate is 0. If A is 1, the output of the AND gate is 1 because both inputs are now 1s. This rule is shown in Figure 4–9, where the lower input is fixed at 1.

FIGURE 4–9
Rule 4: $A \cdot 1 = A$.

$$X = A \cdot 1 = A$$

5. $A + A = A$ A variable ORed with itself is always equal to the variable. If A is 0, then $0 + 0 = 0$; and if A is 1, then $1 + 1 = 1$. This is shown in Figure 4–10, where both inputs are the same variable.

FIGURE 4–10
Rule 5: $A + A = A$.

$$X = A + A = A$$

6. $A + \overline{A} = 1$ A variable ORed with its complement is always equal to 1. If A is 0, then $0 + \overline{0} = 0 + 1 = 1$. If A is 1, then $1 + \overline{1} = 1 + 0 = 1$. See Figure 4–11, where one input is the complement of the other.

FIGURE 4–11
Rule 6: $A + \overline{A} = 1$.

$$X = A + \overline{A} = 1$$

7. $A \cdot A = A$ A variable ANDed with itself is always equal to the variable. If $A = 0$, then $0 \cdot 0 = 0$; and if $A = 1$, then $1 \cdot 1 = 1$. Figure 4–12 illustrates this rule.

FIGURE 4–12
Rule 7: $A \cdot A = A$.

$X = A \cdot A = A$

8. $A \cdot \overline{A} = 0$ A variable ANDed with its complement is always equal to 0. Either A or $\overline{A}$ will always be 0; and when a 0 is applied to the input of an AND gate, the output will be 0 also. Figure 4–13 illustrates this rule.

FIGURE 4–13
Rule 8: $A \cdot \overline{A} = 0$.

$X = A \cdot \overline{A} = 0$

9. $\overline{\overline{A}} = A$ The double complement of a variable is always equal to the variable. If you start with the variable A and complement (invert) it once, you get $\overline{A}$. If you then take $\overline{A}$ and complement (invert) it, you get A, which is the original variable. This rule is shown in Figure 4–14 using inverters.

FIGURE 4–14
Rule 9: $\overline{\overline{A}} = A$.

$\overline{\overline{A}} = A$

10. $A + AB = A$ This rule can be proved by applying the distributive law, rule 2, and rule 4 as follows:

$$A + AB = A(1 + B) \qquad \text{Factoring (distributive law)}$$
$$= A \cdot 1 \qquad \text{Rule 2: } (1 + B) = 1$$
$$= A \qquad \text{Rule 4: } A \cdot 1 = A$$

The proof is shown in Table 4–2, which shows the truth table and the resulting logic circuit simplification.

TABLE 4–2

A	B	AB	A + AB	
0	0	0	0	
0	1	0	0	
1	0	0	1	
1	1	1	1	

11. $A + \overline{A}B = A + B$ This rule can be proved as follows:

$$A + \overline{A}B = (A + AB) + \overline{A}B \qquad \text{Rule 10: } A = A + AB$$
$$= (AA + AB) + \overline{A}B \qquad \text{Rule 7: } A = AA$$
$$= AA + AB + A\overline{A} + \overline{A}B \qquad \text{Rule 8: adding } A\overline{A} = 0$$
$$= (A + \overline{A})(A + B) \qquad \text{Factoring}$$
$$= 1 \cdot (A + B) \qquad \text{Rule 6: } A + \overline{A} = 1$$
$$= A + B \qquad \text{Rule 4: drop the 1}$$

The proof is shown in Table 4–3, which shows the truth table and the resulting logic circuit simplification.

TABLE 4–3

A	B	$\overline{A}B$	$A + \overline{A}B$	$A + B$	
0	0	0	0	0	
0	1	1	1	1	
1	0	0	1	1	
1	1	0	1	1	

equal

12. $(A + B)(A + C) = A + BC$ This rule can be proved as follows:

$$(A + B)(A + C) = AA + AC + AB + BC \qquad \text{Distributive law}$$
$$= A + AC + AB + BC \qquad \text{Rule 7: } AA = A$$
$$= A(1 + C) + AB + BC \qquad \text{Factoring (distributive law)}$$
$$= A \cdot 1 + AB + BC \qquad \text{Rule 2: } 1 + C = 1$$
$$= A(1 + B) + BC \qquad \text{Factoring (distributive law)}$$
$$= A \cdot 1 + BC \qquad \text{Rule 2: } 1 + B = 1$$
$$= A + BC \qquad \text{Rule 4: } A \cdot 1 = A$$

The proof is shown in Table 4–4, which shows the truth table and the resulting logic circuit simplification.

TABLE 4–4

A	B	C	$A + B$	$A + C$	$(A + B)(A + C)$	BC	$A + BC$	
0	0	0	0	0	0	0	0	
0	0	1	0	1	0	0	0	
0	1	0	1	0	0	0	0	
0	1	1	1	1	1	1	1	
1	0	0	1	1	1	0	1	
1	0	1	1	1	1	0	1	
1	1	0	1	1	1	0	1	
1	1	1	1	1	1	1	1	

equal

SECTION 4–2 REVIEW
1. Apply the associative law of addition to the expression $A + (B + C + D)$.
2. Apply the distributive law to the expression $A(B + C + D)$.

4–3 ■ DEMORGAN'S THEOREMS

DeMorgan, a mathematician who knew Boole, proposed two theorems that are an important part of Boolean algebra. In practical terms, DeMorgan's theorems provide mathematical verification of the equivalency of the NAND and negative-OR gates and the equivalency of the NOR and negative-AND gates, which were discussed in Chapter 3.

After completing this section, you should be able to

■ State DeMorgan's theorems ■ Relate DeMorgan's theorems to the equivalency of the NAND and negative-OR gates and to the equivalency of the NOR and negative-AND gates ■ Apply DeMorgan's theorems to the simplification of Boolean expressions

One of DeMorgan's theorems is stated as follows:

> **The complement of a product of variables is equal to the sum of the complements of the variables.**

Stated another way,

> **The complement of two or more variables ANDed is equivalent to the OR of the complements of the individual variables.**

The formula for expressing this theorem for two variables is

$$\overline{XY} = \overline{X} + \overline{Y} \tag{4–6}$$

DeMorgan's second theorem is stated as follows:

> **The complement of a sum of variables is equal to the product of the complements of the variables.**

Stated another way,

> **The complement of two or more variables ORed is equivalent to the AND of the complements of the individual variables.**

The formula for expressing this theorem for two variables is

$$\overline{X + Y} = \overline{X}\,\overline{Y} \tag{4–7}$$

Figure 4–15 shows the gate equivalencies and truth tables for Equations (4–6) and (4–7).

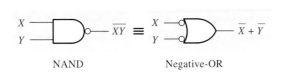

Inputs		Output	
X	Y	$\overline{XY}$	$\overline{X} + \overline{Y}$
0	0	1	1
0	1	1	1
1	0	1	1
1	1	0	0

Inputs		Output	
X	Y	$\overline{X + Y}$	$\overline{X}\,\overline{Y}$
0	0	1	1
0	1	0	0
1	0	0	0
1	1	0	0

FIGURE 4–15

Gate equivalencies and the corresponding truth tables that illustrate DeMorgan's theorems. Notice the equality of the two output columns in each table. This shows that the equivalent gates perform the same logic function.

As stated, DeMorgan's theorems also apply to expressions in which there are more than two variables. The following examples illustrate the application of DeMorgan's theorems to 3-variable and 4-variable expressions.

EXAMPLE 4–3

Apply DeMorgan's theorems to the expressions $\overline{XYZ}$ and $\overline{X + Y + Z}$.

Solution
$$\overline{XYZ} = \overline{X} + \overline{Y} + \overline{Z}$$
$$\overline{X + Y + Z} = \overline{X}\,\overline{Y}\,\overline{Z}$$

Related Problem   Apply DeMorgan's theorem to the expression $\overline{\overline{X} + \overline{Y} + \overline{Z}}$.

EXAMPLE 4–4

Apply DeMorgan's theorems to the expressions $\overline{WXYZ}$ and $\overline{W + X + Y + Z}$.

Solution
$$\overline{WXYZ} = \overline{W} + \overline{X} + \overline{Y} + \overline{Z}$$
$$\overline{W + X + Y + Z} = \overline{W}\,\overline{X}\,\overline{Y}\,\overline{Z}$$

Related Problem   Apply DeMorgan's theorem to the expression $\overline{\overline{W}\,\overline{X}\,\overline{Y}\,\overline{Z}}$.

Each variable in DeMorgan's theorems as stated in Equations (4–6) and (4–7) can also represent a combination of other variables. For example, X can be equal to the term $AB + C$, and Y can be equal to the term $A + BC$. So if you apply DeMorgan's theorem for two variables as stated by $\overline{XY} = \overline{X} + \overline{Y}$ to the expression $\overline{(AB + C)(A + BC)}$, you get the following result:

$$\overline{(AB + C)(A + BC)} = \overline{(AB + C)} + \overline{(A + BC)}$$

Notice that in the preceding result you have two terms, $\overline{AB + C}$ and $\overline{A + BC}$, to each of which you can again apply DeMorgan's theorem $\overline{X + Y} = \overline{X}\,\overline{Y}$ individually, as follows:

$$\overline{(AB + C)} + \overline{(A + BC)} = (\overline{AB})\overline{C} + \overline{A}(\overline{BC})$$

Notice that you still have two terms in the expression to which DeMorgan's theorem can again be applied. These terms are $\overline{AB}$ and $\overline{BC}$. A final application of DeMorgan's theorem gives the following result:

$$(\overline{AB})\overline{C} + \overline{A}(\overline{BC}) = (\overline{A} + \overline{B})\overline{C} + \overline{A}(\overline{B} + \overline{C})$$

Although this result can be simplified further by the use of Boolean rules and laws, DeMorgan's theorems cannot be used any more.

Applying DeMorgan's Theorems

The following procedure illustrates the application of DeMorgan's theorems and Boolean algebra to the specific expression (brackets are the same as parentheses):

$$\overline{\overline{A + B\overline{C}} + \overline{D(\overline{E + \overline{F}})}}$$

Step 1.   Identify the terms to which you can apply DeMorgan's theorems, and think of each term as a single variable. Let $\overline{A + B\overline{C}} = X$ and $\overline{D(\overline{E + \overline{F}})} = Y$.

Step 2.   Since $\overline{X + Y} = \overline{X}\,\overline{Y}$,

$$\overline{[\overline{A + B\overline{C}}] + [\overline{D(\overline{E + \overline{F}})}]} = [\overline{\overline{A + B\overline{C}}}][\overline{\overline{D(\overline{E + \overline{F}})}}]$$

Step 3. Use rule 9 ($\overline{\overline{A}} = A$) to cancel the double bars over the left term (this is not part of DeMorgan's theorem).

$$\overline{\overline{[A + B\overline{C}]}} \; \overline{\overline{[D(\overline{E + \overline{F}})]}} = [A + B\overline{C}][\overline{D(\overline{E + \overline{F}})}]$$

Step 4. In the right term, let $Z = \overline{E + \overline{F}}$.

$$[A + B\overline{C}][\overline{D(\overline{E + \overline{F}})}] = [A + B\overline{C}][\overline{DZ}]$$

Step 5. Since $\overline{DZ} = \overline{D} + \overline{Z}$,

$$[A + B\overline{C}][\overline{D(\overline{E + \overline{F}})}] = (A + B\overline{C})(\overline{D} + \overline{\overline{(E + \overline{F})}})$$

Step 6. Use rule 9 ($\overline{\overline{A}} = A$) to cancel the double bars over the $E + \overline{F}$ part of the term.

$$(A + B\overline{C})(\overline{D} + \overline{\overline{E + \overline{F}}}) = (A + B\overline{C})(\overline{D} + E + \overline{F})$$

The following three examples will further illustrate how to use DeMorgan's theorems.

EXAMPLE 4–5

Apply DeMorgan's theorems to each of the following expressions:

(a) $\overline{(A + B + C)D}$ (b) $\overline{ABC + DEF}$ (c) $\overline{A\overline{B} + \overline{C}D + EF}$

Solution

(a) Let $A + B + C = X$ and $D = Y$. The expression $\overline{(A + B + C)D}$ is of the form $\overline{XY} = \overline{X} + \overline{Y}$ and can be rewritten as

$$\overline{(A + B + C)D} = \overline{A + B + C} + \overline{D}$$

Next, apply DeMorgan's theorem to the term $\overline{A + B + C}$.

$$\overline{A + B + C} + \overline{D} = \overline{A}\,\overline{B}\,\overline{C} + \overline{D}$$

(b) Let $ABC = X$ and $DEF = Y$. The expression $\overline{ABC + DEF}$ is of the form $\overline{X + Y} = \overline{X}\,\overline{Y}$ and can be rewritten as

$$\overline{ABC + DEF} = (\overline{ABC})(\overline{DEF})$$

Next, apply DeMorgan's theorem to each of the terms $\overline{ABC}$ and $\overline{DEF}$.

$$(\overline{ABC})(\overline{DEF}) = (\overline{A} + \overline{B} + \overline{C})(\overline{D} + \overline{E} + \overline{F})$$

(c) Let $A\overline{B} = X$, $\overline{C}D = Y$, and $EF = Z$. The expression $\overline{A\overline{B} + \overline{C}D + EF}$ is of the form $\overline{X + Y + Z} = \overline{X}\,\overline{Y}\,\overline{Z}$ and can be rewritten as

$$\overline{A\overline{B} + \overline{C}D + EF} = (\overline{A\overline{B}})(\overline{\overline{C}D})(\overline{EF})$$

Next, apply DeMorgan's theorem to each of the terms $\overline{A\overline{B}}$, $\overline{\overline{C}D}$, and $\overline{EF}$.

$$(\overline{A\overline{B}})(\overline{\overline{C}D})(\overline{EF}) = (\overline{A} + B)(C + \overline{D})(\overline{E} + \overline{F})$$

Related Problem Apply DeMorgan's theorems to the expression $\overline{\overline{ABC} + D + E}$.

EXAMPLE 4–6

Apply DeMorgan's theorems to each expression:

(a) $\overline{(\overline{A + B}) + \overline{C}}$ **(b)** $\overline{(\overline{A} + B) + CD}$ **(c)** $\overline{(A + B)\overline{C}\overline{D} + E + \overline{F}}$

Solution

(a) $\overline{(\overline{A + B}) + \overline{C}} = (\overline{\overline{A + B}})\overline{\overline{C}} = (A + B)C$

(b) $\overline{(\overline{A} + B) + CD} = (\overline{\overline{A} + B})\overline{CD} = (\overline{\overline{A}}\overline{B})(\overline{C} + \overline{D}) = A\overline{B}(\overline{C} + \overline{D})$

(c) $\overline{(A + B)\overline{C}\overline{D} + E + \overline{F}} = [\overline{(A + B)\overline{C}\overline{D}}](\overline{E + \overline{F}}) = (\overline{A}\overline{B} + C + D)\overline{E}F$

Related Problem Apply DeMorgan's theorems to the expression $\overline{\overline{A}B(C + \overline{D}) + E}$.

EXAMPLE 4–7

The Boolean expression for an exclusive-OR gate is $A\overline{B} + \overline{A}B$. With this as a starting point, develop an expression for the exclusive-NOR gate, using DeMorgan's theorems and any other rules or laws that are applicable.

Solution Start by complementing the exclusive-OR expression and then applying DeMorgan's theorems as follows:

$$\overline{A\overline{B} + \overline{A}B} = (\overline{A\overline{B}})(\overline{\overline{A}B}) = (\overline{A} + \overline{\overline{B}})(\overline{\overline{A}} + \overline{B}) = (\overline{A} + B)(A + \overline{B})$$

Next, apply the distributive law and rule 8 ($A \cdot \overline{A} = 0$).

$$(\overline{A} + B)(A + \overline{B}) = \overline{A}A + \overline{A}\,\overline{B} + AB + B\overline{B} = \overline{A}\,\overline{B} + AB$$

The final expression for the XNOR is $\overline{A}\,\overline{B} + AB$. Note that this expression equals 1 any time both variables are 0s or both variables are 1s.

Related Problem Starting with the expression for a 4-input NAND gate, use De-Morgan's theorems to develop an expression for a 4-input negative-OR gate.

SECTION 4–3 REVIEW

1. Apply DeMorgan's theorems to the following expressions:

    (a) $\overline{\overline{ABC} + (\overline{D} + E)}$ **(b)** $\overline{(\overline{A + B})C}$ **(c)** $\overline{A + B + C + \overline{DE}}$

4–4 ■ BOOLEAN ANALYSIS OF LOGIC CIRCUITS

Boolean algebra provides a concise way to express the operation of a logic circuit formed by a combination of logic gates so that the output can be determined for various combinations of input values.

After completing this section, you should be able to

■ Determine the Boolean expression for a combination of gates   ■ Evaluate the logic operation of a circuit from the Boolean expression   ■ Construct a truth table

The Boolean Expression for a Logic Circuit

To derive the Boolean expression of a given logic circuit, begin at the left-most inputs and work toward the final output, writing the expression for each gate. For the example circuit in Figure 4–16, the Boolean expression is determined as follows:

1. The expression for the left-most AND gate with inputs C and D is CD.
2. The output of the left-most AND gate is one of the inputs to the OR gate and B is the other input. Therefore, the expression for the OR gate is $B + CD$.
3. The output of the OR gate is one of the inputs to the right-most AND gate and A is the other input. Therefore, the expression for this AND gate is $A(B + CD)$, which is the final output expression for the entire circuit.

FIGURE 4–16
A logic circuit showing the development of the Boolean expression for the output.

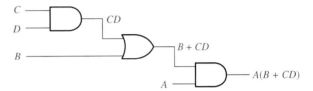

Constructing the Truth Table for a Logic Circuit

Once the Boolean expression for a given logic circuit has been determined, a truth table that shows the output for all possible values of the input variables can be developed. The procedure requires that you evaluate the Boolean expression for all possible combinations of values for the input variables. In the case of the circuit in Figure 4–16, there are four input variables (A, B, C, and D) and therefore sixteen ($2^4 = 16$) combinations of values are possible.

Evaluating the Expression To evaluate the expression $A(B + CD)$, first find the values of the variables that make the expression equal to 1, using the rules for Boolean addition and multiplication. In this case, the expression equals 1 only if $A = 1$ and $B + CD = 1$ because

$$A(B + CD) = 1 \cdot 1 = 1$$

Now determine when the $B + CD$ term equals 1. The term $B + CD = 1$ if either $B = 1$ or $CD = 1$ or if both B and CD equal 1 because

$$B + CD = 1 + 0 = 1$$
$$B + CD = 0 + 1 = 1$$
$$B + CD = 1 + 1 = 1$$

The term $CD = 1$ only if $C = 1$ and $D = 1$.

To summarize, the expression $A(B + CD) = 1$ when $A = 1$ and $B = 1$ regardless of the values of C and D or when $A = 1$ and $C = 1$ and $D = 1$, regardless of the value of B. The expression $A(B + CD) = 0$ for all other value combinations of the variables.

Putting the Results in Truth Table Format The first step is to list the sixteen input variable combinations of 1s and 0s in a binary sequence as shown in Table 4–5. Next, place a 1 in the output column for each combination of input variables that was determined in the evaluation. Finally, place a 0 in the output column for all other combinations of input variables. These results are shown in the truth table in Table 4–5.

TABLE 4–5

Truth table for the logic circuit in Figure 4–16.

Inputs				Output
A	B	C	D	A(B + CD)
0	0	0	0	0
0	0	0	1	0
0	0	1	0	0
0	0	1	1	0
0	1	0	0	0
0	1	0	1	0
0	1	1	0	0
0	1	1	1	0
1	0	0	0	0
1	0	0	1	0
1	0	1	0	0
1	0	1	1	1
1	1	0	0	1
1	1	0	1	1
1	1	1	0	1
1	1	1	1	1

SECTION 4–4 REVIEW

1. Replace the AND gates with OR gates and the OR gate with an AND gate in Figure 4–16 and determine the Boolean expression for the output.
2. Construct a truth table for the circuit in Question 1.

4–5 ■ SIMPLIFICATION USING BOOLEAN ALGEBRA

Many times in the application of Boolean algebra, you have to reduce a particular expression to its simplest form or change its form to a more convenient one to implement the expression most efficiently. The approach taken in this section is to use the basic laws, rules, and theorems of Boolean algebra to manipulate and simplify an expression. This method depends on a thorough knowledge of Boolean algebra and considerable practice in its application, not to mention a little ingenuity and cleverness.

After completing this section, you should be able to

■ Apply the laws, rules, and theorems of Boolean algebra to simplify general expressions

A simplified Boolean expression uses the fewest gates possible to implement a given expression. Four examples follow to illustrate Boolean simplification step by step.

EXAMPLE 4–8

Using Boolean algebra techniques, simplify this expression:

$$AB + A(B + C) + B(B + C)$$

Solution The following is not necessarily the only approach.

Step 1. Apply the distributive law to the second and third terms in the expression, as follows:

$$AB + AB + AC + BB + BC$$

Step 2. Apply rule 7 ($BB = B$) to the fourth term.

$$AB + AB + AC + B + BC$$

Step 3. Apply rule 5 ($AB + AB = AB$) to the first two terms.

$$AB + AC + B + BC$$

Step 4. Apply rule 10 ($B + BC = B$) to the last two terms.

$$AB + AC + B$$

Step 5. Apply rule 10 ($AB + B = B$) to the first and third terms.

$$B + AC$$

At this point the expression is simplified as much as possible. Once you gain experience in applying Boolean algebra, you can combine many individual steps.

Related Problem Simplify the Boolean expression $A\overline{B} + A(\overline{B + C}) + B(\overline{B + C})$.

Figure 4–17 shows that the simplification process in Example 4–8 has significantly reduced the number of logic gates required to implement the expression. Part (a) shows that five gates are required to implement the expression in its original form; however, only two gates are needed for the simplified expression, shown in part (b). It is important to realize that these two gate networks are equivalent. That is, for any combination of levels on the A, B, and C inputs, you get the same output from either circuit.

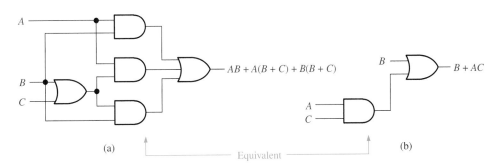

FIGURE 4–17
Gate networks for Example 4–8.

EXAMPLE 4–9

Simplify the following Boolean expression:

$$[A\overline{B}(C + BD) + \overline{A}\,\overline{B}]C$$

Note that brackets and parentheses mean the same thing: the term inside is multiplied (ANDed) with the term outside.

Solution
Step 1. Apply the distributive law to the terms within the brackets.

$$(A\overline{B}C + A\overline{B}BD + \overline{A}\,\overline{B})C$$

Step 2. Apply rule 8 ($\overline{B}B = 0$) to the second term within the parentheses.

$$(A\overline{B}C + A \cdot 0 \cdot D + \overline{A}\,\overline{B})C$$

Step 3. Apply rule 3 ($A \cdot 0 \cdot D = 0$) to the second term within the parentheses.

$$(A\overline{B}C + 0 + \overline{A}\,\overline{B})C$$

Step 4. Apply rule 1 (drop the 0) within the parentheses.

$$(A\overline{B}C + \overline{A}\,\overline{B})C$$

Step 5. Apply the distributive law.

$$A\overline{B}CC + \overline{A}\,\overline{B}C$$

Step 6. Apply rule 7 ($CC = C$) to the first term.

$$A\overline{B}C + \overline{A}\,\overline{B}C$$

Step 7. Factor out $\overline{B}C$.

$$\overline{B}C(A + \overline{A})$$

Step 8. Apply rule 6 ($A + \overline{A} = 1$).

$$\overline{B}C \cdot 1$$

Step 9. Apply rule 4 (drop the 1).

$$\overline{B}C$$

Related Problem   Simplify the Boolean expression $[AB(C + \overline{BD}) + \overline{AB}]CD$.

EXAMPLE 4–10   Simplify the following Boolean expression:

$$\overline{A}BC + A\overline{B}\,\overline{C} + \overline{A}\,\overline{B}\,\overline{C} + A\overline{B}C + ABC$$

Solution

Step 1. Factor BC out of the first and last terms.

$$BC(\overline{A} + A) + A\overline{B}\,\overline{C} + \overline{A}\,\overline{B}\,\overline{C} + A\overline{B}C$$

Step 2. Apply rule 6 ($\overline{A} + A = 1$) to the term in parentheses, and factor $A\overline{B}$ from the second and last terms.

$$BC \cdot 1 + A\overline{B}(\overline{C} + C) + \overline{A}\,\overline{B}\,\overline{C}$$

Step 3. Apply rule 4 (drop the 1) to the first term and rule 6 ($\overline{C} + C = 1$) to the term in parentheses.

$$BC + A\overline{B} \cdot 1 + \overline{A}\,\overline{B}\,\overline{C}$$

Step 4. Apply rule 4 (drop the 1) to the second term.

$$BC + A\overline{B} + \overline{A}\,\overline{B}\,\overline{C}$$

Step 5. Factor $\overline{B}$ from the second and third terms.

$$BC + \overline{B}(A + \overline{A}\,\overline{C})$$

Step 6. Apply rule 11 ($A + \overline{A}\,\overline{C} = A + \overline{C}$) to the term in parentheses.

$$BC + \overline{B}(A + \overline{C})$$

Step 7. Use the distributive and commutative laws to get the following expression:

$$BC + A\overline{B} + \overline{B}\,\overline{C}$$

Related Problem   Simplify the Boolean expression $AB\overline{C} + \overline{A}\,\overline{B}C + \overline{A}BC + \overline{A}\,\overline{B}\,\overline{C}$.

EXAMPLE 4–11   Simplify the following Boolean expression:

$$\overline{AB + AC} + \overline{A}\,\overline{B}C$$

Solution
Step 1.   Apply DeMorgan's theorem to the first term.

$$(\overline{AB})(\overline{AC}) + \overline{A}\,\overline{B}C$$

Step 2.   Apply DeMorgan's theorem to each term in parentheses.

$$(\overline{A} + \overline{B})(\overline{A} + \overline{C}) + \overline{A}\,\overline{B}C$$

Step 3.   Apply the distributive law to the two terms in parentheses.

$$\overline{A}\,\overline{A} + \overline{A}\,\overline{C} + \overline{A}\,\overline{B} + \overline{B}\,\overline{C} + \overline{A}\,\overline{B}C$$

Step 4.   Apply rule 7 $(\overline{A}\,\overline{A} = \overline{A})$ to the first term, and apply rule 10 $[\overline{A}\,\overline{B} + \overline{A}\,\overline{B}C = \overline{A}\,\overline{B}(1 + C) = \overline{A}\,\overline{B}]$ to the third and last terms.

$$\overline{A} + \overline{A}\,\overline{C} + \overline{A}\,\overline{B} + \overline{B}\,\overline{C}$$

Step 5.   Apply rule 10 $[\overline{A} + \overline{A}\,\overline{C} = \overline{A}(1 + \overline{C}) = \overline{A}]$ to the first and second terms.

$$\overline{A} + \overline{A}\,\overline{B} + \overline{B}\,\overline{C}$$

Step 6.   Apply rule 10 $[\overline{A} + \overline{A}\,\overline{B} = \overline{A}(1 + \overline{B}) = \overline{A}]$ to the first and second terms.

$$\overline{A} + \overline{B}\,\overline{C}$$

Related Problem   Simplify the Boolean expression $\overline{AB} + \overline{AC} + \overline{A}\,\overline{B}\,\overline{C}$.

SECTION 4–5
REVIEW

1. Simplify the following Boolean expressions if possible:
   (a) $A + AB + A\overline{B}C$ **(b)** $(\overline{A} + B)C + ABC$ **(c)** $A\overline{B}C(BD + CDE) + A\overline{C}$
2. Implement each expression in Question 1 as originally stated with the appropriate logic gates. Then implement the simplified expression, and compare the number of gates.

4–6 ■ STANDARD FORMS OF BOOLEAN EXPRESSIONS

All Boolean expressions, regardless of their form, can be converted into either of two standard forms: the sum-of-products form or the product-of-sums form. Standardization makes the evaluation, simplification, and implementation of Boolean expressions much more systematic and easier.

After completing this section, you should be able to

■ Identify a sum-of-products expression   ■ Determine the domain of a Boolean expression   ■ Convert any sum-of-products expression to a standard form   ■ Evaluate a standard sum-of-products expression in terms of binary values   ■ Identify a product-of-sums expression   ■ Convert any product-of-sums expression to a standard form   ■ Evaluate a standard product-of-sums expression in terms of binary values   ■ Convert from one standard form to the other

The Sum-of-Products (SOP) Form

A product term was defined in Section 4–1 as a term consisting of the product (Boolean multiplication) of literals (variables or their complements). When two or more product terms are summed by Boolean addition, the resulting expression is a **sum-of-products** (SOP). Some examples are

$$AB + ABC$$

$$ABC + CDE + \overline{B}C\overline{D}$$

$$\overline{A}B + \overline{A}B\overline{C} + AC$$

Also, an SOP expression can contain a single-variable term, as in $A + \overline{A}\,\overline{B}C + BC\overline{D}$. Refer to the simplification examples in the last section, and you will see that each of the final expressions was either a single product term or in SOP form. In an SOP expression, a single overbar cannot extend over more than one variable, although more than one variable in a term can have an overbar. For example, an SOP expression can have the term $\overline{A}\,\overline{B}\,\overline{C}$ but not $\overline{ABC}$.

Domain of a Boolean Expression   The **domain** of a general Boolean expression is the set of variables contained in the expression in either complemented or uncomplemented form. For example, the domain of the expression $\overline{A}B + A\overline{B}C$ is the set of variables A, B, C and the domain of the expression $AB\overline{C} + C\overline{D}E + \overline{B}C\overline{D}$ is the set of variables A, B, C, D, E.

Implementation of an SOP Expression   Implementing an SOP expression simply requires ORing the outputs of two or more AND gates. A product term is produced by an AND operation, and the sum (addition) of two or more product terms is produced by an OR operation. Therefore, an SOP expression can be implemented by AND-OR logic in which the outputs of a number (equal to the number of product terms in the expression) of AND gates connect to the inputs of an OR gate, as shown in Figure 4–18 for the expression $AB + BCD + AC$. The output X of the OR gate equals the SOP expression.

FIGURE 4–18
Implementation of the SOP expression AB + BCD + AC.

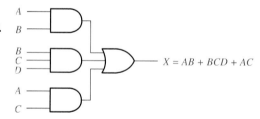

$$X = AB + BCD + AC$$

Conversion of a General Expression to SOP Form

Any logic expression can be changed into SOP form by applying Boolean algebra techniques. For example, the expression $A(B + CD)$ can be converted to SOP form by applying the distributive law:

$$A(B + CD) = AB + ACD$$

EXAMPLE 4–12   Convert each of the following Boolean expressions to SOP form:

(a) $AB + B(CD + EF)$ **(b)** $(A + B)(B + C + D)$ **(c)** $\overline{(A + B) + C}$

Solution
(a) $AB + B(CD + EF) = AB + BCD + BEF$
(b) $(A + B)(B + C + D) = AB + AC + AD + BB + BC + BD$
(c) $\overline{(A + B) + C} = \overline{(A + B)}\overline{C} = (A + B)\overline{C} = A\overline{C} + B\overline{C}$

Related Problem   Convert $\overline{A}B\overline{C} + (A + \overline{B})(B + \overline{C} + A\overline{B})$ to SOP form.

The Standard SOP Form

So far, you have seen SOP expressions in which some of the product terms do not contain all of the variables in the domain of the expression. For example, the expression $\overline{A}B\overline{C} + A\overline{B}D + \overline{A}B\overline{C}D$ has a domain made up of the variables A, B, C, and D. Notice that the complete set of variables in the domain is not represented in the first two terms of the expression; that is, D or $\overline{D}$ is missing from the first term and C or $\overline{C}$ is missing from the second term.

A standard SOP expression is one in which *all* the variables in the domain appear in each product term in the expression. For example, $A\overline{B}CD + \overline{A}\,\overline{B}C\overline{D} + AB\overline{C}D$ is a standard SOP expression. Standard SOP expressions are important in constructing truth tables, covered in Section 4–7, and in the Karnaugh map simplification method, which is covered in Section 4–8. Any nonstandard SOP expression (referred to simply as SOP) can be converted to the standard form using Boolean algebra.

Converting Product Terms to Standard SOP Each product term in an SOP expression that does not contain all the variables in the domain can be expanded to standard form to include all variables in the domain and their complements. As stated in the following steps, a nonstandard SOP expression is converted into standard form using Boolean algebra rule 6 ($A + \overline{A} = 1$) from Table 4–1: A variable added to its complement equals 1.

Step 1. Multiply each nonstandard product term by a term made up of the sum of a missing variable and its complement. This results in two product terms. As you know, you can multiply anything by 1 without changing its value.

Step 2. Repeat Step 1 until all resulting product terms contain all variables in the domain in either complemented or uncomplemented form. In converting a product term to standard form, the number of product terms is doubled for each missing variable, as Example 4–13 shows.

EXAMPLE 4–13

Convert the following Boolean expression into standard SOP form:

$$A\overline{B}C + \overline{A}\,\overline{B} + AB\overline{C}D$$

Solution The domain of this SOP expression is A, B, C, D. Take one term at a time. The first term, $A\overline{B}C$, is missing variable D or $\overline{D}$, so multiply the first term by $D + \overline{D}$ as follows:

$$A\overline{B}C = A\overline{B}C(D + \overline{D}) = A\overline{B}CD + A\overline{B}C\overline{D}$$

In this case, two standard product terms are the result.

The second term, $\overline{A}\,\overline{B}$, is missing variables C or $\overline{C}$ and D or $\overline{D}$, so first multiply the second term by $C + \overline{C}$ as follows:

$$\overline{A}\,\overline{B} = \overline{A}\,\overline{B}(C + \overline{C}) = \overline{A}\,\overline{B}C + \overline{A}\,\overline{B}\,\overline{C}$$

The two resulting terms are missing variable D or $\overline{D}$, so multiply both terms by $D + \overline{D}$ as follows:

$$\overline{A}\,\overline{B} = \overline{A}\,\overline{B}C + \overline{A}\,\overline{B}\,\overline{C} = \overline{A}\,\overline{B}C(D + \overline{D}) + \overline{A}\,\overline{B}\,\overline{C}(D + \overline{D})$$
$$= \overline{A}\,\overline{B}CD + \overline{A}\,\overline{B}C\overline{D} + \overline{A}\,\overline{B}\,\overline{C}D + \overline{A}\,\overline{B}\,\overline{C}\,\overline{D}$$

In this case, four standard product terms are the result.

The third term, $AB\overline{C}D$, is already in standard form. The standard SOP form of the original expression is as follows:

$$A\overline{B}C + \overline{A}\,\overline{B} + AB\overline{C}D = A\overline{B}CD + A\overline{B}C\overline{D} + \overline{A}\,\overline{B}CD + \overline{A}\,\overline{B}C\overline{D} + \overline{A}\,\overline{B}\,\overline{C}D + \overline{A}\,\overline{B}\,\overline{C}\,\overline{D} + AB\overline{C}D$$

Related Problem Convert the expression $W\overline{X}Y + \overline{X}Y\overline{Z} + WX\overline{Y}$ to standard SOP form.

Binary Representation of a Standard Product Term A standard product term is equal to 1 for only one combination of variable values. For example, the product term $A\overline{B}C\overline{D}$ is equal to 1 when $A = 1$, $B = 0$, $C = 1$, and $D = 0$, as shown below, and is 0 for all other combinations of values for the variables.

$$A\overline{B}C\overline{D} = 1 \cdot \overline{0} \cdot 1 \cdot \overline{0} = 1 \cdot 1 \cdot 1 \cdot 1 = 1$$

In this case, the product term has a binary value of 1010 (decimal ten).

Remember, a product term is implemented with an AND gate whose output is 1 only if each of its inputs is 1. Inverters are used to produce the complements of the variables as required.

An SOP expression is equal to 1 only if one or more of the product terms in the expression is equal to 1.

EXAMPLE 4–14

Determine the binary values for which the following standard SOP expression is equal to 1:

$$ABCD + A\overline{B}\,\overline{C}D + \overline{A}\,\overline{B}\,\overline{C}\,\overline{D}$$

Solution The term $ABCD$ is equal to 1 when $A = 1$, $B = 1$, $C = 1$, and $D = 1$.

$$ABCD = 1 \cdot 1 \cdot 1 \cdot 1 = 1$$

The term $A\overline{B}\,\overline{C}D$ is equal to 1 when $A = 1$, $B = 0$, $C = 0$, and $D = 1$.

$$A\overline{B}\,\overline{C}D = 1 \cdot \overline{0} \cdot \overline{0} \cdot 1 = 1 \cdot 1 \cdot 1 \cdot 1 = 1$$

The term $\overline{A}\,\overline{B}\,\overline{C}\,\overline{D}$ is equal to 1 when $A = 0$, $B = 0$, $C = 0$, and $D = 0$.

$$\overline{A}\,\overline{B}\,\overline{C}\,\overline{D} = \overline{0} \cdot \overline{0} \cdot \overline{0} \cdot \overline{0} = 1 \cdot 1 \cdot 1 \cdot 1 = 1$$

The SOP expression equals 1 when any or all of the three product terms is 1.

Related Problem Determine the binary values for which the following SOP expression is equal to 1:

$$\overline{X}YZ + X\overline{Y}Z + XY\overline{Z} + \overline{X}Y\overline{Z} + XYZ$$

Is this a standard SOP expression?

The Product-of-Sums (POS) Form

A sum term was defined in Section 4–1 as a term consisting of the sum (Boolean addition) of literals (variables or their complements). When two or more sum terms are multiplied, the resulting expression is a **product-of-sums** (POS). Some examples are

$$(\overline{A} + B)(A + \overline{B} + C)$$
$$(\overline{A} + \overline{B} + \overline{C})(C + \overline{D} + E)(\overline{B} + C + D)$$
$$(A + B)(A + \overline{B} + C)(\overline{A} + C)$$

A POS expression can contain a single-variable term, as in $\overline{A}(A + \overline{B} + C)(\overline{B} + \overline{C} + D)$. In a POS expression, a single overbar cannot extend over more than one variable, although more than one variable in a term can have an overbar. For example, a POS expression can have the term $\overline{A} + \overline{B} + \overline{C}$ but not $\overline{A + B + C}$.

Implementation of a POS Expression Implementing a POS expression simply requires ANDing the outputs of two or more OR gates. A sum term is produced by an OR operation, and the product of two or more sum terms is produced by an AND operation.

Therefore, a POS expression can be implemented by logic in which the outputs of a number (equal to the number of sum terms in the expression) of OR gates connect to the inputs of an AND gate, as Figure 4–19 shows for the expression $(A + B)(B + C + D)(A + C)$. The output X of the AND gate equals the POS expression.

FIGURE 4–19

Implementation of the POS expression $(A + B)(B + C + D)(A + C)$.

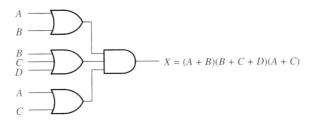

$$X = (A + B)(B + C + D)(A + C)$$

The Standard POS Form

So far, you have seen POS expressions in which some of the sum terms do not contain all of the variables in the domain of the expression. For example, the expression

$$(A + \overline{B} + C)(A + B + \overline{D})(A + \overline{B} + \overline{C} + D)$$

has a domain made up of the variables A, B, C, and D. Notice that the complete set of variables in the domain is not represented in the first two terms of the expression; that is, D or $\overline{D}$ is missing from the first term and C or $\overline{C}$ is missing from the second term.

A standard POS expression is one in which *all* the variables in the domain appear in each sum term in the expression. For example,

$$(\overline{A} + \overline{B} + \overline{C} + \overline{D})(A + \overline{B} + C + D)(A + B + \overline{C} + D)$$

is a standard POS expression. Any nonstandard POS expression (referred to simply as POS) can be converted to the standard form using Boolean algebra.

Converting a Sum Term to Standard POS   Each sum term in a POS expression that does not contain all the variables in the domain can be expanded to standard form to include all variables in the domain and their complements. As stated in the following steps, a nonstandard POS expression is converted into standard form using Boolean algebra rule 8 ($A \cdot \overline{A} = 0$) from Table 4–1: A variable multiplied by its complement equals 0.

Step 1.   Add to each nonstandard product term a term made up of the product of a missing variable and its complement. This results in two sum terms. As you know, you can add 0 to anything without changing its value.

Step 2.   Apply rule 12 from Table 4–1: $A + BC = (A + B)(A + C)$

Step 3.   Repeat Step 1 until all resulting sum terms contain all variables in the domain in either complemented or uncomplemented form.

EXAMPLE 4–15   Convert the following Boolean expression into standard POS form:

$$(A + \overline{B} + C)(\overline{B} + C + \overline{D})(A + \overline{B} + \overline{C} + D)$$

Solution   The domain of this POS expression is A, B, C, D. Take one term at a time. The first term, $A + \overline{B} + C$, is missing variable D or $\overline{D}$, so add $D\overline{D}$ and apply rule 12 as follows:

$$A + \overline{B} + C = A + \overline{B} + C + D\overline{D} = (A + \overline{B} + C + D)(A + \overline{B} + C + \overline{D})$$

The second term, $\overline{B} + C + \overline{D}$, is missing variable A or $\overline{A}$, so add $A\overline{A}$ and apply rule 12 as follows:

$$\overline{B} + C + \overline{D} = \overline{B} + C + \overline{D} + A\overline{A} = (A + \overline{B} + C + \overline{D})(\overline{A} + \overline{B} + C + \overline{D})$$

The third term, $A + \overline{B} + \overline{C} + D$, is already in standard form. The standard POS form of the original expression is as follows:

$$(A + \overline{B} + C)(\overline{B} + C + \overline{D})(A + \overline{B} + \overline{C} + D) =$$

$$(A + \overline{B} + C + D)(A + \overline{B} + C + \overline{D})(A + \overline{B} + C + \overline{D})(\overline{A} + \overline{B} + C + \overline{D})(A + \overline{B} + \overline{C} + D)$$

Related Problem Convert the expression $(A + \overline{B})(B + C)$ to standard POS form.

Binary Representation of a Standard Sum Term A standard sum term is equal to 0 for only one combination of variable values. For example, the sum term $A + \overline{B} + C + \overline{D}$ is 0 when $A = 0, B = 1, C = 0$, and $D = 1$, as shown below, and is 1 for all other combinations of values for the variables.

$$A + \overline{B} + C + \overline{D} = 0 + \overline{1} + 0 + \overline{1} = 0 + 0 + 0 + 0 = 0$$

In this case, the sum term has a binary value of 0101 (decimal 5). Remember, a sum term is implemented with an OR gate whose output is 0 only if each of its inputs is 0. Inverters are used to produce the complements of the variables as required.

A POS expression is equal to 0 only if one or more of the sum terms in the expression is equal to 0.

EXAMPLE 4–16

Determine the binary values of the variables for which the following standard POS expression is equal to 0:

$$(A + B + C + D)(A + \overline{B} + \overline{C} + D)(\overline{A} + \overline{B} + \overline{C} + \overline{D})$$

Solution The term $A + B + C + D$ is equal to 0 when $A = 0, B = 0, C = 0$, and $D = 0$.

$$A + B + C + D = 0 + 0 + 0 + 0 = 0$$

The term $A + \overline{B} + \overline{C} + D$ is equal to 0 when $A = 0, B = 1, C = 1$, and $D = 0$.

$$A + \overline{B} + \overline{C} + D = 0 + \overline{1} + \overline{1} + 0 = 0 + 0 + 0 + 0 = 0$$

The term $\overline{A} + \overline{B} + \overline{C} + \overline{D}$ is equal to 0 when $A = 1, B = 1, C = 1$, and $D = 1$.

$$\overline{A} + \overline{B} + \overline{C} + \overline{D} = \overline{1} + \overline{1} + \overline{1} + \overline{1} = 0 + 0 + 0 + 0 = 0$$

The POS expression equals 0 when any of the three sum terms equals 0.

Related Problem Determine the binary values for which the following POS expression is equal to 0:

$$(X + \overline{Y} + Z)(\overline{X} + Y + Z)(X + Y + \overline{Z})(\overline{X} + \overline{Y} + \overline{Z})(X + \overline{Y} + \overline{Z})$$

Is this a standard POS expression?

Converting Standard SOP to Standard POS

The binary values of the product terms in a given standard SOP expression are not present in the equivalent standard POS expression. Also, the binary values that are not represented in the SOP expression are present in the equivalent POS expression. Therefore, to convert from standard SOP to standard POS, the following steps are taken:

Step 1. Evaluate each product term in the SOP expression. That is, determine the binary numbers which represent the product terms.

Step 2. Determine all of the binary numbers not included in the evaluation in Step 1.

Step 3. Write the equivalent sum term for each binary number from Step 2 and express in POS form.

Using a similar procedure, you can go from POS to SOP.

EXAMPLE 4–17

Convert the following SOP expression to an equivalent POS expression:

$$\overline{A}\,\overline{B}\,\overline{C} + \overline{A}B\overline{C} + \overline{A}BC + A\overline{B}C + ABC$$

Solution   The evaluation is as follows:

$$000 + 010 + 011 + 101 + 111$$

Since there are three variables in the domain of this expression, there are a total of eight (2^3) possible combinations. The SOP expression contains five of these combinations, so the POS must contain the other three which are 001, 100, and 110. Remember, these are the binary values that make the sum term 0. The equivalent POS expression is

$$(A + B + \overline{C})(\overline{A} + B + C)(\overline{A} + \overline{B} + C)$$

Related Problem   Verify that the SOP and POS expressions in this example are equivalent by substituting binary values into each.

**SECTION 4–6
REVIEW**

1. Identify each of the following expressions as SOP, standard SOP, POS, or standard POS:
   (a) $AB + \overline{A}BD + \overline{A}C\overline{D}$ **(b)** $(A + \overline{B} + C)(A + B + \overline{C})$
   (c) $\overline{A}BC + AB\overline{C}$ **(d)** $A(A + \overline{C})(A + B)$
2. Convert each SOP expression in Question 1 to standard form.
3. Convert each POS expression in Question 1 to standard form.

4–7 ■ BOOLEAN EXPRESSIONS AND TRUTH TABLES

All standard Boolean expressions can be easily converted into truth table format using binary values for each term in the expression. The truth table is a common way of presenting, in a concise format, the logical operation of a circuit. Also, standard SOP or POS expressions can be determined from a truth table. You will find truth tables in data sheets and other literature related to the operation of digital circuits.

After completing this section, you should be able to

■ Convert a standard SOP expression into truth table format   ■ Convert a standard POS expression into truth table format   ■ Derive a standard expression from a truth table   ■ Properly interpret truth table data

Converting SOP Expressions to Truth Table Format

Recall from Section 4–6 that an SOP expression is equal to 1 only if at least one of the product terms is equal to 1. A truth table is simply a list of the possible combinations of input variable values and the corresponding output values (1 or 0). For an expression with a domain of two variables, there are four different combinations of those variables $(2^2 = 4)$. For an expression with a domain of three variables, there are eight different combinations of those variables ($2^3 = 8$). For an expression with a domain of four variables, there are sixteen different combinations of those variables ($2^4 = 16$), and so on.

The first step in constructing a truth table is to list all possible combinations of binary values of the variables in the expression. Next, convert the SOP expression to standard form if it is not already. Finally, place a 1 in the output column (X) for each binary value that makes the standard SOP expression a 1 and place a 0 for all the remaining binary values. This procedure is illustrated in Example 4–18.

EXAMPLE 4–18

Develop a truth table for the standard SOP expression $\overline{A}\,\overline{B}C + A\overline{B}\,\overline{C} + ABC$.

Solution There are three variables in the domain, so there are eight possible combinations of binary values of the variables as listed in the left three columns of Table 4–6. The binary values that make the product terms in the expression equal to 1 are $\overline{A}\,\overline{B}C$: 001; $A\overline{B}\,\overline{C}$: 100; and ABC: 111. For each of these binary values, a 1 is placed in the output column as shown in the table. For each of the remaining binary combinations, a 0 is placed in the output column.

TABLE 4–6

Inputs			Output
A	**B**	**C**	**X**
0	0	0	0
0	0	1	1
0	1	0	0
0	1	1	0
1	0	0	1
1	0	1	0
1	1	0	0
1	1	1	1

Related Problem Create a truth table for the standard SOP expression $\overline{A}B\overline{C} + A\overline{B}C$.

Converting POS Expressions to Truth Table Format

Recall that a POS expression is equal to 0 only if at least one of the sum terms is equal to 0. To construct a truth table from a POS expression, list all the possible combinations of binary values of the variables just as was done for the SOP expression. Next, convert the POS expression to standard form if it is not already. Finally, place a 0 in the output column (X) for each binary value that makes the expression a 0 and place a 1 for all the remaining binary values. This procedure is illustrated in Example 4–19.

EXAMPLE 4–19

Determine the truth table for the following standard POS expression:

$$(A + B + C)(A + \overline{B} + C)(A + \overline{B} + \overline{C})(\overline{A} + B + \overline{C})(\overline{A} + \overline{B} + C)$$

Solution There are three variables in the domain and the eight possible binary values are listed in the left three columns of Table 4–7. The binary values that make the sum terms in the expression equal to 0 are $A + B + C$: 000; $A + \overline{B} + C$: 010; $A + \overline{B} + \overline{C}$: 011; $\overline{A} + B + \overline{C}$: 101; and $\overline{A} + \overline{B} + C$: 110. For each of these binary values, a 0 is placed in the output column as shown in the table. For each of the remaining binary combinations, a 1 is placed in the output column.

TABLE 4–7

	Inputs		Output
A	B	C	X
0	0	0	0
0	0	1	1
0	1	0	0
0	1	1	0
1	0	0	1
1	0	1	0
1	1	0	0
1	1	1	1

Notice that the truth table in this example is the same as the one in Example 4–18. This means that the SOP expression in the previous example and the POS expression in this example are equivalent.

Related Problem Develop a truth table for the following standard POS expression:

$$(A + \overline{B} + C)(A + B + \overline{C})(\overline{A} + \overline{B} + \overline{C})$$

Determining Standard Expressions from a Truth Table

To determine the standard SOP expression represented by a truth table, list the binary values of the input variables for which the output is 1. Convert each binary value to the corresponding product term by replacing each 1 with the corresponding variable and each 0 with the corresponding variable complement. For example, the binary value 1010 is converted to a product term as follows:

$$1010 \longrightarrow A\overline{B}C\overline{D}$$

To determine the standard POS expression represented by a truth table, list the binary values for which the output is 0. Convert each binary value to the corresponding sum term by replacing each 1 with the corresponding variable complement and each 0 with the corresponding variable. For example, the binary value 1001 is converted to a sum term as follows:

$$1001 \longrightarrow \overline{A} + B + C + \overline{D}$$

EXAMPLE 4–20

From the truth table in Table 4–8, determine the standard SOP expression and the equivalent standard POS expression.

TABLE 4–8

	Inputs		Output
A	B	C	X
0	0	0	0
0	0	1	0
0	1	0	0
0	1	1	1
1	0	0	1
1	0	1	0
1	1	0	1
1	1	1	1

Solution There are four 1s in the output column and the corresponding binary values are 011, 100, 110, and 111. These binary values are converted to product terms as follows:

$$011 \longrightarrow \overline{A}BC$$

$$100 \longrightarrow A\overline{B}\,\overline{C}$$

$$110 \longrightarrow AB\overline{C}$$

$$111 \longrightarrow ABC$$

The resulting standard SOP expression for the output X is

$$X = \overline{A}BC + A\overline{B}\,\overline{C} + AB\overline{C} + ABC$$

For the POS expression, the output is 0 for binary values 000, 001, 010, and 101. These binary values are converted to sum terms as follows:

$$000 \longrightarrow A + B + C$$

$$001 \longrightarrow A + B + \overline{C}$$

$$010 \longrightarrow A + \overline{B} + C$$

$$101 \longrightarrow \overline{A} + B + \overline{C}$$

The resulting standard POS expression for the output X is

$$X = (A + B + C)(A + B + \overline{C})(A + \overline{B} + C)(\overline{A} + B + \overline{C})$$

Related Problem By substitution of binary values, show that the SOP and the POS expressions derived in this example are equivalent; that is, for any binary value they should either both be 1 or both be 0, depending on the binary value.

SECTION 4–7 REVIEW

1. If a certain Boolean expression has a domain of five variables, how many binary values will be in its truth table?
2. In a certain truth table, the output is a 1 for the binary value 0110. Convert this binary value to the corresponding product term using variables W, X, Y, and Z.
3. In a certain truth table, the output is a 0 for the binary value 1100. Convert this binary value to the corresponding sum term using variables W, X, Y, and Z.

4–8 ■ THE KARNAUGH MAP

The Karnaugh map provides a systematic method for simplifying Boolean expressions and, if properly used, will produce the simplest SOP or POS expression possible. As you have seen, the effectiveness of algebraic simplification depends on your familiarity with all the laws, rules, and theorems of Boolean algebra and on your ability to apply them. The Karnaugh map, on the other hand, basically provides a "cookbook" method for simplification.

After completing this section, you should be able to

■ Construct a Karnaugh map for three or four variables   ■ Determine the binary value of each cell in a Karnaugh map   ■ Determine the standard product term represented by each cell in a Karnaugh map   ■ Explain cell adjacency and identify adjacent cells

A **Karnaugh map** is similar to a truth table because it presents all of the possible values of input variables and the resulting output for each value. Instead of being organized into columns and rows like a truth table, the Karnaugh map is an array of **cells** in which each cell represents a binary value of the input variables. The cells are arranged in a way so that simplification of a given expression is simply a matter of properly grouping the cells. Karnaugh maps can be used for expressions with two, three, four, and five variables, but we will discuss only 3-variable and 4-variable situations to illustrate the principles. Section 4–11 deals with five variables using a 32-cell Karnaugh map. Another method, which is beyond the scope of this book, called the Quine-McClusky method can be used for higher numbers of variables.

The number of cells in a Karnaugh map is equal to the total number of possible input variable combinations as is the number of rows in a truth table. For three variables, the number of cells is $2^3 = 8$. For four variables, the number of cells is $2^4 = 16$.

The 3-Variable Karnaugh Map

The 3-variable Karnaugh map is an array of eight cells, as shown in Figure 4–20(a). In this case, A, B, and C are used for the variables although other letters could be used. Binary values of A and B are along the left side (notice the sequence) and the values of C are across the top.

FIGURE 4–20
A 3-variable Karnaugh map.

(a)   (b)

The value of a given cell is the binary values of A and B at the left in the same row combined with the value of C at the top in the same column. For example, the cell in the upper left corner has a binary value of 000 and the cell in the lower right corner has a binary value of 101. Figure 4–20(b) shows the standard product terms that are represented by each cell in the Karnaugh map.

The 4-Variable Karnaugh Map

The 4-variable Karnaugh map is an array of sixteen cells, as shown in Figure 4–21(a). Binary values of A and B are along the left side and the values of C and D are across the top.

The value of a given cell is the binary values of A and B at the left in the same row combined with the binary values of C and D at the top in the same column. For example, the cell in the upper right corner has a binary value of 0010 and the cell in the lower right corner has a binary value of 1010. Figure 4–21(b) shows the standard product terms that are represented by each cell in the 4-variable Karnaugh map.

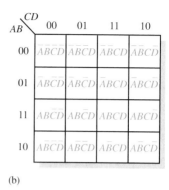

FIGURE 4–21
A 4-variable Karnaugh map.

Cell Adjacency

The cells in a Karnaugh map are arranged so that there is only a single-variable change between adjacent cells. **Adjacency** is defined by a single-variable change. Cells that differ by only one variable are adjacent. For example, in the 3-variable map the 010 cell is adjacent to the 000 cell, the 011 cell, and the 110 cell. Cells with values that differ by more than one variable are not adjacent. For example, the 010 cell is not adjacent to the 001 cell, the 111 cell, the 100 cell, or the 101 cell.

Physically, each cell is adjacent to the cells that are immediately next to it on any of its four sides. A cell is not adjacent to the cells that diagonally touch any of its corners. Also, the cells in the top row are adjacent to the corresponding cells in the bottom row and the cells in the outer left column are adjacent to the corresponding cells in the outer right column. This is called "wrap-around" adjacency because you can think of the map as wrapping around from top to bottom to form a cylinder or from left to right to form a cylinder. Figure 4–22 illustrates the cell adjacencies with a 4-variable map, although the same rules for adjacency apply to Karnaugh maps with any number of cells.

FIGURE 4–22
Cell adjacencies on a Karnaugh map. Arrows point between adjacent cells.

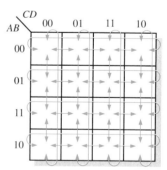

**SECTION 4–8
REVIEW**

1. In a 3-variable Karnaugh map, what is the binary value for the cell in each of the following locations:
   (a) upper left corner   **(b)** lower right corner
   (c) lower left corner   **(d)** upper right corner
2. What is the standard product term for each cell in Question 1 for variables X, Y, and Z?
3. Repeat Question 1 for a 4-variable map.
4. Repeat Question 2 for a 4-variable map using variables W, X, Y, and Z.

4–9 ■ KARNAUGH MAP SOP MINIMIZATION

As stated in the last section, the Karnaugh map is used for simplifying Boolean expressions to their minimum form. A minimized SOP expression contains the fewest possible terms with the fewest possible variables per term. Generally, a minimum SOP expression can be implemented with fewer logic gates than a standard expression and this is the basic purpose in the simplification process.

After completing this section, you should be able to

■ Map a standard SOP expression on a Karnaugh map ■ Combine the 1s on the map into maximum groups ■ Determine the minimum product term for each group on the map ■ Combine the minimum product terms to form a minimum SOP expression ■ Convert a truth table into a Karnaugh map for simplification of the represented expression ■ Use "don't care" conditions on a Karnaugh map

Mapping a Standard SOP Expression

For an SOP expression in standard form, a 1 is placed on the Karnaugh map for each product term in the expression. Each 1 is placed in a cell corresponding to the value of a product term. For example, for the product term $A\overline{B}C$, a 1 goes in the 101 cell on a 3-variable map.

When an SOP expression is completely mapped, there will be a number of 1s on the Karnaugh map equal to the number of product terms in the standard SOP expression. The cells that do not have a 1 are the cells for which the expression is 0. Usually, when working with SOP expressions, the 0s are left off the map. The following steps and the illustration in Figure 4–23 show the mapping process.

Step 1. Determine the binary value of each product term in the standard SOP expression. After some practice, you can usually do the evaluation of terms mentally.

Step 2. As each product term is evaluated, place a 1 on the Karnaugh map in the cell having the same value as the product term.

FIGURE 4–23
Example of mapping a standard SOP expression.

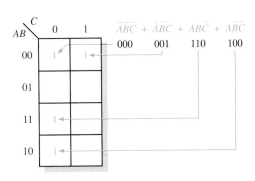

The following two examples will further illustrate the mapping process.

EXAMPLE 4–21   Map the following standard SOP expression on a Karnaugh map:

$$\overline{A}\,\overline{B}C + \overline{A}B\overline{C} + AB\overline{C} + ABC$$

Solution   The expression is evaluated as shown below. A 1 is placed on the 3-variable Karnaugh map in Figure 4–24 for each standard product term in the expression.

$$\overline{A}\,\overline{B}C + \overline{A}B\overline{C} + AB\overline{C} + ABC$$
$$001 \quad 010 \quad 110 \quad 111$$

FIGURE 4–24

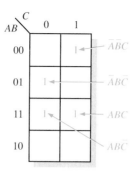

Related Problem Map the standard SOP expression $ABC + A\overline{B}C + ABC$ on a Karnaugh map.

EXAMPLE 4–22 Map the following standard SOP expression on a Karnaugh map:

$$\overline{A}\,\overline{B}CD + \overline{A}B\overline{C}\,\overline{D} + AB\overline{C}D + ABCD + AB\overline{C}\,\overline{D} + \overline{A}\,\overline{B}\,\overline{C}D + A\overline{B}C\overline{D}$$

Solution The expression is evaluated as shown below. A 1 is placed on the 4-variable Karnaugh map in Figure 4–25 for each standard product term in the expression.

$$\overline{A}\,\overline{B}CD + \overline{A}B\overline{C}\,\overline{D} + AB\overline{C}D + ABCD + AB\overline{C}\,\overline{D} + \overline{A}\,\overline{B}\,\overline{C}D + A\overline{B}C\overline{D}$$
$$0011 \quad 0100 \quad 1101 \quad 1111 \quad 1100 \quad 0001 \quad 1010$$

FIGURE 4–25

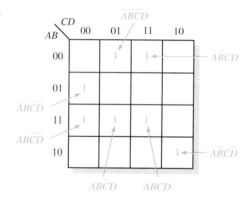

Related Problem Map the following standard SOP expression on a Karnaugh map:

$$\overline{A}BC\overline{D} + AB\overline{C}\,\overline{D} + AB\overline{C}\,\overline{D} + ABCD$$

Mapping a Nonstandard SOP Expression

A Boolean expression must first be in standard form before using the Karnaugh map. If an expression is not in standard form, then it must be converted to standard form by the procedure covered in Section 4–6 or by numerical expansion. Since an expression should be evaluated before mapping anyway, numerical expansion is probably the most efficient approach.

Numerical Expansion of a Nonstandard Product Term   Recall that a nonstandard product term has one or more missing variables. For example, assume that one of the product terms in a certain 3-variable SOP expression is $A\overline{B}$. This term can be expanded numerically to standard form as follows. First, write the binary value of the two variables and attach a 0 for the missing variable $\overline{C}$: 100. Next, write the binary value of the two variables and attach a 1 for the missing variable C: 101. The two resulting binary numbers are the values of the standard SOP terms $A\overline{B}\,\overline{C}$ and $A\overline{B}C$.

As another example, assume that one of the product terms in a 3-variable expression is B (remember that a single variable counts as a product term in an SOP expression). This term can be expanded numerically to standard form as follows. Write the binary value of the variable; then attach all possible values for the missing variables A and C as follows:

$$B$$
$$010$$
$$011$$
$$110$$
$$111$$

The four resulting binary numbers are the values of the standard SOP terms $\overline{A}B\overline{C}$, $\overline{A}BC$, $AB\overline{C}$, and ABC.

EXAMPLE 4–23

Map the following SOP expression on a Karnaugh map: $\overline{A} + A\overline{B} + AB\overline{C}$.

Solution   The SOP expression is obviously not in standard form because each product term does not have three variables. The first term is missing two variables, the second term is missing one variable, and the third term is standard. First expand the terms numerically as follows:

$$\overline{A} \quad + A\overline{B} + AB\overline{C}$$
$$000 \qquad 100 \quad 110$$
$$001 \qquad 101$$
$$010$$
$$011$$

Each of the resulting binary values is mapped by placing a 1 in the appropriate cell of the 3-variable Karnaugh map in Figure 4–26.

FIGURE 4–26

Related Problem   Map the SOP expression $BC + \overline{A}\,\overline{C}$ on a Karnaugh map.

EXAMPLE 4–24   Map the following SOP expression on a Karnaugh map:

$$\overline{B}\,\overline{C} + A\overline{B} + AB\overline{C} + A\overline{B}C\overline{D} + \overline{A}\,\overline{B}\,\overline{C}D + A\overline{B}CD$$

Solution   The SOP expression is obviously not in standard form because each product term does not have four variables. The first and second terms are both missing two variables, the third term is missing one variable, and the rest of the terms are standard. First expand the terms numerically as follows:

$$\overline{B}\,\overline{C} + A\overline{B} \quad + AB\overline{C} \quad + A\overline{B}C\overline{D} + \overline{A}\,\overline{B}\,\overline{C}D + A\overline{B}CD$$

00 00	1 000	1 1 00	1 0 1 0	0 0 0 1	1 0 1 1
00 01	1 001	1 1 01			
1 0 00	1 010				
1 0 01	1 011				

Each of the resulting binary values is mapped by placing a 1 in the appropriate cell of the 4-variable Karnaugh map in Figure 4–27. Notice that some of the values in the expanded expression are redundant.

FIGURE 4–27

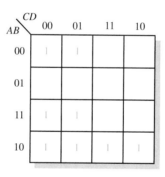

Related Problem   Map the expression $A + \overline{C}D + AC\overline{D} + \overline{A}BCD$ on a Karnaugh map.

Karnaugh Map Simplification of SOP Expressions

The process that results in an expression containing the fewest possible terms with the fewest possible variables is called **minimization.** After an SOP expression has been mapped, there are three steps in the process of obtaining a minimum SOP expression: grouping the 1s, determining the product term for each group, and summing the resulting product terms.

Grouping the 1s   You can group 1s on the Karnaugh map according to the following rules by enclosing those adjacent cells containing 1s. The goal is to maximize the size of the groups and to minimize the number of groups.

1. A group must contain either 1, 2, 4, 8, or 16 cells. In the case of a 3-variable map, eight cells is the maximum group.
2. Each cell in a group must be adjacent to one or more cells in that same group, but all cells in the group do not have to be adjacent to each other.
3. Always include the largest possible number of 1s in a group in accordance with rule 1.
4. Each 1 on the map must be included in at least one group. The 1s already in a group can be included in another group as long as the overlapping groups include noncommon 1s.

EXAMPLE 4–25 Group the 1s in each of the Karnaugh maps in Figure 4–28.

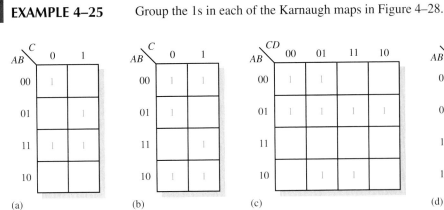

FIGURE 4–28

Solution The groupings are shown in Figure 4–29. In some cases, there may be more than one way to group the 1s to form maximum groupings.

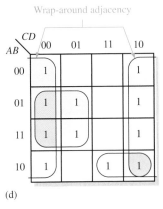

FIGURE 4–29

Related Problem Determine if there are other ways to group the 1s in Figure 4–29 to obtain a minimum number of maximum groupings.

Determining the Minimum SOP Expression from the Map When all the 1s representing the standard product terms in an expression are properly mapped and grouped, the process of determining the resulting minimum SOP expression begins. The following rules are applied to find the minimum product terms and the minimum SOP expression:

1. Group the cells that have 1s. Each group of cells containing 1s creates one product term composed of all variables that occur in only one form (either uncomplemented or complemented) within the group. Variables that occur both uncomplemented and complemented within the group are eliminated. These are called *contradictory variables*.

2. Determine the minimum product term for each group.
   (a) For a 3-variable map:
      (1) A 1-cell group yields a 3-variable product term
      (2) A 2-cell group yields a 2-variable product term
      (3) A 4-cell group yields a 1-variable term
      (4) An 8-cell group yields a value of 1 for the expression

(b) For a 4-variable map:

   (1) A 1-cell group yields a 4-variable product term

   (2) A 2-cell group yields a 3-variable product term

   (3) A 4-cell group yields a 2-variable product term

   (4) An 8-cell group yields a 1-variable term

   (5) A 16-cell group yields a value of 1 for the expression

3. When all the minimum product terms are derived from the Karnaugh map, they are summed to form the minimum SOP expression.

EXAMPLE 4–26

Determine the product terms for the Karnaugh map in Figure 4–30 and write the resulting minimum SOP expression.

FIGURE 4–30

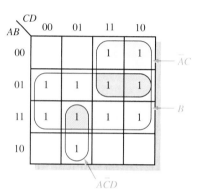

Solution   In Figure 4–30, the product term for the 8-cell group is B because the cells within that group contain both A and $\overline{A}$, C and $\overline{C}$, and D and $\overline{D}$, so these variables are eliminated. The 4-cell group contains $B, \overline{B}, D$, and $\overline{D}$, leaving the product term $\overline{A}C$. The 2-cell group contains B and $\overline{B}$, leaving $A\overline{C}D$ as the product term. Notice how overlapping is used to maximize the size of the groups. The resulting minimum SOP expression is the sum of these product terms:

$$B + \overline{A}C + A\overline{C}D$$

Related Problem   For the Karnaugh map in Figure 4–30, add a 1 in the lower right cell (1010) and determine the resulting SOP expression.

EXAMPLE 4–27

Determine the product terms for each of the Karnaugh maps in Figure 4–31 and write the resulting minimum SOP expression.

Solution   The resulting minimum product term for each group is shown in Figure 4–31. The minimum SOP expressions for each of the Karnaugh maps in the figure are

(a) $AB + BC + \overline{A}\,\overline{B}\,\overline{C}$ **(b)** $\overline{B} + \overline{A}\,\overline{C} + AC$

(c) $\overline{A}B + \overline{A}\,\overline{C} + A\overline{B}D$ **(d)** $\overline{D} + A\overline{B}C + B\overline{C}$

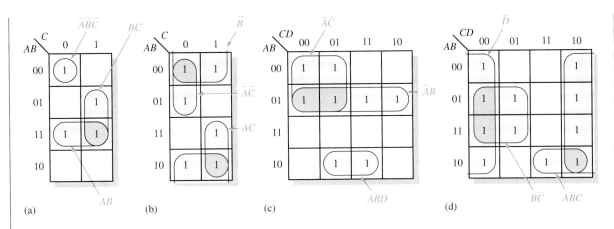

FIGURE 4–31

Related Problem For the Karnaugh map in Figure 4–31(d), add a 1 in the 0111 cell and determine the resulting SOP expression.

EXAMPLE 4–28 Use a Karnaugh map to minimize the following standard SOP expression:

$$A\overline{B}C + \overline{A}BC + \overline{A}\,\overline{B}C + \overline{A}\,\overline{B}\,\overline{C} + A\overline{B}\,\overline{C}$$

Solution The binary values of the expression are

$$101 + 011 + 011 + 000 + 100$$

The standard SOP expression is mapped and the cells are grouped as shown in Figure 4–32.

FIGURE 4–32

Notice the "wrap around" 4-cell group that includes the top row and the bottom row of 1s. The remaining 1 is absorbed in an overlapping group of two cells. The group of four 1s produces a single variable term, $\overline{B}$. This is determined by observing that within the group, $\overline{B}$ is the only variable that does not change from cell to cell. The group of two 1s produces a 2-variable term $\overline{A}C$. This is determined by observing that within the group, $\overline{A}$ and C do not change from one cell to the next. The product term for each group is shown and the resulting minimum SOP expression is

$$\overline{B} + \overline{A}C$$

Keep in mind that this minimum expression is equivalent to the original standard expression.

Related Problem Use a Karnaugh map to simplify the following standard SOP expression:

$$\overline{X}\,\overline{Y}Z + XY\overline{Z} + \overline{X}YZ + \overline{X}Y\overline{Z} + X\overline{Y}\overline{Z} + XYZ$$

EXAMPLE 4–29   Use a Karnaugh map to minimize the following SOP expression:

$$\overline{B}\,\overline{C}\,\overline{D} + \overline{A}B\overline{C}\,\overline{D} + AB\overline{C}\,\overline{D} + \overline{A}\,\overline{B}CD + A\overline{B}C\overline{D} + \overline{A}\,\overline{B}C\overline{D} + \overline{A}B\overline{C}\,\overline{D} + ABC\overline{D} + A\overline{B}CD$$

Solution The first term $\overline{B}\,\overline{C}\,\overline{D}$ must be expanded into $A\overline{B}\,\overline{C}\,\overline{D}$ and $\overline{A}\,\overline{B}\,\overline{C}\,\overline{D}$ to get a standard SOP expression, which is then mapped; and the cells are grouped as shown in Figure 4–33.

FIGURE 4–33

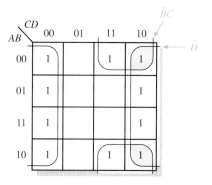

Notice that both groups exhibit "wrap around" adjacency. The group of eight is formed because the cells in the outer columns are adjacent. The group of four is formed to pick up the remaining two 1s because the top and bottom cells are adjacent. The product term for each group is shown and the resulting minimum SOP expression is

$$\overline{D} + \overline{B}C$$

Keep in mind that this minimum expression is equivalent to the original standard expression.

Related Problem Use a Karnaugh map to simplify the following SOP expression:

$$\overline{W}\,\overline{X}\,\overline{Y}\,\overline{Z} + W\overline{X}YZ + W\overline{X}\,\overline{Y}\,\overline{Z} + \overline{W}YZ + W\overline{X}\,\overline{Y}\,\overline{Z}$$

Mapping Directly from a Truth Table

You have seen how to map a Boolean expression; now you will learn how to go directly from a truth table to a Karnaugh map. Recall that a truth table gives the output of a Boolean expression for all possible input variable combinations. An example of a Boolean expression and its truth table representation is shown in Figure 4–34. Notice in the truth table that the output X is 1 for four different input variable combinations. The 1s in the output column of the truth table are mapped directly onto a Karnaugh map into the cells corresponding to the values of the associated input variable combinations, as shown in Figure 4–34. In the figure you can see that the Boolean expression, the truth table, and the Karnaugh map are simply different ways to represent a logic function.

FIGURE 4–34
Example of mapping directly from a truth table to a Karnaugh map.

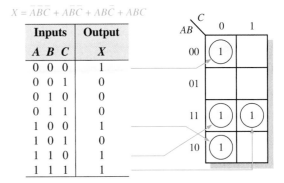

$X = \overline{A}\overline{B}\overline{C} + A\overline{B}\overline{C} + AB\overline{C} + ABC$

Inputs	Output
A B C	X
0 0 0	1
0 0 1	0
0 1 0	0
0 1 1	0
1 0 0	1
1 0 1	0
1 1 0	1
1 1 1	1

"Don't Care" Conditions

Sometimes a situation arises in which some input variable combinations are not allowed. For example, recall that in the BCD code covered in Chapter 2, there are six invalid combinations: 1010, 1011, 1100, 1101, 1110, and 1111. Since these unallowed states will never occur in an application involving the BCD code, they can be treated as **"don't care"** terms with respect to their effect on the output. That is, for these "don't care" terms either a 1 or a 0 may be assigned to the output; it really does not matter since they will never occur.

The "don't care" terms can be used to advantage on the Karnaugh map. Figure 4–35 shows that for each "don't care" term, an X is placed in the cell. When grouping the 1s, the Xs can be treated as 1s to make a larger grouping or as 0s if they cannot be used to advantage. The larger a group, the simpler the resulting term will be.

The truth table in Figure 4–35(a) describes a logic function that has a 1 output only when the BCD code for 7, 8, or 9 is present on the inputs. If the "don't cares" are used as 1s, the resulting expression for the function is $A + BCD$, as indicated in part (b). If the "don't cares" are not used as 1s, the resulting expression is $A\overline{B}\,\overline{C} + \overline{A}BCD$. So you can see the advantage of using "don't care" terms to get the simplest expression.

Inputs	Output
A B C D	Y
0 0 0 0	0
0 0 0 1	0
0 0 1 0	0
0 0 1 1	0
0 1 0 0	0
0 1 0 1	0
0 1 1 0	0
0 1 1 1	1
1 0 0 0	1
1 0 0 1	1
1 0 1 0	X
1 0 1 1	X
1 1 0 0	X
1 1 0 1	X
1 1 1 0	X
1 1 1 1	X

Don't cares

(a) Truth table

(b) Without "don't cares" $Y = A\overline{B}\,\overline{C} + \overline{A}BCD$
     With "don't cares" $Y = A + BCD$

FIGURE 4–35
Example of the use of "don't care" conditions to simplify an expression.

SECTION 4–9 REVIEW

1. Lay out Karnaugh maps for three and four variables.
2. Group the 1s and write the simplified SOP expression for the Karnaugh map in Figure 4–24.
3. Write the original standard SOP expressions for each of the Karnaugh maps in Figure 4–31.

4–10 ■ KARNAUGH MAP POS MINIMIZATION

In the last section, you studied the minimization of an SOP expression using a Karnaugh map. In this section, we will focus on POS expressions. The approaches are much the same except that with POS expressions, 0s representing the standard sum terms are placed on the Karnaugh map instead of 1s. This section can be omitted without impact on future material.

After completing this section, you should be able to

■ Map a standard POS expression on a Karnaugh map   ■ Combine the 0s on the map into maximum groups   ■ Determine the minimum sum term for each group on the map   ■ Combine the minimum sum terms to form a minimum POS expression   ■ Use the Karnaugh map to convert between POS and SOP

Mapping a Standard POS Expression

For a POS expression in standard form, a 0 is placed on the Karnaugh map for each sum term in the expression. Each 0 is placed in a cell corresponding to the value of a sum term. For example, for the sum term $A + \overline{B} + C$, a 0 goes in the 010 cell on a 3-variable map.

When a POS expression is completely mapped, there will be a number of 0s on the Karnaugh map equal to the number of sum terms in the standard POS expression. The cells that do not have a 0 are the cells for which the expression is 1. Usually, when working with POS expressions, the 1s are left off. The following steps and the illustration in Figure 4–36 show the mapping process.

Step 1. Determine the binary value of each sum term in the standard POS expression. This is the binary value that makes the term equal to 0.

Step 2. As each sum term is evaluated, place a 0 on the Karnaugh map in the corresponding cell.

Example 4–30 will further illustrate the mapping process.

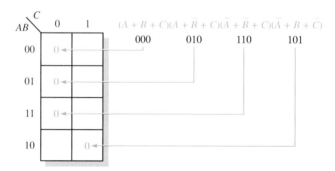

FIGURE 4–36
Example of mapping a standard POS expression.

EXAMPLE 4–30   Map the following standard POS expression on a Karnaugh map:

$$(\overline{A} + \overline{B} + C + D)(\overline{A} + B + \overline{C} + \overline{D})(A + B + \overline{C} + D)(\overline{A} + \overline{B} + \overline{C} + \overline{D})(A + B + \overline{C} + \overline{D})$$

Solution   The expression is evaluated as shown below and a 0 is placed on the 4-variable Karnaugh map in Figure 4–37 for each standard sum term in the expression.

$$(\overline{A} + \overline{B} + C + D)(\overline{A} + B + \overline{C} + \overline{D})(A + B + \overline{C} + D)(\overline{A} + \overline{B} + \overline{C} + \overline{D})(A + B + \overline{C} + \overline{D})$$
$$\quad 1100 \qquad\qquad 1011 \qquad\qquad 0010 \qquad\qquad 1111 \qquad\qquad 0011$$

FIGURE 4–37

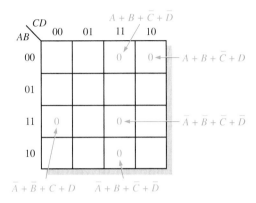

Related Problem   Map the following standard POS expression on a Karnaugh map:

$$(A + \overline{B} + \overline{C} + D)(A + B + C + \overline{D})(A + B + C + D)(\overline{A} + B + \overline{C} + D)$$

Karnaugh Map Simplification of POS Expressions

The process for minimizing a POS expression is basically the same as for an SOP expression except that you group 0s to produce minimum sum terms instead of grouping 1s to produce minimum product terms. The rules for grouping the 0s are the same as those for grouping the 1s that you learned in Section 4–9.

EXAMPLE 4–31   Use a Karnaugh map to minimize the following standard POS expression:

$$(A + B + C)(A + B + \overline{C})(A + \overline{B} + C)(A + \overline{B} + \overline{C})(\overline{A} + \overline{B} + C)$$

Solution   The combinations of binary values of the expression are

$$(0 + 0 + 0)(0 + 0 + 1)(0 + 1 + 0)(0 + 1 + 1)(1 + 1 + 0)$$

The standard POS expression is mapped and the cells are grouped as shown in Figure 4–38.

FIGURE 4–38

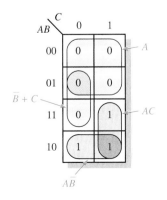

Notice how the 0 in the 110 cell is included into a 2-cell group by utilizing the 0 in the 4-cell group. The sum term for each group is shown in the figure and the resulting minimum POS expression is

$$A(\overline{B} + C)$$

Keep in mind that this minimum POS expression is equivalent to the original standard POS expression.

Grouping the 1s as shown by the gray areas yields an SOP expression that is equivalent to grouping the 0s.

$$AC + A\overline{B} = A(\overline{B} + C)$$

Related Problem   Use a Karnaugh map to simplify the following standard POS expression:

$$(X + \overline{Y} + Z)(X + \overline{Y} + \overline{Z})(\overline{X} + \overline{Y} + Z)(\overline{X} + Y + Z)$$

EXAMPLE 4–32   Use a Karnaugh map to minimize the following POS expression:

$$(B + C + D)(A + B + \overline{C} + D)(\overline{A} + B + C + \overline{D})(A + \overline{B} + C + D)(\overline{A} + \overline{B} + C + D)$$

Solution   The first term must be expanded into $\overline{A} + B + C + D$ and $A + B + C + D$ to get a standard POS expression, which is then mapped; and the cells are grouped as shown in Figure 4–39. The sum term for each group is shown and the resulting minimum POS expression is

$$(C + D)(A + B + D)(\overline{A} + B + C)$$

Keep in mind that this minimum POS expression is equivalent to the original standard POS expression.

FIGURE 4–39

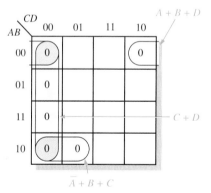

Related Problem   Use a Karnaugh map to simplify the following POS expression:

$$(W + \overline{X} + Y + \overline{Z})(W + X + Y + Z)(W + \overline{X} + \overline{Y} + Z)(\overline{W} + \overline{X} + Z)$$

Converting Between POS and SOP Using the Karnaugh Map

When a POS expression is mapped, it can easily be converted to the equivalent SOP form directly from the Karnaugh map. Also, given a mapped SOP expression, an equivalent POS expression can be derived directly from the map. This provides a good way to compare both minimum forms of an expression to determine if one of them can be implemented with fewer gates than the other.

For a POS expression, all the cells that do not contain 0s contain 1s, from which the SOP expression is derived. Likewise, for an SOP expression, all the cells that do not contain 1s contain 0s, from which the POS expression is derived. Example 4–33 illustrates this conversion.

EXAMPLE 4–33

Using a Karnaugh map, convert the following standard POS expression into a minimum POS expression, a standard SOP expression, and a minimum SOP expression.

$$(\overline{A} + \overline{B} + C + D)(A + \overline{B} + C + D)(A + B + C + \overline{D})$$
$$(A + B + \overline{C} + D)(\overline{A} + B + C + \overline{D})(A + B + \overline{C} + D)$$

Solution The 0s for the standard POS expression are mapped and grouped to obtain the minimum POS expression in Figure 4–40(a). In Figure 4–40(b), 1s are added to the cells that do not contain 0s. From each cell containing a 1, a standard product term is obtained as indicated. These product terms form the standard SOP expression. In Figure 4–40(c), the 1s are grouped and a minimum SOP expression is obtained.

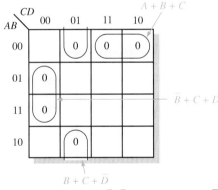

(a) Minimum POS: $(A + B + \overline{C})(\overline{B} + C + D)(B + C + \overline{D})$

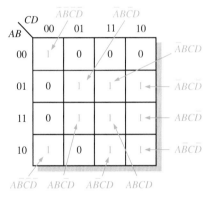

(b) Standard SOP:
$\overline{A}\overline{B}\overline{C}\overline{D} + \overline{A}B\overline{C}D + \overline{A}BCD + \overline{A}BC\overline{D} + AB\overline{C}\overline{D}$
$+ \overline{A}BC\overline{D} + A\overline{B}\overline{C}D + AB\overline{C}D + A\overline{B}CD + ABCD$

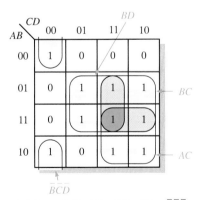

(c) Minimum SOP: $AC + BC + BD + \overline{B}\overline{C}\overline{D}$

FIGURE 4–40

Related Problem   Use a Karnaugh map to convert the following expression to minimum SOP form:

$$(W + \overline{X} + Y + \overline{Z})(\overline{W} + X + \overline{Y} + \overline{Z})(\overline{W} + \overline{X} + \overline{Y} + Z)(\overline{W} + \overline{X} + \overline{Z})$$

<table>
<tr><td>

**SECTION 4–10
REVIEW**

</td><td>

1. What is the difference in mapping a POS expression and an SOP expression?
2. What is the standard sum term for a 0 in cell 1011?
3. What is the standard product term for a 1 in cell 0010?

</td></tr>
</table>

4–11 ■ FIVE-VARIABLE KARNAUGH MAPS

Boolean functions with five variables can be simplified using a 32-cell Karnaugh map. Actually, two 4-variable maps (16 cells each) are used to construct a 5-variable map. You already know the cell adjacencies within each of the 4-variable maps and how to form groups of cells containing 1s to simplify an SOP expression. All you need to learn for five variables is the cell adjacencies between the two 4-variable maps and how to group those adjacent 1s. This section can be omitted without affecting future material.

After completing this section, you should be able to

■ Determine cell adjacencies in a 5-variable map   ■ Form maximum cell groupings in a 5-variable map   ■ Minimize 5-variable Boolean expressions using the Karnaugh map

A Karnaugh map for five variables (*ABCDE*) can be constructed using two 4-variable maps with which you are already familiar. Each map contains 16 cells with all combinations of variables *B*, *C*, *D*, and *E*. One map is for *A* = 0 and the other is for *A* = 1, as shown in Figure 4–41.

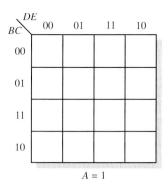

FIGURE 4–41
A 5-variable Karnaugh map.

Cell Adjacencies

You already know how to determine adjacent cells within the 4-variable map. The best way to visualize cell adjacencies between the two 16-cell maps is to imagine that the *A* = 0 map is placed on top of the *A* = 1 map. Each cell in the *A* = 0 map is adjacent to the cell directly below it in the *A* = 1 map.

To illustrate, an example with four groups is shown in Figure 4–42 with the maps in a 3-dimensional arrangement. The 1s in the yellow cells form an 8-bit group (four in the $A = 0$ map combined with four in the $A = 1$ map). The 1s in the orange cells form a 4-bit group. The 1s in the light red cells form a 4-bit group only in the $A = 0$ map. The 1 in the gray cell in the $A = 1$ map is grouped with the 1 in the lower right light red cell in the $A = 0$ map to form a 2-bit group.

FIGURE 4–42
Illustration of groupings of 1s in adjacent cells of a 5-variable map.

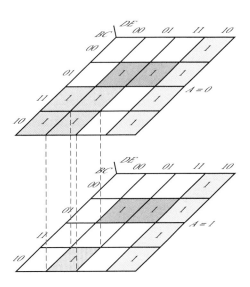

The Simplified Boolean Expression

The original SOP Boolean expression that is plotted on the Karnaugh map in Figure 4–42 contains seventeen 5-variable terms because there are seventeen 1s on the map. As you know, only the variables that do not change within a group remain in the expression for that group. The simplified expression taken from the map is developed as follows:

■ The term for the yellow group is $D\overline{E}$.
■ The term for the orange group is $\overline{B}CE$.
■ The term for the light red group is $\overline{A}B\overline{D}$.
■ The term for the gray cell grouped with the red cell is $B\overline{C}\,\overline{D}E$.

Combining these terms into the simplified SOP expression yields

$$X = D\overline{E} + \overline{B}CE + \overline{A}B\overline{D} + B\overline{C}\,\overline{D}E$$

EXAMPLE 4–34

Use a Karnaugh map to minimize the following standard SOP 5-variable expression:

$$X = \overline{A}\,\overline{B}\,\overline{C}\,\overline{D}\,\overline{E} + \overline{A}\,\overline{B}C\overline{D}\,\overline{E} + \overline{A}BC\overline{D}\,\overline{E} + \overline{A}B\overline{C}\,\overline{D}\,\overline{E} + \overline{A}\,\overline{B}\,\overline{C}\overline{D}E + \overline{A}BC\overline{D}E$$
$$+ \overline{A}BCDE + A\overline{B}\,\overline{C}\,\overline{D}\,\overline{E} + A\overline{B}\,\overline{C}DE + ABC\overline{D}E + ABCDE + A\overline{B}CDE$$

Solution The SOP expression is mapped in Figure 4–43 and the groupings and their corresponding terms are indicated. Combining the terms yields the following minimized SOP expression:

$$X = \overline{A}\,\overline{D}\,\overline{E} + \overline{B}\,\overline{C}\,\overline{D} + BCE + ACDE$$

FIGURE 4–43

Related Problem Minimize the following expression:

$$Y = \overline{A}\,\overline{B}\,\overline{C}\,\overline{D}\,\overline{E} + \overline{A}\,\overline{B}C\overline{D}\overline{E} + \overline{A}BC\overline{D}\overline{E} + \overline{A}\overline{B}\,C\overline{D}\overline{E} + A\overline{B}\,\overline{C}\overline{D}\overline{E} + A\overline{B}C\overline{D}\overline{E} + ABC\overline{D}\overline{E} + AB\overline{C}\overline{D}\overline{E}$$
$$+ \overline{A}\overline{B}\,\overline{C}D\overline{E} + \overline{A}\overline{B}CD\overline{E} + \overline{A}BCD\overline{E} + \overline{A}B\,\overline{C}D\overline{E} + A\overline{B}\,\overline{C}D\overline{E} + A\overline{B}CD\overline{E} + ABCD\overline{E} + AB\overline{C}D\overline{E}$$

SECTION 4–11 REVIEW

1. Why does a 5-variable Karnaugh map require 32 cells?
2. What is the expression represented by a 5-variable Karnaugh map in which each cell contains a 1?

4–12 ■ DIGITAL SYSTEM APPLICATION

Seven-segment displays are used in everything from automobile instruments to Z-meters. The tablet-counting and control system that has been the focus of the system applications in the previous chapters, has two 7-segment displays. These displays are used with logic circuits that decode a binary coded decimal (BCD) number and activate the appropriate digits on the display. In this section, we will focus on a minimum-gate design using decoder logic to illustrate application of Boolean expressions and the Karnaugh map.

After completing this section, you should be able to

■ Appreciate the approach to the design of logic circuits ■ Explain the operation of a 7-segment display ■ Translate a specified requirement into a Boolean expression and/or truth table and then into a minimum-gate logic circuit using a Karnaugh map

The 7-Segment Display

Figure 4–44 shows a common display format composed of seven elements or segments. Energizing certain combinations of these segments can cause each of the ten decimal digits to be displayed. Figure 4–45 illustrates this method of digital display for each of the ten digits by using a red segment to represent one that is energized. To produce a 1, segments *b* and *c* are energized; to produce a 2, segments *a*, *b*, *g*, *e*, and *d* are used; and so on.

FIGURE 4–44
Seven-segment display format showing arrangement of segments.

FIGURE 4–45
Display of decimal digits with a 7-segment device.

LED Displays   One common type of 7-segment display consists of light-emitting diodes (**LED**) arranged as shown in Figure 4–46. Each segment is an LED that emits light when there is current through it. In Figure 4–46(a) the common-anode arrangement requires the driving circuit to provide a low-level voltage in order to activate a given segment. When a LOW is applied to a segment input, the LED is turned on, and there is current through it. In Figure 4–46(b) the common-cathode arrangement requires the driver to provide a high-level voltage to activate a segment. When a HIGH is applied to a segment input, the LED is turned on and there is current through it.

FIGURE 4–46
Arrangements of 7-segment LED displays.

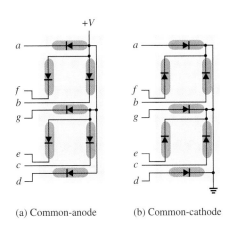

(a) Common-anode   (b) Common-cathode

LCD Displays   Another common type of 7-segment display is the liquid crystal display (**LCD**). LCDs operate by polarizing light so that a nonactivated segment reflects incident light and thus appears invisible against its background. An activated segment does not reflect incident light and thus appears dark. LCDs consume much less power than LEDs but cannot be seen in the dark, while LEDs can.

LCDs operate from a low-frequency signal voltage (30 Hz to 60 Hz) applied between the segment and a common element called the *backplane* (bp). The basic operation is as follows. Figure 4–47 shows a square wave used as the source signal. Each segment in the display is driven by an exclusive-OR gate with one input connected to an output of the 7-segment decoder/driver and the other input connected to the signal source. When the decoder/driver output is HIGH (1), the exclusive-OR acts as an inverter to the source signal, as shown in Figure 4–47(a). You can verify this by reviewing the truth table operation of the exclusive-OR. The resulting voltage between the LCD segment and the backplane is also a square wave because when $V_{seg} = 1$, $V_{bp} = 0$, and vice versa. The voltage difference turns the segment on.

(a) Segment activated (on)

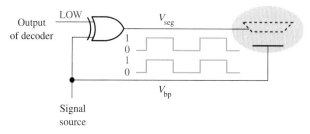

(b) Segment not activated (off)

FIGURE 4–47
Basic operation of an LCD.

When the decoder/driver output is LOW (0), the exclusive-OR gate passes the source signal uninverted, as shown in Figure 4–47(b). The resulting voltage difference between the segment and the backplane is 0 because $V_{seg} = V_{bp}$. This condition turns the segment off.

For driving LCDs, TTL is not recommended. Its low-level voltage is typically a few tenths of a volt, thus creating a dc component across the LCD, which degrades its performance. Therefore, CMOS is used in LCD applications.

Segment Decoding Logic

Each segment is used for various decimal digits, but no one segment is used for all ten digits. Therefore, each segment must be activated by its own decoding circuit that detects the occurrence of any of the numbers in which the segment is used. From Figures 4–44 and 4–45, the segments that are required to be activated for each displayed digit are determined and listed in Table 4–9.

TABLE 4–9
Active segments for each decimal digit.

Digit	Segments Activated
0	a, b, c, d, e, f
1	b, c
2	a, b, d, e, g
3	a, b, c, d, g
4	b, c, f, g
5	a, c, d, f, g
6	a, c, d, e, f, g
7	a, b, c
8	a, b, c, d, e, f, g
9	a, b, c, d, f, g

Truth Table for the Segment Logic   The segment decoding logic requires four binary coded decimal (BCD) inputs and seven outputs, one for each segment in the display, as indicated in the block diagram of Figure 4–48. The multiple-output truth table, shown in Table 4–10, is actually seven truth tables in one and could be separated into a separate table for each segment. A 1 in the segment output columns of the table indicates an activated segment.

FIGURE 4–48
Block diagram of 7-segment logic and display.

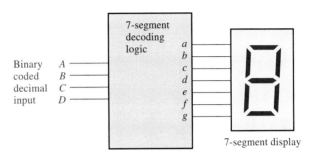

TABLE 4–10
Truth table for 7-segment logic.

Decimal Digit	Inputs				Segment Outputs						
	D	**C**	**B**	**A**	**a**	**b**	**c**	**d**	**e**	**f**	**g**
0	0	0	0	0	1	1	1	1	1	1	0
1	0	0	0	1	0	1	1	0	0	0	0
2	0	0	1	0	1	1	0	1	1	0	1
3	0	0	1	1	1	1	1	1	0	0	1
4	0	1	0	0	0	1	1	0	0	1	1
5	0	1	0	1	1	0	1	1	0	1	1
6	0	1	1	0	1	0	1	1	1	1	1
7	0	1	1	1	1	1	1	0	0	0	0
8	1	0	0	0	1	1	1	1	1	1	1
9	1	0	0	1	1	1	1	1	0	1	1
10	1	0	1	0	X	X	X	X	X	X	X
11	1	0	1	1	X	X	X	X	X	X	X
12	1	1	0	0	X	X	X	X	X	X	X
13	1	1	0	1	X	X	X	X	X	X	X
14	1	1	1	0	X	X	X	X	X	X	X
15	1	1	1	1	X	X	X	X	X	X	X

Output = 1 means segment is activated (on)

Output = 0 means segment is not activated (off)

Output = X means "don't care"

Since the BCD code does not include the binary values 1010, 1011, 1100, 1101, 1110, and 1111, these combinations will never appear on the inputs and can therefore be treated as "don't care" (X) conditions, as indicated in the truth table. To conform with the practice of most IC manufacturers, A represents the least significant bit and D represents the most significant bit in this particular application.

Boolean Expressions for the Segment Logic   From the truth table, a standard SOP or POS expression can be written for each segment. For example, the standard SOP expression for segment a is

$$a = \overline{D}\,\overline{C}\,\overline{B}\,\overline{A} + \overline{D}\,\overline{C}B\overline{A} + \overline{D}\,\overline{C}BA + \overline{D}C\overline{B}A + \overline{D}CB\overline{A} + \overline{D}CBA + D\overline{C}\,\overline{B}\,\overline{A} + D\overline{C}\,\overline{B}A$$

and the standard SOP expression for segment e is

$$e = \overline{D}\,\overline{C}\,\overline{B}\,\overline{A} + \overline{D}\,\overline{C}B\overline{A} + \overline{D}CB\overline{A} + D\overline{C}\,\overline{B}\,\overline{A}$$

Expressions for the other segments can be similarly developed. As you can see, the expression for segment a has eight product terms and the expression for segment e has four product terms representing each of the BCD inputs that activate that segment. This means that the standard SOP implementation of segment-a logic requires an AND-OR circuit consisting of eight 4-input AND gates and one 8-input OR gate. The implementation of segment-e logic requires four 4-input AND gates and one 4-input OR gate. In both cases, four inverters are required to produce the complement of each variable.

Karnaugh Map Minimization of the Segment Logic   Let's begin by obtaining a minimum SOP expression for segment a. A Karnaugh map for segment a is shown in Figure 4–49 and the following steps are carried out:

Step 1.   The 1s are mapped directly from Table 4–10.
Step 2.   All of the "don't cares" (X) are placed on the map.
Step 3.   The 1s are grouped as shown. "Don't cares" and overlapping of cells are utilized to form the largest groups possible.
Step 4.   Write the minimum product term for each group and sum the terms to form the minimum SOP expression.

Keep in mind that "don't cares" do not have to be included in a group, but in this case all of them are used. Also, notice that the 1s in the corner cells are grouped with a "don't care" using the "wrap around" adjacency of the corner cells.

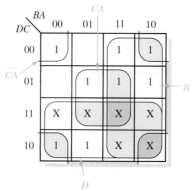

Standard SOP expression:
$$\bar{D}\bar{C}\bar{B}\bar{A} + \bar{D}\bar{C}B\bar{A} + \bar{D}\bar{C}BA + \bar{D}C\bar{B}A + \bar{D}CB\bar{A} + \bar{D}CBA + D\bar{C}\bar{B}\bar{A} + D\bar{C}\bar{B}A$$

Minimum SOP expression: $D + B + CA + \bar{C}\bar{A}$

FIGURE 4–49
Karnaugh map minimization of the segment-a logic expression.

Minimum Implementation of Segment-a Logic   The minimum SOP expression taken from the Karnaugh map in Figure 4–49 for the segment-a logic is

$$D + B + CA + \bar{C}\bar{A}$$

This expression can be implemented with two 2-input AND gates, one 4-input OR gate and two inverters as shown in Figure 4–50. Compare this to the standard SOP implementation for segment-a logic discussed earlier; you'll see that the number of gates and inverters has been reduced from thirteen to five and, as a result, the number of interconnections has been significantly reduced.

The minimum logic for each of the remaining six segments (b, c, d, e, f, and g) can be obtained with a similar approach.

■ SYSTEM ASSIGNMENT

■ **Activity 1:** Determine the minimum logic for segment *b*.

■ **Activity 2:** Determine the minimum logic for segment *c*.

■ **Activity 3:** Determine the minimum logic for segment *d*.

■ **Activity 4:** Determine the minimum logic for segment *e*.

■ **Activity 5:** Determine the minimum logic for segment *f*.

■ **Activity 6:** Determine the minimum logic for segment *g*.

■ **Activity 7:** Determine if there are any duplicate product terms in the 7-segment logic expressions

and, if so, eliminate all but one of the duplicate AND gates from the overall implementation.

■ **Activity 8:** Draw a logic diagram for the complete 7-segment decoding logic by combining all of the individual segment logic circuits and eliminating any duplicate gates or inverters.

■ **Activity 9:** Specify by type number the specific integrated circuits required to implement the logic diagram in Activity 8 using only those that have been introduced up to this point. Utilize all available gates in a package where possible to keep the number of ICs to a minimum.

 The segment decoding logic is in file SA04.EWB on your EWB CD. Use EWB to simulate a test of this logic.

FIGURE 4–50
The minimum logic implementation for segment a of the 7-segment display.

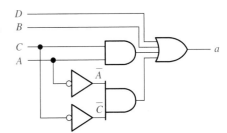

Further Multiple-Output Logic Simplification Once the minimum logic for each individual segment has been developed, there is a possibility of additional simplification by elimination of any duplicate product terms or complemented single-variable terms. For example, if the minimum SOP for segment *a* has a product term that is also found in, say, the minimum SOP for segment *b*, then that duplicate term can be implemented with only one AND gate with its output commonly connected to the inputs of the OR gates for the two segments. Likewise, the complement of a variable needs only to be implemented with one inverter and then the output of that inverter can be connected where it is needed provided that its fan-out is not exceeded.

SECTION 4–12 REVIEW

1. What is the advantage of using "don't care" conditions in a given application?
2. How many duplicate terms did you find in the segment logic in Activity 7?
3. How many gates and inverters were you able to eliminate as a result of duplication in Activity 8?

■ SUMMARY

■ Gate symbols and Boolean expressions for the outputs of an inverter and 2-input gates are shown in Figure 4–51.

$$A \rightarrow\!\!\!\!\triangleright\!\!\circ\! - \overline{A} \qquad \begin{matrix} A \\ B \end{matrix}\!\!\Big]\!\!\!\!\!\Big)\! - AB \qquad \begin{matrix} A \\ B \end{matrix}\!\!\Big]\!\!\!\!\!\Big)\!\!\circ\! - \overline{AB} \qquad \begin{matrix} A \\ B \end{matrix}\!\!\Big]\!\!\!\!\!\Big)\! - A + B \qquad \begin{matrix} A \\ B \end{matrix}\!\!\Big]\!\!\!\!\!\Big)\!\!\circ\! - \overline{A + B}$$

FIGURE 4–51

■ Commutative laws: $A + B = B + A$
$$AB = BA$$

■ Associative laws: $A + (B + C) = (A + B) + C$
$$A(BC) = (AB)C$$

■ Distributive law: $A(B + C) = AB + AC$

■ Boolean rules:
1. $A + 0 = A$	**7.** $A \cdot A = A$
2. $A + 1 = 1$	**8.** $A \cdot \overline{A} = 0$
3. $A \cdot 0 = 0$	**9.** $\overline{\overline{A}} = A$
4. $A \cdot 1 = A$	**10.** $A + AB = A$
5. $A + A = A$	**11.** $A + \overline{A}B = A + B$
6. $A + \overline{A} = 1$	**12.** $(A + B)(A + C) = A + BC$

■ DeMorgan's theorems:

1. The complement of a product is equal to the sum of the complements of the terms in the product.

$$\overline{XY} = \overline{X} + \overline{Y}$$

2. The complement of a sum is equal to the product of the complements of the terms in the sum.

$$\overline{X + Y} = \overline{X}\,\overline{Y}$$

■ Karnaugh maps for 3 and 4 variables are shown in Figure 4–52. A 5-variable map is formed from two 4-variable maps.

FIGURE 4–52

3-variable

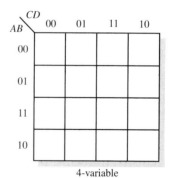

4-variable

■ KEY TERMS

These terms are also in the end-of-book glossary.

Adjacency Characteristic of cells in a Karnaugh map in which there is a single-variable change from one cell to another cell next to it on any of its four sides.

Associative law In addition (ORing) and multiplication (ANDing) of three or more variables, the order in which the variables are grouped makes no difference.

Boolean expression An arrangement of variables and logical operators used to express the operation of a logic circuit.

Cell An area on a Karnaugh map that represents a unique combination of variables in product form.

Commutative law In addition (ORing) and multiplication (ANDing) of two variables, the order in which the variables are ORed or ANDed makes no difference.

Distributive law ORing several variables and then ANDing the result with a single variable is equivalent to ANDing the single variable with each of the several variables and then ORing the products.

Domain All of the variables in a Boolean expression.

"Don't care" A combination of input literals that cannot occur and can be used as a 1 or a 0 on a Karnaugh map.

Karnaugh map An arrangement of cells representing the combinations of literals in a Boolean expression and used for a systematic simplification of the expression.

Literal A variable or the complement of a variable.

Product-of-sums (POS) A form of Boolean expression that is basically the ANDing of ORed terms.

Product term The Boolean product of two or more literals equivalent to an AND operation.

Sum-of-products A form of Boolean expression that is basically the ORing of ANDed terms.

Sum term The Boolean sum of two or more literals equivalent to an OR operation.

Variable A symbol used to represent a logical quantity that can have a value of 1 or 0, usually designated by an italic letter.

■ **SELF-TEST**

1. The complement of a variable is always
   (a) 0   (b) 1   (c) equal to the variable   (d) the inverse of the variable

2. The Boolean expression $A + \overline{B} + C$ is
   (a) a sum term   (b) a literal term   (c) a product term   (d) a complemented term

3. The Boolean expression $A\overline{B}C\overline{D}$ is
   (a) a sum term   (b) a product term   (c) a literal term   (d) always 1

4. The domain of the expression $A\overline{B}CD + A\overline{B} + \overline{C}D + B$ is
   (a) A and D (b) B only   (c) $A, B, C,$ and D (d) none of these

5. According to the commutative law of addition,
   (a) $AB = BA$ (b) $A = A + A$
   (c) $A + (B + C) = (A + B) + C$ (d) $A + B = B + A$

6. According to the associative law of multiplication,
   (a) $B = BB$ (b) $A(BC) = (AB)C$ (c) $A + B = B + A$ (d) $B + B(B + 0)$

7. According to the distributive law,
   (a) $A(B + C) = AB + AC$ (b) $A(BC) = ABC$ (c) $A(A + 1) = A$ (d) $A + AB = A$

8. Which one of the following is *not* a valid rule of Boolean algebra?
   (a) $A + 1 = 1$ (b) $A = \overline{A}$ (c) $AA = A$ (d) $A + 0 = A$

9. Which of the following rules states that if one input of an AND gate is always 1, the output is equal to the other input?
   (a) $A + 1 = 1$ (b) $A + A = A$ (c) $A \cdot A = A$ (d) $A \cdot 1 = A$

10. According to DeMorgan's theorems, the following equality(s) are correct:
    (a) $\overline{AB} = \overline{A} + \overline{B}$ (b) $\overline{XYZ} = \overline{X} + \overline{Y} + \overline{Z}$
    (c) $\overline{A + B + C} = \overline{A}\,\overline{B}\,\overline{C}$ (d) all of these

11. The Boolean expression $X = AB + CD$ represents
    (a) two ORs ANDed together   (b) a 4-input AND gate
    (c) two ANDs ORed together   (d) an exclusive-OR

12. An example of a sum-of-products expression is
    (a) $A + B(C + D)$ (b) $\overline{A}B + A\overline{C} + AB\overline{C}$
    (c) $(\overline{A} + B + C)(A + \overline{B} + C)$ (d) both answers (a) and (b)

13. An example of a product-of-sums expression is
    (a) $A(B + C) + A\overline{C}$ (b) $(A + B)(\overline{A} + B + \overline{C})$
    (c) $\overline{A} + \overline{B} + BC$ (d) both answers (a) and (b)

14. An example of a standard SOP expression is
    (a) $\overline{A}B + A\overline{B}C + AB\overline{D}$ (b) $A\overline{B}C + A\overline{C}D$
    (c) $A\overline{B} + \overline{A}B + AB$ (d) $A\overline{B}C\overline{D} + \overline{A}B + \overline{A}$

15. A 3-variable Karnaugh map has
    (a) eight cells   (b) three cells   (c) sixteen cells   (d) four cells

16. In a 4-variable Karnaugh map, a 2-variable product term is produced by
    (a) a 2-cell group of 1s   (b) an 8-cell group of 1s
    (c) a 4-cell group of 1s   (d) a 4-cell group of 0s

17. On a Karnaugh map, grouping the 0s produces
    (a) a product-of-sums expression   (b) a sum-of-products expression
    (c) a "don't care" condition   (d) AND-OR logic

18. A 5-variable Karnaugh map has
    (a) sixteen cells   (b) thirty-two cells   (c) sixty-four cells

■ PROBLEMS

SECTION 4–1 Boolean Operations and Expressions

1. Using Boolean notation, write an expression that is a 1 whenever one or more of its variables (A, B, C, and D) are 1s.

2. Write an expression that is a 1 only if all of its variables (A, B, C, D, and E) are 1s.

3. Write an expression that is a 1 when one or more of its variables (A, B, and C) are 0s.

4. Evaluate the following operations:

   (a) $0 + 0 + 1$ (b) $1 + 1 + 1$ (c) $1 \cdot 0 \cdot 0$

   (d) $1 \cdot 1 \cdot 1$ (e) $1 \cdot 0 \cdot 1$ (f) $1 \cdot 1 + 0 \cdot 1 \cdot 1$

5. Find the values of the variables that make each product term 1 and each sum term 0.

   (a) AB (b) $A\overline{B}C$ (c) $A + B$ (d) $\overline{A} + B + \overline{C}$

   (e) $\overline{A} + \overline{B} + C$ (f) $\overline{A} + B$ (g) $A\overline{B}\,\overline{C}$

6. Find the value of X for all possible values of the variables.

   (a) $X = (A + B)C + B$ (b) $X = (\overline{A + B})C$ (c) $X = A\overline{B}C + AB$

   (d) $X = (A + B)(\overline{A} + B)$ (e) $X = (A + BC)(\overline{B} + \overline{C})$

SECTION 4–2 Laws and Rules of Boolean Algebra

7. Identify the law of Boolean algebra upon which each of the following equalities is based:

   (a) $A\overline{B} + CD + A\overline{C}D + B = B + A\overline{B} + A\overline{C}D + CD$

   (b) $AB\overline{C}D + \overline{ABC} = D\overline{C}BA + \overline{CBA}$

   (c) $AB(CD + E\overline{F} + GH) = ABCD + ABE\overline{F} + ABGH$

8. Identify the Boolean rule(s) on which each of the following equalities is based:

   (a) $\overline{AB + CD} + \overline{EF} = AB + CD + \overline{EF}$ (b) $A\overline{A}B + AB\overline{C} + AB\overline{B} = AB\overline{C}$

   (c) $A(BC + BC) + AC = A(BC) + AC$ (d) $AB(C + \overline{C}) + AC = AB + AC$

   (e) $A\overline{B} + A\overline{B}C = A\overline{B}$ (f) $ABC + \overline{AB} + \overline{AB}CD = ABC + \overline{AB} + D$

SECTION 4–3 DeMorgan's Theorems

9. Apply DeMorgan's theorems to each expression:

   (a) $\overline{A + \overline{B}}$ (b) $\overline{\overline{A}B}$ (c) $\overline{A + B + C}$ (d) $\overline{ABC}$

   (e) $\overline{A(B + C)}$ (f) $\overline{AB + CD}$ (g) $\overline{AB + CD}$ (h) $\overline{(A + B)(\overline{C} + D)}$

10. Apply DeMorgan's theorems to each expression:

   (a) $\overline{A\overline{B}(C + \overline{D})}$ (b) $\overline{AB(CD + EF)}$

   (c) $\overline{(A + \overline{B} + C + \overline{D}) + ABC\overline{D}}$ (d) $\overline{(\overline{A} + B + C + D)(\overline{ABCD})}$

   (e) $\overline{AB(CD + \overline{EF})(\overline{AB} + \overline{CD})}$

11. Apply DeMorgan's theorems to the following:

   (a) $\overline{\overline{(ABC)(EFG)} + \overline{(HIJ)(KLM)}}$ (b) $\overline{(A + B\overline{C} + CD) + \overline{B}\overline{C}}$

   (c) $\overline{\overline{(A + B)(\overline{C} + D)(E + \overline{F})(G + H)}}$

SECTION 4–4 Boolean Analysis of Logic Circuits

12. Write the Boolean expression for each of the logic gates in Figure 4–53.

(a)

(b)

(c)

(d)

FIGURE 4–53

13. Write the Boolean expression for each of the logic circuits in Figure 4–54.

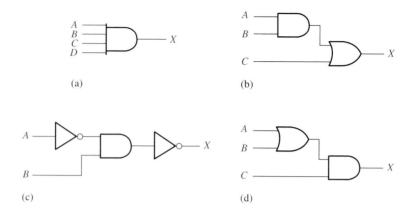

(a) (b)

(c) (d)

FIGURE 4–54

14. Draw the logic circuit represented by each of the following expressions:

   (a) $A + B + C$ (b) ABC (c) $AB + C$ (d) $AB + CD$

15. Draw the logic circuit represented by each expression:

   (a) $A\overline{B} + \overline{A}B$ (b) $AB + \overline{A}\,\overline{B} + \overline{A}BC$

   (c) $\overline{A}B(C + \overline{D})$ (d) $A + B[C + D(B + \overline{C})]$

16. Construct a truth table for each of the following Boolean expressions:

   (a) $A + B$ (b) AB (c) $AB + BC$

   (d) $(A + B)C$ (e) $(A + B)(\overline{B} + C)$

SECTION 4–5   Simplification Using Boolean Algebra

17. Using Boolean algebra techniques, simplify the following expressions as much as possible:

   (a) $A(A + B)$ (b) $A(\overline{A} + AB)$ (c) $BC + \overline{B}C$

   (d) $A(A + \overline{A}B)$ (e) $A\overline{B}C + \overline{A}BC + \overline{A}\,\overline{B}C$

18. Using Boolean algebra, simplify the following expressions:

   (a) $(A + \overline{B})(A + C)$ (b) $\overline{A}B + \overline{A}B\overline{C} + \overline{A}BCD + \overline{A}B\overline{C}\,\overline{D}E$

   (c) $AB + \overline{A}BC + A$ (d) $(A + \overline{A})(AB + AB\overline{C})$

   (e) $AB + (\overline{A} + \overline{B})C + AB$

19. Using Boolean algebra, simplify each expression:

   (a) $BD + B(D + E) + \overline{D}(D + F)$ (b) $\overline{A}\,\overline{B}C + (\overline{A + B + \overline{C}}) + \overline{A}\,\overline{B}\,\overline{C}D$

   (c) $(B + BC)(B + \overline{B}C)(B + D)$ (d) $ABCD + AB(\overline{CD}) + (\overline{AB})CD$

   (e) $ABC[AB + \overline{C}(BC + AC)]$

20. Determine which of the logic circuits in Figure 4–55 are equivalent.

SECTION 4–6   Standard Forms of Boolean Expressions

21. Convert the following expressions to sum-of-product (SOP) forms:

   (a) $(A + B)(C + \overline{B})$ (b) $(A + \overline{B}C)C$ (c) $(A + C)(AB + AC)$

22. Convert the following expressions to sum-of-product (SOP) forms:

   (a) $AB + CD(\overline{AB} + CD)$ (b) $AB(\overline{B}\,\overline{C} + BD)$ (c) $A + B[AC + (B + \overline{C})D]$

23. Define the domain of each SOP expression in Problem 21 and convert the expression to standard SOP form.

24. Convert each SOP expression in Problem 22 to standard SOP form.

25. Determine the binary value of each term in the standard SOP expressions from Problem 23.

26. Determine the binary value of each term in the standard SOP expressions from Problem 24.

27. Convert each standard SOP expression in Problem 23 to standard POS form.

28. Convert each standard SOP expression in Problem 24 to standard POS form.

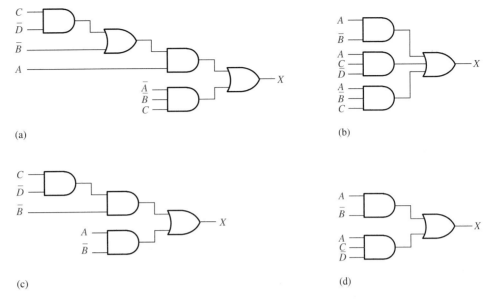

(a)

(b)

(c)

(d)

FIGURE 4–55

SECTION 4–7 Boolean Expressions and Truth Tables

29. Develop a truth table for each of the following standard SOP expressions:

   (a) $A\overline{B}C + \overline{A}B\overline{C} + ABC$ **(b)** $\overline{XYZ} + \overline{X}\overline{Y}Z + XY\overline{Z} + X\overline{Y}Z + \overline{X}YZ$

30. Develop a truth table for each of the following standard SOP expressions:

   (a) $\overline{A}B\overline{C}D + \overline{A}BC\overline{D} + A\overline{B}\overline{C}D + \overline{ABCD}$

   (b) $WXYZ + WX\overline{Y}\overline{Z} + \overline{W}XYZ + W\overline{X}YZ + WX\overline{Y}Z$

31. Develop a truth table for each of the SOP expressions:

   (a) $\overline{A}B + AB\overline{C} + \overline{AC} + A\overline{B}C$ **(b)** $\overline{X} + Y\overline{Z} + WZ + X\overline{Y}Z$

32. Develop a truth table for each of the standard POS expressions:

   (a) $(\overline{A} + \overline{B} + \overline{C})(A + B + C)(A + \overline{B} + C)$

   (b) $(\overline{A} + B + \overline{C} + D)(A + \overline{B} + C + \overline{D})(A + \overline{B} + \overline{C} + D)(\overline{A} + B + C + \overline{D})$

33. Develop a truth table for each of the standard POS expressions:

   (a) $(A + B)(A + C)(A + B + C)$ **(b)** $(A + \overline{B})(A + \overline{B} + \overline{C})(B + C + \overline{D})(\overline{A} + B + \overline{C} + D)$

34. For each truth table in Figure 4–56, derive a standard SOP and a standard POS expression.

FIGURE 4–56

ABC	X
0 0 0	0
0 0 1	1
0 1 0	0
0 1 1	0
1 0 0	1
1 0 1	1
1 1 0	0
1 1 1	1

(a)

ABC	X
0 0 0	0
0 0 1	0
0 1 0	0
0 1 1	0
1 0 0	0
1 0 1	1
1 1 0	1
1 1 1	1

(b)

ABCD	X
0 0 0 0	1
0 0 0 1	1
0 0 1 0	0
0 0 1 1	1
0 1 0 0	0
0 1 0 1	1
0 1 1 0	1
0 1 1 1	0
1 0 0 0	0
1 0 0 1	1
1 0 1 0	0
1 0 1 1	0
1 1 0 0	1
1 1 0 1	0
1 1 1 0	0
1 1 1 1	0

(c)

ABCD	X
0 0 0 0	0
0 0 0 1	0
0 0 1 0	1
0 0 1 1	0
0 1 0 0	1
0 1 0 1	1
0 1 1 0	0
0 1 1 1	1
1 0 0 0	0
1 0 0 1	0
1 0 1 0	0
1 0 1 1	1
1 1 0 0	1
1 1 0 1	0
1 1 1 0	0
1 1 1 1	1

(d)

SECTION 4–8   The Karnaugh Map

35. Sketch a 3-variable Karnaugh map and label each cell according to its binary value.

36. Sketch a 4-variable Karnaugh map and label each cell according to its binary value.

37. Write the standard problem term for each cell in a 3-variable Karnaugh map.

SECTION 4–9   Karnaugh Map SOP Minimization

38. Use a Karnaugh map to find the minimum SOP form for each expression:

(a) $\overline{A}\,\overline{B}\,\overline{C} + \overline{A}\,\overline{B}C + A\overline{B}C$ (b) $AC(\overline{B} + C)$

(c) $\overline{A}(BC + B\overline{C}) + A(BC + B\overline{C})$ (d) $\overline{A}\,\overline{B}\,\overline{C} + A\overline{B}\,\overline{C} + \overline{A}BC + AB\overline{C}$

39. Use a Karnaugh map to simplify each expression to a minimum SOP form:

(a) $\overline{A}\,\overline{B}\,\overline{C} + A\overline{B}C + \overline{A}BC + AB\overline{C}$ (b) $AC[\overline{B} + B(B + \overline{C})]$

(c) $DE\overline{F} + \overline{D}E\overline{F} + \overline{D}\,\overline{E}\,\overline{F}$

40. Expand each expression to a standard SOP form:

(a) $AB + A\overline{B}C + ABC$ (b) $A + BC$

(c) $A\overline{B}\,\overline{C}D + AC\overline{D} + \overline{B}CD + \overline{A}BC\overline{D}$ (d) $A\overline{B} + A\overline{B}CD + CD + B\overline{C}D + ABCD$

41. Minimize each expression in Problem 40 with a Karnaugh map.

42. Use a Karnaugh map to reduce each expression to a minimum SOP form:

(a) $A + B\overline{C} + CD$

(b) $\overline{A}\,\overline{B}\,\overline{C}\,\overline{D} + \overline{A}\,\overline{B}\,\overline{C}D + ABCD + ABC\overline{D}$

(c) $\overline{A}B(\overline{C}\,\overline{D} + \overline{C}D) + AB(\overline{C}\,\overline{D} + \overline{C}D) + A\overline{B}\,\overline{C}D$

(d) $(\overline{A}\,\overline{B} + A\overline{B})(CD + C\overline{D})$

(e) $\overline{A}\,\overline{B} + A\overline{B} + \overline{C}\,\overline{D} + C\overline{D}$

43. Reduce the function specified in the truth table in Figure 4–57 to its minimum SOP form by using a Karnaugh map.

44. Use the Karnaugh map method to implement in the minimum SOP form the logic function specified in the truth table in Figure 4–58.

45. Solve Problem 44 for a situation in which the last six binary combinations are not allowed.

Inputs				Output
A	B	C	D	X
0	0	0	0	0
0	0	0	1	1
0	0	1	0	1
0	0	1	1	0
0	1	0	0	0
0	1	0	1	0
0	1	1	0	1
0	1	1	1	1
1	0	0	0	1
1	0	0	1	0
1	0	1	0	1
1	0	1	1	0
1	1	0	0	1
1	1	0	1	1
1	1	1	0	0
1	1	1	1	1

Inputs			Output
A	B	C	X
0	0	0	1
0	0	1	1
0	1	0	0
0	1	1	1
1	0	0	1
1	0	1	1
1	1	0	0
1	1	1	1

FIGURE 4–57   **FIGURE 4–58**

SECTION 4–10 Karnaugh Map POS Minimization

46. Use a Karnaugh map to find the minimum POS for each expression:

(a) $(A + B + C)(\overline{A} + \overline{B} + \overline{C})(A + \overline{B} + C)$

(b) $(X + \overline{Y})(\overline{X} + Z)(X + \overline{Y} + \overline{Z})(\overline{X} + \overline{Y} + Z)$

(c) $A(B + \overline{C})(\overline{A} + C)(A + \overline{B} + C)(\overline{A} + B + \overline{C})$

47. Use a Karnaugh map to simplify each expression to minimum POS form:

(a) $(A + \overline{B} + C + \overline{D})(\overline{A} + B + \overline{C} + D)(\overline{A} + \overline{B} + \overline{C} + \overline{D})$

(b) $(X + \overline{Y})(W + \overline{Z})(\overline{X} + \overline{Y} + \overline{Z})(W + X + Y + Z)$

48. For the function specified in the truth table of Figure 4–57, determine the minimum POS expression using a Karnaugh map.

49. Determine the minimum POS expression for the function in the truth table of Figure 4–58.

50. Convert each of the following POS expressions to minimum SOP expressions using a Karnaugh map:

(a) $(A + \overline{B})(A + \overline{C})(\overline{A} + \overline{B} + C)$

(b) $(\overline{A} + B)(\overline{A} + \overline{B} + \overline{C})(B + \overline{C} + D)(A + \overline{B} + C + \overline{D})$

SECTION 4–11 Five-Variable Karnaugh Maps

51. Minimize the following SOP expression using a Karnaugh map:

$$X = \overline{A}B\overline{C}D\overline{E} + \overline{A}\,\overline{B}\,\overline{C}DE + A\overline{B}\,\overline{C}DE + AB\overline{C}\,\overline{D}E + \overline{A}BCD\overline{E} + \overline{A}BC\overline{D}E$$
$$+ \overline{A}\,\overline{B}\,\overline{C}\,\overline{D}E + \overline{A}\,\overline{B}CDE + AB\overline{C}D\overline{E} + AB\overline{C}DE$$

52. Apply the Karnaugh map method to minimize the following SOP expression:

$$A = \overline{V}WXYZ + V\overline{W}XYZ + VW\overline{X}YZ + VWX\overline{Y}Z + VWXY\overline{Z} + \overline{V}\,\overline{W}\,\overline{X}\,\overline{Y}\,\overline{Z}$$
$$+ \overline{V}\,\overline{W}\,\overline{X}Y\overline{Z} + \overline{V}\,\overline{W}X\overline{Y}\,\overline{Z} + \overline{V}W\overline{X}\,\overline{Y}\,\overline{Z}$$

SECTION 4–12 Digital System Application

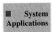

53. If you are required to choose a type of digital display for low light conditions, will you select LED or LCD 7-segment displays? Why?

54. Explain why the codes 1010, 1011, 1100, 1101, 1110, and 1111 fall into the "don't care" category in 7-segment display applications.

55. For segment b, how many fewer gates and inverters does it take to implement the minimum SOP expression than the standard SOP expression?

56. Repeat Problem 55 for the logic for segments c through g.

Special Design Problems

57. The logic for segment a in Figure 4–50 produces a HIGH output to activate the segment and so do the circuits for each of the other segments. If a type of 7-segment display is used that requires a LOW to activate each segment, modify the segment logic accordingly.

58. Redesign the logic for segment a using a minimum POS approach. Which is simpler, minimum POS or the minimum SOP?

59. Repeat Problem 58 for segments b through g.

60. Summarize the results of your redesign effort in Problems 58 and 59 and recommend the best design based on fewer ICs. Specify the types of ICs.

Electronics Workbench Troubleshooting

61. Open file PR04_61.EWB and observe the operation of the logic circuit using the existing test set-up. Determine whether or not a fault exists.

62. Open file PR04_62.EWB and observe the operation of the logic circuit using the existing test set-up. Determine whether or not a fault exists.

63. Open file PR04_63.EWB and observe the operation of the logic circuit using the existing test set-up. Determine whether or not a fault exists.

■ ANSWERS TO SECTION REVIEWS

SECTION 4–1

1. $\overline{A} = \overline{0} = 1$ **2.** $A = 1, B = 1, C = 0; \overline{A} + \overline{B} + C = \overline{1} + \overline{1} + 0 = 0 + 0 + 0 = 0$

3. $A = 1, B = 0, C = 1; A\overline{B}C = 1 \cdot \overline{0} \cdot 1 = 1 \cdot 1 \cdot 1 = 1$

SECTION 4–2

1. $A + (B + C + D) = (A + B + C) + D$ **2.** $A(B + C + D) = AB + AC + AD$

SECTION 4–3

1. (a) $\overline{ABC} + \overline{(D + E)} = \overline{A} + \overline{B} + \overline{C} + D\overline{E}$ **(b)** $\overline{(A + B)C} = \overline{A}B + \overline{C}$

(c) $\overline{A + B + C} + \overline{\overline{DE}} = \overline{A}\,\overline{B}\,\overline{C} + D + E$

SECTION 4–4

1. $(C + D)B + A$

2. Abbreviated truth table: The expression is a 1 when A is 1 or when B and C are 1s or when B and D are 1s. The expression is 0 for all other variable combinations.

SECTION 4–5

1. (a) $A + AB + A\overline{B}C = A$ **(b)** $(\overline{A} + B)C + ABC = C(\overline{A} + B)$

(c) $A\overline{B}C(BD + CDE) + A\overline{C} = A(\overline{C} + \overline{B}DE)$

2. (a) *Original:* 2 AND gates, 1 OR gate, 1 inverter; *Simplified:* No gates (straight connection)

(b) *Original:* 2 OR gates, 2 AND gates, 1 inverter; *Simplified:* 1 OR gate, 1 AND gate, 1 inverter

(c) *Original:* 5 AND gates, 2 OR gates, 2 inverters; *Simplified:* 2 AND gates, 1 OR gate, 2 inverters

SECTION 4–6

1. (a) SOP   **(b)** standard POS   **(c)** standard SOP   **(d)** POS

2. (a) $AB\overline{C}\,\overline{D} + AB\overline{C}D + ABC\overline{D} + ABCD + \overline{A}B\overline{C}D + \overline{A}BCD + \overline{A}\,\overline{B}C\overline{D} + \overline{A}BC\overline{D}$

(c) Already standard

3. (b) Already standard

(d) $(A + \overline{B} + \overline{C})(A + \overline{B} + C)(A + B + \overline{C})(A + B + C)$

SECTION 4–7

1. $2^5 = 32$ **2.** $0110 \longrightarrow \overline{W}XY\overline{Z}$ **3.** $1100 \longrightarrow \overline{W} + \overline{X} + Y + Z$

SECTION 4–8

1. (a) upper left cell: 000   **(b)** lower right cell: 101   **(c)** lower left cell: 100

(d) upper right cell: 001

2. (a) upper left cell: $\overline{X}\,\overline{Y}\,\overline{Z}$ **(b)** lower right cell: $X\overline{Y}Z$ **(c)** lower left cell: $X\overline{Y}\,\overline{Z}$

(d) upper right cell: $\overline{X}\,\overline{Y}Z$

3. (a) upper left cell: 0000   **(b)** lower right cell: 1010   **(c)** lower left cell: 1000

(d) upper right cell: 0010

4. (a) upper left cell: $\overline{W}\,\overline{X}\,\overline{Y}\,\overline{Z}$ **(b)** lower right cell: $W\overline{X}Y\overline{Z}$ **(c)** lower left cell: $W\overline{X}\,\overline{Y}\,\overline{Z}$

(d) upper right cell: $\overline{W}\,\overline{X}Y\overline{Z}$

SECTION 4–9

1. 8-cell map for 3 variables; 16-cell map for 4 variables
2. $AB + B\overline{C} + \overline{A}\,\overline{B}C$
3. (a) $\overline{A}\,\overline{B}\,\overline{C} + \overline{A}BC + ABC + AB\overline{C}$
   (b) $\overline{A}\,\overline{B}\,\overline{C} + \overline{A}\,\overline{B}C + \overline{A}B\overline{C} + ABC + A\overline{B}\,\overline{C} + A\overline{B}C$
   (c) $\overline{A}\,\overline{B}\,\overline{C}\,\overline{D} + \overline{A}\,\overline{B}\overline{C}D + \overline{A}B\overline{C}\,\overline{D} + \overline{A}B\overline{C}D + \overline{A}BCD + A\overline{B}\,\overline{C}D + A\overline{B}\,\overline{C}D + A\overline{B}CD$
   (d) $\overline{A}\,\overline{B}\,\overline{C}\,\overline{D} + \overline{A}\,\overline{B}C\overline{D} + A\overline{B}C\overline{D} + A\overline{B}\,\overline{C}\,\overline{D} + \overline{A}\,\overline{B}\overline{C}D + AB\overline{C}D + A\overline{B}CD + \overline{A}\,\overline{B}CD + \overline{A}BC\overline{D} + ABC\overline{D} + A\overline{B}CD$

SECTION 4–10

1. In mapping a POS expression, 0s are placed in cells whose value makes the standard sum term zero; and in mapping an SOP expression 1s are placed in cells having the same values as the product terms.
2. 0 in the 1011 cell: $\overline{A} + B + \overline{C} + \overline{D}$ 3. 1 in the 0010 cell: $\overline{A}\,\overline{B}C\overline{D}$

SECTION 4–11

1. There are 32 combinations of 5 variables ($2^5 = 32$).
2. $X = 1$ because the function is 1 for all possible combinations of 5 variables.

SECTION 4–12

1. "Don't cares" simplify a logic expression.
2. You should have found seven duplicate terms in the segment logic.
3. Eight AND gates and eleven inverters were eliminated.

■ **ANSWERS TO RELATED PROBLEMS FOR EXAMPLES**

4–1 $\overline{A} + B = 0$ when $A = 1$ and $B = 0$.   **4–2** $\overline{A}B = 1$ when $A = 0$ and $B = 0$.   **4–3** XYZ

4–4 $W + X + Y + Z$ **4–5** $ABC\overline{D}E$ **4–6** $(A + \overline{B} + \overline{C}D)\overline{E}$

4–7 $\overline{ABCD} = \overline{A} + \overline{B} + \overline{C} + \overline{D}$ **4–8** $A\overline{B}$ **4–9** CD **4–10** $AB\overline{C} + \overline{A}C + \overline{A}B$

4–11 $\overline{A} + \overline{B} + \overline{C}$ **4–12** $\overline{A}B\overline{C} + AB + A\overline{C} + A\overline{B} + \overline{B}C$

4–13 $W\overline{X}YZ + W\overline{X}Y\overline{Z} + W\overline{X}Y\overline{Z} + \overline{W}\,\overline{X}Y\overline{Z} + WX\overline{Y}Z + WX\overline{Y}Z$

4–14 011, 101, 110, 010, 111. Yes

4–15 $(A + \overline{B} + C)(A + \overline{B} + \overline{C})(A + B + C)(\overline{A} + B + C)$

4–16 010, 100, 001, 111, 011. Yes   **4–17** SOP and POS expressions are equivalent.

4–18 See Table 4–11.   **4–19** See Table 4–12.

TABLE 4–11

A	B	C	X
0	0	0	0
0	0	1	0
0	1	0	1
0	1	1	0
1	0	0	0
1	0	1	1
1	1	0	0
1	1	1	0

TABLE 4–12

A	B	C	X
0	0	0	1
0	0	1	0
0	1	0	0
0	1	1	1
1	0	0	1
1	0	1	1
1	1	0	1
1	1	1	0

4–20 The SOP and POS expressions are equivalent.   **4–21** See Figure 4–59.
4–22 See Figure 4–60.   **4–23** See Figure 4–61.   **4–24** See Figure 4–62.

FIGURE 4–59

FIGURE 4–60

FIGURE 4–61

FIGURE 4–62

4–25 No other ways   **4–26** $X = B + \overline{A}C + A\overline{C}D + C\overline{D}$ **4–27** $X = \overline{D} + A\overline{B}C + B\overline{C} + \overline{A}B$
4–28 $Q = X + Y$ **4–29** $Q = \overline{X}\,\overline{Y}\,\overline{Z} + W\overline{X}Z + \overline{W}YZ$ **4–30** See Figure 4–63.

FIGURE 4–63

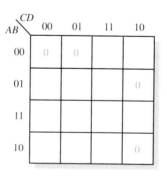

4–31 $Q = (X + \overline{Y})(X + \overline{Z})(\overline{X} + Y + Z)$
4–32 $Q = (\overline{X} + \overline{Y} + Z)(\overline{W} + \overline{X} + Z)(W + X + Y + Z)(W + \overline{X} + Y + \overline{Z})$
4–33 $Q = \overline{Y}\,\overline{Z} + \overline{X}\,\overline{Z} + \overline{W}Y + \overline{X}\,\overline{Y}Z$
4–34 $Y = \overline{D}\,\overline{E} + \overline{A}\,\overline{E} + \overline{B}\,\overline{C}\,\overline{E}$

■ **ANSWERS**
TO SELF-TEST

1. (d)   **2.** (a)   **3.** (b)   **4.** (c)   **5.** (d)   **6.** (b)   **7.** (a)   **8.** (b)
9. (d)   **10.** (d)   **11.** (c)   **12.** (b)   **13.** (b)   **14.** (c)   **15.** (a)   **16.** (c)
17. (a)   **18.** (b)

5

COMBINATIONAL LOGIC

 Computer Simulation
of Digital Circuits:
Electronics Workbench (EWB) Tutorial at:
http://www.prenhall.com/floyd

■ CHAPTER OBJECTIVES

❑ Analyze basic combinational logic circuits, such as AND-OR, AND-OR-Invert, exclusive-OR, exclusive-NOR, and other general combinational networks

❑ Use AND-OR and AND-OR-Invert circuits to implement sum-of-products (SOP) and product-of-sums (POS) expressions

❑ Write the Boolean output expression for any combinational logic circuit

❑ Develop a truth table from the output expression for a combinational logic circuit

❑ Use the Karnaugh map to expand an output expressions containing suppressed variable terms into a full SOP form

❑ Design a combinational logic circuit for a given Boolean output expression

❑ Design a combinational logic circuit for a given truth table

❑ Simplify a combinational logic circuit to its minimum form

❑ Use NAND gates to implement any combinational logic function

❑ Use NOR gates to implement any combinational logic function

❑ Troubleshoot faulty logic circuits

❑ Troubleshoot logic circuits by using signal tracing and waveform analysis

■ CHAPTER OVERVIEW

In Chapters 3 and 4, logic gates were discussed on an individual basis and in simple combinations. You were introduced to SOP and POS implementations, which are basic forms of combinational logic. When logic gates are connected together to produce a specified output for certain specified combinations of input variables, with no storage involved, the resulting circuit is in the category of **combinational logic.** In combinational logic, the output level is at all times dependent on the combination of input levels. This chapter expands on the material introduced in earlier chapters with a coverage of the analysis, design, and troubleshooting of various combinational logic circuits.

 SPECIFIC DEVICES

74HC58   74LS51   74LS54

■ DIGITAL SYSTEM APPLICATION

This Digital System Application illustrates the concepts taught in this chapter by demonstrating how combinational logic can be used for a specific purpose in a practical application. In Section 5–7, a digital circuit is used to control a lumber-processing operation in a furniture factory. The digital circuit operates from four switch inputs to control four motors that run the system. Based on the switch settings, the logic circuit determines which motors should be on and which should be off and shuts the system down if the switches are set improperly.

Motors

Input switches
and
motor interface

5–1 ■ BASIC COMBINATIONAL LOGIC CIRCUITS

In Chapter 4, you learned that SOP expressions are implemented with an AND gate for each product term and one OR gate for summing all of the product terms. This SOP implementation is called AND-OR logic and is the basic form for realizing standard Boolean functions. In this section, the AND-OR and the AND-OR-Invert are examined; and the exclusive-OR and exclusive-NOR gates, which are actually a form of AND-OR logic, are covered.

After completing this section, you should be able to

- Analyze and apply AND-OR circuits
- Analyze and apply AND-OR-Invert circuits
- Analyze and apply exclusive-OR gates
- Analyze and apply exclusive-NOR gates

AND-OR Logic

Figure 5–1(a) shows an AND-OR circuit consisting of two 2-input AND gates and one 2-input OR gate; Figure 5–1(b) is the ANSI standard rectangular outline symbol. The Boolean expressions for the AND gate outputs and the resulting SOP expression for the output X are shown on the diagram. In general, an AND-OR circuit can have any number of AND gates each with any number of inputs.

The truth table for the 4-input AND-OR logic circuit is shown in Table 5–1. The intermediate AND gate outputs (the AB and CD columns) are also shown in the table.

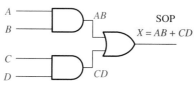

(a) Logic diagram (ANSI standard distinctive shape symbols)

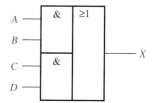

(b) ANSI standard rectangular outline symbol

FIGURE 5–1
An example of AND-OR logic.

TABLE 5–1
Truth table for the AND-OR logic in Figure 5–1.

	Inputs					Output
A	B	C	D	AB	CD	X
0	0	0	0	0	0	0
0	0	0	1	0	0	0
0	0	1	0	0	0	0
0	0	1	1	0	1	1
0	1	0	0	0	0	0
0	1	0	1	0	0	0
0	1	1	0	0	0	0
0	1	1	1	0	1	1
1	0	0	0	0	0	0
1	0	0	1	0	0	0
1	0	1	0	0	0	0
1	0	1	1	0	1	1
1	1	0	0	1	0	1
1	1	0	1	1	0	1
1	1	1	0	1	0	1
1	1	1	1	1	1	1

An AND-OR circuit directly implements an SOP expression, assuming the comple-ments (if any) of the variables are available. The logical operation of the AND-OR circuit in Figure 5–1 is stated as follows:

For a 4-input AND-OR logic circuit, the output *X* is HIGH (1) if both input *A* and input *B* are HIGH (1) or both input *C* and input *D* are HIGH (1).

EXAMPLE 5–1

In a certain chemical-processing plant, a liquid chemical is used in a manufacturing process. The chemical is stored in three different tanks. A level sensor in each tank produces a HIGH voltage when the level of chemical in the tank drops below a specified point.

Design a circuit that monitors the chemical level in each tank and indicates when the level in any two of the tanks drops below the specified point.

Solution   The AND-OR circuit in Figure 5–2 has inputs from the sensors on tanks *A*, *B*, and *C* as shown. The AND gate G_1 checks the levels in tanks *A* and *B*, gate G_2 checks tanks *A* and *C*, and gate G_3 checks tanks *B* and *C*. When the chemical level in any two of the tanks gets too low, one of the AND gates will have HIGHs on both of its inputs causing its output to be HIGH, and so the final output *X* from the OR gate is HIGH. This HIGH input is then used to activate an indicator such as a lamp or audible alarm, as shown in the figure.

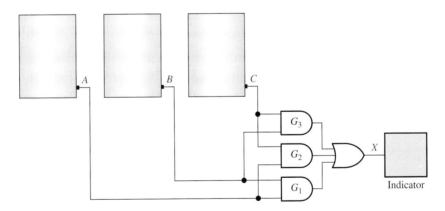

FIGURE 5–2

Related Problem   Write the Boolean SOP expression for the AND-OR logic in Figure 5–2.

  A Specific AND-OR Integrated Circuit   The 74HC58 is an example of AND-OR logic in an IC. This is a CMOS device that includes two separate AND-OR circuits in a single package. One circuit has two 2-input AND gates and the other circuit has two 3-input AND gates, as shown in Figure 5–3(a). Pin numbers for dual-in-line or SOIC packages are shown. The ANSI standard rectangular outline symbol is given in part (b).

(a) Logic diagram   (b) Rectangular outline symbol

FIGURE 5–3
The 74HC58 dual AND-OR.

AND-OR-Invert Logic

When the output of an AND-OR circuit is complemented (inverted), it results in an AND-OR-Invert circuit. Recall that AND-OR logic directly implements SOP expressions. POS expressions can be implemented with AND-OR-Invert logic. This is illustrated as follows starting with a POS expression and developing the corresponding AND-OR-Invert expression.

$$X = (\overline{A} + \overline{B})(\overline{C} + \overline{D}) = (\overline{AB})(\overline{CD}) = \overline{(\overline{AB})(\overline{CD})} = \overline{\overline{AB}} + \overline{\overline{CD}} = \overline{AB + CD}$$

The logic diagram in Figure 5–4(a) shows an AND-OR-Invert circuit and the development of the POS output expression. The ANSI standard rectangular outline symbol is shown in part (b). In general, an AND-OR Invert circuit can have any number of AND gates each with any number of inputs.

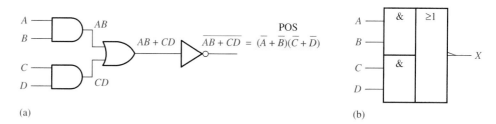

(a)   (b)

FIGURE 5–4
An AND-OR-Invert circuit produces a POS output.

The logical operation of the AND-OR-Invert circuit in Figure 5–4 is stated as follows:

> **For a 4-input AND-OR-Invert logic circuit, the output *X* is LOW (0) if both input *A* and input *B* are HIGH (1) or both input *C* and input *D* are HIGH (1).**

A truth table can be developed from the AND-OR truth table in Table 5–1 by simply changing all 1s to 0s and all 0s to 1s in the output column.

EXAMPLE 5–2

The sensors in the chemical tanks of Example 5–1 are being replaced by a new model that produces a LOW voltage instead of a HIGH voltage when the level of the chemical in the tank drops below a critical point.

Modify the circuit in Figure 5–2 to operate with the different input levels and still produce a HIGH output to activate the indicator when the level in any two of the tanks drops below the critical point. Show the logic diagram.

Solution The AND-OR-Invert circuit in Figure 5–5 has inputs from the sensors on tanks A, B, and C as shown. The AND gate G_1 checks the levels in tanks A and B, gate G_2 checks tanks A and C, and gate G_3 checks tanks B and C. When the chemical level in any two of the tanks gets too low, each AND gate will have a LOW on at least one input causing its output to be LOW and, thus, the final output X from the inverter is HIGH. This HIGH output is then used to activate an indicator.

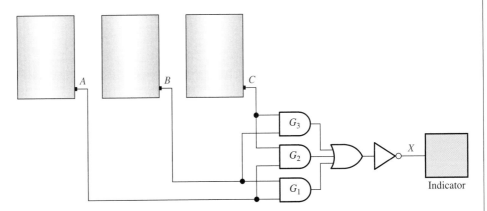

FIGURE 5–5

Related Problem Write the Boolean expression for the AND-OR-Invert logic in Figure 5–5 and show that the output is HIGH (1) when any two of the inputs A, B, and C are LOW (0).

Specific AND-OR-Invert Integrated Circuits The 74LS51 and the 74LS54 are examples of AND-OR-Invert logic. Both are TTL low-power Schottky devices but are available in other logic families as well. The 74LS51 includes two separate AND-OR-Invert circuits in a single package. One circuit has two 2-input AND gates and the other circuit has two 3-input AND gates as shown in Figure 5–6(a). The 74LS54 is a single AND-OR-Invert circuit that has two 2-input AND gates and two 3-input AND gates as shown in Figure 5–6(b). Notice that the inversion is indicated by a bubble on the output of the OR gates in each case, showing that the OR-Invert part of the circuit is effectively a NOR gate.

Exclusive-OR Logic

The exclusive-OR gate and the 74XX86 IC were introduced in Chapter 3. Although, because of its importance, this circuit is considered a type of logic gate with its own unique symbol, it is actually a combination of two AND gates, one OR gate, and two inverters as shown in Figure 5–7(a). The two ANSI standard logic symbols are shown in parts (b) and (c).

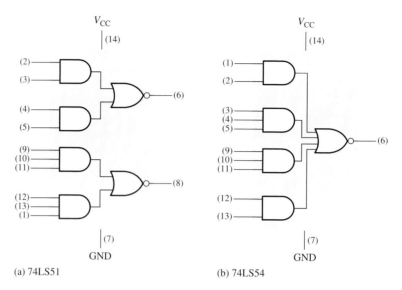

FIGURE 5–6
The 74LS51 dual 2-wide AND-OR-Invert and the 74LS54 single 4-wide AND-OR-Invert. The term 2-wide means there are two AND gates per device and 4-wide means there are four AND gates per device.

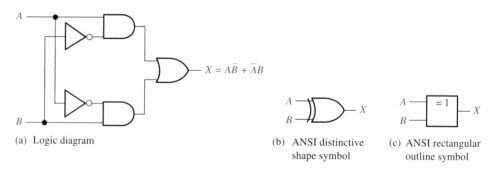

(a) Logic diagram

(b) ANSI distinctive shape symbol

(c) ANSI rectangular outline symbol

FIGURE 5–7
Exclusive-OR logic diagram and symbols.

The output expression for the circuit in Figure 5–7 is

$$X = A\overline{B} + \overline{A}B$$

Evaluation of this expression results in the truth table in Table 5–2. Notice that the output is HIGH only when the two inputs are at opposite levels. A special exclusive-OR operator $\oplus$ is often used, so the expression $X = A\overline{B} + \overline{A}B$ can be stated as "X is equal to A exclusive-OR B" and can be written as

$$X = A \oplus B$$

TABLE 5–2
Truth table for an exclusive-OR.

A	B	X
0	0	0
0	1	1
1	0	1
1	1	0

Exclusive-NOR Logic

As you know, the complement of the exclusive-OR function is the exclusive-NOR, which is derived as follows:

$$X = \overline{A\overline{B} + \overline{A}B} = \overline{(A\overline{B})(\overline{A}B)} = (\overline{A} + B)(A + \overline{B}) = \overline{A}\,\overline{B} + AB$$

Notice that the output X is HIGH only when the two inputs, A and B, are at the same level.

The exclusive-NOR can be implemented by simply inverting the output of an exclusive-OR, as shown in Figure 5–8(a), or by directly implementing the expression $\overline{A}\,\overline{B} + AB$, as shown in part (b).

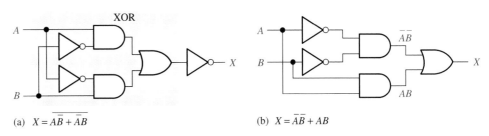

(a) $X = A\overline{B} + \overline{A}B$ (b) $X = \overline{A}\,\overline{B} + AB$

FIGURE 5–8
Two equivalent ways of implementing the exclusive-NOR.

SECTION 5–1 REVIEW

1. Determine the output (1 or 0) of a 4-variable AND-OR-Invert circuit for each of the following input conditions:
   (a) $A = 1, B = 0, C = 1, D = 0$ **(b)** $A = 1, B = 1, C = 0, D = 1$
   (c) $A = 0, B = 1, C = 1, D = 1$
2. Determine the output (1 or 0) of an exclusive-OR gate for each of the following input conditions:
   (a) $A = 1, B = 0$ **(b)** $A = 1, B = 1$
   (c) $A = 0, B = 1$ **(d)** $A = 0, B = 0$
3. Develop the truth table for a certain 3-input logic circuit with the output expression $X = A\overline{B}C + \overline{A}BC + \overline{A}\,\overline{B}\,\overline{C} + AB\overline{C} + ABC$.
4. Draw the logic diagram for an exclusive-NOR circuit.

5–2 ■ IMPLEMENTING COMBINATIONAL LOGIC

In this section, examples are used to illustrate how to implement a logic circuit from a Boolean expression or a truth table. Minimization of a logic circuit using the methods covered in Chapter 4 is also discussed.

After completing this section, you should be able to

■ Implement a logic circuit from a Boolean expression   ■ Implement a logic circuit from a truth table   ■ Minimize a logic circuit

From a Boolean Expression to a Logic Circuit

Let's examine the following Boolean expression:

$$X = AB + CDE$$

A brief inspection shows that this expression is composed of two terms, AB and CDE, with a domain of five variables. The first term is formed by ANDing A with B, and the second term is formed by ANDing C, D, and E. The two terms are then ORed to form the output X. These operations are indicated in the structure of the expression as follows:

$$X = \quad AB \quad + \quad CDE$$

Note that in this particular expression, the AND operations forming the two individual terms, AB and CDE, must be performed *before* the terms can be ORed.

To implement this Boolean expression, a 2-input AND gate is required to form the term AB, and a 3-input AND gate is needed to form the term CDE. A 2-input OR gate is then required to combine the two AND terms. The resulting logic circuit is shown in Figure 5–9.

FIGURE 5–9
Logic circuit for X = AB + CDE.

As another example, let's implement the following expression:

$$X = AB(C\overline{D} + EF)$$

A breakdown of this expression shows that the terms A, B, and $C\overline{D} + EF$ are ANDed. The term $C\overline{D} + EF$ is formed by first ANDing C and $\overline{D}$ and ANDing E and F, and then ORing these two terms. This structure is indicated in relation to the expression as follows:

$$X = AB(C\overline{D} + EF)$$

AND
NOT
OR
AND

Before the expression can be formed, you must have the term $C\overline{D} + EF$; but before you can get this term, you must have the terms $C\overline{D}$ and EF; but before you can get the term $C\overline{D}$, you must have $\overline{D}$. So, as you can see, the logic operations must be done in the proper order.

The logic gates required to implement $X = AB(C\overline{D} + EF)$ are as follows:

1. One inverter to form $\overline{D}$
2. Two 2-input AND gates to form $C\overline{D}$ and EF
3. One 2-input OR gate to form $C\overline{D} + EF$
4. One 3-input AND gate to form X

The logic circuit for this expression is shown in Figure 5–10(a). Notice that there is a maximum of three gates and an inverter between an input and output in this circuit (from input D to output). Often the total propagation delay time through a logic circuit is a major consideration. Propagation delays are additive, so the more gates or inverters between input and output, the greater the propagation delay time.

COMPUTER NOTE

Many control programs require logic operations to be performed by a computer. A driver program is a control program that is used with computer peripherals. For example, a mouse driver requires logic tests to determine if a button has been pressed and further logic operations to determine if it has moved, either horizontally or vertically. Within the heart of a microprocessor is the arithmetic logic unit (ALU) which performs these logic operations as directed by program instructions. All of the logic described in this chapter can also be performed by the ALU, given the proper instructions.

(a)

(b) Sum-of-products implementation of the circuit in part (a)

FIGURE 5–10
Logic circuits for X = AB(CD̄ + EF) = ABCD̄ + ABEF.

Unless an intermediate term, such as $C\overline{D} + EF$ in Figure 5–10(a), is required as an output for some other purpose, it is usually best to reduce a circuit to its SOP form. The expression is converted to SOP as follows, and the resulting circuit is shown in Figure 5–10(b).

$$AB(C\overline{D} + EF) = ABC\overline{D} + ABEF$$

From a Truth Table to a Logic Circuit

If you begin with a truth table instead of an expression, you can write the SOP expression from the truth table and then implement the logic circuit. Table 5–3 specifies a logic function.

TABLE 5–3

	Inputs		Output	
A	**B**	**C**	**X**	
0	0	0	0	
0	0	1	0	
0	1	0	0	
0	1	1	1	$\longrightarrow \overline{A}BC$
1	0	0	1	$\longrightarrow A\overline{B}\,\overline{C}$
1	0	1	0	
1	1	0	0	
1	1	1	0	

The Boolean SOP expression obtained from the truth table by ORing the product terms for which $X = 1$ (the blue rows) is

$$X = \overline{A}BC + A\overline{B}\,\overline{C}$$

The first term in the expression is formed by ANDing the three variables $\overline{A}$, B, and C. The second term is formed by ANDing the three variables A, $\overline{B}$, and $\overline{C}$.

The logic gates required to implement this expression are as follows: three inverters to form the $\overline{A}$, $\overline{B}$, and $\overline{C}$ variables; two 3-input AND gates to form the terms $\overline{A}BC$ and $A\overline{B}\,\overline{C}$; and one 2-input OR gate to form the final output function, $\overline{A}BC + A\overline{B}\,\overline{C}$.

The implementation of this logic function is illustrated in Figure 5–11.

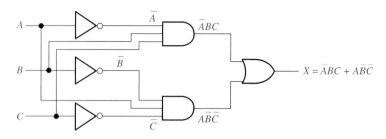

FIGURE 5–11
Logic circuit for $X = \overline{A}BC + A\overline{B}\,\overline{C}$.

EXAMPLE 5–3

Design a logic circuit to implement the operation specified in the truth table of Table 5–4.

TABLE 5–4

Inputs			Output
A	B	C	X
0	0	0	0
0	0	1	0
0	1	0	0
0	1	1	1
1	0	0	0
1	0	1	1
1	1	0	1
1	1	1	0

Solution   Notice that $X = 1$ for only three of the input conditions. Therefore, the logic expression is

$$X = \overline{A}BC + A\overline{B}C + AB\overline{C}$$

The logic gates required are three inverters, three 3-input AND gates and one 3-input OR gate. The logic circuit is shown in Figure 5–12.

FIGURE 5–12

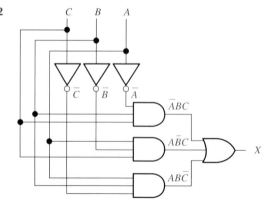

Related Problem   Determine if the logic circuit of Figure 5–12 can be simplified.

EXAMPLE 5–4

Develop a logic circuit with four input variables that will only produce a 1 output when exactly three input variables are 1s.

Solution Out of sixteen possible combinations of four variables, the combinations in which there are exactly three 1s are listed in Table 5–5, along with the corresponding product term for each.

TABLE 5–5

A	B	C	D	
0	1	1	1	$\longrightarrow \overline{A}BCD$
1	0	1	1	$\longrightarrow A\overline{B}CD$
1	1	0	1	$\longrightarrow AB\overline{C}D$
1	1	1	0	$\longrightarrow ABC\overline{D}$

The product terms are ORed to get the following expression:

$$X = \overline{A}BCD + A\overline{B}CD + AB\overline{C}D + ABC\overline{D}$$

This expression is implemented in Figure 5–13 with AND-OR logic.

FIGURE 5–13

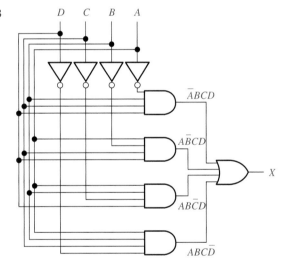

Related Problem Determine if the logic circuit of Figure 5–13 can be simplified.

EXAMPLE 5–5

Reduce the combinational logic circuit in Figure 5–14 to a minimum form.

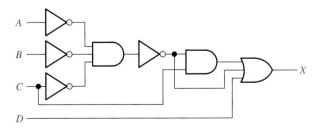

FIGURE 5–14

Solution The expression for the output of the circuit is

$$X = (\overline{\overline{A}\,\overline{BC}})C + \overline{\overline{A}\,\overline{BC}} + D$$

Applying DeMorgan's theorem and Boolean algebra,

$$X = (\overline{\overline{A}} + \overline{\overline{B}} + \overline{\overline{C}})C + \overline{\overline{A}} + \overline{\overline{B}} + \overline{\overline{C}} + D$$
$$= AC + BC + CC + A + B + C + D$$
$$= AC + BC + C + A + B + \cancel{C} + D$$
$$= C(A + B + 1) + A + B + D$$
$$X = A + B + C + D$$

The simplified circuit is a 4-input OR gate as shown in Figure 5–15.

FIGURE 5–15

A
B
C
D
— X

Related Problem Verify the minimized expression $A + B + C + D$ using a Karnaugh map.

EXAMPLE 5–6

Minimize the combinational logic circuit in Figure 5–16. Inverters for the complemented variables are not shown.

FIGURE 5–16

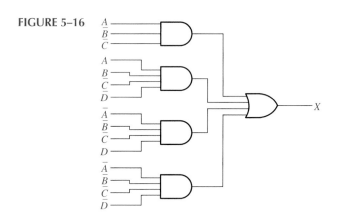

Solution The output expression is

$$X = A\overline{B}\,\overline{C} + AB\overline{C}\,\overline{D} + \overline{A}\,\overline{B}\,C\overline{D} + \overline{A}\,B\overline{C}\,\overline{D}$$

Expanding the first term to include the missing variables D and $\overline{D}$,

$$X = A\overline{B}\,\overline{C}(D + \overline{D}) + AB\overline{C}\,\overline{D} + \overline{A}\,\overline{B}\,C\overline{D} + \overline{A}\,B\overline{C}\,\overline{D}$$
$$= A\overline{B}\,\overline{C}D + A\overline{B}\,\overline{C}\,\overline{D} + AB\overline{C}\,\overline{D} + \overline{A}\,\overline{B}\,C\overline{D} + \overline{A}\,B\overline{C}\,\overline{D}$$

This expanded SOP expression is mapped and simplified on the Karnaugh map in Figure 5–17(a). The simplified implementation is shown in part (b). Inverters are not shown.

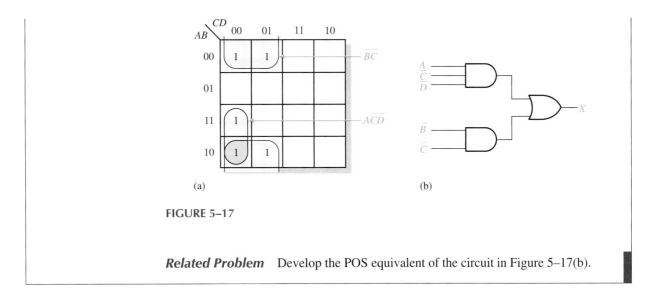

FIGURE 5–17

Related Problem Develop the POS equivalent of the circuit in Figure 5–17(b).

SECTION 5–2 REVIEW

1. Implement the following Boolean expressions as they are stated:
   (a) $X = ABC + AB + AC$ (b) $X = AB(C + DE)$
2. Develop a logic circuit that will produce a 1 on its output only when all three inputs are 1s or when all three inputs are 0s.
3. Reduce the circuits in Question 1 to minimum SOP form.

5–3 ■ THE UNIVERSAL PROPERTY OF NAND AND NOR GATES

Up to this point, combinational circuits implemented with AND gates, OR gates, and inverters have been studied. In this section, the universal property of the NAND gate and the NOR gate is discussed. The universality of the NAND gate means that it can be used as an inverter and that combinations of NAND gates can be used to implement the AND, OR, and NOR operations. Similarly, the NOR gate can be used to implement the inverter, AND, OR, and NAND operations.

After completing this section, you should be able to

■ Use NAND gates to implement the inverter, the AND gate, the OR gate, and the NOR gate ■ Use NOR gates to implement the inverter, the AND gate, the OR gate, and the NAND gate

The NAND Gate as a Universal Logic Element

The NAND gate is a **universal gate** because it can be used to produce the NOT operation, the AND operation, the OR operation, and the NOR operation. An inverter can be made from a NAND gate by connecting all of the inputs together and creating, in effect, a single input, as shown in Figure 5–18(a) for a 2-input gate. An AND operation can be generated by the use of NAND gates alone, as shown in Figure 5–18(b). An OR operation can be produced with NAND gates, as illustrated in part (c). Finally, a NOR operation is produced, as shown in part (d).

(a) A NAND gate used as an inverter

(b) Two NAND gates used as an AND gate

(c) Three NAND gates used as an OR gate

(d) Four NAND gates used as a NOR gate

FIGURE 5–18
Universal application of NAND gates.

In Figure 5–18(b), a NAND gate is used to invert (complement) a NAND output to form the AND operation, as indicated in the following equation:

$$X = \overline{\overline{AB}} = AB$$

In Figure 5–18(c), NAND gates G_1 and G_2 are used to invert the two input variables before they are applied to NAND gate G_3. The final OR output is derived as follows by application of DeMorgan's theorem:

$$X = \overline{\overline{A}\,\overline{B}} = A + B$$

In Figure 5–18(d), NAND gate G_4 is used as an inverter connected to the circuit of part (c) to produce the NOR operation $\overline{A + B}$.

The NOR Gate as a Universal Logic Element

Like the NAND gate, the NOR gate can be used to produce the NOT, AND, OR, and NAND operations. A NOT circuit, or inverter, can be made from a NOR gate by connecting all of the inputs together to effectively create a single input, as shown in Figure 5–19(a) with a 2-input example. Also, an OR gate can be produced from NOR gates, as illustrated in Figure 5–19(b). An AND gate can be constructed by the use of NOR gates, as shown in Figure 5–19(c). In this case the NOR gates G_1 and G_2 are used as inverters, and the final output is derived by the use of DeMorgan's theorem as follows:

$$X = \overline{\overline{A} + \overline{B}} = AB$$

Figure 5–19(d) shows how NOR gates are used to form a NAND operation.

(a) A NOR gate used as an inverter

(b) Two NOR gates used as an OR gate

(c) Three NOR gates used as an AND gate

(d) Four NOR gates used as a NAND gate

FIGURE 5–19
Universal application of NOR gates.

SECTION 5–3 REVIEW

1. Use NAND gates to implement each expression:
   (a) $X = \overline{A} + B$ (b) $X = A\overline{B}$
2. Use NOR gates to implement each expression:
   (a) $X = \overline{A} + B$ (b) $X = A\overline{B}$

5–4 ■ COMBINATIONAL LOGIC USING NAND AND NOR GATES

In this section, you will see how NAND and NOR gates can be used to implement a logic function. Recall from Chapter 3 that the NAND gate also exhibits an equivalent operation called the negative-OR and that the NOR gate exhibits an equivalent operation called the negative-AND. You will see how the use of the appropriate symbols to represent the equivalent operations makes "reading" a logic diagram easier.

After completing this section, you should be able to

■ Use NAND gates to implement a logic function   ■ Use NOR gates to implement a logic function   ■ Use the appropriate dual symbol in a logic diagram.

NAND Logic

As you have learned, a NAND gate can function as either a NAND or a negative-OR because, by DeMorgan's theorem,

$$\overline{AB} = \overline{A} + \overline{B}$$

$$\text{NAND} \underline{\qquad} \uparrow \qquad \uparrow \underline{\qquad} \text{negative-OR}$$

Consider the NAND logic in Figure 5–20. The output expression is developed in the following steps:

$$X = \overline{(\overline{AB})(\overline{CD})}$$

$$= \overline{(\overline{A} + \overline{B})(\overline{C} + \overline{D})}$$

$$= \overline{(\overline{A} + \overline{B})} + \overline{(\overline{C} + \overline{D})}$$

$$= \overline{\overline{A}}\,\overline{\overline{B}} + \overline{\overline{C}}\,\overline{\overline{D}}$$

$$= AB + CD$$

FIGURE 5–20
NAND logic for X = AB + CD.

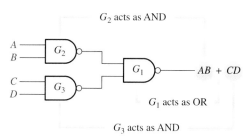

Notice that in the last step of the development of the output expression, $AB + CD$ is in the form of two AND terms ORed together. This form of the expression shows that gates G_2 and G_3 in Figure 5–20 act as AND gates and that gate G_1 acts as an OR gate, as illustrated in Figure 5–21(a). This circuit is redrawn in part (b) with NAND symbols for gates G_2 and G_3 and a negative-OR symbol for gate G_1.

Notice in Figure 5–21(b) the bubble-to-bubble connections between the outputs of gates G_2 and G_3 and the inputs of gate G_1. *Since a bubble represents an inversion, two*

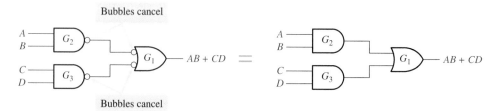

(a) Original NAND logic diagram showing effective
gate operation relative to the output expression

(b) Equivalent NAND/Negative-OR logic diagram   (c) AND-OR equivalent

FIGURE 5–21
Development of the AND-OR equivalent of the circuit in Figure 5–20.

connected bubbles represent a double inversion and therefore cancel each other. This inversion cancellation can be seen in the previous development of the output expression $AB + CD$ and is indicated by the absence of barred terms in the output expression. Thus, the circuit in Figure 5–21(b) is *effectively* an AND-OR circuit as shown in Figure 5–21(c).

NAND Logic Diagrams

All logic diagrams using NAND gates should be drawn with each gate represented by either a NAND symbol or the equivalent negative-OR symbol to reflect the operation of the gate within the logic circuit. The NAND symbol and the negative-OR symbol are called **dual symbols.**

If we begin by representing the output gate with a negative-OR symbol, then we will use the NAND symbol for the level of gates right before the output gate and will alternate the symbols for successive levels of gates as we move away from the output. Always use the gate symbols in such a way that every connection between a gate output and a gate input is either bubble-to-bubble or nonbubble-to-nonbubble. A bubble output should not be connected to a nonbubble input or vice versa in a logic diagram.

Figure 5–22 shows an arrangement of gates to illustrate the procedure of using the appropriate dual symbols for a NAND circuit with several gate levels. Although using all NAND symbols as in Figure 5–22(a) is correct and there is nothing wrong with it, the diagram in part (b) is much easier to "read" and is the preferred method. The shape of the gate indicates the way its inputs will appear in the output expression and thus shows how the gate functions within the logic circuit. For a NAND symbol the inputs appear ANDed in the output expression, and for a negative-OR symbol the inputs appear ORed in the output expression, as Figure 5–22(b) illustrates. You can see that the dual-symbol diagram in part (b) makes it much easier to determine the output expression directly from the logic diagram. This is because *each gate symbol indicates the relationship of its input variables as they appear in the output expression.*

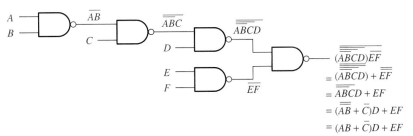

(a) Several Boolean steps are required to arrive at final output expression.

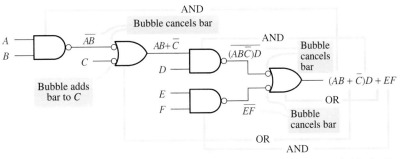

(b) Output expression can be obtained directly from the function of each gate symbol in the diagram.

FIGURE 5–22

Illustration of the use of the appropriate dual symbols in a NAND logic diagram.

EXAMPLE 5–7

Redraw the logic diagram and develop the output expression for the circuit in Figure 5–23, using the appropriate dual symbols.

FIGURE 5–23

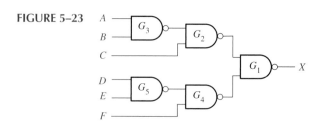

Solution Redraw the logic diagram in Figure 5–23 with the use of equivalent negative-OR symbols as shown in Figure 5–24. Writing the expression for X directly from the indicated logic operation of each gate gives $X = (\overline{A} + \overline{B})C + (\overline{D} + \overline{E})F$.

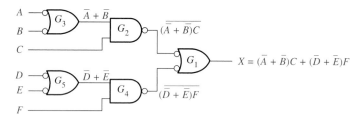

FIGURE 5–24

Related Problem Derive the output expression from Figure 5–23 and show it is equivalent to the expression in the solution.

EXAMPLE 5–8

Implement each expression with NAND logic:
(a) $ABC + DE$ **(b)** $ABC + \overline{D} + \overline{E}$

Solution See Figure 5–25.

FIGURE 5–25

Related Problem Convert the NAND circuits in Figure 5–25(a) and (b) to equivalent AND-OR logic.

NOR Logic

A NOR gate can function as either a NOR or a negative-AND, as shown by DeMorgan's theorem:

$$\overline{A + B} = \overline{A}\,\overline{B}$$

NOR ———↑   ↑——— negative-AND

FIGURE 5–26
NOR logic for X = (A + B)(C + D).

A — G_2
B —
C — G_3
D —
G_1 — $X = (A + B)(C + D)$

Consider the NOR logic in Figure 5–26. The output expression is developed as follows:

$$X = \overline{\overline{A + B} + \overline{C + D}} = (\overline{\overline{A + B}})(\overline{\overline{C + D}}) = (A + B)(C + D)$$

Notice that the expression $(A + B)(C + D)$ consists of two OR terms ANDed together. This shows that gates G_2 and G_3 act as OR gates and gate G_1 acts as an AND gate, as illustrated in Figure 5–27(a). This circuit is redrawn in part (b) with a negative-AND symbol for gate G_1.

As with NAND logic, the purpose for using the dual symbols is to make the logic diagram easier to read and analyze, as illustrated in the NOR logic circuit in Figure 5–28. When the circuit in part (a) is redrawn with dual symbols in part (b), notice that all

(a)   (b)

FIGURE 5–27

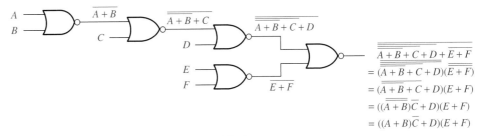

(a) Final output expression is obtained after several Boolean steps.

(b) Output expression can be obtained directly from the function of each gate symbol in the diagram.

FIGURE 5–28
Illustration of the use of the appropriate dual symbols in a NOR logic diagram.

output-to-input connections between gates are bubble-to-bubble or nonbubble-to-nonbubble. Again, you can see that the shape of each gate symbol indicates the type of term (AND or OR) that it produces in the output expression, thus making the output expression easier to determine and the logic diagram easier to analyze.

EXAMPLE 5–9

Using appropriate dual symbols, redraw the logic diagram and develop the output expression for the circuit in Figure 5–29.

FIGURE 5–29

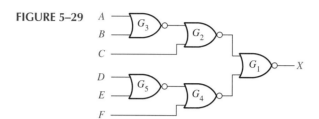

Solution Redraw the logic diagram with the equivalent negative-AND symbols as shown in Figure 5–30. Writing the expression for X directly from the indicated logic operation of each gate,

$$X = (\overline{A}\,\overline{B} + C)(\overline{D}\,\overline{E} + F)$$

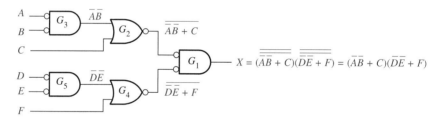

FIGURE 5–30

Related Problem Prove that the output of the NOR circuit in Figure 5–29 is the same as for the circuit in Figure 5–30.

SECTION 5–4 REVIEW

1. Implement the expression $X = \overline{(\overline{A} + \overline{B} + \overline{C})DE}$ by using NAND logic.
2. Implement the expression $X = \overline{\overline{A}\,\overline{B}\,\overline{C} + (D + E)}$ with NOR logic.

5–5 ■ OPERATION WITH PULSE WAVEFORMS

Several examples of general combinational logic circuits with pulse waveform inputs are examined in this section. Keep in mind that the logical operation of each gate is the same for pulse inputs as for constant-level inputs. The output of a logic circuit at any given time depends on the inputs at that particular time, so the relationship of the time-varying inputs is of primary importance.

After completing this section, you should be able to

■ Analyze combinational logic circuits with pulse waveform inputs ■ Develop a timing diagram for any given combinational logic circuit with specified inputs

The logical operation of any gate is the same regardless of whether its inputs are pulsed or constant levels. The nature of the inputs (pulsed or constant levels) does not alter the truth table of a circuit. The examples in this section illustrate the analysis of combinational logic circuits with pulsed inputs.

The following is a review of the logical operation of individual gates for use in analyzing combinational circuits with pulse waveform inputs:

1. The output of an AND gate is HIGH only when all inputs are HIGH at the same time.
2. The output of an OR gate is HIGH only when at least one of its inputs is HIGH.
3. The output of a NAND gate is LOW only when all inputs are HIGH at the same time.
4. The output of a NOR gate is LOW only when at least one of its inputs is HIGH.

EXAMPLE 5–10

Determine the final output waveform X for the circuit in Figure 5–31, with input waveforms A, B, and C as shown.

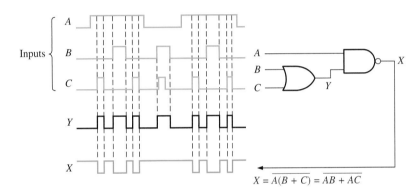

$$X = \overline{A(B + C)} = \overline{AB + AC}$$

FIGURE 5–31

Solution The output expression, $\overline{AB + AC}$, indicates that the output X is LOW when both A and B are HIGH or when both A and C are HIGH or when all inputs are HIGH. The output waveform X is shown in the timing diagram of Figure 5–31. The intermediate waveform Y at the output of the OR gate is also shown.

Related Problem Determine the output waveform if input A is a constant HIGH level.

EXAMPLE 5–11

Show the timing diagram for the circuit in Figure 5–32 showing the outputs of G_1, G_2, and G_3 with the input waveforms, A and B, as indicated.

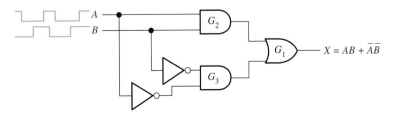

$X = AB + \overline{A}\overline{B}$

FIGURE 5–32

Solution When both inputs are HIGH or when both inputs are LOW, the output X is HIGH as shown in Figure 5–33. Notice that this is an exclusive-NOR circuit. The intermediate outputs of gates G_2 and G_3 are also shown in Figure 5–33.

FIGURE 5–33

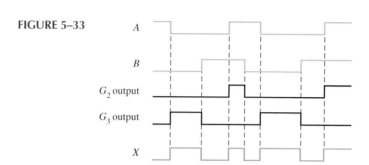

Related Problem Determine the output X in Figure 5–32 if input B is inverted.

EXAMPLE 5–12

Determine the output waveform X for the logic circuit in Figure 5–34(a) by first finding the intermediate waveform at each of points Y_1, Y_2, Y_3, and Y_4. The input waveforms are shown in Figure 5–34(b).

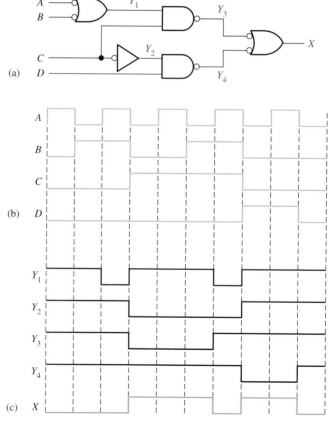

FIGURE 5–34

Solution All the intermediate waveforms and the final output waveform are shown in the timing diagram of Figure 5–34(c).

Related Problem Determine the waveforms Y_1 Y_2, Y_3, Y_4 and X if input waveform A is inverted.

EXAMPLE 5–13

Determine the output waveform X for the circuit in Example 5–12, Figure 5–34(a), directly from the output expression.

Solution The output expression for the circuit is developed in Figure 5–35. The SOP form indicates that the output is HIGH when A is LOW and C is HIGH or when B is LOW and C is HIGH or when C is LOW and D is HIGH.

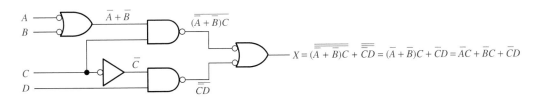

FIGURE 5–35

The result is shown in Figure 5–36 and is the same as the one obtained by the intermediate-waveform method in Example 5–12. The corresponding product terms for each waveform condition that results in a HIGH output are indicated.

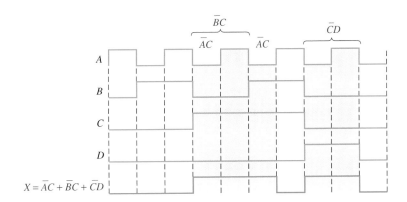

FIGURE 5–36

Related Problem Repeat this example if all the input waveforms are inverted.

**SECTION 5–5
REVIEW**

1. One pulse with $t_W = 50\ \mu s$ is applied to one of the inputs of an exclusive-OR circuit. A second positive pulse with $t_W = 10\ \mu s$ is applied to the other input beginning 15 μs after the leading edge of the first pulse. Show the output in relation to the inputs.
2. The pulse waveforms A and B in Figure 5–31 are applied to the exclusive-NOR circuit in Figure 5–32. Develop a complete timing diagram.

5–6 ■ TROUBLESHOOTING

The preceding sections have given you some insight into the operation of combinational logic circuits and the relationships of inputs and outputs. This type of understanding is essential when you troubleshoot digital circuits because you must know what logic levels or waveforms to look for throughout the circuit for a given set of input conditions.

In this section, the logic pulser, the probe, and the current tracer are used to troubleshoot a logic circuit when a gate output is connected to several gate inputs. Also, an example of signal tracing and waveform analysis methods is presented using a logic analyzer for locating a fault in a combinational logic circuit.

After completing this section, you should be able to

■ Define a circuit node ■ Use a logic probe to find a faulty circuit node ■ Use a pulser and probe to find an open gate output ■ Use a pulser and current tracer to find a shorted gate input or output ■ Use an oscilloscope or a logic analyzer for signal tracing in a combinational logic circuit

In a combinational logic circuit, the output of one gate may be connected to two or more gate inputs as shown in Figure 5–37. The interconnecting paths share a common electrical point known as a **node.**

FIGURE 5–37
Illustration of a node in a logic circuit.

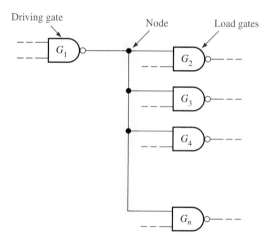

Gate G_1 in Figure 5–37 is driving the node, and the other gates represent loads connected to the node. A driving gate can drive a number of load gate inputs up to its specified fan-out. Several types of failures are possible in this situation. Some of these failure modes are difficult to isolate to a single bad gate because all the gates connected to the node are affected. Common types of failures are the following:

1. *Open output in driving gate.* This failure will cause a loss of signal to all load gates.
2. *Open input in a load gate.* This failure will not affect the operation of any of the other gates connected to the node, but it will result in loss of signal output from the faulty gate.
3. *Shorted output in driving gate.* This failure can cause the node to be stuck in the LOW state (short to ground) or in the HIGH state (short to V_{CC}).
4. *Shorted input in a load gate.* This failure can also cause the node to be stuck in the LOW state (short to ground) or in the HIGH state (short to V_{CC}).

When troubleshooting logic circuits, begin with a visual check, looking for obvious problems. In addition to components, visual inspection should include connectors. Edge connectors are frequently used to bring power, ground, and signals to a circuit board. The mating surfaces of the connector need to be clean and have a good mechanical fit. A dirty connector can cause intermittent or complete failure of the circuit. Edge connectors can be cleaned with a common pencil eraser and wiped clean with a Q-tip soaked in alcohol. Also, all connectors should be checked for loose-fitting pins.

Troubleshooting Common Faults

Open Output in Driving Gate   In this situation there is no pulse activity on the node. With circuit power on, an open node will normally result in a "floating" level and may be indicated by a dim lamp on the logic probe, as illustrated in Figure 5–38.

FIGURE 5–38
Open output in driving gate.

Open Input in a Load Gate   If the check for an open driver output is negative, then a check for an open input in a load gate should be performed. Apply the logic pulser tip to the node with all nonpulsed inputs HIGH. Then check the output of each gate for pulse activity with the logic probe, as illustrated in Figure 5–39. If one of the inputs that is normally connected to the node is open, no pulses will be detected on that gate's output.

Output Shorted to Ground in Driving Gate   This fault can cause the node to be stuck LOW, as previously mentioned. A quick check with a pulser and a logic probe will indicate this, as shown in Figure 5–40(a). A short to ground in the driving gate's output or in any load gate input will cause this symptom, and further checks must therefore be made to isolate the short to a particular gate.

If the driving gate's output is internally shorted to ground, then, essentially, no current activity will be present on any of the connections to the node. Thus, a current tracer will indicate no activity with circuit power on, as illustrated in Figure 5–40(b).

To further verify a shorted output, a pulser and a current tracer can be used with the circuit power off, as shown in Figure 5–40(c). When current pulses are applied to the node

Pin 4 input of this gate in IC2 is open

74AHC00 pin diagram from data sheet

Pulse the node.   Probe   HIGH   Probe   HIGH

HIGH   OFF   Probe

Check the output pin of each gate connected to the node with other gate inputs HIGH. No pulse activity on an output indicates an open gate input or open gate output. A HIGH on the gate inputs are produced by connecting a 1.0 kΩ resistor to V_{CC}.

FIGURE 5–39
Open input in a load gate.

Pulse the gate input with other input HIGH.   Probe indicates LOW level at all points connected to the node.

Output of this gate in IC1 is shorted

HIGH   OFF

(a) Node is "stuck" LOW.

Pulse the gate input with other input HIGH.   Current tracer indicates no current in node interconnections.

HIGH   OFF

(b) No current in any node interconnection

Tracer OFF—no current
Pulse the node.   in any of these paths

HIGH   OFF

Current tracer indicates current only into shorted output.

(c) There is current from point of pulser application directly to short. There is no current in other paths connected to node.

FIGURE 5–40
Shorted output in driving gate.

with the pulser, all of the current will be into the shorted output, and none through the circuit paths into the load gate inputs.

Input Shorted to Ground in a Load Gate If one of the load gate inputs is internally shorted to ground, the node will be stuck in the LOW state. Again, as in the case of a shorted output, the logic pulser and current tracer can be used to isolate the faulty gate.

When the node is pulsed with circuit power off, essentially all the current will be into the shorted input, and tracing its path with the current tracer will lead to the shorted input, as illustrated in Figure 5–41.

FIGURE 5–41
Shorted input in a load gate.

COMPUTER NOTE
Testing of newly manufactured circuits has become more difficult due to increased complexity and reduced size of components causing restricted access to pins. A method that has become popular for testing high-density circuits relies on computer testing a logic board using ICs that have additional built-in logic, called boundary scan logic. The concept is to add test points (boundary scan cells) inside the IC at the I/O points in a logic device. Each device has additional pins to support the boundary scan logic. States of the device can be read and controlled through a serial input to a computer that controls the test and can isolate faults to the pin where the fault lies.

Signal Tracing and Waveform Analysis

Although the methods of isolating an open or a short at a node point are very useful from time to time, a more general troubleshooting technique called **signal tracing** is of great value to the technician or technologist in just about every troubleshooting situation. Waveform measurement is accomplished with an oscilloscope or a logic analyzer.

Basically, the signal tracing method requires that you observe the waveforms and their time relationships at all accessible points in the logic circuit. You can begin at the inputs and, from an analysis of the waveform timing diagram for each point, determine where an incorrect waveform first occurs. With this procedure you can usually isolate the fault to a specific gate. A procedure beginning at the output and working back toward the inputs can also be used.

The general procedure for signal tracing starting at the inputs is outlined as follows:

■ Within a system, define the section of logic that is suspected of being faulty.
■ Start at the inputs to the section of logic under examination. We assume, for this discussion, that the input waveforms coming from other sections of the system have been found to be correct.
■ For each gate, beginning at the input and working toward the output of the logic circuit, observe the output waveform of the gate and compare it with the input waveforms by using the oscilloscope or the logic analyzer.
■ Determine if the output waveform is correct, using your knowledge of the logical operation of the gate.
■ If the output is incorrect, the gate under test may be faulty. Pull the IC containing the gate that is suspected of being faulty, and test it out-of-circuit. If the gate is found to be faulty, replace the IC. If it works correctly, the fault is in the external circuitry or in another IC to which the tested one is connected.
■ If the output is correct, go to the next gate. Continue checking each gate until an incorrect waveform is observed.

Figure 5–42 is an example that illustrates the general procedure for a specific logic circuit in the following steps:

Step 1. Observe the output of gate G_1 (test point 5) relative to the inputs. If it is correct, check the inverter next. If the output is not correct, the gate or its connections are bad; or, if the output is LOW, the input to gate G_2 may be shorted.

Step 2. Observe the output of the inverter (TP6) relative to the input. If it is correct, check gate G_2 next. If the output is not correct, the inverter or its connections are bad; or, if the output is LOW, the input to gate G_3 may be shorted.

Step 3. Observe the output of gate G_2 (TP7) relative to the inputs. If it is correct, check gate G_3 next. If the output is not correct, the gate or its connections are bad; of, if the output is LOW, the input to gate G_4 may be shorted.

FIGURE 5–42

Example of signal tracing and waveform analysis in a portion of a digital system. TP indicates test point.

Step 4. Observe the output of gate G_3 (TP8) relative to the inputs. If it is correct, check gate G_4 next. If the output is not correct, the gate or its connections are bad; or, if the output is LOW, the input to gate G_4 (TP9) may be shorted.

Step 5. Observe the output of gate G_4 relative to the inputs. If it is correct, the circuit is okay. If the output is not correct, the gate or its connections are bad.

EXAMPLE 5–14

Determine the fault in the logic circuit of Figure 5–43(a) by using waveform analysis. The red waveforms are correct, and you have observed the waveforms shown in green in Figure 5–43(b).

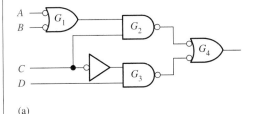

(a)

(b)

FIGURE 5–43

Solution

1. Determine what the correct waveform should be for each gate. The correct waveforms are shown in red, superimposed on the actual measured waveforms, in Figure 5–43(b).
2. Compare waveforms gate by gate until you find a measured waveform that does not match the correct waveform.

In this example, everything tested is correct until gate G_3 is checked. The output of this gate is not correct as the differences in the waveforms indicate. An analysis of the waveforms indicates that if the D input to gate G_3 is open and acting as a HIGH, you will get the output waveform measured (shown in red). Notice that the output of G_4 is correct for the inputs measured, although the input from G_3 is incorrect.

Replace the IC containing G_3, and check the circuit's operation again.

Related Problem For the inputs in Figure 5–43(b), determine the output waveform for the logic circuit (output of G_4) if the inverter has an open output.

As you know, testing and troubleshooting logic circuits often require observing and comparing two digital waveforms simultaneously, such as an input and the output of a gate, on a two-channel oscilloscope. For digital waveforms, the scope should always be set to DC coupling on each channel input to avoid "shifting" the ground level. You should determine where the 0 V level is on the screen for both channels.

To compare the timing of the waveforms, the scope should be triggered from only one channel (don't use vertical mode or composite triggering). The channel selected for triggering should always be the one that has the lowest frequency waveform, if possible.

SECTION 5–6
REVIEW

1. List four common internal failures in logic gates.
2. One input of a NOR gate is externally shorted to $+V_{CC}$. How does this condition affect the gate operation?
3. Determine the output of gate G_4 in Figure 5–43(a), with inputs as shown in part (b), for the following faults:
   (a) one input to G_1 shorted to ground
   (b) the inverter input shorted to ground
   (c) an open output in G_3

5–7 ■ DIGITAL SYSTEM APPLICATION

In this system application, a digital control system for a portion of a lumber-processing operation in a furniture factory is introduced. This system controls four motors that run a conveyor, its lubrication pump, and two saws.

After completing this section, you should be able to

■ Describe the system operation   ■ Design the control logic for a portion of the system   ■ Troubleshoot the system board   ■ Develop a test procedure for the control logic

Basic System Operation

The system uses four manual on/off switches, control logic, and motor drive interface to control the conveyor lubrication pump motor, the conveyor motor, the cross-cut saw motor, and the band saw motor, as indicated by the block diagram in Figure 5–44. This system application focuses on the control logic portion of the system.

Operational Requirements   Switch input S_1 controls the lubrication pump motor (output M_1). Switch input S_2 controls the conveyor motor (output M_2). Switch input S_3 controls the band saw motor (output M_3). Switch input S_4 controls the cross-cut saw motor (output M_4).

The motor that provides for lubrication of the conveyor must be on ($M_1 = 1$) when the conveyor is on. The motor that drives the conveyor is on ($M_2 = 1$) only when both switch 1 and switch 2 are on ($S_1 = 1$ and $S_2 = 1$). The band saw motor is on ($M_3 = 1$) when switch 3 is on ($S_3 = 1$), and the cross-cut saw motor is on ($M_4 = 1$) when switch 4 is on ($S_4 = 1$). The band saw and cross-cut saw motors require no lubrication, but they must never be on at the same time. If switches 3 and 4 are both turned on at the same time, the

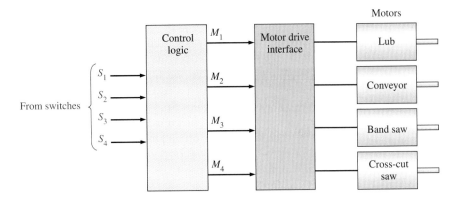

FIGURE 5–44
Basic block diagram for the control system.

system must be completely shut down, including the conveyor and lubrication motors. Also, the cross-cut saw and the conveyor cannot be on at the same time. The control logic controls the motors to prevent any of the unallowed conditions from occurring should the switches be improperly operated.

Truth Table for the Control Logic

Table 5–6 is a truth table that shows the states of the switches and motors based on the conditions previously described. A 1 indicates the ON state and a 0 indicates the OFF state. Since there are four switches, there are sixteen possible ON/OFF combinations of the switches and, as a result, sixteen possible ON/OFF combinations for the motors. The states of the switches are the input variables and the states of the motors are the output variables. The unallowed conditions in this case are not treated as "don't cares." They are switch input conditions that should not occur but if they do, the system must be shut down with all motors off.

TABLE 5–6
Truth table for the control logic.

Inputs				Outputs				
S_1	S_2	S_3	S_4	M_1 Lub	M_2 Conv	M_3 Band	M_4 Cross	Comments
0	0	0	0	0	0	0	0	Shut down
0	0	0	1	0	0	0	1	OK
0	0	1	0	0	0	1	0	OK
0	0	1	1	0	0	0	0	Unallowed (2 saws on). Shut down
0	1	0	0	0	0	0	0	Unallowed (No lub with conveyor). Shut down
0	1	0	1	0	0	0	0	Unallowed (conveyor and cross cut). Shut down
0	1	1	0	0	0	0	0	Unallowed (No lub with conveyor). Shut down
0	1	1	1	0	0	0	0	Unallowed (2 saws and conveyor). Shut down
1	0	0	0	1	0	0	0	OK
1	0	0	1	1	0	0	1	OK
1	0	1	0	1	0	1	0	OK
1	0	1	1	0	0	0	0	Unallowed (2 saws). Shut down
1	1	0	0	1	1	0	0	OK
1	1	0	1	0	0	0	0	Unallowed (conveyor and cross cut). Shut down
1	1	1	0	1	1	1	0	OK
1	1	1	1	0	0	0	0	Unallowed (all on). Shut down

Design of the Control Logic

There are four separate logic circuits, one for each of the motors. Let's begin by designing the logic circuit for the lubrication pump motor (output M_1). The first step is to transfer the data from the truth table to a Karnaugh map and develop an SOP expression.

The switch variables S_1, S_2, S_3, and S_4 are map variables and the states of M_1 are plotted and grouped as shown in Figure 5–45(a). The 0s on the map are for switch conditions when the motor is off and the 1s are for switch conditions when the motor is on. The resulting SOP expression for the lubrication pump motor logic results in the NAND implementation shown in part (b).

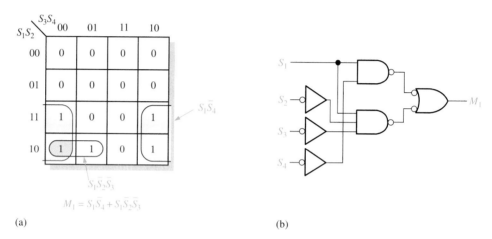

(a)

(b)

FIGURE 5–45
Karnaugh map simplification and implementation for the lubrication motor logic.

■ SYSTEM ASSIGNMENTS

The lubrication motor logic has been designed and now its your turn to complete the control logic design for the remaining three motors. When you have completed the design, it should agree with the logic on the system printed circuit (PC) board. All of the connections to the V_{CC} and ground pads are on the back side of the board and not shown for simplicity. Circuit interconnections on the back side of the board feed through to the pads on the component side. Figure 5–46 shows the pin diagrams for the ICs on the board.

(a) 74AHC04 hex inverters

(b) 74AHC00 quad 2-input NAND

(c) 74AHC10 triple 3-input NAND

FIGURE 5–46
IC pin diagrams.

■ SYSTEM ASSIGNMENT 1:   Design

- ***Activity 1:***   Using Table 5–6 and the Karnaugh map method, design the logic for controlling the conveyor motor using NAND gates and inverters.

- ***Activity 2:***   Design the logic for controlling the band saw motor.

- ***Activity 3:***   Design the logic for controlling the cross-cut saw motor.

- ***Activity 4:***   Combine the logic for each of the four motor controls into a complete logic diagram for the control logic. Specify the specific devices to be used.

- ***Activity 5:***   Trace out the PC board and draw the resulting logic diagram with the aid of the IC logic pin diagrams in Figure 5–46. The logic that you designed should agree with the logic on the circuit board (with the possible exception of some pin numbers).

- ***Activity 6:***   Label the inputs and outputs on the PC board using the variables in the truth table.

■ SYSTEM ASSIGNMENT 2:   Testing and Troubleshooting

■ Assume there are four identical control logic boards to be tested. To test each board, a device (binary counter) that produces a sequence of binary numbers for 0 through 15 is connected to the board to simulate the switch inputs according to Table 5–6. A logic analyzer is used to observe the output waveforms. For each board a set of output waveforms is produced. For each display deter-

mine if the board is operating properly and, if not, specify the possible fault or faults.

The control logic is in file SA05.EWB on your EWB CD. Use the EWB word generator and logic analyzer to simulate this test for each fault.

SECTION 5–7 REVIEW

1. State the purpose of the control logic in this application.
2. What happens if switches 1 and 3 are on at the same time with switch 2 and switch 4 off?
3. What happens if switches 2 and 4 are on at the same time with switch 1 and switch 3 off?

 All or portions of Chapter 15, "Integrated Circuit Technologies," may be covered at this point. The coverage in Chapter 15 is not a prerequisite to any other material in this text, and its omission or partial omission will not affect any other material.

■ SUMMARY

- AND-OR logic produces an output expression in SOP form.
- AND-OR-Invert logic produces a complemented SOP form, which is actually a POS form.
- The operational symbol for exclusive-OR is $\oplus$. An exclusive-OR expression can be stated in two equivalent ways:

$$A\overline{B} + \overline{A}B = A \oplus B$$

- To do an analysis of a logic circuit, start with the logic circuit, and develop the Boolean output expression or the truth table or both.
- Implementation of a logic circuit is the process in which you start with the Boolean output expressions or the truth table and develop a logic circuit that produces the output function.
- All NAND or NOR logic diagrams should be drawn using appropriate dual symbols so that bubble outputs are connected to bubble inputs and nonbubble outputs are connected to nonbubble inputs.
- When two negation indicators (bubbles) are connected, they effectively cancel each other.

■ KEY TERMS

These terms are also in the end-of-book glossary.

Dual symbols The negative-AND is the dual symbol for the NOR gate, and the negative-OR is the dual symbol for the NAND gate.

Node A common connection point in a circuit in which a gate output is connected to one or more gate inputs.

Signal tracing A troubleshooting technique in which waveforms are observed in a step-by-step manner beginning at the input and working toward the output or vice versa. At each point the observed waveform is compared with the correct signal for that point.

Universal gate Either a NAND gate or a NOR gate. The term *universal* refers to the property of a gate that permits any logic function to be implemented by that gate or by a combination of gates of that kind.

■ SELF-TEST

1. The output expression for an AND-OR circuit having one AND gate with inputs A, B, C, and D and one AND gate with inputs E and F is

   (a) $ABCDEF$
   (b) $A + B + C + D + E + F$
   (c) $(A + B + C + D)(E + F)$
   (d) $ABCD + EF$

2. A logic circuit with an output $X = A\overline{B}C + A\overline{C}$ consists of

   (a) two AND gates and one OR gate
   (b) two AND gates, one OR gate, and two inverters
   (c) two OR gates, one AND gate, and two inverters
   (d) two AND gates, one OR gate, and one inverter

3. To implement the expression $\overline{A}BCD + A\overline{B}CD + ABC\overline{D}$, it takes one OR gate and
   (a) one AND gate
   (b) three AND gates
   (c) three AND gates and four inverters
   (d) three AND gates and three inverters

4. The expression $\overline{A}BCD + ABC\overline{D} + A\overline{B}\,\overline{C}D$
   (a) cannot be simplified
   (b) can be simplified to $\overline{A}BC + A\overline{B}$
   (c) can be simplified to $ABC\overline{D} + \overline{A}B\overline{C}$
   (d) None of these answers is correct.

5. The output expression for an AND-OR-Invert circuit having one AND gate with inputs A, B, C, and D and one AND gate with inputs E and F is
   (a) $ABCD + EF$
   (b) $\overline{A} + \overline{B} + \overline{C} + \overline{D} + \overline{E} + \overline{F}$
   (c) $\overline{(A + B + C + D)(E + F)}$
   (d) $(\overline{A} + \overline{B} + \overline{C} + \overline{D})(\overline{E} + \overline{F})$

6. An exclusive-OR function is expressed as
   (a) $\overline{A}\,\overline{B} + AB$
   (b) $\overline{A}B + A\overline{B}$
   (c) $(\overline{A} + B)(A + \overline{B})$
   (d) $(\overline{A} + \overline{B}) + (A + B)$

7. The AND operation can be produced with
   (a) two NAND gates
   (b) three NAND gates
   (c) one NOR gate
   (d) three NOR gates

8. The OR operation can be produced with
   (a) two NOR gates
   (b) three NAND gates
   (c) four NAND gates
   (d) both answers (a) and (b)

9. When using dual symbols in a logic diagram,
   (a) bubble outputs are connected to bubble inputs
   (b) the NAND symbols produce the AND operations
   (c) the negative-OR symbols produce the OR operations
   (d) All of these answers are true.
   (e) None of these answers is true.

10. All Boolean expressions can be implemented with
    (a) NAND gates only
    (b) NOR gates only
    (c) combinations of NAND and NOR gates
    (d) combinations of AND gates, OR gates, and inverters
    (e) any of these

■ PROBLEMS

SECTION 5–1 Basic Combinational Logic Circuits

1. Sketch the ANSI distinctive shape logic diagram for a 3-wide, 4-input AND-OR-Invert circuit. Also draw the ANSI standard rectangular outline symbol.

2. Write the output expression for each circuit in Figure 5–47.

(a)

(b)

FIGURE 5–47

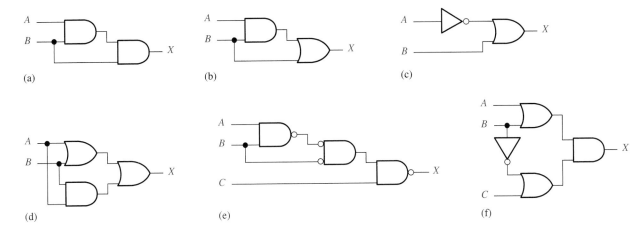

FIGURE 5–48

3. Write the output expression for each circuit as it appears in Figure 5–48.
4. Write the output expression for each circuit as it appears in Figure 5–49 and then change each circuit to an equivalent AND-OR configuration.
5. Develop the truth table for each circuit in Figure 5–48.
6. Develop the truth table for each circuit in Figure 5–49.
7. Show that an exclusive-NOR circuit produces a POS output.

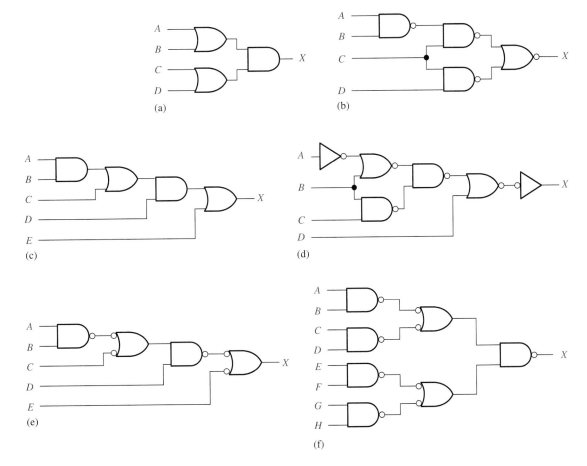

FIGURE 5–49

SECTION 5–2   Implementing Combinational Logic

8. Use AND gates, OR gates, or combinations of both to implement the following logic expressions as stated:

(a) $X = AB$ **(b)** $X = A + B$ **(c)** $X = AB + C$

(d) $X = ABC + D$ **(e)** $X = A + B + C$ **(f)** $X = ABCD$

(g) $X = A(CD + B)$ **(h)** $X = AB(C + DEF) + CE(A + B + F)$

9. Use AND gates, OR gates, and inverters as needed to implement the following logic expressions as stated:

(a) $X = AB + \overline{B}C$ **(b)** $X = A(B + \overline{C})$

(c) $X = A\overline{B} + AB$ **(d)** $X = \overline{ABC} + B(EF + \overline{G})$

(e) $X = A[BC(A + B + C + D)]$ **(f)** $X = B(C\overline{D}E + \overline{E}FG)(\overline{AB} + C)$

10. Use NAND gates, NOR gates, or combinations of both to implement the following logic expressions as stated:

(a) $X = \overline{A}B + CD + (\overline{A + B})(ACD + \overline{BE})$ **(b)** $X = AB\overline{C}\,\overline{D} + D\overline{E}F + \overline{AF}$

(c) $X = \overline{A}[B + \overline{C}(D + E)]$

11. Implement a logic circuit for the truth table in Table 5–7.

TABLE 5–7

Inputs			Output
A	B	C	X
0	0	0	1
0	0	1	0
0	1	0	1
0	1	1	0
1	0	0	1
1	0	1	0
1	1	0	1
1	1	1	1

12. Implement a logic circuit for the truth table in Table 5–8.

TABLE 5–8

Inputs				Output
A	B	C	D	X
0	0	0	0	0
0	0	0	1	0
0	0	1	0	1
0	0	1	1	1
0	1	0	0	1
0	1	0	1	0
0	1	1	0	0
0	1	1	1	0
1	0	0	0	1
1	0	0	1	1
1	0	1	0	1
1	0	1	1	1
1	1	0	0	0
1	1	0	1	0
1	1	1	0	0
1	1	1	1	1

13. Simplify the circuit in Figure 5–50 as much as possible, and verify that the simplified circuit is equivalent to the original by showing that the truth tables are identical.

14. Repeat Problem 13 for the circuit in Figure 5–51.

15. Minimize the gates required to implement the functions in each part of Problem 9 in SOP form.

16. Minimize the gates required to implement the functions in each part of Problem 10 in SOP form.

17. Minimize the gates required to implement the function of the circuit in each part of Figure 5–49 in SOP form.

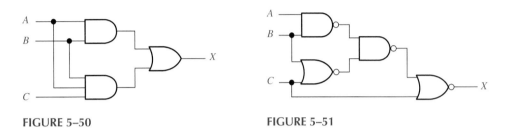

FIGURE 5–50   **FIGURE 5–51**

SECTION 5–3   The Universal Property of NAND and NOR Gates

18. Implement the logic circuits in Figure 5–47 using only NAND gates.

19. Implement the logic circuits in Figure 5–51 using only NAND gates.

20. Repeat Problem 18 using only NOR gates.

21. Repeat Problem 19 using only NOR gates.

SECTION 5–4   Combinational Logic Using NAND and NOR Gates

22. Show how the following expressions can be implemented as stated using only NOR gates:

(a) $X = ABC$ (b) $X = \overline{ABC}$ (c) $X = A + B$

(d) $X = A + B + \overline{C}$ (e) $X = \overline{AB} + \overline{CD}$ (f) $X = (A + B)(C + D)$

(g) $X = AB[C(\overline{DE} + \overline{AB}) + \overline{BCE}]$

23. Repeat Problem 22 using only NAND gates.

24. Implement each function in Problem 8 by using only NAND gates.

25. Implement each function in Problem 9 by using only NAND gates.

SECTION 5–5   Operation with Pulse Waveforms

26. Given the logic circuit and the input waveforms in Figure 5–52, sketch the output waveform.

27. For the logic circuit in Figure 5–53, sketch the output waveform in proper relationship to the inputs.

FIGURE 5–52

FIGURE 5–53

28. For the input waveforms in Figure 5–54, what logic circuit will generate the output waveform shown?

29. Repeat Problem 28 for the waveforms in Figure 5–55.

FIGURE 5–54

FIGURE 5–55

30. For the circuit in Figure 5–56, sketch the waveforms at the numbered points in the proper relationship to each other.

31. Assuming a propagation delay through each gate of 10 nanoseconds (ns), determine if the *desired* output waveform X in Figure 5–57 (a pulse with a minimum $t_W = 25$ ns positioned as shown) will be generated properly with the given inputs.

FIGURE 5–56

FIGURE 5–57

SECTION 5–6   Troubleshooting

32. For the logic circuit and the input waveforms in Figure 5–58, the indicated output waveform is observed. Determine if this is the correct output waveform.

33. The output waveform in Figure 5–59 is incorrect for the inputs that are applied to the circuit. Assuming that one gate in the circuit has failed, with its output either an apparent constant HIGH or a constant LOW, determine the faulty gate and the type of failure (output open or shorted).

FIGURE 5–58

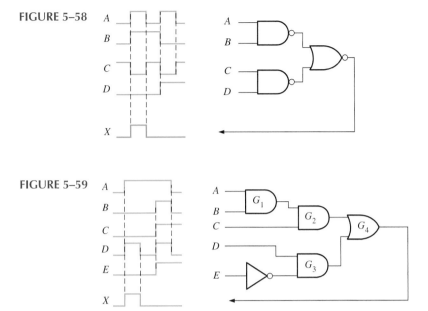

FIGURE 5–59

34. Repeat Problem 33 for the circuit in Figure 5–60, with input and output waveforms as shown.

35. The node in Figure 5–61 has been found to be stuck in the LOW state. A pulser and current tracer yield the series of indications shown in the diagram. From these observations, what do you conclude?

FIGURE 5–60

FIGURE 5–61

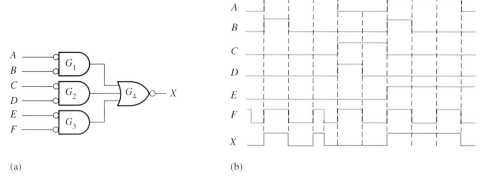

(a)

(b)

FIGURE 5–62

36. Figure 5–62(a) is a logic circuit under test. Figure 5–62(b) shows the waveforms as observed on a logic analyzer. The output waveform is incorrect for the inputs that are applied to the circuit. Assuming that one gate in the circuit has failed, with its output either an apparent constant HIGH or a constant LOW, determine the faulty gate and the type of failure.

37. The logic circuit in Figure 5–63 has the input waveforms shown.

   (a) Determine the correct output waveform in relation to the inputs.

   (b) Determine the output waveform if the output of gate G_3 is open.

   (c) Determine the output waveform if the upper input to gate G_5 is shorted to ground.

38. The logic circuit in Figure 5–64 has only one intermediate test point available besides the output, as indicated. For the inputs shown, you observe the indicated waveform at the test point. Is this waveform correct? If not, what are the possible faults that would cause it to appear as it does?

FIGURE 5–63

FIGURE 5–64

SECTION 5–7 Digital System Application

39. Implement the lubrication motor logic in Figure 5–45 with NOR gates and inverters.

40. Repeat Activities 1 through 3 in System Application 1 using NOR gates and inverters.

41. Modify the motor control logic for the condition $S_1S_2S_3S_4 = 0101$ to shut down the conveyor but keep the cross-cut saw running.

Special Design Problems

42. Design a logic circuit to produce a HIGH output only if the input, represented by a 4-bit binary number, is greater than twelve or less than three. First develop the truth table and then draw the logic diagram.

43. Develop the logic circuit necessary to meet the following requirements:
A battery-powered lamp in a room is to be operated from two switches, one at the back door and one at the front door. The lamp is to be on if the front switch is on and the back switch is off, or if the front switch is off and the back switch is on. The lamp is to be off if both switches are off or if both switches are on. Let a HIGH output represent the on condition and a LOW output represent the off condition.

44. Design a circuit to add to the control logic developed in System Application 1 that will indicate a shut-down condition with a red light.

45. Develop the NAND logic for a hexadecimal keypad encoder that will convert each key closure to binary.

Electronics Workbench Troubleshooting

46. Open file PR05_46.EWB and test the logic circuit to determine if there is a fault. If there is a fault, identify it if possible.

47. Open file PR05_47.EWB and test the logic circuit to determine if there is a fault. If there is a fault, identify it if possible.

48. Open file PR05_48.EWB and test the logic circuit to determine if there is a fault. If there is a fault, identify it if possible.

49. Open file PR05_49.EWB and test the logic circuit to determine if there is a fault. If there is a fault, identify it if possible.

■ **ANSWERS TO SECTION REVIEWS**

SECTION 5–1

1. (a) $\overline{AB + CD} = \overline{1 \cdot 0 + 1 \cdot 0} = 1$ **(b)** $\overline{AB + CD} = \overline{1 \cdot 1 + 0 \cdot 1} = 0$
(c) $\overline{AB + CD} = \overline{0 \cdot 1 + 1 \cdot 1} = 0$

2. (a) $A\overline{B} + \overline{A}B = 1 \cdot \overline{0} + \overline{1} \cdot 0 = 1$ **(b)** $A\overline{B} + \overline{A}B = 1 \cdot \overline{1} + \overline{1} \cdot 1 = 0$
(c) $A\overline{B} + \overline{A}B = 0 \cdot \overline{1} + \overline{0} \cdot 1 = 1$ **(d)** $A\overline{B} + \overline{A}B = 0 \cdot \overline{0} + \overline{0} \cdot 0 = 0$

3. $X = 1$ when $ABC = 000, 011, 101, 110,$ and 111; $X = 0$ when $ABC = 001, 010,$ and 100

4. $X = AB + \overline{A}\overline{B}$; the circuit consists of two AND gates, one OR gate, and two inverters. See Figure 5–8(b) for diagram.

SECTION 5–2

1. (a) $X = ABC + AB + AC$: three AND gates, one OR gate
(b) $X = AB(C + DE)$: three AND gates, one OR gate

2. $X = ABC + \overline{A}\,\overline{B}\,\overline{C}$; two AND gates, one OR gate, and three inverters

3. (a) $X = AB(C + 1) + AC = AB + AC$ **(b)** $X = AB(C + DE) = ABC + ABDE$

SECTION 5–3

1. (a) $X = \overline{A} + B$: a 2-input NAND gate with A and $\overline{B}$ on its inputs.
(b) $X = A\overline{B}$: a 2-input NAND with A and $\overline{B}$ on its inputs, followed by one NAND used as an inverter

2. (a) $X = \overline{A} + B$: a 2-input NOR with inputs $\overline{A}$ and B, followed by one NOR used as an inverter.
(b) $X = A\overline{B}$: a 2-input NOR with $\overline{A}$ and B on its inputs

SECTION 5–4

1. $X = \overline{(\overline{A} + \overline{B} + \overline{C})DE}$: a 3-input NAND with inputs, A, B, and C, with its output connected to a second 3-input NAND with two other inputs, D and E

2. $X = \overline{\overline{ABC} + (D + E)}$: a 3-input NOR with inputs A, B, and C, with its output connected to a second 3-input NOR with two other inputs, D and E

SECTION 5–5

1. The exclusive-OR output is a 15 μs pulse followed by a 25 μs pulse, with a separation of 10 μs between the pulses.

2. The output of the exclusive-NOR is HIGH when both inputs are HIGH or when both inputs are LOW.

SECTION 5–6

1. Common gate failures are input or output open; input or output shorted to ground.

2. Input shorted to V_{CC} causes output to be stuck LOW.

3. **(a)** G_4 output is HIGH until rising edge of seventh pulse, then it goes LOW.

   (b) G_4 output is the same as input D

   (c) G_4 output is the inverse of the G_2 output shown in Figure 5–43(b).

SECTION 5–7

1. The control logic decodes switch inputs and produces motor control outputs.

2. Condition is allowed. The bandsaw runs.

3. Unallowed condition because the conveyor and the cross-cut saw can't be on at same time (shutdown) and also because the conveyor is on with no lubrication.

■ **ANSWERS TO RELATED PROBLEMS FOR EXAMPLES**

5–1 $Y = AB + AC + BC$

5–2 $Y = \overline{AB + AC + BC}$

 If $A = 0$ and $B = 0$, $Y = \overline{0 \cdot 0 + 0 \cdot 1 + 0 \cdot 1} = \overline{0} = 1$

 If $A = 0$ and $C = 0$, $Y = \overline{0 \cdot 1 + 0 \cdot 0 + 1 \cdot 0} = \overline{0} = 1$

 If $B = 0$ and $C = 0$, $Y = \overline{1 \cdot 0 + 1 \cdot 0 + 0 \cdot 0} = \overline{0} = 1$

5–3 Cannot be simplified   **5–4** Cannot be simplified   **5–5** $X = A + B + C + D$ is valid.

5–6 See Figure 5–65.   **5–7** $X = \overline{(\overline{ABC})(\overline{DEF})} = (\overline{AB})C + (\overline{DE})F = (\overline{A} + \overline{B})C + (\overline{D} + \overline{E})F$

5–8 See Figure 5–66.

5–9 $X = \overline{(\overline{\overline{A + B} + C}) + (\overline{\overline{D + E} + F})} = (\overline{A + B} + C)(\overline{D + E} + F) = (\overline{A}\overline{B} + C)(\overline{D}\overline{E} + F)$

5–10 See Figure 5–67.   **5–11** See Figure 5–68.

5–12 See Figure 5–69.   **5–13** See Figure 5–70.   **5–14** See Figure 5–71.

FIGURE 5–65

$X = \overline{C}(A + \overline{B})(\overline{B} + \overline{D})$

(a)   (b)

FIGURE 5–66

FIGURE 5–67

FIGURE 5–68

FIGURE 5–69

FIGURE 5–70

FIGURE 5–71

■ **ANSWERS TO SELF-TEST**

1. (d) **2.** (b) **3.** (c) **4.** (a) **5.** (d) **6.** (b) **7.** (a) **8.** (d)

9. (d) **10.** (e)

6

FUNCTIONS OF COMBINATIONAL LOGIC

Computer Simulation
of Digital Circuits:
Electronics Workbench (EWB) tutorial at
http://www.prenhall.com/floyd

■ CHAPTER OBJECTIVES

- ❏ Distinguish between half-adders and full-adders
- ❏ Use full-adders to implement multibit parallel binary adders
- ❏ Explain the differences between ripple carry and look-ahead carry parallel adders
- ❏ Use the magnitude comparator to determine the relationship between two binary numbers and use cascaded comparators to handle the comparison of larger numbers
- ❏ Implement a basic binary decoder
- ❏ Use BCD-to-7-segment decoders in display systems
- ❏ Apply a decimal-to-BCD priority encoder in a simple keyboard application
- ❏ Convert from BCD to binary, binary to Gray code, and Gray code to binary by using logic devices
- ❏ Apply multiplexers in data selection, multiplexed displays, logic function generation, and simple communications systems
- ❏ Use decoders as demultiplexers
- ❏ Explain the meaning of parity
- ❏ Use parity generators and checkers to detect bit errors in digital systems
- ❏ Implement a simple data communications system
- ❏ Identify glitches, common bugs in digital systems

CHAPTER OVERVIEW

In this chapter, several types of MSI combinational logic circuits are introduced including adders, comparators, decoders, encoders, code converters, multiplexers (data selectors), demultiplexers, and parity generators/checkers. Examples of system applications of many of these devices are included to demonstrate how the devices can be used in practical situations.

SPECIFIC DEVICES

74HC42	74LS47	74LS83A
74HC85	74HC138	74LS139
74HC147	74F148	74HC151
74HC154	74HC157A	74184
74185	74LS280	74LS283

DIGITAL SYSTEM APPLICATION

This Digital System Application illustrates concepts taught in this chapter and deals with one portion of a traffic light control system. The system application in the next chapter presents another approach, while the system applications in Chapters 8, 9, and 11 focus on other parts of the system. Ba-sically, this system controls the traffic light at the intersection of a busy street and a lightly traveled side street. The system includes a combinational logic portion to which the topics in this chapter and Chapter 5 apply, a timing circuit portion to which Chapter 8 applies, and a sequential logic portion to which both Chapters 8 and 9 apply.

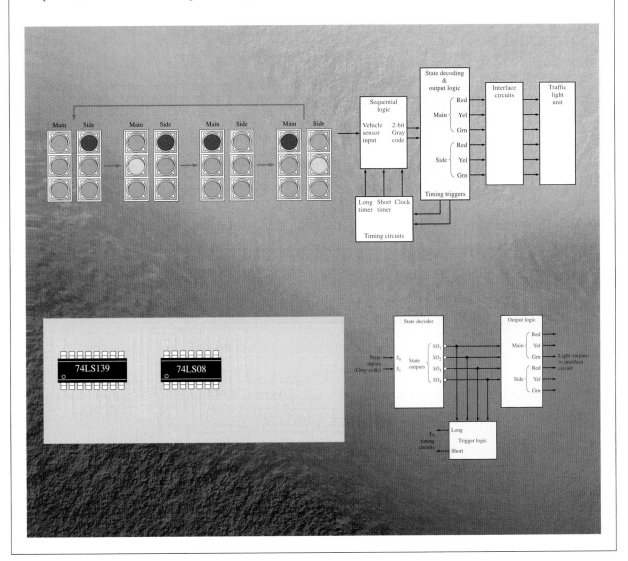

6–1 ■ BASIC ADDERS

Adders are important not only in computers, but in many types of digital systems in which numerical data are processed. An understanding of the basic adder operation is fundamental to the study of digital systems. In this section, the half-adder and the full-adder are introduced.

After completing this section, you should be able to

- Describe the function of a half-adder ■ Draw a half-adder logic diagram
- Describe the function of the full-adder ■ Draw a full-adder logic diagram using half-adders ■ Implement a full-adder using AND-OR logic

The Half-Adder

Recall the basic rules for binary addition as stated in Chapter 2.

$$0 + 0 = 0$$
$$0 + 1 = 1$$
$$1 + 0 = 1$$
$$1 + 1 = 10$$

The operations are performed by a logic circuit called a **half-adder.**

> **The half-adder accepts two binary digits on its inputs and produces two binary digits on its outputs, a sum bit and a carry bit.**

A half-adder is represented by the logic symbol in Figure 6–1.

FIGURE 6–1
Logic symbol for a half-adder.

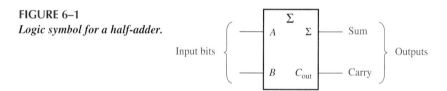

Half-Adder Logic From the logical operation of the half-adder as stated in Table 6–1, expressions can be derived for the sum and the output carry as functions of the inputs. Notice that the output carry (C_{out}) is a 1 only when both A and B are 1s; therefore, C_{out} can be expressed as the AND of the input variables.

$$C_{out} = AB \qquad\qquad (6\text{–}1)$$

Now observe that the sum output (Σ) is a 1 only if the input variables, A and B, are not equal. The sum can therefore be expressed as the exclusive-OR of the input variables.

$$\Sigma = A \oplus B \qquad\qquad (6\text{–}2)$$

TABLE 6–1
Half-adder truth table.

A	B	C_{out}	Σ
0	0	0	0
0	1	0	1
1	0	0	1
1	1	1	0

Σ = sum

C_{out} = output carry

A and B = input variables (operands)

From Equations (6–1) and (6–2), the logic implementation required for the half-adder function can be developed. The output carry is produced with an AND gate with A and B on the inputs, and the sum output is generated with an exclusive-OR gate, as shown in Figure 6–2.

FIGURE 6–2
Half-adder logic diagram.

$\Sigma = A \oplus B$

$C_{\text{out}} = AB$

The Full-Adder

The second category of adder is the **full-adder.**

> **The full-adder accepts two input bits and an input carry and generates a sum output and an output carry.**

The basic difference between a full-adder and a half-adder is that the full-adder accepts an input carry. A logic symbol for a full-adder is shown in Figure 6–3, and the truth table in Table 6–2 shows the operation of a full-adder.

FIGURE 6–3
Logic symbol for a full-adder.

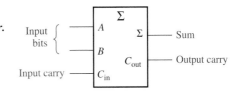

TABLE 6–2
Full-adder truth table.

A	B	C_{in}	C_{out}	Σ
0	0	0	0	0
0	0	1	0	1
0	1	0	0	1
0	1	1	1	0
1	0	0	0	1
1	0	1	1	0
1	1	0	1	0
1	1	1	1	1

C_{in} = input carry, sometimes designated as *CI*
C_{out} = output carry, sometimes designated as *CO*
Σ = sum
A and *B* = input variables (operands)

Full-Adder Logic The full-adder must add the two input bits and the input carry. From the half-adder you know that the sum of the input bits A and B is the exclusive-OR of those two variables, $A \oplus B$. For the input carry (C_{in}) to be added to the input bits, it must be exclusive-ORed with $A \oplus B$, yielding the equation for the sum output of the full-adder.

$$\Sigma = (A \oplus B) \oplus C_{\text{in}} \tag{6–3}$$

This means that to implement the full-adder sum function, two 2-input exclusive-OR gates can be used. The first must generate the term $A \oplus B$, and the second has as its inputs the output of the first XOR gate and the input carry, as illustrated in Figure 6–4(a).

The output carry is a 1 when both inputs to the first XOR gate are 1s or when both inputs to the second XOR gate are 1s. You can verify this fact by studying Table 6–2. The output carry of the full-adder is therefore produced by the inputs A ANDed with B and

(a) Logic required to form the sum of three bits

(b) Complete logic circuit for a full-adder (each half-adder is enclosed by a shaded area)

FIGURE 6–4
Full-adder logic.

$A \oplus B$ ANDed with C_{in}. These two terms are ORed, as expressed in Equation (6–4). This function is implemented and combined with the sum logic to form a complete full-adder circuit, as shown in Figure 6–4(b).

$$C_{out} = AB + (A \oplus B)C_{in} \tag{6–4}$$

Notice in Figure 6–4(b) there are two half-adders, connected as shown in the block diagram of Figure 6–5(a), with their output carries ORed. The logic symbol shown in Figure 6–5(b) will normally be used to represent the full-adder.

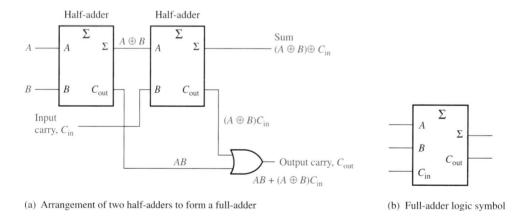

(a) Arrangement of two half-adders to form a full-adder

(b) Full-adder logic symbol

FIGURE 6–5
Full-adder implemented with half-adders.

EXAMPLE 6–1

For each of the three full-adders in Figure 6–6, determine the outputs for the inputs shown.

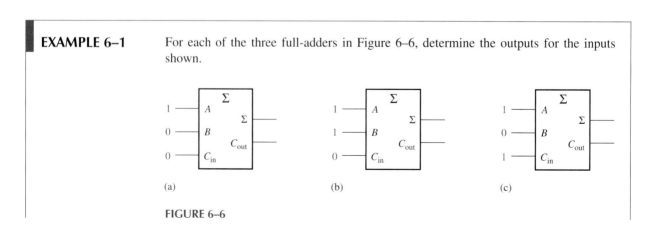

(a)

(b)

(c)

FIGURE 6–6

Solution

(a) The input bits are $A = 1$, $B = 0$, and $C_{in} = 0$.

$$1 + 0 + 0 = 1 \text{ with no carry}$$

Therefore, $\Sigma = 1$ and $C_{out} = 0$.

(b) The input bits are $A = 1$, $B = 1$, and $C_{in} = 0$.

$$1 + 1 + 0 = 0 \text{ with a carry of } 1$$

Therefore, $\Sigma = 0$ and $C_{out} = 1$.

(c) The input bits are $A = 1$, $B = 0$, and $C_{in} = 1$.

$$1 + 0 + 1 = 0 \text{ with a carry of } 1$$

Therefore, $\Sigma = 0$ and $C_{out} = 1$.

Related Problem   What are the full-adder outputs for $A = 1$, $B = 1$, and $C_{in} = 1$?

**SECTION 6–1
REVIEW**

1. Determine the sum (Σ) and the output carry (C_{out}) of a half-adder for each set of input bits:
   (a) 01   (b) 00   (c) 10   (d) 11
2. A full-adder has $C_{in} = 1$. What are the sum (Σ) and the output carry (C_{out}) when $A = 1$ and $B = 1$?

6–2 ■ PARALLEL BINARY ADDERS

Two or more full-adders are connected to form parallel binary adders. In this section, you will learn the basics of this type of adder so that you will understand all the necessary input and output functions when working with these devices.

After completing this section, you should be able to

■ Use full-adders to implement a parallel binary adder   ■ Explain the addition process in a parallel binary adder   ■ Use the truth table for a 4-bit parallel adder   ■ Apply the 74LS83A and the 74LS283 for the addition of two 4-bit numbers   ■ Expand the 4-bit adder to accommodate 8-bit or 16-bit addition

COMPUTER NOTE
Addition is performed by computers on two numbers at a time, called *operands*. The *source operand* is a number that is to be added to an existing number called the *destination operand*, which is held in an ALU register, such as the accumulator. The sum of the two numbers is then stored back in the accumulator. Addition is performed on integer numbers or floating-point numbers using ADD or FADD instructions respectively.

As you saw in Section 6–1, a single full-adder is capable of adding two 1-bit numbers and an input carry. To add binary numbers with more than one bit, additional full-adders must be used. When one binary number is added to another, each column generates a sum bit and a 1 or 0 carry bit to the next column to the left, as illustrated here with 2-bit numbers.

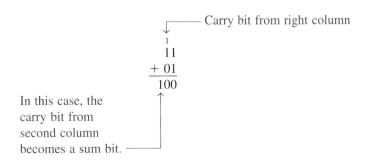

To add two binary numbers, a full-adder is required for each bit in the numbers. So for 2-bit numbers, two adders are needed; for 4-bit numbers, four adders are used; and so on. The carry output of each adder is connected to the carry input of the next higher-order adder, as shown in Figure 6–7 for a 2-bit adder. Notice that either a half-adder can be used for the least significant position or the carry input of a full-adder can be made 0 (grounded) because there is no carry input to the least significant bit position.

FIGURE 6–7

Block diagram of a basic 2-bit parallel adder using two full-adders.

In Figure 6–7 the least significant bits (LSB) of the two numbers are represented by A_1 and B_1. The next higher-order bits are represented by A_2 and B_2. The three sum bits are Σ_1, Σ_2, and Σ_3. Notice that the output carry from the left-most full-adder becomes the most significant bit (MSB) in the sum, Σ_3.

EXAMPLE 6–2

Determine the sum generated by the 3-bit parallel adder in Figure 6–8 and show the intermediate carries when the binary numbers 101 and 011 are being added.

FIGURE 6–8

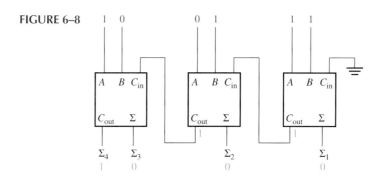

Solution   The LSBs of the two numbers are added in the right-most full-adder. The sum bits and the intermediate carries are indicated in blue in Figure 6–8.

Related Problem   What are the sum outputs when 111 and 101 are added by the 3-bit parallel adder?

Four-Bit Parallel Adders

A group of four bits is called a **nibble**. A basic 4-bit parallel adder is implemented with four full-adder stages as shown in Figure 6–9. Again, the LSBs (A_1 and B_1) in each number being added go into the right-most full-adder; the higher-order bits are applied as

(a) Block diagram

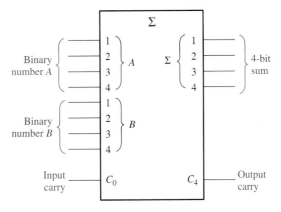

(b) Logic symbol

FIGURE 6–9
A 4-bit parallel adder.

shown to the successively higher-order adders, with the MSBs (A_4 and B_4) in each number being applied to the left-most full-adder. The carry output of each adder is connected to the carry input of the next higher-order adder as indicated. These are called *internal carries.*

In keeping with most manufacturers' data sheets, the input labeled C_0 is the input carry to the least significant bit adder; C_4, in the case of four bits, is the output carry of the most significant bit adder; and Σ_1 (LSB) through Σ_4 (MSB) are the sum outputs. The logic symbol is shown in Figure 6–9(b).

In terms of the method used to handle carries in a parallel adder, there are two types: the *ripple carry* adder and the *carry look-ahead* adder. A **ripple carry** adder is one in which the carry output of each full-adder is connected to the carry input of the next higher-order stage (a stage is one full-adder). The sum and the output carry of any stage cannot be produced until the input carry occurs; this causes a time delay in the addition process. The carry propagation delay for each full-adder is the time from the application of the input carry until the output carry occurs, assuming that the A and B inputs are already present.

A method of speeding up the addition process by eliminating this ripple carry delay is called **look-ahead carry** addition. The look-ahead carry adder anticipates the output carry of each stage, and based on the input bits of each stage, produces the output carry by either *carry generation* or *carry propagation.*

Carry generation occurs when an output carry is produced (generated) internally by the full-adder. A carry is generated only when both input bits are 1s. The generated carry, C_g, is expressed as the AND function of the two input bits, A and B.

$$C_g = AB$$

Carry propagation occurs when the input carry is rippled to become the output carry. An input carry may be propagated by the full-adder when either or both of the input bits are 1s. The propagated carry, C_p, is expressed as the OR function of the input bits.

$$C_p = A + B$$

Truth Table for a 4-Bit Parallel Adder

Table 6–3 is the truth table for a 4-bit adder. On some data sheets, truth tables may be called *function tables* or *functional truth tables*. The subscript n represents the adder bits and can be 1, 2, 3, or 4 for the 4-bit adder. C_{n-1} is the carry from the previous adder. Carries C_1, C_2, and C_3 are generated internally. C_0 is an external carry input and C_4 is an output. Example 6–3 illustrates how to use Table 6–3.

TABLE 6–3

Truth table for each stage of a 4-bit parallel adder.

C_{n-1}	A_n	B_n	Σ_n	C_n
0	0	0	0	0
0	0	1	1	0
0	1	0	1	0
0	1	1	0	1
1	0	0	1	0
1	0	1	0	1
1	1	0	0	1
1	1	1	1	1

EXAMPLE 6–3

Use the 4-bit parallel adder truth table (Table 6–3) to find the sum and output carry for the addition of the following two 4-bit numbers if the input carry (C_{n-1}) is 0:

$$A_4A_3A_2A_1 = 1100 \quad \text{and} \quad B_4B_3B_2B_1 = 1100$$

Solution For $n = 1$: $A_1 = 0$, $B_1 = 0$, and $C_{n-1} = 0$. From the 1st row of the table,

$$\Sigma_1 = \mathbf{0} \quad \text{and} \quad C_1 = 0$$

For $n = 2$: $A_2 = 0$, $B_2 = 0$, and $C_{n-1} = 0$. From the 1st row of the table,

$$\Sigma_2 = \mathbf{0} \quad \text{and} \quad C_2 = 0$$

For $n = 3$: $A_3 = 1$, $B_3 = 1$, and $C_{n-1} = 0$. From the 4th row of the table,

$$\Sigma_3 = \mathbf{0} \quad \text{and} \quad C_3 = 1$$

For $n = 4$: $A_4 = 1$, $B_4 = 1$, and $C_{n-1} = 1$. From the last row of the table,

$$\Sigma_4 = \mathbf{1} \quad \text{and} \quad C_4 = \mathbf{1}$$

C_4 becomes the output carry; the sum of 1100 and 1100 is 11000.

Related Problem Use the truth table (Table 6–3) to find the result of adding the binary numbers 1011 and 1010.

MSI Adders

Examples of 4-bit parallel adders that are available as medium-scale integrated (MSI) circuits are the 74LS83A and the 74LS283 low-power Schottky TTL devices. The 74LS83A and the 74LS283 are functionally identical to each other but not pin compatible; that is, the pin numbers for the inputs and outputs are different due to different power and ground pin connections. For the 74LS83A, V_{CC} is pin 5 and ground is pin 12 on the 16-pin package. For the 74LS283, V_{CC} is pin 16 and ground is pin 8, which is a more standard configuration. Pin diagrams and logic symbols for both of these devices are shown, with pin numbers in parentheses on the logic symbols, in Figure 6–10.

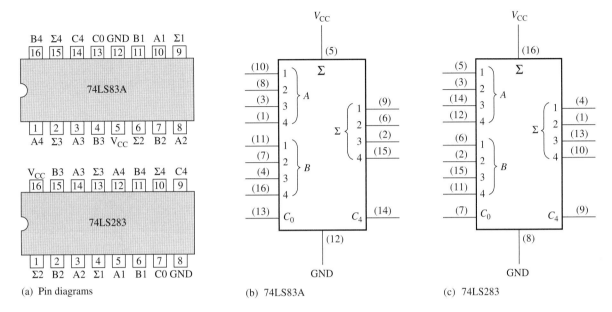

(a) Pin diagrams (b) 74LS83A (c) 74LS283

FIGURE 6–10
MSI 4-bit parallel adders.

AC Data Sheet Characteristics   Recall that logic gates have one specified propagation delay time, t_P, from an input to the output. For MSI logic, there may be several different specifications for t_P. The 4-bit parallel adder has four t_P specifications as shown in Figure 6–11, which is part of a 74LS283 data sheet.

Adder Expansion

The 4-bit parallel adder can be expanded to handle the addition of two 8-bit numbers by using two 4-bit adders and connecting the carry input of the low-order adder (C_0) to ground because there is no carry into the least significant bit position and by connecting the carry

Symbol	Parameter	Limits			Unit
		Min	Typ	Max	
t_{PLH} t_{PHL}	Propagation delay, C_0 input to any Σ output		16 15	24 24	ns
t_{PLH} t_{PHL}	Propagation delay, any A or B input to Σ outputs		15 15	24 24	ns
t_{PLH} t_{PHL}	Propagation delay, C_0 input to C_4 output		11 11	17 22	ns
t_{PLH} t_{PHL}	Propagation delay, any A or B input to C_4 output		11 12	17 17	ns

FIGURE 6–11
AC characteristics for the 74LS283.

output of the low-order adder to the carry input of the high-order adder as shown in Figure 6–12(a). This process is known as **cascading.** Notice that, in this case, the output carry is designated C_8 because it is generated from the eighth bit position. The low-order adder is the one that adds the lower or less significant four bits in the numbers and the high-order adder is the one that adds the higher or more significant four bits in the 8-bit numbers.

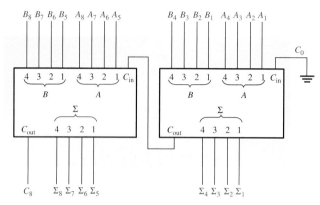

(a) Cascading of 4-bit adders to form an 8-bit adder

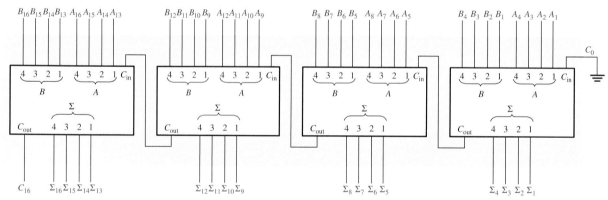

(b) Cascading of 4-bit adders to form a 16-bit adder

FIGURE 6–12
Examples of adder expansion.

Similarly, four 4-bit adders can be cascaded to handle two 16-bit numbers as shown in Figure 6–12(b). Notice that the output carry is designated C_{16} because it is generated from the sixteenth bit position.

EXAMPLE 6–4

Show how two 74LS83A adders can be connected to form an 8-bit parallel adder. Show output bits for the following 8-bit input numbers:

$$A_8A_7A_6A_5A_4A_3A_2A_1 = 10111001 \quad \text{and} \quad B_8B_7B_6B_5B_4B_3B_2B_1 = 10011110$$

Solution Two 74LS83A 4-bit parallel adders are used to implement the 8-bit adder. The only connection between the two 74LS83As is the carry output (pin 14) of the low-order adder to the carry input (pin 13) of the high-order adder, as shown in Figure 6–13. Pin 13 of the low-order adder is grounded (no carry input).

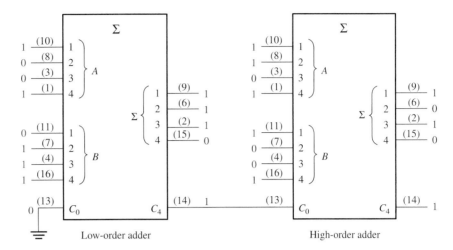

FIGURE 6–13

Two 74LS83A adders connected as an 8-bit parallel adder (pin numbers are in parentheses).

Related Problem   Use 74LS283 adders to implement a 12-bit parallel adder.

Application Example

An example of full-adder and parallel adder application is a simple voting system that can be used to simultaneously provide the number of "yes" votes and the number of "no" votes. For example, this type of system can be used where a group of people are assembled and there is a need for immediately determining opinions (for or against), making decisions, or voting on certain issues or other matters.

In its simplest form, the system includes a switch for "yes" or "no" selection at each position in the assembly and a digital display for the number of yes votes and one for the number of no votes. The basic system is shown in Figure 6–14 on page 282 for a 6-position setup, but it can be expanded to any number of positions with additional 6-position modules and additional parallel adder and display circuits.

In Figure 6–14 each full-adder can produce the sum of up to three votes. The sum and output carry of each full-adder then goes to the two lower-order inputs of a parallel binary adder. The two higher-order inputs of the parallel adder are connected to ground (0) because there is never a case where the binary input exceeds 0011 (decimal 3). For this basic 6-position system, the outputs of the parallel adder go to a BCD-to-7-segment decoder that drives the 7-segment display. As mentioned, additional circuits must be included when the system is expanded.

The resistors from the inputs of each full-adder to ground assure that each input is LOW when the switch is in the neutral position (CMOS logic is used). When a switch is moved to the "yes" or to the "no" position, a HIGH level (V_{CC}) is applied to the associated full-adder input.

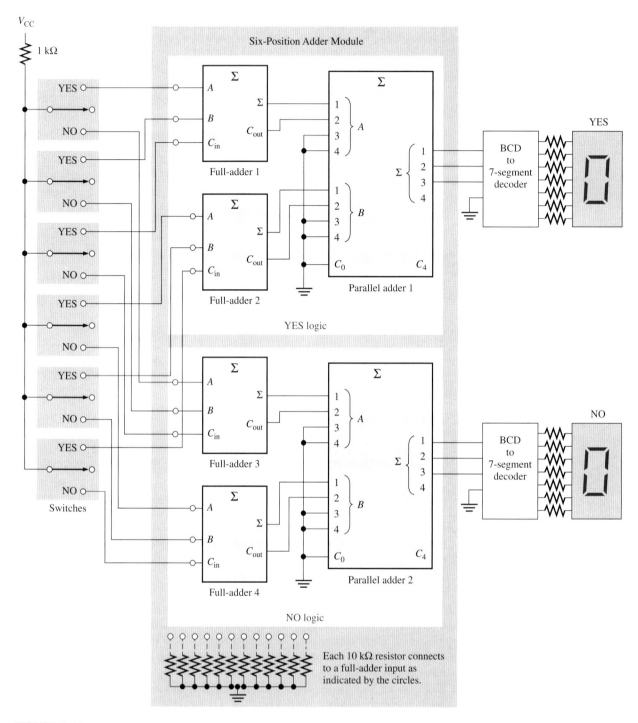

FIGURE 6–14
A voting system using full-adders and parallel binary adders.

SECTION 6–2 REVIEW

1. Two 4-bit numbers (1101 and 1011) are applied to a 4-bit parallel adder. The input carry is 1. Determine the sum (Σ) and the output carry.

2. How many 74LS283 adders would be required to add two binary numbers each representing decimal numbers up through 1000_{10}?

6–3 ■ COMPARATORS

The basic function of a comparator is to compare the magnitudes of two binary quantities to determine the relationship of those quantities. In its simplest form, a comparator circuit determines whether two numbers are equal.

After completing this section, you should be able to

■ Use the exclusive-OR gate as a basic comparator ■ Analyze the internal logic of a magnitude comparator that has both equality and inequality outputs ■ Apply the 74HC85 magnitude comparator to the comparison of two 4-bit numbers ■ Cascade 74HC85s to expand a comparator to eight or more bits

Equality

As you learned in Chapter 3, the exclusive-OR gate can be used as a basic comparator because its output is a 1 if the two input bits are not equal and a 0 if the input bits are equal. Figure 6–15 shows the exclusive-OR gate as a 2-bit comparator.

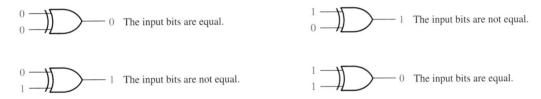

FIGURE 6–15
Basic comparator operation.

In order to compare binary numbers containing two bits each, an additional exclusive-OR gate is necessary. The two least significant bits (LSBs) of the two numbers are compared by gate G_1, and the two most significant bits (MSBs) are compared by gate G_2, as shown in Figure 6–16. If the two numbers are equal, their corresponding bits are the same, and the output of each exclusive-OR gate is a 0. If the corresponding sets of bits are not equal, a 1 occurs on that exclusive-OR gate output.

In order to produce a single output indicating an equality or inequality of two numbers, two inverters and an AND gate can be used, as shown in Figure 6–16. The output of each exclusive-OR gate is inverted and applied to the AND gate input. When the two input bits for each exclusive-OR are equal, the corresponding bits of the numbers are equal, producing a 1 on both inputs to the AND gate and thus a 1 on the output. When the two numbers are not equal, one or both sets of corresponding bits are unequal, and a 0 appears on at least one input to the AND gate to produce a 0 on its output. Thus, the output of the AND gate indicates equality (1) or inequality (0) of the two numbers.

General format: Binary number $A \rightarrow A_1A_0$
Binary number $B \rightarrow B_1B_0$

FIGURE 6–16
Logic diagram for equality comparison of two 2-bit numbers.

Example 6–5 illustrates this operation for two specific cases. The exclusive-OR gate and inverter are replaced by an exclusive-NOR symbol. Recall that in the system application for Chapter 3, you used this basic type of circuit to compare two 8-bit numbers.

EXAMPLE 6–5

Apply each of the following sets of binary numbers to the comparator inputs in Figure 6–17, and determine the output by following the logic levels through the circuit.
(a) 10 and 10   **(b)** 11 and 10

(a)

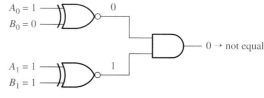

(b)

FIGURE 6–17

Solution
(a) The output is **1** for inputs 10 and 10, as shown in Figure 6–17(a).
(b) The output is **0** for inputs 11 and 10, as shown in Figure 6–17(b).

Related Problem   Repeat the process for binary inputs of 01 and 10.

COMPUTER NOTE
In a computer, the *cache* is a very fast intermediate memory between the central processing unit (CPU) and the slower main memory. The CPU requests data by sending out its *address* (unique location) in memory. Part of this address is called a *tag*. The *tag address comparator* compares the tag from the CPU with the tag from the cache directory. If the two agree, the addressed data is already in the cache and is retrieved very quickly. If the tags disagree, the data must be retrieved from the main memory at a much slower rate.

As you know from Chapter 3, the basic comparator can be expanded to any number of bits. The AND gate sets the condition that all corresponding bits of the two numbers must be equal if the two numbers themselves are equal.

Inequality

In addition to the equality output, many integrated circuit comparators provide additional outputs that indicate which of the two binary numbers being compared is the larger. That is, there is an output that indicates when number A is greater than number B ($A > B$) and an output that indicates when number A is less than number B ($A < B$), as shown in the logic symbol for a 4-bit comparator in Figure 6–18.

FIGURE 6–18
Logic symbol for a 4-bit comparator with in-equality indication.

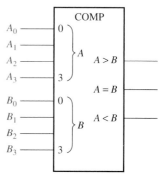

To determine an inequality of binary numbers A and B, you first examine the highest-order bit in each number. The following conditions are possible:

1. If $A_3 = 1$ and $B_3 = 0$, number A is greater than number B.
2. If $A_3 = 0$ and $B_3 = 1$, number A is less than number B.
3. If $A_3 = B_3$, then you must examine the next lower bit position for an inequality.

These three operations are valid for each bit position in the numbers. The general procedure used in a comparator is to check for an inequality in a bit position, starting with the highest order bits (MSBs). When such an inequality is found, the relationship of the two numbers is established, and any other inequalities in lower-order bit positions must be ignored because it is possible for an opposite indication to occur; *the highest-order indication must take precedence.*

EXAMPLE 6–6

Determine the $A = B$, $A > B$, and $A < B$ outputs for the input numbers shown on the comparator in Figure 6–19.

FIGURE 6–19

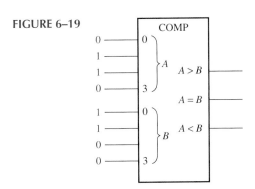

Solution The number on the A inputs is 0110 and the number on the B inputs is 0011. The $A > B$ **output is HIGH and the other outputs are LOW.**

Related Problem What are the comparator outputs when $A_3A_2A_1A_0 = 1001$ and $B_3B_2B_1B_0 = 1010$?

An MSI 4-Bit Magnitude Comparator

The 74HC85 is an MSI comparator that is also available in other IC families. The pin diagram and logic symbol are shown in Figure 6–20. Notice that this device has all the inputs and outputs of the generalized comparator previously discussed and, in addition, has three cascading inputs: $A < B$, $A = B$, $A > B$. These inputs allow several comparators to be cascaded for comparison of any number of bits greater than four. To expand the comparator, the $A < B$, $A = B$, and $A > B$ outputs of the lower-order comparator are connected to the corresponding cascading inputs of the next higher-order comparator. The lowest-order comparator must have a HIGH on the $A = B$ input and LOWs on the $A < B$ and $A > B$ inputs.

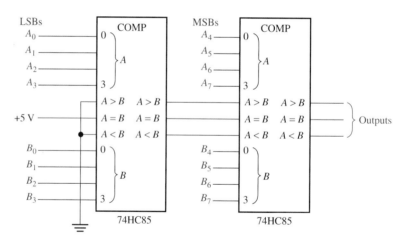

FIGURE 6–20
Pin diagram and logic symbol for the 74HC85 4-bit magnitude comparator (pin numbers are in parentheses).

EXAMPLE 6–7

Use 74HC85 comparators to compare the magnitudes of two 8-bit numbers. Show the comparators with proper interconnections.

Solution Two 74HC85s are required to compare two 8-bit numbers. They are connected as shown in Figure 6–21, in a cascaded arrangement.

FIGURE 6–21
An 8-bit magnitude comparator using two 74HC85s.

Related Problem Expand the circuit in Figure 6–21 to a 16-bit comparator.

Most CMOS devices contain protection circuitry to guard against damage from high static voltages or electric fields. However, precautions must be taken to avoid applications of any voltages higher than maximum rated voltages. For proper operation, input and output voltages should be between ground and V_{CC}. Also, remember that unused inputs must always be connected to an appropriate logic level (ground or V_{CC}). Unused outputs may be left open.

SECTION 6–3 REVIEW

1. The binary numbers $A = 1011$ and $B = 1010$ are applied to the inputs of a 74HC85. Determine the outputs.
2. The binary numbers $A = 11001011$ and $B = 11010100$ are applied to the 8-bit comparator in Figure 6–21. Determine the states of output pins 5, 6, and 7 on each 74HC85.

6–4 ■ DECODERS

The basic function of a decoder is to detect the presence of a specified combination of bits (code) on its inputs and to indicate the presence of that code by a specified output level. In its general form, a decoder has n input lines to handle n bits and from one to 2^n output lines to indicate the presence of one or more n-bit combinations. In this section, several decoders are introduced. The basic principles can be extended to other types of decoders.

After completing this section, you should be able to

- Define *decoder* ■ Design a logic circuit to decode any combination of bits
- Describe the 74HC154 binary-to-decimal decoder ■ Describe the 74HC42 BCD-to-decimal decoder ■ Expand decoders to accommodate larger numbers of bits in a code ■ Describe the 74LS47 BCD-to-7-segment decoder ■ Discuss zero suppression in 7-segment displays ■ Apply decoders to specific applications

COMPUTER NOTE

An *instruction* tells the computer what operation to perform. Instructions are in machine code (1s and 0s) and, in order for the computer to carry out an instruction, the instruction must be decoded. Instruction decoding is one the steps in instruction *pipelining* which are as follows: Instruction is read from the memory (instruction fetch), instruction is decoded, operand(s) read from memory (operand fetch), instruction is executed, result is written back to memory. Basically, pipelining allows the next instruction to begin processing before the current one is completed.

The Basic Binary Decoder

Suppose you need to determine when a binary 1001 occurs on the inputs of a digital circuit. An AND gate can be used as the basic decoding element because it produces a HIGH output only when all of its inputs are HIGH. Therefore, you must make sure that all of the inputs to the AND gate are HIGH when the binary number 1001 occurs; this can be done by inverting the two middle bits (the 0s), as shown in Figure 6–22.

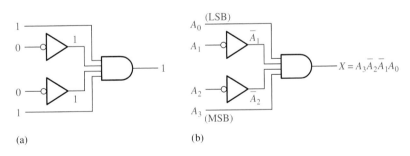

FIGURE 6–22
Decoding logic for the binary code 1001 with an active-HIGH output.

The logic equation for the **decoder** of Figure 6–22(a) is developed as illustrated in Figure 6–22(b). You should verify that the output is 0 except when $A_0 = 1$, $A_1 = 0$, $A_2 = 0$, and $A_3 = 1$ are applied to the inputs. A_0 is the LSB and A_3 is the MSB. *In the representation of a binary number or other weighted code in this book, the LSB is the right-most bit in a horizontal arrangement and the topmost bit in a vertical arrangement, unless specified otherwise.*

If a NAND gate is used in place of the AND gate in Figure 6–22, a LOW output will indicate the presence of the proper binary code, which is 1001 in this case.

EXAMPLE 6–8

Determine the logic required to decode the binary number 1011 by producing a HIGH level on the output.

Solution   The decoding function can be formed by complementing only the variables that appear as 0 in the desired binary number, as follows:

$$X = A_3\overline{A}_2A_1A_0 \qquad (1011)$$

This function can be implemented by connecting the true (uncomplemented) variables A_0, A_1, and A_3 directly to the inputs of an AND gate, and inverting the variable A_2 before applying it to the AND gate input. The decoding logic is shown in Figure 6–23.

FIGURE 6–23
Decoding logic for producing a HIGH output when 1011 is on the inputs.

Related Problem   Develop the logic required to detect the binary code 10010 and produce an active-LOW output.

The 4-Bit Decoder

In order to decode all possible combinations of four bits, sixteen decoding gates are required ($2^4 = 16$). This type of decoder is commonly called a *4-line-to-16-line decoder* because there are four inputs and sixteen outputs or a *1-of-16 decoder* because for any given code on the inputs, one of the sixteen outputs is activated. A list of the sixteen binary codes and their corresponding decoding functions is given in Table 6–4.

TABLE 6–4
Decoding functions and truth table for a 4-line-to-16-line decoder with active-LOW outputs.

Decimal Digit	Binary Inputs				Decoding Function	Outputs																
	A_3	A_2	A_1	A_0		0	1	2	3	4	5	6	7	8	9	10	11	12	13	14	15
0	0	0	0	0	$\overline{A}_3\overline{A}_2\overline{A}_1\overline{A}_0$	0	1	1	1	1	1	1	1	1	1	1	1	1	1	1	1
1	0	0	0	1	$\overline{A}_3\overline{A}_2\overline{A}_1A_0$	1	0	1	1	1	1	1	1	1	1	1	1	1	1	1	1
2	0	0	1	0	$\overline{A}_3\overline{A}_2A_1\overline{A}_0$	1	1	0	1	1	1	1	1	1	1	1	1	1	1	1	1
3	0	0	1	1	$\overline{A}_3\overline{A}_2A_1A_0$	1	1	1	0	1	1	1	1	1	1	1	1	1	1	1	1
4	0	1	0	0	$\overline{A}_3A_2\overline{A}_1\overline{A}_0$	1	1	1	1	0	1	1	1	1	1	1	1	1	1	1	1
5	0	1	0	1	$\overline{A}_3A_2\overline{A}_1A_0$	1	1	1	1	1	0	1	1	1	1	1	1	1	1	1	1
6	0	1	1	0	$\overline{A}_3A_2A_1\overline{A}_0$	1	1	1	1	1	1	0	1	1	1	1	1	1	1	1	1
7	0	1	1	1	$\overline{A}_3A_2A_1A_0$	1	1	1	1	1	1	1	0	1	1	1	1	1	1	1	1
8	1	0	0	0	$A_3\overline{A}_2\overline{A}_1\overline{A}_0$	1	1	1	1	1	1	1	1	0	1	1	1	1	1	1	1
9	1	0	0	1	$A_3\overline{A}_2\overline{A}_1A_0$	1	1	1	1	1	1	1	1	1	0	1	1	1	1	1	1
10	1	0	1	0	$A_3\overline{A}_2A_1\overline{A}_0$	1	1	1	1	1	1	1	1	1	1	0	1	1	1	1	1
11	1	0	1	1	$A_3\overline{A}_2A_1A_0$	1	1	1	1	1	1	1	1	1	1	1	0	1	1	1	1
12	1	1	0	0	$A_3A_2\overline{A}_1\overline{A}_0$	1	1	1	1	1	1	1	1	1	1	1	1	0	1	1	1
13	1	1	0	1	$A_3A_2\overline{A}_1A_0$	1	1	1	1	1	1	1	1	1	1	1	1	1	0	1	1
14	1	1	1	0	$A_3A_2A_1\overline{A}_0$	1	1	1	1	1	1	1	1	1	1	1	1	1	1	0	1
15	1	1	1	1	$A_3A_2A_1A_0$	1	1	1	1	1	1	1	1	1	1	1	1	1	1	1	0

If an active-LOW output is required for each decoded number, the entire decoder can be implemented with NAND gates and inverters. In order to decode each of the sixteen binary codes, sixteen NAND gates are required (AND gates can be used to produce active-HIGH outputs).

A logic symbol for a 4-line-to-16-line (1-of-16) decoder with active-LOW outputs is shown in Figure 6–24. The BIN/DEC label indicates that a binary input makes the corresponding decimal output active. The input labels 8, 4, 2, and 1 represent the binary weights of the input bits ($2^3 2^2 2^1 2^0$).

FIGURE 6–24
Logic symbol for a 4-line-to-16-line (1-of-16) decoder.

An MSI 1-of-16 Decoder

The 74HC154 is a good example of an MSI decoder. The logic symbol is shown in Figure 6–25. There is an enable function (*EN*) provided on this device, which is implemented with a NOR gate used as a negative-AND. A LOW level on each chip select input, $\overline{CS}_1$ and $\overline{CS}_2$, is required in order to make the enable gate output (*EN*) HIGH. The enable gate output

FIGURE 6–25
Pin diagram and logic symbol for the 74HC154 1-of-16 decoder.

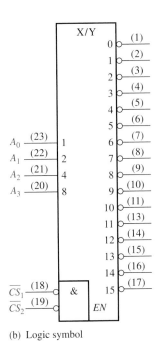

(a) Pin diagram   (b) Logic symbol

is connected to an input of *each* NAND gate in the decoder, so it must be HIGH for the NAND gates to be enabled. If the enable gate is not activated by a LOW on both inputs, then all sixteen decoder outputs (Y) will be HIGH regardless of the states of the four input variables, A_0, A_1, A_2, and A_3.

EXAMPLE 6–9

A certain application requires that a 5-bit number be decoded. Use 74HC154 decoders to implement the logic. The binary number is represented by the format $A_4A_3A_2A_1A_0$.

Solution Since the 74HC154 can handle only four bits, two decoders must be used to decode five bits. The fifth bit, A_4, is connected to the chip select inputs, $\overline{CS}_1$ and $\overline{CS}_2$, of one decoder, and $\overline{A}_4$ is connected to the enable inputs of the other decoder, as shown in Figure 6–26. When the decimal number is 15 or less, $A_4 = 0$, and the low-order decoder is enabled and the high-order decoder is disabled. When the decimal number is greater than 15, $A_4 = 1$ so $\overline{A}_4 = 0$, and the high-order decoder is enabled and the low-order decoder is disabled.

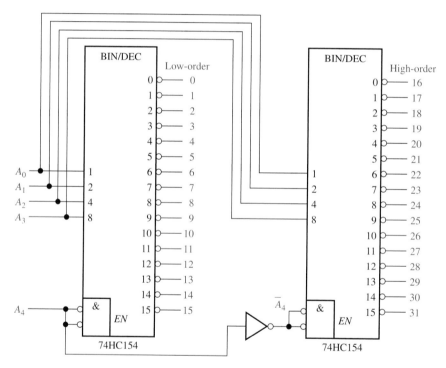

FIGURE 6–26
A 5-bit decoder using 74HC154s.

Related Problem Determine the output in Figure 6–26 that is activated for the binary input 10110.

Application Example

Decoders are used in many types of applications. One example is in computers for input/output selection as depicted in the general diagram of Figure 6–27.

Computers must communicate with a variety of external devices called *peripherals* by sending and/or receiving data through what is known as input/output (I/O) ports. These

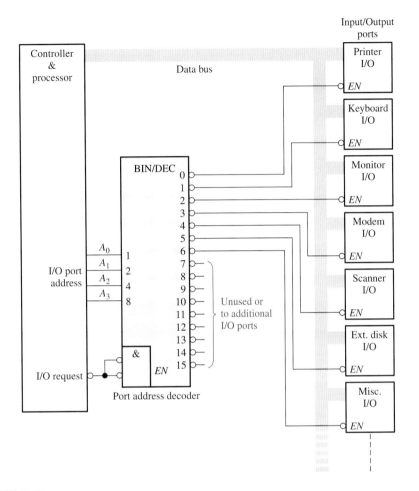

FIGURE 6–27

A simplified computer I/O port system with a port address decoder with only four address lines shown.

external devices include printers, modems, scanners, external disk drives, keyboard, video monitors, and other computers. As indicated in Figure 6–27, a decoder is used to select the I/O port as determined by the computer so that data can be sent or received from a specific external device.

Each I/O port has a number, called an address, which uniquely identifies it. When the computer wants to communicate with a particular device, it issues the appropriate address code for the I/O port to which that particular device is connected. This binary port address is decoded and the appropriate decoder output is activated to enable the I/O port.

As shown in Figure 6–27, binary data are transferred within the computer on a data bus, which is a set of parallel lines. For example, an 8-bit bus consists of eight parallel lines that can carry one byte of data at a time. The data bus goes to all of the I/O ports, but any data coming in or going out will only pass through the port that is enabled by the port address decoder.

The BCD-to-Decimal Decoder

The BCD-to-decimal decoder converts each BCD code (8421 code) into one of ten possible decimal digit indications. It is frequently referred as a *4-line-to-10-line decoder* or a *1-of-10 decoder.*

The method of implementation is the same as for the 1-of-16 decoder previously discussed, except that only ten decoding gates are required because the BCD code represents only the ten decimal digits 0 through 9. A list of the ten BCD codes and their corresponding decoding functions is given in Table 6–5. Each of these decoding functions is implemented with NAND gates to provide active-LOW outputs. If an active-HIGH output is required, AND gates are used for decoding. The logic is identical to that of the first ten decoding gates in the 1-of-16 decoder (see Table 6–4).

TABLE 6–5
BCD decoding functions.

Decimal Digit	BCD Code				Decoding Function
	A_3	A_2	A_1	A_0	
0	0	0	0	0	$\overline{A_3}\overline{A_2}\overline{A_1}\overline{A_0}$
1	0	0	0	1	$\overline{A_3}\overline{A_2}\overline{A_1}A_0$
2	0	0	1	0	$\overline{A_3}\overline{A_2}A_1\overline{A_0}$
3	0	0	1	1	$\overline{A_3}\overline{A_2}A_1A_0$
4	0	1	0	0	$\overline{A_3}A_2\overline{A_1}\overline{A_0}$
5	0	1	0	1	$\overline{A_3}A_2\overline{A_1}A_0$
6	0	1	1	0	$\overline{A_3}A_2A_1\overline{A_0}$
7	0	1	1	1	$\overline{A_3}A_2A_1A_0$
8	1	0	0	0	$A_3\overline{A_2}\overline{A_1}\overline{A_0}$
9	1	0	0	1	$A_3\overline{A_2}\overline{A_1}A_0$

EXAMPLE 6–10

The 74HC42 is an integrated circuit BCD-to-decimal decoder. The logic symbol is shown in Figure 6–28. If the input waveforms in Figure 6–29(a) are applied to the inputs of the 74HC42, show the output waveforms.

FIGURE 6–28
The 74HC42 BCD-to-decimal decoder.

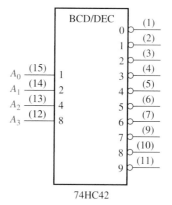

74HC42

Solution   The output waveforms are shown in Figure 6–29(b). As you can see, the inputs are sequenced through the BCD for digits 0 through 9. The output waveforms in the timing diagram indicate that sequence.

Related Problem   Construct a timing diagram showing input and output waveforms for the case where the binary inputs sequence through the decimal numbers as follows: 0, 2, 4, 6, 8, 1, 3, 5, and 9.

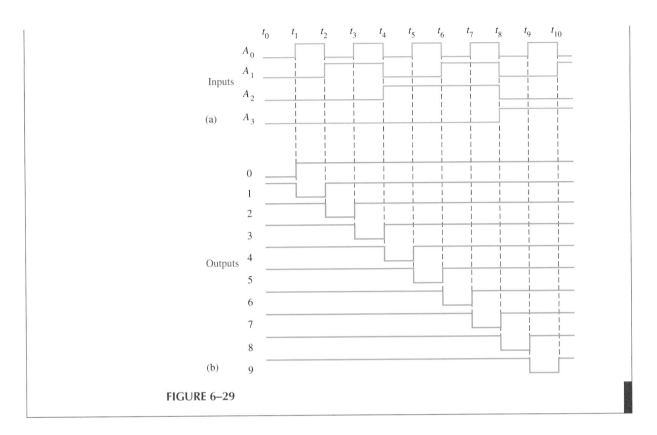

FIGURE 6–29

The BCD-to-7-Segment Decoder

As you learned in the system application of Chapter 4, the BCD-to-7-segment decoder accepts the BCD code on its inputs and provides outputs to drive 7-segment display devices to produce a decimal readout. The logic diagram for a basic 7-segment decoder is shown in Figure 6–30.

FIGURE 6–30
Logic symbol for a BCD-to-7-segment decoder/driver with active-LOW outputs.

An MSI BCD-to-7-Segment Decoder/Driver

The 74LS47 is an example of an MSI device that decodes a BCD input and drives a 7-segment display. In addition to its decoding and segment drive capability, the 74LS47 has several additional features as indicated by the $\overline{LT}$, $\overline{RBI}$, $\overline{BI/RBO}$ functions in the logic symbol of Figure 6–31. As indicated by the bubbles on the logic symbol, all of the outputs (*a* through *g*) are active-LOW as are the $\overline{LT}$ (lamp test), *RBI* (ripple blanking input), and $\overline{BI/RBO}$ (blanking input/ripple blanking output) functions. The outputs can drive a common-anode 7-segment display directly. Recall that 7-segment displays were discussed in Chapter 4.

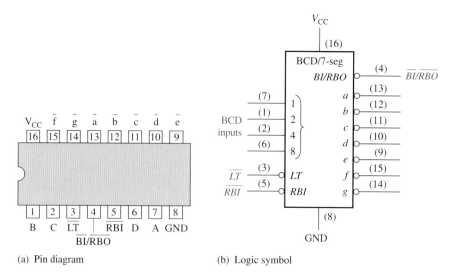

FIGURE 6–31
Pin diagram and logic symbol for the 74LS47 BCD-to-7-segment decoder/driver.

In addition to decoding a BCD input and producing the appropriate 7-segment outputs, the 74LS47 has lamp test and zero suppression capability.

Lamp Test When a LOW is applied to the $\overline{LT}$ input and the $\overline{BI}/\overline{RBO}$ is HIGH, all of the 7 segments in the display are turned on. Lamp test is used to verify that no segments are burned out.

Zero Suppression **Zero suppression** is a feature used for multidigit displays to blank out unnecessary zeros. For example, in a 6-digit display the number 6.4 may be displayed as 006.400 if the zeros are not blanked out. Blanking the zeros at the front of a number is called *leading zero suppression* and blanking the zeros at the back of the number is called *trailing zero suppression.* Keep in mind that only nonessential zeros are blanked. With zero suppression, the number 030.080 will be displayed as 30.08 (the essential zeros remain).

Zero suppression in the 74LS47 is accomplished using the $\overline{RBI}$ and $\overline{BI}/\overline{RBO}$ functions. $\overline{RBI}$ is the ripple blanking input and $\overline{RBO}$ is the ripple blanking output on the 74LS47; these are used for zero suppression. $\overline{BI}$ is the blanking input which shares the same pin with $\overline{RBO}$; in other words, the $\overline{BI}/\overline{RBO}$ pin can be used as an input or an output. When used as a $\overline{BI}$ (blanking input), all segment outputs are HIGH (nonactive) when $\overline{BI}$ is LOW, which overrides all other inputs. The $\overline{BI}$ function is not part of the zero suppression capability of the device.

All of the segment outputs of the decoder are nonactive (HIGH) if a zero code (0000) is on its BCD inputs and if its $\overline{RBI}$ is LOW. This causes the display to be blank and produces a LOW $\overline{RBO}$.

The logic diagram in Figure 6–32(a) illustrates leading zero suppression for a whole number. The highest-order digit position (left-most) is always blanked if a zero code is on its BCD inputs because the $\overline{RBI}$ of the most-significant decoder is made LOW by connecting it to ground. The $\overline{RBO}$ of each decoder is connected to the $\overline{RBI}$ of the next lowest-order decoder so that all zeros to the left of the first nonzero digit are blanked. For example, in part (a) of the figure the two highest-order digits are zeros and therefore are blanked. The remaining two digits, 3 and 9 are displayed.

The logic diagram in Figure 6–32(b) illustrates trailing zero suppression for a fractional number. The lowest-order digit (right-most) is always blanked if a zero code is on its

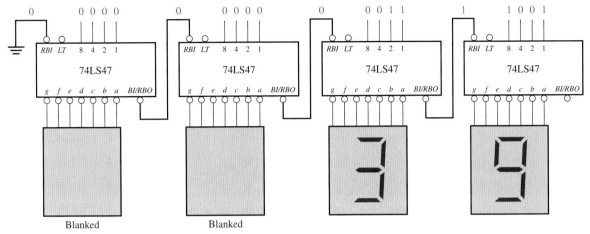

(a) Illustration of leading zero suppression

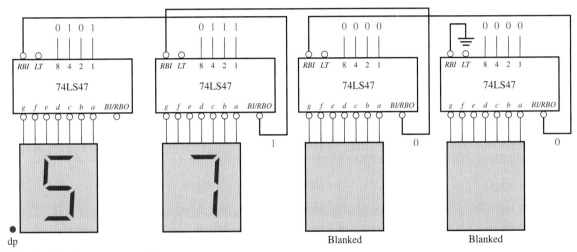

(b) Illustration of trailing zero suppression

FIGURE 6–32

Examples of zero suppression using the 74LS47 BCD to 7-segment decoder/driver.

BCD inputs because the $\overline{RBI}$ is connected to ground. The $\overline{RBO}$ of each decoder is connected to the $\overline{RBI}$ of the next highest-order decoder so that all zeros to the right of the first nonzero digit are blanked. In part (b) of the figure, the two lowest-order digits are zeros and therefore are blanked. The remaining two digits, 5 and 7 are displayed. To combine both leading and trailing zero suppression in one display and to have decimal point capability, additional logic is required.

SECTION 6–4 REVIEW

1. A 3-line-to-8-line decoder can be used for octal-to-decimal decoding. When a binary 101 is on the inputs, which output line is activated?
2. How many 74HC154 1-of-16 decoders are necessary to decode a 6-bit binary number?
3. Would you select a decoder/driver with active-HIGH or active-LOW outputs to drive a common-cathode 7-segment LED display?

6–5 ■ ENCODERS

An encoder is a combinational logic circuit that essentially performs a "reverse" decoder function. An encoder accepts an active level on one of its inputs representing a digit, such as a decimal or octal digit, and converts it to a coded output, such as BCD or binary. Encoders can also be devised to encode various symbols and alphabetic characters. The process of converting from familiar symbols or numbers to a coded format is called encoding.

After completing this section, you should be able to

■ Determine the logic for a decimal encoder   ■ Explain the purpose of the priority feature in encoders   ■ Describe the 74HC147 decimal-to-BCD priority encoder   ■ Describe the 74F148 octal-to-binary priority encoder   ■ Expand an encoder   ■ Apply the encoder to a specific application

COMPUTER NOTE

An *assembler* can be thought of as a software encoder because it interprets the mnemonic instructions with which a program is written and carries out the applicable encoding to convert each mnemonic to a machine code instruction (series of 1s and 0s) which the computer can understand. Examples of mnemonic instructions are ADD, MOV (move data), MUL (multiply), XOR, JMP (jump), and OUT (output to a port).

The Decimal-to-BCD Encoder

This type of **encoder** has ten inputs—one for each decimal digit—and four outputs corresponding to the BCD code, as shown in Figure 6–33. This is a basic 10-line-to-4-line encoder.

FIGURE 6–33
Logic symbol for a decimal-to-BCD encoder.

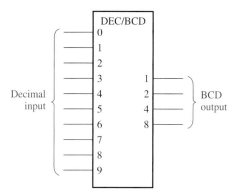

The BCD (8421) code is listed in Table 6–6. From this table you can determine the relationship between each BCD bit and the decimal digits in order to analyze the logic. For instance, the most significant bit of the BCD code, A_3, is always a 1 for decimal digit 8 or 9. An OR expression for bit A_3 in terms of the decimal digits can therefore be written as

$$A_3 = 8 + 9$$

Bit A_2 is always a 1 for decimal digit 4, 5, 6, or 7 and can be expressed as an OR function as follows:

$$A_2 = 4 + 5 + 6 + 7$$

Bit A_1 is always a 1 for decimal digit 2, 3, 6, or 7 and can be expressed as

$$A_1 = 2 + 3 + 6 + 7$$

Finally, A_0 is always a 1 for decimal digit 1, 3, 5, 7, or 9. The expression for A_0 is

$$A_0 = 1 + 3 + 5 + 7 + 9$$

TABLE 6–6

Decimal Digit	BCD Code			
	A_3	A_2	A_1	A_0
0	0	0	0	0
1	0	0	0	1
2	0	0	1	0
3	0	0	1	1
4	0	1	0	0
5	0	1	0	1
6	0	1	1	0
7	0	1	1	1
8	1	0	0	0
9	1	0	0	1

Now let's implement the logic circuitry required for encoding each decimal digit to a BCD code by using the logic expressions just developed. It is simply a matter of ORing the appropriate decimal digit input lines to form each BCD output. The basic encoder logic resulting from these expressions is shown in Figure 6–34.

The basic operation of the circuit in Figure 6–34 is as follows: When a HIGH appears on *one* of the decimal digit input lines, the appropriate levels occur on the four BCD output lines. For instance, if input line 9 is HIGH (assuming all other input lines are LOW), this condition will produce a HIGH on outputs A_0 and A_3 and LOWs on outputs A_1 and A_2, which is the BCD code (1001) for decimal 9.

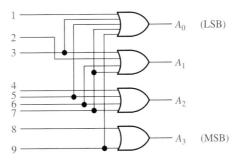

FIGURE 6–34
Basic logic diagram of a decimal-to-BCD encoder. A 0-digit input is not needed because the BCD outputs are all LOW when there are no HIGH inputs.

The Decimal-to-BCD Priority Encoder This type of encoder performs the same basic encoding function as previously discussed. It also offers additional flexibility in that it can be used in applications that require priority detection. The priority function means that the encoder will produce a BCD output corresponding to the *highest-order decimal digit* input that is active and will ignore any other lower-order active inputs. For instance, if the 6 and the 3 inputs are both active, the BCD output is 0110 (which represents decimal 6).

An MSI Decimal-to-BCD Encoder

The 74HC147 is a **priority encoder** with active-LOW inputs (0) for decimal digits 1 through 9 and active-LOW BCD outputs as indicated in the logic symbol in Figure 6–35. A BCD zero output is represented when none of the inputs is active. The device pin numbers are in parentheses.

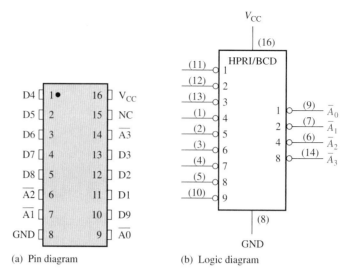

(a) Pin diagram   (b) Logic diagram

FIGURE 6–35
Pin diagram and logic symbol for the 74HC147 decimal-to-BCD priority encoder (HPRI means highest value input has priority).

EXAMPLE 6–11

If LOW levels appear on pins 1, 4, and 13 of the 74HC147 shown in Figure 6–35, indicate the state of the four outputs. All other inputs are HIGH.

Solution Pin 4 is the highest-order decimal digit input having a LOW level and represents decimal 7. Therefore, the output levels indicate the BCD code for decimal 7 where $\overline{A}_0$ is the LSB and $\overline{A}_3$ is the MSB. Output $\overline{A}_0$ is LOW, $\overline{A}_1$ is LOW, $\overline{A}_2$ is LOW, and $\overline{A}_3$ is HIGH.

Related Problem What are the outputs of the 74HC147 if all its inputs are LOW? If all its inputs are HIGH?

An MSI 8-Line-to-3-Line Encoder

The 74F148 is a priority encoder that has eight active-LOW inputs and three active-LOW binary outputs, as shown in Figure 6–36. This device can be used for converting octal inputs (recall that the octal digits are 0 through 7) to a 3-bit binary code. To enable the device, the *EI* (enable input) must be LOW. It also has the *EO* (enable output) and *GS* output for expansion purposes. The *EO* is LOW when the *EI* is LOW and none of the inputs (0 through 7) is active. *GS* is LOW when *EI* is LOW and any of the inputs is active.

The 74F148 can be expanded to a 16-line-to-4-line encoder by connecting the *EO* of the higher-order encoder to the *EI* of the lower-order encoder and negative-ORing the corresponding binary outputs as shown in Figure 6–37. The *EO* is used as the fourth and most-significant bit. This particular configuration produces active-HIGH outputs for the 4-bit binary number.

Application Example

A classic application example is a keyboard encoder. The ten decimal digits on the keyboard of a computer, for example, must be encoded for processing by the logic circuitry. When one of the keys is pressed, the decimal digit is encoded to the corresponding BCD

FIGURE 6–36
Logic symbol for the 74F148 8-line-to-3-line encoder.

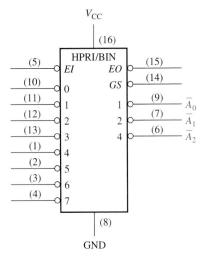

FIGURE 6–37
A 16-line-to-4 line encoder using 74F148s and external logic.

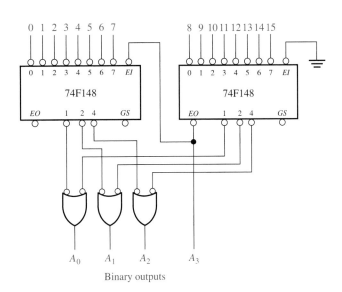

code. Figure 6–38 shows a simple keyboard encoder arrangement using a 74HC147 priority encoder. The keys are represented by ten push-button switches, each with a **pull-up resistor** to $+V$. The pull-up resistor ensures that the line is HIGH when a key is not depressed. When a key is depressed, the line is connected to ground, and a LOW is applied to the corresponding encoder input. The zero key is not connected because the BCD output represents zero when none of the other keys is depressed.

The BCD complement output of the encoder goes into a storage device, and each successive BCD code is stored until the entire number has been entered. Methods of storing BCD numbers and binary data are covered in later chapters.

FIGURE 6–38
A simplified keyboard encoder.

SECTION 6–5 REVIEW

1. Suppose that HIGH levels are applied to the 2 input and the 9 input of the circuit in Figure 6–34.
   (a) What are the states of the output lines?
   (b) Does this represent a valid BCD code?
   (c) What is the restriction on the encoder logic in Figure 6–34?
2. (a) What is the $\overline{A}_3\overline{A}_2\overline{A}_1\overline{A}_0$ output when LOWs are applied to pins 1 and 5 of the 74HC147 in Figure 6–35?
   (b) What does this output represent?

6–6 ■ CODE CONVERTERS

In this section, we will examine some methods of using combinational logic circuits to convert from one code to another.

After completing this section, you should be able to

■ Explain the process for converting BCD to binary ■ Apply the 74184 and the 74185 to BCD-to-binary and binary-to-BCD conversion, respectively ■ Expand the 74184 and the 74185 to handle larger binary and BCD numbers ■ Use exclusive-OR gates for conversions between binary and Gray codes

BCD-to-Binary Conversion

One method of BCD-to-binary code conversion uses adder circuits. The basic conversion process is as follows:

1. The value, or weight, of each bit in the BCD number is represented by a binary number.
2. All of the binary representations of the weights of bits that are 1s in the BCD number are added.
3. The result of this addition is the binary equivalent of the BCD number.

A more concise statement of this operation is

The binary numbers representing the weights of the BCD bits are summed to produce the total binary number.

Let's examine an 8-bit BCD code (one that represents a 2-digit decimal number) to understand the relationship between BCD and binary. For instance, you already know that the decimal number 87 can be expressed in BCD as

$$\underbrace{1000}_{8} \quad \underbrace{0111}_{7}$$

The left-most 4-bit group represents 80, and the right-most 4-bit group represents 7. That is, the left-most group has a weight of 10, and the right-most group has a weight of 1. Within each group, the binary weight of each bit is as follows:

	Tens Digit				Units Digit			
Weight:	80	40	20	10	8	4	2	1
Bit designation:	B_3	B_2	B_1	B_0	A_3	A_2	A_1	A_0

The binary equivalent of each BCD bit is a binary number representing the weight of that bit within the total BCD number. This representation is given in Table 6–7.

TABLE 6–7
Binary representations of BCD bit weights.

BCD Bit	BCD Weight	(MSB) 64	32	Binary Representation 16	8	4	2	(LSB) 1
A_0	1	0	0	0	0	0	0	1
A_1	2	0	0	0	0	0	1	0
A_2	4	0	0	0	0	1	0	0
A_3	8	0	0	0	1	0	0	0
B_0	10	0	0	0	1	0	1	0
B_1	20	0	0	1	0	1	0	0
B_2	40	0	1	0	1	0	0	0
B_3	80	1	0	1	0	0	0	0

If the binary representations for the weights of all the 1s in the BCD number are added, the result is the binary number that corresponds to the BCD number. Example 6–12 illustrates this.

EXAMPLE 6–12

Convert the BCD numbers 00100111 (decimal 27) and 10011000 (decimal 98) to binary.

Solution Write the binary representations of the weights of all 1s appearing in the numbers, and then add them together.

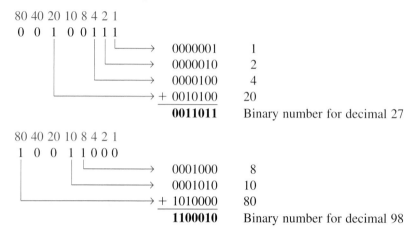

```
80 40 20 10 8 4 2 1
 0  0  1  0 0 1 1 1
                    →   0000001        1
                    →   0000010        2
                    →   0000100        4
                    → + 0010100       20
                        0011011       Binary number for decimal 27

80 40 20 10 8 4 2 1
 1  0  0  1 1 0 0 0
                    →   0001000        8
                    →   0001010       10
                    → + 1010000       80
                        1100010       Binary number for decimal 98
```

Related Problem Show the process of converting 01000001 in BCD to binary.

With this basic procedure in mind, let's determine how the process can be implemented with logic circuits. Once the binary representation for each 1 in the BCD number is determined, adder circuits can be used to add the 1s in each column of the binary representation. The 1s occur in a given column only when the corresponding BCD bit is a 1. The occurrence of a BCD 1 can therefore be used to generate the proper binary 1 in the appropriate column of the adder structure. To handle a two-decimal-digit (two-decade) BCD code, eight BCD input lines and seven binary outputs are required. (It takes seven bits to represent binary numbers through ninety-nine.)

MSI BCD-to-Binary and Binary-to-BCD Converters

MSI code converters are generally implemented with programmable logic devices (PLDs) which are introduced in Chapter 7. Examples are the 74184, a PLD preprogrammed as a BCD-to-binary converter, and the 74185, a PLD preprogrammed as a binary-to-BCD converter. Logic symbols for these devices used as 6-bit converters are shown in Figure 6–39. Figure 6–40 and Figure 6–41 show how these devices can be expanded for conversions involving larger numbers of bits.

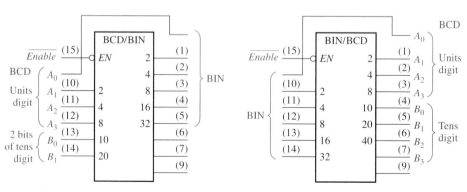

(a) The 74184 6-bit BCD-to-binary converter   (b) The 74185 6-bit binary-to-BCD converter

FIGURE 6–39

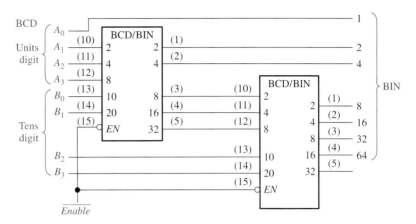

FIGURE 6–40
74184s expanded for conversion of two BCD digits to binary.

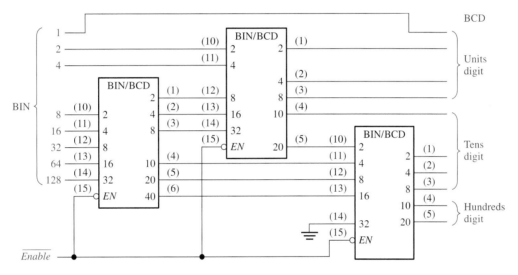

FIGURE 6–41
74185s expanded for 8-bit binary-to-BCD conversion.

Binary-to-Gray and Gray-to-Binary Conversion

The basic process for Gray-binary conversions was covered in Chapter 2. Exclusive-OR gates can be used for these conversions. Programmable logic devices (PLDs) can also be programmed for these code conversions. Figure 6–42 shows a 4-bit binary-to-Gray code converter, and Figure 6–43 illustrates a 4-bit Gray-to-binary converter.

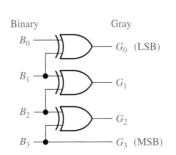

FIGURE 6–42
Four-bit binary-to-Gray conversion logic.

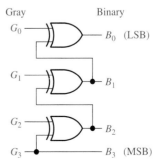

FIGURE 6–43
Four-bit Gray-to-binary conversion logic.

EXAMPLE 6–13 (a) Convert the binary number 0101 to Gray code with exclusive-OR gates.
(b) Convert the Gray code 1011 to binary with exclusive-OR gates.

Solution
(a) 0101_2 is 0111 Gray. See Figure 6–44(a).
(b) 1011 Gray is 1101_2. See Figure 6–44(b).

FIGURE 6–44

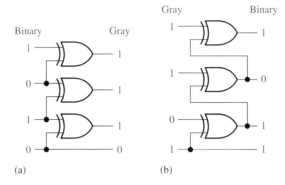

(a)   (b)

Related Problem How many exclusive-OR gates are required to convert 8-bit binary to Gray?

**SECTION 6–6
REVIEW**

1. Convert the BCD number 10000101 to binary.
2. Draw the logic diagram for converting an 8-bit binary number to Gray code.

6–7 ■ MULTIPLEXERS (DATA SELECTORS)

A multiplexer (MUX) is a device that allows digital information from several sources to be routed onto a single line for transmission over that line to a common destination. The basic multiplexer has several data-input lines and a single output line. It also has data-select inputs, which permit digital data on any one of the inputs to be switched to the output line. Multiplexers are also known as data selectors.

After completing this section, you should be able to

■ Explain the basic operation of a multiplexer ■ Describe the 74HC151 and the 74HC157A MSI multiplexers ■ Expand a multiplexer to handle more data inputs ■ Use the multiplexer as a logic function generator

A logic symbol for a 4-input **multiplexer (MUX)** is shown in Figure 6–45. Notice that there are two data-select lines because with two select bits, any one of the four data-input lines can be selected.

In Figure 6–45, a 2-bit code on the data-select (S) inputs will allow the data on the selected data input to pass through to the data output. If a binary 0 ($S_1 = 0$ and $S_0 = 0$) is applied to the data-select lines, the data on input D_0 appear on the data-output line. If a binary 1 ($S_1 = 0$ and $S_0 = 1$) is applied to the data-select lines, the data on input D_1 appear on the data output. If a binary 2 ($S_1 = 1$ and $S_0 = 0$) is applied, the data on D_2 appear on the output. If a binary 3 ($S_1 = 1$ and $S_0 = 1$) is applied, the data on D_3 are switched to the output line. A summary of this operation is given in Table 6–8.

FIGURE 6–45
Logic symbol for a 1-of-4 data selector/multiplexer.

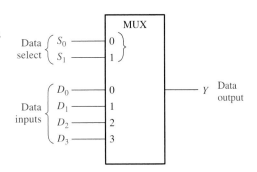

TABLE 6–8
Data selection for a 1-of-4 multiplexer.

Data-select Inputs		Input Selected
S_1	S_0	
0	0	D_0
0	1	D_1
1	0	D_2
1	1	D_3

COMPUTER NOTE
A *bus* is an internal pathway along which electrical signals are sent from one part of a computer to another. In computer networks, a *shared bus* is one that is connected to all the microprocessors in the system in order to exchange data. A shared bus may contain memory and input/output devices that can be accessed by all the microprocessors in the system. Access to the shared bus is controlled by a *bus arbiter* (a multiplexer of sorts) that allows only one microprocessor at a time to use the system's shared bus.

Now let's look at the logic circuitry required to perform this multiplexing operation. The data output is equal to the state of the *selected* data input. You can therefore, derive a logic expression for the output in terms of the data input and the select inputs.

The data output is equal to D_0 only if $S_1 = 0$ and $S_0 = 0$: $Y = D_0\overline{S}_1\overline{S}_0$.

The data output is equal to D_1 only if $S_1 = 0$ and $S_0 = 1$: $Y = D_1\overline{S}_1 S_0$.

The data output is equal to D_2 only if $S_1 = 1$ and $S_0 = 0$: $Y = D_2 S_1\overline{S}_0$.

The data output is equal to D_3 only if $S_1 = 1$ and $S_0 = 1$: $Y = D_3 S_1 S_0$.

When these terms are ORed, the total expression for the data output is

$$Y = D_0\overline{S}_1\overline{S}_0 + D_1\overline{S}_1 S_0 + D_2 S_1\overline{S}_0 + D_3 S_1 S_0$$

The implementation of this equation requires four 3-input AND gates, a 4-input OR gate, and two inverters to generate the complements of S_1 and S_0, as shown in Figure 6–46. Because data can be selected from any one of the input lines, this circuit is also referred to as a **data selector.**

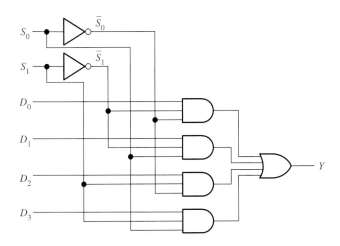

FIGURE 6–46
Logic diagram for a 4-input multiplexer.

EXAMPLE 6–14   The data-input and data-select waveforms in Figure 6–47(a) are applied to the multiplexer in Figure 6–46. Determine the output waveform in relation to the inputs.

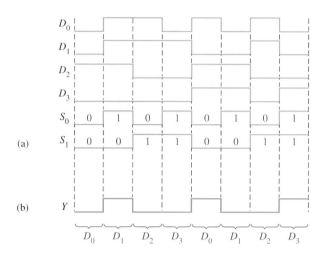

FIGURE 6–47

Solution   The binary state of the data-select inputs during each interval determines which data input is selected. Notice that the data-select inputs go through a repetitive binary sequence 00, 01, 10, 11, 00, 01, 10, 11, and so on. The resulting output waveform is shown in Figure 6–47(b).

Related Problem   Construct a timing diagram showing all inputs and the output if the S_0 and S_1 waveforms in Figure 6–47 are interchanged.

An MSI Quad 2-Input Data Selector/Multiplexer

The 74HC157A, as well as its LS version, consists of four separate 2-input multiplexers. Each of the four multiplexers shares a common data-select line and a common *Enable*. Because there are only two inputs to be selected in each multiplexer, a single data-select input is sufficient.

A LOW on the $\overline{Enable}$ input allows the selected input data to pass through to the output. A HIGH on the $\overline{Enable}$ input prevents data from going through to the output; that is, it disables the multiplexers.

The ANSI/IEEE Logic Symbol   The pin diagram for the 74HC157A is shown in Figure 6–48(a). The ANSI/IEEE logic symbol for the 74HC157A is shown in Figure 6–48(b). Notice that the four multiplexers are indicated by the partitioned outline and that the inputs common to all four multiplexers are indicated as inputs to the notched block at the top, which is called the *common control block*. All labels within the upper MUX block apply to the other blocks below it.

Notice the 1 and $\overline{1}$ labels in the MUX blocks and the G1 label in the common control block. These labels are an example of the **dependency notation** system specified in the ANSI/IEEE Standard 91–1984. In this case G1 indicates an AND relationship between the data-select input and the data inputs with 1 or $\overline{1}$ labels. (The $\overline{1}$ means that the AND relationship applies to the complement of the G1 input.) In other words, when

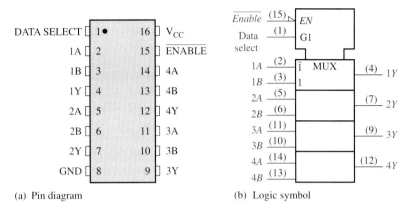

FIGURE 6–48

Pin diagram and logic symbol for the 74HC157A quadruple 2-input data selector/multiplexer.

the data-select input is HIGH, the *B* inputs of the multiplexers are selected; and when the data-select input is LOW, the *A* inputs are selected. A "G" is always used to denote AND dependency. Other aspects of dependency notation are introduced as appropriate throughout the book.

An MSI 8-Input Data Selector/Multiplexer

The 74HC151 has eight data inputs (D_0–D_7) and, therefore, three data-select or address input lines (S_0–S_2). Three bits are required to select any one of the eight data inputs ($2^3 = 8$). A LOW on the $\overline{Enable}$ input allows the selected input data to pass through to the output. Notice that the data output and its complement are both available. The pin diagram is shown in Figure 6–49(a), and the ANSI/IEEE logic symbol is shown in part (b). In this case there is no need for a common control block on the logic symbol because there is only one multiplexer to be controlled, not four as in the 74HC157A. The $G\frac{0}{7}$ label within the logic symbol indicates the AND relationship between the data-select inputs and each of the data inputs 0 through 7.

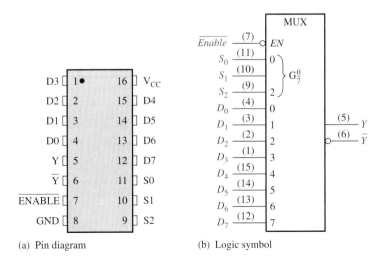

FIGURE 6–49

Pin diagram and logic symbol for the 74HC151 8-input data selector/multiplexer.

EXAMPLE 6–15

Use 74HC151s and any other logic necessary to multiplex 16 data lines onto a single data-output line.

Solution   An implementation of this system is shown in Figure 6–50. Four bits are required to select one of 16 data inputs ($2^4 = 16$). In this application the $\overline{Enable}$ input is used as the most significant data-select bit. When the MSB in the data-select code is LOW, the left 74HC151 is enabled, and one of the data inputs (D_0 through D_7) is selected by the other three data-select bits. When the data-select MSB is HIGH, the right 74HC151 is enabled, and one of the data inputs (D_8 through D_{15}) is selected. The selected input data are then passed through to the negative-OR gate and onto the single output line.

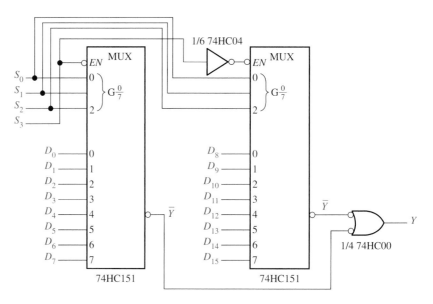

FIGURE 6–50
A 16-input multiplexer.

Related Problem   Determine the codes on the select inputs required to select each of the following data inputs: D_0, D_4, D_8, and D_{13}.

Application Examples

A 7-Segment Display Multiplexer   Figure 6–51 shows a simplified method of multiplexing BCD numbers to a 7-segment display. In this example, 2-digit numbers are displayed on the 7-segment readout by the use of a single BCD-to-7-segment decoder. This basic method of display multiplexing can be extended to displays with any number of digits. The basic operation is as follows.

Two BCD digits ($A_3A_2A_1A_0$ and $B_3B_2B_1B_0$) are applied to the multiplexer inputs. A square wave is applied to the data-select line, and when it is LOW, the A bits ($A_3A_2A_1A_0$) are passed through to the inputs of the 74LS48 BCD-to-7-segment decoder. The LOW on the data-select also puts a LOW on the 1 input of the 74LS139 2-line-to-4-line decoder, thus activating its 0 output and enabling the A-digit display by effectively connecting its common terminal to ground. The A digit is now on and the B digit is off.

FIGURE 6–51
Simplified 7-segment display multiplexing logic.

When the data-select line goes HIGH, the B bits $(B_3B_2B_1B_0)$ are passed through to the inputs of the BCD-to-7-segment decoder. Also, the 74LS139 decoder's 1 output is activated, thus enabling the B-digit display. The B digit is now on and the A digit is off. The cycle repeats at the frequency of the data-select square wave. This frequency must be high enough (about 30 Hz) to prevent visual flicker as the digit displays are multiplexed.

A Logic Function Generator A useful application of the data selector/multiplexer is in the generation of combinational logic functions in sum-of-products form. When used in this way, the device can replace discrete gates, can often greatly reduce the number of ICs, and can make design changes much easier.

To illustrate, a 74HC151 8-input data selector/multiplexer can be used to implement any specified 3-variable logic function if the variables are connected to the data-select inputs and each data input is set to the logic level required in the truth table for that function. For example, if the function is a 1 when the variable combination is $\overline{A}_2A_1\overline{A}_0$, the 2 input (selected by 010) is connected to a HIGH. This HIGH is passed through to the output when this particular combination of variables occurs on the data-select lines. An example will help clarify this application.

EXAMPLE 6–16

Implement the logic function specified in Table 6–9 by using a 74HC151 8-input data selector/multiplexer. Compare this method with a discrete logic gate implementation.

TABLE 6–9

	Inputs		Output
A_2	A_1	A_0	Y
0	0	0	0
0	0	1	1
0	1	0	0
0	1	1	1
1	0	0	0
1	0	1	1
1	1	0	1
1	1	1	0

Solution   Notice from the truth table that Y is a 1 for the following input variable combinations: 001, 011, 101, and 110. For all other combinations, Y is 0. For this function to be implemented with the data selector, the data input selected by each of the above-mentioned combinations must be connected to a HIGH (5 V). All the other data inputs must be connected to a LOW (ground), as shown in Figure 6–52.

The implementation of this function with logic gates would require four 3-input AND gates, one 4-input OR gate, and three inverters unless the expression can be simplified.

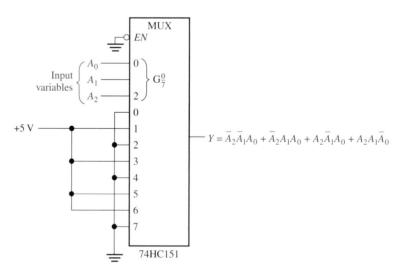

$$Y = \bar{A}_2\bar{A}_1A_0 + \bar{A}_2A_1A_0 + A_2\bar{A}_1A_0 + A_2A_1\bar{A}_0$$

FIGURE 6–52
Data selector/multiplexer connected as a 3-variable logic function generator.

Related Problem   Use the 74HC151 to implement the following expression:

$$Y = \bar{A}_2\bar{A}_1\bar{A}_0 + A_2\bar{A}_1\bar{A}_0 + \bar{A}_2A_1\bar{A}_0$$

Example 6–16 illustrated how the 8-input data selector can be used as a logic function generator for three variables. Actually, this device can be also used as a 4-variable logic function generator by the utilization of one of the bits (A_0) in conjunction with the data inputs.

A 4-variable truth table has sixteen combinations of input variables. When an 8-bit data selector is used, each input is selected twice: the first time when A_0 is 0 and the second time when A_0 is 1. With this in mind, the following rules can be applied (Y is the output, and A_0 is the least significant bit):

1. If $Y = 0$ both times a given data input is selected by a certain combination of the input variables, $A_3A_2A_1$, connect that data input to ground (0).
2. If $Y = 1$ both times a given data input is selected by a certain combination of the input variables, $A_3A_2A_1$, connect the data input to $+V$ (1).
3. If Y is different the two times a given data input is selected by a certain combination of the input variables, $A_3A_2A_1$, and if $Y = A_0$, connect that data input to A_0.
4. If Y is different the two times a given data input is selected by a certain combination of the input variables, $A_3A_2A_1$, and if $Y = \overline{A}_0$, connect that data input to $\overline{A}_0$.

The following example illustrates this method.

EXAMPLE 6–17

Implement the logic function in Table 6–10 by using a 74HC151 8-input data selector/multiplexer. Compare this method with a discrete logic gate implementation.

TABLE 6–10

Decimal Digit	A_3	A_2	A_1	A_0	Output Y
0	0	0	0	0	0
1	0	0	0	1	1
2	0	0	1	0	1
3	0	0	1	1	0
4	0	1	0	0	0
5	0	1	0	1	1
6	0	1	1	0	1
7	0	1	1	1	1
8	1	0	0	0	1
9	1	0	0	1	0
10	1	0	1	0	1
11	1	0	1	1	0
12	1	1	0	0	1
13	1	1	0	1	1
14	1	1	1	0	0
15	1	1	1	1	1

(The header "Inputs" spans columns A_3, A_2, A_1, A_0.)

Solution The data-select inputs are $A_3A_2A_1$. In the first row of the table, $A_3A_2A_1 = 000$ and $Y = A_0$. In the second row, where $A_3A_2A_1$ again is 000, $Y = A_0$. Thus, A_0 is connected to the 0 input. In the third row of the table, $A_3A_2A_1 = 001$ and $Y = \overline{A}_0$. Also, in the fourth row, when $A_3A_2A_1$ again is 001, $Y = \overline{A}_0$. Thus, A_0 is inverted and connected to the 1 input. This analysis is continued until each input is properly connected according to the specified rules. The implementation is shown in Figure 6–53.

If implemented with logic gates, the function would require as many as ten 4-input AND gates, one 10-input OR gate, and four inverters, although possible simplification would reduce this requirement.

$$Y = \bar{A}_3\bar{A}_2\bar{A}_1A_0 + \bar{A}_3\bar{A}_2A_1\bar{A}_0 + \bar{A}_3A_2\bar{A}_1A_0$$
$$+ \bar{A}_3A_2A_1\bar{A}_0 + \bar{A}_3A_2A_1A_0 + A_3\bar{A}_2\bar{A}_1\bar{A}_0$$
$$+ A_3\bar{A}_2A_1\bar{A}_0 + A_3A_2\bar{A}_1\bar{A}_0 + A_3A_2\bar{A}_1A_0$$
$$+ A_3A_2A_1A_0$$

FIGURE 6–53
Data selector/multiplexer connected as a 4-variable logic function generator.

Related Problem   In Table 6–10, if $Y = 0$ when the inputs are all zeros and is alternately a 1 and a 0 for the remaining rows in the table, use a 74HC151 to implement the resulting logic function.

SECTION 6–7 REVIEW

1. In Figure 6–46, $D_0 = 1$, $D_1 = 0$, $D_2 = 1$, $D_3 = 0$, $S_0 = 1$, and $S_1 = 0$. What is the output?
2. Identify each MSI device.
   (a) 74LS157   **(b)** 74LS151
3. A 74HC151 has alternating LOW and HIGH levels on its data inputs beginning with $D_0 = 0$. The data-select lines are sequenced through a binary count (000, 001, 010, and so on) at a frequency of 1 kHz. The enable input is LOW. Describe the data output waveform.
4. Briefly describe the purpose of each of the following devices in Figure 6–51:
   (a) 74HC157A   **(b)** 74LS48   **(c)** 74LS139

6–8 ■ DEMULTIPLEXERS

A demultiplexer (DEMUX) basically reverses the multiplexing function. It takes data from one line and distributes them to a given number of output lines. For this reason, the demultiplexer is also known as a data distributor. As you will learn, decoders can also be used as demultiplexers.

After completing this section, you should be able to

■ Explain the basic operation of a demultiplexer   ■ Describe how the 74HC154 4-line-to-16-line decoder can be used as a demultiplexer   ■ Develop the timing diagram for a demultiplexer with specified data and data selection inputs

---

Figure 6–54 shows a 1-line-to-4-line **demultiplexer (DEMUX)** circuit. The data-input line goes to all of the AND gates. The two data-select lines enable only one gate at a time,

FIGURE 6–54
A 1-line-to-4-line demultiplexer.

and the data appearing on the data-input line will pass through the selected gate to the associated data-output line.

EXAMPLE 6–18

The serial data-input waveform (Data in) and data-select inputs (S_0 and S_1) are shown in Figure 6–55. Determine the data-output waveforms on D_0 through D_3 for the demultiplexer in Figure 6–54.

FIGURE 6–55

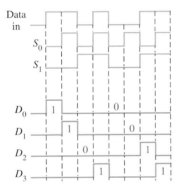

Solution Notice that the select lines go through a binary sequence so that each successive input bit is routed to D_0, D_1, D_2, and D_3 in sequence, as shown by the output waveforms in Figure 6–55.

Related Problem Develop the timing diagram for the demultiplexer if the S_0 and S_1 waveforms are both inverted.

An MSI Demultiplexer

We have already discussed the 74HC154 in its application as a 4-line-to-16-line decoder (Section 6–4). This device and other decoders can also be used in demultiplexing applications. The logic symbol for this device when used as a demultiplexer is shown in Figure 6–56. In demultiplexer applications, the input lines are used as the data-select lines. One of the chip select inputs is used as the data-input line, with the other chip select input held LOW to enable the internal negative-AND gate at the bottom of the diagram.

FIGURE 6–56
The 74HC154 used as a demultiplexer.

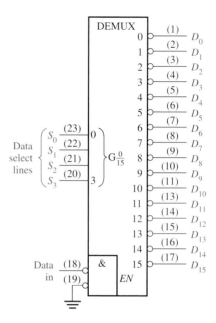

<table>
<tr><td>SECTION 6–8
REVIEW</td><td>1. Generally, how can an MSI decoder be used as a demultiplexer?
2. The 74HC154 demultiplexer in Figure 6–56 has a binary code of 1010 on the data-select lines, and the data-input line is LOW. What are the states of the output lines?</td></tr>
</table>

6–9 ■ PARITY GENERATORS/CHECKERS

Errors can occur as digital codes are being transferred from one point to another within a digital system or while codes are being transmitted from one system to another. The errors take the form of undesired changes in the bits that make up the coded information; that is, a 1 can change to a 0, or a 0 to a 1, because of component malfunctions or electrical noise. In most digital systems, the probability that even a single bit error will occur is very small, and the likelihood that more than one will occur is even smaller. Nevertheless, when an error occurs undetected, it can cause serious problems in a digital system.

After completing this section, you should be able to

■ Explain the concept of parity ■ Implement a basic parity circuit with exclusive-OR gates ■ Describe the operation of basic parity generating and checking logic ■ Discuss the 74LS280 9-bit parity generator/checker ■ Discuss how error detection can be implemented in a data transmission

The parity method of error detection in which a **parity bit** is attached to a group of information bits in order to make the total number of 1s either even or odd (depending on the system) was covered in Chapter 2. In addition to parity bits, several specific codes also provide inherent error detection.

Basic Parity Logic

In order to check for or to generate the proper parity in a given code, a very basic principle can be used:

The sum (disregarding carries) of an even number of 1s is always 0, and the sum of an odd number of 1s is always 1.

Therefore, to determine if a given code has **even parity** or **odd parity,** all the bits in that code are summed. As you know, the sum of two bits can be generated by an exclusive-OR gate, as shown in Figure 6–57(a); the sum of four bits can be formed by three exclusive-OR gates connected as shown in Figure 6–57(b); and so on. When the number of 1s on the inputs is even, the output X is 0 (LOW). When the number of 1s is odd, the output X is 1 (HIGH).

FIGURE 6–57

(a) Summing of two bits   (b) Summing of four bits

An MSI 9-Bit Parity Generator/Checker

The logic symbol and function table for a 74LS280 are shown in Figure 6–58. This particular MSI device can be used to check for odd or even parity on a 9-bit code (eight data bits and one parity bit) or it can be used to generate a parity bit for a binary code with up to nine bits. The inputs are A through I; when there is an even number of 1s on the inputs, the Σ Even output is HIGH and the Σ Odd output is LOW.

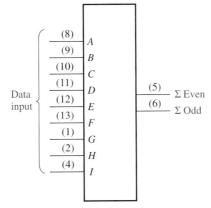

(a) Traditional logic symbol

Number of Inputs A Thru I That Are HIGH	Outputs	
	Σ Even	Σ Odd
0, 2, 4, 6, 8	H	L
1, 3, 5, 7, 9	L	H

(b) Function table

FIGURE 6–58
The 74LS280 9-bit parity generator/checker.

Parity Checker When this device is used as an even parity checker, the number of input bits should always be even; and when a parity error occurs, the Σ Even output goes LOW and the Σ Odd output goes HIGH. When it is used as an odd parity checker, the number of input bits should always be odd; and when a parity error occurs, the Σ Odd output goes LOW and the Σ Even output goes HIGH.

Parity Generator If this device is used as an even parity generator, the parity bit is taken at the Σ Odd output because this output is a 0 if there is an even number of input bits and it is a 1 if there is an odd number. When used as an odd parity generator, the parity bit is taken at the Σ Even output because it is a 0 when the number of inputs bits is odd.

A Data Transmission System with Error Detection

A simplified data transmission system is shown in Figure 6–59 to illustrate an application of parity generators/checkers, as well as multiplexers and demultiplexers, and to illustrate the need for data storage in some applications.

FIGURE 6–59

Simplified data transmission system with error detection.

In this application, digital data from seven sources are multiplexed onto a single line for transmission to a distant point. The seven data bits (D_0 through D_6) are applied to the multiplexer data inputs and, at the same time, to the even parity generator inputs. The Σ Odd output of the parity generator is used as the even parity bit. This bit is 0 if the number of 1s on the inputs A through I is even and is a 1 if the number of 1s on A through I is odd. This bit is D_7 of the transmitted code.

The data-select inputs are repeatedly cycled through a binary sequence, and each data bit, beginning with D_0, is serially passed through and onto the transmission line ($\overline{Y}$). In this example, the transmission line consists of four conductors: one carries the serial data and three carry the timing signals (data selects). There are more sophisticated ways of sending the timing information, but we are using this direct method to illustrate a basic principle.

At the demultiplexer end of the system, the data-select signals and the serial data stream are applied to the demultiplexer. The data bits are distributed by the demultiplexer onto the output lines in the order in which they occurred on the multiplexer inputs. That is,

D_0 comes out on the D_0 output, D_1 comes out on the D_1 output, and so on. The parity bit comes out on the D_7 output. These eight bits are temporarily stored and applied to the even parity checker. Not all of the bits are present on the parity checker inputs until the parity bit D_7 comes out and is stored. At this time, the error gate is enabled by the data-select code 111. If the parity is correct, a 0 appears on the Σ Even output, keeping the ERROR output at 0. If the parity is incorrect, all 1s appear on the error gate inputs, and a 1 on the ERROR output results.

This particular application has demonstrated the need for data storage so that you will be better able to appreciate the usefulness of the storage devices introduced in Chapter 8 and used in other later chapters.

The timing diagram in Figure 6–60 illustrates a specific case in which two 8-bit words are transmitted, one with correct parity and one with an error.

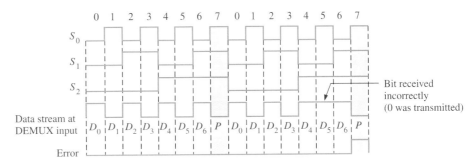

FIGURE 6–60
Example of data transmission with and without error for the system in Figure 6–59.

SECTION 6–9 REVIEW

1. Add an even parity bit to each of the following codes:
   (a) 110100   (b) 01100011
2. Add an odd parity bit to each of the following codes:
   (a) 1010101   (b) 1000001
3. Check each of the even parity codes for an error.
   (a) 100010101   (b) 1110111001

6–10 ■ TROUBLESHOOTING

In this section, the problem of decoder glitches is introduced and examined from a troubleshooting standpoint. A glitch is any undesired voltage or current spike (pulse) of very short duration. A glitch can be interpreted as a valid signal by a logic circuit and may cause improper operation.

After completing this section, you should be able to

■ Explain what a glitch is   ■ Determine the cause of glitches in a decoder application
■ Use the method of output strobing to eliminate glitches

The 74LS138 was used as a DEMUX in the data transmission system in Figure 6–59. Now the 74HC138 is used as a 3-line-to-8-line decoder (binary-to-octal) in Figure 6–61 to illustrate how **glitches** occur and how to identify their cause. The $A_2A_1A_0$ inputs of the decoder are sequenced through a binary count, and the resulting waveforms of the inputs and outputs can be displayed on the screen of a logic analyzer as shown in Figure 6–61.

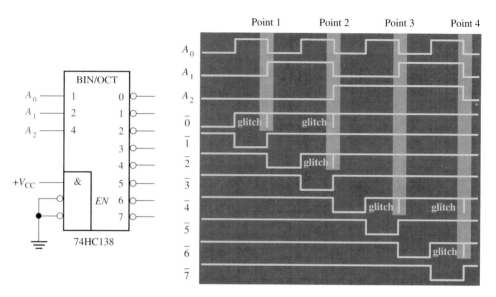

FIGURE 6–61
Decoder waveforms with output glitches.

A_2 transitions are delayed from A_1 transitions and A_1 transitions are delayed from A_0 transitions. This commonly occurs when waveforms are generated by a binary counter, as you will learn in Chapter 9.

The output waveforms are correct except for the glitches that occur on some of the output signals. A logic analyzer or an oscilloscope can be used to display glitches, which are normally very difficult to see. Generally, the logic analyzer is preferred, especially for low repetition rates (less than 10 kHz) and/or irregular occurrence because most logic analyzers have a *glitch capture* capability. Oscilloscopes can be used to observe glitches with reasonable success particularly if the glitches occur at a regular high repetition rate (greater than 10 kHz).

The points of interest indicated by the highlighted areas on the input waveforms in Figure 6–61 are displayed as shown in Figure 6–62. At point 1 there is a transitional state of 000 due to delay differences in the waveforms. This causes the first glitch on the $\overline{0}$ output of the decoder. At point 2 there are two transitional states, 010 and 000. These cause the glitch on the $\overline{2}$ output of the decoder and the second glitch on the $\overline{0}$ output, respectively. At point 3 the transitional state is 100, which causes the first glitch on the $\overline{4}$ output of the decoder. At point 4 the two transitional states, 110 and 100, result in the glitch on the $\overline{6}$ output and the second glitch on the $\overline{4}$ output, respectively.

One way to eliminate the glitch problem is a method called **strobing,** in which the decoder is enabled by a strobe pulse only during the times when the waveforms are not in transition. This method is illustrated in Figure 6–63.

In addition to glitches that are the result of propagation delays, as you have seen in the case of a decoder, other types of unwanted noise spikes can also be a problem. Current and voltage spikes on the V_{CC} and ground lines are caused by the fast switching waveforms in digital circuits. This problem can be minimized by proper printed circuit board layout. Switching spikes can be absorbed by decoupling the circuit board with a 1 μF capacitor from V_{CC} to ground. Also, smaller decoupling capacitors (0.022 μF to 0.1 μF) should be distributed at various points between V_{CC} and ground over the circuit board. Decoupling should be done especially near devices that are switching at higher rates or driving more loads such as oscillators, counters, buffers, and bus drivers.

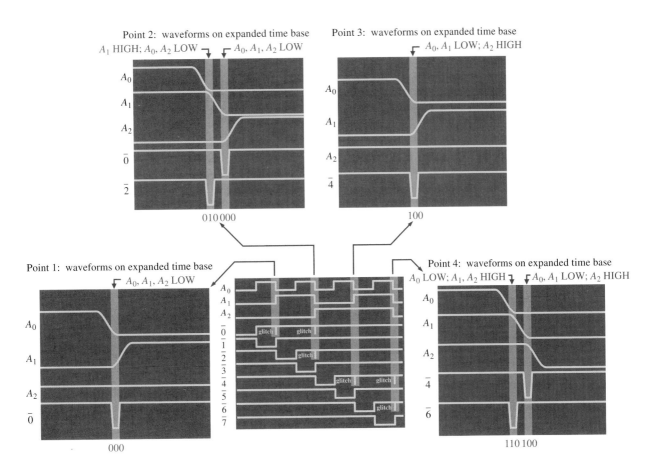

FIGURE 6–62
Decoder waveform displays showing how transitional input states produce glitches in the output waveforms.

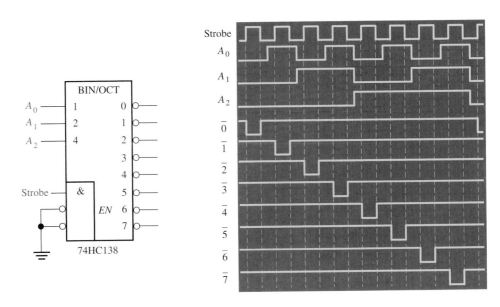

FIGURE 6–63
Application of a strobe waveform to eliminate glitches on decoder outputs.

1. Define the term *glitch*.
2. Explain the basic cause of glitches in decoder logic.
3. Define the term *strobe*.

6–11 ■ DIGITAL SYSTEM APPLICATION

In this system application, you will begin working with a traffic light control system. In this section, the system requirements are established, a general block diagram is developed, and a state diagram is created to help define the sequence of operation. A portion of the system involving combinational logic is designed and methods of testing are considered. The sequential and timing portions of the system will be dealt with in Chapters 8 and 9.

After completing this section, you should be able to

■ Determine system specifications ■ Develop a system block diagram from basic specifications ■ Explain the purpose of a state diagram ■ Use a state diagram and system specifications to design and implement the combinational logic portions of the system ■ Develop a test procedure for testing the breadboarded combinational logic ■ Specify a simple test setup

General System Requirements

A digital controller is required to control a traffic light at the intersection of a busy main street and an occasionally used side street. The main street is to have a green light for a minimum of 25 s or as long as there is no vehicle on the side street. The side street is to have a green light until there is no vehicle on the side street or for a maximum of 25 s. There is to be a 4 s caution light (yellow) between changes from green to red on both the main street and on the side street. These requirements are illustrated in the pictorial diagram in Figure 6–64.

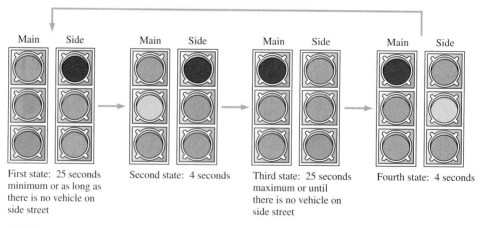

First state: 25 seconds minimum or as long as there is no vehicle on side street

Second state: 4 seconds

Third state: 25 seconds maximum or until there is no vehicle on side street

Fourth state: 4 seconds

FIGURE 6–64
Requirements for the traffic light sequence.

Developing a Block Diagram of the System

From the requirements, you can develop a block diagram of the system. First, you know that the system must control six different pairs of lights. These are the red, yellow, and green lights for both directions on the main street and the red, yellow, and green lights for both directions on the side street. Also, you know that there is one external input (other than power) from a side street vehicle sensor. Figure 6–65 is a minimal block diagram showing these requirements.

FIGURE 6–65
A minimal system block diagram.

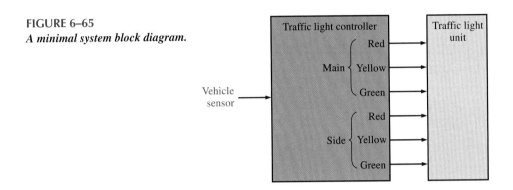

Using the minimal system block diagram, you can begin to fill in the details. The system has four states, as indicated in Figure 6–64, so a logic circuit is needed to control the sequence of states (sequential logic). Also, circuits are needed to generate the proper time intervals of 25 s and 4 s that are required in the system and to generate a clock signal for cycling the system (timing circuits). The time intervals (long and short) and the vehicle sensor are inputs to the sequential logic because the sequencing of states is a function of these variables. Logic circuits are also needed to determine which of the four states the system is in at any given time, to generate the proper outputs to the lights (state decoding and output logic), and to initiate the long and short time intervals. Finally, an interface circuit is needed to convert the logic levels of the decoding and output circuit to the voltages and currents required to turn on each of the lights. Figure 6–66 is a more detailed block diagram showing these essential elements.

The State Diagram

A state diagram graphically shows the sequence of states in a system and the conditions for each state and for transitions from one state to the next. Actually, Figure 6–64 is a form of state diagram because it shows the sequence of states and the conditions.

Definition of Variables Before a traditional state diagram can be developed, the variables that determine how the system sequences through its states must be defined. These variables and their symbols are listed as follows:

- Vehicle present on side street $= V_S$
- 25 s timer (long timer) is *on* $= T_L$
- 4 s timer (short timer) is *on* $= T_S$

The use of complemented variables indicates the opposite conditions. For example, $\overline{V}_S$ indicates that there is no vehicle on the side street, $\overline{T}_L$ indicates the long timer is *off*, and $\overline{T}_S$ indicates the short timer is *off*.

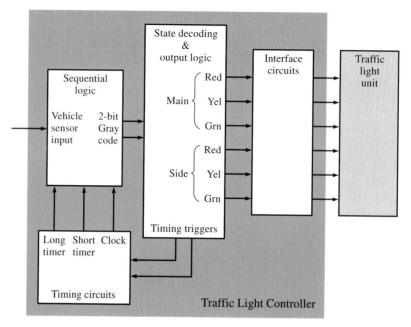

FIGURE 6–66
System block diagram showing essential elements.

Description of the State Diagram A state diagram is shown in Figure 6–67. Each of the four states is labeled according to the 2-bit Gray code sequence, as indicated by the circles. The looping arrow at each state indicates that the system remains in that state under the condition defined by the associated variable or expression. Each of the arrows going from one state to the next indicates a state transition under the condition defined by the associated variable or expression.

First state The Gray code for this state is 00. The main street light is green and the side street light is red. The system remains in this state for at least 25 s when the long timer is *on* or as long as there is no vehicle on the side street ($T_L + \overline{V}_s$). The system goes to the next state when the 25 s timer is *off* and there is a vehicle on the side street ($\overline{T}_L V_s$).

Second state The Gray code for this state is 01. The main street light is yellow (caution) and the side street light is red. The system remains in this state for 4 s when the short timer is *on* (T_S) and goes to the next state when the short timer goes *off* ($\overline{T}_S$).

Third state The Gray code for this state is 11. The main street light is red and the side street light is green. The system remains in this state when the long timer is *on* and there is a vehicle on the side street ($T_L V_s$). The system goes to the next state when the 25 s have elapsed or when there is no vehicle on the side street, whichever comes first ($\overline{T}_L + \overline{V}_s$).

Fourth state The Gray code for this state is 10. The main street light is red and the side street light is yellow. The system remains in this state for 4 s when the short timer is *on* (T_S) and goes back to the first state when the short timer goes *off* ($\overline{T}_S$).

The State Decoding and Output Logic

The focus in this chapter's system application is the state decoding and output logic portion of the block diagram of Figure 6–66. The timing and the sequential logic circuits will be the subjects of the system application sections in Chapters 8 and 9.

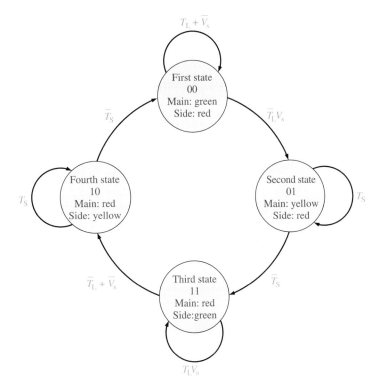

FIGURE 6–67
State diagram for the traffic light control system showing the Gray code sequence.

A block diagram for just the state decoding and output logic portion of the system is developed as a first step in designing the logic. The three functions that this logic must perform are defined as follows, and the resulting diagram with a block for each of the three functions is shown in Figure 6–68:

■ **State Decoder** Decodes the 2-bit Gray code from the sequential logic to determine which of the four states the system is in.
■ **Output Logic** Uses the decoded state to activate (through the interface circuits) the appropriate two pairs of traffic lights out of the six pairs of traffic lights in the light unit.
■ **Trigger Logic** Uses the decoded states to produce signals for properly initiating (triggering) the long timer and the short timer.

Implementing the Logic The *state decoder* has two inputs (2-bit Gray code) and must have an output for each of the four states. A 2-line-to-4-line decoder is needed and the 74LS139 should do the job nicely. The 74LS139 contains two separate decoders but only one of them will be used.

The *output logic* must take the four active-LOW state outputs from the 74LS139 decoder and produce six outputs for activating the traffic lights. A truth table for the decoding and output logic is given in Table 6–11. The state outputs ($\overline{SO}_1$ through $\overline{SO}_4$) are active-LOW (0), and it is required that the light outputs from the output logic be active-LOW (0).

The *trigger logic* produces two outputs. The long output produces a LOW-to-HIGH transition to trigger the 25 s timing circuit (long timer) when the system goes into the first or third states. The short output produces is a LOW-to-HIGH transition to trigger the 4 s timing circuit (short timer) when the system goes into the second or fourth states. This

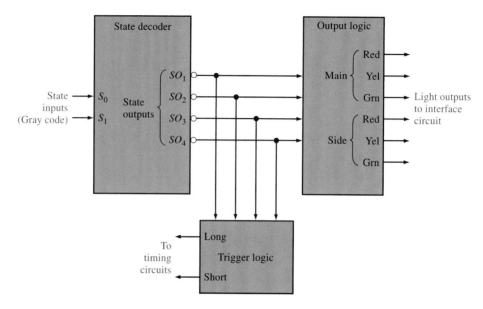

FIGURE 6–68
Block diagram of the state decoding and output logic.

TABLE 6–11
Truth table for the decoding and output logic.

State Inputs		State Outputs				Light Outputs						Trigger Outputs	
S_1	S_0	$\overline{SO}_1$	$\overline{SO}_2$	$\overline{SO}_3$	$\overline{SO}_4$	$\overline{MR}$	$\overline{MY}$	$\overline{MG}$	$\overline{SR}$	$\overline{SY}$	$\overline{SG}$	*Long*	*Short*
0	0	0	1	1	1	1	1	0	0	1	1	1	0
0	1	1	0	1	1	1	0	1	0	1	1	0	1
1	1	1	1	0	1	0	1	1	1	1	0	1	0
1	0	1	1	1	0	0	1	1	1	0	1	0	1

State outputs are active-LOW and light outputs are active-LOW. *MR* stands for main street red, *SG* for side street green, etc.

requirement can be seen by examining the system requirements and the state diagram in Figure 6–67.

■ SYSTEM ASSIGNMENTS

System Assignment 1 involves the design activities for the state decoder and output logic portion of the system. Make every effort to design the logic diagram required for this part of the system and then compare your results with the portion of the printed circuit board shown in System Assignment 2. Only the portion of the circuit board containing the state decoder and output logic is shown. Keep in mind that you will be adding to your logic diagram as we progress through the rest of this system in Chapters 8 and 9.

In System Assignment 2, all of the connections to the V_{CC} and ground pads are on the back side of the board. Circuit interconnections on the back side of the board feed through to the pads on the component side.

■ SYSTEM ASSIGNMENT 1:   Design

■ *Activity 1:* Using the information from Table 6–11 and the pin diagram from a 74LS08 data sheet shown below, design the output logic required to generate the six light outputs to the interface circuit. Draw the logic diagram. *Hint: An AND gate can be treated as a negative-NOR gate where a LOW on either or both inputs produces a LOW output as shown below.*

■ *Activity 2:* Using the state diagram and the 74LS08, design the trigger logic to meet the previously stated requirements. Add this to your logic diagram from Activity 1.

74LS08

■ *Activity 3:* Combine your circuit from Activities 1 and 2 with the 74LS139 decoder for which the pin diagram and truth table from the data sheet are shown below to create a complete logic diagram

for the state decoder and output portion of the system. Make sure your diagram includes pin numbers.

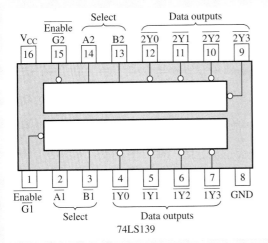

74LS139

Inputs		Outputs			
Enable	**Select**				
$\overline{G}$	B	A	$\overline{Y}_0$	$\overline{Y}_1$	$\overline{Y}_2$	$\overline{Y}_3$
H	X	X	H	H	H	H
L	L	L	L	H	H	H
L	L	H	H	L	H	H
L	H	L	H	H	L	H
L	H	H	H	H	H	L

NOTE: H = High Level, L = Low Level, X = Don't Care

■ SYSTEM ASSIGNMENT 2:   Verification and Testing

■ **Activity 1:**   Compare the logic diagram that you developed in System Assignment 1 with the portion of the system PC board shown below and make sure they agree.

■ **Activity 2:**   Identify the function of each of the points on the PC board labeled with a circled number.

■ **Activity 3:**   Use the switches connected as shown to simulate the Gray code input from the sequential portion of the system to develop a test procedure to check out this portion of the system board.

■ **Activity 4:**   When the switches are thrown into each of the positions shown in the table, determine from the indications at the test points represented by the circled numbers if the circuit is operating properly for each case and, if not, what the most likely problem is. Take each case separately. Each case is independent of all the other cases in the table.

 The PC board logic for the state decoder and output logic is in file SA06.EWB on your EWB CD. Use EWB to simulate a test of this circuit.

S_1 / S_0	Case	③	④	⑤	⑥	⑦	⑧	⑨	⑩	⑪
HIGH / HIGH	1	LOW	HIGH	HIGH	LOW	HIGH	LOW	HIGH	HIGH	HIGH
	2	LOW	HIGH	HIGH	LOW	HIGH	HIGH	HIGH	HIGH	HIGH
HIGH / HIGH	3	HIGH	HIGH	LOW	HIGH	HIGH	LOW	HIGH	HIGH	LOW
	4	HIGH	HIGH	HIGH	HIGH	HIGH	LOW	HIGH	HIGH	LOW
HIGH / HIGH	5	HIGH	LOW	HIGH	LOW	LOW	HIGH	HIGH	LOW	HIGH
	6	HIGH	HIGH	HIGH	HIGH	HIGH	HIGH	HIGH	HIGH	HIGH
HIGH / HIGH	7	HIGH	HIGH	LOW	HIGH	LOW	HIGH	LOW	HIGH	HIGH
	8	HIGH	HIGH	LOW	HIGH	HIGH	HIGH	LOW	HIGH	HIGH

SECTION 6–11 REVIEW

1. List the sequence of states on the state decoder inputs.
2. Explain how the NAND gates used for the output logic and the trigger logic on the breadboard in the System Assignment act as negative-OR gates.
3. When do the timing trigger output transitions occur for the long timer? For the short timer?

All or portions of Chapter 15, "Integrated Circuit Technology," may be covered at this point. The coverage in Chapter 15 is not a prerequisite to any other material in this text, and its omission or partial omission will not affect any other material.

■ **SUMMARY**

■ Half-adder and full-adder operations are summarized in Figure 6–69.
■ Logic symbols with pin numbers for the ICs used in this chapter are shown in Figure 6–70, which continues on the next page. Pin designations may differ from some manufacturers' data sheets.

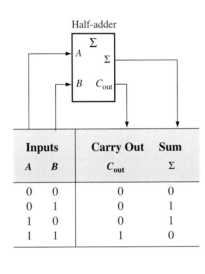

Half-adder

Inputs		Carry Out	Sum
A	B	C_{out}	Σ
0	0	0	0
0	1	0	1
1	0	0	1
1	1	1	0

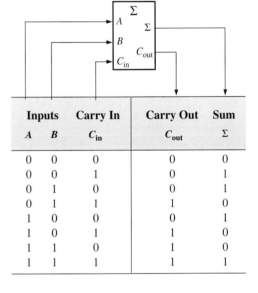

Full-adder

Inputs		Carry In	Carry Out	Sum
A	B	C_{in}	C_{out}	Σ
0	0	0	0	0
0	0	1	0	1
0	1	0	0	1
0	1	1	1	0
1	0	0	0	1
1	0	1	1	0
1	1	0	1	0
1	1	1	1	1

FIGURE 6–69

74HC42 BCD-to-decimal decoder   74LS47/74LS48 BCD-to-7-segment decoder/driver   74LS83A 4-bit adder   74HC85 4-bit magnitude comparator

FIGURE 6–70

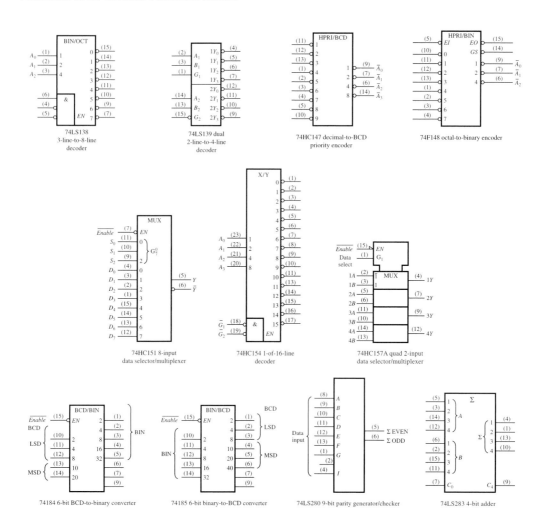

FIGURE 6–70
(Continued)

■ KEY TERMS

These terms are also in the end-of-book glossary.

Carry generation The process of producing an output carry in a full-adder when both input bits are 1s.

Carry propagation The process of rippling an input carry to become an output carry in a full-adder when either or both of the input bits are 1s and the input carry is a 1.

Cascading Connecting the output of one device to the input of a similar device, allowing one device to drive another in order to expand the operational capability.

Code converter A digital device that converts one type of coded information into another coded form.

Data selector A circuit that selects data from several inputs one at a time in a sequence and places them on the output; also called a multiplexer.

Decoder A digital circuit that converts coded information into a familiar or noncoded form.

Demultiplexer (DEMUX) A circuit that switches digital data from one input line to several output lines in a specified time sequence.

Dependency notation A notational system for logic symbols that specifies input and output relationships, thus fully defining a given function; an integral part of ANSI/IEEE Std. 91-1984.

Encoder A digital circuit that converts information to a coded form.

Even parity The condition of having an even number of 1s in every group of bits.

Full-adder A digital circuit that adds two bits and an input carry to produce a sum and an output carry.

Glitch A voltage or current spike of short duration, usually unintentionally produced and unwanted.

Half-adder A digital circuit that adds two bits and produces a sum and an output carry. It cannot handle input carries.

Look-ahead carry A method of binary addition whereby carries from preceding adder stages are anticipated, thus eliminating carry propagation delays.

Multiplexer (MUX) A circuit that switches digital data from several input lines onto a single output line in a specified time sequence.

Nibble A group of four bits.

Odd parity The condition of having an odd number of 1s in every group of bits.

Parity bit A bit attached to each group of information bits to make the total number of 1s odd or even for every group of bits.

Priority encoder An encoder in which only the highest value input digit is encoded and any other active input is ignored.

Pull-up resistor A resistor with one end connected to the dc supply voltage used to keep a given point in a circuit HIGH when in the inactive state.

Ripple carry A method of binary addition in which the output carry from each adder becomes the input carry of the next higher-order adder.

Strobing A process of using a pulse to sample the occurrence of an event at a specified time in relation to the event.

Zero suppression The process of blanking out leading or trailing zeros in a digital display.

■ SELF-TEST

1. A half-adder is characterized by
   - (a) two inputs and two outputs
   - (b) three inputs and two outputs
   - (c) two inputs and three outputs
   - (d) two inputs and one output

2. A full-adder is characterized by
   - (a) two inputs and two outputs
   - (b) three inputs and two outputs
   - (c) two inputs and three outputs
   - (d) two inputs and one output

3. The inputs to a full-adder are $A = 1$, $B = 1$ and $C_{in} = 0$. The outputs are
   - (a) $\Sigma = 1$, $C_{out} = 1$
   - (b) $\Sigma = 1$, $C_{out} = 0$
   - (c) $\Sigma = 0$, $C_{out} = 1$
   - (d) $\Sigma = 0$, $C_{out} = 0$

4. A 4-bit parallel adder can add
   - (a) two 4-bit binary numbers
   - (b) two 2-bit binary numbers
   - (c) four bits at a time
   - (d) four bits in sequence

5. The 74LS83A is an example of a 4-bit parallel adder. To expand this device to an 8-bit adder, you must
   - (a) use four adders with no interconnections
   - (b) use two adders and connect the sum outputs of one to the bit inputs of the other
   - (c) use eight adders with no interconnections
   - (d) use two adders with the carry output of one connected to the carry input of the other

6. If a 74HC85 magnitude comparator has $A = 1011$ and $B = 1001$ on its inputs, the outputs are
   (a) $A > B = 0, A < B = 1, A = B = 0$ (b) $A > B = 1, A < B = 0, A = B = 0$
   (c) $A > B = 1, A < B = 1, A = B = 0$ (d) $A > B = 0, A < B = 0, A = B = 1$

7. If a 1-of-16 decoder with active-LOW outputs exhibits a LOW on the decimal 12 output, what are the inputs?
   (a) $A_3A_2A_1A_0 = 1010$ (b) $A_3A_2A_1A_0 = 1110$
   (c) $A_3A_2A_1A_0 = 1100$ (d) $A_3A_2A_1A_0 = 0100$

8. A BCD-to-7-segment decoder has 0100 on its inputs. The active outputs are
   (a) a, c, f, g (b) b, c, f, g (c) b, c, e, f (d) b, d, e, g

9. If an octal-to-binary priority encoder has its 0, 2, 5, and 6 inputs at the active level, the active-HIGH binary output is
   (a) 110   (b) 010   (c) 101   (d) 000

10. In general, a multiplexer has
    (a) one data input, several data outputs, and selection inputs
    (b) one data input, one data output, and one selection input
    (c) several data inputs, several data outputs, and selection inputs
    (d) several data inputs, one data output, and selection inputs

11. Data selectors are basically the same as
    (a) decoders   (b) demultiplexers
    (c) multiplexers   (d) encoders

12. Which of the following codes exhibit even parity?
    (a) 10011000   (b) 01111000   (c) 11111111
    (d) 11010101   (e) all   (f) both answers (b) and (c)

■ PROBLEMS

SECTION 6–1 Basic Adders

1. For the full-adder of Figure 6–4, determine the logic state (1 or 0) at each gate output for the following inputs:
   (a) $A = 1, B = 1, C_{in} = 1$ (b) $A = 0, B = 1, C_{in} = 1$ (c) $A = 0, B = 1, C_{in} = 0$

2. What are the full-adder inputs that will produce each of the following outputs:
   (a) $\Sigma = 0, C_{out} = 0$ (b) $\Sigma = 1, C_{out} = 0$
   (c) $\Sigma = 1, C_{out} = 1$ (d) $\Sigma = 0, C_{out} = 1$

3. Determine the outputs of a full-adder for each of the following inputs:
   (a) $A = 1, B = 0, C_{in} = 0$ (b) $A = 0, B = 0, C_{in} = 1$
   (c) $A = 0, B = 1, C_{in} = 1$ (d) $A = 1, B = 1, C_{in} = 1$

SECTION 6–2 Parallel Binary Adders

4. For the parallel adder in Figure 6–71, determine the complete sum by analysis of the logical operation of the circuit. Verify your result by longhand addition of the two input numbers.

FIGURE 6–71

FIGURE 6–72

FIGURE 6–73
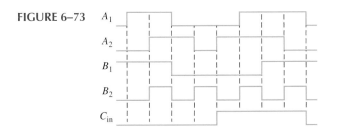

5. Repeat Problem 4 for the circuit and input conditions in Figure 6–72.

6. The input waveforms in Figure 6–73 are applied to a 2-bit adder. Determine the waveforms for the sum and the output carry in relation to the inputs by constructing a timing diagram.

7. The following sequences of bits (right-most bit first) appear on the inputs to a 74LS83A adder. Determine the resulting sequence of bits on each output.

$$
\begin{array}{ll}
A_1 & 1001 \\
A_2 & 1110 \\
A_3 & 0000 \\
A_4 & 1011 \\
B_1 & 1111 \\
B_2 & 1100 \\
B_3 & 1010 \\
B_4 & 0010 \\
\end{array}
$$

8. In the process of checking a 74LS83A 4-bit full-adder, the following voltage levels are observed on its pins: 1-LOW, 2-HIGH, 3-HIGH, 4-HIGH, 6-HIGH, 7-HIGH, 8-LOW, 9-LOW, 10-LOW, 11-LOW, 13-LOW, 14-HIGH, 15-LOW, and 16-HIGH. Determine if the IC is functioning properly.

SECTION 6–3 Comparators

9. The waveforms in Figure 6–74 are applied to the comparator as shown. Determine the output $(A = B)$ waveform.

FIGURE 6–74
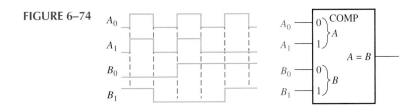

10. For the 4-bit comparator in Figure 6–75, plot each output waveform for the inputs shown. The outputs are active-HIGH.

11. For each set of binary numbers, determine the output states for the comparator of Figure 6–18.

    (a) $A_3A_2A_1A_0 = 1100$ **(b)** $A_3A_2A_1A_0 = 1000$ **(c)** $A_3A_2A_1A_0 = 0100$
        $B_3B_2B_1B_0 = 1001$ $B_3B_2B_1B_0 = 1011$ $B_3B_2B_1B_0 = 0100$

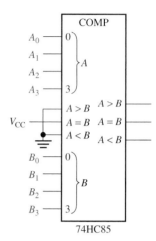

FIGURE 6–75

SECTION 6–4 Decoders

12. When a HIGH is on the output of each of the decoding gates in Figure 6–76, what is the binary code appearing on the inputs? The MSB is A_3.

13. Show the decoding logic for each of the following codes if an active-HIGH (1) output is required:

    (a) 1101   **(b)** 1000   **(c)** 11011   **(d)** 11100
    (e) 101010   **(f)** 111110   **(g)** 000101   **(h)** 1110110

14. Solve Problem 13, given that an active-LOW (0) output is required.

15. You wish to detect only the presence of the codes 1010, 1100, 0001, and 1011. An active-HIGH output is required to indicate their presence. Develop the minimum decoding logic with a single output that will indicate when any one of these codes is on the inputs. For any other code, the output must be LOW.

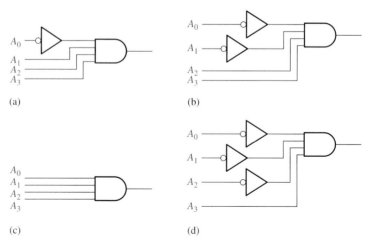

FIGURE 6–76

16. If the input waveforms are applied to the decoding logic as indicated in Figure 6–77, sketch the output waveform in proper relation to the inputs.

17. BCD numbers are applied sequentially to the BCD-to-decimal decoder in Figure 6–78. Draw a timing diagram, showing each output in the proper relationship with the others and with the inputs.

18. A 7-segment decoder/driver drives the display in Figure 6–79. If the waveforms are applied as indicated, determine the sequence of digits that appears on the display.

FIGURE 6–77

FIGURE 6–78

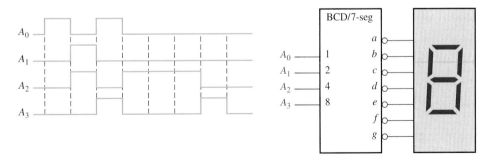

FIGURE 6–79

SECTION 6–5   Encoders

19. For the decimal-to-BCD encoder logic of figure 6–34, assume that the 9 input and the 3 input are both HIGH. What is the output code? Is it a valid BCD (8421) code?

20. A 74HC147 encoder has LOW levels on pins 2, 5, and 12. What BCD code appears on the outputs if all the other inputs are HIGH?

SECTION 6–6   Code Converters

21. Convert the following decimal numbers first to BCD. Then, show the BCD digits applied to the BCD-to-binary converter in Figure 6–39(a) with the resulting binary outputs.

(a) 2   (b) 8   (c) 13   (d) 26   (e) 33

22. Show the logic required to convert a 10-bit binary number to Gray code, and use that logic to convert the following binary numbers to Gray code:

(a) 1010101010   (b) 1111100000   (c) 0000001110   (d) 1111111111

23. Show the logic required to convert a 10-bit Gray code to binary, and use that logic to convert the following Gray code words to binary:

(a) 1010000000   (b) 0011001100   (c) 1111000111   (d) 0000000001

SECTION 6–7   Multiplexers (Data Selectors)

24. For the multiplexer in Figure 6–80, determine the output for the following input states: $D_0 = 0$, $D_1 = 1$, $D_2 = 1$, $D_3 = 0$, $S_0 = 1$, $S_1 = 0$.

25. If the data-select inputs to the multiplexer in Figure 6–80 are sequenced as shown by the waveforms in Figure 6–81, determine the output waveform with the data inputs specified in Problem 24.

26. The waveforms in Figure 6–82 are observed on the inputs of a 74HC151 8-input multiplexer. Sketch the Y output waveform.

FIGURE 6–80

FIGURE 6–81

FIGURE 6–82

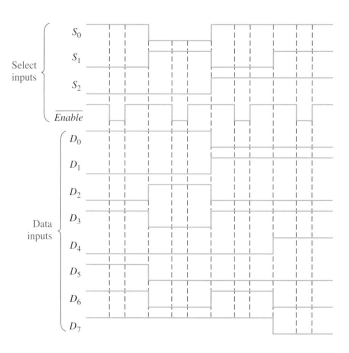

SECTION 6–8 Demultiplexers

27. Develop the total timing diagram (inputs and outputs) for a 74HC154 used in a demultiplexing application in which the inputs are as follows: The data-select inputs are repetitively sequenced through a straight binary count beginning with 0000, and the data input is a serial data stream carrying BCD data representing the decimal number 2468. The least significant digit (8) is first in the sequence, with its LSB first, and it should appear in the first 4-bit positions of the output.

SECTION 6–9 Parity Generators/Checkers

28. The waveforms in Figure 6–83 are applied to the 4-bit parity logic. Determine the output waveform in proper relation to the inputs. For how many bit times does even parity occur, and how is it indicated? The timing diagram includes eight bit times.

29. Determining the Σ Even and the Σ Odd outputs of a 74LS280 9-bit parity generator/checker for the inputs in Figure 6–84. Refer to the truth table in Figure 6–58.

FIGURE 6–83

FIGURE 6–84

SECTION 6–10 Troubleshooting

30. The full-adder in Figure 6–85 is tested under all input conditions with the input waveforms shown. From your observation of the Σ and C_{out} waveforms, is it operating properly, and if not, what is the most likely fault?

FIGURE 6–85

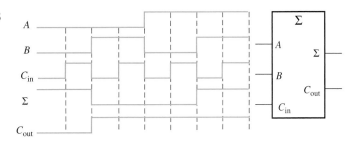

31. List the possible faults for each decoder/display in Figure 6–86.

32. Develop a systematic test procedure to check out the complete operation of the keyboard encoder in Figure 6–38.

33. You are testing a BCD-to-binary converter consisting of 4-bit adders as shown in Figure 6–87. First verify that the circuit converts BCD to binary. The test procedure calls for applying BCD numbers in sequential order beginning with 0_{10} and checking for the correct binary output. What symptom or symptoms will appear on the binary outputs in the event of each of the following faults? For what BCD number is each fault *first* detected?

(a) The A_1 input is open (top adder).

(b) The C_{out} is open (top adder).

(c) The Σ_4 output is shorted to ground (top adder).

(d) The 32 output is shorted to ground (bottom adder).

(a)

(b)

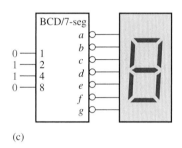
(c)

FIGURE 6–86

FIGURE 6–87

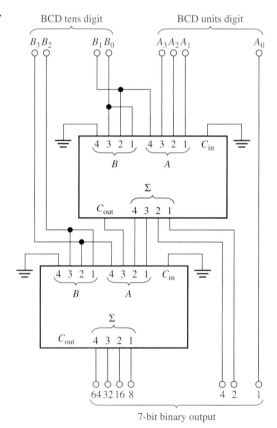

7-bit binary output

34. For the display multiplexing system in Figure 6–51, determine the most likely cause or causes for each of the following symptoms:

(a) The B-digit (MSD) display does not turn on at all.

(b) Neither 7-segment display turns on.

(c) The f-segment of both displays appears to be on all the time.

(d) There is a visible flicker on the displays.

35. Develop a systematic procedure to fully test the 74HC151 data selector IC.

36. During the testing of the data transmission system in Figure 6–59, a code is applied to the D_0 through D_6 inputs that contains an odd number of 1s. A single bit error is deliberately introduced on the serial data transmission line between the MUX and the DEMUX, but the system does not indicate an error (error output = 0). After some investigation, you check the inputs to the even parity checker and find that D_0 through D_6 contain an even number of 1s, as you would expect. Also, you find that the D_7 parity bit is a 1. What are the possible reasons for the system not indicating the error?

37. In general, describe how you would fully test the data transmission system in Figure 6–59, and specify a method for the introduction of parity errors.

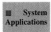

SECTION 6–11 Digital System Application

38. The output logic block is implemented on the workbench using a 74LS08 with the AND gates operating as negative-NOR gates. Use a 74LS00 (quad NAND gates) and any other device that may be required to produce the active-HIGH outputs.

39. Implement the output logic if active-LOW outputs are required.

Special Design Problems

40. Modify the design of the 7-segment display multiplexing system in Figure 6–52 to accommodate two additional digits.

41. Using Table 6–2, write the SOP expressions for the Σ and C_{out} of a full-adder. Use a Karnaugh map to minimize the expressions and then implement them with inverters and AND-OR logic. Show how you can replace the AND-OR logic with 74HC151 data selectors.

42. Implement the logic function specified in Table 6–12 by using a 74HC151 data selector.

TABLE 6–12

	Inputs			Output
A_3	A_2	A_1	A_0	Y
0	0	0	0	0
0	0	0	1	0
0	0	1	0	1
0	0	1	1	1
0	1	0	0	0
0	1	0	1	0
0	1	1	0	1
0	1	1	1	1
1	0	0	0	1
1	0	0	1	0
1	0	1	0	1
1	0	1	1	1
1	1	0	0	0
1	1	0	1	1
1	1	1	0	0
1	1	1	1	1

FIGURE 6–88

43. Using two of the 6-position adder modules from Figure 6–14, design a 12-position voting system.

44. The adder block in the tablet-counting and control system in Figure 6–88 performs the addition of the 8-bit binary number from the counter and the 16-bit binary number from Register B. The result from the adder goes back into Register B. Use 74LS283s to implement this function and draw a complete logic diagram including pin numbers. Refer to Chapters 1 and 2 to review the system operation.

45. The comparator block in the tablet-counting and control system in Figure 6–88 was implemented in Chapter 3 using discrete gates. Now you are prepared to use MSI devices to simplify the design. The comparator compares the 8-bit binary number (actually only seven bits are required) from the BCD-to-binary converter with the 8-bit binary number from the counter. Use 74HC85s to implement this function and draw a complete logic diagram including pin numbers.

46. Two BCD-to-7-segment decoders are used in the tablet-counting and control system in Figure 6–88. One is required to drive the 2-digit *tablets/bottle* display and the other to drive the 5-digit *total tablets bottled* display. Use 74LS47s to implement each decoder and draw a complete logic diagram including pin numbers.

47. The encoder shown in the system block diagram of Figure 6–88 encodes each decimal key closure and converts it to BCD. Use a 74HC147 to implement this function and draw a complete logic diagram including pin numbers.

48. The system in Figure 6–88 requires two code converters. The BCD-to-binary converter changes the 2-digit BCD number in Register A to an 8-bit binary code (actually only 7 bits are required because the MSB is always 0). Use appropriate MSI code converters to implement the BCD-to-binary converter function and draw a complete logic diagram including pin numbers.

Electronics Workbench Troubleshooting

49. Open file PR06_49.EWB and test the logic circuit to determine if there is a fault. If there is a fault, identify it if possible.

50. Open file PR06_50.EWB and test the logic circuit to determine if there is a fault. If there is a fault, identify it if possible.

51. Open file PR06_51.EWB and test the logic circuit to determine if there is a fault. If there is a fault, identify it if possible.

52. Open file PR06_52.EWB and test the logic circuit to determine if there is a fault. If there is a fault, identify it if possible.

■ **ANSWERS TO SECTION REVIEWS**

SECTION 6–1

1. (a) $\Sigma = 1, C_{out} = 0$ **(b)** $\Sigma = 0, C_{out} = 0$
   (c) $\Sigma = 1, C_{out} = 0$ **(d)** $\Sigma = 0, C_{out} = 1$

2. $\Sigma = 1, C_{out} = 1$

SECTION 6–2

1. $C_{out}\Sigma_4\Sigma_3\Sigma_2\Sigma_1 = 11001$ **2.** Three 74LS283s are required to add two 10-bit numbers.

SECTION 6–3

1. $A > B = 1, A < B = 0, A = B = 0$ when $A = 1011$ and $B = 1010$

2. Right comparator: pin 7: $A < B = 1$; pin 6: $A = B = 0$; pin 5: $A > B = 0$
Left comparator: pin 7: $A < B = 0$; pin 6: $A = B = 0$; pin 5: $A > B = 1$

SECTION 6–4

1. Output 5 is active when 101 is on the inputs.

2. Four 74HC154s are used to decode a 6-bit binary number.

3. Active-LOW output drives a common-cathode LED display.

SECTION 6–5

1. (a) $A_0 = 1, A_1 = 1, A_2 = 0, A_3 = 1$

   (b) No, this is not a valid BCD code.

   (c) Only one input can be active for a valid output.

2. (a) $\overline{A}_3 = 0, \overline{A}_2 = 1, \overline{A}_1 = 1, \overline{A}_0 = 1$

   (b) The output is 0111 which is the complement of 1000 (8).

SECTION 6–6

1. 10000101 (BCD) = 1010101_2

2. An 8-bit binary-to-Gray converter consists of seven exclusive-OR gates in an arrangement like that in Figure 6–42.

SECTION 6–7

1. The output is 0.

2. (a) 74LS157: Quad 2-input data selector

   (b) 74LS151: 8-input data selector

3. The data output alternates between LOW and HIGH as the data-select inputs sequence through the binary states.

4. (a) The 74HC157A multiplexes the two BCD codes to the 7-segment decoder.

   (b) The 74LS48 decodes the BCD to energize the display.

   (c) The 74LS139 enables the 7-segment displays alternately.

SECTION 6–8

1. A decoder can be used as a multiplexer by using the input lines for data selection and an Enable line for data input.

2. The outputs are all HIGH except D_{10}, which is LOW.

SECTION 6–9

1. (a) Even parity: $\underline{1}110100$ **(b)** Even parity: 001100011

2. (a) Odd parity: $\underline{1}1010101$ **(b)** Odd parity: $\underline{1}1000001$

3. (a) Code is correct, four 1s.

   (b) Code is in error, seven 1s

SECTION 6–10

1. A glitch is a very short-duration voltage spike (usually unwanted).

2. Glitches are caused by transition states.

3. Strobe is the enabling of a device for a specified period of time when the device is not in transition.

SECTION 6–11

1. Decoder input state sequence is 00, 01, 11, 10.

2. Negative-OR: $\overline{A} + \overline{B} = \overline{AB}$

3. Long timer triggers when going into the first or third states.
Short timer triggers when going into the second or fourth states.

■ **ANSWERS TO RELATED PROBLEMS FOR EXAMPLES**

6–1 $\Sigma = 1$, $C_{out} = 1$

6–2 $\Sigma_1 = 0$
$\Sigma_2 = 0$
$\Sigma_3 = 1$
$\Sigma_4 = 1$

6–3 $1011 + 1010 = 10101$ **6–4** See Figure 6–89.   **6–5** See Figure 6–90.

6–6 $A > B = 0, A = B = 0, A < B = 1$

6–7 See Figure 6–91.   **6–8** See Figure 6–92.

6–9 Output 22   **6–10** See Figure 6–93.

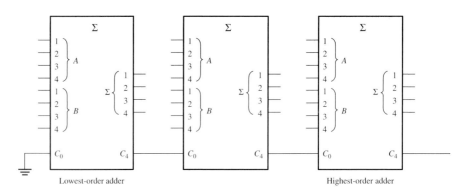

Lowest-order adder   Highest-order adder

FIGURE 6–89

FIGURE 6–90

$A_0 = 1$
$B_0 = 0$ → 0

$A_1 = 0$
$B_1 = 1$ → 0

$0 \rightarrow$ not equal

FIGURE 6–91

FIGURE 6–92

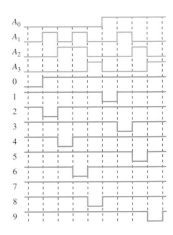

FIGURE 6–93

6–11 All inputs LOW: $\overline{A}_0 = 0, \overline{A}_1 = 1, \overline{A}_2 = 1, \overline{A}_3 = 0$
All inputs HIGH: All outputs HIGH.

6–12 BCD 01000001

$$\begin{array}{ll} \longrightarrow 00000001 & 1 \\ \longrightarrow 00101000 & 40 \\ \hline \text{Binary} \quad 00101001 & 41 \end{array}$$

6–13 Seven exclusive-OR gates   **6–14** See Figure 6–94.

6–15 D_0: $S_3 = 0, S_2 = 0, S_1 = 0, S_0 = 0$
D_4: $S_3 = 0, S_2 = 1, S_1 = 0, S_0 = 0$
D_8: $S_3 = 1, S_2 = 0, S_1 = 0, S_0 = 0$
D_{13}: $S_3 = 1, S_2 = 1, S_1 = 0, S_0 = 1$

6–16 See Figure 6–95.

FIGURE 6–94

FIGURE 6–95

$$Y = \overline{A}_2\overline{A}_1\overline{A}_0 + A_2\overline{A}_1\overline{A}_0 + \overline{A}_2 A_1\overline{A}_0$$

74HC151

6–17 See Figure 6–96.

6–18 See Figure 6–97.

FIGURE 6–96

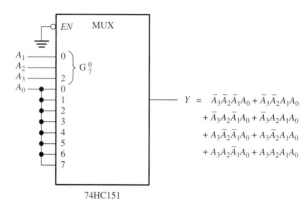

$$Y = \bar{A}_3\bar{A}_2\bar{A}_1A_0 + \bar{A}_3\bar{A}_2A_1A_0$$
$$+ \bar{A}_3A_2\bar{A}_1A_0 + \bar{A}_3A_2A_1A_0$$
$$+ A_3\bar{A}_2\bar{A}_1A_0 + A_3\bar{A}_2A_1A_0$$
$$+ A_3A_2\bar{A}_1A_0 + A_3A_2A_1A_0$$

74HC151

FIGURE 6–97

■ **ANSWERS TO SELF-TEST**

1. (a) **2.** (b) **3.** (c) **4.** (a) **5.** (d) **6.** (b) **7.** (c) **8.** (b)

9. (a) **10.** (d) **11.** (c) **12.** (f)

7

INTRODUCTION TO PROGRAMMABLE LOGIC DEVICES

■ CHAPTER OBJECTIVES

❑ Discuss programmable arrays and describe the major types of programmable logic devices (PLDs)

❑ Describe the basic PAL structure and explain how programming the AND array produces standard logic functions

❑ Describe the basic GAL structure and explain how programming the AND array produces standard logic functions

❑ Describe the basic GAL22V10 architecture

❑ Describe the basic GAL16V8 architecture

❑ Discuss the basic procedure of PLD programming

❑ Write simple ABEL programs to implement combinational logic in a PLD

❑ Implement simple combinational logic using a GAL

■ CHAPTER OVERVIEW

Programmable logic devices (PLDs) are introduced in this chapter. The emphasis is on using PLDs to implement combinational logic. Sequential logic applications of PLDs are discussed in Chapter 11. PLDs, particularly programmable array logic (PAL) and generic array logic (GAL) devices, can be used to replace SSI and MSI logic devices in most types of applications, resulting in fewer parts and lower cost.

     Programming software, such as ABEL and CUPL, are used to implement logic designs in PLDs. In this book, ABEL is used to illustrate basic programming principles. Once you are familiar with one PLD programming language, the others are easily learned because of the similarities. The coverage of ABEL in this text is limited and is not intended as a comprehensive treatment. Materials from the manufacturer and from other sources are recommended to supplement this coverage. This chapter may be postponed until after Chapter 10 or completely omitted without affecting any other topics except the material in Chapter 11.

SPECIFIC DEVICES

PAL16L8   GAL22V10   GAL16V8
GAL20V8

■ DIGITAL SYSTEM APPLICATION

This Digital System Application illustrates concepts taught in this chapter. The combinational logic portion of the traffic light control system from Chapter 6 is redesigned using a programmable logic device. A program is developed using ABEL to implement the state decoding and output logic.

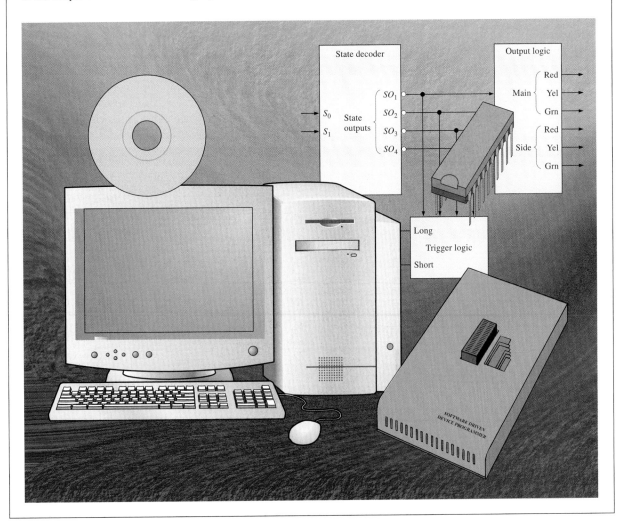

7–1 ■ PLD ARRAYS AND CLASSIFICATIONS

Programmable logic devices (PLDs) are used in many applications to replace SSI and MSI circuits; they save space and reduce the actual number and cost of devices in a given design. A PLD consists of a large array of AND gates and OR gates that can be programmed to achieve specified logic functions. Four types of devices that are classified as PLDs are the programmable read-only memory (PROM), the programmable logic array (PLA), the programmable array logic (PAL), and the generic array logic (GAL).

After completing this section you should be able to

■ Explain a basic OR array ■ Explain a basic AND array ■ Describe the structure of a PROM ■ Describe the structure of a PLA ■ Describe the basic PAL ■ Describe the basic GAL ■ Discuss the differences between a PAL and a GAL

Programmable Arrays

All **PLDs** consist of programmable arrays. A programmable **array** is essentially a grid of conductors that form rows and columns with a fusible link at each cross point. Arrays can be either fixed or programmable. The earliest type of programmable array, dating back to the 1960s, was a matrix with a diode at each cross point of the matrix.

The OR Array The original diode array evolved into the integrated OR array, which consists of an array of OR gates connected to a programmable matrix with fusible links at each cross point of a row and column, as shown in Figure 7–1(a). The array is programmed by blowing **fuses** to eliminate selected variables from the output functions, as illustrated in part (b) for a specific case. For each input to an OR gate, only one fuse is left intact in order to connect the desired variable to the gate input. Once a fuse is blown, it cannot be reconnected.

The AND Array This type of array consists of AND gates connected to a programmable matrix with fusible links at each cross point, as shown in Figure 7–2(a). Like the OR array, the AND array is programmed by blowing fuses to eliminate variables from the output function, as illustrated in part (b). For each input to an AND gate, only one fuse is left intact in order to connect the desired variable to the gate input. Also like the OR array, the AND array with fusible links is one-time programmable.

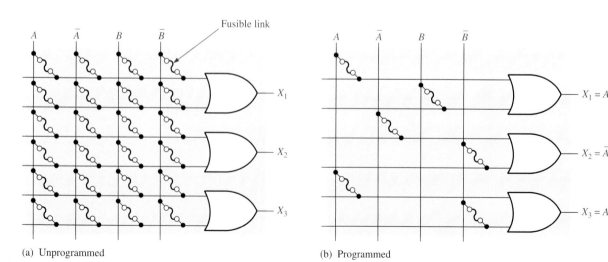

(a) Unprogrammed

(b) Programmed

FIGURE 7–1

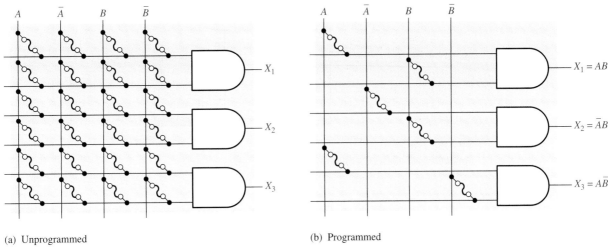

(a) Unprogrammed

(b) Programmed

FIGURE 7–2
An example of a basic AND array.

Classifications of PLDs

PLDs are classified according to their **architecture,** which is basically the internal functional arrangement of the elements that give a device its particular operating characteristic.

Programmable Read-Only Memory   The **PROM** consists of a set of fixed (nonprogrammable) AND gates connected as a decoder and a programmable OR array, as shown in the generalized block diagram of Figure 7–3. The PROM is used primarily as an addressable memory and not as a logic device because of limitations imposed by the fixed AND gates. PROMs are covered in detail in Chapter 12.

COMPUTER NOTE

Most system-level designs incorporate a variety of devices such as RAMs, ROMs, controllers, processors, etc. that are interconnected by a large quantity of general-purpose logic devices often referred to as "glue" logic. In the last several years, PLDs have come to replace many of the older SSI and MSI "glue" devices. The use of PLDs provides a reduction in package count.

 For example, in computer memory systems, PLDs can be used for memory address decoding and to generate memory write signals.

FIGURE 7–3
Block diagram of a PROM (programmable read-only memory).

Programmable Logic Array (PLA)   The **PLA** is a PLD that consists of a programmable AND array and a programmable OR array, as shown in Figure 7–4. The PLA was developed to overcome some of the limitations of the PROM. The PLA is also called an FPLA (field-programmable logic array) because the user in the field, not the manufacturer, programs it.

Programmable Array Logic (PAL)   The **PAL** is a PLD that was developed to overcome certain disadvantages of the PLA, such as longer delays due to the additional fusible links that result from using two programmable arrays and more circuit complexity. The basic PAL consists of a programmable AND array and a fixed OR array with output logic, as shown in Figure 7–5. The PAL is the most common one-time programmable logic device and is implemented with bipolar technology (TTL or ECL).

FIGURE 7–4
Block diagram of a PLA (programmable logic array).

FIGURE 7–5
Block diagram of a PAL (programmable array logic).

Generic Array Logic (GAL) A more recent development in PLDs is the GAL. The **GAL,** like the PAL, has a programmable AND array and a fixed OR array with programmable output logic. The two main differences between GAL and PAL devices are (a) the GAL is reprogrammable and (b) the GAL has programmable output configurations.

   The GAL can be reprogrammed again and again because it uses **E^{2}CMOS** (electrically erasable CMOS) technology instead of bipolar technology and fusible links. The block diagram of a GAL is shown in Figure 7–6.

FIGURE 7–6
Block diagram of a GAL (generic array logic).

**SECTION 7–1
REVIEW**

1. List four PLDs.
2. What is the difference between a PLA and a PAL?
3. Describe the differences between a PAL and a GAL?

7–2 ■ PROGRAMMABLE ARRAY LOGIC (PAL)

The PAL and the GAL are the most common PLDs used for logic implementation. PAL is a designation originally used by Monolithic Memories, Inc. (now a part of Advanced Micro Devices, AMD) and later licensed to other manufacturers. As you learned in the last section, the PAL in its basic form is a PLD with a one-time programmable AND array and a fixed OR array. In this section, you will learn how PALs are used to produce specified combinational logic functions and examine a specific PAL.

After completing this section, you should be able to

■ Describe basic PAL operation ■ Show how a sum-of-products expression is implemented in a PAL ■ Discuss simplified PAL logic diagrams ■ Explain the three basic types of PAL output combinational logic ■ Interpret standard PAL numbers ■ Discuss the PAL16L8

PAL Operation

The PAL consists of a programmable array of AND gates that connects to a fixed array of OR gates. This structure allows any sum-of-products (SOP) logic expression with a defined number of variables to be implemented. Recall from Chapter 4 that any logic function can be expressed in SOP form.

The basic structure of a PAL is illustrated in Figure 7–7 for two input variables and one output although most PALs have many inputs and many outputs. As you know, a programmable array is essentially a grid of conductors forming rows and columns with a fusible link at each cross point. Each fused cross point of a row and column is called a **cell** and is the programmable element of a PAL. Each row is connected to the input of an AND gate and each column is connected to an input variable or its complement. By using the presence or absence of fused connections created by programming, any combination of input variables or complements can be applied to an AND gate to form any desired product term.

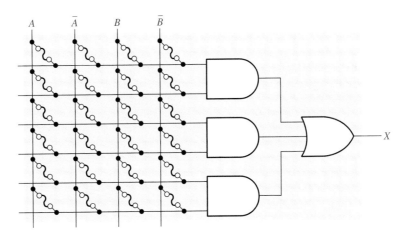

FIGURE 7–7
Basic structure of a PAL.

Implementing a Sum-of-Products Expression In its simplest form, each cell in a basic AND array consists of a fusible link connecting a row and a column as represented in Figure 7–7. When the connection between a row and column is required, the fuse is left intact. When no connection between a row and column is required, the fuse is blown open during the programming process.

As an example, a simple array is programmed as shown in Figure 7–8 so that the product term AB is produced by the top AND gate, $A\overline{B}$ by the middle AND gate, and $\overline{A}\,\overline{B}$ by the bottom AND gate. As you can see, the fusible links are left intact to connect the desired variables or their complements to the appropriate AND gate inputs. The fusible links are opened where a variable or its complement is not used in a given product term. The final output from the OR gate is the SOP expression,

$$X = AB + A\overline{B} + \overline{A}\,\overline{B}$$

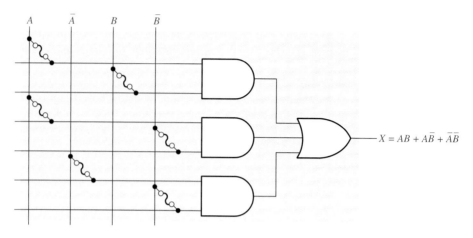

FIGURE 7–8
PAL implementation of a sum-of-products expression.

Simplified Symbols

What you have seen so far represents a small segment of a typical PAL. Actual PALs have many AND gates and many OR gates in addition to other circuitry and are capable of handling many input variables and their complements. Since PALs are very complex integrated circuit devices, manufacturers have adopted a simplified notation for the logic diagrams to keep them from being overwhelmingly complicated.

Input Buffers   The input variables to a PAL are buffered to prevent loading by the large number of AND gate inputs to which a variable or its complement may be connected. An inverting buffer produces the complement of an input variable. The symbol representing the **buffer** circuit that produces both the variable and its complement on its outputs is shown in Figure 7–9 where the bubble output is the complement.

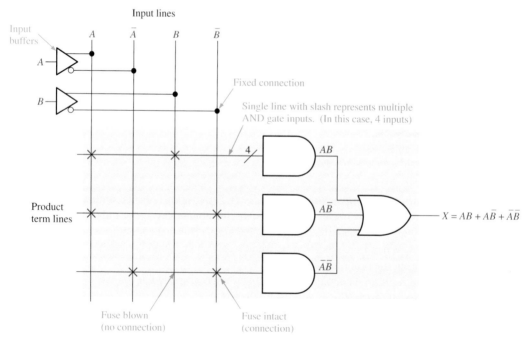

FIGURE 7–9
Simplified diagram of a programmed PAL.

AND Gates A typical PAL AND array has an extremely large number of interconnecting lines, and each AND gate has multiple inputs. PAL logic diagrams show an AND gate that actually has several inputs by using an AND gate symbol with a single input line representing all of its input lines, as indicated Figure 7–9. Also, multiple input lines are sometimes indicated by a slash and the number of lines as shown in the top AND gate for the case of four lines.

PAL Connections To keep a logic diagram as simple as possible, the fusible links in a programmed AND array are indicated by an X at the cross point if the fuse is left intact and by the absence of an X if the fuse is blown, as indicated in Figure 7–9. Fixed connections use the standard dot notation, as also indicated.

EXAMPLE 7–1

Show how a PAL is programmed for the following 3-variable logic function:

$$X = A\overline{B}C + \overline{A}B\overline{C} + \overline{A}\overline{B} + AC$$

Solution The programmed array is shown in Figure 7–10. The intact fusible links are indicated by small red Xs. The absence of an X means that the fuse has been blown.

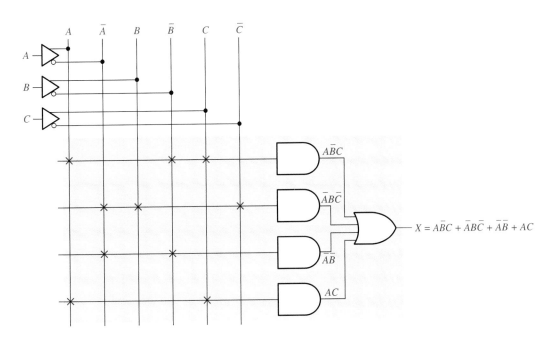

FIGURE 7–10

Related Problem Write the expression for the output if the fusible links connecting input A to the top row and to the bottom row in Figure 7–10 are also blown.

The PAL Block Diagram

A block diagram of a PAL is shown in Figure 7–11. The AND array outputs go to the OR array, and the output of each OR gate goes to its associated output logic. A typical PAL has eight or more inputs to its AND array and up to eight outputs from its output logic as indicated, where $n \geq 8$ and $m \leq 8$. Some PALS provide a combined input and output **(I/O)** pin that can be programmed as either an output or an input. The symbol ◁▷ means that a pin can be either an input or an output.

FIGURE 7–11
Block diagram of a PAL.

PAL Output Combinational Logic

There are several basic types of PAL output logic that allows you to configure the device for a specific application. In this chapter, only the aspects of the output logic related to combinational logic functions are discussed. Additional features are covered in Chapter 11 after you have studied flip-flops, counters, and registers. Figure 7–12 shows three basic types of combinational output logic with tristate outputs and the associated OR gate. The following are types of PAL output logic:

■ *Combinational output*   This output is used for an SOP function and is usually available as either an active-LOW or an active-HIGH output.
■ *Combinational input/output (I/O)*   This output is used when the output function must feed back to be an input to the array or be used to make the I/O pin an input only.
■ *Programmable polarity output*   This output is used for selecting either the output function or its complement by programming the exclusive-OR gate for inversion or noninversion. The fusible link on the exclusive-OR input is blown open for inversion and left intact for noninversion.

(a) Combinational output (active-LOW). An active-HIGH output would be shown without the bubble on the tristate gate symbol.

(b) Combinational input/output (active-LOW)

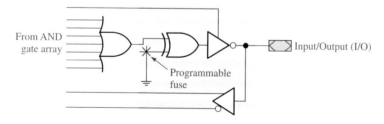

(c) Programmable polarity output

FIGURE 7–12
Basic types of PAL combinational output logic.

Standard PAL Numbering

Standard PALs come in a variety of configurations, each of which is identified by a unique part number. This part number always begins with the prefix PAL. The first two digits following PAL indicate the number of inputs, which includes outputs that can be configured as inputs. The letter following the number of inputs designates the type of output: L—active-LOW, H—active-HIGH, or P—programmable polarity. The one or two digits that follow the output type is the number of outputs. The following number is an example.

PAL10L8

Programmable array logic
Ten inputs
Eight outputs
Active-LOW output

In addition, a PAL part number may carry suffixes that specify speed, package type, and temperature range.

As an example of a PAL configuration, a block diagram of the PAL16L8 is shown in Figure 7–13. This particular device has 10 dedicated inputs (I), 2 dedicated outputs (O), and 6 pins that can be used either as inputs or as outputs. Each of the outputs is active-LOW.

FIGURE 7–13
Block diagram of the PAL16L8.

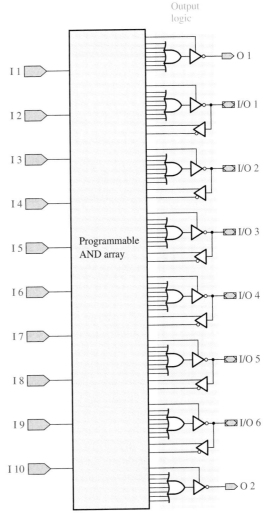

I = input, O = output, I/O = input/output

EXAMPLE 7–2

Determine the number of inputs, the number of outputs, and the type of output for each of the following PAL part numbers:

(a) PAL12H6   (b) PAL16L2   (c) PAL20P8

Solution

(a) 12 inputs, 6 outputs, active-HIGH outputs

(b) 16 inputs, 2 outputs, active-LOW outputs

(c) 20 inputs, 8 outputs, programmable polarity outputs

Related Problem   Describe a PAL with the part number PAL14H4.

SECTION 7–2 REVIEW

1. What is a PAL?
2. Explain the purpose of a fusible link in a PAL.
3. Can any combinational logic function be implemented with a PAL?
4. List three types of PAL output logic.
5. What does the designation PAL12H6 mean?

7–3 ■ GENERIC ARRAY LOGIC (GAL)

GAL is a designation originally used by Lattice Semiconductor and later licensed to other manufacturers. As you learned in Section 7–1, the GAL in its basic form is a PLD with a reprogrammable AND array, a fixed OR array, and programmable output logic. In this section, basic concepts are introduced and in Sections 7–4 and 7–5 specific GALs are examined.

After completing this section, you should be able to

■ Describe basic GAL operation   ■ Show how a sum-of-products expression is implemented in a GAL   ■ Compare the E^2CMOS cell in a GAL with the fusible link cell in a PAL   ■ Define OLMC and explain its purpose   ■ Interpret standard GAL numbers

GAL Operation

The GAL basically consists of a reprogrammable array of AND gates that connects to a fixed array of OR gates. Just as in a PAL, this structure allows any sum-of-products (SOP) logic expression with a defined number of variables to be implemented.

The basic structure of a GAL is illustrated in Figure 7–14 for two input variables and one output although most GALs have many inputs and many outputs. The reprogrammable array is essentially a grid of conductors forming rows and columns with an electrically erasable CMOS (E^2CMOS) cell at each cross point, rather than a fuse as in a PAL. These cells are shown as blocks in the figure.

Each row is connected to the input of an AND gate, and each column is connected to an input variable or its complement. By programming each E^2CMOS cell to be either *on* or *off,* any combination of input variables or complements can be applied to an AND gate to form any desired product term. A cell that is *on* effectively connects its corresponding row and column, and a cell that is *off* disconnects the row and column. The cells can be electrically erased and reprogrammed. A typical E^2CMOS cell can retain its programmed state for 20 years or more. See Chapter 15 for a discussion of how an E^2CMOS cell works.

Implementing a Sum-of-Products Expression   As an example, a simple GAL array is programmed as shown in Figure 7–15 so that the product term $\overline{A}B$ is produced by the top AND gate, AB by the middle AND gate, and $\overline{A}\,\overline{B}$ by the bottom AND gate. As shown, the

FIGURE 7–14
Basic E^2CMOS array structure of a GAL.

FIGURE 7–15
GAL implementation of a sum-of-products expression.

E^2CMOS cells are *on* to connect the desired variables or their complements to the appropriate AND gate inputs. The E^2CMOS cells are *off* where a variable or its complement is not used in a given product term. The final output from the OR gate is an SOP expression.

EXAMPLE 7–3

Show how a GAL is programmed for the following 3-variable logic function:

$$X = \overline{A}B\overline{C} + \overline{A}BC + BC + A\overline{B}$$

Solution   The programmed array using simplified notation is shown in Figure 7–16. Cells that are *on* are indicated by small red Xs. The absence of an X means that the cell is *off*.

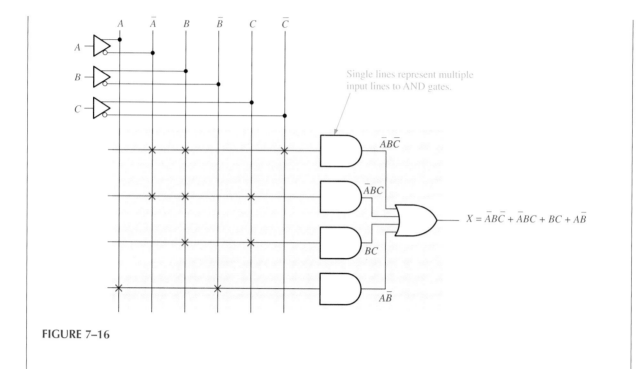

FIGURE 7–16

Related Problem   Write the expression for the output if the cells connecting input $\overline{A}$ to the top row and to the second row in Figure 7–16 are also *off*.

The GAL Block Diagram

A block diagram of a GAL is shown in Figure 7–17. The AND array outputs go to the output logic macrocells (**OLMC**), which contain the OR gates and programmable logic. A typical GAL has eight or more inputs to its AND array and eight or more input/outputs (I/Os) from its OLMCs as indicated, where $n \geq 8$ and $m \geq 8$.

FIGURE 7–17
Block diagram of a GAL.

The OLMC is made up of logic circuits that can be programmed as either *combinational logic* or as *registered logic*. The OLMC provides much more flexibility than the fixed output logic in a PAL. In this chapter, the combinational output configurations of OLMCs are considered; the registered configurations are discussed in Chapter 11 after you have studied flip-flops, counters, and registers.

Standard GAL Numbering

GALs come in a variety of configurations, each of which is identified by a unique part number. This part number always begins with the prefix GAL. The first two digits following the prefix indicate the number of inputs, which includes outputs that can be configured as inputs. The letter V following the number of inputs designates a variable-output configuration. The one or two digits that follow the output type is the number of outputs. The following number is an example.

GAL16V8

Generic array logic — Sixteen inputs — Eight outputs — Variable-output configuration

---

EXAMPLE 7–4

Determine the number of inputs and the number of outputs for each of the following GAL part numbers:
(a) GAL20V8   **(b)** GAL22V10

Solution
(a) 20 inputs, 8 outputs
(b) 22 inputs, 10 outputs

Related Problem   Describe a GAL with the part number GAL18V10.

SECTION 7–3 REVIEW

1. What is a GAL?
2. Explain the purpose of the E^2CMOS cell in a GAL.
3. Can any combinational logic function be implemented with a GAL?
4. What is the OLMC?

7–4 ■ THE GAL22V10

The various GALs all have the same type of programmable array. They differ in the size of the array, in the type of OLMCs, and in operating parameters such as speed and power dissipation. In this section, a popular generic array logic device, the GAL22V10, is discussed in order to illustrate principles that are common to all types of GALs. In the next section, we will look at the GAL16V8.

After completing this section, you should be able to

■ Explain the operation of the OLMC and list the four configurations ■ Describe the combinational mode ■ Discuss the operation and purpose of the tristate output buffer ■ Explain how the OLMC can operate as either an input or an output ■ Describe how the E^2CMOS cells are organized ■ Illustrate how an SOP function is implemented

Block Diagram of the GAL22V10

The GAL22V10 contains twelve dedicated inputs and ten input/outputs (I/Os) as shown in the block diagram of Figure 7–18. This device is available in either a 24-pin DIP (dual in-line package) or a 28-pin PLCC (plastic chip carrier), as shown in Figure 7–19. This device is also available in a low-voltage version, the GAL22LV10.

FIGURE 7–18
GAL22V10 block diagram.

The Output Logic Macrocells (OLMCs)

As stated in the discussion of GALs, an OLMC contains programmable logic circuits that can be configured either for a combinational output or input or for a registered output. In the registered mode (discussed in Chapter 11), the output comes from a flip-flop. In this chapter, the focus is on the combinational mode. OLMC combinational mode configurations are automatically set by programming.

As indicated by the notation in the block diagram of Figure 7–18, of the ten available GAL22V10 OLMCs, two have eight product terms (lines from the AND array to the OR gate), two have ten product terms, two have twelve product terms, two have fourteen product terms, and two have sixteen product terms. Each OLMC can be programmed for either an active-HIGH or an active-LOW output. Also, each OLMC can be programmed as an input.

Logic Diagram A basic logic diagram for the GAL22V10 OLMC is shown in Figure 7–20. The inputs from the AND gates to the OR gate vary from ten to sixteen, as mentioned. The logic inside the blue box consists of a flip-flop (defined in Chapter 1 and covered in Chapter 8 but not used in this chapter) and two multiplexers (covered in Chapter 6).

FIGURE 7–19
GAL22V10 package diagrams.

(a) PLCC (b) DIP

FIGURE 7–20
The GAL22V10 OLMC.

The 1-of-4 multiplexer connects one of its four input lines to the tristate output buffer based on the states of two select inputs, S_0 and S_1. The inputs to the 1-of-4 multiplexer are the OR gate output, the complement of the OR gate output, the flip-flop output, and the complement of the flip-flop output. The 1-of-2 multiplexer connects either the output of the tristate buffer or the flip-flop back through a buffer to the AND array based on the state of S_1. The select bits, S_0 and S_1, for each OLMC are programmed into a dedicated group of cells in the array.

The four OLMC configurations are

■ Combinational mode with active-LOW output
■ Combinational mode with active-HIGH output
■ Registered mode with active-LOW output
■ Registered mode with active-HIGH output

Combinational Mode When $S_0 = 0$ and $S_1 = 1$, the multiplexer selects the OR gate output. The *effective* logic paths of the OLMC are shown in red in Figure 7–21(a) where the output polarity is active-LOW because of the inversion of the tristate output buffer. When $S_0 = 1$ and $S_1 = 1$, the complement of the OR gate output is selected, and the *effective* logic paths of the OLMC are as shown in red in part (b). The output is active-HIGH because of the double inversion (complement of the OR and tristate inversion). The S_0 and

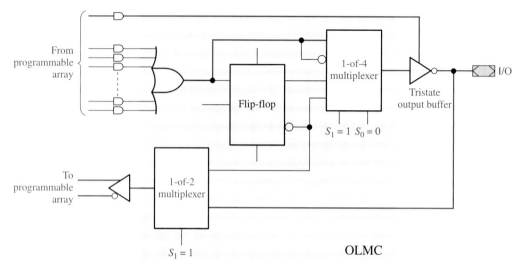

(a) Active-LOW output: $S_1 = 1$, $S_0 = 0$

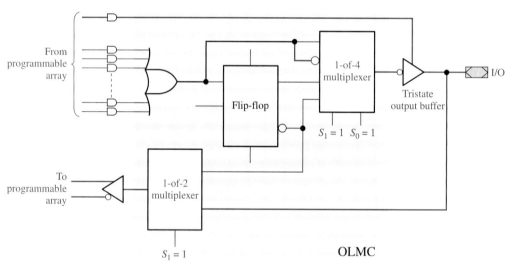

(b) Active-HIGH output: $S_1 = 1$, $S_0 = 1$

FIGURE 7–21
OLMC combinational mode.

S_1 bits for each of the ten OLMCs are set by programming a set of dedicated cells in the array separate from the logic array cells. The OLMC can be configured as an output or an input by controlling the tristate output buffer.

Tristate Output Buffer   The **tristate output buffer,** shown in Figure 7–22(a), is a two-state logic inverter with a control line that permits the output to be disconnected from the input. The three output states are *LOW* (when the input is HIGH), *HIGH* (when the input is LOW), and *high-impedance* (when the output is disconnected from the input). When the control line is HIGH, the buffer is active; and when the control line is LOW, the buffer is in the high-impedance state. These three states are illustrated in Figure 7–22(b). Notice that in the high-impedance state the buffer effectively acts as an open switch.

Output or Input Selection   In the OLMC combinational mode logic of Figure 7–21, you can see that the control line for the tristate buffer comes from an AND gate in the array. When this control line is HIGH, the OLMC produces a combinational output. When this control line is LOW, the buffer is open (high-impedance) and the pin becomes an input. This is indicated in Figure 7–23. The state of the tristate control line is set by programming.

(a)

(b)

FIGURE 7–22
Tristate output buffer operation.

(a) Output (active LOW shown for illustration)

(b) Input

FIGURE 7–23
OLMC combinational output and input configurations.

EXAMPLE 7–5

Determine the output expression of an OLMC in a GAL22V10 for the product terms from the AND array and the select bits as shown in Figure 7–24.

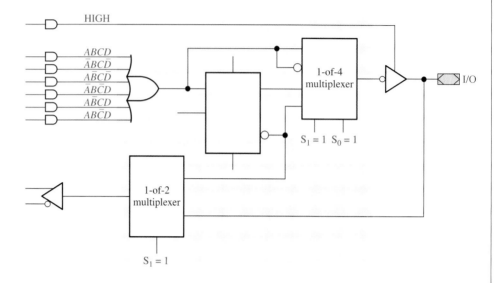

FIGURE 7–24

Solution   The OLMC is configured for an active-HIGH combinational output; therefore, the SOP expression is

$$X = ABCD + \overline{A}BC\overline{D} + A\overline{B}C\overline{D} + \overline{A}B\overline{C}D + A\overline{B}CD + AB\overline{C}D$$

Related Problem   Write the SOP expression for the output if $S_0 = 0$ and $S_1 = 1$.

The GAL22V10 Array

The diagram for the GAL22V10 in Figure 7–25 is similar to the block diagram, shown in Figure 7–18, except that details of the programmable AND array are shown.

The GAL22V10 programmable array is organized as 22 input lines and complements crossing 132 product term lines. There is a programmable E^2CMOS cell at each of the 5808 intersections. Each cell can be programmed in the *on* state to connect an input variable or its complement to a product term line.

A single line represents the product term line in the diagram of Figure 7–25. However, in the GAL22V10 there are actually 44 lines to each AND gate. So, each product term line consists of 44 AND gate inputs, one for each input line and its complement. The numbers shown for each product term line are the beginning cell numbers for the 44 cells associated with that line. The number in each OLMC for S_0 and S_1 are the cell numbers in the array into which the bits are programmed. These twenty special cells (5808 through 5827) are not shown in the array diagram.

Figure 7–26 shows details of the programmable array for the first OLMC section. The OLMC has a capacity for eight product terms that can be used in an SOP function. One product term is reserved for the tristate control, and one product term is used for a reset function in the registered mode for all of the OLMCs. The latter product term is called a *global* function because it is common to all OLMCs.

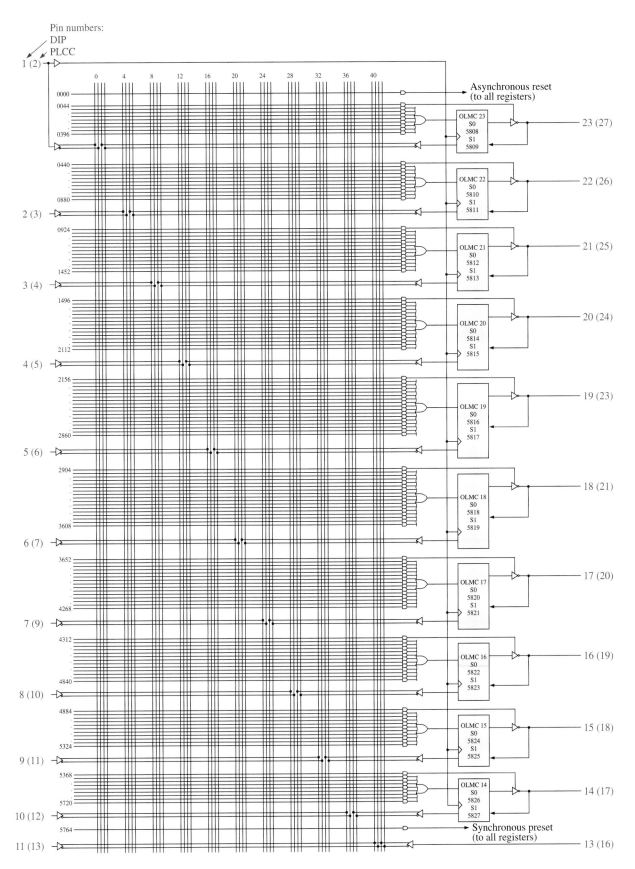

FIGURE 7–25
GAL22V10 array diagram.

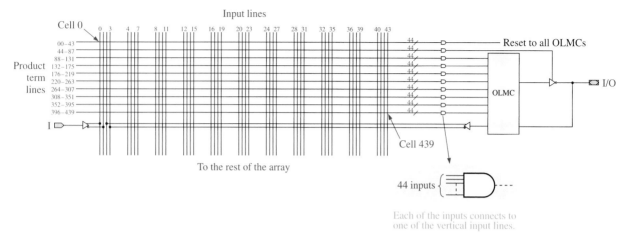

FIGURE 7–26
Organization of the programmable array showing one OLMC portion.

Implementing an SOP Function

As mentioned, a cell is programmed to the *on* state where a variable or its complement is part of a product term. Programming a cell to the *on* state simply connects an input line to a product term line. Each AND gate in the array produces one product term. When all of the input variables that are connected to an AND gate are HIGH, the output of the AND gate is HIGH.

The product terms from all of the AND gates for a given OLMC are ORed to form the sum-of-products function. The OLMC select bits are programmed to route the OR gate output, either complemented or uncomplemented, to the output pin.

With the GAL22V10, up to ten separate SOP functions can be programmed, the largest of which can contain sixteen product terms. If an SOP expression with more than sixteen terms is required, two OLMCs are necessary. The OR gate output of one OLMC is fed back to the array and connected into the other OLMC by programming.

EXAMPLE 7–6   Show how the following 6-variable SOP function is implemented with the GAL22V10.

$$X = ABCDEF + A\overline{B}C\overline{D}E\overline{F} + \overline{A}B\overline{C}D\overline{E}F + \overline{A}BCDE\overline{F} + A\overline{B}\,\overline{C}\,\overline{D}EF + \overline{A}\,\overline{B}\,CDEF + \overline{A}\,\overline{B}\,\overline{C}\,\overline{D}\,\overline{E}\,\overline{F}$$

Solution   The programmed portion of the array with the associated OLMC is shown in Figure 7–27. The rest of the array remains unused in this case. The Xs represent E^2CMOS cells that are programmed to the *on* state.

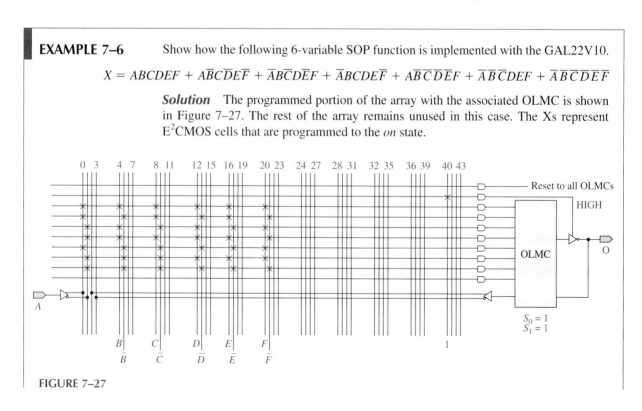

FIGURE 7–27

Related Problem How many more product terms can be added to the SOP expression programmed in Figure 7–27?

**SECTION 7–4
REVIEW**

1. What are the two modes for the GAL22V10?
2. How is the mode determined?
3. How is the polarity of the output set?
4. What is the maximum number of product terms in an SOP expression that can be implemented with a single GAL22V10 OLMC?

7–5 ■ THE GAL16V8

In this section, another type of generic array logic device, the GAL16V8, is discussed. The main feature of the GAL16V8 is that it can be programmed to emulate many types of PALs.

After completing this section, you should be able to

■ List the modes for the OLMC   ■ Discuss the simple mode   ■ Discuss the complex mode   ■ State how the GAL16V8 can replace various types of PALs

Block Diagram of the GAL16V8

The GAL16V8 contains eight dedicated inputs and eight input/outputs (I/Os) as shown in the block diagram of Figure 7–28(a) with the package pin numbers indicated. This device is available in either a 20-pin DIP (dual in-line package) or a 20-pin PLCC (plastic chip carrier) as shown in part (b). This device is also available in a low-voltage version, the GAL16LV8.

The OLMC

The OLMC in the GAL16V8 is somewhat different from that in the GAL22V10. The GAL16V8 is designed to be programmed in one of three available modes to emulate most of the existing PALs; thus, it may replace the PAL for which it is programmed.

PAL Emulation All of the possible OLMC configurations of the GAL16V8 are classified into three basic modes: *simple, complex,* and *registered.* Each of these modes allows the GAL16V8 to be programmed to emulate a number of PALs as specified in Table 7–1.

TABLE 7–1
PALs which can be emulated by each OLMC mode of the GAL16V8.

GAL16V8 Simple Mode	GAL16V8 Complex Mode	GAL16V8 Registered Mode
PAL10L8	PAL16L8	PAL16R8
PAL12L6	PAL16H6	PAL16R6
PAL14L4	PAL16P8	PAL16R4
PAL16L2		PAL16RP8
PAL10H8		PAL16RP6
PAL12H6		PAL16RP4
PAL14H4		
PAL16H2		
PAL10P8		
PAL12P6		
PAL14P4		
PAL16P2		

FIGURE 7–28
GAL16V8 block diagram and packaging.

The simple and complex modes are associated only with combinational outputs and are discussed here. The registered mode is covered in Chapter 11.

Simple Mode In this mode, the OLMCs are configured as dedicated active combinational outputs or as dedicated inputs (limited to six). The three possible configurations in the simple mode are

■ Combinational output
■ Combinational output with feedback to AND array
■ Dedicated input

Each of these configurations is illustrated in Figure 7–29. Notice that the tristate control line is HIGH (V_{CC}) making the buffer active for the combinational output configurations and LOW (ground) making the buffer open (high-impedance) for the dedicated input configuration.

The feedback allows the combinational output to be connected back to the AND array. The *XOR* input to the exclusive-OR gate determines the active state of the output. When *XOR* = 0, the output is active-LOW because the input from the OR gate is not inverted. When *XOR* = 1, the output is active-HIGH because the input from the OR gate is inverted.

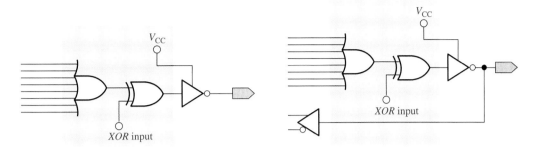

(a) Combinational output

(b) Combinational output with feedback to AND array

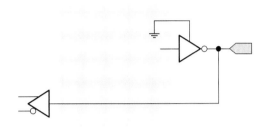

(c) Dedicated input

FIGURE 7–29
OLMC configurations in the simple mode.

The eight inputs to the OR gate come from the AND array and indicate that the OLMC can produce up to eight product terms in an SOP expression.

Complex Mode In this mode, the OLMCs can be configured two ways:

■ Combinational output
■ Combinational input/output (I/O)

These two configurations are illustrated in Figure 7–30. The I/Os are limited to six in this mode. Notice that there are only seven inputs to the OR gate from the AND array because the eighth input is used for the tristate control. This means that an OLMC can produce up to seven product terms in an SOP expression.

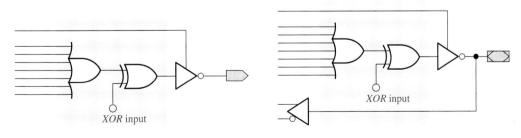

(a) Combinational output

(b) Combinational input/output

FIGURE 7–30
OLMC configurations in the complex mode.

CPLDs and FPGAs

In addition to fairly simple PLDs like the GAL, there are much larger programmable logic devices available. Two basic types are *complex PLDs* (CPLDs) and *field-programmable gate arrays* (FPGAs).

CPLDs are essentially much larger versions of simple PLDs, with a centralized internal interconnect matrix used to connect the device macrocells together. As with the simple PLDs discussed, the interconnect matrix is programmed to selectively connect input signals to a programmable AND layer connected in turn to a fixed OR layer. The outputs of the OR gates are connected in turn to configurable macrocells that allow the user to specify the output polarity, select registered or combinational operation, provide tristating capability, and optionally feed the signal back into the interconnect matrix. Regardless of the logic expression implemented, however, the CPLD pathway is essentially the same for each signal: from an input pin through a buffer or inverter to the interconnect matrix, through the interconnect matrix to the AND-OR logic, and from the AND-OR logic through the output macrocell. In addition, input, output, and clock signals are typically limited to specific pins on the CPLD, so that specific pin types are functionally identical to each other as well. This architecture limits the flexibility of the design to some degree, but it also simplifies design analysis by offering very predictable input to output signal timing.

FPGAs, on the other hand, consist of a large array of simple logic cells (simpler than the CPLD macrocells) with interconnecting horizontal and vertical routing channels. FPGAs do not have a predefined implementation structure as CPLDs do and through which each CPLD signal must pass. Instead they allow the routing channels to connect logic cells together in a great many ways to implement desired logic functions. In fact, depending upon the design constraints and available resources, two identical logic functions might even be implemented in completely different ways inside the same device. This design makes FPGAs extremely flexible and allows designers to implement very complex combinational and sequential logic designs. Some larger FPGAs can even be programmed to function as microprocessors.

SECTION 7–5 REVIEW

1. State the basic differences between a GAL16V8 and a GAL22V10.
2. How does the GAL16V8 emulate a PAL device?
3. List the three OLMC modes.

7–6 ■ PLD PROGRAMMING

As you have learned, PALs are programmed by leaving specified fusible links intact and blowing open all others. GALs are programmed in a similar way except the E^2CMOS cells are turned on or off. The logic function(s) to be implemented determines which cells are affected. In order to program a PAL or GAL, the following items are required: a computer, programming software, and a PLD programmer.

After completing this section, you should be able to

■ Name the components required to program a PLD ■ Describe the programming process in terms of a flow chart ■ Define in-system programming (ISP)

The three components required to program PLDs are

1. Programming software (logic compiler)
2. A computer that meets the software requirements
3. A software-driven programmer

The computer must meet the **software** requirements. The **programmer** is a hardware device that accepts programming data from the computer and implements a specified logic design in the PLD that is inserted in a socket in the programmer. These elements are illustrated in Figure 7–31.

FIGURE 7–31
Components for PLD programming.

Computer

Any computer that meets the software and programmer specifications can be used. These specifications normally include the type of microprocessor on which the computer is based, the amount of memory, and the operating system (DOS, Windows, and Macintosh are examples).

Software

The software packages for PLD programming are called logic **compilers.** Various software packages are available including, but not limited to, the following: ABEL, CUPL, OrCAD-PLD, and LOG/iC. All of these software packages perform similar functions; they process and synthesize the logic design entered by a specified method, convert the entered data into an intermediate file, and then generate a final output file called a **JEDEC file** (also known as a *cell map* or *fuse map*) for the device programmer. Also, with the software, a logic design can be fully simulated and debugged before any hardware is built.

A key feature of PLD software is the method for entering a logic design. Three basic design entry methods are *Boolean equation, truth table,* and *state diagram.* Additional entry methods are schematic, timing waveform, and hardware description. All available software packages include some or all of these design entry methods.

The Programmer

The PLD device is inserted into the programmer socket, which is usually a **ZIF socket** (zero insertion force socket). The programmer has a software driver program that reads the JEDEC file generated by the logic compiler and converts it to instructions for applying required voltages to specified PLD pins to alter the specified cells in the array as directed by the fuse map. All PLD software and programmers, regardless of the manufacturer, conform to a standard for generation of the JEDEC file established by the Joint Electronic Device Engineering Council (JEDEC).

The Programming Process

The general flow chart in Figure 7–32 shows the sequence for programming a PLD. First, the logic circuit is designed and expressed in terms of Boolean equations, truth tables, logic schematic, or other acceptable form, and the software is loaded into the computer.

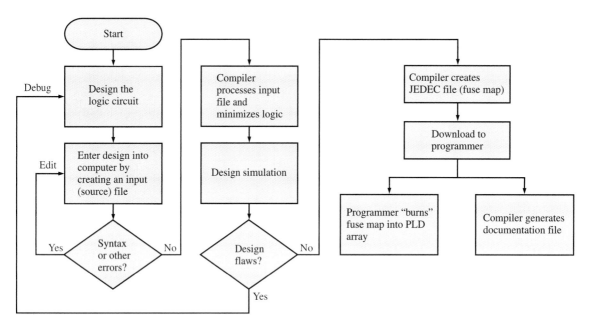

FIGURE 7–32
Flow chart of a PLD programming sequence.

Entering the Design The logic design is entered into the computer by creating an input or source file. Typically, some preliminary information is entered, such as user's name, company, data, and description of the design. Then, the type of PLD device, the input and output pin numbers, and variable labels are entered. Finally, the logic function(s) is entered in the form of a Boolean equation, truth table, schematic diagram, or the like. Any syntax errors or other errors made in entering data into the input file are indicated and must be corrected. **Syntax** is the prescribed format and symbology used in describing a category of functions.

Running the Software The software compiler processes and translates the input file and minimizes the logic. The logic design is then simulated using a set of hypothetical inputs known as *test vectors*. This process effectively "exercises" the design in software to determine if it works properly before the PLD is actually programmed. If any design flaws are discovered during software simulation, the design must be debugged and modified to correct the flaw.

Once the design is finalized, the compiler generates a **documentation file,** which includes the final logic equations, the JEDEC file, and if desired a pinout diagram of the PLD. The documentation file essentially provides a "hard copy" of the final design.

Programming the Device When the design is finalized, the compiler creates a fuse map (JEDEC file) and downloads it to the programmer. The fuse map tells the programmer which fuse links to blow and which to leave intact or, in the case of a GAL, which E^2CMOS cells to turn *on* or *off*. The programmer uses its own software driver program to pattern the PLD device according to the fuse map.

Most PLD programmers provide a zero insertion force (ZIF) socket that will accept a variety of PLD packages. It is vital that the PLD be inserted into the ZIF socket in the correct orientation and at the correct end of the socket. If inserted incorrectly, damage may be done to the programmer. Many programmers provide an insertion diagram adjacent to the socket as a guide for proper insertion of a PLD.

An introduction to PLD programming using ABEL software is provided in Section 7–7.

In-System Programmability Standard PLDs including EEPROMs, PALs, and GALs are programmed using the hardware programmer, as you have just learned. A variety of programmable logic including GALs is also available as in-system programmable (ISP) devices; for example, an in-system programmable version of the GAL22V10 is designated ispGAL22V10. The ISP device is the same as the standard device except that it has additional logic for in-system programming.

An ISP device can be programmed as a standard device prior to being placed on a printed circuit board, or it can be programmed in several ways after it has been installed on a printed circuit board. ISP devices can be programmed in a system as follows:

- Using a programmer before assembly on a printed circuit board
- Using automatic test equipment (ATE) after assembly on a board
- Using a computer after assembly on a board
- Using an in-system microprocessor (embedded) after assembly on a board

In-system programmability (ISP) has several advantages over traditional PLD programming. A few of these advantages are as follows:

1. In-system programming allows logic designs to be programmed and modified without removing the device(s) from the circuit board.
2. With traditional PLDs, manual handling is required during both design/debugging and manufacturing stages. Fragile leads, especially on LSI PLDs, as thin as 0.5 mm can easily bend in the programmer socket causing damage. ISP packages can be soldered onto the printed circuit board before programming and debugging and left there.
3. Multiple designs can be implemented with a generic circuit board layout.
4. Updates can be made in the field by loading a new circuit configuration from disk or downloading it through a modem.

**SECTION 7–6
REVIEW**

1. What is required to program a PLD?
2. What are the software packages for PLDs called?
3. What is a JEDEC file?
4. Define ISP.
5. How does an ispGAL differ from a standard GAL?

7–7 ■ PLD SOFTWARE

As mentioned earlier, there are several software packages for implementing logic designs in PLDs. ABEL and CUPL are two commonly used hardware description languages (HDLs). Because both languages are similar and produce the same results in terms of programming a PLD, usage is often a matter of preference and availability. This section introduces basic concepts of PLD software using the ABEL language and is not intended to provide a comprehensive coverage. **Digital Design Using ABEL,** *by Pellerin and Holley (Prentice Hall), is recommended for a thorough coverage.*

After completing this section, you should be able to

■ List the three formats for design entry ■ Define *test vector* ■ Identify the ABEL logic operator symbols ■ Express logic functions using the ABEL equation format ■ Express logic functions using the ABEL truth table format ■ Develop test vectors for a given logic design ■ Write an ABEL input file for a given logic design ■ Describe a documentation file ■ Discuss the JEDEC file

Introduction to ABEL

ABEL, which is the acronym for *Advanced Boolean Expression Language,* allows logic designs to be implemented in programmable logic devices. ABEL can be used to program any type of PLD and is therefore a device-independent language. As discussed in the previous section, ABEL is run on a computer connected to a language-independent device programmer into which the PLD is inserted.

Logic Design Entry   ABLE provides three different formats for describing and entering a logic design from the computer keyboard: *equations, truth tables,* and *state diagrams.* The equation and truth table formats are used for combinational logic design, which is the focus of this chapter. The state diagram format, as well as the other two formats, can be used for sequential logic design (state machines) and is not discussed until Chapter 11.

Design Simulation   Once a logic circuit design has been entered, its operation can be simulated using **test vectors** to make sure there are no design errors. Basically, a test vectors section lists a sample of input values and the corresponding output values. The software essentially "exercises" the logic design to make sure it works as expected by taking it through various combinations of inputs and checking the resulting outputs.

Logic Synthesis   The software process of converting a circuit description in the form of equations, truth tables, or state diagrams to a standard JEDEC file format required to actually implement the design in a PLD is called logic **synthesis.** Logic synthesis usually includes optimization and minimization of a design.

Boolean Operations

The NOT (inversion), AND, OR, and exclusive-OR operations have special symbols in ABEL as shown in Table 7–2 in order of precedence. As in Boolean algebra, parentheses can be used to alter the order of precedence.

TABLE 7–2
ABEL symbols for the logic operations.

Logic Operation	ABEL Symbol
NOT	!
AND	&
OR	#
XOR	$

Examples of the basic operations in standard Boolean notation and in ABEL notation are as follows:

Standard Boolean	*ABEL*
$\overline{A}$	!A
AB	A & B
$A + B$	A # B
$A \oplus B$	A $ B

Equations

One of the ABEL entry methods uses logic equations. The equal sign ($=$) in ABEL is the same as in standard equations, and any letters or combinations of letters and numbers can be used as variable identifiers. With respect to variable identifiers, ABEL is case-sensitive. For example, uppercase A and lowercase a are two different variables in ABEL.

All equations in ABEL must end with a semicolon (;). For example, the sum-of-products expression $X = AB + \overline{A}B + A\overline{B}$ is written in the ABEL language as

$$X = A \& B \# !A \& B \# A \& !B;$$

EXAMPLE 7–7

Write each of the following logic expressions in ABEL:
(a) $X = A\overline{B}C + \overline{A}\,\overline{B}\,\overline{C} + AB + \overline{B}C$ (b) $Y = (\overline{A} + B + \overline{C} + D)(A + B + C)$

Solution
(a) $X = A \& !B \& C \# !A \& !B \& !C \# A \& B \# !B \& C;$
(b) $Y = (!A \# B \# !C \# D) \& (A \# B \# C);$

Note that the parentheses in part (b) are used to change the order of precedence as defined in Table 7–2. That is, the variables within the parentheses are first ORed and then the two resulting terms are ANDed.

Related Problem Express $W = X + Y$ in the ABEL format, where $X = \overline{A}B\overline{C}$ and $Y = A + \overline{B} + C$.

Sets In some cases, multiple input or output variables may be grouped as a set using brackets in order to simplify an equation. For example, a group of inputs labeled A_3, A_2, A_1, A_0 can be defined as a single variable A using ABEL set notation as follows:

$$A = [A3, A2, A1, A0];$$

EXAMPLE 7–8

Develop the logic equations to describe the 12-input, 4-output multiplexer shown in Figure 7–33(a) using ABEL. The select inputs, S_0 and S_1, select one of the three sets of four data inputs and route it to the four output lines. The function table is shown in part (b).

(a) Block diagram

Select Inputs		Data Outputs			
S_1	S_0	Y_3	Y_2	Y_1	Y_0
0	1	A_3	A_2	A_1	A_0
1	0	B_3	B_2	B_1	B_0
1	1	C_3	C_2	C_1	C_0

(b) Function table

FIGURE 7–33

Solution The sum-of-products equations in standard Boolean form are

$$Y_3 = A_3\overline{S}_1 S_0 + B_3 S_1 \overline{S}_0 + C_3 S_1 S_0$$
$$Y_2 = A_2\overline{S}_1 S_0 + B_2 S_1 \overline{S}_0 + C_2 S_1 S_0$$
$$Y_1 = A_1\overline{S}_1 S_0 + B_1 S_1 \overline{S}_0 + C_1 S_1 S_0$$
$$Y_0 = A_0\overline{S}_1 S_0 + B_0 S_1 \overline{S}_0 + C_0 S_1 S_0$$

The equations in ABEL format are

```
Y3 = A3 & !S1 & S0 # B3 & S1 & !S0 # C3 & S1 & S0;

Y2 = A2 & !S1 & S0 # B2 & S1 & !S0 # C2 & S1 & S0;

Y1 = A1 & !S1 & S0 # B1 & S1 & !S0 # C1 & S1 & S0;

Y0 = A0 & !S1 & S0 # B0 & S1 & !S0 # C0 & S1 & S0;
```

The four equations can be combined into one ABEL equation by grouping the inputs and outputs into sets with the following declaration:

```
A = [A3, A2, A1, A0];

B = [B3, B2, B1, B0];

C = [C3, C2, C1, C0];

Y = [Y3, Y2, Y1, Y0];

S = [S1, S0];
```

The sets are then used as follows to produce a single multiplexer output equation:

```
Y = (S == 1) & A # (S == 2) & B # (S == 3) & C;
```

The symbol == is a relational operator that provides an equality comparison for the sets. For example, the first term in the right side of the equation is (S == 1). This means that the select inputs, S_1 and S_0, are equal to 1 or binary 01, which is expressed in standard Boolean notation as $\overline{S}_1 S_0$.

With the set declarations and the single ABEL equation, the compiler software will implement each of the four original logic equations.

Related Problem Rewrite the single multiplexer ABEL equation if the order of selection in the multiplexer is reversed. That is, the C inputs are selected when the select inputs are equal to binary 01, etc.

Truth Tables

Another ABEL entry method uses truth tables. A specified logic function can be entered in either equation form, as you have seen, or in truth table form. Both methods are equivalent in terms of the result. Truth tables are usually more convenient than equations for describing certain types of circuits such as decoders.

The ABEL truth table format includes a header and the truth table entries. The header format is illustrated by the following sample header:

```
TRUTH_TABLE ([A, B, C, D] -> [X1, X2])
```

where A, B, C, and D are the inputs and X1 and X2 are the outputs. The symbol –> is an operator that indicates the inputs produce combinational outputs. To illustrate how an ABEL truth table is constructed, a simple example of an exclusive-OR gate with inputs *A* and *B* and output *X* is written as follows:

```
TRUTH_TABLE ([A, B] -> [X])
            [1, 0] -> [1];
            [0, 1] -> [1];
            [0, 0] -> [0];
            [1, 1] -> [0];
```

EXAMPLE 7–9

Develop an ABEL truth table for a BCD-to-7-segment decoder. Refer to the truth table in Table 4–10. The block diagram and 7-segment format are shown in Figure 7–34. The outputs are to be active-HIGH.

FIGURE 7–34

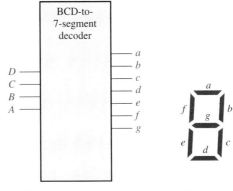

(a) Block diagram   (b) 7-segment display format

Solution   The most significant input is *D* and the least significant input is *A*. The segment outputs are *a* through *g*. An output of 1 means the segment is *on* and 0 means the segment is *off*. The ABEL truth table is

```
TRUTH_TABLE ([D, C, B, A] -> [a, b, c, d, e, f, g])
            [0, 0, 0, 0] -> [1, 1, 1, 1, 1, 1, 0];
            [0, 0, 0, 1] -> [0, 1, 1, 0, 0, 0, 0];
            [0, 0, 1, 0] -> [1, 1, 0, 1, 1, 0, 1];
            [0, 0, 1, 1] -> [1, 1, 1, 1, 0, 0, 1];
            [0, 1, 0, 0] -> [0, 1, 1, 0, 0, 1, 1];
            [0, 1, 0, 1] -> [1, 0, 1, 1, 0, 1, 1];
            [0, 1, 1, 0] -> [1, 0, 1, 1, 1, 1, 1];
            [0, 1, 1, 1] -> [1, 1, 1, 0, 0, 0, 0];
            [1, 0, 0, 0] -> [1, 1, 1, 1, 1, 1, 1];
            [1, 0, 0, 1] -> [1, 1, 1, 1, 0, 1, 1];
```

By using a set declaration for the inputs and decimal numbers for the BCD input values, the table can be simplified.

```
BCD = [D, C, B, A];

TRUTH_TABLE (BCD -> [a, b, c, d, e, f, g])
            0  -> [1, 1, 1, 1, 1, 1, 0];
            1  -> [0, 1, 1, 0, 0, 0, 0];
            2  -> [1, 1, 0, 1, 1, 0, 1];
            3  -> [1, 1, 1, 1, 0, 0, 1];
            4  -> [0, 1, 1, 0, 0, 1, 1];
            5  -> [1, 0, 1, 1, 0, 1, 1];
            6  -> [1, 0, 1, 1, 1, 1, 1];
            7  -> [1, 1, 1, 0, 0, 0, 0];
            8  -> [1, 1, 1, 1, 1, 1, 1];
            9  -> [1, 1, 1, 1, 0, 1, 1];
```

Related Problem   Rework the ABEL truth table for active-low outputs (0).

Test Vectors

As you know, test vectors are used in ABEL to describe input signals and expected outputs for logic simulation. In other words, the test vectors examine a logic design before it is committed to hardware by applying selected input combinations and checking for the proper outputs.

Test vectors are essentially the same as truth tables and are entered in the same format. For example, the test vectors section for the BCD-to-7-segment decoder is

```
TEST_VECTORS ([D, C, B, A] -> [a, b, c, d, e, f, g])
              [0, 0, 0, 0] -> [1, 1, 1, 1, 1, 1, 0];
              [0, 0, 0, 1] -> [0, 1, 1, 0, 0, 0, 0];
              [0, 0, 1, 0] -> [1, 1, 0, 1, 1, 0, 1];
              [0, 0, 1, 1] -> [1, 1, 1, 1, 0, 0, 1];
              [0, 1, 0, 0] -> [0, 1, 1, 0, 0, 1, 1];
              [0, 1, 0, 1] -> [1, 0, 1, 1, 0, 1, 1];
              [0, 1, 1, 0] -> [1, 0, 1, 1, 1, 1, 1];
              [0, 1, 1, 1] -> [1, 1, 1, 0, 0, 0, 0];
              [1, 0, 0, 0] -> [1, 1, 1, 1, 1, 1, 1];
              [1, 0, 0, 1] -> [1, 1, 1, 1, 0, 1, 1];
```

The test vectors can be simplified using sets, the same as truth tables.

The ABEL Input File

When you create an **input** (source) **file** in ABEL, you create a module which contains three basic sections following the module and title statements: *declarations, logic descriptions,* and *test vectors.*

Declarations The declarations section generally includes *device declaration, pin declarations,* and *set declarations.* We used set declarations in Examples 7–8 and 7–9.

Device declarations are used to specify the PLD that is to be programmed, sometimes referred to as a *target device.* An example of a device declaration statement is

```
decoder device 'P22V10';
```

The first word is simply a description and can be anything you wish. The second word is an essential keyword and can appear in the statement in either uppercase or lowercase letters. The specific device must be in the form shown with the prefix P used for both PALs and GALs.

An example of a pin declaration statement is

```
A0, A1, A2, A3, PIN 1, 2, 3, 4;
```

The keyword PIN can also be written in lowercase letters. The numbers following the keyword are pin numbers on the PLD device that is to be programmed. A0 goes to pin 1, A1 goes to pin 2, A2 goes to pin 3, and A3 goes to pin 4.

Logic Descriptions The two methods of entry that we have already discussed, equations and truth tables, are two types of logic descriptions. The third type is the state diagram which is covered in Chapter 11.

Test Vectors As stated before, the test vector section tests a logic design before it is finalized and implemented in hardware. Test vectors have the same form as truth tables.

EXAMPLE 7–10

Create an ABEL input file for implementing the quad 1-of-4 multiplexer shown as a block diagram in Figure 7–35(a). The function table is shown in part (b), and the pin assignment diagram of the GAL20V8 target device is shown in part (c).

Solution The ABEL input file for the multiplexer logic design is as follows:

```
Module    quad_1of4_mux
Title        'Quad 1 of 4 multiplexer in a GAL20V8'
  mux        device        'P20V8';
  A0, A1, A2, A3            pin 1, 2, 3, 4;
  B0, B1, B2, B3            pin 5, 6, 7, 8;
  C0, C1, C2, C3            pin 9, 10, 11, 13;
  D0, D1, D2, D3            pin 14, 15, 16, 17;
  Aout, Bout, Cout, Dout    pin 21, 20, 19, 18;
  S0, S1                    pin 22, 23;
```

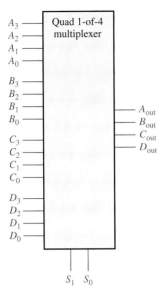

(a) Block diagram

Select Inputs		Data Outputs			
S_1	S_0	D_{out}	C_{out}	B_{out}	A_{out}
0	0	D_0	C_0	B_0	A_0
0	1	D_1	C_1	B_1	A_1
1	0	D_2	C_2	B_2	A_2
1	1	D_3	C_3	B_3	A_3

(b) Function table

(c) Pin assignments

FIGURE 7–35

```
Equations

 Aout =!S1 & !S0 & A0 # !S1 & S0 & A1 # S1 & !S0 & A2 # S1 & S0 & A3;

 Bout =!S1 & !S0 & B0 # !S1 & S0 & B1 # S1 & !S0 & B2 # S1 & S0 & B3;

 Cout =!S1 & !S0 & C0 # !S1 & S0 & C1 # S1 & !S0 & C2 # S1 & S0 & C3;

 Dout =!S1 & !S0 & D0 # !S1 & S0 & D1 # S1 & !S0 & D2 # S1 & S0 & D3;

Test_vectors
([S1,S0,A0,A1,A2,A3,B0,B1,B2,B3,C0,C1,C2,C3,D0,D1,D2,D3] ->
[Aout,Bout,Cout,Dout])

"S  S  A  A  A  A  B  B  B  B  C  C  C  C  D  D  D  D      outputs
"1  0  0  1  2  3  0  1  2  3  0  1  2  3  0  1  2  3    A  B  C  D

[0, 0, 1, 0, 0, 0, 0, 1, 0, 0, 0, 0, 1, 0, 0, 0, 0, 1] ->[1, 0, 0, 0];
[0, 1, 1, 0, 0, 0, 0, 1, 0, 0, 0, 0, 1, 0, 0, 0, 0, 1] ->[0, 1, 0, 0];
[1, 0, 1, 0, 0, 0, 0, 1, 0, 0, 0, 0, 1, 0, 0, 0, 0, 1] ->[0, 0, 1, 0];
[1, 1, 1, 0, 0, 0, 0, 1, 0, 0, 0, 0, 1, 0, 0, 0, 0, 1] ->[0, 0, 0, 1];

[0, 0, 1, 1, 1, 0, 1, 1, 0, 1, 1, 0, 1, 1, 0, 1, 1, 1] ->[1, 1, 1, 0];
[0, 1, 1, 1, 1, 0, 1, 1, 0, 1, 1, 0, 1, 1, 0, 1, 1, 1] ->[1, 1, 0, 1];
[1, 0, 1, 1, 1, 0, 1, 1, 0, 1, 1, 0, 1, 1, 0, 1, 1, 1] ->[1, 0, 1, 1];
[1, 1, 1, 1, 1, 0, 1, 1, 0, 1, 1, 0, 1, 1, 0, 1, 1, 1] ->[0, 1, 1, 1];

END
```

Notice that comment lines can be inserted anyplace but must be preceded by the quotation symbol " as we did in the test vectors section to clarify the inputs and outputs in the entries. Comment lines in general are used to clarify or identify parts of the input file for later use by anyone who reads the file. Comment lines are ignored by the ABEL compiler and do not affect the final design.

Related Problem   Use set declarations to write the logic equations for the multiplexer.

The Documentation File

After an input file is processed, ABEL generates a documentation file. The documentation file provides a hard copy of the final reduced equations, a JEDEC file such as the one shown in Figure 7–36, and a device pin diagram.

```
QP24* QF2706*
L0000
000000000000000000000000000000000000000000
000000000000000000000000000000000000000000
000000000000000000000000000000000000000000
000000000000000000000000000000000000000000
000000000000000000000000000000000000000000
000000000000000000000000000000000000000000
000000000000000000000000000000000000000000
000000000000000000000000000000000000000000
```

A_{out} logic
```
110111011101111111111111111111111111
011111011101111111111111111111111111
111101011110111111111111111111111111
111111010101111111111111111111111111
000000000000000000000000000000000000
000000000000000000000000000000000000
000000000000000000000000000000000000
000000000000000000000000000000000000
```

B_{out} logic
```
111111011100111111111111111111111111
111111011011110111111111111111111111
111111011110111111101111111111111111
111111011110111111111110111111111111
000000000000000000000000000000000000
000000000000000000000000000000000000
000000000000000000000000000000000000
000000000000000000000000000000000000
```

C_{out} logic
```
111111011101111111111111111011111111
111111101101111111111111111101111111
111111011110111111111111111111110111
111111011101111111111111111111111101
000000000000000000000000000000000000
000000000000000000000000000000000000
000000000000000000000000000000000000
000000000000000000000000000000000000
```

D_{out} logic
```
111111011101111111111111111111011111
111111101101111111111111111110111111
111111011110111111111111110111111111
111110111101111111110111111111111111
000000000000000000000000000000000000
000000000000000000000000000000000000
000000000000000000000000000000000000
000000000000000000000000000000000000*
```

Used for configuring OLMCs and other special programming
```
L2560
01111000*
L2568
0000000000000000000000000000000000000000000000000000000000000000*
L2632
1*
L2633
0*
L2634
0*
L2635
0*
L2636
0*
L2637
1*
L2638
1*
L2639
1*
L2640
1111111111111111111111111111111111111111111111111111111111111111*
L2704
10*
```

Test vectors section (not programmed into array)
```
V0001 10000100001N00001LLLHOON*
V0002 10000100001N00001LLHL10N*
V0003 10000100001N00001LHLL01N*
V0004 10000100001N00001HLLL11N*
V0005 11101101101N10111LHHHOON*
V0006 11101101101N10111HLHH10N*
V0007 11101101101N10111HHLH01N*
V0008 11101101101N10111HHHL11N*
C5127*
```

FIGURE 7–36
JEDEC file for the quad 1-of-4 multiplexer from Example 7–10. The red labels are not part of the file.

The JEDEC File As mentioned in Section 7–6, this part of the documentation file is also called a *cell map* or a *fuse map*. The JEDEC file is downloaded to the programmer for programming the target device. JEDEC files are standarized for all programming languages.

The cell map shows which cells in the device array are disconnected (*off* or blown) and which are connected (*on* or intact). A 0 in a cell map indicates a cell that is connected and a 1 indicates a cell that is disconnected.

Figure 7–36 illustrates the cell map for the quad 1-of-4 multiplexer from Example 7–10. The four digits preceded by an "L" correspond to the product line numbers in the

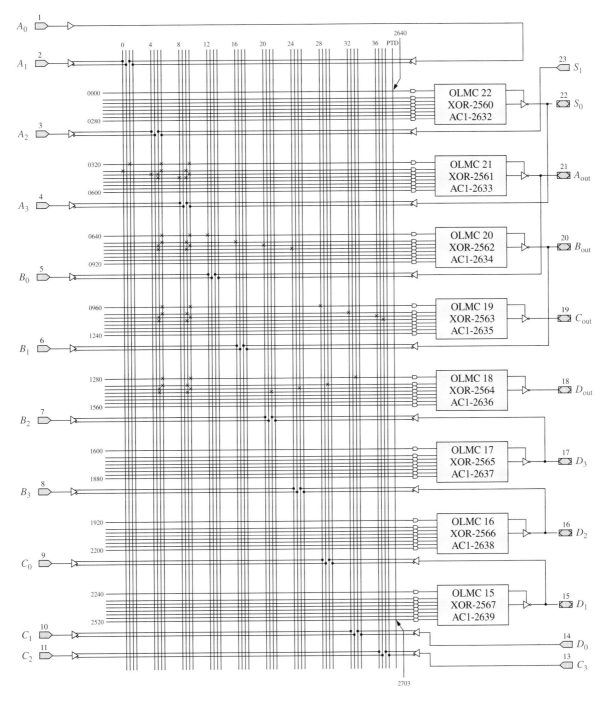

FIGURE 7–37
GAL20V8 for Example 7–11 on the next page.

array. The four sum-of-product equations for multiplexer outputs are implemented in the four sections of the array as indicated. An asterisk indicates the end of the logic implementation. All of the lines that start with an "L" are for special cell programming for user identification, configuring the OLMCs, and other functions. The lines in the cell map beginning with "V" are used for the test vectors, and the last line beginning with a "C" is a checksum.

EXAMPLE 7–11

Show the GAL20V8 logic array with the input and output assignments and the array programming for the quad 1-of-4 multiplexer based on the JEDEC file in Figure 7–36.

Solution   The programmed logic diagram for the GAL20V8 is shown in Figure 7–37 on the preceding page where the red Xs indicate the cells that are *on*.

Related Problem   How many I/O pins in the GAL20V8 in Figure 7–37 are used as inputs and how many are used as outputs?

SECTION 7–7 REVIEW

1. What is ABEL?
2. List the three formats for design entry in ABEL.
3. Write the ABEL symbols for NOT, AND, OR, and XOR.
4. Express the following Boolean equation using ABEL:

$$Y = \overline{D}E\overline{F} + \overline{D}\overline{E} + F$$

5. What are test vectors?
6. List the three parts of an ABEL input file.
7. What types of declarations are usually found in an input file?
8. What is a documentation file?
9. Describe a JEDEC file and give two other names used for it.

7–8 ■ DIGITAL SYSTEM APPLICATION

In the last chapter the combinational logic portion of the traffic light control system was implemented using SSI and MSI devices. You will now apply your knowledge of PLDs and PLD programming to implement the state decoding and output logic with a GAL22V10. You should review Section 6–11 before continuing.

After completing this section, you should be able to

■ Program the GAL22V10   ■ Explain how the GAL replaces the logic previously implemented in Chapter 6

State Decoding and Output Logic

The block diagram of the state decoding and output logic is shown in Figure 7–38. The traffic light sequence and the corresponding state diagram are given in Figure 7–39 for reference.

The state decoding and output logic portion of the system has two inputs, S_1 and S_0, and eight outputs, *MR, MY, MG, SR, SY, SG, Long,* and *Short*. These input and output variables are defined as follows:

S_1 State input, Gray code MSB
S_0 State input, Gray code LSB

MR	Main street red light output
MY	Main street yellow light output
MG	Main street green light output
SR	Side street red light output
SY	Side street yellow light output
SG	Side street green light output
Long	Trigger output to long timer
Short	Trigger output to short timer

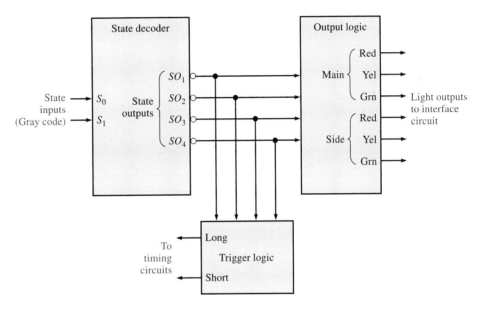

FIGURE 7–38
Block diagram of the state decoding and output logic.

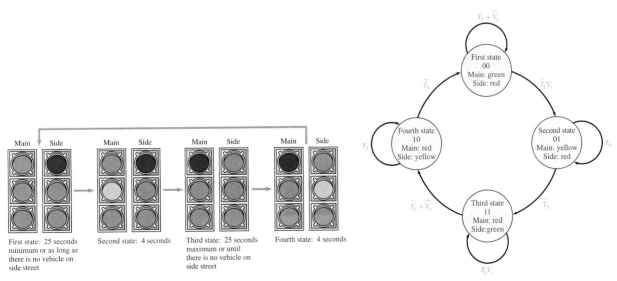

(a) Traffic light sequence

(b) State diagram

FIGURE 7–39
Traffic light sequence and state diagram.

■ SYSTEM ASSIGNMENT 1:   GAL22V10 Programming

■ *Activity 1:*   Using the truth table and the pin assignment diagram for a GAL22V10 shown here, write an ABEL input file to program the device with the state decoding and output logic.

State Inputs		Light Outputs						Trigger Outputs	
S_1	S_0	$\overline{MR}$	$\overline{MY}$	$\overline{MG}$	$\overline{SR}$	$\overline{SY}$	$\overline{SG}$	*Long*	*Short*
0	0	1	1	0	0	1	1	1	0
0	1	1	0	1	0	1	1	0	1
1	1	0	1	1	1	1	0	1	0
1	0	0	1	1	1	0	1	0	1

■ SYSTEM ASSIGNMENT 2:   Verification and Testing

The single GAL22V10 package replaces the 74LS139 and the 74LS08 that were used to implement the same logic in the Chapter 6 system application. However, the GAL can be reprogrammed for additional logic functions required in this system, as you will see in Chapter 11.

■ *Activity 1:*   Identify the function of each of the points labeled with a circled number on the portion of the system PC board shown.

■ *Activity 2:*   Using the switches connected as shown to simulate the Gray code sequence on the inputs, develop a test procedure to check out this portion of the system board.

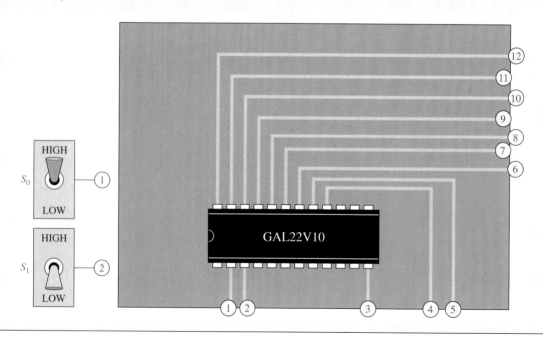

SECTION 7–8 REVIEW

1. After programming the GAL22V10 in this section, how many outputs are still unused and available?
2. How many inputs are unused and available?
3. Compare the GAL implementation to the implementation in Chapter 6 in terms of complexity?

■ SUMMARY

- Four basic types of PLDs are shown in Figure 7–40.
- PALs are one-time programmable.
- GALs are reprogrammable.
- PALs and GALs can produce any sum-of-products logic function limited only by the number of variables.
- GALs can emulate most types of PALs.
- A hardware description language (HDL) such as ABEL or CUPL is used to program PLDs.

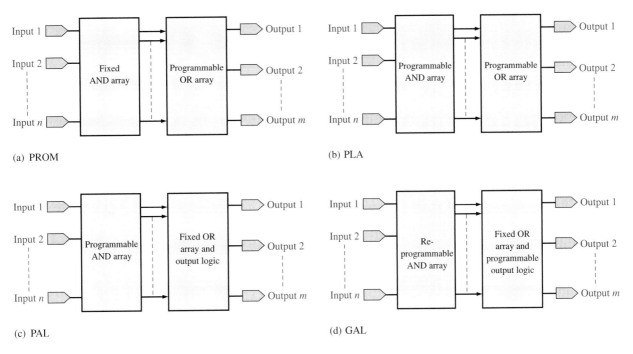

(a) PROM

(b) PLA

(c) PAL

(d) GAL

FIGURE 7–40

■ KEY TERMS

These terms are also in the end-of-book glossary.

ABEL   Advanced Boolean Expression Language. A software compiler language for PLD programming; a type of hardware description language (HDL).

Architecture   The internal functional arrangement of the elements that give a device its particular operating characteristics.

Array   In a PLD, a matrix formed by rows of product-term lines and columns of input lines with a programmable cell at each junction.

Buffer   A circuit that prevents loading of an input or output.

Cell   A fused cross point of a row and column in a PLD.

Compiler   Software that translates from high-level language that uses words or symbols, such as ABEL or CUPL, into low-level machine language (1s and 0s).

Documentation file   The information from a computer that documents the final design after the input file has been processed.

E²CMOS   Electrically erasable CMOS (EECMOS). The circuit technology used for the reprogrammable cells in a GAL.

Fuse   The programmable element in certain types of PLDs; also called a fusible link.

GAL   Generic array logic. A PLD with a reprogrammable AND array, a fixed OR array, and programmable output logic macrocells.

Input file   The information entered in a computer that describes a logic design using a PLD programming language such as ABEL or CUPL.

Input/Output (I/O)   A terminal of a device that can be used as either an input or as an output.

JEDEC file   A Joint Electronic Device Engineering Council standard software file generated from the compiler software that is used by a programming device to implement a logic design in a PLD; also called a fuse map or cell map.

OLMC   Output logic macrocell. The programmable output logic in a GAL.

PAL   Programmable array logic. A PLD with a programmable AND array and a fixed OR array.

PLA   Programmable logic array. A PLD with programmable AND and OR arrays.

PLD   Programmable logic device.

Programmer   An instrument that programs a PLD using a JEDEC file downloaded from a computer running hardware description language software.

PROM   Programmable read-only semiconductor memory. A PLD with a fixed AND array and programmable OR array; used as a memory device and normally not as a logic circuit device.

Software   Computer programs; programs that instruct a computer what to do in order to carry out a given set of tasks.

Synthesis   The software process of converting a circuit description to a standard JEDEC file for PLD programming.

Tristate output buffer   A logic circuit having three output states: HIGH, LOW, and high impedance (open).

ZIF socket   Zero insertion force socket. A type of socket used in most programmers that accepts a PLD package.

■ SELF-TEST

1. The basic types of programmable arrays are
   - **(a)** NOT and XOR   **(b)** NAND and NOR
   - **(c)** AND and OR   **(d)** diode and transistor

2. An array can best be described as
   - **(a)** a group of programmable logic gates
   - **(b)** a matrix of rows and columns with a programmable cell at each intersection
   - **(c)** rows of programmable cells
   - **(d)** columns of fusible links

3. PLDs can be classified as
   - **(a)** combinational and registered
   - **(b)** programmable and nonprogrammable
   - **(c)** one-time programmable and reprogrammable
   - **(d)** PROMs, PLAs, PALs, and GALs
   - **(e)** answers (a) and (c)
   - **(f)** answers (c) and (d)

4. A PLD that has two programmable arrays is

    (a) PAL   **(b)** PROM   **(c)** PLA   **(d)** GAL

5. A PLD that has a programmable AND array and a fixed OR array is

    (a) PROM   **(b)** PLA   **(c)** PAL   **(d)** GAL

6. A PLD that has a reprogrammable AND array is

    (a) PROM   **(b)** PLA   **(c)** PAL   **(d)** GAL

7. The two types of lines in a PLD array are

    (a) vertical and horizontal   **(b)** input and product term

    (c) fusible and nonfusible   **(d)** buffered and tristate

8. A connection between a row and column in a PAL array is made by

    (a) blowing a fusible link

    (b) leaving a fusible link intact

    (c) connecting an input variable to the input line

    (d) connecting an input variable to the product term line

9. The device number PAL14H4 indicates

    (a) a PAL with fourteen active-HIGH outputs and four inputs

    (b) a PAL that implements fourteen AND gates and four OR gates

    (c) a PAL with fourteen inputs and four active-HIGH outputs

    (d) who the manufacturer is

10. A GAL is different from a PAL because

    (a) a GAL has more inputs and outputs

    (b) a GAL is implemented with a different technology

    (c) a GAL can replace several different PALs

    (d) a GAL can be reprogrammed and a PAL cannot

    (e) all of the above answers

    (f) all except answer (a)

    (g) all except answer (c)

11. The reprogrammable cells in a GAL array are

    (a) TTL   **(b)** E^2CMOS **(c)** ECL   **(d)** bipolar fuses

12. OLMC is an acronym for

    (a) Output Logic Main Cell

    (b) Optimum Logic Multiple Channel

    (c) Output Logic Macrocell

    (d) Odd-parity Logic Master Check

13. Two ways in which a GAL output can be configured are

    (a) combinational and I/O

    (b) simple and complex

    (c) fixed and variable

    (d) combinational and registered

14. The device number GAL22V10 means that

    (a) the device has 22 dedicated inputs and 10 dedicated outputs

    (b) the device has 22 inputs including dedicated inputs and I/Os and 10 outputs either dedicated or I/Os

    (c) the device has a variable number of inputs from a maximum of 22 to a minimum of 10

15. To program a PLD, you need only a

    (a) special fixture

    (b) special fixture and a master PLD that has been preprogrammed at the factory

    (c) computer and a programmer

    (d) computer, a programmer, and HDL software

    (e) computer, a programmer, and BASIC software

16. ISP stands for

  (a) In-System Programmable

  (b) Integrated System Program

  (c) Integrated Silicon Programmer

17. ABEL and CUPL are types of

  (a) programmers   **(b)** PLDs

  (c) HDL software   **(d)** disk operating systems

18. The symbols for NOT, AND, OR, and XOR in ABEL are

  (a) !, &, +, \$ **(b)** !, @, #, \$ **(c)** !, &, #, \$ **(d)** *, &, +, !

19. The sum-of-products expression, $X = \overline{A}BC + A\overline{B}C + AB\overline{C}$ is written in ABEL as

  (a) X = !ABC # A!BC # AB!C;

  (b) X == !A & B & C # A & !B & C # A & B & !C;

  (c) X = !A & B & C # A & !B & C # A & B & !C;

  (d) X = (!A & B & C) + (A & !B & C) + (A & B & !C);

20. A logic circuit design to be implemented in a PLD is first described in

  (a) a declaration   **(b)** an input file   **(c)** a JEDEC file   **(d)** a documentation file

---

■ PROBLEMS

SECTION 7–1   PLD Arrays and Classifications

1. Name four types of PLDs and describe each.

2. In the simple programmed AND array with fusible links in Figure 7–41, determine the Boolean output expressions.

3. Determine by row and column number which fusible links must be blown in the programmable AND array of Figure 7–42 to implement each of the following product terms: $X_1 = \overline{A}BC$, $X_2 = AB\overline{C}$, $X_3 = \overline{A}B\overline{C}$.

FIGURE 7–41

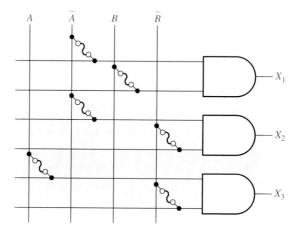

SECTION 7–2   Programmable Array Logic (PAL)

4. Determine the Boolean output expression for the simple PAL shown in Figure 7–43, where the Xs represent intact fusible links.

5. Show how the PAL in Figure 7–43 should be programmed in order to implement each of the following SOP expressions. Use an X to indicate an intact fuse. Simplify the expressions, if necessary, to fit the PAL shown.

  (a) $Y = A\overline{B}C + \overline{A}B\overline{C} + ABC$

  (b) $Y = A\overline{B}C + \overline{A}\,\overline{B}C + A\overline{B}\,\overline{C} + \overline{A}BC$

6. Interpret each of the PAL device numbers.

  (a) PAL16L2   **(b)** PAL12H6   **(c)** PAL10P8   **(d)** PAL16R6

7. Explain how a programmed polarity output in a PAL works.

FIGURE 7–42

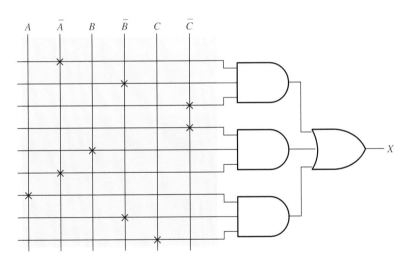

FIGURE 7–43

SECTION 7–3 Generic Array Logic (GAL)

8. Determine the output expression for the GAL array shown in Figure 7–44 on the next page. The Xs represent an *on* cell.

9. Show how the GAL in Figure 7–44 should be reprogrammed in order to implement each of the following expressions. Use Xs to indicate *on* cells. Modify the expressions, if necessary.

(a) $X = A\overline{B}C + \overline{A}B\overline{C} + A\overline{B} + BC$ **(b)** $X = (A + \overline{B} + \overline{C})(\overline{A} + B)$

SECTION 7–4 The GAL22V10

10. Describe a GAL22V10 in terms of

    (a) the number of dedicated inputs

    (b) the number of I/Os

    (c) the number of dedicated inputs plus the number of I/Os

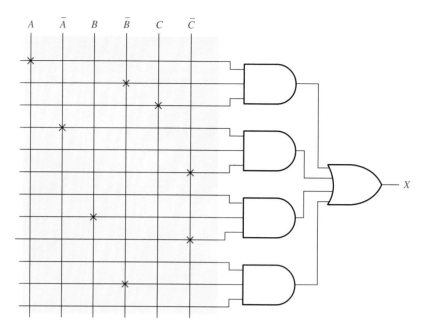

FIGURE 7–44

11. If fifteen inputs are required for a particular GAL22V10 implementation, what is the maximum number of available outputs?

12. What is the largest single SOP expression that can be implemented with a GAL22V10 in terms of the number of variables and the number of product terms? Assume that only one output is used.

13. State the purpose of the select inputs to each OLMC in a GAL22V10.

14. Determine the SOP output of an OLMC in a GAL22V10 for the product terms and the select bits shown in Figure 7–45.

FIGURE 7–45

SECTION 7–5   The GAL16V8

15. Describe a GAL16V8 in terms of

   (a) the number of dedicated inputs

   (b) the number of I/Os

   (c) the number of dedicated inputs plus the number of I/Os

16. If there are ten inputs to a GAL16V8, what is the maximum number of available outputs?

17. What is the largest single SOP expression that can be implemented with a GAL16V8 in terms of the number of variables and the number of product terms? Assume that only one output is used.

18. PAL emulation with a GAL16V8 can be achieved by programming the device in any one of three modes. What are these modes?

SECTION 7–6   PLD Programming

19. List the items you must have before programming a PLD.

20. List the sequence of events that must occur to program a PLD.

21. How does an ispGAL22V10 differ from a GAL22V10?

SECTION 7–7   PLD Software

22. Express each of the following Boolean operations using ABEL symbols:

    (a) NOT   **(b)** AND   **(c)** OR   **(d)** XOR

23. Convert each of the following Boolean expressions to ABEL notation:

    (a) $\overline{A}\,\overline{B}\,\overline{C}$ **(b)** $A + B + \overline{C}$ **(c)** $A(\overline{B + C})$

24. Determine the standard Boolean expression for each of the ABEL equations.

    (a) W = X & Y & Z # !X & !Y & !Z;

    (b) X = !(A & B) # !(!A & B & !C);

    (c) !Y = !(A # A & !B # !A & B);

25. Write each of the logic equations using ABEL:

    (a) $X = A\overline{B}C\overline{D}E + A\overline{B}C + \overline{A}\,\overline{B}$

    (b) $Y = AB\overline{C} + DEF + \overline{G}H\overline{I} + \overline{J}\overline{K}L$

    (c) $Z = (\overline{A} + \overline{B} + C + D)(E + \overline{F}) + G$

26. Write a set declaration for the following ABEL equations and then express the set of equations with a single equation using the net notation.

    ```
    X1 = A1 & P1 & Q1 # B1 & !P1 & Q1 # C1 & P1 & !Q1;
    X2 = A2 & P1 & Q1 # B2 & !P1 & Q1 # C2 & P1 & !Q1;
    X3 = A3 & P1 & Q1 # B3 & !P1 & Q1 # C3 & P1 & !Q1;
    ```

27. Write an ABEL truth table for an exclusive-NOR gate.

28. Write an ABEL truth table for a BCD-to-decimal decoder with active-HIGH outputs.

29. Write the test vectors for the BCD-to-decimal decoder in Problem 28.

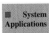

SECTION 7–8   Digital System Application

30. For the traffic light control system, the long time interval is to be increased from 25 s to 45 s and the short time interval from 4 s to 5 s. How does this change affect the truth table for the state decoding and output logic?

31. Can a GAL16V8 be used to implement the state decoding and output logic? Give the reason for your answer.

Special Design Problems

32. Write a complete ABEL input file for a BCD-to-decimal decoder with active-HIGH outputs to be implemented with a GAL22V10.

33. Write an ABEL input file to implement a decimal-to-BCD priority encoder with active-HIGH outputs using a GAL16V8. Assume that only one input at a time will be active.

34. Use a GAL22V10 to replace the three 74185s used for the 8-bit binary-to-BCD converter in Chapter 6 (Figure 6–41). Develop the input file. Discuss any potential problems in this implementation.

▪ ANSWERS TO SECTION REVIEWS

SECTION 7–1

1. Four types of PLDs are PROM, PLA, PAL, and GAL.

2. In a PLA, both arrays are programmable. In a PAL, only the AND array is programmable.

3. A PAL is one-time programmable; a GAL is reprogrammable and has programmable output sections.

SECTION 7–2

1. A PAL is a PLD with a programmable AND array and a fixed OR array.

2. The fusible link is the programmable element in an array.

3. Any combinational logic can be implemented with a PAL within the limitations of available inputs and outputs.

4. Three types of PAL output logic are combinational output, combinational I/O, and programmable polarity output.

5. A PAL12H6 is a PAL that has 12 inputs, including I/Os, and 6 active-HIGH outputs.

SECTION 7–3

1. A GAL is a PLD with a reprogrammable AND array, a fixed OR array, and programmable output logic.

2. An E^2CMOS cell can be programmed repeatedly to connect or disconnect an input line and a product term line.

3. Any combinational logic can be implemented with a GAL within the limitations of available inputs and outputs.

4. OLMC (output logic marocell) is the programmable output logic in a GAL. It can be programmed for several output configurations.

SECTION 7–4

1. The GAL22V10 has a combinational mode and a registered mode.

2. The mode is determined by the programmable select bits.

3. The polarity is set by the programmable select bits.

4. A maximum of sixteen product terms can be implemented in an SOP expression with a GAL22V10 OLMC.

SECTION 7–5

1. A GAL22V10 has more inputs and outputs than a GAL16V8, but the GAL16V8 can be readily programmed to emulate many PALs.

2. PAL emulation is done by programming the OLMCs.

3. GAL16V8 modes are simple, complex, and registered.

SECTION 7–6

1. Programming of a PLD requires a computer, a programmer, and PLD software.

2. The software for PLDs are compilers.

3. A JEDEC (Joint Electronic Device Engineering Council) file is the final output from the PLD software that actually programs the device.

4. ISP means in-system programmable.

5. An ispGAL can be programmed after being connected on a printed circuit board: a standard GAL must be programmed in a programmer.

SECTION 7–7

1. ABEL (Advanced Boolean Expression Language) is a PLD hardware description language.

2. Design entry formats are equations, truth tables, and state diagrams.

3. NOT: !,   AND: &,   OR: #,   XOR: $

4. Y = !D & E & !F # !D & !E # F

5. Test vectors list input and output values for testing a logic design.

6. Three parts of an ABEL input file are declarations, logic description, and test vectors.

7. An input file usually contains a device declaration, pin declarations, and set declarations.

8. A documentation file is the output file containing reduced equations, the JEDEC file, and a pin diagram.

9. A JEDEC file is the file that goes to the programmer to actually program the PLD. It is also called a fuse map or cell map.

SECTION 7-8

1. Two outputs are unused after programming the GAL22V10.

2. Ten inputs are unused and available.

3. The GAL replaces two IC packages.

■ **ANSWERS TO RELATED PROBLEMS FOR EXAMPLES**

7-1 $X = \overline{B}C + \overline{A}B\overline{C} + \overline{A}\,\overline{B} + C$

7-2 A PAL14H4 has 14 inputs and 4 active-HIGH outputs.

7-3 $X = B\overline{C} + BC + BC + A\overline{B} = \overline{B}C + BC + A\overline{B}$

7-4 A GAL18V10 has up to 18 inputs and 10 outputs.

7-5 $X = \overline{\overline{A}BCD + \overline{A}BC\overline{D} + A\overline{B}C\overline{D} + \overline{A}B\overline{C}D + A\overline{B}CD + AB\overline{C}D}$

7-6 One more product term can be added.

7-7 W = !A & B & !C # A # !B # C;

7-8 Y = (S == 3) & A # (S == 2) & B # (S == 1) & C;

7-9 TRUTH_TABLE (BCD -> [a, b, c, d, e, f, g])
```
                0  -> [0, 0, 0, 0, 0, 0, 1];
                1  -> [1, 0, 0, 1, 1, 1, 1];
                2  -> [0, 0, 1, 0, 0, 1, 0];
                3  -> [0, 0, 0, 0, 1, 1, 0];
                4  -> [1, 0, 0, 1, 1, 0, 0];
                5  -> [0, 1, 0, 0, 1, 0, 0];
                6  -> [0, 1, 0, 0, 0, 0, 0];
                7  -> [0, 0, 0, 1, 1, 1, 1];
                8  -> [0, 0, 0, 0, 0, 0, 0];
                9  -> [0, 0, 0, 0, 1, 0, 0];
```

7-10
```
BIT0 = [D0, C0, B0, A0];
BIT1 = [D1, C1, B1, A1];
BIT2 = [D2, C2, B2, A2];
BIT3 = [D3, C3, B3, A3];
   S = [S1, S0];
 OUT = [Dout, Cout, Bout, Aout];
 OUT = (S == 0) & BIT0 # (S == 1) & BIT1 # (S == 2)
                                  & BIT2 # (S == 3) & BIT3;
```

7-11 18 pins are used as inputs. Four pins are used as outputs.

■ **ANSWERS TO SELF-TEST**

1. (c)	**2.** (b)	**3.** (f)	**4.** (c)	**5.** (c)	**6.** (d)	**7.** (b)	**8.** (b)
9. (c)	**10.** (f)	**11.** (b)	**12.** (c)	**13.** (d)	**14.** (b)	**15.** (d)	**16.** (a)
17. (c)	**18.** (c)	**19.** (c)	**20.** (b)				

8

FLIP-FLOPS AND RELATED DEVICES

 Computer Simulation of Digital Circuits:
Electronics Workbench (EWB) Tutorial at http://www.prenhall.com/floyd

■ CHAPTER OBJECTIVES

❑ Use logic gates to construct basic latches
❑ Explain the difference between an S-R latch and a D latch
❑ Recognize the difference between a latch and a flip-flop
❑ Explain how S-R, D, and J-K flip-flops differ
❑ Explain how edge-triggered and master-slave flip-flops differ
❑ Understand the significance of propagation delays, set-up time, hold time, maximum operating frequency, minimum clock pulse widths, and power dissipation in the application of flip-flops
❑ Apply flip-flops in basic applications
❑ Analyze circuits for race conditions and the occurrence of glitches
❑ Explain how retriggerable and nonretriggerable one-shots differ
❑ Connect a 555 timer to operate as either an astable multivibrator or a one-shot
❑ Approach the debugging of a new design
❑ Troubleshoot basic flip-flop and one-shot circuits
❑ Apply one-shots in a system application

■ CHAPTER OVERVIEW

In this chapter, bistable, monostable, and astable logic devices called *multivibrators* are covered. Two categories of bistable devices are the latch and the flip-flop. Bistable devices have two stable states, called SET and RESET; they can retain either of these states indefinitely, making them useful as storage devices. The basic difference between latches and flip-flops is the way in which they are changed from one state to the other. The flip-flop is a basic building block for counters, registers, and other sequential control logic and is used in certain types of memories. The monostable multivibrator, commonly known as the one-shot, has only one

stable state. A one-shot produces a single controlled-width pulse when activated or triggered. The astable multivibrator has no stable state and is used primarily as an oscillator, which is a self-sustained waveform generator. Pulse oscillators are used as the sources for timing waveforms in digital systems.

 ■ SPECIFIC DEVICES

74AHC74	74LS279	74LS122
555	74121	74LS75
74HC112		

■ DIGITAL SYSTEM APPLICATION

This Digital System Application illustrates the concepts taught in this chapter. The system application in Section 8–9 continues with the traffic light control system from Chapter 6. The focus of this chapter is the timing circuit

portion of the system that produces the clock, the long time interval for the red and green lights, and the short time interval for the caution light. The clock is used as the basic system timing signal for advancing the sequential logic through its states. The sequential logic will be developed in Chapter 9.

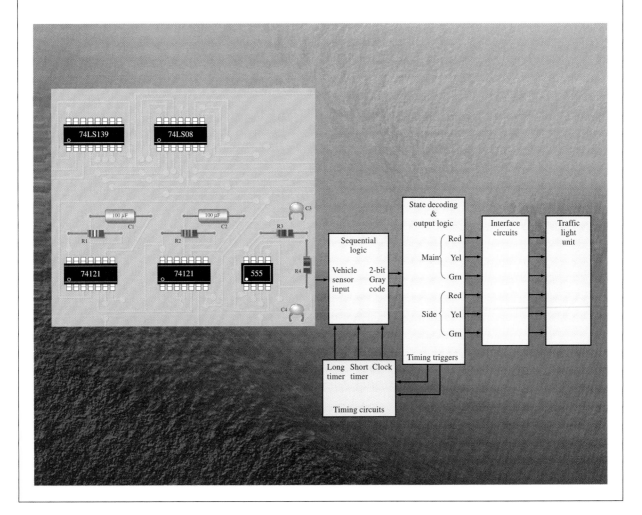

8–1 ■ LATCHES

The latch is a type of temporary storage device that has two stable states (bistable) and is normally placed in a category separate from that of flip-flops. Latches are basically similar to flip-flops because they are bistable devices that can reside in either of two states using a feedback arrangement, in which the outputs are connected back to the opposite inputs. The main difference between latches and flip-flops is in the method used for changing their state.

After completing this section, you should be able to

■ Explain the operation of a basic S-R latch ■ Explain the operation of a gated S-R latch ■ Explain the operation of a gated D latch ■ Implement an S-R or D latch with logic gates ■ Describe the 74LS279 and 74LS75 quad latches

COMPUTER NOTE

Latches are sometimes used in computer systems for multiplexing data onto a bus. For example, data being input to a computer from an external source have to share the data bus with data from other sources. When the data bus becomes unavailable to the external source, the existing data must be stored, and latches placed between the external source and the data bus may be used to do this. When the data bus is unavailable to the external source, the latches must be disconnected from the bus using a method known as tristating. When the data bus becomes available, the external data pass through the latches, thus the term *transparent latch.* The gated D latch performs this function because when it is enabled, the data on its input appear on the output just as though there were a direct connection. Data on the input are stored as soon as the latch is disabled.

The S-R (SET-RESET) Latch

A **latch** is a type of **bistable** logic device or **multivibrator.** An active-HIGH input S-R (SET-RESET) latch is formed with two cross-coupled NOR gates as shown in Figure 8–1(a); an active-LOW input $\overline{S}$-$\overline{R}$ latch is formed with two cross-coupled NAND gates as shown in Figure 8–1(b). Notice that the output of each gate is connected to an input of the opposite gate. This produces the regenerative **feedback** that is characteristic of all latches and flip-flops.

(a) Active-HIGH input S-R latch   (b) Active-LOW input $\overline{S}$-$\overline{R}$ latch

FIGURE 8–1
Two versions of SET-RESET (S-R) latches.

To explain the operation of the latch, we will use the NAND gate $\overline{S}$-$\overline{R}$ latch in Figure 8–1(b). This latch is redrawn in Figure 8–2 with the negative-OR equivalents used for the NAND gates. This is done because LOWs on the $\overline{S}$ and $\overline{R}$ lines are the activating inputs.

The latch in Figure 8–2 has two inputs, $\overline{S}$ and $\overline{R}$, and two outputs, Q and $\overline{Q}$. Let's start by assuming that both inputs and the Q output are HIGH. Since the Q output is connected back to an input of gate G_2, and the $\overline{R}$ input is HIGH, the output of G_2 must be LOW. This LOW output is coupled back to an input of gate G_1, ensuring that its output is HIGH.

FIGURE 8–2
Negative-OR equivalent of the NAND gate $\overline{S}$-$\overline{R}$ latch in Figure 8–1(b).

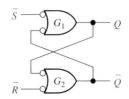

When the Q output is HIGH, the latch is in the **SET** state. It will remain in this state indefinitely until a LOW is temporarily applied to the $\overline{R}$ input. With a LOW on the $\overline{R}$ input and a HIGH on $\overline{S}$, the output of gate G_2 is forced HIGH. This HIGH on the $\overline{Q}$ output is coupled back to an input of G_1, and since the $\overline{S}$ input is HIGH, the output of G_1 goes LOW. This LOW on the Q output is then coupled back to an input of G_2, ensuring that the $\overline{Q}$ output remains HIGH even when the LOW on the $\overline{R}$ input is removed. When the Q output is

LOW, the latch is in the **RESET** state. Now the latch remains indefinitely in the RESET state until a LOW is applied to the $\overline{S}$ input.

In normal operation, the outputs of a latch are always complements of each other.

When Q is HIGH, $\overline{Q}$ is LOW, and when Q is LOW, $\overline{Q}$ is HIGH.

An invalid condition in the operation of an active-LOW input $\overline{S}$-$\overline{R}$ latch occurs when LOWs are applied to both $\overline{S}$ and $\overline{R}$ at the same time. As long as the LOW levels are simultaneously held on the inputs, both the Q and $\overline{Q}$ outputs are forced HIGH, thus violating the basic complementary operation of the outputs. Also, if the LOWs are released simultaneously, both outputs will attempt to go LOW. Since there is always some small difference in the propagation delay time of the gates, one of the gates will dominate in its transition to the LOW output state. This, in turn, forces the output of the slower gate to remain HIGH. In this situation, you cannot reliably predict the next state of the latch.

Figure 8–3 illustrates the active-LOW input $\overline{S}$-$\overline{R}$ latch operation for each of the four possible combinations of levels on the inputs. (The first three combinations are valid, but the last is not.) Table 8–1 summarizes the logic operation in truth table form. Operation of the active-HIGH input NOR gate latch in Figure 8–1(a) is similar but requires the use of opposite logic levels.

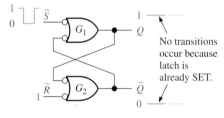

(a) Two possibilities for the SET operation

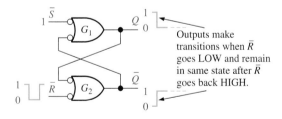

(b) Two possibilities for the RESET operation

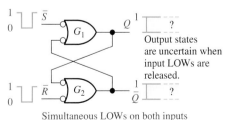

(c) No-change condition

(d) Invalid condition

FIGURE 8–3

The three modes of basic $\overline{S}$-$\overline{R}$ latch operation (SET, RESET, no-change) and the invalid condition.

TABLE 8–1

Truth table for an active-LOW input $\overline{S}$-$\overline{R}$ latch.

Inputs		Outputs		
$\overline{S}$	$\overline{R}$	Q	$\overline{Q}$	Comments
1	1	NC	NC	No change. Latch remains in present state.
0	1	1	0	Latch SET.
1	0	0	1	Latch RESET.
0	0	1	1	Invalid condition

FIGURE 8–4

Logic symbols for the S-R and $\overline{S}$-$\overline{R}$ latch.

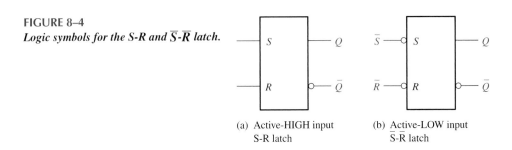

(a) Active-HIGH input
S-R latch

(b) Active-LOW input
$\overline{S}$-$\overline{R}$ latch

Logic symbols for both the active-HIGH input and the active-LOW input latches are shown in Figure 8–4.

Example 8–1 illustrates how an active-LOW input $\overline{S}$-$\overline{R}$ latch responds to conditions on its inputs. LOW levels are pulsed on each input in a certain sequence and the resulting Q output waveform is observed. The $\overline{S} = 0$, $\overline{R} = 0$ condition is avoided because it results in an invalid mode of operation and is a major drawback of any SET-RESET type of latch.

EXAMPLE 8–1

If the $\overline{S}$ and $\overline{R}$ waveforms in Figure 8–5(a) are applied to the inputs of the latch in Figure 8–4(b), determine the waveform that will be observed on the Q output. Assume that Q is initially LOW.

FIGURE 8–5

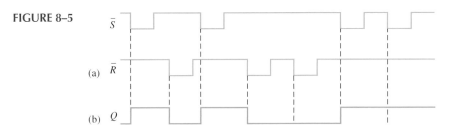

(a) $\overline{S}$ $\overline{R}$

(b) Q

Solution See Figure 8–5(b).

Related Problem Determine the Q output of an active-HIGH input S-R latch if the waveforms in Figure 8–5(a) are inverted and applied to the inputs.

Application Example

The Latch as a Contact-Bounce Eliminator A good example of an application of an $\overline{S}$-$\overline{R}$ latch is in the elimination of mechanical switch contact "bounce." When the pole of a switch strikes the contact upon switch closure, it physically vibrates or bounces several times before finally making a solid contact. Although these bounces are very short in duration, they produce voltage spikes that are often not acceptable in a digital system. This situation is illustrated in Figure 8–6(a).

FIGURE 8–6
The $\overline{S}$-$\overline{R}$ latch used to eliminate switch contact bounce.

An $\overline{S}$-$\overline{R}$ latch can be used to eliminate the effects of switch bounce as shown in Figure 8–6(b). The switch is normally in position 1, keeping the $\overline{R}$ input LOW and the latch RESET. When the switch is thrown to position 2, $\overline{R}$ goes HIGH because of the pull-up resistor to V_{CC}, and $\overline{S}$ goes LOW on the first contact. Although $\overline{S}$ remains LOW for only a very short time before the switch bounces, this is sufficient to set the latch. Any further voltage spikes on the $\overline{S}$ input due to switch bounce do not affect the latch, and it remains SET. Notice that the Q output of the latch provides a clean transition from LOW to HIGH, thus eliminating the voltage spikes caused by contact bounce. Similarly, a clean transition from HIGH to LOW is made when the switch is thrown back to position 1.

An IC SET-RESET Latch

The 74LS279 is a quad $\overline{S}$-$\overline{R}$ latch represented by the logic diagram of Figure 8–7(a) and the pin diagram in part (b). Notice that two of the latches each have two $\overline{S}$ inputs.

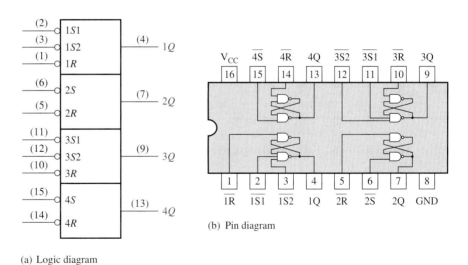

(a) Logic diagram

(b) Pin diagram

FIGURE 8–7
The 74LS279 quad $\overline{S}$-$\overline{R}$ latch.

The Gated S-R Latch

A gated latch requires an enable input, *EN* (*G* is also used to designate an enable input). The logic diagram and logic symbol for a gated S-R latch are shown in Figure 8–8. The *S* and *R* inputs control the state to which the latch will go when a HIGH level is applied to the *EN* input. The latch will not change until *EN* is HIGH, but as long as it remains HIGH, the output is controlled by the state of the *S* and *R* inputs. In this circuit, the invalid state occurs when both *S* and *R* are simultaneously HIGH.

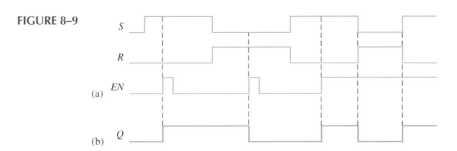

(a) Logic diagram   (b) Logic symbol

FIGURE 8–8
A gated S-R latch.

EXAMPLE 8–2

Determine the Q output waveform if the inputs shown in Figure 8–9(a) are applied to a gated S-R latch that is initially RESET.

FIGURE 8–9

(a)

(b)

Solution The Q waveform is shown in Figure 8–9(b). Anytime S is HIGH and R is LOW, a HIGH on the *EN* input sets the latch. Anytime S is LOW and R is HIGH, a HIGH on the *EN* input resets the latch.

Related Problem Determine the Q output of a gated S-R latch if the S and R inputs in Figure 8–9(a) are inverted.

The Gated D Latch

Another type of gated latch is called the D latch. It differs from the S-R latch because it has only one input in addition to *EN*. This input is called the D (data) input. Figure 8–10 contains a logic diagram and logic symbol of a D latch. When the D input is HIGH and the *EN*

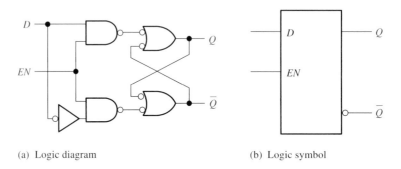

(a) Logic diagram   (b) Logic symbol

FIGURE 8–10
A gated D latch.

input is HIGH, the latch will set. When the *D* input is LOW, and *EN* is HIGH, the latch will reset. Stated another way, the output *Q* follows the input *D* when *EN* is HIGH.

EXAMPLE 8–3

Determine the *Q* output waveform if the inputs shown in Figure 8–11(a) are applied to a gated D latch, which is initially RESET.

FIGURE 8–11

Solution   The *Q* waveform is shown in Figure 8–11(b). Whenever *D* is HIGH and *EN* is HIGH, *Q* goes HIGH. Whenever *D* is LOW and *EN* is HIGH, *Q* goes LOW. When *EN* is LOW, the state of the latch is not affected by the *D* input.

Related Problem   Determine the *Q* output of the gated D latch if the *D* input in Figure 8–11(a) is inverted.

An IC D Latch

An example of an IC gated D latch is the 74LS75 represented by the logic symbol in Figure 8–12(a). This device has four latches. Notice that each active-HIGH *EN* input is shared by two latches and is designated as a control input (*C*). The truth table for each latch is shown in Figure 8–12(b). The X in the truth table represents a "don't care" condition. In this case, when the *EN* input is LOW, it does not matter what the *D* input is because the outputs are unaffected and remain in their prior states.

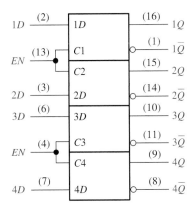

(a) Logic symbol

Inputs		Outputs		
D	*EN*	*Q*	$\bar{Q}$	**Comments**
0	1	0	1	RESET
1	1	1	0	SET
X	0	Q_0	$\bar{Q}_0$	No change

NOTE: Q_0 is the prior output level before the indicated input conditions were established.

(b) Truth table (each latch)

FIGURE 8–12
The 74LS75 quad gated D latches.

SECTION 8–1 REVIEW

1. List three types of latches.
2. Develop the truth table for the active-HIGH input S-R latch in Figure 8–1(a).
3. What is the Q output of a D latch when $EN = 1$ and $D = 1$?

8–2 ■ EDGE-TRIGGERED FLIP-FLOPS

Flip-flops are synchronous bistable devices, also known as bistable multivibrators. In this case, the term synchronous means that the output changes state only at a specified point on a triggering input called the clock (CLK) which is designated as a control input, C; that is, changes in the output occur in synchronization with the clock.

After completing this section, you should be able to

■ Define *clock* ■ Define *edge-triggered flip-flop* ■ Explain the difference between a flip-flop and a latch ■ Identify an edge-triggered flip-flop by its logic symbol ■ Discuss the difference between a positive and a negative edge-triggered flip-flop ■ Discuss and compare the operation of S-R, D, and J-K edge-triggered flip-flops and explain the differences in their truth tables. ■ Discuss the asynchronous inputs of a flip-flop ■ Describe the 74AHC74 and the 74HC112 flip-flops

COMPUTER NOTE
Computer semiconductor memories consist of large numbers of individual cells. Each storage cell holds a 1 or a 0. One type of memory is the Static Random Access Memory or SRAM, which uses flip-flops for the storage cells because a flip-flop will retain either of its two states indefinitely as long as dc power is applied, thus the term *static*. This type of memory is classified as a *volatile* memory because all the stored data are lost when power is turned off. Another type of memory, the Dynamic Random Access Memory or DRAM, uses capacitance rather than flip-flops as the basic storage element and must be periodically refreshed in order to maintain the stored data.

An **edge-triggered flip-flop** changes state either at the positive edge (rising edge) or at the negative edge (falling edge) of the clock pulse and is sensitive to its inputs only at this transition of the clock. Three types of edge-triggered flip-flops are covered in this section: S-R, D, and J-K. Although the S-R flip-flop is not available in IC form, it is the basis for the D and J-K flip-flops and is, therefore, important to cover. The logic symbols for all of these flip-flops are shown in Figure 8–13. Notice that each type can be either positive edge-triggered (no bubble at C input) or negative edge-triggered (bubble at C input). The key to identifying an edge-triggered flip-flop by its logic symbol is the small triangle inside the block at the clock (C) input. This triangle is called the *dynamic input indicator*.

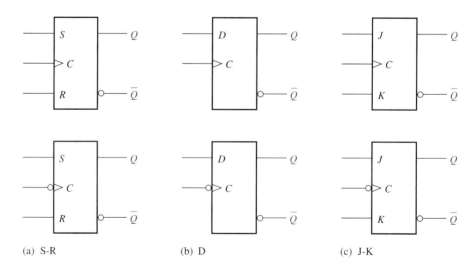

(a) S-R   (b) D   (c) J-K

FIGURE 8–13
Edge-triggered flip-flop logic symbols (top: positive edge-triggered; bottom: negative edge-triggered).

The Edge-Triggered S-R Flip-Flop

The S and R inputs of the **S-R flip-flop** are called **synchronous** inputs because data on these inputs are transferred to the flip-flop's output only on the triggering edge of the clock pulse. When S is HIGH and R is LOW, the Q output goes HIGH on the triggering edge of the clock pulse, and the flip-flop is SET. When S is LOW and R is HIGH, the Q output goes LOW on the triggering edge of the clock pulse, and the flip-flop is RESET. When both S and R are LOW, the output does not change from its prior state. An invalid condition exists when both S and R are HIGH.

This basic operation of a positive edge-triggered flip-flop is illustrated in Figure 8–14, and Table 8–2 is the truth table for this type of flip-flop. Remember, *the flip-flop cannot change state except on the triggering edge of a clock pulse.* The S and R inputs can be changed at any time when the clock input is LOW or HIGH (except for a very short interval around the triggering transition of the clock) without affecting the output.

(a) $S = 1, R = 0$ flip-flop SETS on positive clock edge. (If already SET, it remains SET.)

(b) $S = 0, R = 1$ flip-flop RESETS on positive clock edge. (If already RESET, it remains RESET.)

(c) $S = 0, R = 0$ flip-flop does not change. (If SET, it remains SET; if RESET, it remains RESET.)

FIGURE 8–14
Operation of a positive edge-triggered S-R flip-flop.

TABLE 8–2
Truth table for a positive edge-triggered S-R flip-flop.

Inputs			Outputs		
S	R	CLK	Q	$\overline{Q}$	Comments
0	0	X	Q_0	$\overline{Q}_0$	No change
0	1	↑	0	1	RESET
1	0	↑	1	0	SET
1	1	↑	?	?	Invalid

↑ = clock transition LOW to HIGH

X = irrelevant ("don't care")

Q_0 = output level prior to clock transition

The operation and truth table for a negative edge-triggered S-R flip-flop are the same as those for a positive edge-triggered device except that the falling edge of the clock pulse is the triggering edge.

EXAMPLE 8–4

Determine the Q and $\overline{Q}$ output waveforms of the flip-flop in Figure 8–15 for the S, R, and CLK inputs in Figure 8–16(a). Assume that the positive edge-triggered flip-flop is initially RESET.

FIGURE 8–15

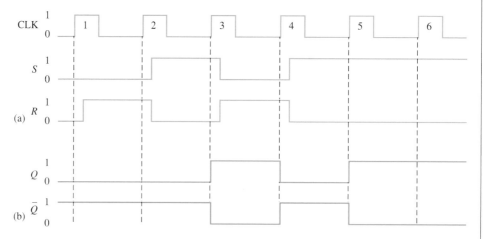

FIGURE 8–16

Solution
1. At clock pulse 1, S is LOW and R is LOW, so Q does not change.
2. At clock pulse 2, S is LOW and R is HIGH, so Q remains LOW (RESET).
3. At clock pulse 3, S is HIGH and R is LOW, so Q goes HIGH (SET).
4. At clock pulse 4, S is LOW and R is HIGH, so Q goes LOW (RESET).
5. At clock pulse 5, S is HIGH and R is LOW, so Q goes HIGH (SET).
6. At clock pulse 6, S is HIGH and R is LOW, so Q stays HIGH.

Once Q is determined, $\overline{Q}$ is easily found since it is simply the complement of Q. The resulting waveforms for Q and $\overline{Q}$ are shown in Figure 8–16(b) for the input waveforms in part (a).

Related Problem Determine Q and $\overline{Q}$ for the S and R inputs in Figure 8–16(a) if the flip-flop is a negative edge-triggered device.

A Method of Edge-Triggering

A simplified implementation of an edge-triggered S-R flip-flop is illustrated in Figure 8–17(a) and is used to demonstrate the concept of edge-triggering. This coverage of the S-R flip-flop does not imply that it is the most important type. Actually, the D flip-flop and the J-K flip-flop are available in IC form and more widely used than the S-R type. However, understanding the S-R is important because both the D and the J-K flip-flops are derived from the S-R flip-flop. Notice that the S-R flip-flop differs from the gated

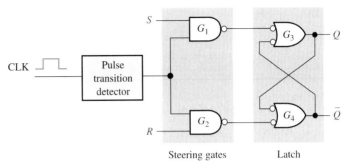

(a) A simplified logic diagram for a positive edge-triggered S-R flip-flop

(b) A type of pulse transition detector

FIGURE 8–17
Edge triggering.

S-R latch only in that it has a pulse transition detector. This circuit produces a very short-duration spike on the positive-going transition of the clock pulse.

One basic type of pulse transition detector is shown in Figure 8–17(b). As you can see, there is a small delay on one input to the NAND gate so that the inverted clock pulse arrives at the gate input a few nanoseconds after the true clock pulse. This produces an output spike with a duration of only a few nanoseconds. In a negative edge-triggered flip-flop the clock pulse is inverted first, thus producing a narrow spike on the negative-going edge.

Notice that the circuit in Figure 8–17 is partitioned into two sections, one labeled Steering gates, and the other labeled Latch. The steering gates direct, or steer, the clock spike either to the input to gate G_3 or to the input to gate G_4, depending on the state of the S and R inputs. To understand the operation of this flip-flop, begin with the assumptions that it is in the RESET state ($Q = 0$) and that the S, R, and CLK inputs are all LOW. For this condition, the outputs of gate G_1 and gate G_2 are both HIGH. The LOW on the Q output is coupled back into one input of gate G_4, making the $\overline{Q}$ output HIGH. Because $\overline{Q}$ is HIGH, both inputs to gate G_3 are HIGH (remember, the output of gate G_1 is HIGH), holding the Q output LOW. If a pulse is applied to the CLK input, the outputs of gates G_1 and G_2 remain HIGH because they are disabled by the LOWs on the S input and the R input; therefore, there is no change in the state of the flip-flop—it remains RESET.

Let's now make S HIGH, leave R LOW and apply a clock pulse. Because the S input to gate G_1 is now HIGH, the output of gate G_1 goes LOW for a very short time (spike) when CLK goes HIGH, causing the Q output to go HIGH. Both inputs to gate G_4 are now HIGH (remember, gate G_2 output is HIGH because R is LOW), forcing the $\overline{Q}$ output LOW. This LOW on $\overline{Q}$ is coupled back into one input of gate G_3, ensuring that the Q output will remain HIGH. The flip-flop is now in the SET state. Figure 8–18 illustrates the logic level transitions that take place within the flip-flop for this condition.

Next, let's make S LOW and R HIGH and apply a clock pulse. Because the R input is now HIGH, the positive-going edge of the clock produces a negative-going spike on the output of gate G_2, causing the $\overline{Q}$ output to go HIGH. Because of this HIGH on $\overline{Q}$, both inputs to gate G_3 are now HIGH (remember, the output of gate G_1 is HIGH because of the LOW on S), forcing the Q output to go LOW. This LOW on Q is coupled back into one input of gate G_4, ensuring that $\overline{Q}$ will remain HIGH. The flip-flop is now in the RESET state. Figure 8–19 illustrates the logic level transitions that occur within the flip-flop for this condition. As with the gated latch, an invalid condition exists if a clock pulse occurs when both S and R are HIGH at the same time. This is the major drawback of the S-R flip-flop.

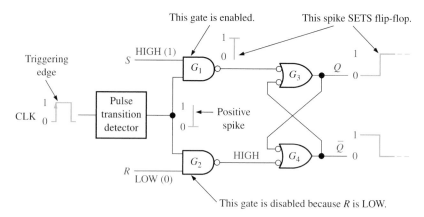

FIGURE 8–18

Flip-flop making a transition from the RESET state to the SET state on the positive-going edge of the clock pulse.

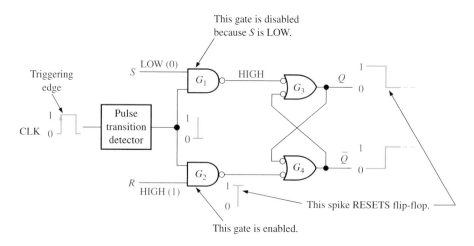

FIGURE 8–19

Flip-flop making a transition from the SET state to the RESET state on the positive-going edge of the clock pulse.

The Edge-Triggered D Flip-Flop

The **D flip-flop** is useful when a single data bit (1 or 0) is to be stored. The addition of an inverter to an S-R flip-flop creates a basic D flip-flop, as in Figure 8–20, where a positive edge-triggered type is shown.

Notice that the flip-flop in Figure 8–20 has only one input, the *D* input, in addition to the clock. If there is a HIGH on the *D* input when a clock pulse is applied, the flip-flop will set, and the HIGH on the *D* input is stored by the flip-flop on the positive-going edge of the clock pulse. If there is a LOW on the *D* input when the clock pulse is applied, the flip-flop

FIGURE 8–20

A positive edge-triggered D flip-flop formed with an S-R flip-flop and an inverter.

will reset, and the LOW on the D input is stored by the flip-flop on the leading edge of the clock pulse. In the SET state the flip-flop is storing a 1, and in the RESET state it is storing a 0.

The logic operation of the positive edge-triggered D flip-flop is summarized in Table 8–3. The operation of a negative edge-triggered device is, of course, the same, except that triggering occurs on the falling edge of the clock pulse. *Remember, Q follows D at the active or triggering clock edge.*

TABLE 8–3
Truth table for a positive edge-triggered D flip-flop.

Inputs		Outputs		
D	CLK	Q	$\overline{Q}$	Comments
1	↑	1	0	SET (stores a 1)
0	↑	0	1	RESET (stores a 0)

↑ = clock transition LOW to HIGH

EXAMPLE 8–5

Given the waveforms in Figure 8–21(a) for the D input and the clock, determine the Q output waveform if the flip-flop starts out RESET.

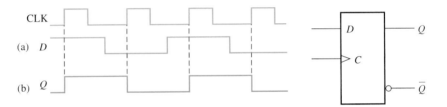

FIGURE 8–21

Solution The Q output goes to the state of the D input at the time of the positive-going clock edge. The resultant output is shown in Figure 8–21(b).

Related Problem Determine the Q output for the D flip-flop if the D input in Figure 8–21(a) is inverted.

The Edge-Triggered J-K Flip-Flop

The **J-K flip-flop** is versatile and is a widely used type of flip-flop. The J and K designations for the inputs have no known significance except that they are adjacent letters in the alphabet.

The functioning of the J-K flip-flop is identical to that of the S-R flip-flop in the SET, RESET, and no-change conditions of operation. *The difference is that the J-K flip-flop has no invalid state as does the S-R flip-flop.*

Figure 8–22 shows the basic internal logic for a positive edge-triggered J-K flip-flop. Notice that it differs from the S-R edge-triggered flip-flop in that the Q output is connected back to the input of gate G_2, and the $\overline{Q}$ output is connected back to the input of gate G_1. The two inputs are labeled J and K. A J-K flip-flop can also be of the negative edge-triggered type, in which case the clock input is inverted.

Let's assume that the flip-flop in Figure 8–23 is RESET and that the J input is HIGH and the K input is LOW rather than as shown. When a clock pulse occurs, a leading-edge spike indicated by ① is passed through gate G_1 because $\overline{Q}$ is HIGH and J is HIGH. This will cause the latch portion of the flip-flop to change to the SET state.

FIGURE 8–22
A simplified logic diagram for a positive edge-triggered J-K flip-flop.

FIGURE 8–23
Transitions illustrating the toggle operation when J = 1 and K = 1.

The flip-flop is now SET. If you now make J LOW and K HIGH, the next clock spike indicated by ② will pass through gate G_2 because Q is HIGH and K is HIGH. This will cause the latch portion of the flip-flop to change to the RESET state.

Now if a LOW is applied to both the J and K inputs, the flip-flop will stay in its present state when a clock pulse occurs. So, a LOW on both J and K results in a *no-change* condition.

So far, the logical operation of the J-K flip-flop is the same as that of the S-R type in the SET, RESET, and no-change modes. The difference in operation occurs when both the J and K inputs are HIGH. To see this, assume that the flip-flop is reset. The HIGH on the $\overline{Q}$ enables gate G_1, so the clock spike indicated by ③ passes through to set the flip-flop. Now there is a HIGH on Q, which allows the next clock spike to pass through gate G_2 and reset the flip-flop.

As you can see, on each successive clock spike, the flip-flop changes to the opposite state. This mode is called **toggle** operation. Figure 8–23 illustrates the transitions when the flip-flop is in the toggle mode.

Table 8–4 summarizes the logic operation of the edge-triggered J-K flip-flop in truth table form. *Notice that there is no invalid state as there is with an S-R flip-flop.* The truth table for a negative edge-triggered device is identical except that it is triggered on the falling edge of the clock pulse.

TABLE 8–4
Truth table for a positive edge-triggered J-K flip-flop.

Inputs			Outputs		
J	**K**	**CLK**	**Q**	**$\overline{Q}$**	**Comments**
0	0	↑	Q_0	$\overline{Q}_0$	No change
0	1	↑	0	1	RESET
1	0	↑	1	0	SET
1	1	↑	$\overline{Q}_0$	Q_0	Toggle

↑ = clock transition LOW to HIGH

Q_0 = output level prior to clock transition

EXAMPLE 8–6

The waveforms in Figure 8–24(a) are applied to the J, K, and clock inputs as indicated. Determine the Q output, assuming that the flip-flop is initially RESET.

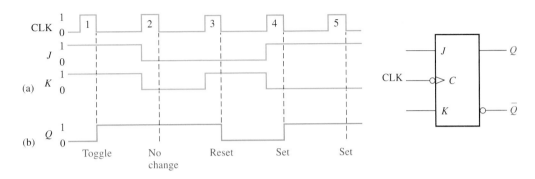

FIGURE 8–24

Solution

1. First, since this is a negative edge-triggered flip-flop, as indicated by the "bubble" at the clock input, the Q output will change only on the negative-going edge of the clock pulse.
2. At the first clock pulse, both J and K are HIGH; and because this is a toggle condition, Q goes HIGH.
3. At clock pulse 2, a no-change condition exists on the inputs, keeping Q at a HIGH level.
4. When clock pulse 3 occurs, J is LOW and K is HIGH, resulting in a RESET condition; Q goes LOW.
5. At clock pulse 4, J is HIGH and K is LOW, resulting in a SET condition; Q goes HIGH.
6. A SET condition still exists on J and K when clock pulse 5 occurs, so Q will remain HIGH.

The resulting Q waveform is indicated in Figure 8–24(b).

Related Problem Determine the Q output of the J-K flip-flop if the J and K inputs in Figure 8–24(a) are inverted.

EXAMPLE 8–7

The waveforms in Figure 8–25(a) are applied to the flip-flop as shown. Determine the Q output, starting in the RESET state.

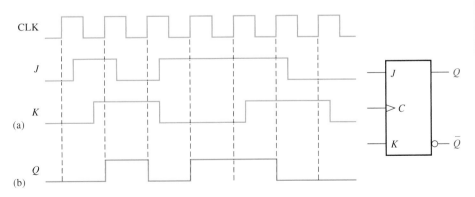

FIGURE 8–25

Solution The *Q* output assumes the state determined by the states of the *J* and *K* inputs at the positive-going edge (triggering edge) of the clock pulse. A change in *J* or *K* after the triggering edge of the clock has no effect on the output, as shown in Figure 8–25(b).

Related Problem Interchange the *J* and *K* inputs and determine the resulting *Q* output.

Asynchronous Preset and Clear Inputs

For the flip-flops just discussed, the *S-R*, *D*, and *J-K* inputs are called *synchronous inputs* because data on these inputs are transferred to the flip-flop's output only on the triggering edge of the clock pulse; that is, the data are transferred synchronously with the clock.

   Most integrated circuit flip-flops also have **asynchronous** inputs. These are inputs that affect the state of the flip-flop *independent of the clock*. They are normally labeled **preset** (*PRE*) and **clear** (*CLR*), or *direct set* (S_D) and *direct reset* (R_D) by some manufacturers. An active level on the preset input will set the flip-flop, and an active level on the clear input will reset it. A logic symbol for a J-K flip-flop with preset and clear inputs is shown in Figure 8–26. These inputs are active-LOW, as indicated by the bubbles. These preset and clear inputs must both be kept HIGH for synchronous operation.

FIGURE 8–26
Logic symbol for a J-K flip-flop with active-LOW preset and clear inputs.

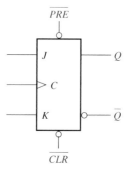

   Figure 8–27 shows the logic diagram for an edge-triggered J-K flip-flop with active-LOW preset ($\overline{PRE}$) and clear ($\overline{CLR}$) inputs. This figure illustrates basically how these inputs work. As you can see, they are connected so that they override the effect of the synchronous inputs, *J*, *K*, and the clock.

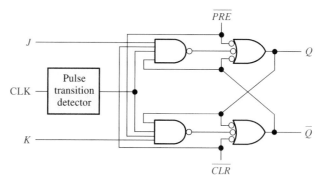

FIGURE 8–27
Logic diagram for a basic J-K flip-flop with active-LOW preset and clear.

EXAMPLE 8–8

For the positive edge-triggered J-K flip-flop with preset and clear inputs in Figure 8–28(a), determine the Q output for the inputs shown in the timing diagram if Q is initially LOW.

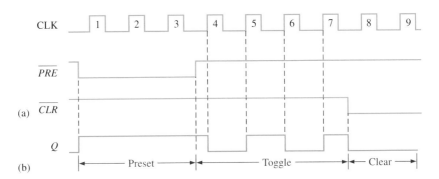

(a)

(b)

FIGURE 8–28

Solution

1. During clock pulses 1, 2, and 3, the preset ($\overline{PRE}$) is LOW, keeping the flip-flop SET regardless of the synchronous J and K inputs.
2. For clock pulses 4, 5, 6, and 7, toggle operation occurs because J is HIGH, K is HIGH, and both $\overline{PRE}$ and $\overline{CLR}$ are HIGH.
3. For clock pulses 8 and 9, the clear ($\overline{CLR}$) input is LOW, keeping the flip-flop RESET regardless of the synchronous inputs. The resulting Q output is shown in Figure 8–27(b).

Related Problem If you interchange the $\overline{PRE}$ and $\overline{CLR}$ waveforms in Figure 8–28(a), what will the Q output look like?

IC Flip-FLops

Let's look at two specific IC edge-triggered flip-flops. They are representative of the various types of flip-flops available in IC form and, like most other devices, are available in CMOS and in TTL.

74AHC74 Dual D Flip-Flops The CMOS device contains two identical flip-flops that are independent of each other except for sharing V_{CC} and ground. The flip-flops are positive edge-triggered and have active-LOW asynchronous preset and clear inputs. The logic

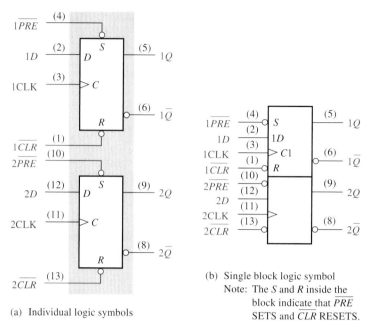

(a) Individual logic symbols

(b) Single block logic symbol
Note: The S and R inside the block indicate that $\overline{PRE}$ SETS and $\overline{CLR}$ RESETS.

FIGURE 8–29
Logic symbols for the 74AHC74 dual positive edge-triggered D flip-flops.

symbols for the individual flip-flops within the package are shown in Figure 8–29(a), and an ANSI/IEEE standard single block symbol representing the entire device is shown in part (b) of the figure. The pin numbers are shown in parentheses.

74HC112 Dual J-K Flip-Flops This CMOS device also has two identical flip-flops that are negative edge-triggered and have active-LOW asynchronous preset and clear inputs. The logic symbols are shown in Figure 8–30.

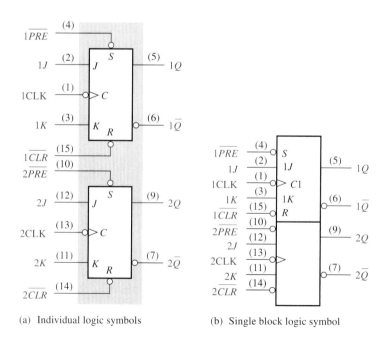

(a) Individual logic symbols

(b) Single block logic symbol

FIGURE 8–30
Logic symbols for the 74HC112 dual negative edge-triggered J-K flip-flops.

EXAMPLE 8–9

The $1J$, $1K$, $1CLK$, $1\overline{PRE}$, and $1\overline{CLR}$ waveforms in Figure 8–31(a) are applied to one of the negative edge-triggered flip-flops in a 74HC112 package. Determine the $1Q$ output waveform.

FIGURE 8–31

Solution   The resulting $1Q$ waveform is shown in Figure 8–31(b). Notice that each time a LOW is applied to the $1\overline{PRE}$ or $1\overline{CLR}$, the flip-flop is set or reset regardless of the states of the other inputs.

Related Problem   Determine the $1Q$ output waveform if the waveforms for $1\overline{PRE}$ and $1\overline{CLR}$ are interchanged.

SECTION 8–2 REVIEW

1. Describe the main difference between a gated S-R latch and an edge-triggered S-R flip-flop.
2. How does a J-K flip-flop differ from an S-R flip-flop in its basic operation?
3. Assume that the flip-flop in Figure 8–21 is negative edge-triggered. Describe the output waveform for the same CLK and D waveforms.

8–3 ■ MASTER-SLAVE FLIP-FLOPS

Another class of flip-flop is the pulse-triggered master-slave, which has largely been replaced by the edge-triggered devices. Although master-slave flip-flops are essentially becoming obsolete, you may encounter this type of flip-flop in some existing equipment. Data are entered into the flip-flop at the leading edge of the clock pulses, but the output does not reflect the input state until the trailing edge. The pulse-triggered master-slave flip-flop does not allow data to change while the clock pulse is active.

After completing this section, you should be able to

■ Identify a master-slave flip-flop by its logic symbol   ■ Explain how the master-slave flip-flop differs from the edge-triggered devices   ■ Discuss the basic operation of the pulse-triggered master-slave flip-flop

The Pulse-Triggered Master-Slave J-K Flip-Flop

A basic **master-slave** J-K flip-flop is shown in Figure 8–32. The truth table operation is the same as that for the edge-triggered J-K flip-flop except for the way it is clocked. Internally, though, it is quite different.

FIGURE 8–32
Basic logic diagram for a master-slave J-K flip-flop.

This type of flip-flop is composed of two sections, the master section and the slave section. The master section is basically a gated latch, and the slave section is the same except that it is clocked on the inverted clock pulse and is controlled by the outputs of the master section rather than by the external *J-K* inputs.

The master section will assume the state determined by the J and K inputs beginning at the leading (positive-going) edge of the clock pulse. The state of the master section is then transferred to the slave section on the trailing edge of the clock pulse because the outputs of the master are applied to the inputs of the slave and the clock pulse to the slave is inverted. At the trailing edge of the clock pulse, the state of the slave then appears on the Q and $\overline{Q}$ outputs. The Q output is connected back to an input of gate G_2 and the $\overline{Q}$ output is connected back to an input of gate G_1 to produce the characteristic toggle operation when $J = 1$ and $K = 1$. The logic operation is summarized in Table 8–5. One limitation to master-slave operation is that the inputs (J and K) cannot change while the clock pulse is active because the state of the master latch can change during this time.

TABLE 8–5
Truth table for the master-slave J-K flip-flop.

Inputs			Outputs		
J	**K**	**CLK**	**Q**	**$\overline{Q}$**	**Comments**
0	0	⊓	Q_0	$\overline{Q}_0$	No change
0	1	⊓	0	1	RESET
1	0	⊓	1	0	SET
1	1	⊓	$\overline{Q}_0$	Q_0	Toggle

⊓ = clock pulse
Q_0 = output level before clock pulse

The logic symbols for the J-K master-slave flip-flop are shown in Figure 8–33. The key to identifying a pulse-triggered (master-slave) flip-flop by its logic symbol is the ANSI/IEEE *postponed output* symbol (⅂) at the outputs. This symbol means that the output does not reflect the *J-K* input data until the occurrence of the clock edge (either positive-going or negative-going) following the triggering edge. Notice that there is no dynamic input indicator (▷) at the clock (*C*) input as there is for an edge-triggered flip-flop.

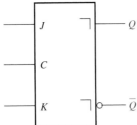

(a) Active-HIGH clock: Data are
clocked in on positive-going edge
of clock pulse and transferred to
output on the following negative-
going edge.

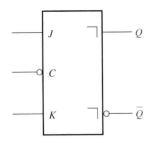

(b) Active-LOW clock: Data are
clocked in on negative-going
edge of clock pulse and
transferred to output on the
following positive-going edge.

FIGURE 8–33
Pulse-triggered (master-slave) J-K flip-flop logic symbols.

EXAMPLE 8–10   Determine the Q output of the master-slave J-K flip-flop for the input waveforms shown
in Figure 8–34(a). The flip-flop starts out RESET and the clock is active-LOW.

FIGURE 8–34

Solution   The Q waveform is shown in Figure 8–34(b). The input states and events at
the beginning and end of each clock pulse are labeled to demonstrate the operation. NC
means no change.

Related Problem   What would the Q output look like if the J and K waveforms were
inverted?

SECTION 8–3 REVIEW

1. Describe the basic difference between pulse-triggered and edge-triggered flip-flops.
2. Suppose that the *D* input of a flip-flop changes from LOW to HIGH in the middle of a positive-going clock pulse.
   (a) Describe what happens if the flip-flop is a positive edge-triggered type.
   (b) Describe what happens if the flip-flop is a pulse-triggered master-slave type.

8–4 ■ FLIP-FLOP OPERATING CHARACTERISTICS

The performance, operating requirements, and limitations of flip-flops are specified by several operating characteristics or parameters found on the data sheet for the device. Generally, the specifications are applicable to all CMOS and TTL flip-flops.

After completing this section, you should be able to

■ Define *propagation delay time* ■ Explain the various propagation delay time specifications ■ Define *set-up time* and discuss how it limits flip-flop operation ■ Define *hold time* and discuss how it limits flip-flop operation ■ Discuss the significance of maximum clock frequency ■ Discuss the various pulse width specifications ■ Define *power dissipation* and calculate its value for a specific device ■ Compare various series of flip-flops in terms of their operating parameters

Propagation Delay Times

A **propagation delay time** is the interval of time required after an input signal has been applied for the resulting output change to occur. Four categories of propagation delay are important in the operation of a flip-flop:

1. Propagation delay t_{PLH} as measured from the triggering edge of the clock pulse to the LOW-to-HIGH transition of the output. This delay is illustrated in Figure 8–35(a).
2. Propagation delay t_{PHL} as measured from the triggering edge of the clock pulse to the HIGH-to-LOW transition of the output. This delay is illustrated in Figure 8–35(b).
3. Propagation delay t_{PLH} as measured from the preset input to the LOW-to-HIGH transition of the output. This delay is illustrated in Figure 8–36(a) for an active-LOW preset input.

(a)   (b)

FIGURE 8–35
Propagation delays, clock to output.

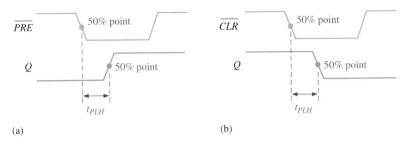

FIGURE 8–36
Propagation delays, preset input to output and clear input to output.

4. Propagation delay t_{PHL} as measured from the clear input to the HIGH-to-LOW transition of the output. This delay is illustrated in Figure 8–35(b) for an active-LOW clear input.

Set-up Time

The **set-up time** (t_s) is the minimum interval required for the logic levels to be maintained constantly on the inputs (J and K, or S and R, or D) prior to the triggering edge of the clock pulse in order for the levels to be reliably clocked into the flip-flop. This interval is illustrated in Figure 8–37 for a D flip-flop.

FIGURE 8–37
Set-up time (t_s).

Hold Time

The **hold time** (t_h) is the minimum interval required for the logic levels to remain on the inputs after the triggering edge of the clock pulse in order for the levels to be reliably clocked into the flip-flop. This is illustrated in Figure 8–38 for a D flip-flop.

FIGURE 8–38
Hold time (t_h).

Maximum Clock Frequency

The maximum clock frequency (f_{max}) is the highest rate at which a flip-flop can be reliably triggered. At clock frequencies above the maximum, the flip-flop would be unable to respond quickly enough, and its operation would be impaired.

Pulse Widths

Minimum pulse widths (t_W) for reliable operation are usually specified by the manufacturer for the clock, preset, and clear inputs. Typically, the clock is specified by its minimum HIGH time and its minimum LOW time.

Power Dissipation

The **power dissipation** of any digital circuit is the total power consumption of the device. For example, if the flip-flop operates on a +5 V dc source and draws 5 mA of current, the power dissipation is

$$P = V_{CC} \times I_{CC} = 5 \text{ V} \times 5 \text{ mA} = 25 \text{ mW}$$

The power dissipation is very important in most applications in which the capacity of the dc supply is a concern. As an example, let's assume that you have a digital system that requires a total of ten flip-flops, and each flip-flop dissipates 25 mW of power. The total power requirement is

$$P_T = 10 \times 25 \text{ mW} = 250 \text{ mW} = 0.25 \text{ W}$$

This tells you the output capacity required of the dc supply. If the flip-flops operate on +5 V dc, then the amount of current that the supply must provide is as follows:

$$I = \frac{250 \text{ mW}}{5 \text{ V}} = 50 \text{ mA}$$

You must use a +5 V dc supply that is capable of providing at least 50 mA of current.

Comparison of Specific Flip-Flops

Table 8–6 provides a comparison, in terms of the operating parameters discussed in this section, of four CMOS and TTL flip-flops of the same type.

TABLE 8–6

Comparison of operating parameters for four IC families of flip-flops of the same type at 25°C.

Parameter (Times in ns)	CMOS		TTL	
	74HC74A	74AHC74	74LS74A	74F74
t_{PHL} (CLK to Q)	17	4.6	40	6.8
t_{PLH} (CLK to Q)	17	4.6	25	8.0
t_{PHL} ($\overline{CLR}$ to Q)	18	4.8	40	9.0
t_{PLH} ($\overline{PRE}$ to Q)	18	4.8	25	6.1
t_s (set-up time)	14	5.0	20	2.0
t_h (hold time)	3.0	0.5	5	1.0
t_W (CLK HIGH)	10	5.0	25	4.0
t_W (CLK LOW)	10	5.0	25	5.0
t_W ($\overline{CLR/PRE}$)	10	5.0	25	4.0
f_{max} (MHz)	35	170	25	100
Power (mW), quiescent	0.012	1.1		
Power (mW), 50% duty cycle			44	88

An advantage of CMOS is that it can operate over a wider range of dc supply voltages (typically 2 V to 6 V) than TTL and, therefore, less expensive power supplies that do not have precise regulation can be used. Also, batteries can be used as secondary or primary sources for CMOS circuits. In addition, lower voltages mean that the IC dissipates less power. The drawback is that the performance of CMOS is degraded with lower supply voltages. For example, the guaranteed maximum clock frequency of a CMOS flip-flop is much less at $V_{CC} = 2$ V than at $V_{CC} = 6$ V.

**SECTION 8–4
REVIEW**

1. Define the following:
   (a) set-up time (b) hold time
2. Which specific flip-flop in Table 8–6 can be operated at the highest frequency?

8–5 ■ FLIP-FLOP APPLICATIONS

In this section, three general applications of flip-flops are discussed to give you an idea of how they can be used. In Chapters 9 and 10 there are detailed coverages of flip-flop applications in counters and registers.

After completing this section, you should be able to

■ Discuss the application of flip-flops in data storage ■ Describe how flip-flops are used for frequency division ■ Explain how flip-flops are used in basic counter applications

Application Examples

Parallel Data Storage   A common requirement in digital systems is to store several bits of data from parallel lines simultaneously in a group of flip-flops. This operation is illustrated in Figure 8–39(a) using four flip-flops. Each of the four parallel data lines is connected to the D input of a flip-flop. The clock inputs of the flip-flops are connected together, so that each flip-flop is triggered by the same clock pulse. In this example, positive edge-triggered flip-flops are used, so the data on the D inputs are stored simultaneously by the flip-flops on the positive edge of the clock, as indicated in the timing diagram in Figure 8–39(b). Also, the asynchronous reset (R) inputs are connected to a common $\overline{CLR}$ line, which initially resets all the flip-flops.

This group of four flip-flops is an example of a basic register used for data storage. In digital systems, data are normally stored in groups of bits (usually eight or multiples thereof) that represent numbers, codes, or other information. Registers are covered in detail in Chapter 10.

Frequency Division   Another application of a flip-flop is dividing (reducing) the frequency of a periodic waveform. When a pulse waveform is applied to the clock input of a J-K flip-flop that is connected to toggle ($J = K = 1$), the Q output is a square wave with one-half the frequency of the clock input. Thus, a single flip-flop can be applied as a divide-by-2 device, as is illustrated in Figure 8–40. As you can see, the flip-flop changes state on each triggering clock edge (positive edge-triggered in this case). This results in an output that changes at half the frequency of the clock waveform.

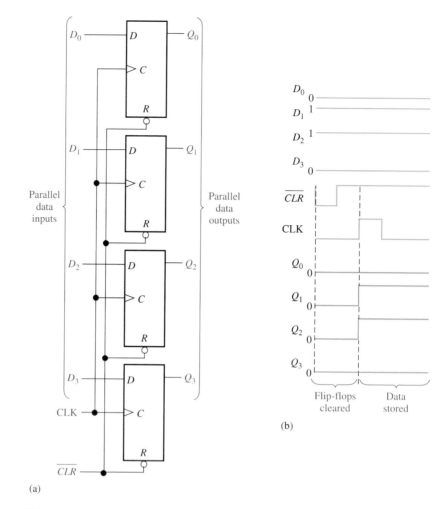

FIGURE 8–39
Example of flip-flops used in a basic register for parallel data storage.

FIGURE 8–40
The J-K flip-flop as a divide-by-2 device.

Further division of a clock frequency can be achieved by using the output of one flip-flop as the clock input to a second flip-flop, as shown in Figure 8–41. The frequency of the Q_A output is divided by 2 by flip-flop B. The Q_B output is, therefore, one-fourth the frequency of the original clock input. Propagation delay times are not shown on the timing diagrams.

By connecting flip-flops in this way, a frequency division of 2^n is achieved, where n is the number of flip-flops. For example, three flip-flops divide the clock frequency by $2^3 = 8$; four flip-flops divide the clock frequency by $2^4 = 16$; and so on.

FIGURE 8–41
Example of two J-K flip-flops used to divide the clock frequency by 4.

EXAMPLE 8–11

Develop the f_{out} waveform for the circuit in Figure 8–42 when an 8 kHz square wave input is applied to the clock input of flip-flop A.

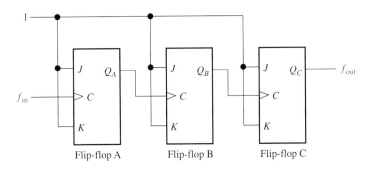

FIGURE 8–42

Solution The three flip-flops are connected to divide the input frequency by eight ($2^3 = 8$) and the f_{out} waveform is shown in Figure 8–43. Since these are positive edge-triggered flip-flops, the outputs change on the positive-going clock edge. There is one output pulse for every eight input pulses, so the output frequency is 1 kHz. Waveforms of Q_A and Q_B are also shown.

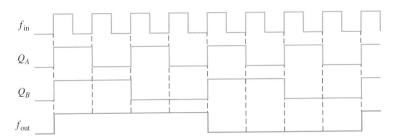

FIGURE 8–43

Related Problem How many flip-flops are required to divide a frequency by thirty-two?

Counting   Another important application of flip-flops is in digital counters which are covered in detail in the next chapter. The concept is illustrated in Figure 8–44. The flip-flops are negative edge-triggered J-Ks. Both flip-flops are initially RESET. Flip-flop A toggles on the negative-going transition of each clock pulse. The Q output of flip-flop A clocks flip-flop B, so each time Q_A makes a HIGH-to-LOW transition, flip-flop B toggles. The resulting Q_A and Q_B waveforms are shown in the figure.

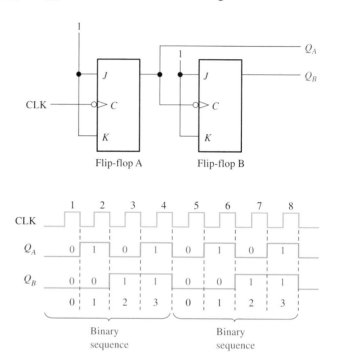

FIGURE 8–44
Flip-flops used to generate a binary count sequence. Two repetitions (00, 01, 10, 11) are shown.

Observe the sequence of Q_A and Q_B in Figure 8–44. Prior to clock pulse 1, $Q_A = 0$ and $Q_B = 0$; after clock pulse 1, $Q_A = 1$ and $Q_B = 0$; after clock pulse 2, $Q_A = 0$ and $Q_B = 1$; and after clock pulse 3, $Q_A = 1$ and $Q_B = 1$. If we take Q_A as the least significant bit, a 2-bit sequence is produced as the flip-flops are clocked. This binary sequence repeats every four clock pulses, as shown in the timing diagram of Figure 8–44. Thus, the flip-flops are counting in sequence from 0 to 3 (00, 01, 10, 11) and then recycling back to 0 to begin the sequence again.

EXAMPLE 8–12

Determine the output waveforms in relation to the clock for Q_A, Q_B, and Q_C in the circuit of Figure 8–45 and show the binary sequence represented by these waveforms.

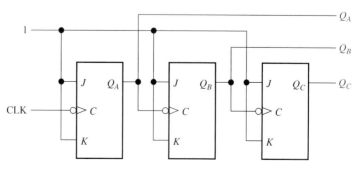

FIGURE 8–45

Solution The output timing diagram is shown in Figure 8–46. Notice that the outputs change on the negative-going edge of the clock pulses. The outputs go through the binary sequence 000, 001, 010, 011, 100, 101, 110, and 111 as indicated.

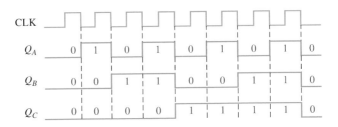

FIGURE 8–46

Related Problem How many flip-flops are required to produce a binary sequence representing decimal numbers 0 through 15?

SECTION 8–5 REVIEW

1. A group of flip-flops used for data storage is called a _____.
2. How must a J-K flip-flop be connected to function as a divide-by-2 device?
3. How many flip-flops are required to produce a divide-by-64 device?

8–6 ■ ONE-SHOTS

The one-shot is a monostable multivibrator, a device with only one stable state. A one-shot is normally in its stable state and will change to its unstable state only when triggered. Once it is triggered, the one-shot remains in its unstable state for a predetermined length of time and then automatically returns to its stable state. The time that the device stays in its unstable state determines the pulse width of its output.

After completing this section, you should be able to

■ Describe the basic operation of a one-shot ■ Explain how a nonretriggerable one-shot works ■ Explain how a retriggerable one-shot works ■ Set up the 74121 and the 74LS122 one-shots to obtain a specified output pulse width ■ Recognize a Schmitt trigger symbol and explain basically what it means

Figure 8–47 shows a basic **one-shot (monostable multivibrator)** that is composed of a logic gate and an inverter. When a pulse is applied to the **trigger** input, the output of gate

FIGURE 8–47
A simple one-shot circuit.

G_1 goes LOW. This HIGH-to-LOW transition is coupled through the capacitor to the input of inverter G_2. The apparent LOW on G_2 makes its output go HIGH. This HIGH is connected back into G_1, keeping its output LOW. Up to this point the trigger pulse has caused the output of the one-shot, Q, to go HIGH.

The capacitor immediately begins to charge through R toward the high voltage level. The rate at which it charges is determined by the RC time constant. When the capacitor charges to a certain level, which appears as a HIGH to G_2, the output goes back LOW.

To summarize, the output of inverter G_2 goes HIGH in response to the trigger input. It remains HIGH for a time set by the RC time constant. At the end of this time, it goes LOW. So *a single narrow trigger pulse produces a single output pulse whose time duration is controlled by the RC time constant.* This operation is illustrated in Figure 8–47.

A typical one-shot logic symbol is shown in Figure 8–48(a), and the same symbol with an external R and C is shown in Figure 8–48(b). The two basic types of IC one-shots are nonretriggerable and retriggerable.

FIGURE 8–48
Basic one-shot logic symbols. CX and RX stand for external components.

(a)   (b)

Nonretriggerable One-Shots

A nonretriggerable one-shot will not respond to any additional trigger pulses from the time it is triggered into its unstable state (fired) until it returns to its stable state. In other words, it will ignore any trigger pulses occurring before it times out. The time that the one-shot remains in its unstable state is the pulse width of the output.

Figure 8–49 shows the nonretriggerable one-shot being triggered at intervals greater than its pulse width and at intervals less than the pulse width. Notice that in the second case, the additional pulses are ignored.

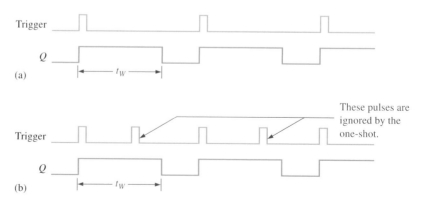

FIGURE 8–49
Nonretriggerable one-shot action.

A Nonretriggerable One-Shot The 74121 is an example of a nonretriggerable IC one-shot. It has provisions for external R and C, as shown in Figure 8–50. The inputs labeled A_1, A_2, and B are gated trigger inputs. The R_{INT} input connects to a 2 kΩ internal timing resistor.

(a) Traditional logic symbol

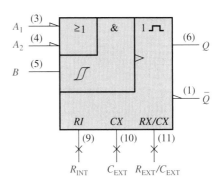

(b) ANSI/IEEE std. 91–1984 logic symbol ($\times$ = nonlogic connection). "1 ⊓" is the qualifying symbol for a nonretriggerable one-shot.

FIGURE 8–50
Logic symbols for the 74121 nonretriggerable one-shot.

Setting the Pulse Width A typical pulse width of about 30 ns is produced when no external timing components are used and the internal timing resistor (R_{INT}) is connected to V_{CC}, as shown in Figure 8–51(a). The pulse width can be set anywhere between about 30 ns and 28 s by the use of external components. Figure 8–51(b) shows connection of the internal resistor (2 kΩ) and an external capacitor. Part (c) illustrates connection of an external resistor and an external capacitor. The output pulse width is set by the values of the resistor (R_{INT} = 2 kΩ, and R_{EXT} is selected) and the capacitor according to the following formula:

$$t_W = 0.7RC_{EXT} \tag{8–1}$$

where R is either R_{INT} or R_{EXT}. When R is in kilohms (kΩ) and C_{EXT} is in picofarads (pF), the output pulse width t_W is in nanoseconds (ns).

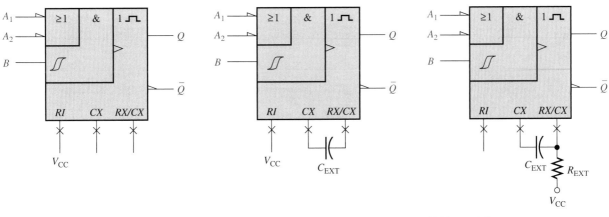

(a) No external components
R_{INT} to V_{CC}
$t_W \cong 30$ ns

(b) C_{EXT} and R_{INT} to V_{CC}
$t_W = 0.7(2 \text{ k}\Omega)C_{EXT}$

(c) R_{EXT} and C_{EXT}
$t_W = 0.7R_{EXT}C_{EXT}$

FIGURE 8–51
Three ways to set the pulse width of a 74121.

The Schmitt-Trigger Symbol   The symbol $\int$ indicates a Schmitt-trigger input. This type of input uses a special threshold circuit that produces **hysteresis,** a characteristic that prevents erratic switching between states when a slow-changing trigger voltage hovers around the critical input level. This allows reliable triggering to occur even when the input is changing as slowly as 1 volt/second.

EXAMPLE 8–13

A certain application requires a one-shot with a pulse width of approximately 100 ms. Using a 74121, show the connections and the component values.

Solution   Arbitrarily select $R_{EXT} = $ **39 kΩ** and calculate the necessary capacitance.

$$t_W = 0.7R_{EXT}C_{EXT}$$

$$C_{EXT} = \frac{t_W}{0.7R_{EXT}}$$

where C_{EXT} is in pF, R_{EXT} is in kΩ, and t_W is in ns. Since 100 ms $= 1 \times 10^8$ ns,

$$C_{EXT} = \frac{1 \times 10^8 \text{ ns}}{0.7(39 \text{ k}\Omega)} = 3.66 \times 10^6 \text{ pF} = \textbf{3.66 } \boldsymbol{\mu}\textbf{F}$$

A standard 3.3 μF capacitor will give an output pulse width of 91 ms. The proper connections are shown in Figure 8–52. To achieve a pulse width closer to 100 ms, other combinations of values for R_{EXT} and C_{EXT} can be tried. For example, $R_{EXT} = 68$ kΩ and $C_{EXT} = 2.2$ μF gives a pulse width of 105 ms.

FIGURE 8–52

Related Problem   Use an external capacitor in conjunction with R_{INT} to produce an output pulse width of 10 μs from the 74121.

Retriggerable One-Shots

A retriggerable one-shot can be triggered before it times out. The result of retriggering is an extension of the pulse width as illustrated in Figure 8–53.

A Retriggerable One-Shot   The 74LS122 is an example of a retriggerable IC one-shot with a clear input. It also has provisions for external R and C, as shown in Figure 8–54. The inputs labeled A_1, A_2, B_1, and B_2 are the gated trigger inputs.

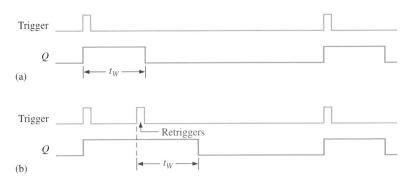

FIGURE 8–53
Retriggerable one-shot action.

(a) Traditional logic symbol

(b) ANSI/IEEE std. 91–1984 logic symbol
($\times$ = nonlogic connection). ⌐⌐ is the
qualifying symbol for a retriggerable
one-shot.

FIGURE 8–54
Logic symbol for the 74LS122 retriggerable one-shot.

A minimum pulse width of approximately 45 ns is obtained with no external components. Wider pulse widths are achieved by using external components. A general formula for calculating the values of these components for a specified pulse width (t_W) is

$$t_W = 0.32RC_{EXT}\left(1 + \frac{0.7}{R}\right) \tag{8–2}$$

where 0.32 is a constant determined by the particular type of one-shot, R is in kΩ and is either the internal or the external resistor, C_{EXT} is in pF, and t_W is in ns. The internal resistance is 10 kΩ and can be used instead of an external resistor. (Notice the difference between this formula and that for the 74121, shown in Equation (8–1).)

EXAMPLE 8–14

Determine the values of R_{EXT} and C_{EXT} that will produce a pulse width of 1 μs when connected to a 74LS122.

Solution Assume a value of C_{EXT} = **560 pF** and then solve for R_{EXT}. The pulse width must be expressed in ns and C_{EXT} in pF. R_{EXT} will be in kΩ.

$$t_W = 0.32R_{EXT}C_{EXT}\left(\frac{1 + 0.7}{R_{EXT}}\right) = 0.32R_{EXT}C_{EXT} + 0.7\left(\frac{0.32R_{EXT}C_{EXT}}{R_{EXT}}\right)$$

$$= 0.32R_{EXT}C_{EXT} + (0.7)(0.32)C_{EXT}$$

$$R_{EXT} = \frac{t_W - (0.7)(0.32)C_{EXT}}{0.32C_{EXT}} = \frac{t_W}{0.32C_{EXT}} - 0.7$$

$$= \frac{1000 \text{ ns}}{(0.32)560 \text{ pF}} - 0.7 = \textbf{4.88 k}\Omega$$

Use a standard value of **4.7 kΩ**.

Related Problem   Show the connections and component values for a 74LS122 one-shot with an output pulse width of 5 μs. Assume $C_{EXT} = 560$ pF.

Application Example

One practical one-shot application is a sequential timer that can be used to illuminate a series of lights. This type of circuit can be used, for example, in a lane change directional indicator for highway construction projects or in sequential turn signals on automobiles.

Figure 8–55 shows three 74LS122 one-shots connected as a sequential timer. This particular circuit produces a sequence of three 1 s pulses. The first one-shot is triggered by a switch closure or a low-frequency pulse input producing a 1 s output pulse. When the first one-shot (OS 1) times out and the 1 s pulse goes LOW, the second one-shot (OS 2) is triggered also producing a 1 s output pulse. When this second pulse goes LOW, the third one-shot (OS 3) is triggered and the third 1 s pulse is produced. The output timing is illustrated in the figure. Variations of this basic arrangement can be used to produce a variety of timed outputs.

FIGURE 8–55
A sequential timing circuit using three 74LS122 one-shots.

SECTION 8–6 REVIEW

1. Describe the difference between a nonretriggerable and a retriggerable one-shot.
2. How is the output pulse width set in most IC one-shots?

8–7 ■ THE 555 TIMER

The 555 timer is a versatile and widely used device because it can be configured in two different modes as either a monostable multivibrator (one-shot) or as an astable multivibrator (oscillator). An astable multivibrator has no stable states and therefore changes back and forth (oscillates) between two unstable states without any external triggering.

After completing this section, you should be able to

■ Describe the basic elements in a 555 timer ■ Set up a 555 timer as a one-shot ■ Set up a 555 timer as an oscillator

Basic Operation

A functional diagram showing the internal components of a 555 **timer** is given in Figure 8–56. The comparators are devices whose outputs are HIGH when the voltage on the positive (+) input is greater than the voltage on the negative (−) input and LOW when the − input voltage is greater than the + input voltage. The voltage divider consisting of three $5 \text{ k}\Omega$ resistors provides a trigger level of $\frac{1}{3}V_{CC}$ and a threshold level of $\frac{2}{3}V_{CC}$. The control voltage input (pin 5) can be used to externally adjust the trigger and threshold levels to other values if necessary. When the normally HIGH trigger input momentarily goes below $\frac{1}{3}V_{CC}$, the output of comparator B switches from LOW to HIGH and sets the S-R latch, causing the output (pin 3) to go HIGH and turning the discharge transistor Q_1 off. The output will stay HIGH until the normally LOW threshold input goes above $\frac{2}{3}V_{CC}$ and causes the output of comparator A to switch from LOW to HIGH. This resets the latch, causing the output to go back LOW and turning the discharge transistor on. The external reset input can be used to reset the latch independent of the threshold circuit. The trigger and threshold inputs (pins 2 and 6) are controlled by external components connected to produce either monostable or astable action.

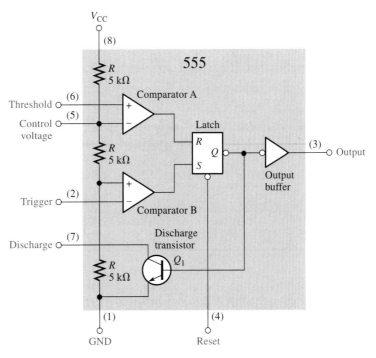

FIGURE 8–56
Internal functional diagram of a 555 timer (pin numbers are in parentheses).

Monostable (One-Shot) Operation

An external resistor and capacitor connected as shown in Figure 8–57 are used to set up the 555 timer as a nonretriggerable one-shot. The pulse width of the output is determined by the time constant of R_1 and C_1 according to the following formula:

$$t_W = 1.1R_1C_1 \tag{8–3}$$

The control voltage input is not used and is connected to a decoupling capacitor C_2 to prevent noise from affecting the trigger and threshold levels.

FIGURE 8–57
The 555 timer connected as a one-shot.

Before a trigger pulse is applied, the output is LOW and the discharge transistor Q_1 is *on,* keeping C_1 discharged as shown in Figure 8–58(a). When a negative-going trigger pulse is applied at t_0, the output goes HIGH and the discharge transistor turns *off,* allowing capacitor C_1 to begin charging through R_1 as shown in part (b). When C_1 charges to ⅓V_{CC}, the output goes back LOW at t_1 and Q_1 turns *on* immediately, discharging C_1 as shown in part (c). As you can see, the charging rate of C_1 determines how long the output is HIGH.

EXAMPLE 8–15

What is the output pulse width for a 555 monostable circuit with $R_1 = 2.2$ kΩ and $C_1 = 0.01$ μF?

Solution   From Equation (8–3) the pulse width is

$$t_W = 1.1R_1C_1 = 1.1(2.2 \text{ k}\Omega)(0.01 \text{ μF}) = \textbf{24.2 μs}$$

Related Problem   For $C_1 = 0.01$ μF, determine the value of R_1 for a pulse width of 1 ms.

Astable Operation

A 555 timer connected to operate as an **astable** multivibrator, which is a nonsinusoidal **oscillator,** is shown in Figure 8–59. Notice that the threshold input *(THRESH)* is now connected to the trigger input *(TRIG).* The external components R_1, R_2, and C_1 form the timing network that sets the frequency of oscillation. The 0.01 μF capacitor C_2 connected to the control *(CONT)* input is strictly for decoupling and has no effect on the operation; in some cases it can be left off.

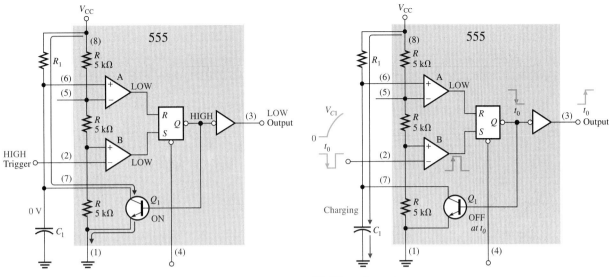

(a) Prior to triggering   (b) When triggered

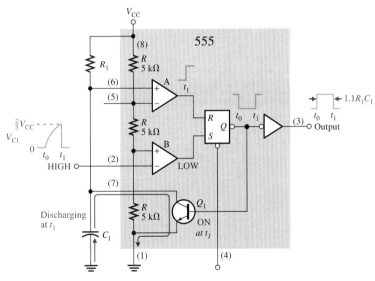

(c) At end of charging interval

FIGURE 8–58
One-shot operation of the 555 timer.

FIGURE 8–59
The 555 timer connected as an astable multivibrator (oscillator).

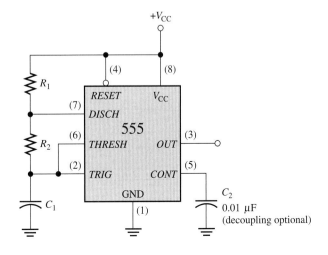

FIGURE 8–60

Operation of the 555 timer in the astable mode.

Initially, when the power is turned on, the capacitor (C_1) is uncharged and thus the trigger voltage (pin 2) is at 0 V. This causes the output of comparator B to be HIGH and the output of comparator A to be LOW, forcing the output of the latch, and thus the base of Q_1, LOW and keeping the transistor off. Now, C_1 begins charging through R_1 and R_2 as indicated in Figure 8–60. When the capacitor voltage reaches $\frac{1}{3}V_{CC}$, comparator B switches to its LOW output state, and when the capacitor voltage reaches $\frac{2}{3}V_{CC}$, comparator A switches to its HIGH output state. This resets the latch, causing the base of Q_1 to go HIGH and turning on the transistor. This sequence creates a discharge path for the capacitor through R_2 and the transistor, as indicated. The capacitor now begins to discharge, causing comparator A to go LOW. At the point where the capacitor discharges down to $\frac{1}{3}V_{CC}$, comparator B switches HIGH; this sets the latch, making the base of Q_1 LOW and turning off the transistor. Another charging cycle begins, and the entire process repeats. The result is a rectangular wave output whose duty cycle depends on the values of R_1 and R_2. The frequency of oscillation is given by the following formula, or it can be found using the graph in Figure 8–61.

$$f = \frac{1.44}{(R_1 + 2R_2)C_1} \tag{8–4}$$

By selecting R_1 and R_2, the duty cycle of the output can be adjusted. Since C_1 charges through $R_1 + R_2$ and discharges only through R_2, duty cycles approaching a minimum of 50 percent can be achieved if $R_2 \gg R_1$ so that the charging and discharging times are approximately equal.

An expression for the duty cycle is developed as follows. The time that the output is HIGH (t_H) is how long it takes C_1 to charge from $\frac{1}{3}V_{CC}$ to $\frac{2}{3}V_{CC}$. It is expressed as

$$t_H = 0.7(R_1 + R_2)C_1 \tag{8–5}$$

FIGURE 8–61
Frequency of oscillation as a function of C_1 and $R_1 + 2R_2$. The sloped lines are values of $R_1 + 2R_2$.

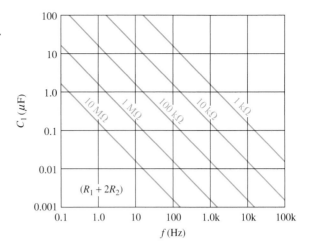

The time that the output is LOW (t_L) is how long it takes C_1 to discharge from $\frac{1}{3}V_{CC}$ to $\frac{2}{3}V_{CC}$. It is expressed as

$$t_L = 0.7R_2C_1 \qquad (8\text{--}6)$$

The period, T, of the output waveform is the sum of t_H and t_L.

$$T = t_H + t_L = 0.7(R_1 + 2R_2)C_1$$

This is the reciprocal of f in Equation (8–4). Finally, the duty cycle is

$$\text{Duty cycle} = \frac{t_H}{T} = \frac{t_H}{t_H + t_L}$$

$$\text{Duty cycle} = \left(\frac{R_1 + R_2}{R_1 + 2R_2}\right)100\% \qquad (8\text{--}7)$$

To achieve duty cycles of less than 50 percent, the circuit in Figure 8–59 can be modified so that C_1 charges through only R_1 and discharges through R_2. This is achieved with a diode D_1 placed as shown in Figure 8–62. The duty cycle can be made less than 50 percent by making R_1 less than R_2. Under this condition, the expression for the duty cycle is

$$\text{Duty cycle} = \left(\frac{R_1}{R_1 + R_2}\right)100\% \qquad (8\text{--}8)$$

FIGURE 8–62
The addition of diode D_1 allows the duty cycle of the output to be adjusted to less than 50 percent by making $R_1 < R_2$.

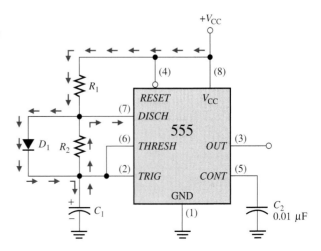

EXAMPLE 8–16   A 555 timer configured to run in the astable mode (oscillator) is shown in Figure 8–63. Determine the frequency of the output and the duty cycle.

FIGURE 8–63

Solution   Use Equations (8–4) and (8–7).

$$f = \frac{1.44}{(R_1 + 2R_2)C_1} = \frac{1.44}{(2.2 \text{ k}\Omega + 9.4 \text{ k}\Omega)0.022 \; \mu\text{F}} = \textbf{5.64 kHz}$$

$$\text{Duty cycle} = \left(\frac{R_1 + R_2}{R_1 + 2R_2}\right)100\% = \left(\frac{2.2 \text{ k}\Omega + 4.7 \text{ k}\Omega}{2.2 \text{ k}\Omega + 9.4 \text{ k}\Omega}\right)100\% = \textbf{59.5\%}$$

Related Problem   Determine the duty cycle in Figure 8–63 if a diode is connected across R_2 as indicated in Figure 8–62.

SECTION 8–7 REVIEW

1. Explain the difference in operation between an astable multivibrator and a monostable multivibrator.
2. For a certain astable multivibrator, $t_H = 15$ ms and $T = 20$ ms. What is the duty cycle of the output?

8–8 ■ TROUBLESHOOTING

It is standard practice to test a new circuit design to be sure that it is operating as specified. New designs are usually simulated on a computer and then "breadboarded" and tested before the design is finalized. The term breadboard refers to a method of temporarily hooking up a circuit so that its operation can be verified and any faults (bugs) worked out before a prototype unit is built. In this section, we will consider an example case.

After completing this section, you should be able to

■ Describe how the timing of a circuit can produce erroneous glitches   ■ Approach the debugging of a new design with greater insight and awareness of potential problems

(a)

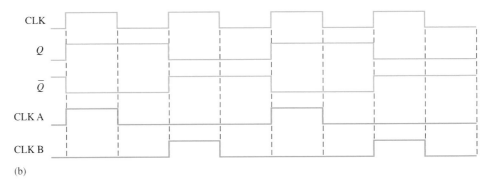

(b)

FIGURE 8–64
Two-phase clock generator with ideal waveforms.

The circuit shown in Figure 8–64(a) generates two clock waveforms (CLK A and CLK B) having an alternating occurrence of pulses. Each waveform is to be one-half the frequency of the original clock (CLK), as shown in the ideal timing diagram in part (b).

When the circuit is tested, the CLK A and CLK B waveforms appear on the oscilloscope or logic analyzer as shown in Figure 8–65(a). Since glitches occur on both waveforms, something is wrong with the circuit either in its basic design or in the way it is connected. Further investigation reveals that the glitches are caused by a **race** condition

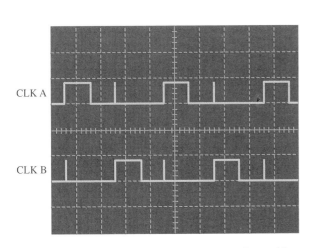

(a) Logic analyzer display of CLK A and CLK B waveforms with glitches indicated by the "spikes".

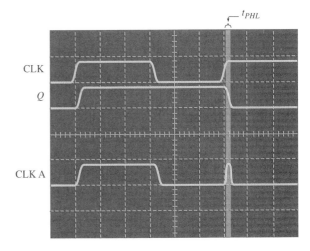

(b) Logic analyzer display showing propagation delay that creates glitch on CLK A waveform

FIGURE 8–65
Logic analyzer displays for the circuit in Figure 8–64.

between the CLK signal and the Q and $\overline{Q}$ signals at the inputs of the AND gates. As displayed in Figure 8–65(b), the propagation delays between CLK and Q and $\overline{Q}$ create a short-duration coincidence of HIGH levels at the leading edges of alternate clock pulses. Thus, there is a basic design flaw.

The problem can be corrected by using a negative edge-triggered flip-flop in place of the positive edge-triggered device, as shown in Figure 8–66(a). Although the propagation delays between CLK and Q and $\overline{Q}$ still exist, they are initiated on the trailing edges of the clock (CLK), thus eliminating the glitches, as shown in the timing diagram of Figure 8–66(b).

(a)

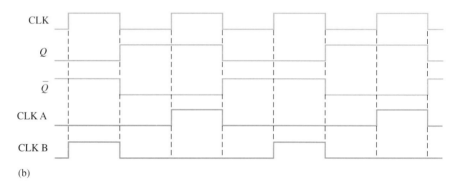

(b)

FIGURE 8–66
Two-phase clock generator using negative edge-triggered flip-flop to eliminate glitches.

Glitches that occur in digital systems are very fast (extremely short in duration) and can be difficult to see on an oscilloscope, particularly at lower sweep rates. A logic analyzer, however, can show a glitch easily. To look for glitches using a logic analyzer, select "latch" mode or (if available) transitional sampling. In the latch mode, the analyzer looks for a voltage level change. When a change occurs, even if it is of extremely short duration (a few nanoseconds), the information is "latched" into the analyzer's memory as another sampled data point. When the data are displayed, the glitch will show as an obvious change in the sampled data, making it easy to identify.

**SECTION 8–8
REVIEW**

1. Can a negative edge-triggered D flip-flop be used in the circuit of Figure 8–66?
2. What device can be used to provide the clock for the circuit in Figure 8–66?

8–9 ■ DIGITAL SYSTEM APPLICATION

In this system application, you will continue working with the traffic light control system that was started in Chapter 6. You should review Section 6–11 before continuing. In this chapter, the focus is on the timing circuits that produce the 4 s interval for the caution light, the 25 s interval for the red and green lights, and the clock.

After completing this section, you should be able to

■ Develop a block diagram for the timing circuits portion of the system ■ Implement the short and long timers to meet specifications ■ Implement the clock oscillator to meet specifications ■ Develop a schematic from the printed circuit board ■ Develop a test procedure and troubleshoot the circuit board

General Requirements for the Timing Circuits

The overall system block diagram was discussed in the system application for Chapter 6 and is shown in Figure 8–67 again for reference. The state decoding and output logic block has been developed, and now we are going to concentrate on the timing circuits that must produce a 4 s positive pulse or a 25 s positive pulse when triggered from the state decoding and output logic and a 10 kHz clock signal. All three of these outputs are used by the sequential logic that will be developed in Chapter 9.

FIGURE 8–67
Traffic light control system block diagram.

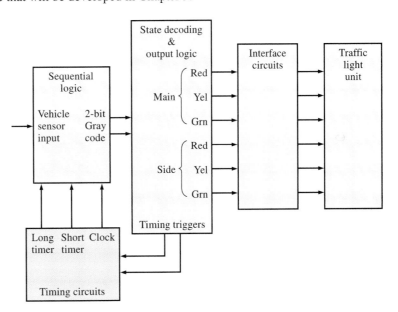

The timing circuits consist of the long timer, the short timer, and the clock oscillator as shown in Figure 8–68. The long timer and the short timer are implemented with one-shots and the clock oscillator is implemented with a 555 timer connected in the astable mode.

FIGURE 8–68
Block diagram of the timing circuits.

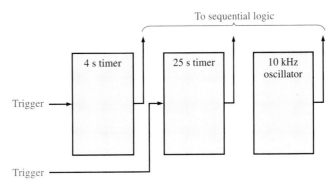

■ SYSTEM ASSIGNMENTS

In the Chapter 6 system application you focused on the portion of the traffic light controller board containing the state decoder and output logic. In this application, the portion of the board containing the timing circuits is now considered. As before, all of the connections to the V_{CC} and ground pads are on the back side of the board. Circuit interconnections on the back side of the board feed through to the pads on the component side.

■ SYSTEM ASSIGNMENT 1: Circuit Verification and Design

■ **Activity 1:** Using the PC board, add the timing circuit portion of the system to the schematic that you began in Section 6–11.

■ **Activity 2:** Verify that the timing circuit portion of the PC board is correct and properly connected to the state decoder and output logic.

■ **Activity 3:** Identify the function of each of the points on the PC board to the right of each of the three circled numbers at the bottom of the board.

■ **Activity 4:** Determine a value for each resistor and capacitor associated with the one-shots on the board that will produce a 4 s positive pulse, a 25 s positive pulse, and a 10 kHz pulse waveform (duty cycle is not critical).

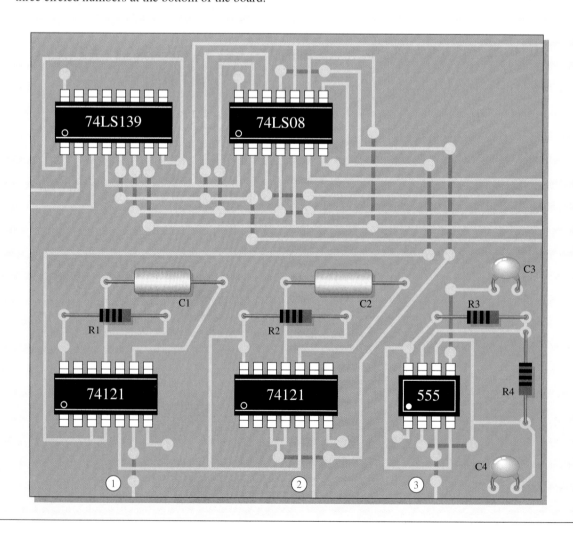

■ SYSTEM ASSIGNMENT 2:   Testing and Troubleshooting

■ *Activity 1:*   Using the switches connected as shown to simulate the Gray code inputs from the sequential portion of the system, develop a test procedure to check out this portion of the system board.

■ *Activity 2:*   When the switches are thrown into each of the positions shown in the table, determine from the indications at the test points represented

by the circled numbers if the circuit is operating properly for each case and, if not, what the most likely problem is.

The PC board logic for the state decoding and output logic and the timing circuits is in file SA08.EWB on your EWB CD. Use EWB to simulate a test of this circuit.

S_1	S_0	Case 1 ①	Case 1 ②	Case 1 ③	Case 2 ①	Case 2 ②	Case 2 ③	Case 3 ①	Case 3 ②	Case 3 ③	Case 4 ①	Case 4 ②	Case 4 ③
HIGH / LOW	HIGH / LOW	HIGH 25 s	LOW	10 kHz	LOW	HIGH 25 s	10 kHz	HIGH 25 s	LOW	10 kHz	LOW	LOW	LOW
HIGH / LOW	HIGH / LOW	LOW	HIGH 4 s	10 kHz	HIGH 4 s	LOW	10 kHz	LOW	LOW	10 kHz	LOW	LOW	LOW
HIGH / LOW	HIGH / LOW	HIGH 25 s	LOW	10 kHz	LOW	HIGH 25 s	10 kHz	HIGH 25 s	LOW	10 kHz	LOW	LOW	LOW
HIGH / LOW	HIGH / LOW	LOW	HIGH 4 s	10 kHz	HIGH 4 s	LOW	10 kHz	LOW	LOW	10 kHz	LOW	LOW	LOW

SECTION 8–9
REVIEW

1. Explain the purpose of the two switches in System Assignment 2.
2. Why are the 4 s one-shot and the 25 s one-shot triggered separately?
3. Upon what other parts of the system does the operation of the clock oscillator depend?

All or portions of Chapter 15, "Integrated Circuit Technologies," may be covered at this point. The coverage in Chapter 15 is not a prerequisite to any other material in this text, and its omission or partial omission will not affect any other material.

■ **SUMMARY**

- Symbols for latches and flip-flops are shown in Figure 8–69.
- Latches are bistable devices whose state normally depends on asynchronous inputs.
- Edge-triggered flip-flops are bistable devices with synchronous inputs whose state depends on the inputs only at the triggering transition of a clock pulse. Changes in the outputs occur at the triggering transition of the clock.

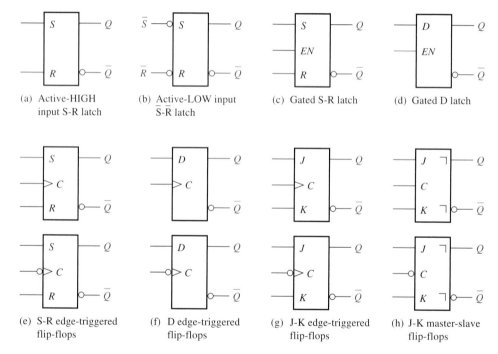

FIGURE 8–69

- Pulse-triggered master-slave flip-flops are bistable devices with synchronous inputs whose state depends on the inputs at the leading edge of the clock pulse, but whose output is postponed and does not reflect the internal state until the trailing edge of the clock pulse. The synchronous inputs should not be allowed to change while the clock is HIGH.
- Monostable multivibrators (one-shots) have one stable state. When the one-shot is triggered, the output goes to its unstable state for a time determined by an RC circuit.
- Astable multivibrators have no stable states and are used as oscillators to generate timing waveforms in digital systems.

■ KEY TERMS

These terms are also in the end-of-book glossary.

Astable   Having no stable state. An astable multivibrator oscillates between two quasi-stable states.

Asynchronous   Having no fixed time relationship.

Bistable   Having two stable states. Flip-flops and latches are bistable multivibrators.

Clear   An asynchronous input used to reset a flip-flop (make the Q output 0).

D flip-flop   A type of bistable multivibrator in which the output assumes the state of the D input on the triggering edge of a clock pulse.

Edge-triggered flip-flop   A type of flip-flop in which the data are entered and appear on the output on the same clock edge.

Feedback   The output voltage or a portion of it that is connected back to the input of a circuit.

Hold time   The time interval required for the control levels to remain on the inputs to a flip-flop after the triggering edge of the clock in order to reliably activate the device.

Hysteresis   A characteristic of a threshold-triggered circuit, such as the Schmitt trigger, where the device turns on and off at different input levels.

J-K flip-flop   A type of flip-flop that can operate in the SET, RESET, no-change, and toggle modes.

Latch   A bistable digital circuit used for storing a bit.

Master-slave flip-flop   A type of flip-flop in which the input data are entered into the device on the leading edges of clock pulses and appear at the output on trailing edges. Master-slave flip-flops have, for the most part, been replaced by edge-triggered types.

Monostable   Having only one stable state. A monostable multivibrator, commonly called a one-shot, produces a single pulse in response to a triggering input.

One-shot   A monostable multivibrator.

Preset   An asynchronous input used to set a flip-flop (make the Q output 1)

RESET   The state of a flip-flop or latch when the output is 0; the action of producing a RESET state.

SET   The state of a flip-flop or latch when the output is 1; the action of producing a SET state.

Set-up time   The time interval required for the control levels to be on the inputs to a digital circuit, such as a flip-flop, prior to the triggering edge of a clock pulse.

S-R flip-flop   A SET-RESET flip-flop.

Synchronous   Having a fixed time relationship.

Timer   A circuit that can be used as a one-shot or as an oscillator.

Toggle   The action of a flip-flop when it changes state on each clock pulse.

■ **SELF-TEST**

1. If an S-R latch has a 1 on the S input and a 0 on the R input and then the S input goes to 0, the latch will be

   (a) set (b) reset (c) invalid (d) clear

2. The invalid state of an S-R latch occurs when

   (a) $S = 1, R = 0$ (b) $S = 0, R = 1$
   (c) $S = 1, R = 1$ (d) $S = 0, R = 0$

3. For a gated D latch, the Q output always equals the D input

   (a) before the enable pulse (b) during the enable pulse
   (c) immediately after the enable pulse (d) answers (b) and (c)

4. Like the latch, the flip-flop belongs to a category of logic circuits known as

   (a) monostable multivibrators (b) bistable multivibrators
   (c) astable multivibrators (d) one-shots

5. The purpose of the clock input to a flip-flop is to

   (a) clear the device
   (b) set the device
   (c) always cause the output to change states
   (d) cause the output to assume a state dependent on the controlling (S-R, J-K, or D) inputs.

6. For an edge-triggered D flip-flop,

   (a) a change in the state of the flip-flop can occur only at a clock pulse edge
   (b) the state that the flip-flop goes to depends on the D input
   (c) the output follows the input at each clock pulse
   (d) all of these answers

7. A feature that distinguishes the J-K flip-flop from the S-R flip-flop is the

   (a) toggle condition (b) preset input
   (c) type of clock (d) clear input

8. A flip-flop is in the toggle condition when

   (a) $J = 1, K = 0$ (b) $J = 1, K = 1$
   (c) $J = 0, K = 0$ (d) $J = 0, K = 1$

9. A J-K flip-flop with $J = 1$ and $K = 1$ has a 10 kHz clock input. The Q output is

   (a) constantly HIGH (b) constantly LOW
   (c) a 10 kHz square wave (d) a 5 kHz square wave

10. A one-shot is a type of

    (a) monostable multivibrator (b) astable multivibrator
    (c) timer (d) answers (a) and (c)
    (e) answers (b) and (c)

11. The output pulse width of a nonretriggerable one-shot depends on

    (a) the trigger intervals (b) the supply voltage
    (c) a resistor and capacitor (d) the threshold voltage

12. An astable multivibrator

    (a) requires a periodic trigger input (b) has no stable state
    (c) is an oscillator (d) produces a periodic pulse output
    (e) answers (a), (b), (c), and (d) (f) answers (b), (c), and (d) only

■ **PROBLEMS**

SECTION 8–1 Latches

1. If the waveforms in Figure 8–70 are applied to an active-LOW input S-R latch, sketch the resulting Q output waveform in relation to the inputs. Assume that Q starts LOW.

2. Solve Problem 1 for the input waveforms in Figure 8–71 applied to an active-HIGH S-R latch.

3. Solve Problem 1 for the input waveforms in Figure 8–72.

FIGURE 8-70

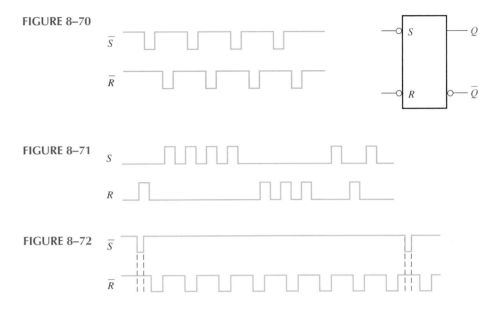

FIGURE 8-71

FIGURE 8-72

4. For a gated S-R latch, determine the Q and $\overline{Q}$ outputs for the inputs in Figure 8–73. Show them in proper relation to the enable input. Assume that Q starts LOW.

5. Solve Problem 4 for the inputs in Figure 8–74.

FIGURE 8-73

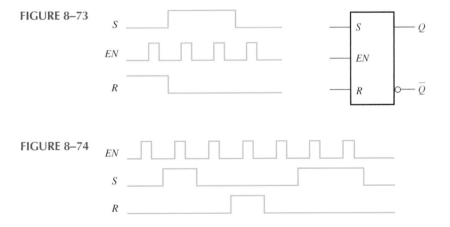

FIGURE 8-74

6. Solve Problem 4 for the inputs in Figure 8–75.

7. For a gated D latch, the waveforms shown in Figure 8–76 are observed on its inputs. Sketch the timing diagram showing the output waveform you would expect to see at Q if the latch is initially RESET.

FIGURE 8-75

FIGURE 8-76

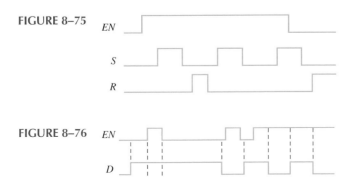

SECTION 8–2 Edge-Triggered Flip-Flops

8. Two edge-triggered S-R flip-flops are shown in Figure 8–77. If the inputs are as shown, sketch the Q output of each flip-flop relative to the clock, and explain the difference between the two. The flip-flops are initially RESET.

9. The Q output of an edge-triggered S-R flip-flop is shown in relation to the clock signal in Figure 8–78. Determine the input waveforms on the S and R inputs that are required to produce this output if the flip-flop is a positive edge-triggered type.

FIGURE 8–77

FIGURE 8–78

10. Sketch the Q output relative to the clock for a D flip-flop with the inputs as shown in Figure 8–79. Assume positive edge-triggering and Q initially LOW.

11. Solve Problem 10 for the inputs in Figure 8–80.

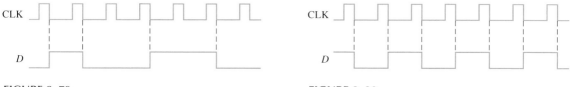

FIGURE 8–79   **FIGURE 8–80**

12. For a positive edge-triggered J-K flip-flop with inputs as shown in Figure 8–81, determine the Q output relative to the clock. Assume that Q starts LOW.

13. Solve Problem 12 for the inputs in Figure 8–82.

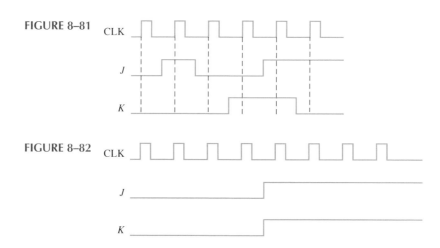

FIGURE 8–81

FIGURE 8–82

14. Determine the Q waveform relative to the clock if the signals shown in Figure 8–83 are applied to the inputs of the J-K flip-flop. Assume that Q is initially low.

15. For a negative edge-triggered J-K flip-flop with the inputs in Figure 8–84, develop the Q output waveform relative to the clock. Assume that Q is initially LOW.

FIGURE 8–83

FIGURE 8–84

16. The following serial data are applied to the flip-flop through the AND gates as indicated in Figure 8–85. Determine the resulting serial data that appear on the Q output. There is one clock pulse for each bit time. Assume that Q is initially 0 and that $\overline{PRE}$ and $\overline{CLR}$ are HIGH. Rightmost bits are applied first.

J_1: 1 0 1 0 0 1 1
J_2: 0 1 1 1 0 1 0
J_3: 1 1 1 1 0 0 0
K_1: 0 0 0 1 1 1 0
K_2: 1 1 0 1 1 0 0
K_3: 1 0 1 0 1 0 1

FIGURE 8–85

17. For the circuit in Figure 8–85, complete the timing diagram in Figure 8–86 by showing the Q output (which is initially LOW). Assume $\overline{PRE}$ and $\overline{CLR}$ remain HIGH.

FIGURE 8–86

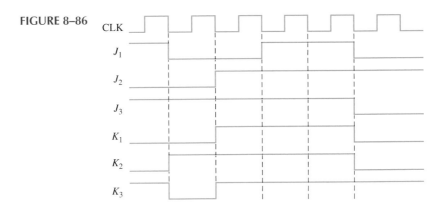

18. Solve Problem 17 with the same J and K inputs but with the $\overline{PRE}$ and $\overline{CLR}$ inputs as shown in Figure 8–87 in relation to the clock.

FIGURE 8–87

SECTION 8–3 Master-Slave Flip-Flops

19. Determine the Q output in Figure 8–88. Assume that Q is initially LOW.

20. Show the Q output of flip-flop B in Figure 8–89 in proper relation to the clock. The flip-flops are initially RESET.

FIGURE 8–88

FIGURE 8–89

21. Show the Q output of flip-flop B in Figure 8–90 in proper relation to the clock. The flip-flops are initially RESET.

FIGURE 8–90

22. For a master-slave J-K flip-flop with the inputs in Figure 8–84, develop the timing diagram showing Q starting in the LOW state. Assume the clock is active-HIGH.

23. The D input and a single clock pulse are shown in Figure 8–91. Compare the resulting Q outputs for positive edge-triggered, negative edge-triggered, and pulse-triggered master-slave flip-flops. The flip-flops are initially RESET.

FIGURE 8–91

SECTION 8–4   Flip-Flop Operating Characteristics

24. Typically, a manufacturer's data sheet specifies four different propagation delay times associated with a flip-flop. Name and describe each one.

25. The data sheet of a certain flip-flop specifies that the minimum HIGH time for the clock pulse is 30 ns and the minimum LOW time is 37 ns. What is the maximum operating frequency?

26. The flip-flop in Figure 8–92 is initially RESET. Show the relation between the Q output and the clock pulse if propagation delay t_{PLH} (clock to Q) is 8 ns.

FIGURE 8–92

27. The direct current required by a particular flip-flop that operates on a +5 V dc source is found to be 10 mA. A certain digital device uses 15 of these flip-flops. Determine the current capacity required for the +5 V dc supply and the total power dissipation of the system.

28. For the circuit in Figure 8–93, determine the maximum frequency of the clock signal for reliable operation if the set-up time for each flip-flop is 2 ns and the propagation delays (t_{PLH} and t_{PHL}) from clock to output are 5 ns for each flip-flop.

FIGURE 8–93

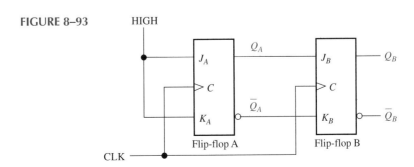

SECTION 8–5   Flip-Flop Applications

29. A D flip-flop is connected as shown in Figure 8–94. Determine the Q output in relation to the clock. What specific function does this device perform?

30. For the circuit in Figure 8–93, develop a timing diagram for eight clock pulses, showing the Q_A and Q_B outputs in relation to the clock.

FIGURE 8–94

SECTION 8–6   One-Shots

31. Determine the pulse width of a 74121 one-shot if the external resistor is 3.3 kΩ and the external capacitor is 2000 pF.

32. An output pulse of 5 μs duration is to be generated by a 74LS122 one-shot. Using a capacitor of 10,000 pF, determine the value of external resistance required.

SECTION 8–7   The 555 Timer

33. Create a one-shot, using a 555 timer that will produce a 0.25 s output pulse.

34. A 555 timer is configured to run as an astable multivibrator as shown in Figure 8–95. Determine its frequency.

FIGURE 8–95

35. Determine the values of the external resistors for a 555 timer used as an astable multivibrator with an output frequency of 20 kHz, if the external capacitor C is 0.002 μF and the duty cycle is to be approximately 75%.

 ## SECTION 8–8   Troubleshooting

36. The flip-flop in Figure 8–96 is tested under all input conditions as shown. Is it operating properly? If not, what is the most likely fault?

37. A 74HC00 quad NAND gate IC is used to construct a gated S-R latch on a protoboard in the lab as shown in Figure 8–97. The schematic in part (a) is used to connect the circuit in part (b). When you try to operate the latch, you find that the Q output stays HIGH no matter what the inputs are. Determine the problem.

38. Determine if the flip-flop in Figure 8–98 is operating properly, and if not, identify the most probable fault.

(a)

(b)

(c)

(d)

FIGURE 8–96

(a)

(b)

FIGURE 8–97

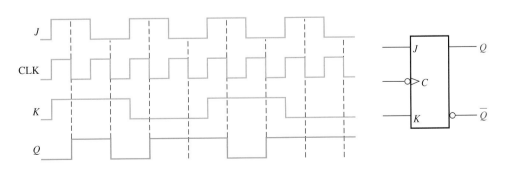

FIGURE 8–98

39. The parallel data storage circuit in Figure 8–39 does not operate properly. To check it out, you first make sure that V_{CC} and ground are connected, and then you apply LOW levels to all the D inputs and pulse the clock line. You check the Q outputs and find them all to be LOW; so far, so good. Next you apply HIGHs to all the D inputs and again pulse the clock line. When you check the Q outputs, they are still all LOW. What is the problem, and what procedure will you use to isolate the fault to a single device?

40. The flip-flop circuit in Figure 8–99(a) is used to generate a binary count sequence. The gates form a decoder that is supposed to produce a HIGH when a binary zero or a binary three state occurs (00 or 11). When you check the Q_A and Q_B outputs, you get the display shown in part (b), which reveals glitches on the decoder output (X) in addition to the correct pulses. What is causing these glitches, and how can you eliminate them?

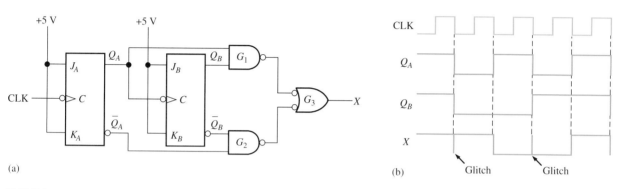

(a)

(b)

FIGURE 8–99

41. Determine the Q_A, Q_B, and X outputs over six clock pulses in Figure 8–99(a) for each of the following faults in the TTL circuits. Start with both Q_A and Q_B LOW.

    (a) J_A input open   **(b)** K_B input open

    (c) Q_B output open   **(d)** clock input to flip-flop B shorted

    (e) gate G_2 output open

42. Two 74121 one-shots are connected on a circuit board as shown in Figure 8–100. After observing the oscilloscope display, do you conclude that the circuit is operating properly? If not, what is the most likely problem?

FIGURE 8–100

SECTION 8–9 Digital System Application

43. Use 555 timers to implement the 4 s and the 25 s one-shots for the timing circuits portion of the system. The trigger input to the 555 cannot stay LOW after its negative-going transition, so you will have to develop a circuit to produce very short negative-going pulses to trigger the long and short timers when the system goes into each state.

44. Modify the 10 kHz oscillator circuit in the timing portion of the traffic light control system to produce an output with a 25% duty cycle.

Special Design Problems

45. Devise a basic counting circuit that produces a binary sequence from zero through seven by using negative edge-triggered J-K flip-flops.

46. In the shipping department of a softball factory, the balls roll down a conveyor and through a chute single file into boxes for shipment. Each ball passing through the chute activates a switch circuit that produces an electrical pulse. The capacity of each box is 32 balls. Design a logic circuit to indicate when a box is full so that an empty box can be moved into position.

47. List the design changes that would be necessary in the traffic light control system to add a 15 s left turn arrow for the main street. The turn arrow will occur after the red light and prior to the green light. Modify the state diagram from Chapter 6 (Figure 6–67) to show these changes.

Electronics Workbench Troubleshooting

48. Open file PR08_48.EWB and test the latches to determine which one is faulty.

49. Open file PR08_49.EWB and test the J-K flip-flops to determine which one is faulty.

50. Open file PR08_50.EWB and test the D flip-flops to determine which one is faulty.

51. Open file PR08_51.EWB and test the one-shots to determine which one is faulty.

52. Open file PR08_52.EWB and test the divide-by-four circuit to determine if there is a fault. If there is a fault, identify it if possible.

■ ANSWERS TO SECTION REVIEWS

SECTION 8–1

1. Three types of latches are S-R, gated S-R, and gated D.

2. $SR = 00$, NC; $SR = 01$, $Q = 0$; $SR = 10$, $Q = 1$; $SR = 11$, invalid

3. $Q = 1$

SECTION 8–2

1. The output of a gated S-R latch can change any time the gate enable (EN) input is active. The output of an edge-triggered S-R flip-flop can change only on the triggering edge of a clock pulse.

2. The J-K flip-flop does not have an invalid state as does the S-R flip-flop.

3. Output Q goes HIGH on the trailing edge of the first clock pulse, LOW on the trailing edge of the second pulse, HIGH on the trailing edge of the third pulse, and LOW on the trailing edge of the fourth pulse.

SECTION 8–3

1. In the pulse-triggered flip-flop (master-slave), a data bit goes into the master section on the leading edge of a clock pulse and is transferred to the output (slave) on the trailing edge. In the edge-triggered flip-flop, a data bit goes into the flip-flop and appears on the output on the same clock edge.

2. **(a)** Nothing happens for the positive edge-triggered flip-flop.

(b) The output of a pulse-triggered master-slave flip-flop will change.

SECTION 8–4

1. **(a)** Set-up time is the time required for input data to be present before the triggering edge of the clock pulse.

   (b) Hold time is the time required for data to remain on input after the triggering edge of the clock pulse.

2. The 74AHC74 can be operated at the highest frequency, according to Table 8–6.

SECTION 8–5

1. A group of data storage flip-flops is a register.

2. For divide-by-2 operation, the flip-flop must toggle ($J = 1$, $K = 1$).

3. Six flip-flops are used in a divide-by-64 device.

SECTION 8–6

1. A nonretriggerable one-shot times out before it can respond to another trigger input. A retriggerable one-shot responds to each trigger input.

2. Pulse width is set with external R and C components.

SECTION 8–7

1. An astable has no stable state. A monostable has one stable state.

2. Duty cycle = (15 ms/20 ms)100% = 75%

SECTION 8–8

1. Yes, a negative edge-triggered D flip-flop can be used.

2. An astable multivibrator using a 555 timer can be used to provide the clock.

SECTION 8–9

1. The switches are used to simulate the 2-bit Gray code input.

2. The 4 s interval and the 25 s interval occur at different states.

3. none

■ **ANSWERS TO RELATED PROBLEMS FOR EXAMPLES**

8–1 The Q output is the same as shown in Figure 8–5(b).

8–2 See Figure 8–101.   **8–3** See Figure 8–102.

8–4 See Figure 8–103.   **8–5** See Figure 8–104.

8–6 See Figure 8–105.   **8–7** See Figure 8–106.

8–8 See Figure 8–107.   **8–9** See Figure 8–108.

FIGURE 8–101

FIGURE 8–102

FIGURE 8–103

FIGURE 8–104

FIGURE 8–105

FIGURE 8–106

FIGURE 8–107

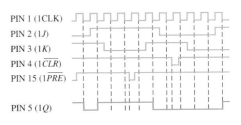

FIGURE 8–108

8–10 See Figure 8–109.

8–11 $2^5 = 32$. Five flip-flops are required.

8–12 Sixteen states require four flip-flops ($2^4 = 16$).

8–13 $C_{EXT} = 7143$ pF (6800 pF ∥ 330 pF) connected from CX to RX/CX of the 74121.

8–14 $C_{EXT} = 560$ pF, $R_{EXT} = 27$ kΩ. See Figure 8–110.

8–15 $R_1 = 91$ kΩ **8–16** Duty cycle $\cong 32\%$

FIGURE 8–109

FIGURE 8–110

1. (a) **2.** (c) **3.** (d) **4.** (b) **5.** (d) **6.** (d) **7.** (a) **8.** (b)

9. (d) **10.** (d) **11.** (c) **12.** (f)

9

COUNTERS

Computer Simulation
of Digital Circuits:
Electronics Workbench (EWB) Tutorial at
http://www.prenhall.com/floyd

■ CHAPTER OBJECTIVES

❑ Describe the difference between an asynchronous
  and a synchronous counter
❑ Analyze counter timing diagrams
❑ Analyze counter circuits
❑ Explain how propagation delays affect the opera-
  tion of a counter
❑ Determine the modulus of a counter
❑ Modify the modulus of a counter
❑ Recognize the difference between a 4-bit binary
  counter and a decade counter
❑ Use an up/down counter to generate forward and re-
  verse binary sequences
❑ Determine the sequence of a counter
❑ Use IC counters in various applications
❑ Design a counter that will have any specified se-
  quence of states
❑ Use cascaded counters to achieve a higher modulus
❑ Use logic gates to decode any given state of a
  counter
❑ Eliminate glitches in counter decoding
❑ Explain how a digital clock operates
❑ Troubleshoot counters for various types of faults
❑ Interpret counter logic symbols that use dependency
  notation
❑ Apply a counter in a system application

■ CHAPTER OVERVIEW

As you learned in Chapter 8, flip-flops can be connected together to perform counting operations. Such a group of flip-flops is a counter. The number of flip-flops used and the way in which they are connected determine the number of states (called the modulus) and also the specific sequence of states that the counter goes through during each complete cycle.

Counters are classified into two broad categories according to the way they are clocked: asynchronous and synchronous. In asynchronous counters, commonly called *ripple counters*, the first flip-flop is clocked by the external clock pulse and then each successive flip-flop is clocked by the output of the preceding flip-flop. In synchronous counters, the clock input is connected to all of the flip-flops so that they are clocked simultaneously. Within each of these two categories, counters are classified primarily by the type of sequence, the number of states, or the number of flip-flops in the counter.

 SPECIFIC DEVICES

74LS93A	74HC160	74HC161
74HC163	74LS190	74LS247

■ DIGITAL SYSTEM APPLICATION

This Digital System Application illustrates the concepts taught in this chapter. The system application in Section 9–10 continues the traffic light control system from the last three chapters. The focus of this chapter is the sequential logic portion of the system that produces the traffic light sequence based on inputs from the timing circuits and the vehicle sensor.

9–1 ■ ASYNCHRONOUS COUNTER OPERATION

The term asynchronous refers to events that do not have a fixed time relationship with each other and, generally, do not occur at the same time. An asynchronous counter is one in which the flip-flops (FF) within the counter do not change states at exactly the same time because they do not have a common clock pulse.

After completing this section, you should be able to

■ Describe the operation of a 2-bit asynchronous binary counter ■ Describe the operation of a 3-bit asynchronous binary counter ■ Define *ripple* in relation to counters ■ Describe the operation of an asynchronous decade counter ■ Develop counter timing diagrams ■ Discuss the 74LS93A 4-bit asynchronous binary counter

A 2-Bit Asynchronous Binary Counter

Figure 9–1 shows a 2-bit counter connected for **asynchronous** operation. Notice that the clock (CLK) is applied to the clock input (C) of only the first flop-flop, FF0. The second flip-flop, FF1, is triggered by the $\overline{Q}_0$ output of FF0. FF0 changes state at the positive-going edge of each clock pulse, but FF1 changes only when triggered by a positive-going transition of the $\overline{Q}_0$ output of FF0. Because of the inherent propagation delay time through a flip-flop, a transition of the input clock pulse (CLK) and a transition of the $\overline{Q}_0$ output of FF0 can never occur at exactly the same time. Therefore, the two flip-flops are never simultaneously triggered, so the counter operation is asynchronous.

FIGURE 9–1
A 2-bit asynchronous binary counter.

The Timing Diagram   Let's examine the basic operation of the **asynchronous counter** of Figure 9–1 by applying four clock pulses to FF0 and observing the Q output of each flip-flop. Figure 9–2 illustrates the changes in the state of the flip-flop outputs in response to the clock pulses. Both flip-flops are connected for toggle operation ($J = 1, K = 1$) and are assumed to be initially RESET (Q LOW).

The positive-going edge of CLK1 (clock pulse 1) causes the Q_0 output of FF0 to go HIGH, as shown in Figure 9–2. At the same time the $\overline{Q}_0$ output goes LOW, but it has no

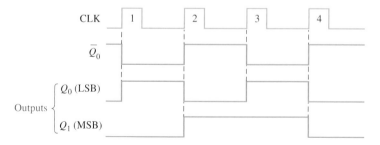

FIGURE 9–2
Timing diagram for the counter of Figure 9–1. As in previous chapters, output waveforms are shown in green.

effect on FF1 because a positive-going transition must occur to trigger the flip-flop. After the leading edge of CLK1, $Q_0 = 1$ and $Q_1 = 0$. The positive-going edge of CLK2 causes Q_0 to go LOW. Output $\overline{Q}_0$ goes HIGH and triggers FF1, causing Q_1 to go HIGH. After the leading edge of CLK2, $Q_0 = 0$ and $Q_1 = 1$. The positive-going edge of CLK3 causes Q_0 to go HIGH again. Output $\overline{Q}_0$ goes LOW and has no effect on FF1. Thus, after the leading edge of CLK3, $Q_0 = 1$ and $Q_1 = 1$. The positive-going edge of CLK4 causes Q_0 to go LOW, while $\overline{Q}_0$ goes HIGH and triggers FF1, causing Q_1 to go LOW. After the leading edge of CLK4, $Q_0 = 0$ and $Q_1 = 0$. The counter has now recycled to its original state (both flip-flops are RESET).

In the timing diagram, the waveforms of the Q_0 and Q_1 outputs are shown relative to the clock pulses as illustrated in Figure 9–2. For simplicity, the transitions of Q_0, Q_1, and the clock pulses are shown as simultaneous even though this is an asynchronous counter. There is, of course, some small delay between the CLK and the Q_0 transition and between the $\overline{Q}_0$ transition and the Q_1 transition.

Note in Figure 9–2 that the 2-bit counter exhibits four different states, as you would expect with two flip-flops ($2^2 = 4$). Also, notice that if Q_0 represents the least significant bit (LSB) and Q_1 represents the most significant bit (MSB), the sequence of counter states represents a sequence of binary numbers as listed in Table 9–1. *In digital logic, Q_0 is always the LSB unless otherwise specified.*

TABLE 9–1
Binary state sequence for the counter in Figure 9–1.

Clock Pulse	Q_1	Q_0
Initially	0	0
1	0	1
2	1	0
3	1	1
4 (recycles)	0	0

Since it goes through a binary **sequence,** the counter in Figure 9–1 is a binary counter. It actually counts the number of clock pulses up to three, and on the fourth pulse it recycles to its original state ($Q_0 = 0$, $Q_1 = 0$). The term **recycle** is commonly applied to counter operation; it refers to the transition of the counter from its *final state* back to its *original state.*

A 3-Bit Asynchronous Binary Counter

A 3-bit asynchronous binary counter is shown in Figure 9–3(a). The basic operation is the same as that of the 2-bit counter just discussed, except that the 3-bit counter has eight states, due to its three flip-flops. A timing diagram is shown in Figure 9–3(b) for eight clock pulses. Notice that the counter in Figure 9–3 progresses through a binary count of zero through seven and then recycles to the zero state. This counter sequence is listed in Table 9–2.

TABLE 9–2
Binary state sequence for a 3-bit binary counter.

Clock Pulse	Q_2	Q_1	Q_0
Initially	0	0	0
1	0	0	1
2	0	1	0
3	0	1	1
4	1	0	0
5	1	0	1
6	1	1	0
7	1	1	1
8 (recycles)	0	0	0

(a)

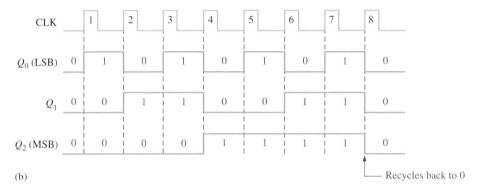

(b)

└─ Recycles back to 0

FIGURE 9–3
Three-bit asynchronous binary counter and its timing diagram for one cycle.

Propagation Delay Asynchronous counters are commonly referred to as **ripple coun-ters** for the following reason: The effect of the input clock pulse is first "felt" by FF0. This effect cannot get to FF1 immediately because of the propagation delay through FF0. Then there is the propagation delay through FF1 before FF2 can be triggered. Thus, the effect of an input clock pulse "ripples" through the counter, taking some time, due to propagation delays, to reach the last flip-flop.

To illustrate, notice that all three flip-flops in the counter of Figure 9–3 change state on the leading edge of CLK4. This ripple clocking effect is shown in Figure 9–4 for the first four clock pulses, with the propagation delays indicated. The HIGH-to-LOW transi-tion of Q_0 occurs one delay time (t_{PHL}) after the positive-going transition of the clock pulse. The HIGH-to-LOW transition of Q_1 occurs one delay time (t_{PHL}) after the positive-going transition of $\overline{Q}_0$. The LOW-to-HIGH transition of Q_2 occurs one delay time (t_{PHL}) after the positive-going transition of $\overline{Q}_1$. As you can see, FF2 is not triggered until two delay times after the positive-going edge of the clock pulse, CLK4. Thus, it takes three

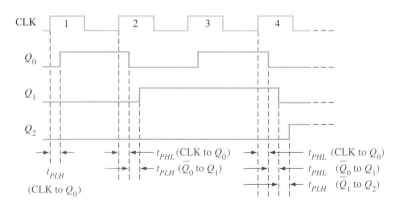

FIGURE 9–4
Propagation delays in a 3-bit asynchronous (ripple-clocked) binary counter.

propagation delay times for the effect of the clock pulse, CLK4, to ripple through the counter and change Q_2 from LOW to HIGH.

This cumulative delay of an asynchronous counter is a major disadvantage in many applications because it limits the rate at which the counter can be clocked and creates decoding problems. The maximum cumulative delay in a counter must be less than the period of the clock waveform.

EXAMPLE 9–1

A 4-bit asynchronous binary counter is shown in Figure 9–5(a). Each flip-flop is negative edge-triggered and has a propagation delay of 10 nanoseconds (ns). Develop a timing diagram showing the Q output of each flip-flop, and determine the total propagation delay time from the triggering edge of a clock pulse until a corresponding change can occur in the state of Q_3. Also determine the maximum clock frequency at which the counter can be operated.

(a)

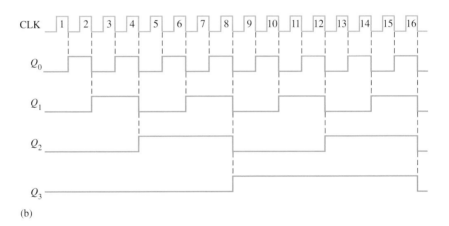

(b)

FIGURE 9–5
Four-bit asynchronous binary counter and its timing diagram.

Solution The timing diagram with delays omitted is as shown in Figure 9–5(b). For the total delay time, the effect of CLK8 or CLK16 must propagate through four flip-flops before Q_3 changes, so

$$t_{p(tot)} = 4 \times 10 \text{ ns} = \textbf{40 ns}$$

The maximum clock frequency is

$$f_{\max} = \frac{1}{t_{p(tot)}} = \frac{1}{40 \text{ ns}} = \textbf{25 MHz}$$

Related Problem Show the timing diagram if all of the flip-flops in Figure 9–5(a) are positive edge-triggered.

Asynchronous Decade Counters

The **modulus** of a counter is the number of unique states that the counter will sequence through. The maximum possible number of states (maximum modulus) of a counter is 2^n, where n is the number of flip-flops in the counter.

Counters can also be designed to have a number of states in their sequence that is less than the maximum of 2^n. The resulting sequence is called a **truncated sequence.**

One common modulus for counters with **truncated** sequences is ten (called MOD10). Counters with ten states in their sequence are called **decade counters.** A **decade** counter with a count sequence of zero (0000) through nine (1001) is a BCD decade counter because its ten-state sequence produces the BCD code. This type of counter is useful in display applications in which BCD is required for conversion to a decimal readout.

To obtain a truncated sequence, it is necessary to force the counter to recycle before going through all of its possible states. For example, the BCD decade counter must recycle back to the 0000 state after the 1001 state. A decade counter requires four flip-flops (three flip-flops are insufficient because $2^3 = 8$).

Let's use a 4-bit asynchronous counter such as the one in Figure 9–5(a) and modify its sequence to illustrate the principle of truncated counters. One way to make the counter recycle after the count of nine (1001) is to decode count ten (1010) with a NAND gate and connect the output of the NAND gate to the clear ($\overline{CLR}$) inputs of the flip-flops, as shown in Figure 9–6(a).

(a)

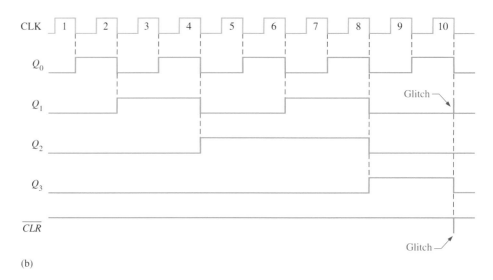

(b)

FIGURE 9–6

An asynchronously clocked decade counter with asynchronous recycling.

Partial Decoding Notice in Figure 9–6(a) that only Q_1 and Q_3 are connected to the NAND gate inputs. This arrangement is an example of *partial decoding,* in which the two unique states ($Q_1 = 1$ and $Q_3 = 1$) are sufficient to decode the count of ten because none of the other states (zero through nine) have both Q_1 and Q_3 HIGH at the same time. When the counter goes into count ten (1010), the decoding gate output goes LOW and asynchronously resets all the flip-flops.

The resulting timing diagram is shown in Figure 9–6(b). Notice that there is a glitch on the Q_1 waveform. The reason for this glitch is that Q_1 must first go HIGH before the count of ten can be decoded. Not until several nanoseconds after the counter goes to the count of ten does the output of the decoding gate go LOW (both inputs are HIGH). Thus, the counter is in the 1010 state for a short time before it is reset to 0000, thus producing the glitch on Q_1 and the resulting glitch on the $\overline{CLR}$ line that resets the counter.

Other truncated sequences can be implemented in a similar way, as Example 9–2 shows.

EXAMPLE 9–2

Show how an asynchronous counter can be implemented having a modulus of twelve with a straight binary sequence from 0000 through 1011.

Solution Since three flip-flops can produce a maximum of eight states, four flip-flops are required to produce any modulus greater than eight but less than or equal to sixteen.

When the counter gets to its last state, 1011, it must recycle back to 0000 rather than going to its normal next state of 1100, as illustrated in the following sequence chart:

Q_3	Q_2	Q_1	Q_0	
0	0	0	0	←
.	.	.	.	
.	.	.	.	Recycles
.	.	.	.	
1	0	1	1	
1	1	0	0	← Normal next state

Observe that Q_0 and Q_1 both go to 0 anyway, but Q_2 and Q_3 must be forced to 0 on the twelfth clock pulse. Figure 9–7(a) shows the modulus-12 counter. The NAND gate partially decodes count twelve (1100) and resets flip-flop 2 and flip-flop 3. Thus, on the twelfth clock pulse, the counter is forced to recycle from count eleven to count zero, as shown in the timing diagram of Figure 9–7(b) on the next page. (It is in count twelve for only a few nanoseconds before it is reset by the glitch on $\overline{CLR}$.)

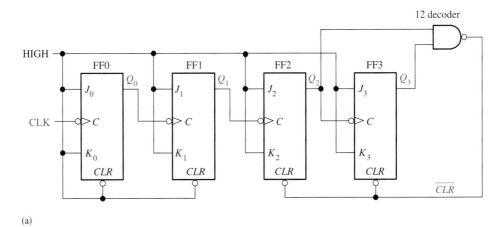

(a)

FIGURE 9–7
Asynchronously clocked modulus-12 counter with asynchronous recycling.

FIGURE 9–7
(Continued)

Related Problem How can the counter in Figure 9–7(a) be modified to make it a modulus-13 counter?

A 4-Bit Asynchronous Binary Counter

The 74LS93A is an example of a specific integrated circuit asynchronous counter. As the logic diagram in Figure 9–8 shows, this device actually consists of a single flip-flop and a 3-bit asynchronous counter. This arrangement is for flexibility. It can be used as a divide-by-2 device if only the single flip-flop is used, or it can be used as a modulus-8 counter if only the 3-bit counter portion is used. This device also provides gated reset inputs, *RO(1)* and *RO(2)*. When both of these inputs are HIGH, the counter is reset to the 0000 state by $\overline{CLR}$.

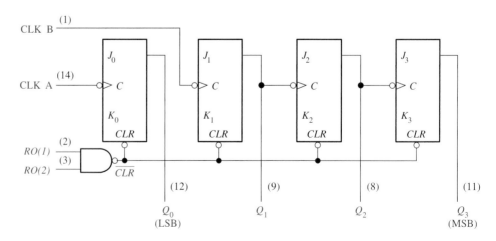

FIGURE 9–8
The 74LS93A 4-bit asynchronous binary counter logic diagram. (Pin numbers are in parentheses, and all J and K inputs are internally connected HIGH.)

Additionally, the 74LS93A can be used as a 4-bit modulus-16 counter (counts 0 through 15) by connecting the Q_0 output to the CLK B input as shown in Figure 9–9(a). It can also be configured as a decade counter (counts 0 through 9) with asynchronous recycling by using the gated reset inputs for partial decoding of count ten, as shown in Figure 9–9(b).

(a) 74LS93A connected as a modulus-16 counter

(b) 74LS93A connected as a decade counter

FIGURE 9–9

Two configurations of the 74LS93A asynchronous counter. (The qualifying label, CTR DIV n, indicates a counter with n states.)

EXAMPLE 9–3

Show how the 74LS93A can be used as a modulus-12 counter.

Solution Use the gated reset inputs, *RO(1)* and *RO(2)*, to partially decode count 12 (remember, there is an internal NAND gate associated with these inputs). The count-12 decoding is accomplished by connecting Q_3 to *RO(1)* and Q_2 to *RO(2)*, as shown in Figure 9–10. Output Q_0 is connected to CLK B to create a 4-bit counter.

FIGURE 9–10

74LS93A connected as a modulus-12 counter.

Immediately after the counter goes to count 12 (1100), it is reset to 0000. The recycling, however, results in a glitch on Q_2 because the counter must go into the 1100 state for several nanoseconds before recycling.

Related Problem Show how the 74LS93A can be connected as a modulus-13 counter.

SECTION 9–1 REVIEW

1. What does the term *asynchronous* mean in relation to counters?
2. How many states does a modulus-14 counter have? What is the minimum number of flip-flops required?

9–2 ■ SYNCHRONOUS COUNTER OPERATION

The term synchronous refers to events that have a fixed time relationship with each other. With respect to counter operation, synchronous means that all the flip-flops in the counter are clocked at the same time by a common clock pulse.

After completing this section, you should be able to

■ Describe the operation of a 2-bit synchronous binary counter ■ Describe the operation of a 3-bit synchronous binary counter ■ Describe the operation of a 4-bit synchronous binary counter ■ Describe the operation of a synchronous decade counter ■ Develop counter timing diagrams ■ Discuss the 74HC163 4-bit binary counter and the 74HC160 BCD decade counter

A 2-Bit Synchronous Binary Counter

Figure 9–11 shows a 2-bit **synchronous** binary counter. Notice that an arrangement different from that for the asynchronous counter must be used for the J_1 and K_1 inputs of FF1 in order to achieve a binary sequence.

FIGURE 9–11
A 2-bit synchronous binary counter.

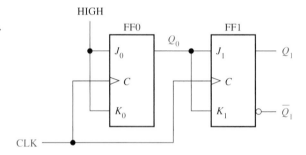

The operation of this **synchronous counter** is as follows: First, assume that the counter is initially in the binary 0 state; that is, both flip-flops are RESET. When the positive edge of the first clock pulse is applied, FF0 will toggle and Q_0 will therefore go HIGH. What happens to FF1 at the positive-going edge of CLK1? To find out, let's look at the input conditions of FF1. Inputs J_1 and K_1 are both LOW because Q_0, to which they are connected, has not yet gone HIGH. Remember, there is a propagation delay from the triggering edge of the clock pulse until the Q output actually makes a transition. So, $J = 0$ and $K = 0$ when the leading edge of the first clock pulse is applied. This is a no-change condition, and therefore FF1 does not change state. A timing detail of this portion of the counter operation is shown in Figure 9–12(a).

After CLK1, $Q_0 = 1$ and $Q_1 = 0$ (which is the binary 1 state). When the leading edge of CLK2 occurs, FF0 will toggle and Q_0 will go LOW. Since FF1 has a HIGH ($Q_0 = 1$) on its J_1 and K_1 inputs at the triggering edge of this clock pulse, the flip-flop toggles and Q_1 goes HIGH. Thus, after CLK2, $Q_0 = 0$ and $Q_1 = 1$ (which is a binary 2 state). The timing detail for this condition is shown in Figure 9–12(b).

When the leading edge of CLK3 occurs, FF0 again toggles to the SET state ($Q_0 = 1$), and FF1 remains SET ($Q_1 = 1$) because its J_1 and K_1 inputs are both LOW ($Q_0 = 0$). After this triggering edge, $Q_0 = 1$ and $Q_1 = 1$ (which is a binary 3 state). The timing detail is shown in Figure 9–12(c).

Finally, at the leading edge of CLK4, Q_0 and Q_1 go LOW because they both have a toggle condition on their J and K inputs. The timing detail is shown in Figure 9–12(d). The counter has now recycled to its original state, binary 0.

The complete timing diagram for the counter in Figure 9–11 is shown in Figure 9–13. Notice that all the waveform transitions appear coincident; that is, the propagation

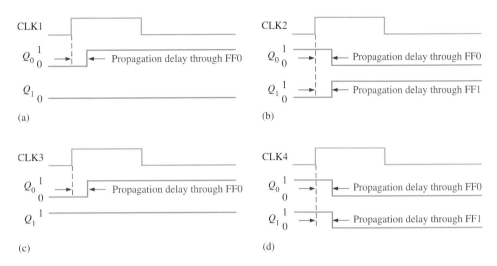

FIGURE 9–12

Timing details for the 2-bit synchronous counter operation (the propagation delays of both flip-flops are assumed to be equal).

FIGURE 9–13

Timing diagram for the counter of Figure 9–11.

delays are not indicated. Although the delays are an important factor in the synchronous counter operation, in an overall timing diagram they are normally omitted for simplicity. Major waveform relationships resulting from the normal operation of a circuit can be conveyed completely without showing small delay and timing differences. However, in high-speed digital circuits, these small delays are an important consideration in design and troubleshooting.

A 3-Bit Synchronous Binary Counter

A 3-bit synchronous binary counter is shown in Figure 9–14, and its timing diagram is shown in Figure 9–15. You can understand this counter operation by examining its sequence of states as shown in Table 9–3.

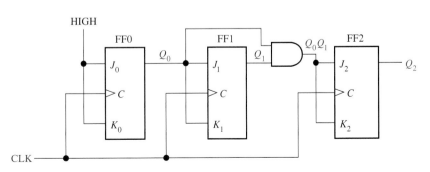

FIGURE 9–14

A 3-bit synchronous binary counter.

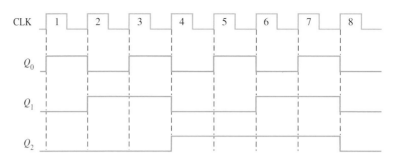

FIGURE 9–15
Timing diagram for the counter of Figure 9–14.

TABLE 9–3
Binary state sequence for a 3-bit binary counter.

Clock Pulse	Q_2	Q_1	Q_0
Initially	0	0	0
1	0	0	1
2	0	1	0
3	0	1	1
4	1	0	0
5	1	0	1
6	1	1	0
7	1	1	1
8 (recycles)	0	0	0

First, let's look at Q_0. Notice that Q_0 changes on each clock pulse as the counter progresses from its original state to its final state and then back to its original state. To produce this operation, FF0 must be held in the toggle mode by constant HIGHs on its J_0 and K_0 inputs. Notice that Q_1 goes to the opposite state following each time Q_0 is a 1. This change occurs at CLK2, CLK4, CLK6, and CLK8. The CLK8 pulse causes the counter to recycle. To produce this operation, Q_0 is connected to the J_1 and K_1 inputs of FF1. When Q_0 is a 1 and a clock pulse occurs, FF1 is in the toggle mode and therefore changes state. The other times, when Q_0 is a 0, FF1 is in the no-change mode and remains in its present state.

Next, let's see how FF2 is made to change at the proper times according to the binary sequence. Notice that both times Q_2 changes state, it is preceded by the unique condition in which both Q_0 and Q_1 are HIGH. This condition is detected by the AND gate and applied to the J_2 and K_2 inputs of FF2. Whenever both Q_0 and Q_1 are HIGH, the output of the AND gate makes the J_2 and K_2 inputs of FF2 HIGH, and FF2 toggles on the following clock pulse. At all other times, the J_2 and K_2 inputs of FF2 are held LOW by the AND gate output, and FF2 does not change state.

A 4-Bit Synchronous Binary Counter

Figure 9–16(a) shows a 4-bit synchronous binary counter, and Figure 9–16(b) shows its timing diagram. This particular counter is implemented with negative edge-triggered flip-flops. The reasoning behind the J and K input control for the first three flip-flops is the same as previously discussed for the 3-bit counter. The fourth stage, FF3, changes only twice in the sequence. Notice that both of these transitions occur following the times that Q_0, Q_1, and Q_2 are all HIGH. This condition is decoded by AND gate G_2 so that when a clock pulse occurs, FF3 will change state. For all other times the J_3 and K_3 inputs of FF3 are LOW, and it is in a no-change condition.

(a)

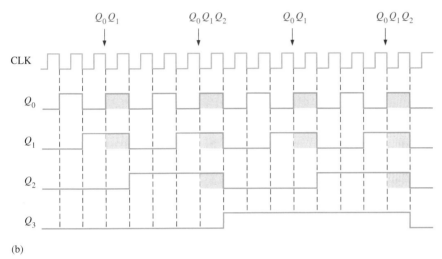

(b)

FIGURE 9–16

A 4-bit synchronous binary counter and timing diagram. Points where the AND gate outputs are HIGH are indicated by the shaded areas.

A 4-Bit Synchronous Decade Counter

As you know, a BCD decade counter exhibits a truncated binary sequence and goes from 0000 through the 1001 state. Rather than going from the 1001 state to the 1010 state, it re-cycles to the 0000 state. A synchronous BCD decade counter is shown in Figure 9–17. The timing diagram for the decade counter is shown in Figure 9–18.

The counter operation can be understood by examining the sequence of states in Table 9–4 and by following the implementation in Figure 9–17. First, notice that FF0 (Q_0) toggles on each clock pulse, so the logic equation for its J_0 and K_0 inputs is

$$J_0 = K_0 = 1$$

This equation is implemented by connecting J_0 and K_0 to a constant HIGH level.

Next, notice in Table 9–4 that FF1 (Q_1) changes on the next clock pulse each time $Q_0 = 1$ and $Q_3 = 0$, so the logic equation for the J_1 and K_1 inputs is

$$J_1 = K_1 = Q_0\overline{Q}_3$$

This equation is implemented by ANDing Q_0 and $\overline{Q}_3$ and connecting the gate output to the J_1 and K_1 inputs of FF1.

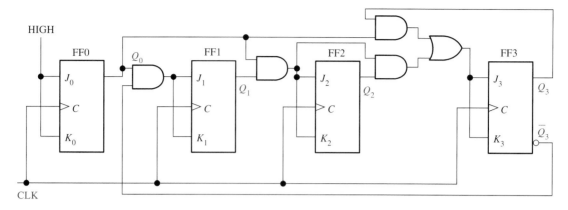

FIGURE 9–17
A synchronous BCD decade counter.

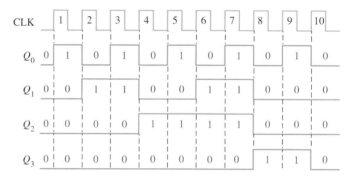

FIGURE 9–18
Timing diagram for the BCD decade counter (Q_0 is the LSB).

TABLE 9–4
States of a BCD decade counter.

Clock Pulse	Q_3	Q_2	Q_1	Q_0
Initially	0	0	0	0
1	0	0	0	1
2	0	0	1	0
3	0	0	1	1
4	0	1	0	0
5	0	1	0	1
6	0	1	1	0
7	0	1	1	1
8	1	0	0	0
9	1	0	0	1
10 (recycles)	0	0	0	0

Flip-flop 2 (Q_2) changes on the next clock pulse each time both $Q_0 = 1$ and $Q_1 = 1$. This requires an input logic equation as follows:

$$J_2 = K_2 = Q_0Q_1$$

This equation is implemented by ANDing Q_0 and Q_1 and connecting the gate output to the J_2 and K_2 inputs of FF2.

Finally, FF3 (Q_3) changes to the opposite state on the next clock pulse each time $Q_0 = 1$, $Q_1 = 1$, and $Q_2 = 1$ (state 7), or when $Q_0 = 1$ and $Q_3 = 1$ (state 9). The equation for this is as follows:

$$J_3 = K_3 = Q_0Q_1Q_2 + Q_0Q_3$$

This function is implemented with the AND/OR logic connected to the J_3 and K_3 inputs of FF3 as shown in the logic diagram in Figure 9–17. Notice that the only difference between this decade counter and the modulus-16 binary counter in Figure 9–16 is the Q_0Q_3 AND gate and the OR gate; this arrangement detects the occurrence of the 1001 state and causes the counter to recycle properly on the next clock pulse.

A 4-Bit IC Synchronous Binary Counter

The 74HC163 is an example of an integrated circuit 4-bit synchronous binary counter. A logic symbol is shown in Figure 9–19 with pin numbers in parentheses. This counter has several features in addition to the basic functions previously discussed for the general synchronous binary counter.

FIGURE 9–19

The 74HC163 4-bit synchronous binary counter. (The qualifying label CTR DIV 16 indicates a counter with sixteen states.)

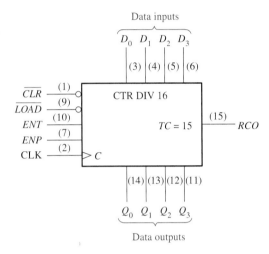

First, the counter can be synchronously preset to any 4-bit binary number by applying the proper levels to the parallel data inputs. When a LOW is applied to the $\overline{LOAD}$ input, the counter will assume the state of the data inputs on the next clock pulse. Thus, the counter sequence can be started with any 4-bit binary number.

Also, there is an active-LOW clear input ($\overline{CLR}$), which synchronously resets all four flip-flops in the counter. There are two enable inputs, *ENP* and *ENT*. These inputs must both be HIGH for the counter to sequence through its binary states. When at least one input is LOW, the counter is disabled. The ripple clock output (*RCO*) goes HIGH when the counter reaches a **terminal count** of fifteen (*TC* = 15). This output, in conjunction with the enable inputs, allows these counters to be cascaded for higher count sequences, as will be discussed later.

Figure 9–20 shows a timing diagram of this counter being preset to twelve (1100) and then counting up to its terminal count, fifteen (1111). Input D_0 is the least significant input bit, and Q_0 is the least significant output bit.

Let's examine this timing diagram in detail. This will aid you in interpreting timing diagrams found later in this chapter or on manufacturers' data sheets. To begin, the LOW level pulse on the $\overline{CLR}$ input causes all the outputs (Q_0, Q_1, Q_2, and Q_3) to go LOW.

Next, the LOW level pulse on the $\overline{LOAD}$ input synchronously enters the data on the data inputs (D_0, D_1, D_2, and D_3) into the counter. These data appear on the *Q* outputs at the time of the first positive-going clock edge after $\overline{LOAD}$ goes LOW. This is the preset operation. In this particular example, Q_0 is LOW, Q_1 is LOW, Q_2 is HIGH, and Q_3 is HIGH. This, of course, is a binary 12 (Q_0 is the LSB).

The counter now advances through states 13, 14, and 15 on the next three positive-going clock edges. It then recycles to 0, 1, 2 on the following clock pulses. Notice that both

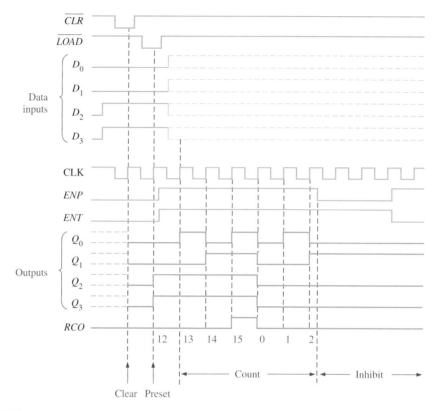

FIGURE 9–20
Timing example for a 74HC163.

ENP and *ENT* inputs are HIGH during the state sequence. When *ENP* goes LOW, the counter is inhibited and remains in the binary 2 state.

A Synchronous BCD Decade Counter

The 74HC160 has the same inputs and outputs as the 74HC163 binary counter previously discussed. It can be preset to any BCD count by the use of the data inputs and a LOW on the $\overline{LOAD}$ input. A LOW on the asynchronous $\overline{CLR}$ will reset the counter. The enable inputs *ENP* and *ENT* must both be HIGH for the counter to advance through its sequence of states in response to a positive transition on the CLK input. As in the 74HC163, the enable inputs in conjunction with the ripple clock output *RCO* (terminal count of 1001) provide for cascading several decade counters. Figure 9–21 shows the logic symbol for the

FIGURE 9–21
The 74HC160 synchronous BCD decade counter. (The qualifying label CTR DIV 10 indicates a counter with ten states.)

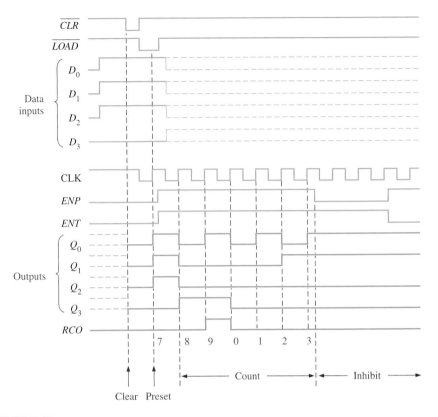

FIGURE 9–22
Timing example for a 74HC160.

74HC160 counter, and Figure 9–22 is a timing diagram showing the counter being preset to count 7 (0111). Cascaded counters will be discussed in Section 9–5.

SECTION 9–2 REVIEW

1. How does a synchronous counter differ from an asynchronous counter?
2. Explain the function of the preset feature of counters such as the 74HC160 and the 74HC163.
3. Describe the purpose of the *ENP* and *ENT* inputs and the *RCO* output for the two specific counters introduced in this section.

9–3 ■ UP/DOWN SYNCHRONOUS COUNTERS

An up/down counter is one that is capable of progressing in either direction through a certain sequence. An up/down counter, sometimes called a bidirectional counter, can have any specified sequence of states. A 3-bit binary counter that advances upward through its sequence (0, 1, 2, 3, 4, 5, 6, 7) and then can be reversed so that it goes through the sequence in the opposite direction (7, 6, 5, 4, 3, 2, 1, 0) is an illustration of up/down sequential operation.

After completing this section, you should be able to

■ Explain the basic operation of an up/down counter   ■ Discuss the 74LS190 up/down decade counter

In general, most **up/down counters** can be reversed at any point in their sequence. For instance, the 3-bit binary counter can be made to go through the following sequence:

$$\overbrace{0, 1, 2, 3, 4, 5,}^{\text{UP}} \underbrace{4, 3, 2,}_{\text{DOWN}} \overbrace{3, 4, 5, 6, 7,}^{\text{UP}} \underbrace{6, 5,}_{\text{DOWN}} \text{etc.}$$

Table 9–5 shows the complete up/down sequence for a 3-bit binary counter. The arrows indicate the state-to-state movement of the counter for both its UP and its DOWN modes of operation. An examination of Q_0 for both the up and down sequences shows that FF0 toggles on each clock pulse. So the J_0 and K_0 inputs of FF0 are

$$J_0 = K_0 = 1$$

For the up sequence, Q_1 changes state on the next clock pulse when $Q_0 = 1$. For the down sequence, Q_1 changes on the next clock pulse when $Q_0 = 0$. Thus, the J_1 and K_1 inputs of FF1 must equal 1 under the conditions expressed by the following equation:

$$J_1 = K_1 = (Q_0 \cdot \text{UP}) + (\overline{Q}_0 \cdot \text{DOWN})$$

For the up sequence, Q_2 changes state on the next clock pulse when $Q_0 = Q_1 = 1$. For the down sequence, Q_2 changes on the next clock pulse when $Q_0 = Q_1 = 0$. Thus, the J_2 and K_2 inputs of FF2 must equal 1 under the conditions expressed by the following equation:

$$J_2 = K_2 = (Q_0 \cdot Q_1 \cdot \text{UP}) + (\overline{Q}_0 \cdot \overline{Q}_1 \cdot \text{DOWN})$$

Each of the conditions for the J and K inputs of each flip-flop produces a toggle at the appropriate point in the counter sequence.

TABLE 9–5
Up/Down sequence for a 3-bit binary counter.

Clock Pulse	UP	Q_2	Q_1	Q_0	DOWN
0		0	0	0	
1		0	0	1	
2		0	1	0	
3		0	1	1	
4		1	0	0	
5		1	0	1	
6		1	1	0	
7		1	1	1	

FIGURE 9–23
A basic 3-bit up/down synchronous counter.

Figure 9–23 shows a basic implementation of a 3-bit up/down binary counter using the logic equations just developed for the J and K inputs of each flip-flop. Notice that the $UP/\overline{DOWN}$ control input is HIGH for UP and LOW for DOWN.

EXAMPLE 9–4

Show the timing diagram and determine the sequence of a 4-bit synchronous binary up/down counter if the clock and $UP/\overline{DOWN}$ control inputs have waveforms as shown in Figure 9–24(a). The counter starts in the all 0s state and is positive edge-triggered.

FIGURE 9–24

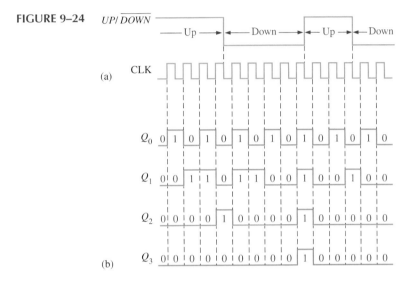

(a)

(b)

Solution   The timing diagram showing the Q outputs is shown in Figure 9–24(b). From these waveforms, the counter sequence is as shown in Table 9–6.

TABLE 9–6

Q_3	Q_2	Q_1	Q_0	
0	0	0	0	
0	0	0	1	
0	0	1	0	UP
0	0	1	1	
0	1	0	0	
0	0	1	1	
0	0	1	0	
0	0	0	1	DOWN
0	0	0	0	
1	1	1	1	
0	0	0	0	
0	0	0	1	UP
0	0	1	0	
0	0	0	1	DOWN
0	0	0	0	

Related Problem   Show the timing diagram if the $UP/\overline{DOWN}$ control waveform in Figure 9–24(a) is inverted.

An Up/Down Decade Counter

Figure 9–25 shows a logic diagram for the 74LS190, an example of an integrated circuit up/down synchronous counter. The direction of the count is determined by the level of the up/down input ($D/\overline{U}$). When this input is HIGH, the counter counts down; when it is LOW, the counter counts up. Also, this device can be preset to any desired BCD digit as determined by the states of the data inputs when the $\overline{LOAD}$ input is LOW.

FIGURE 9–25
The 74LS190 up/down synchronous decade counter.

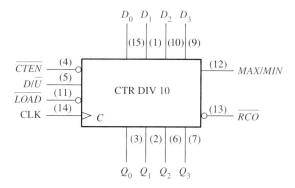

The *MAX/MIN* output produces a HIGH pulse when the terminal count nine (1001) is reached in the UP mode or when the terminal count zero (0000) is reached in the DOWN mode. This *MAX/MIN* output, along with the ripple clock output ($\overline{RCO}$) and the count enable input ($\overline{CTEN}$), is used when cascading counters. (Cascaded counters are discussed in Section 9–5.)

Figure 9–26 is an example timing diagram showing the 74LS190 counter preset to seven (0111) and then going through a count-up sequence followed by a count-down sequence. The *MAX/MIN* output is HIGH when the counter is in either the all-0s state (*MIN*) or the 1001 state (*MAX*).

FIGURE 9–26
Timing example for a 74LS190.

**SECTION 9–3
REVIEW**

1. A 4-bit up/down binary counter is in the DOWN mode and in the 1010 state. On the next clock pulse, to what state does the counter go?
2. What is the terminal count of a 4-bit binary counter in the UP mode? In the DOWN mode? What is the next state after the terminal count in the DOWN mode?

9–4 ■ DESIGN OF SYNCHRONOUS COUNTERS*

This section is recommended for those who want an introduction to counter design or to state machine design in general.

After completing this section, you should be able to

- Describe a general sequential circuit in terms of its basic parts and its input and outputs
- Develop a state diagram for a given sequence ■ Develop a next-state table for a specified counter sequence ■ Create a flip-flop transition table ■ Use the Karnaugh map method to derive the logic requirements for a synchronous counter ■ Implement a counter to produce a specified sequence of states

General Model of a Sequential Circuit

Before proceeding with a specific counter design technique, let's begin with a general definition of a **sequential circuit (state machine)**: A general sequential circuit consists of a combinational logic section and a memory section (flip-flops), as shown in Figure 9–27. In a clocked sequential circuit, there is a clock input to the memory section as indicated.

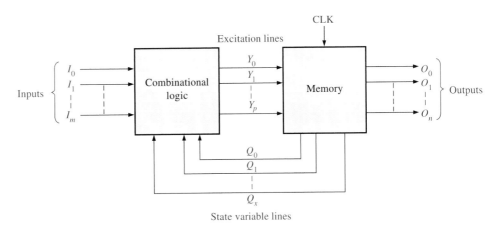

FIGURE 9–27
General clocked sequential circuit.

The information stored in the memory section, as well as the inputs to the combinational logic ($I_0, I_1, . . . , I_m$), is required for proper operation of the circuit. At any given time, the memory is in a state called the *present state* and will advance to a *next state* on a clock pulse as determined by conditions on the excitation lines ($Y_0, Y_1, . . . , Y_p$). The present state of the memory is represented by the state variables ($Q_0, Q_1, . . . , Q_x$). These state variables, along with the inputs ($I_0, I_1, . . . , I_m$), determine the system outputs ($O_0, O_1, . . . , O_n$).

---

*This section is optional and may be omitted without affecting the flow of material in the remainder of the book.

Not all sequential circuits have input and output variables as in the general model just discussed. However, all have excitation variables and state variables. Counters are a special case of clocked sequential circuits. In this section, a general design procedure for sequential circuits is applied to synchronous counters in a series of steps.

Step 1: State Diagram

A counter is first described by a **state diagram,** which shows the progression of states through which the counter advances when it is clocked. As an example, Figure 9–28 is a state diagram for a basic 3-bit Gray code counter. This particular circuit has no inputs other than the clock and no outputs other than the outputs taken off each flip-flop in the counter. You may wish to review the coverage of the Gray code in Chapter 2 at this time.

FIGURE 9–28
State diagram for a 3-bit Gray code counter.

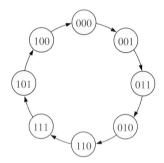

Step 2: Next-State Table

Once the sequential circuit is defined by a state diagram, the second step is to derive a next-state table, which lists each state of the counter (present state) along with the corresponding next state. *The next state is the state that the counter goes to from its present state upon application of a clock pulse.* The next-state table is derived from the state diagram and is shown in Table 9–7 for the 3-bit Gray code counter. Q_0 is the least significant bit.

TABLE 9–7
Next-state table for 3-bit Gray code counter.

Present State			Next State		
Q_2	Q_1	Q_0	Q_2	Q_1	Q_0
0	0	0	0	0	1
0	0	1	0	1	1
0	1	1	0	1	0
0	1	0	1	1	0
1	1	0	1	1	1
1	1	1	1	0	1
1	0	1	1	0	0
1	0	0	0	0	0

Step 3: Flip-Flop Transition Table

Table 9–8 is a transition table for the J-K flip-flop. All possible output transitions are listed by showing the Q output of the flip-flop going from present states to next states. Q_N is the present state of the flip-flop (before a clock pulse) and Q_{N+1} is the next state (after a clock pulse). For each output transition, the J and K inputs that will cause the transition to occur are listed. The Xs indicate a "don't care" (the input can be either a 1 or a 0).

To design the counter, the transition table is applied to each of the flip-flops in the counter, based on the next-state table (Table 9–7). For example, for the present state 000, Q_0 goes from a present state of 0 to a next state of 1. To make this happen, J_0 must be a 1 and you don't care what K_0 is ($J_0 = 1$, $K_0 = X$), as you can see in the transition table

TABLE 9–8
Transition table for a J-K flip-flop.

Output Transitions Q_N		Q_{N+1}	Flip-Flop Inputs J	K
0	$\longrightarrow$	0	0	X
0	$\longrightarrow$	1	1	X
1	$\longrightarrow$	0	X	1
1	$\longrightarrow$	1	X	0

Q_N: present state
Q_{N+1}: next state
X: "don't care"

(Table 9–8). Next, Q_1 is 0 in the present state and remains a 0 in the next state. For this transition, $J_1 = 0$ and $K_1 = X$. Finally, Q_2 is 0 in the present state and remains a 0 in the next state. Therefore, $J_2 = 0$ and $K_2 = X$. This analysis is repeated for each present state in Table 9–7.

Step 4: Karnaugh Maps

Karnaugh maps can be used to determine the logic required for the J and K inputs of each flip-flop in the counter. There is a Karnaugh map for the J input and a Karnaugh map for the K input of each flip-flop. In this design procedure, each cell in a Karnaugh map represents one of the present states in the counter sequence listed in Table 9–7.

From the J and K states in the transition table (Table 9–8) a 1, 0, or X is entered into each present state cell on the maps depending on the transition of the Q output for a particular flip-flop. To illustrate this procedure, two sample entries are shown for the J_0 and the K_0 inputs to the least significant flip-flop (Q_0) in Figure 9–29.

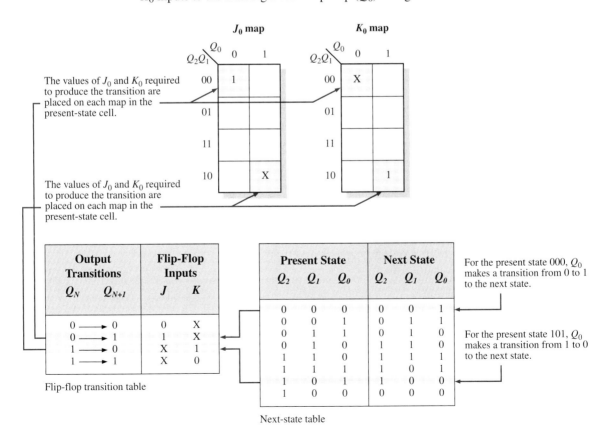

FIGURE 9–29
Examples of the mapping procedure for the counter sequence represented in Table 9–7 and Table 9–8.

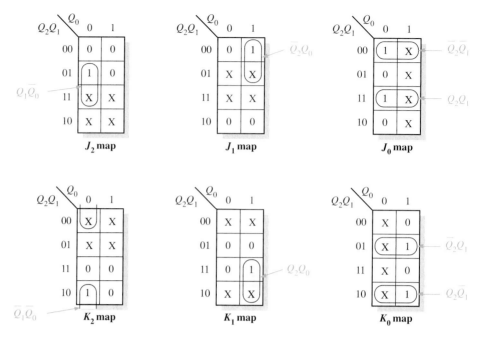

FIGURE 9–30
Karnaugh maps for present-state J and K inputs.

The completed Karnaugh maps for all three flip-flops in the counter are shown in Figure 9–30. The cells are grouped as indicated and the corresponding Boolean expressions for each group are derived.

Step 5: Logic Expressions for Flip-Flop Inputs

From the Karnaugh maps of Figure 9–30 you obtain the following expressions for the J and K inputs of each flip-flop:

$$J_0 = Q_2Q_1 + \overline{Q}_2\overline{Q}_1 = \overline{Q_2 \oplus Q_1}$$
$$K_0 = Q_2\overline{Q}_1 + \overline{Q}_2Q_1 = Q_2 \oplus Q_1$$
$$J_1 = \overline{Q}_2Q_0$$
$$K_1 = Q_2Q_0$$
$$J_2 = Q_1\overline{Q}_0$$
$$K_2 = \overline{Q}_1\overline{Q}_0$$

Step 6: Counter Implementation

The final step is to implement the combinational logic from the expressions for the J and K inputs and connect the flip-flops to form the complete 3-bit Gray code counter as shown in Figure 9–31.

A summary of steps used in the design of this counter follows. In general, these steps can be applied to any sequential circuit.

1. Specify the counter sequence and draw a state diagram.
2. Derive a next-state table from the state diagram.
3. Develop a transition table showing the flip-flop inputs required for each transition. The transition table is always the same for a given type of flip-flop.
4. Transfer the J and K states from the transition table to Karnaugh maps. There is a Karnaugh map for each input of each flip-flop.

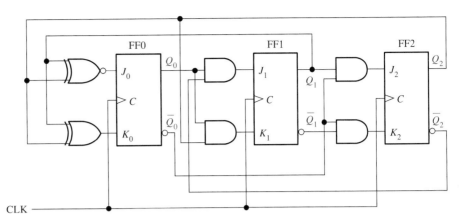

FIGURE 9–31
Three-bit Gray code counter.

5. Factor the maps to generate a logic expression for each flip-flop input.
6. Implement the expressions with combinational logic, and combine with the flip-flops to create the counter.

This procedure is now applied to the design of other synchronous counters in Examples 9–5 and 9–6.

EXAMPLE 9–5

Design a counter with the irregular binary count sequence shown in the state diagram of Figure 9–32. Use J-K flip-flops.

FIGURE 9–32

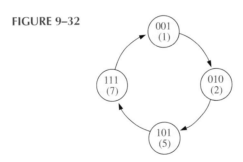

Solution

Step 1. The state diagram is as shown. Although there are only four states, a 3-bit counter is required to implement this sequence because the maximum binary count is seven. Since the required sequence does not include all the possible binary states, the invalid states (0, 3, 4, and 6) can be treated as "don't cares" in the design. However, if the counter should erroneously get into an invalid state, you must make sure that it goes back to a valid state.

Step 2. The next-state table is developed from the state diagram and is given in Table 9–9.

TABLE 9–9
Next-state table.

Present State			Next State		
Q_2	Q_1	Q_0	Q_2	Q_1	Q_0
0	0	1	0	1	0
0	1	0	1	0	1
1	0	1	1	1	1
1	1	1	0	0	1

Step 3. The transition table for the J-K flip-flop is repeated in Table 9–10.

TABLE 9–10
Transition table for a J-K flip-flop.

Output Transitions		Flip-Flop Inputs	
Q_N	Q_{N+1}	J	K
0 $\longrightarrow$ 0		0	X
0 $\longrightarrow$ 1		1	X
1 $\longrightarrow$ 0		X	1
1 $\longrightarrow$ 1		X	0

Step 4. The J and K inputs are plotted on the present-state Karnaugh maps in Figure 9–33. Also "don't cares" can be placed in the cells corresponding to the invalid states of 000, 011, 100, and 110, as indicated by the red Xs.

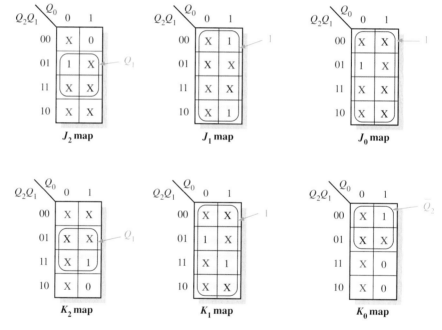

FIGURE 9–33

Step 5. Group the 1s, taking advantage of as many of the "don't care" states as possible for maximum simplification, as shown in Figure 9–33. Notice that when *all* cells in a map are grouped, the expression is simply equal to 1. The expression for each J and K input taken from the maps is as follows:

$$J_0 = 1, K_0 = \overline{Q}_2$$
$$J_1 = K_1 = 1$$
$$J_2 = K_2 = Q_1$$

Step 6. The implementation of the counter is shown in Figure 9–34.

An analysis shows that if the counter, by accident, gets into one of the invalid states (0, 3, 4, 6), it will always return to a valid state according to the following sequences: $0 \rightarrow 3 \rightarrow 4 \rightarrow 7$, and $6 \rightarrow 1$.

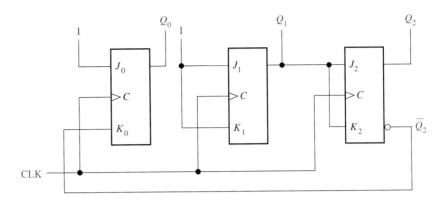

FIGURE 9–34

Related Problem Verify the analysis that proves the counter will always return (eventually) to a valid state from an invalid state.

EXAMPLE 9–6

Develop a synchronous 3-bit up/down counter with a Gray code sequence. The counter should count up when an UP/$\overline{\text{DOWN}}$ control input is 1 and count down when the control input is 0.

Solution

Step 1. The state diagram is shown in Figure 9–35. The 1 or 0 beside each arrow indicates the state of the UP/$\overline{\text{DOWN}}$ control input, Y.

FIGURE 9–35
State diagram for a 3-bit up/down Gray code counter.

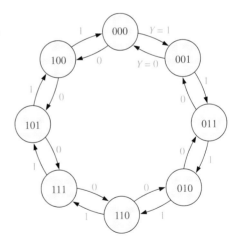

Step 2. The next-state table is derived from the state diagram and is shown in Table 9–11. Notice that for each present state there are two possible next states, depending on the UP/$\overline{\text{DOWN}}$ control variable, Y.

Step 3. The transition table for the J-K flip-flops is repeated in Table 9–12.

Step 4. The Karnaugh maps for the J and K inputs of the flip-flops are shown in Figure 9–36. The UP/$\overline{\text{DOWN}}$ control input, Y, is considered one of the state variables along with Q_0, Q_1, and Q_2. Using the next-state table, the information in the "Flip-Flop Inputs" column of Table 9–12 is transferred onto the maps as indicated for each present state of the counter.

TABLE 9–11

Next-state table for 3-bit up/down Gray code counter.

Present State			Next State					
			Y = 0 (DOWN)			Y = 1 (UP)		
Q_2	Q_1	Q_0	Q_2	Q_1	Q_0	Q_2	Q_1	Q_0
0	0	0	1	0	0	0	0	1
0	0	1	0	0	0	0	1	1
0	1	1	0	0	1	0	1	0
0	1	0	0	1	1	1	1	0
1	1	0	0	1	0	1	1	1
1	1	1	1	1	0	1	0	1
1	0	1	1	1	1	1	0	0
1	0	0	1	0	1	0	0	0

$Y = UP/\overline{DOWN}$ control input.

TABLE 9–12

Transition table for a J-K flip-flop.

Output Transitions			Flip-Flop Inputs	
Q_N		Q_{N+1}	J	K
0	$\longrightarrow$	0	0	X
0	$\longrightarrow$	1	1	X
1	$\longrightarrow$	0	X	1
1	$\longrightarrow$	1	X	0

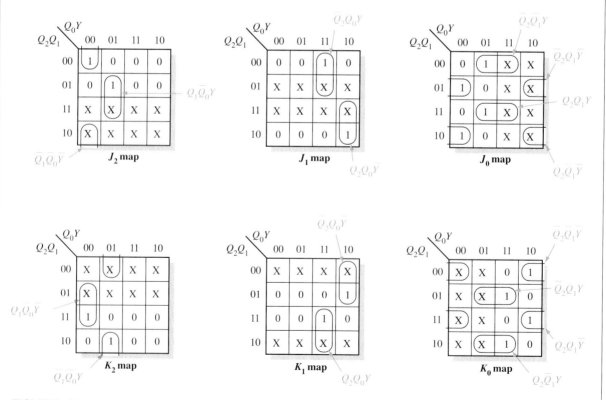

FIGURE 9–36

J and K maps for Table 9–11. The UP/$\overline{DOWN}$ control input, Y, is treated as a fourth variable.

Step 5. The 1s are combined in the largest possible groupings, with "don't cares" (Xs) used where possible. The groups are factored, and the expressions for the J and K inputs are as follows:

$$J_0 = Q_2Q_1Y + Q_2\overline{Q_1}\,\overline{Y} + \overline{Q_2}Q_1\overline{Y} + \overline{Q_2}\,\overline{Q_1}Y \qquad K_0 = \overline{Q_2}\,\overline{Q_1}\overline{Y} + \overline{Q_2}Q_1Y + Q_2Q_1\overline{Y} + Q_2\overline{Q_1}Y$$

$$J_1 = \overline{Q_2}Q_0Y + Q_2Q_0\overline{Y} \qquad\qquad\qquad K_1 = \overline{Q_2}Q_0\overline{Y} + Q_2Q_0Y$$

$$J_2 = Q_1\overline{Q_0}Y + \overline{Q_1}\,\overline{Q_0}\,\overline{Y} \qquad\qquad\qquad K_2 = Q_1\overline{Q_0}\,\overline{Y} + \overline{Q_1}\,\overline{Q_0}Y$$

Step 6. The J and K equations are implemented with combinational logic, and the complete counter is shown in Figure 9–37. This is a good candidate for PLD implementation.

FIGURE 9–37
Three-bit up/down Gray code counter.

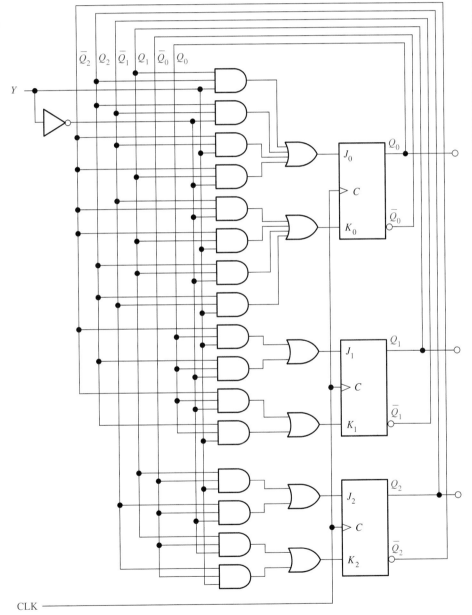

Related Problem Verify that the logic in Figure 9–37 agrees with the expressions in Step 5.

In this section, you have seen how sequential circuit design techniques can be applied specifically to counter design. In general, sequential circuits can be classified into two types: (1) those in which the output or outputs depend only on the present internal state (called *Moore circuits*) and (2) those in which the output or outputs depend on both the present state and the input or inputs (called *Mealy circuits*).

SECTION 9–4 REVIEW

1. A flip-flop is presently in the RESET state and must go to the SET state on the next clock pulse. What must J and K be?
2. A flip-flop is presently in the SET state and must remain SET on the next clock pulse. What must J and K be?
3. A binary counter is in the $Q_3Q_2Q_1Q_0 = 1010$ state.
   (a) What is its next state?
   (b) What condition must exist on each flip-flop input to ensure that it goes to the proper next state on the clock pulse?

9–5 ■ CASCADED COUNTERS

Counters can be connected in cascade to achieve higher-modulus operation. In essence, cascading means that the last-stage output of one counter drives the input of the next counter.

After completing this section, you should be able to

■ Determine the overall modulus of cascaded counters ■ Analyze the timing diagram of a cascaded counter configuration ■ Use cascaded counters as a frequency divider ■ Use cascaded counters to achieve specified truncated sequences

COMPUTER NOTE
The time stamp counter (TSC), mentioned in the last computer note, is a 64-bit counter. It is interesting to observe that if this counter (or any full-modulus 64-bit counter) is clocked at a frequency of 100 MHz, it will take 5,849 years for it to go through all of its states and reach its terminal count. In contrast, a 32-bit full-modulus counter will exhaust all of its states in approximately 43 seconds when clocked at 100 MHz.

An example of two counters connected in **cascade** is shown for a 2-bit and a 3-bit ripple counter in Figure 9–38. The timing diagram is shown in Figure 9–39.

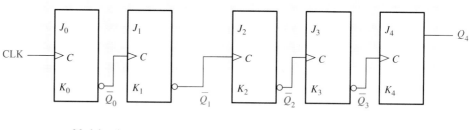

Modulus-4 counter   Modulus-8 counter

FIGURE 9–38
Two cascaded counters (all J and K inputs are HIGH).

Notice that in the timing diagram in Figure 9–39, the final output of the modulus-8 counter, Q_4, occurs once for every 32 input clock pulses. The overall modulus of the cascaded counters is 32; that is, they act as a divide-by-32 counter. In general, the overall modulus of cascaded counters is equal to the product of the individual moduli. For instance, for the counter in Figure 9–38, the overall modulus is $4 \times 8 = 32$.

When operating synchronous counters in a cascaded configuration, it is necessary to use the count enable and the terminal count functions to achieve higher-modulus

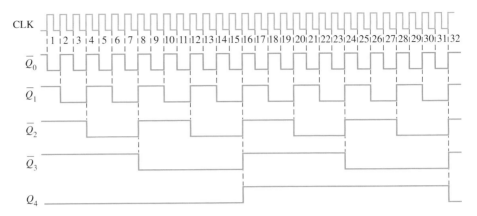

FIGURE 9–39
Timing diagram for the cascaded counter configuration of Figure 9–38.

operation. On some devices the count enable is labeled simply *CTEN* or some other designation, and terminal count (*TC*) is analogous to ripple clock output (*RCO*) on some IC counters.

Figure 9–40 shows two decade counters connected in cascade. The terminal count (*TC*) output of counter 1 is connected to the count enable (*CTEN*) input of counter 2. Counter 2 is inhibited by the LOW on its *CTEN* input until counter 1 reaches its last, or terminal, state and its terminal count output goes HIGH. This HIGH now enables counter 2, so that when the first clock pulse after counter 1 reaches its terminal count (CLK10), counter 2 goes from its initial state to its second state. Upon completion of the entire second cycle of counter 1 (when counter 1 reaches terminal count the second time), counter 2 is again enabled and advances to its next state. This sequence continues. Since these are decade counters, counter 1 must go through ten complete cycles before counter 2 completes its first cycle. In other words, for every ten cycles of counter 1, counter 2 goes through one cycle. Thus, counter 2 will complete one cycle after one hundred clock pulses. The overall modulus of these two cascaded counters is $10 \times 10 = 100$.

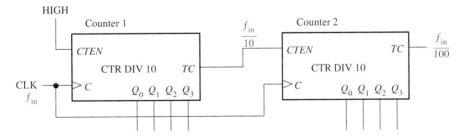

FIGURE 9–40
A modulus-100 counter using two cascaded decade counters.

When viewed as a frequency divider, the circuit of Figure 9–40 divides the input clock frequency by 100. Cascaded counters are often used to divide a high-frequency clock signal to obtain highly accurate pulse frequencies. Cascaded counter configurations used for such purposes are sometimes called *countdown chains.*

For example, suppose that you have a basic clock frequency of 1 MHz and you wish to obtain 100 kHz, 10 kHz, and 1 kHz; a series of cascaded decade counters can be used. If the 1 MHz signal is divided by 10, the output is 100 kHz. Then if the

100 kHz signal is divided by 10, the output is 10 kHz. Another division by 10 produces the 1 kHz frequency. The general implementation of this countdown chain is shown in Figure 9–41.

FIGURE 9–41
Three cascaded decade counters forming a divide-by-1000 frequency divider with intermediate divide-by-10 and divide-by-100 outputs.

EXAMPLE 9–7   Determine the overall modulus of the two cascaded counter configurations in Figure 9–42.

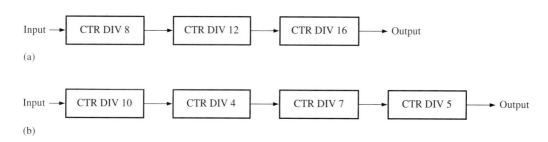

FIGURE 9–42

Solution   In Figure 9–42(a), the overall modulus for the 3-counter configuration is

$$8 \times 12 \times 16 = \mathbf{1536}$$

In Figure 9–42(b), the overall modulus for the 4 counter configuration is

$$10 \times 4 \times 7 \times 5 = \mathbf{1400}$$

Related Problem   How many cascaded decade counters are required to divide a clock frequency by 100,000?

EXAMPLE 9–8   Use 74HC160 counters to obtain a 10 kHz waveform from a 1 MHz clock. Show the logic diagram.

Solution   To obtain 10 kHz from a 1 MHz clock requires a division factor of 100. Two 74HC160 counters must be cascaded as shown in Figure 9–43. The left counter produces an *RCO* pulse for every 10 clock pulses. The right counter produces an *RCO* pulse for every 100 clock pulses.

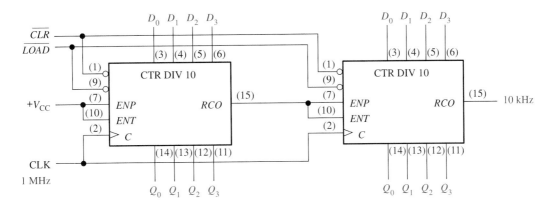

FIGURE 9–43
A divide-by-100 counter using two 74HC160 decade counters.

Related Problem Determine the frequency of the waveform at the Q_0 output of the second counter (the one on the right) in Figure 9–43.

Cascaded IC Counters with Truncated Sequences

The preceding discussion has shown how to achieve an overall modulus (divide-by-factor) that is the product of the individual moduli of all the cascaded counters. This can be considered *full-modulus cascading*.

Often an application requires an overall modulus that is less than that achieved by full-modulus cascading. That is, a truncated sequence must be implemented with cascaded counters. To illustrate this method, the cascaded counter configuration in Figure 9–44 is used. This particular circuit uses four 74HC161 4-bit synchronous binary counters. If these four counters (sixteen bits total) were cascaded in a full-modulus arrangement, the modulus would be

$$2^{16} = 65,536$$

Let's assume that a certain application requires a divide-by-40,000 counter (modulus 40,000). The difference between 65,536 and 40,000 is 25,536, which is the number of states that must be *deleted* from the full-modulus sequence. The technique used in the circuit of Figure 9–44 is to preset the cascaded counter to 25,536 (63C0 in hexadecimal) each

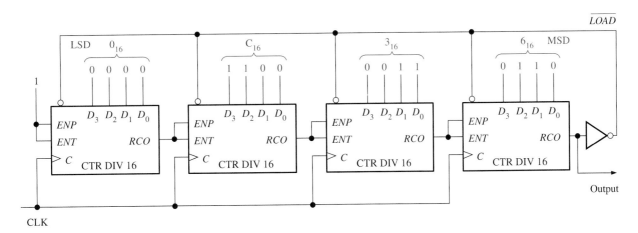

FIGURE 9–44
A divide-by-40,000 counter using 74HC161 4-bit binary counters. Note that each of the parallel data inputs is shown in binary order (the right-most bit D_0 is the LSB in each counter).

time it recycles, so that it will count from 25,536 up to 65,535 on each full cycle. Therefore, each full cycle of the counter consists of 40,000 states.

Notice in Figure 9–44 that the *RCO* output of the right-most counter is inverted and applied to the $\overline{LOAD}$ input of each 4-bit counter. Each time the count reaches its terminal value of 65,535, which is 1111111111111111_2, *RCO* goes HIGH and causes the number on the parallel data inputs ($63C0_{16}$) to be loaded into the counter synchronous with the clock pulse. Thus, there is one *RCO* pulse from the right-most 4-bit counter for every 40,000 clock pulses.

With this technique any modulus can be achieved by synchronous loading of the counter to the appropriate initial state on each cycle.

SECTION 9–5 REVIEW

1. How many decade counters are necessary to implement a divide-by-1000 (modulus-1000) counter? A divide-by-10,000?
2. Show with general block diagrams how to achieve each of the following, using a flip-flop, a decade counter, and a 4-bit binary counter, or any combination of these:
   (a) Divide-by-20 counter   **(b)** Divide-by-32 counter
   (c) Divide-by-160 counter   **(d)** Divide-by-320 counter

9–6 ■ COUNTER DECODING

In many applications, it is necessary that some or all of the counter states be decoded. The decoding of a counter involves using decoders or logic gates to determine when the counter is in a certain binary state in its sequence. For instance, the terminal count function previously discussed is a single decoded state (the last state) in the counter sequence.

After completing this section, you should be able to

■ Implement the decoding logic for any given state in a counter sequence   ■ Explain why glitches occur in counter decoding logic   ■ Use the method of strobing to eliminate decoding glitches

Suppose that you wish to decode binary state 6 (110) of a 3-bit binary counter. When $Q_2 = 1$, $Q_1 = 1$, and $Q_0 = 0$, a HIGH appears on the output of the decoding gate, indicating that the counter is at state 6. This can be done as shown in Figure 9–45. This is called *active-HIGH decoding*. Replacing the AND gate with a NAND gate provides active-LOW decoding.

FIGURE 9–45
Decoding of state 6 (110).

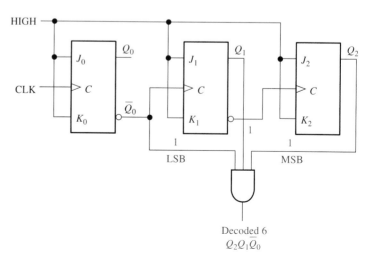

Decoded 6
$Q_2 Q_1 \overline{Q_0}$

EXAMPLE 9–9

Implement the decoding of binary state 2 and binary state 7 of a 3-bit synchronous counter. Show the entire counter timing diagram and the output waveforms of the decoding gates. Binary $2 = \overline{Q}_2Q_1\overline{Q}_0$, and binary $7 = Q_2Q_1Q_0$.

Solution See Figure 9–46. The 3-bit counter was originally discussed in Section 9–2 (Figure 9–14).

FIGURE 9–46
A 3-bit counter with active-HIGH decoding of count 2 and count 7.

Related Problem Show the logic for decoding state 5 in the 3-bit counter.

Decoding Glitches

The problem of glitches produced by the decoding process was introduced in Chapter 6. As you have learned, the propagation delays due to the ripple effect in asynchronous counters create transitional states in which the counter outputs are changing at slightly different times. These transitional states produce undesired voltage spikes of short duration (glitches) on the outputs of a decoder connected to the counter. The glitch problem can also occur to some degree with synchronous counters because the propagation delays from clock to Q outputs of each flip-flop in a counter can vary slightly.

Figure 9–47 shows a basic asynchronous BCD decade counter connected to a BCD-to-decimal decoder. To see what happens in this case, let's look at a timing diagram in which the propagation delays are taken into account, as shown in Figure 9–48. Notice that these delays cause false states of short duration. The value of the false binary state at each

FIGURE 9–47
A basic decade (BCD) counter and decoder.

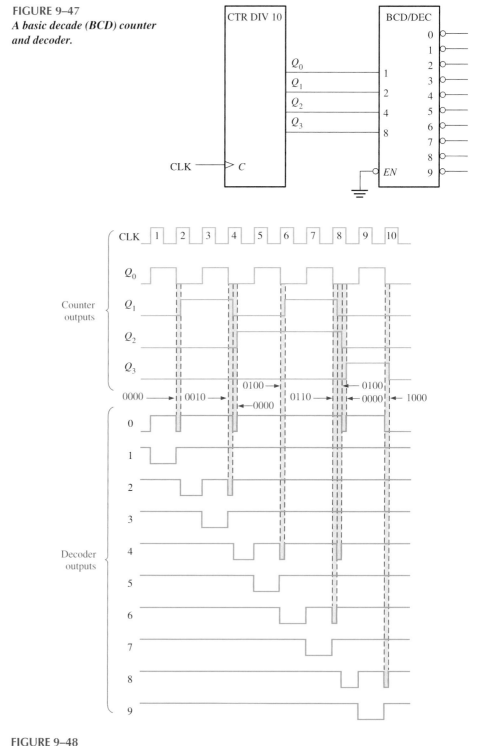

FIGURE 9–48
Outputs with glitches from the decoder in Figure 9–47. Glitch widths are exaggerated for illustration and are usually only a few nanoseconds wide.

critical transition is indicated on the diagram. The resulting glitches can be seen on the decoder outputs.

One way to eliminate the glitches is to enable the decoded outputs at a time after the glitches have had time to disappear. This method is known as *strobing* and can be accomplished in the case of an active-HIGH clock by using the LOW level of the clock to enable the decoder, as shown in Figure 9–49. The resulting improved timing diagram is shown in Figure 9–50.

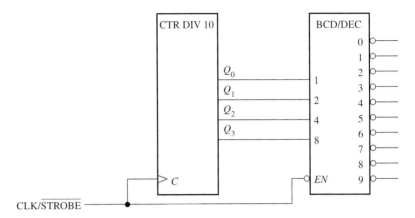

FIGURE 9–49
The basic decade counter and decoder with strobing to eliminate glitches.

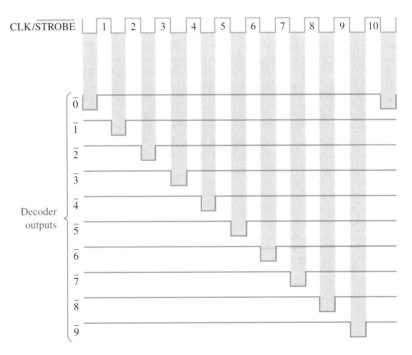

FIGURE 9–50
Strobed decoder outputs for the circuit of Figure 9–49.

**SECTION 9–6
REVIEW**

1. What transitional states are possible when a 4-bit asynchronous binary counter changes from
   (a) count 2 to count 3   (b) count 3 to count 4
   (c) count 10_{10} to count 11_{10} (d) count 15 to count 0

9–7 ■ COUNTER APPLICATIONS

The digital counter is a useful and versatile device that is found in many applications. In this section, some representative counter applications are presented.

After completing this section, you should be able to

■ Describe how counters are used in a basic digital clock system   ■ Explain how a divide-by-60 counter is implemented and how it is used in a digital clock   ■ Explain how the hours counter is implemented   ■ Discuss the application of a counter in an automobile parking control system   ■ Describe how a counter is used in the process of parallel-to-serial data conversion

Application Examples

The Digital Clock   A common example of a counter application is in timekeeping systems. Figure 9–51 is a simplified logic diagram of a digital clock that displays seconds, minutes, and hours. First, a 60 Hz sinusoidal ac voltage is converted to a 60 Hz pulse waveform and divided down to a 1 Hz pulse waveform by a divide-by-60 counter formed by a divide-by-10 counter followed by a divide-by-6 counter. Both the *seconds* and *minutes* counts are also produced by divide-by-60 counters, the details of which are shown in Figure 9–52. These counters count from 0 to 59 and then recycle to 0; synchronous decade counters are used in this particular implementation. Notice that the divide-by-6 portion is

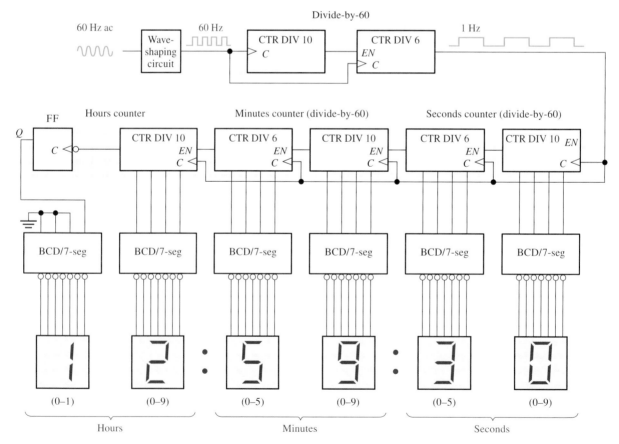

FIGURE 9–51

Simplified logic diagram for a 12-hour digital clock. Logic details using specific devices are shown in Figures 9–52 and 9–53.

FIGURE 9–52
Logic diagram of typical divide-by-60 counter using 74LS160A synchronous decade counters.
Note that the outputs are in binary order (the right-most bit is the LSB).

formed with a decade counter with a truncated sequence achieved by using the decoder count 6 to asynchronously clear the counter. The terminal count, 59, is also decoded to enable the next counter in the chain.

The *hours* counter is implemented with a decade counter and a flip-flop as shown in Figure 9–53. Consider that initially both the decade counter and the flip-flop are RESET, and the decode-12 gate and decode-9 gate outputs are HIGH. The decade counter advances through all of its states from zero to nine, and on the clock pulse that recycles it from nine back to zero, the flip-flop goes to the SET state ($J = 1$, $K = 0$). This illuminates a 1 on the

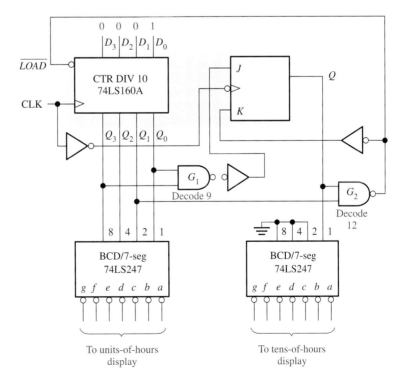

FIGURE 9–53
Logic diagram for hours counter and decoders. Note that on the counter inputs and outputs, the right-most bit is the LSB.

tens-of-hours display. The total count is now ten (the decade counter is in the zero state and the flip-flop is SET).

Next, the total count advances to eleven and then to twelve. In state 12 the Q_2 output of the decade counter is HIGH, the flip-flop is still SET, and thus the decode-12 gate output is LOW. This activates the $\overline{LOAD}$ input of the decade counter. On the next clock pulse, the decade counter is preset to state 1 by the data inputs, and the flip-flop is RESET ($J = 0$, $K = 1$). As you can see, this logic always causes the counter to recycle from twelve back to one rather than back to zero.

Automobile Parking Control   A simple application example illustrates the use of an up/down counter to solve an everyday problem. The problem is to devise a means of monitoring available spaces in a one-hundred-space parking garage and provide for an indication of a full condition by illuminating a display sign and lowering a gate bar at the entrance.

A system that solves this problem consists of (1) optoelectronic sensors at the entrance and exit of the garage, (2) an up/down counter and associated circuitry, and (3) an interface circuit that uses the counter output to turn the FULL sign on or off as required and lower or raise the gate bar at the entrance. A general block diagram of this system is shown in Figure 9–54.

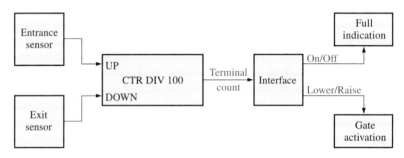

FIGURE 9–54
Functional block diagram for parking garage control.

A logic diagram of the up/down counter is shown in Figure 9–55. It consists of two cascaded 74LS190 up/down decade counters. The operation is described in the following paragraphs.

The counter is initially preset to 0 using the parallel data inputs which are not shown. Each automobile entering the garage breaks a light beam, activating a sensor that produces an electrical pulse. This positive pulse sets the S-R latch on its leading edge. The LOW on the $\overline{Q}$ output of the latch puts the counter in the UP mode. Also, the sensor pulse

FIGURE 9–55
Logic diagram for modulus-100 up/down counter for automobile parking control.

goes through the NOR gate and clocks the counter on the LOW-to-HIGH transition of its trailing edge. Each time an automobile enters the garage, the counter is advanced by one **(incremented).** When the one-hundredth automobile enters, the counter goes to its last state (100_{10}). The *MAX/MIN* output goes HIGH and activates the interface circuit (no detail), which lights the FULL sign and lowers the gate bar to prevent further entry.

When an automobile exits, an optoelectronic sensor produces a positive pulse, which resets the S-R latch and puts the counter in the DOWN mode. The trailing edge of the clock decreases the count by one **(decrements).** If the garage is full and an automobile leaves, the *MAX/MIN* output of the counter goes LOW, turning off the FULL sign and raising the gate.

Parallel-to-Serial Data Conversion (Multiplexing) A simplified example of data transmission using multiplexing and demultiplexing techniques was introduced in Chapter 6. Essentially, the parallel data bits on the multiplexer inputs are converted to serial data bits on the single transmission line. A group of bits appearing simultaneously on parallel lines is called *parallel data.* A group of bits appearing on a single line in a time sequence is called *serial data.*

Parallel-to-serial conversion is normally accomplished by the use of a counter to provide a binary sequence for the data-select inputs of a data selector/multiplexer, as illustrated in Figure 9–56. The Q outputs of the modulus-8 counter are connected to the data-select inputs of an 8-bit multiplexer.

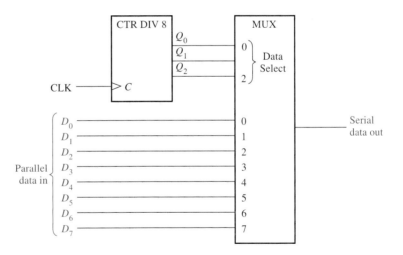

FIGURE 9–56
Parallel-to-serial data conversion logic.

Figure 9–57 is a timing diagram illustrating the operation of this circuit. The first byte (eight-bit group) of parallel data is applied to the multiplexer inputs. As the counter goes through a binary sequence from zero to seven, each bit, beginning with D_0, is sequentially selected and passed through the multiplexer to the output line. After eight clock pulses the data byte has been converted to a serial format and sent out on the transmission line. When the counter recycles back to 0, the next byte is applied to the data inputs and is sequentially converted to serial form as the counter cycles through its eight states. This process continues repeatedly as each parallel byte is converted to a serial byte. An end-of-chapter problem (Problem 52) requires that this circuit be implemented with specific devices.

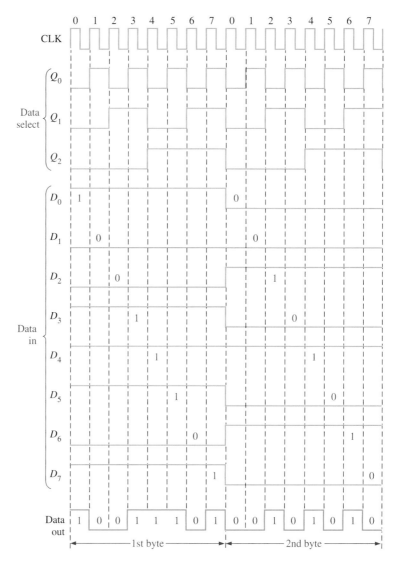

FIGURE 9–57
Example of parallel-to-serial conversion timing for the circuit in Figure 9–56.

1. Explain the purpose of each NAND gate in Figure 9–53.
2. Identify the two recycle conditions for the hours counter in Figure 9–51, and explain the reason for each.

9–8 ■ TROUBLESHOOTING

The troubleshooting of counters can be simple or quite involved, depending on the type of counter and the type of fault. This section will give you some good practical insight into how to approach the troubleshooting of sequential circuits.

After completing this section, you should be able to

■ Detect a faulty counter   ■ Isolate faults in maximum-modulus cascaded counters
■ Isolate faults in cascaded counters with truncated sequences   ■ Determine faults in counters implemented with individual flip-flops

IC Counters

For an IC counter with a straightforward sequence that is not controlled by external logic, about the only thing to check (other than V_{CC} and ground) is the possibility of open or shorted inputs or outputs. An IC counter almost never alters its sequence of states because of an internal fault, so you need only check for pulse activity on the Q outputs to detect the existence of an open or a short. The absence of pulse activity on one of the Q outputs indicates an internal open or a short on the line, which may be internal or external to the IC. Absence of pulse activity on all the Q outputs indicates that the clock input is faulty or the clear input is stuck in its active state.

To check the clear input, apply a constant active level while the counter is clocked. You will observe a LOW on each of the Q outputs if it is functioning properly.

A synchronous parallel load feature on a counter can be checked by activating the $\overline{LOAD}$ input and exercising each state as follows: Apply LOWs to the parallel data inputs, pulse the clock input once, and check for LOWs on all the Q outputs. Next, apply HIGHs to all the parallel data inputs, pulse the clock input once, and check for HIGHs on all the Q outputs.

Cascaded IC Counters with Maximum Modulus

A failure in one of the counters in a chain of cascaded counters can affect all the counters that follow it. For example, if a count enable input opens, it effectively acts as a HIGH (for TTL), and the counter is always enabled. This type of failure in one of the counters will cause that counter to run at the full clock rate and will also cause all the succeeding counters to run at higher than normal rates. This is illustrated in Figure 9–58 for a divide-by-1000 cascaded counter arrangement where an open enable (*CTEN*) input acts as a TTL HIGH and continuously enables the second counter. Other faults that can affect "downstream" counter stages are open or shorted clock inputs or terminal count outputs. In some of these situations, pulse activity can be observed, but it may be at the wrong frequency. Exact frequency or frequency ratio measurements must be made.

(a) Normal operation

(b) Count Enable (*CTEN*) input of second counter open

FIGURE 9–58
Example of a failure that affects following counters in a cascaded arrangement.

Cascaded Counters with Truncated Sequences

The count sequence of a cascaded counter with a truncated sequence, such as that in Figure 9–59, can be affected by other types of faults in addition to those mentioned for maximum-modulus cascaded counters. For example, a failure in one of the parallel data inputs, the $\overline{LOAD}$ input, or the inverter can alter the preset count and thus change the modulus of the counter.

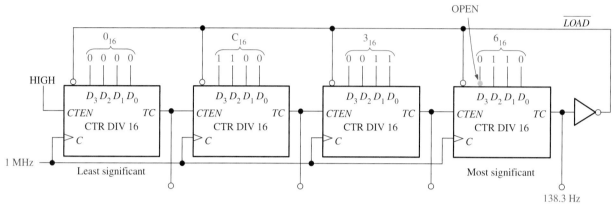

FIGURE 9–59
Example of a failure in a cascaded counter with a truncated sequence.

For example, suppose the D_3 input of the most significant counter in Figure 9–59 is open and acts as a HIGH. Instead of 6_{16} (0110) being preset into the counter, E_{16} (1110) is preset in. So, instead of beginning with $63C0_{16}$ ($25,536_{10}$) each time the counter recycles, the sequence will begin with $E3C0_{16}$ ($58,304_{10}$). This changes the modulus of the counter from 40,000 to $65,536 - 58,304 = 7232$.

To check this counter, apply a known clock frequency, for example 1 MHz, and measure the output frequency at the final terminal count output. If the counter is operating properly, the output frequency is

$$f_{out} = \frac{f_{in}}{modulus} = \frac{1 \text{ MHz}}{40,000} = 25 \text{ Hz}$$

In this case, the specific failure described in the preceding paragraph will cause the output frequency to be

$$f_{out} = \frac{f_{in}}{modulus} = \frac{1 \text{ MHz}}{7232} = 138.3 \text{ Hz}$$

EXAMPLE 9–10

Frequency measurements are made on the truncated counter in Figure 9–60 as indicated. Determine if the counter is working properly, and if not, isolate the fault.

Solution Check to see if the frequency measured at TC 4 is correct. If it is, the counter is working properly.

$$\text{truncated modulus} = \text{full modulus} - \text{preset count}$$
$$= 16^4 - 82C0_{16}$$
$$= 65,536 - 33,472 = 32,064$$

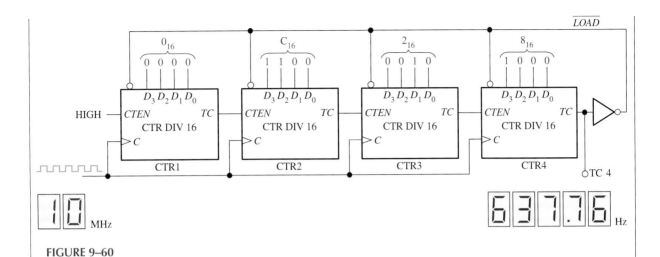

FIGURE 9–60

The correct frequency at TC 4 is

$$f_4 = \frac{10 \text{ MHz}}{32,064} = 311.88 \text{ Hz}$$

Uh oh! There is a problem. The measured frequency of 637.76 Hz does not agree with the correct calculated frequency of 311.88 Hz.

To find the faulty counter, determine the actual truncated modulus as follows:

$$\text{modulus} = \frac{f_{\text{out}}}{f_{\text{in}}} = \frac{10 \text{ MHz}}{637.76 \text{ Hz}} = 15,680$$

Because the truncated modulus should be 32,064, most likely the counter is being preset to the wrong count when it recycles. The actual preset count is determined as follows:

$$\text{truncated modulus} = \text{full modulus} - \text{preset count}$$
$$\text{preset count} = \text{full modulus} - \text{truncated modulus}$$
$$= 65,536 - 15,680$$
$$= 49,856$$
$$= \text{C2C0}_{16}$$

This shows that the counter is being preset to C2C0_{16} instead of 82C0_{16} each time it recycles.

Counters 1, 2, and 3 are being preset properly but counter 4 is not. Since $\text{C}_{16} = 1100_2$, the D_2 input to counter 4 is HIGH when it should be LOW. This is most likely caused by an **open input.** Check for an external open caused by a bad solder connection, a broken conductor, or a bent pin on the IC. If none can be found, replace the IC and the counter should work properly.

Related Problem Determine what the output frequency at TC 4 would be if the D_3 input of counter 3 were open.

Counters Implemented with Individual Flip-Flops

Counters implemented with individual flip-flop and gate ICs are sometimes more difficult to troubleshoot because there are many more inputs and outputs with external connections than there are in an IC counter. The sequence of a counter can be altered by a single open or short on an input or output, as Example 9–11 illustrates.

EXAMPLE 9–11

Suppose that you observe the output waveforms that are indicated for the counter in Figure 9–61. Determine if there is a problem with the counter.

FIGURE 9–61

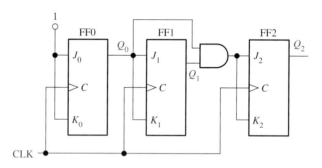

Solution The Q_2 waveform is incorrect. The correct waveform is shown as a red dashed line. You can see that the Q_2 waveform looks exactly like the Q_1 waveform. So whatever is causing FF1 to toggle appears to also be controlling FF2.

Checking the J and K inputs to FF2, you find a waveform that looks like Q_0. This result indicates that Q_0 is somehow getting through the AND gate. The only way this can happen is if the Q_1 input to the AND gate is always HIGH. But you have seen that Q_1 has a correct waveform. This observation leads to the conclusion that the lower input to the AND gate must be internally open and acting as a HIGH. Replace the AND gate and retest the circuit.

Related Problem Describe the Q_2 output of the counter in Figure 9–61 if the Q_1 output of FF1 is open.

When observing the time relationship between two digital signals with a dual-trace oscilloscope, the proper way to trigger the scope is with the slower of the two signals. The reason for this is that the slower signal has fewer possible trigger points than the faster signal and there will be no ambiguity for starting the sweep. Vertical mode triggering uses a composite of both channels and should never be used for determining absolute time information. Since clock signals are usually the fastest signal in a digital system, they should not be used for triggering.

SECTION 9–8 REVIEW

1. What failures can cause the counter in Figure 9–58 to have no pulse activity on any of the *TC* outputs?
2. What happens if the inverter in Figure 9–60 develops an open output?

9–9 ■ LOGIC SYMBOLS WITH DEPENDENCY NOTATION*

Up to this point, the logic symbols with dependency notation specified in ANSI/IEEE Standard 91–1984 have been introduced on a limited basis. In many cases, the new symbols do not deviate greatly from the traditional symbols. A significant departure from what we are accustomed to does occur, however, for some devices, including counters and other more complex devices. Although we will continue to use primarily the more traditional and familiar symbols throughout this book, a brief coverage of logic symbols with dependency notation is provided.

After completing this section, you should be able to

■ Interpret logic symbols that include dependency notation ■ Identify the common block and the individual elements of a counter symbol ■ Interpret the qualifying symbol ■ Discuss control dependency ■ Discuss mode dependency ■ Discuss AND dependency

Dependency notation is fundamental to the ANSI/IEEE standard. Dependency notation is used in conjunction with the logic symbols to specify the relationships of inputs and outputs so that the logical operation of a given device can be determined entirely from its logic symbol without a prior knowledge of the details of its internal structure and without a detailed logic diagram for reference.

The 74HC163 4-bit synchronous binary counter is used for illustration. For comparison, Figure 9–62 shows a traditional block symbol and the ANSI/IEEE symbol with dependency notation. Basic descriptions of the symbol and the dependency notation follow.

(a) Traditional block symbol

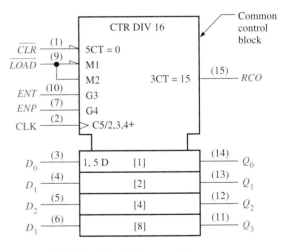

(b) ANSI/IEEE Std. 91–1984 logic symbol

FIGURE 9–62
The 74HC163 4-bit synchronous counter.

Common Control Block The upper block with notched corners in Figure 9–62(b) has inputs and an output that are considered common to all elements in the device and not unique to any one of the elements.

Individual Elements The lower block in Figure 9–62(b), which is partitioned into four abutted sections, represents the four storage elements (D flip-flops) in the counter, with inputs D_0, D_1, D_2, and D_3 and outputs Q_0, Q_1, Q_2, and Q_3.

*This section is optional, and its omission will not affect the remainder of the book.

Qualifying Symbol   The label "CTR DIV 16" in Figure 9–62(b) identifies the device as a counter (CTR) with sixteen states (DIV 16).

Control Dependency (C)   As shown in Figure 9–62(b), the letter C denotes control dependency. Control inputs usually enable or disable the data inputs (D, J, K, S, and R) of a storage element. The C input is usually the clock input. In this case the digit 5 following C (C5/2,3,4+) indicates that the inputs labeled with a 5 prefix are dependent on the clock (synchronous with the clock). For example, $5CT = 0$ on the $\overline{CLR}$ input indicates that the clear function is dependent on the clock; that is, it is a synchronous clear. When the $\overline{CLR}$ input is LOW (0), the counter is reset to zero ($CT = 0$) on the triggering edge of the clock pulse. Also, the 5 D label at the input of storage element [1] indicates that the data storage is dependent on (synchronous with) the clock. All labels in the [1] storage element apply to the [2], [4], and [8] elements below it since they are not labeled differently.

Mode Dependency (M)   As shown in Figure 9–62(b), the letter M denotes mode dependency. This label is used to indicate how the functions of various inputs or outputs depend on the mode in which the device is operating. In this case the device has two modes of operation. When the $\overline{LOAD}$ input is LOW (0), as indicated by the triangle input, the counter is in a preset mode (M1) in which the input data (D_0, D_1, D_2, and D_3) are synchronously loaded into the four flip-flops. The digit 1 following M in M1 and the 1 in the label 1, 5 D show a dependency relationship and indicate that input data are stored only when the device is in the preset mode (M1), in which $\overline{LOAD} = 0$. When the $\overline{LOAD}$ input is HIGH (1), the counter advances through its normal binary sequence, as indicated by M2 and the 2 in C5/2,3,4+.

AND Dependency (G)   As shown in Figure 9–62(b), the letter G denotes AND dependency, indicating that an input designated with G followed by a digit is ANDed with any other input or output having the same digit as a prefix in its label. In this particular example, the G3 at the ENT input and the $3CT = 15$ at the RCO output are related, as indicated by the 3, and that relationship is an AND dependency, indicated by the G. This tells us that ENT must be HIGH (no triangle on the input) *and* the count must be fifteen ($CT = 15$) for the RCO output to be HIGH.

Also, the digits 2, 3, and 4 in the label C5/2,3,4+ indicate that the counter advances through its states when $\overline{LOAD} = 1$, as indicated by the mode dependency label M2, and when $ENT = 1$ and $ENP = 1$, as indicated by the AND dependency labels G3 and G4. The + indicates that the counter advances by one count when these conditions exist.

This coverage of a specific logic symbol with dependency notation is intended to aid in the interpretation of other such symbols that you may encounter in the future.

SECTION 9–9 REVIEW

1. In dependency notation, what do the letters C, M, and G stand for?
2. By what letter is data storage denoted?

9–10 ■ DIGITAL SYSTEM APPLICATION

In this system application, you will continue working with the traffic light control system. You should review Section 6–11 and Section 8–9 before continuing. In this section, the focus is on the sequential logic portion of the system which is the last block of the system to be developed.

After completing this section, you should be able to

■ Develop a block diagram for the sequential logic portion of the system   ■ Implement the counter and input logic and complete the system logic diagram   ■ Compare the logic diagram to the printed circuit board   ■ Develop a test procedure and troubleshoot the circuit board

General Requirements for the Sequential Logic

The overall system block diagram with the sequential logic block highlighted in green and the light sequence is shown in Figure 9–63. The state decoding and output logic block and the timing circuits block have been developed in Chapters 6 and 8. In this section, you are going to concentrate on the sequential logic that controls the sequencing of the traffic lights based on inputs from the timing circuits and from the vehicle sensor. The sequential logic produces a 2-bit Gray code sequence for the four states of the system that are indicated in Figure 9–63.

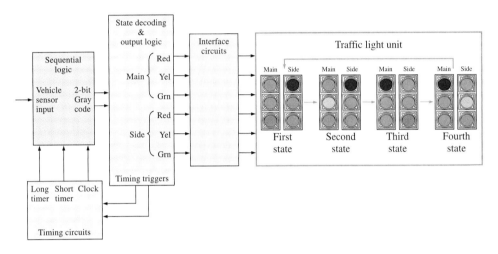

FIGURE 9–63
Traffic light control system block diagram and light sequence.

A Block Diagram of the Sequential Logic

The sequential logic consists of a 2-bit Gray code counter and associated input logic as shown in Figure 9–64. The counter produces a sequence of four states. Transitions from one state to the next are determined by the 4 s interval (short timer) and the 25 s interval (long timer) produced by the timing circuits and the vehicle sensor input. The clock for the counter is the 10 kHz signal produced by the 555 timer in the timing circuits portion of the system.

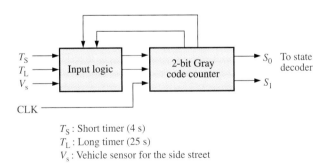

T_S : Short timer (4 s)
T_L : Long timer (25 s)
V_s : Vehicle sensor for the side street

FIGURE 9–64
Block diagram of the sequential logic.

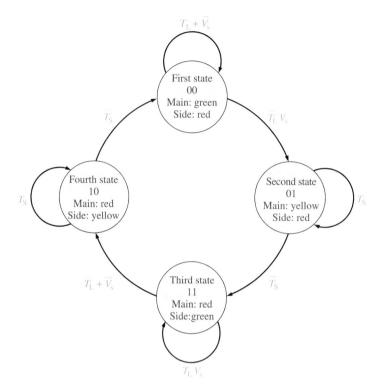

FIGURE 9–65
State diagram showing the 2-bit Gray code sequence.

The State Diagram and the Sequential Operation

The state diagram for the system was introduced in Chapter 6 and is shown again in Figure 9–65. Based on this state diagram, the sequential Gray code operation is described as follows.

First state: The Gray code for this state is 00. The main street light is green and the side street light is red. The system remains in this state for at least 25 s when the long timer is *on* or as long as there is no vehicle on the side street ($T_L + \overline{V}_s$). The system goes to the next state when the 25 s timer is *off* and there is a vehicle on the side street ($\overline{T}_L V_s$).

Second state: The Gray code for this state is 01. The main street light is yellow (caution) and the side street light is red. The system remains in this state for 4 s when the short timer is *on* (T_S) and goes to the next state when the short timer goes *off* ($\overline{T}_S$).

Third state: The Gray code for this state is 11. The main street light is red and the side street light is green. The system remains in this state when the long timer is *on* and there is a vehicle on the side street ($T_L V_s$). The system goes to the next state when the 25 s have elapsed or when there is no vehicle on the side street, whichever comes first ($\overline{T}_L + \overline{V}_s$).

Fourth state: The Gray code for this state is 10. The main street light is red and the side street light is yellow. The system remains in this state for 4 s when the short timer is *on* (T_S) and goes back to the first state when the short timer goes *off* ($\overline{T}_S$).

Implementing the Logic

The diagram in Figure 9–66 further defines the sequential logic. The counter consists of two synchronously clocked D flip-flops for which the D inputs come from the input logic. The clock is the 10 kHz output from the 555 timer. The input logic consists of five input variables: Q_1, Q_0, T_L, T_S, and V_s.

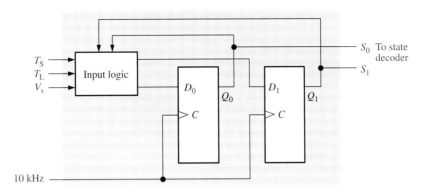

FIGURE 9–66
Sequential logic.

TABLE 9–13
Next-state table for the sequential logic transition.

Present State		Next State		Input Conditions
Q_1	Q_0	Q_1	Q_0	
0	0	0	0	$T_L + \overline{V}_s$
0	0	0	1	$\overline{T}_L V_s$
0	1	0	1	T_S
0	1	1	1	$\overline{T}_S$
1	1	1	1	$T_L V_s$
1	1	1	0	$\overline{T}_L + \overline{V}_s$
1	0	1	0	T_S
1	0	0	0	$\overline{T}_S$

TABLE 9–14
D flip-flop transition table.

Output Transitions			Flip-Flop Input
Q_N		Q_{N+1}	D
0	$\longrightarrow$	0	0
0	$\longrightarrow$	1	1
1	$\longrightarrow$	0	0
1	$\longrightarrow$	1	1

From the state diagram, a next-state table can be developed as shown in Table 9–13. The input conditions for T_L, T_S, and V_s for each present-state/next-state combination are also listed in the table. The D flip-flop transition table is shown in Table 9–14.

■ SYSTEM ASSIGNMENTS

In the Chapter 6 system assignment you worked with the portion of the traffic light controller board containing the state decoder and output logic. In Chapter 8, the timing circuits were developed. In this section, the entire board including the sequential logic portion is completed. As before, all of the connections to the V_{CC} and ground pads are on the back side of the board and are omitted for simplicity. Circuit interconnections on the back side of the board, shown as darker traces, feed through to the pads on the component side.

■ SYSTEM ASSIGNMENT 1:   Design

From Table 9–13 (next-state table) and Table 9–14 (D flip-flop transition table), you can determine the logic conditions required for each flip-flop to have a next state of 1. For example, Q_0 goes from 0 to 1 when the present state is 00 and the input condition is $\overline{T}_L V_s$, as indicated on the second row of the table shown here. D_0 must be a 1 to make Q_0 go to a 1 or to remain a 1. This condition for D_0 can be written as a logic expression as follows:

$$D_0 = \overline{Q}_1 \overline{Q}_0 \overline{T}_L V_s$$

| Present State | | Next State | | Input Conditions | Flip-Flop Inputs | |
Q_1	Q_0	Q_1	Q_0		D_1	D_0
0	0	0	0	$T_L + \overline{V}_s$		
0	0	0	1	$\overline{T}_L V_s$	0	1
0	1	0	1	T_S		
0	1	1	1	$\overline{T}_S$		
1	1	1	1	$T_L V_s$		
1	1	1	0	$\overline{T}_L + \overline{V}_s$		
1	0	1	0	T_S		
1	0	0	0	$\overline{T}_S$		

■ *Activity 1:* Complete the *flop-flop inputs* part of the table. Determine the rest of the logic conditions that make $D_0 = 1$ and develop a logic expression for D_0. Keep in mind that you are looking for conditions for each present state that make the flip-flop go to a 1 or stay a 1 in the next state. The resulting expression should be a sum-of-products.

■ *Activity 2:* Determine all of the logic conditions that make $D_1 = 1$ and develop a logic expression for D_1. Keep in mind that you are looking for conditions for each present state that make the flip-flop go to a 1 or stay a 1 in the next state. The resulting expression should be a sum-of-products.

■ *Activity 3:* Using the SOP logic expressions developed for D_0 and D_1, implement the sequential logic based on the block diagram in Figure 9–64. Refer to the following data sheets:

■ 74LS74A dual D flip-flops
■ 74LS08 quad AND gates
■ 74LS153 data selector (multiplexer)

Use these devices to implement the logic. The 74LS153 is a dual 4-input data selector to be used as a logic function generator. This topic was discussed in Section 6–7.

■ SYSTEM ASSIGNMENT 2: Verification

■ *Activity 1:* The complete system board shown here includes the sequential logic, state decoding and output logic, and the timing circuits. As the system block diagram shows, the outputs from this board go to the interface circuits which are a separate part of the system. Develop the complete system logic diagram from this board.

■ *Activity 2:* Verify that your sequential logic design from System Assignment 1 agrees with the circuitry on this board.

■ *Activity 3:* Identify and label the outputs and input on the circuit board connector.

■ SYSTEM ASSIGNMENT 3:   Troubleshooting

■ *Activity 1:*   Develop a test procedure for testing the system board in the laboratory by simulating the vehicle sensor input, the interface circuit, and the traffic lights with available components.

■ *Activity 2:*   The total system is shown in the diagram above with the correct sequence of lights indicated. For each of the following cases, determine if there is a problem, and if so, identify the most likely fault or faults on the system circuit board. Assume that the interface circuits and the lights are working properly so that any problem is confined to the system board.

The complete system PC board logic is in file SA09.EWB on your EWB CD. Use EWB to test this system.

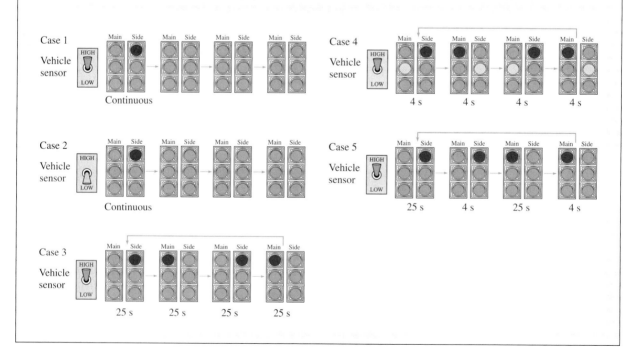

SECTION 9–10 REVIEW

1. Explain the purpose of the sequential logic portion of the system.
2. What controls the transitions of the sequential logic from one state to the next?
3. Explain why the frequency of the clock could be greater or less than 10 kHz and not visually affect the light sequence.

 All or portions of Chapter 15, "Integrated Circuit Technologies," may be covered at this point. The coverage in Chapter 15 is not a prerequisite to any other material in this text, and its omission or partial omission will not affect any other material.

■ SUMMARY

■ Asynchronous and synchronous counters differ only in the way in which they are clocked, as shown in Figure 9–67. Synchronous counters can run at faster clock rates than asynchronous counters.

■ Connection diagrams for the IC counters introduced in this chapter are shown in Figure 9–68.

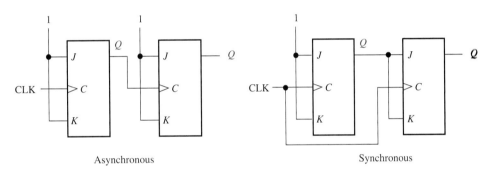

FIGURE 9–67
Comparison of asynchronous and synchronous counters.

74LS93A 4-bit asynchronous
binary counter

74HC160 synchronous decade
counter with asynchronous clear

74HC161 4-bit synchronous binary
counter with asynchronous clear

74HC163 4-bit synchronous binary
counter with synchronous clear

74LS190 synchronous
up/down decade counter

FIGURE 9–68
Note that the labels (names of inputs and outputs) are consistent with text but may differ from the particular manufacturer's data book you are using. The devices shown are functionally the same and pin compatible with the same device types in other available IC families.

■ The maximum modulus of a counter is the maximum number of possible states and is a function of the number of stages (flip-flops). Thus,

$$\text{Maximum modulus} = 2^n$$

where n is the number of stages in the counter. The modulus of a counter is the *actual* number of states in its sequence and can be equal to or less than the maximum modulus.

■ The overall modulus of cascaded counters is equal to the product of the moduli of the individual counters.

■ KEY TERMS

These terms are also in the end-of-book glossary.

Asynchronous counter   A type of counter in which each stage is clocked from the output of the preceding stage.

Cascade   To connect "end-to-end" as when several counters are connected from the terminal count output of one counter to the enable input of the next counter.

Decade   Characterized by ten states or values.

Decade counter   A digital counter having ten states.

Recycle   To undergo transition (as in a counter) from the final or terminal state back to the initial state.

Ripple counter   An asynchronous counter.

Sequence   The order in which several things occur in a specified time relationship.

Sequential circuit   A digital circuit whose logic states follow on a specified time sequence.

State diagram   A graphic depiction of a sequence of states or values.

State machine   A logic system exhibiting a sequence of states conditioned by internal logic and external inputs; any sequential circuit exhibiting a specified sequence of states.

Synchronous counter   A type of counter in which each stage is clocked by the same pulse.

Terminal count   The final state in a counter's sequence.

Truncated   Shortened.

Truncated sequence   A sequence that does not include all of the possible states of a counter.

Up/Down counter   A counter that can progress in either direction through a certain sequence.

■ SELF-TEST

1. Asynchronous counters are known as
   (a) ripple counters   (b) multiple clock counters
   (c) decade counters   (d) modulus counters
2. An asynchronous counter differs from a synchronous counter in
   (a) the number of states in its sequence   (b) the method of clocking
   (c) the type of flip-flops used   (d) the value of the modulus
3. The modulus of a counter is
   (a) the number of flip-flops
   (b) the actual number of states in its sequence
   (c) the number of times it recycles in a second
   (d) the maximum possible number of states
4. A 3-bit binary counter has a maximum modulus of
   (a) 3   (b) 6   (c) 8   (d) 16

5. A 4-bit binary counter has a maximum modulus of
   (a) 16   (b) 32   (c) 8   (d) 4

6. A modulus-12 counter must have
   (a) 12 flip-flops   (b) 3 flip-flops
   (c) 4 flip-flops   (d) synchronous clocking

7. Which one of the following is an example of a counter with a truncated modulus?
   (a) Modulus 8   (b) Modulus 14
   (c) Modulus 16   (d) Modulus 32

8. A 4-bit ripple counter consists of flip-flops that each have a propagation delay from clock to Q output of 12 ns. For the counter to recycle from 1111 to 0000, it takes a total of
   (a) 12 ns   (b) 24 ns   (c) 48 ns   (d) 36 ns

9. A BCD counter is an example of
   (a) a full-modulus counter   (b) a decade counter
   (c) a truncated-modulus counter   (d) answers (b) and (c)

10. Which of the following is an invalid state in an 8421 BCD counter?
    (a) 1100   (b) 0010   (c) 0101   (d) 1000

11. Three cascaded modulus-10 counters have an overall modulus of
    (a) 30   (b) 100   (c) 1000   (d) 10,000

12. A 10 MHz clock frequency is applied to a cascaded counter consisting of a modulus-5 counter, a modulus-8 counter, and two modulus-10 counters. The lowest output frequency possible is
    (a) 10 kHz   (b) 2.5 kHz   (c) 5 kHz   (d) 25 kHz

13. A 4-bit binary up/down counter is in the binary state of zero. The next state in the DOWN mode is
    (a) 0001   (b) 1111   (c) 1000   (d) 1110

14. The terminal count of a modulus-13 binary counter is
    (a) 0000   (b) 1111   (c) 1101   (d) 1100

■ **PROBLEMS**

SECTION 9–1   Asynchronous Counter Operation

1. For the ripple counter shown in Figure 9–69, show the complete timing diagram for eight clock pulses, showing the clock, Q_0, and Q_1 waveforms.

2. For the ripple counter in Figure 9–70, show the complete timing diagram for sixteen clock pulses. Show the clock, Q_0, Q_1, and Q_2 waveforms.

FIGURE 9–69

FIGURE 9–70

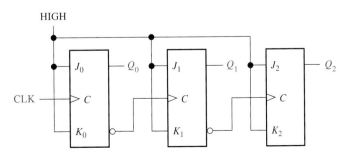

3. In the counter of Problem 2, assume that each flip-flop has a propagation delay from the triggering edge of the clock to a change in the Q output of 8 ns. Determine the worst-case (longest) delay time from a clock pulse to the arrival of the counter in a given state. Specify the state or states for which this worst-case delay occurs.

4. Show how to connect a 74LS93A 4-bit asynchronous counter for each of the following moduli:
   (a) 9   (b) 11   (c) 13   (d) 14   (e) 15

SECTION 9–2   Synchronous Counter Operation

5. If the counter of Problem 3 were synchronous rather than asynchronous, what would be the longest delay time?

6. Show the complete timing diagram for the 5-stage synchronous binary counter in Figure 9–71. Verify that the waveforms of the Q outputs represent the proper binary number after each clock pulse.

7. By analyzing the J and K inputs to each flip-flop prior to each clock pulse, prove that the decade counter in Figure 9–72 progresses through a BCD sequence. Explain how these conditions in each case cause the counter to go to the next proper state.

FIGURE 9–71

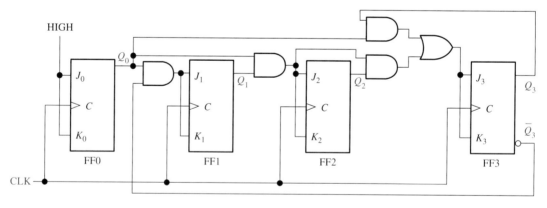

FIGURE 9–72

8. The waveforms in Figure 9–73 are applied to the count enable, clear, and clock inputs as indicated. Show the counter output waveforms in proper relation to these inputs. The clear input is asynchronous.

9. A BCD decade counter is shown in Figure 9–74. The waveforms are applied to the clock and clear inputs as indicated. Determine the waveforms for each of the counter outputs (Q_0, Q_1, Q_2, and Q_3). The clear is synchronous, and the counter is initially in the binary 1000 state.

10. The waveforms in Figure 9–75 are applied to a 74HC163 counter. Determine the Q outputs and the RCO. The inputs are $D_0 = 1$, $D_1 = 1$, $D_2 = 0$, and $D_3 = 1$.

FIGURE 9–73

FIGURE 9–74

FIGURE 9–75

11. The waveforms in Figure 9–75 are applied to a 74HC160 counter. Determine the Q outputs and the RCO. The inputs are $D_0 = 1$, $D_1 = 0$, $D_2 = 0$, and $D_3 = 1$.

SECTION 9–3 Up/Down Synchronous Counters

12. Show a complete timing diagram for a 3-bit up/down counter that goes through the following sequence. Indicate when the counter is in the UP mode and when it is in the DOWN mode. Assume positive edge-triggering.

$$0, 1, 2, 3, 2, 1, 2, 3, 4, 5, 6, 5, 4, 3, 2, 1, 0$$

13. Develop the Q output waveforms for a 74LS190 up/down counter with the input waveforms shown in Figure 9–76. A binary 0 is on the data inputs. Start with a count of 0000.

FIGURE 9–76

SECTION 9–4 Design of Synchronous Counters

14. Determine the sequence of the counter in Figure 9–77.

15. Determine the sequence of the counter in Figure 9–78. Begin with the counter cleared.

FIGURE 9–77

FIGURE 9–78

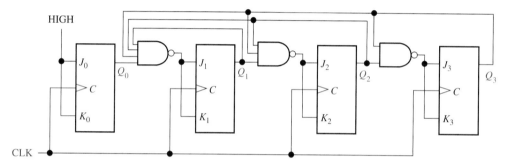

16. Design a counter to produce the following sequence. Use J-K flip-flops.

$$00, 10, 01, 11, 00, \ldots$$

17. Design a counter to produce the following binary sequence. Use J-K flip-flops.

$$1, 4, 3, 5, 7, 6, 2, 1, \ldots$$

18. Design a counter to produce the following binary sequence. Use J-K flip-flops.

$$0, 9, 1, 8, 2, 7, 3, 6, 4, 5, 0, \ldots$$

19. Design a binary counter with the sequence shown in the state diagram of Figure 9–79.

FIGURE 9–79

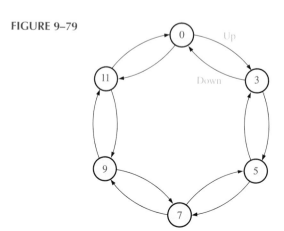

SECTION 9–5   Cascaded Counters

20. For each of the cascaded counter configurations in Figure 9–80, determine the frequency of the waveform at each point indicated by a circled number, and determine the overall modulus.

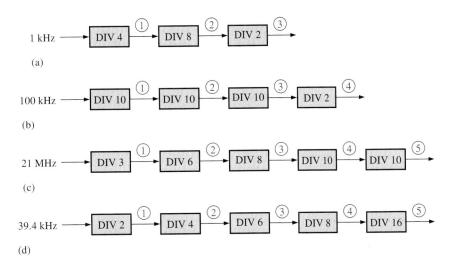

(a)

(b)

(c)

(d)

FIGURE 9–80

21. Expand the counter in Figure 9–41 to create a divide-by-10,000 counter and a divide-by-100,000 counter.

22. With general block diagrams, show how to obtain the following frequencies from a 10 MHz clock by using single flip-flops, modulus-5 counters, and decade counters:

    (a) 5 MHz   **(b)** 2.5 MHz   **(c)** 2 MHz   **(d)** 1 MHz   **(e)** 500 kHz

    (f) 250 kHz   **(g)** 62.5 kHz   **(h)** 40 kHz   **(i)** 10 kHz   **(j)** 1 kHz

SECTION 9–6   Counter Decoding

23. Given a BCD decade counter with only the Q outputs available, show what decoding logic is required to decode each of the following states and how it should be connected to the counter. A HIGH output indication is required for each decoded state. The MSB is to the left.

    (a) 0001   **(b)** 0011   **(c)** 0101   **(d)** 0111   **(e)** 1000

24. For the 4-bit binary counter connected to the decoder in Figure 9–81, determine each of the decoder output waveforms in relation to the clock pulses.

FIGURE 9–81

25. If the counter in Figure 9–81 is asynchronous, determine where the decoding glitches occur on the decoder output waveforms.

26. Modify the circuit in Figure 9–81 to eliminate decoding glitches.

27. Analyze the counter in Figure 9–45 for the occurrence of glitches on the decode gate output. If glitches occur, suggest a way to eliminate them.

28. Analyze the counter in Figure 9–46 for the occurrence of glitches on the outputs of the decoding gates. If glitches occur, make a design change that will eliminate them.

SECTION 9–7   Counter Applications

29. Assume that the digital clock of Figure 9–51 is initially reset to 12 o'clock. Determine the binary state of each counter after sixty-two 60 Hz pulses have occurred.

30. What is the output frequency of each counter in the digital clock circuit of Figure 9–51?

31. For the automobile parking control system in Figure 9–54 a pattern of entrance and exit sensor pulses during a given 24-hour period are shown in Figure 9–82. If there were 53 cars already in the garage at the beginning of the period, what is the state of the counter at the end of the 24 hours?

FIGURE 9–82

32. The binary number for decimal 57 appears on the parallel data inputs of the parallel-to-serial converter in Figure 9–56 (D_0 is the LSB). The counter initially contains all zeros and a 10 kHz clock is applied. Develop the timing diagram showing the clock, the counter outputs, and the serial data output.

SECTION 9–8   Troubleshooting

33. For the counter in Figure 9–1 show the timing diagram for the Q_0 and Q_1 waveforms for each of the following faults (assume Q_0 and Q_1 are initially LOW):

  (a) clock input to FF0 shorted to ground

  (b) Q_0 output open

  (c) clock input to FF1 open

  (d) J input to FF0 open

  (e) K input to FF1 shorted to ground

34. Solve Problem 33 for the counter in Figure 9–11.

35. Isolate the fault in the counter in Figure 9–3 by analyzing the waveforms in Figure 9–83.

FIGURE 9–83

36. From the waveform diagram in Figure 9–84, determine the most likely fault in the counter of Figure 9–14.

37. Solve Problem 36 if the Q_2 output has the waveform observed in Figure 9–85. Outputs Q_0 and Q_1 are the same as in Figure 9–84.

FIGURE 9–84

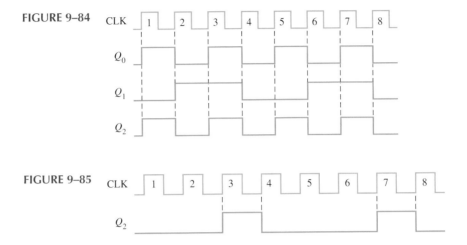

38. You apply a 5 MHz clock to the cascaded counter in Figure 9–44 and measure a frequency of 76.2939 Hz at the last *RCO* output. Is this correct, and if not, what is the most likely problem?

39. Develop a table for use in testing the counter in Figure 9–44 that will show the frequency at the final *RCO* output for all possible open failures of the parallel data inputs (D_0, D_1, D_2, and D_3) taken one at a time. Use 10 MHz as the test frequency for the clock.

40. The tens-of-hours 7-segment display in the digital clock system of Figure 9–51 continuously displays a 1. All the other digits work properly. What could be the problem?

41. What would be the visual indication of an open Q_1 output in the tens portion of the minutes counter in Figure 9–51? Also see Figure 9–52.

42. One day (perhaps a Monday) complaints begin flooding in from patrons of a parking garage that has the control system depicted in Figures 9–54 and 9–55. The patrons say that they enter the garage because the gate is up and the FULL sign is off but that, once in, they can find no empty space. As the technician in charge of this facility, what do you think the problem is, and how will you troubleshoot and repair the system as quickly as possible?

SECTION 9–10   Digital System Application

43. Implement the input logic in the sequential circuit portion of the traffic light control system using only NAND gates.

44. Replace the D flip-flops in the 2-bit Gray code state counter in Figure 9–66 with specific J-K flip-flops.

45. Specify how you would change the time interval for the green light from 25 s to 60 s.

Special Design Problems

46. Design a modulus-1000 counter by using 74HC160 decade counters.

47. Modify the design of the counter in Figure 9–44 to achieve a modulus of 30,000.

48. Repeat Problem 47 for a modulus of 50,000.

49. Modify the digital clock in Figures 9–51, 9–52, and 9–53 so that it can be preset to any desired time.

50. Design an alarm circuit for the digital clock that can detect a predetermined time (hours and minutes only) and produce a signal to activate an audio alarm.

51. Modify the design of the circuit in Figure 9–55 for a 1000-space parking garage and a 3000-space parking garage.

52. Implement the parallel-to-serial data conversion logic in Figure 9–56 with specific devices.

53. In Problem 15 you found that the counter locks up and alternates between two states. It turns out that this operation is the result of a design flaw. Redesign the counter so that when it goes into the second of the lock-up states, it will recycle to the all-0s state on the next clock pulse.

54. Modify the block diagram of the traffic light control system in Figure 9–63 to reflect the addition of a 15 s left turn signal on the main street immediately preceding the green light.

Electronics Workbench Troubleshooting

55. Open file PR09_55.EWB and test the 4-bit asynchronous counter to determine if there is a fault. If there is a fault, identify it if possible.

56. Open file PR09_56.EWB and test the 3-bit synchronous counter to determine if there is a fault. If there is a fault, identify it if possible.

57. Open file PR09_57.EWB and test the BCD counter to determine if there is a fault. If there is a fault, identify it if possible.

58. Open file PR09_58.EWB and test the 74163 4-bit binary counter to determine if there is a fault. If there is a fault, identify it if possible.

59. Open file PR09_59.EWB and test the 74190 Up/Down decade counter to determine if there is a fault. If there is a fault, identify it if possible.

■ ANSWERS TO SECTION REVIEWS

SECTION 9–1

1. Asynchronous means that each flip-flop after the first one is enabled by the output of the preceding flip-flop.

2. A modulus-14 counter has fourteen states requiring four flip-flops.

SECTION 9–2

1. All flip-flops in a synchronous counter are clocked simultaneously.

2. The counter can be preset (initialized) to any given state.

3. Counter is enabled when *ENP* and *ENT* are both HIGH; *RCO* goes HIGH when final state in sequence is reached.

SECTION 9–3

1. The counter goes to 1001. **2.** UP: 1111; DOWN: 0000; the next state is 1111.

SECTION 9–4

1. $J = 1, K = X$ ("don't care") **2.** $J = X$ ("don't care"), $K = 0$

3. **(a)** The next state is 1011.

(b) Q_3 (MSB): no-change or SET; Q_2: no-change or RESET; Q_1: no change or SET; Q_0 (LSB): SET or toggle

SECTION 9–5

1. Three decade counters produce ÷1000; 4 decade counters produce ÷10,000.

2. **(a)** ÷20: flip-flop and DIV 10 **(b)** ÷32: flip-flop and DIV 16

(c) ÷160: DIV 16 and DIV 10 **(d)** ÷320: DIV 16 and DIV 10 and flip-flop

SECTION 9–6

1. **(a)** No transitional states because there is a single bit change

(b) 0000, 0001, 0010, 0101, 0110, 0111

(c) No transitional states because there is a single bit change

(d) 0001, 0010, 0011, 0100, 0101, 0110, 0111, 1000, 1001, 1010, 1011, 1100, 1101, 1110

SECTION 9–7

1. Gate G_1 resets flip-flop on first clock pulse after count 12. Gate G_2 decodes count 12 to preset counter to 0001.

2. The hours decade counter advances through each state from zero to nine, and as it recycles from nine back to zero, the flip-flop is toggled to the SET state. This produces a ten (10) on the display. When the hours decade counter is in state 12, the decode NAND gate causes the counter to recycle to state 1 on the next clock pulse. The flip-flop resets. This results in a one (01) on the display.

SECTION 9–8

1. No pulses on TC outputs: $CTEN$ of first counter shorted to ground or to a LOW; clock input of first counter open; clock line shorted to ground or to a LOW; TC output of first counter shorted to ground or to a LOW

2. With inverter output open, the counter does not recycle at the preset count but acts as a full-modulus counter.

SECTION 9–9

1. C: control, usually clock; M: mode; G: AND   2. D indicates data storage.

SECTION 9–10

1. The sequential logic produces the proper sequence of states to control the lights.

2. The long timer, the short timer, and/or the vehicle sensor controls sequential logic transitions.

3. The shortest light interval of 4 s is much greater than the period of a 10 kHz clock.

■ **ANSWERS TO RELATED PROBLEMS FOR EXAMPLES**

9–1 See Figure 9–86.

9–2 Connect Q_0 to the NAND gate as a third input (Q_2 and Q_3 are two of the inputs). Connect the $\overline{CLR}$ line to the $\overline{CLR}$ input of FF0 as well as FF2 and FF3.

9–3 See Figure 9–87.   9–4 See Figure 9–88.

FIGURE 9–86

FIGURE 9–87

FIGURE 9–88

TABLE 9–15

Present Invalid State			J-K Inputs						Next State			
Q_2	Q_1	Q_0	J_2	K_2	J_1	K_1	J_0	K_0	Q_2	Q_1	Q_0	
0	0	0	0	0	1	1	1	1	0	1	1	
0	1	1	1	1	1	1	1	1	1	0	0	
1	0	0	0	0	1	1	1	0	1	1	1	valid state
1	1	0	1	1	1	1	1	0	0	0	1	valid state

9–5 See Table 9–15.

9–6 Application of Boolean algebra to the logic in Figure 9–37 shows that the output of each OR gate agrees with the expression in step 5.

9–7 Five decade counters are required. $10^5 = 100,000$

9–8 $f_{Q0} = 1 \text{ MHz}/[(10)(2)] = 50 \text{ kHz}$

9–9 See Figure 9–89.

9–10 $8AC0_{16}$ would be loaded. $16^4 - 8AC0_{16} = 65,536 - 32,520 = 30,016$
$f_{TC4} = 10 \text{ MHz}/30,016 = 333.2 \text{ Hz}$

9–11 See Figure 9–90.

FIGURE 9–89

FIGURE 9–90

■ **ANSWERS TO SELF-TEST**

1. (a)	**2.** (b)	**3.** (b)	**4.** (c)	**5.** (a)	**6.** (c)	**7.** (b)	**8.** (c)
9. (d)	**10.** (a)	**11.** (c)	**12.** (b)	**13.** (b)	**14.** (d)		

10

SHIFT REGISTERS

 Computer Simulation
of Digital Circuits:
Electronics Workbench (EWB) Tutorial at
http://www.prenhall.com/floyd

■ CHAPTER OBJECTIVES

❏ Identify the basic forms of data movement in shift registers

❏ Explain how serial in/serial out, serial in/parallel out, parallel in/serial out, and parallel in/parallel out shift registers operate

❏ Describe how a bidirectional shift register operates

❏ Determine the sequence of a Johnson counter

❏ Set up a ring counter to produce a specified sequence

❏ Construct a ring counter from a shift register

❏ Use a shift register as a time-delay device

❏ Use a shift register to implement a serial-to-parallel data converter

❏ Implement a basic shift-register-controlled keyboard encoder

❏ Troubleshoot digital systems by "exercising" the system with a known test pattern

❏ Interpret ANSI/IEEE Standard 91–1984 shift register symbols with dependency notation

❏ Use shift registers in a system application

■ CHAPTER OVERVIEW

Shift registers are a type of sequential logic circuit closely related to digital counters. Registers are used primarily for the storage of digital data and typically do not possess a characteristic internal sequence of states as do counters. There are exceptions, however, and these are covered in Section 10–7.

In this chapter, the basic types of shift registers are studied and several applications are presented.

Also, an important troubleshooting method is introduced.

■ SPECIFIC DEVICES

| 74HC164 | 74HC165 | 74HC174A |
| 74HC194 | 74HC195 | |

■ DIGITAL SYSTEM APPLICATION

This Digital System Application illustrates the concepts taught in this chapter. The system application in Section 10–11 introduces a security entry system for controlling the alarms in a building. This system uses two types of shift registers as well as other types of devices covered in previous chapters. The system also includes a memory that will be the focus in Chapter 12.

10–1 ■ BASIC SHIFT REGISTER FUNCTIONS

Shift registers consist of an arrangement of flip-flops and are important in applications involving the storage and transfer of data in a digital system. The basic difference between a register and a counter is that a register has no specified sequence of states, except in certain very specialized applications. A register, in general, is used solely for storing and shifting data (1s and 0s) entered into it from an external source and typically possesses no characteristic internal sequence of states.

After completing this section, you should be able to

■ Explain how a flip-flop stores a data bit ■ Define the storage capacity of a shift register ■ Define the shifting capability of a register

A **register** is a digital circuit with two basic functions: data storage and data movement. The storage capability of a register makes it an important type of *memory* device. Figure 10–1 illustrates the concept of storing a 1 or a 0 in a D flip-flop. A 1 is applied to the data input as shown, and a clock pulse is applied that stores the 1 by *setting* the flip-flop. When the 1 on the input is removed, the flip-flop remains in the SET state, thereby storing the 1. A similar procedure applies to the storage of a 0 by *resetting* the flip-flop, as also illustrated in Figure 10–1.

FIGURE 10–1
The flip-flop as a storage element.

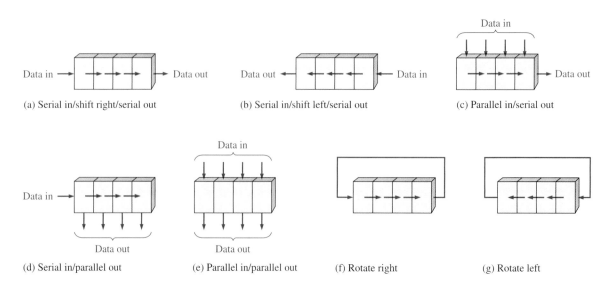

FIGURE 10–2
Basic data movement in shift registers (four bits are used for illustration).

The *storage capacity* of a register is the total number of bits (1s and 0s) of digital data it can retain. Each **stage** (flip-flop) in a **shift** register represents one bit of storage capacity; therefore, the number of stages in a register determines its storage capacity.

The shifting capability of a register permits the movement of data from stage to stage within the register or into or out of the register upon application of clock pulses. Figure 10–2 illustrates the types of data movement in shift registers. The block represents any arbitrary 4-bit register, and the arrows indicate the direction of data movement.

<table>
<tr><td>SECTION 10–1
REVIEW</td><td>1. Generally, what is the difference between a counter and a shift register?
2. What two principal functions are performed by a shift register?</td></tr>
</table>

10–2 ■ SERIAL IN/SERIAL OUT SHIFT REGISTERS

The serial in/serial out shift register accepts data serially—that is, one bit at a time on a single line. It produces the stored information on its output also in serial form.

After completing this section, you should be able to

■ Explain how data bits are serially entered into a shift register ■ Describe how data bits are shifted through the register ■ Explain how data bits are serially taken out of a shift register ■ Develop and analyze timing diagrams for serial in/serial out registers

Let's first look at the serial entry of data into a typical shift register. Figure 10–3 shows a 4-bit device implemented with D flip-flops. With four stages, this register can store up to four bits of data.

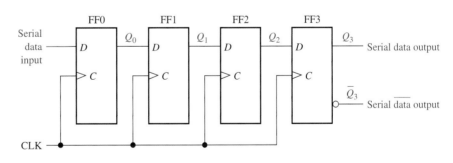

FIGURE 10–3
Serial in/serial out shift register.

Figure 10–4 illustrates entry of the four bits 1010 into the register, beginning with the right-most bit. The register is initially clear. The 0 is put onto the data input line, making $D = 0$ for FF0. When the first clock pulse is applied, FF0 is reset, thus storing the 0.

Next the second bit, which is a 1, is applied to the data input, making $D = 1$ for FF0 and $D = 0$ for FF1 because the D input of FF1 is connected to the Q_0 output. When the second clock pulse occurs, the 1 on the data input is shifted into FF0, causing FF0 to set; and the 0 that was in FF0 is shifted into FF1.

The third bit, a 0, is now put onto the data-input line, and a clock pulse is applied. The 0 is entered into FF0, the 1 stored in FF0 is shifted into FF1, and the 0 stored in FF1 is shifted into FF2.

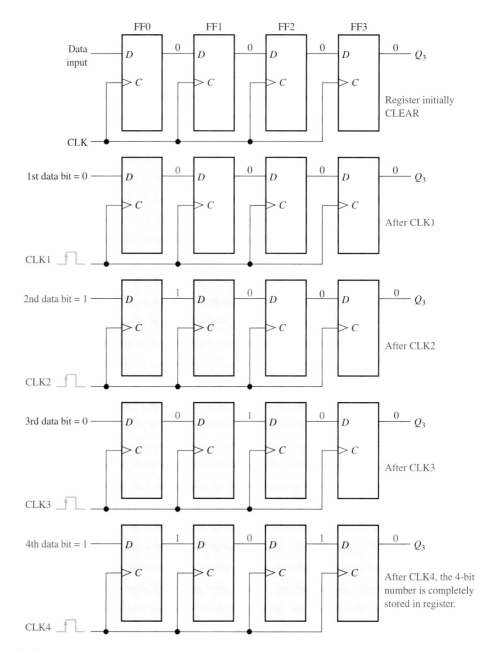

FIGURE 10–4
Four bits (1010) being entered serially into the register.

The last bit, a 1, is now applied to the data input, and a clock pulse is applied. This time the 1 is entered into FF0, the 0 stored in FF0 is shifted into FF1, the 1 stored in FF1 is shifted into FF2, and the 0 stored in FF2 is shifted into FF3. This completes the serial entry of the four bits into the shift register, where they can be stored for any length of time as long as the flip-flops have dc power.

If you want to get the data out of the register, the bits must be shifted out serially and taken off the Q_3 output, as Figure 10–5 illustrates. After CLK4 in the data-entry operation just described, the right-most bit, 0, appears on the Q_3 output. When clock pulse CLK5 is applied, the second bit appears on the Q_3 output. Clock pulse CLK6 shifts the third bit to the output, and CLK7 shifts the fourth bit to the output. Notice that while the original four bits are being shifted out, more bits can be shifted in. All zeros are shown being shifted in.

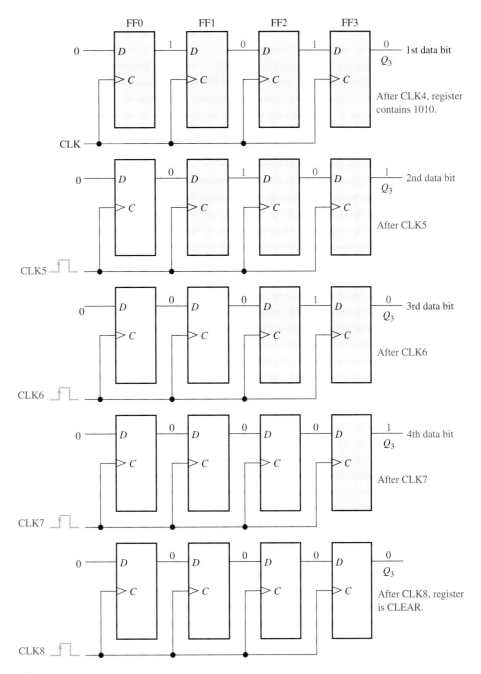

FIGURE 10–5
Four bits (1010) being serially shifted out of the register and replaced by all zeros.

EXAMPLE 10–1

Show the states of the 5-bit register in Figure 10–6(a) for the specified data input and clock waveforms. Assume that the register is initially cleared (all 0s).

Solution The first data bit (1) is entered into the register on the first clock pulse and then shifted from left to right as the remaining bits are entered and shifted. The register contains $Q_4 Q_3 Q_2 Q_1 Q_0 = 11010$ after five clock pulses. See Figure 10–6(b).

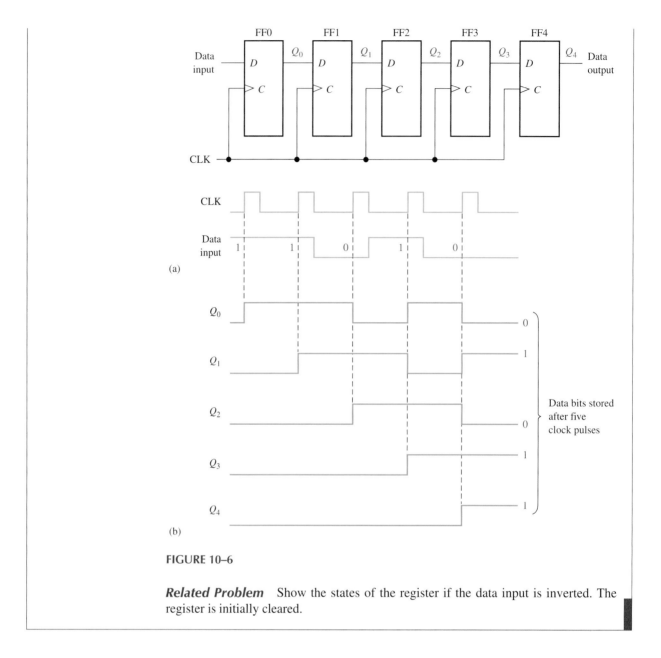

FIGURE 10–6

Related Problem Show the states of the register if the data input is inverted. The register is initially cleared.

A traditional logic block symbol for an 8-bit serial in/serial out shift register is shown in Figure 10–7. The "SRG 8" designation means a shift register (SRG) with an 8-bit capacity.

FIGURE 10–7
Logic symbol for an 8-bit serial in/serial out shift register.

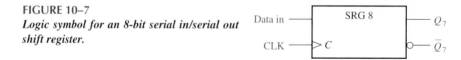

SECTION 10–2 REVIEW

1. Develop the logic diagram for the shift register in Figure 10–3, using J-K flip-flops to replace the D flip-flops.
2. How many clock pulses are required to enter a byte of data serially into an 8-bit shift register?

10–3 ■ SERIAL IN/PARALLEL OUT SHIFT REGISTERS

Data bits are entered serially (right-most bit first) into this type of register in the same manner as discussed in Section 10–2. The difference is the way in which the data bits are taken out of the register; in the parallel output register, the output of each stage is available. Once the data are stored, each bit appears on its respective output line, and all bits are available simultaneously, rather than on a bit-by-bit basis as with the serial output.

After completing this section, you should be able to

■ Explain how data bits are taken out of a shift register in parallel ■ Compare serial output to parallel output ■ Discuss the 74HC164 8-bit shift register ■ Develop and analyze timing diagrams for serial in/parallel out registers

Figure 10–8 shows a 4-bit serial in/parallel out shift register and its logic block symbol.

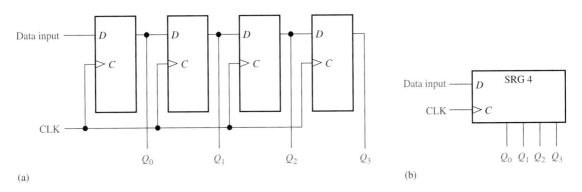

FIGURE 10–8
A serial in/parallel out shift register.

EXAMPLE 10–2

Show the states of the 4-bit register (SRG 4) for the data input and clock waveforms in Figure 10–9(a). The register initially contains all 1s.

Solution   The register contains 0110 after four clock pulses. See Figure 10–9(b).

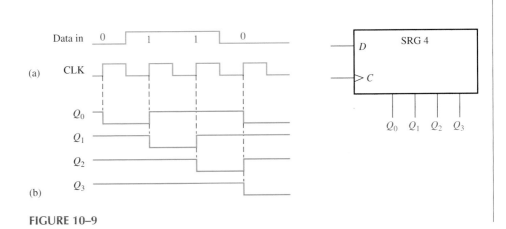

FIGURE 10–9

> ***Related Problem***   If the data input remains 0 after the fourth clock pulse, what is the state of the register after three additional clock pulses?

An 8-Bit Serial In/Parallel Out Shift Register

The 74HC164 is an example of an IC shift register having serial in/parallel out operation. The logic diagram is shown in Figure 10–10(a), and a typical logic block symbol is shown in part (b). Notice that this device has two gated serial inputs, A and B, and a clear ($\overline{CLR}$) input that is active-LOW. The parallel outputs are Q_0 through Q_7.

(a) Logic diagram

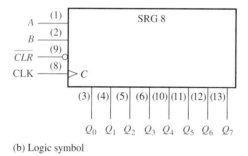

(b) Logic symbol

FIGURE 10–10
The 74HC164 8-bit serial in/parallel out shift register.

A sample timing diagram for the 74HC164 is shown in Figure 10–11. Notice that the serial input data on input A are shifted into and through the register after input B goes HIGH.

SECTION 10–3 REVIEW

1. The bit sequence 1101 is serially entered (right-most bit first) into a 4-bit parallel out shift register that is initially clear. What are the Q outputs after two clock pulses?
2. How can a serial in/parallel out register be used as a serial in/serial out register?

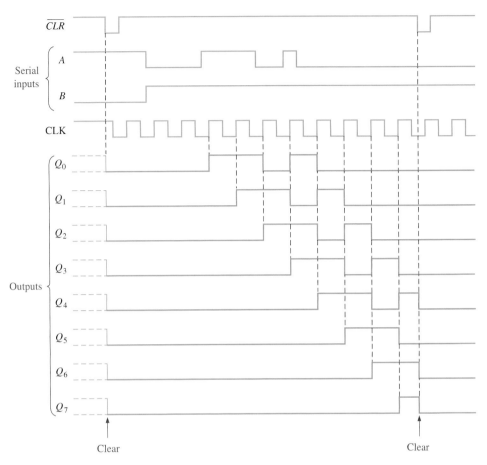

FIGURE 10–11
Sample timing diagram for a 74HC164 shift register.

10–4 ■ PARALLEL IN/SERIAL OUT SHIFT REGISTERS

For a register with parallel data inputs, the bits are entered simultaneously into their respective stages on parallel lines rather than on a bit-by-bit basis on one line as with serial data inputs. The serial output is the same as described in Section 10–2, once the data are completely stored in the register.

After completing this section, you should be able to

■ Explain how data bits are entered into a shift register in parallel ■ Compare serial input to parallel input ■ Discuss the 74HC165 8-bit parallel-load shift register ■ Develop and analyze timing diagrams for parallel in/serial out registers

Figure 10–12 illustrates a 4-bit parallel in/serial out shift register and a typical logic symbol. Notice that there are four data-input lines, D_0, D_1, D_2, and D_3, and a $SHIFT/\overline{LOAD}$ input, which allows four bits of data to be **loaded** in parallel into the register. When $SHIFT/\overline{LOAD}$ is LOW, gates G_1 through G_3 are enabled, allowing each data bit to be applied to the D input of its respective flip-flop. When a clock pulse is applied, the flip-flops with $D = 1$ will set and those with $D = 0$ will reset, thereby storing all four bits simultaneously.

When $SHIFT/\overline{LOAD}$ is HIGH, gates G_1 through G_3 are disabled and gates G_4 through G_6 are enabled, allowing the data bits to shift right from one stage to the next. The OR gates allow either the normal shifting operation or the parallel data-entry operation, depending on which AND gates are enabled by the level on the $SHIFT/\overline{LOAD}$ input.

(a) Logic diagram

(b) Logic symbol

FIGURE 10–12
*A **4-bit parallel in/serial out** shift register.*

EXAMPLE 10–3   Show the data-output waveform for a 4-bit register with the parallel input data and the clock and *SHIFT/$\overline{LOAD}$* waveforms given in Figure 10–13(a). Refer to Figure 10–12(a) for the logic diagram.

FIGURE 10–13

Solution On clock pulse 1, the parallel data ($D_0D_1D_2D_3 = 1010$) are loaded into the register, making Q_3 a 0. On clock pulse 2 the 1 from Q_2 is shifted onto Q_3; on clock pulse 3 the 0 is shifted onto Q_3; on clock pulse 4 the last data bit (1) is shifted onto Q_3; and on clock pulse 5, all data bits have been shifted out, and only 1s remain in the register (assuming the D input remains a 1). See Figure 10–13(b).

Related Problem Show the data-output waveform for the clock and *SHIFT/$\overline{LOAD}$* inputs shown in Figure 10–13(a) if the parallel data are $D_0D_1D_2D_3 = 0101$.

An 8-Bit Parallel Load Shift Register

The 74HC165 is an example of an IC shift register that has a parallel in/serial out operation (it can also be operated as serial in/serial out). Figure 10–14(a) shows the internal logic diagram for this device, and part (b) shows a typical logic block symbol. A LOW on the *SHIFT/$\overline{LOAD}$* input (*SH/$\overline{LD}$*) enables all the NAND gates for parallel loading. When an

(a) Logic diagram

(b) Logic symbol

FIGURE 10–14
The 74HC165 8-bit parallel load shift register.

input data bit is a 1, the flip-flop is asynchronously set by a LOW out of the upper gate. When an input data bit is a 0, the flip-flop is asynchronously reset by a LOW out of the lower gate. Additionally, data can be entered serially on the *SER* input. Also, the clock can be inhibited anytime with a HIGH on the *CLK INH* input. The serial data outputs of the register are Q_7 and its complement $\overline{Q}_7$. This implementation is different from the synchronous method of parallel loading previously discussed, demonstrating that there are usually several ways to accomplish the same function.

Figure 10–15 is a timing diagram showing an example of the operation of a 74HC165 shift register.

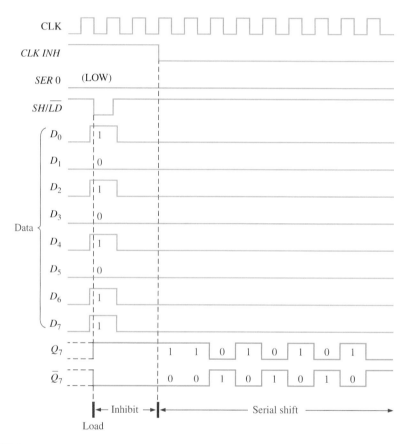

FIGURE 10–15
Sample timing diagram for a 74HC165 shift register.

SECTION 10–4 REVIEW
1. Explain the function of the *SHIFT/$\overline{LOAD}$* input.
2. Is the parallel load operation in a 74HC165 shift register synchronous or asynchronous? What does this mean?

10–5 ■ PARALLEL IN/PARALLEL OUT SHIFT REGISTERS

Parallel entry of data was described in Section 10–4, and parallel output of data has also been discussed previously. The parallel in/parallel out register employs both methods. Immediately following the simultaneous entry of all data bits, the bits appear on the parallel outputs.

■ Discuss the 74HC195 4-bit parallel-access shift register ■ Develop and analyze timing diagrams for parallel in/parallel out registers

Figure 10–16 shows a parallel in/parallel out register.

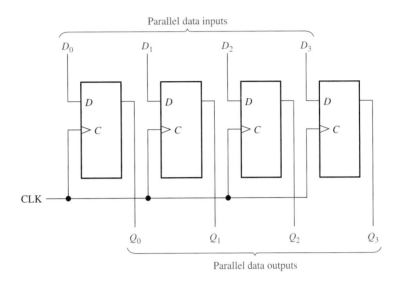

FIGURE 10–16
A parallel in/parallel out register.

A 4-Bit Parallel-Access Shift Register

The 74HC195 can be used for parallel in/parallel out operation. Because it also has a serial input, it can be used for serial in/serial out and serial in/parallel out operations. It can be used for parallel in/serial out operation by using Q_3 as the output. A typical logic block symbol is shown in Figure 10–17.

When the *SHIFT/$\overline{LOAD}$* input (*SH/$\overline{LD}$*) is LOW, the data on the parallel inputs are entered synchronously on the positive transition of the clock. When *SH/$\overline{LD}$* is HIGH, stored data will shift right (Q_0 to Q_3) synchronously with the clock. Inputs J and $\overline{K}$ are the serial data inputs to the first stage of the register (Q_0); Q_3 can be used for serial output data. The active-LOW clear input is asynchronous.

The timing diagram in Figure 10–18 illustrates the operation of this register.

FIGURE 10–17
The 74LS195A 4-bit parallel access shift register.

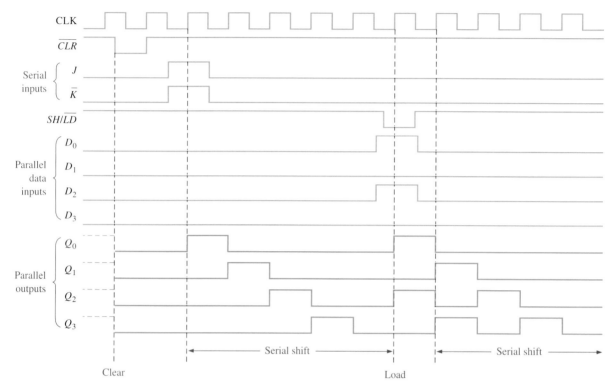

FIGURE 10–18
Sample timing diagram for a 74HC195 shift register.

**SECTION 10–5
REVIEW**

1. In Figure 10–16, $D_0 = 1$, $D_1 = 0$, $D_2 = 0$, and $D_3 = 1$. After three clock pulses, what are the data outputs?
2. For a 74HC195, $SH/\overline{LD} = 1$, $J = 1$, and $\overline{K} = 1$. What is Q_0 after one clock pulse?

10–6 ■ BIDIRECTIONAL SHIFT REGISTERS

A bidirectional shift register is one in which the data can be shifted either left or right. It can be implemented by using gating logic that enables the transfer of a data bit from one stage to the next stage to the right or to the left, depending on the level of a control line.

After completing this section, you should be able to

■ Explain the operation of a bidirectional shift register   ■ Discuss the 74HC194 4-bit bidirectional universal shift register   ■ Develop and analyze timing diagrams for bidirectional shift registers

A 4-bit **bidirectional** shift register is shown in Figure 10–19. A HIGH on the $RIGHT/\overline{LEFT}$ control input allows data bits inside the register to be shifted to the right, and a LOW enables data bits inside the register to be shifted to the left. An examination of the gating logic will make the operation apparent. When the $RIGHT/\overline{LEFT}$ control input is HIGH, gates G_1 through G_4 are enabled, and the state of the Q output of each flip-flop is passed through to the D input of the *following* flip-flop. When a clock pulse occurs, the

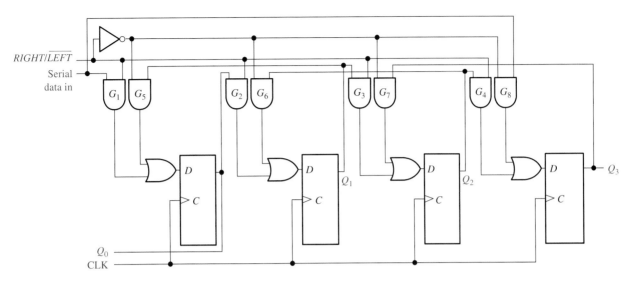

FIGURE 10–19
Four-bit bidirectional shift register.

data bits are shifted one place to the *right*. When the $RIGHT/\overline{LEFT}$ control input is LOW, gates G_5 through G_8 are enabled, and the Q output of each flip-flop is passed through to the D input of the *preceding* flip-flop. When a clock pulse occurs, the data bits are then shifted one place to the *left*.

EXAMPLE 10–4

Determine the state of the shift register of Figure 10–19 after each clock pulse for the given $RIGHT/\overline{LEFT}$ control input waveform in Figure 10–20(a). Assume that $Q_0 = 1$, $Q_1 = 1$, $Q_2 = 0$, and $Q_3 = 1$ and that the serial data-input line is LOW.

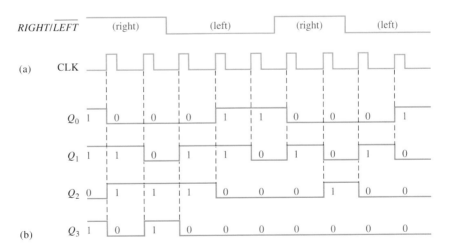

FIGURE 10–20

Solution See Figure 10–20(b).

Related Problem Invert the $RIGHT/\overline{LEFT}$ waveform, and determine the state of the shift register in Figure 10–19 after each clock pulse.

A 4-Bit Bidirectional Universal Shift Register

The 74HC194 is an example of a universal bidirectional shift register in integrated circuit form. A **universal shift register** has both serial and parallel input and output capability. A logic block symbol is shown in Figure 10–21, and a sample timing diagram is shown in Figure 10–22.

Parallel loading, which is synchronous with a positive transition of the clock, is accomplished by applying the four bits of data to the parallel inputs and a HIGH to the S_0 and

FIGURE 10–21
The 74HC194 4-bit bidirectional universal shift register.

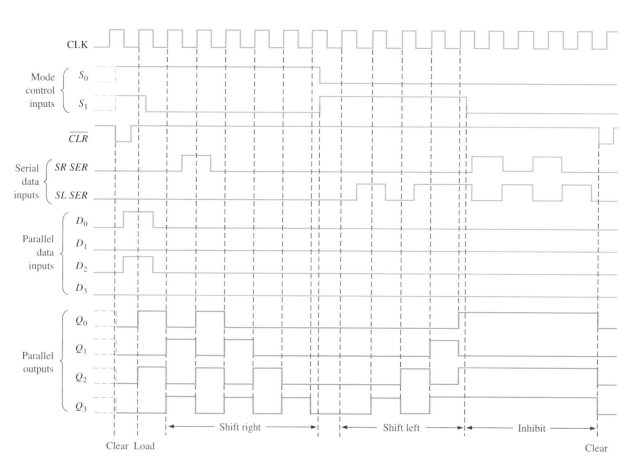

FIGURE 10–22
Sample timing diagram for a 74HC194 shift register.

S_1 inputs. Shift right is accomplished synchronously with the positive edge of the clock when S_0 is HIGH and S_1 is LOW. Serial data in this mode are entered at the shift-right serial input (*SR SER*). When S_0 is LOW and S_1 is HIGH, data bits shift left synchronously with the clock, and new data are entered at the shift-left serial input (*SL SER*). Input *SR SER* goes into the Q_0 stage, and *SL SER* goes into the Q_3 stage.

SECTION 10–6 REVIEW

1. Assume that the 4-bit bidirectional shift register in Figure 10–19 has the following contents: $Q_0 = 1$, $Q_1 = 1$, $Q_2 = 0$, and $Q_3 = 0$. There is a 1 on the serial data-input line. If *RIGHT/LEFT* is HIGH for three clock pulses and LOW for two more clock pulses, what are the contents after the fifth clock pulse?

10–7 ▪ SHIFT REGISTER COUNTERS

A shift register counter is basically a shift register with the serial output connected back to the serial input to produce special sequences. These devices are often classified as counters because they exhibit a specified sequence of states. Two of the most common types of shift register counters, the Johnson counter and the ring counter, are introduced in this section.

After completing this section, you should be able to

▪ Discuss how a shift register counter differs from a basic shift register ▪ Explain the operation of a Johnson counter ▪ Specify a Johnson sequence for any number of bits ▪ Explain the operation of a ring counter and determine the sequence of any specific ring counter

The Johnson Counter

In a **Johnson counter** the complement of the output of the last flip-flop is connected back to the D input of the first flip-flop (it can be implemented with other types of flip-flops as well). This feedback arrangement produces a characteristic sequence of states, as shown in Table 10–1 for a 4-bit device and in Table 10–2 for a 5-bit device. Notice that the 4-bit sequence has a total of eight states, or bit patterns, and that the 5-bit sequence has a total of ten states. In general, a Johnson counter will produce a modulus of $2n$, where n is the number of stages in the counter.

The implementations of the 4-stage and 5-stage Johnson counters are shown in Figure 10–23. The implementation of a Johnson counter is very straightfoward and is the

TABLE 10–1
Four-bit Johnson sequence.

Clock Pulse	Q_0	Q_1	Q_2	Q_3
0	0	0	0	0
1	1	0	0	0
2	1	1	0	0
3	1	1	1	0
4	1	1	1	1
5	0	1	1	1
6	0	0	1	1
7	0	0	0	1

TABLE 10–2
Five-bit Johnson sequence.

Clock Pulse	Q_0	Q_1	Q_2	Q_3	Q_4
0	0	0	0	0	0
1	1	0	0	0	0
2	1	1	0	0	0
3	1	1	1	0	0
4	1	1	1	1	0
5	1	1	1	1	1
6	0	1	1	1	1
7	0	0	1	1	1
8	0	0	0	1	1
9	0	0	0	0	1

(a) Four-bit Johnson counter

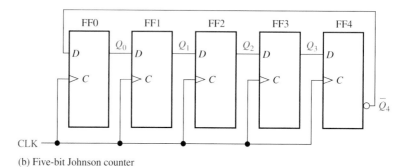

(b) Five-bit Johnson counter

FIGURE 10–23
Four-bit and 5-bit Johnson counters.

same regardless of the number of stages. The Q output of each stage is connected to the D input of the next stage (assuming that D flip-flops are used). The single exception is that the $\overline{Q}$ output of the last stage is connected back to the D input of the first stage. As the sequences in Table 10–1 and 10–2 show, the counter will "fill up" with 1s from left to right, and then it will "fill up" with 0s again.

Diagrams of the timing operations of the 4-bit and 5-bit counters are shown in Figures 10–24 and 10–25, respectively.

The Ring Counter

The **ring counter** utilizes one flip-flop for each state in its sequence. It has the advantage that decoding gates are not required. In the case of a 10-bit ring counter, there is a unique output for each decimal digit.

A logic diagram for a 10-bit ring counter is shown in Figure 10–26. The sequence for this ring counter is given in Table 10–3 on page 540. Initially, a 1 is preset into the first

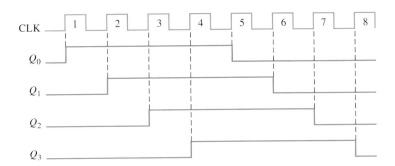

FIGURE 10–24
Timing sequence for a 4-bit Johnson counter.

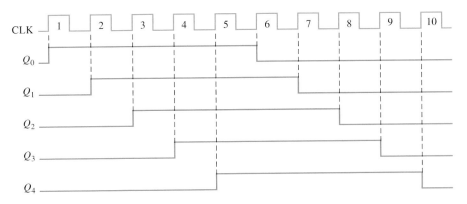

FIGURE 10–25
Timing sequence for a 5-bit Johnson counter.

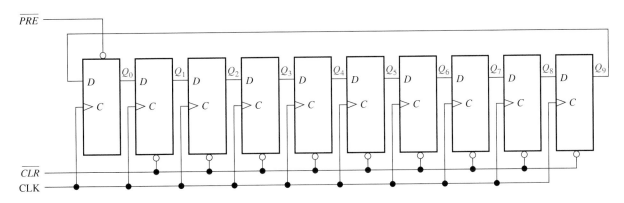

FIGURE 10–26
A 10-bit ring counter.

flip-flop, and the rest of the flip-flops are cleared. Notice that the interstage connections are the same as those for a Johnson counter, except that Q rather than $\overline{Q}$ is fed back from the last stage. The ten outputs of the counter indicate directly the decimal count of the clock pulse. For instance, a 1 on Q_0 represents a zero, a 1 on Q_1 represents a one, a 1 on Q_2 represents a two, a 1 on Q_3 represents a three, and so on. You should verify for yourself that the 1 is always retained in the counter and simply shifted "around the ring," advancing one stage for each clock pulse.

Modified sequences can be achieved by having more than a single 1 in the counter, as illustrated in Example 10–5.

TABLE 10–3

Ten-bit ring counter sequence.

Clock Pulse	Q_0	Q_1	Q_2	Q_3	Q_4	Q_5	Q_6	Q_7	Q_8	Q_9
0	1	0	0	0	0	0	0	0	0	0
1	0	1	0	0	0	0	0	0	0	0
2	0	0	1	0	0	0	0	0	0	0
3	0	0	0	1	0	0	0	0	0	0
4	0	0	0	0	1	0	0	0	0	0
5	0	0	0	0	0	1	0	0	0	0
6	0	0	0	0	0	0	1	0	0	0
7	0	0	0	0	0	0	0	1	0	0
8	0	0	0	0	0	0	0	0	1	0
9	0	0	0	0	0	0	0	0	0	1

EXAMPLE 10–5

If a 10-bit ring counter similar to Figure 10–26 has the initial state 1010000000, determine the waveform for each of the Q outputs.

Solution   See Figure 10–27.

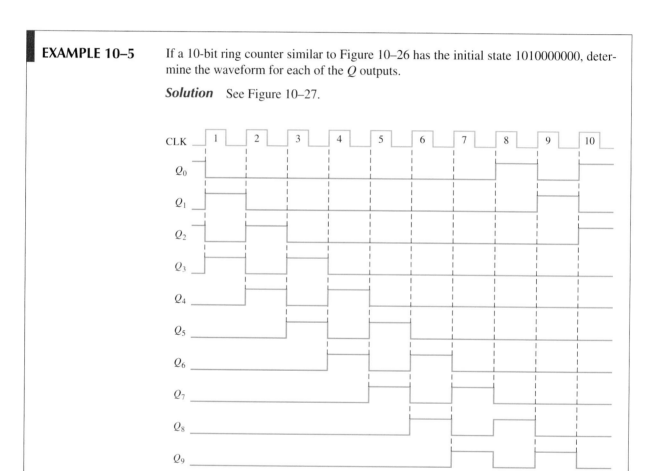

FIGURE 10–27

Related Problem   If a 10-bit ring counter has an initial state 0101001111, determine the waveform for each Q output.

**SECTION 10–7
REVIEW**

1. How many states are there in an 8-bit Johnson counter sequence?
2. Write the sequence of states for a 3-bit Johnson counter starting with 000.

10–8 ■ SHIFT REGISTER APPLICATIONS

Shift registers are found in many types of applications, a few of which are presented in this section.

After completing this section, you should be able to

■ Use a shift register to generate a time delay ■ Implement a specified ring counter sequence using a 74HC195 shift register ■ Discuss how shift registers are used for serial-to-parallel conversion of data ■ Define *UART* ■ Explain the operation of a keyboard encoder and how registers are used in this application

COMPUTER NOTE
The general-purpose registers in the Pentium are all 32-bit registers that can be used for temporary data storage as well as specific uses. Four of these registers are as follows. The *accumulator* (EAX) is used mainly for temporary storage of data and instruction operands. The *base register* (EBX) is used to store a value temporarily. The *count register* (ECX) is mainly used to determine the number of repetitions in certain loop, string, shift, or rotate operations. The *data register* (EDX), is normally used for the temporary storage of data.

Time Delay

The serial in/serial out shift register can be used to provide a time delay from input to output that is a function of both the number of stages (n) in the register and the clock frequency.

When a data pulse is applied to the serial input in Figure 10–28 (A and B connected together), it enters the first stage on the triggering edge of the clock pulse. It is then shifted from stage to stage on each successive clock pulse until it appears on the serial output n clock periods later. This time-delay operation is illustrated in Figure 10–28, in which an 8-bit serial in/serial out shift register is used with a clock frequency of 1 MHz to achieve a time delay (t_d) of 8 μs ($8 \times 1 \ \mu$s). This time can be adjusted up or down by changing the clock frequency. The time delay can also be increased by cascading shift registers and decreased by taking the outputs from successively lower stages in the register if the outputs are available, as illustrated in Example 10–6.

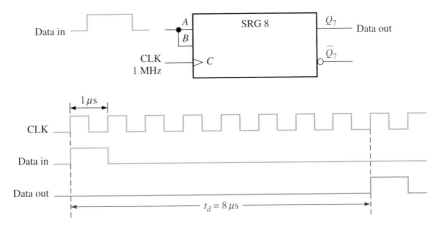

FIGURE 10–28
The shift register as a time-delay device.

EXAMPLE 10–6

Determine the amount of time delay between the serial input and each output in Figure 10–29. Show a timing diagram to illustrate.

FIGURE 10–29

* Data shifts from Q_0 toward Q_7.

Solution   The clock period is 2 μs. Thus, the time delay can be increased or decreased in 2 μs increments from a minimum of 2 μs to a maximum of 16 μs, as illustrated in Figure 10–30.

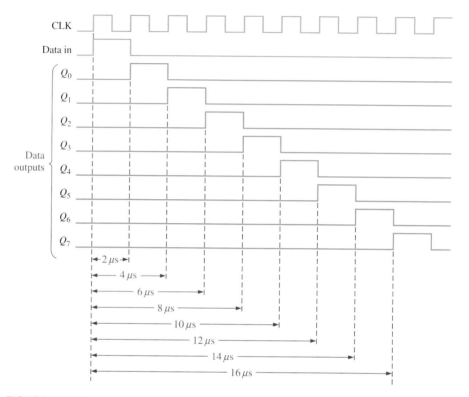

FIGURE 10–30
Timing diagram showing time delays for the register in Figure 10–29.

Related Problem   Determine the clock frequency required to obtain a time delay of 24 μs to the Q_7 output in Figure 10–29.

A Ring Counter Using a 74HC195 Shift Register

If the output is connected back to the serial input, a shift register can be used as a ring counter. Figure 10–31 illustrates this application with a 74HC195 4-bit shift register.

Initially, a bit pattern of 1000 (or any other pattern) can be synchronously preset into the counter by applying the bit pattern to the parallel data inputs, taking the $SH/\overline{LD}$ input LOW, and applying a clock pulse. After this initialization, the 1 continues to circulate through the ring counter, as the timing diagram in Figure 10–32 shows.

FIGURE 10–31
74HC195 connected as a ring counter.

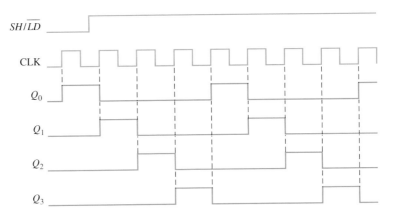

FIGURE 10–32
Timing diagram showing two complete cycles of the ring counter in Figure 10–31 when it is initially preset to 1000.

Serial-to-Parallel Data Converter

Serial data transmission from one digital system to another is commonly used to reduce the number of wires in the transmission line. For example, eight bits can be sent serially over one wire, but it takes eight wires to send the same data in parallel.

A computer or microprocessor-based system commonly requires incoming data to be in parallel format, thus the requirement for serial-to-parallel conversion. A simplified serial-to-parallel data converter, in which two types of shift registers are used, is shown in Figure 10–33.

To illustrate the operation of this serial-to-parallel converter, the serial data format shown in Figure 10–34 is used. It consists of eleven bits. The first bit (start bit) is always 0 and always begins with a HIGH-to-LOW transition. The next eight bits (D_7 through D_0) are the data bits (one of the bits can be parity), and the last two bits (stop bits) are always 1s. When no data are being sent, there is a continuous 1 on the serial data line.

The HIGH-to-LOW transition of the start bit sets the control flip-flop, which enables the clock generator. After a fixed delay time, the clock generator begins producing a pulse

FIGURE 10–33
Simplified logic diagram of a serial-to-parallel converter.

Start bit (0)	D_7	D_6	D_5	D_4	D_3	D_2	D_1	D_0	Stop bit (1)	Stop bit (1)

FIGURE 10–34
Serial data format.

waveform, which is applied to the data-input register and to the divide-by-8 counter. The clock has a frequency precisely equal to that of the incoming serial data, and the first clock pulse after the start bit occurs simultaneously with the first data bit.

The timing diagram in Figure 10–35 illustrates the following basic operation: The eight data bits (D_7 through D_0) are serially shifted into the data-input register. After the eighth clock pulse, a HIGH-to-LOW transition of the terminal count (TC) output of the counter ANDed with the clock (TC·CLK) loads the eight bits that are in the data-

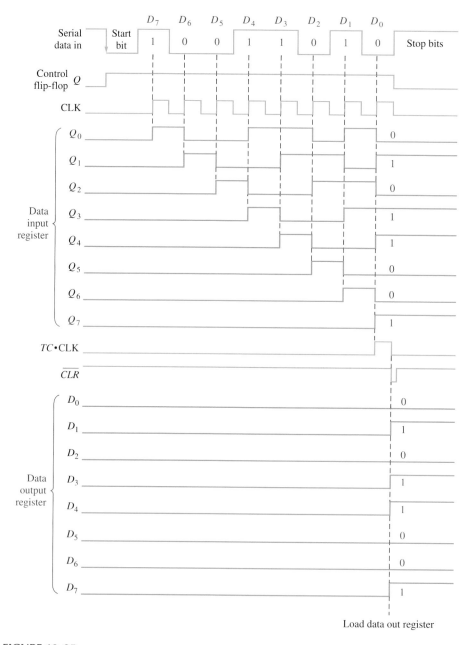

FIGURE 10–35
Timing diagram illustrating the operation of the serial-to-parallel data converter in Figure 10–33.

input register into the data-output register. This same transition also triggers the one-shot, which produces a short-duration pulse to clear the counter and reset the control flip-flop and thus disable the clock generator. The system is now ready for the next group of eleven bits, and it waits for the next HIGH-to-LOW transition at the beginning of the start bit.

By reversing the process just stated, parallel-to-serial data conversion can be accomplished. However, since the serial data format must be produced, additional requirements must be taken into consideration. Problem 43 at the end of the chapter relates to the design of an 8-bit parallel-to-serial data converter.

Universal Asynchronous Receiver Transmitter (UART)

As mentioned, computers and microprocessor-based systems often send and receive data in a parallel format. Frequently, these systems must communicate with external devices that send and/or receive serial data. An interfacing device used to accomplish these conversions is the UART. Figure 10–36 illustrates the UART in a general microprocessor-based system application.

A UART includes a serial-to-parallel data converter such as we have discussed and a parallel-to-serial converter, as shown in Figure 10–37. The data bus is basically a set of parallel conductors along which data move between the UART and the microprocessor system. Buffers interface the data registers with the data bus.

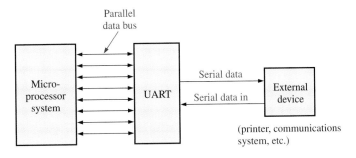

FIGURE 10–36
UART interface.

FIGURE 10–37
Basic UART block diagram.

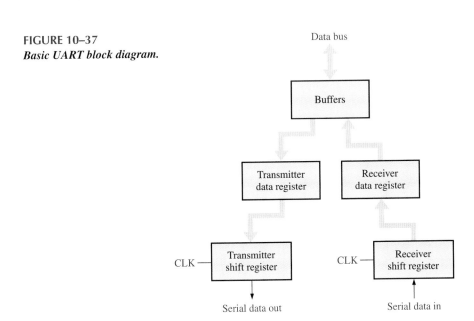

The UART receives data in serial format, converts the data to parallel format, and places them on the data bus. The UART also accepts parallel data from the data bus, converts the data to serial format, and transmits them to an external device.

Keyboard Encoder

The keyboard encoder is a good example of the application of a shift register used as a ring counter in conjunction with other devices. Recall that a simplified computer keyboard encoder without data storage was presented in Chapter 6.

Figure 10–38 shows a simplified keyboard encoder for encoding a key closure in a 64-key matrix organized in eight rows and eight columns. Two 74HC195 4-bit shift regis-

FIGURE 10–38
Simplified keyboard encoding circuit.

ters are connected as an 8-bit ring counter with a fixed bit pattern of seven 1s and one 0 preset into it when the power is turned on. Two 74HC147 priority encoders (introduced in Chapter 6) are used as eight-line-to-three-line encoders (9 input HIGH, 8 output unused) to encode the ROW and COLUMN lines of the keyboard matrix. The 74HC174A (hex flip-flops) is used as a parallel in/parallel out register in which the ROW/COLUMN code from the priority encoders is stored.

The basic operation of the keyboard encoder in Figure 10–38 is as follows: The ring counter "scans" the rows for a key closure as the clock signal shifts the 0 around the counter at a 5 kHz rate. The 0 (LOW) is sequentially applied to each ROW line, while all other ROW lines are HIGH. All the ROW lines are connected to the ROW encoder inputs, so the 3-bit output of the ROW encoder at any time is the binary representation of the ROW line that is LOW. When there is a key closure, one COLUMN line is connected to one ROW line. When the ROW line is taken LOW by the ring counter, that particular COLUMN line is also pulled LOW. The COLUMN encoder produces a binary output corresponding to the COLUMN in which the key is closed. The 3-bit ROW code plus the 3-bit COLUMN code uniquely identifies the key that is closed. This 6-bit code is applied to the inputs of the key code register. When a key is closed, the two one-shots produce a delayed clock pulse to parallel-load the 6-bit code into the key code register. This delay allows the contact bounce to die out. Also, the first one-shot output inhibits the ring counter to prevent it from scanning while the data are being loaded into the key code register.

The 6-bit code in the key code register is now applied to a ROM (read-only memory) to be converted to an appropriate alphanumeric code that identifies the keyboard character. ROMs are studied in Chapter 12.

SECTION 10–8 REVIEW

1. In the keyboard encoder, how many times per second does the ring counter scan the keyboard?
2. What is the 6-bit ROW/COLUMN code (key code) for the top row and the left-most column in the keyboard encoder?
3. What is the purpose of the diodes in the keyboard encoder? What is the purpose of the resistors?

10–9 ■ TROUBLESHOOTING

One basic method of troubleshooting sequential logic and other more complex digital systems uses a procedure of "exercising" the circuit under test with a known input waveform (stimulus) and then observing the output for the proper bit pattern.

After completing this section, you should be able to

■ Explain the procedure of "exercising" as a troubleshooting technique ■ Discuss exercising of a serial-to-parallel converter ■ Discuss the basics of signature analysis

The serial-to-parallel data converter in Figure 10–33 is used to illustrate the "exercising" procedure. The main objective in exercising the circuit is to force all elements (flip-flops and gates) into all of their states to be certain that nothing is stuck in a given state as a result of a fault. The input test pattern, in this case, must be designed to force each flip-flop in the registers into both states, to clock the counter through all of its eight states, and to take the control flip-flop, the clock generator, the one-shot, and the AND gate through their paces.

The input test pattern that accomplishes this objective for the serial-to-parallel data converter is based on the serial data format in Figure 10–34. It consists of the pattern

FIGURE 10–39
Sample test pattern.

10101010 in one serial group of data bits followed by 01010101 in the next group, as shown in Figure 10–39. These patterns are generated on a repetitive basis by a special test-pattern generator. The basic test setup is shown in Figure 10–40.

After both patterns have been run through the circuit under test, all the flip-flops in the data-input register and in the data-output register have resided in both SET and RESET states, the counter has gone through its sequence (once for each bit pattern), and all the other devices have been exercised.

To check for proper operation, each of the parallel data outputs is observed for an alternating pattern of 1s and 0s as the input test patterns are repetitively shifted into the data-input register and then loaded into the data-output register. The proper timing diagram is shown in Figure 10–41. The outputs can be observed in pairs with a dual-trace oscilloscope, or all eight outputs can be observed simultaneously with a logic analyzer configured for timing analysis.

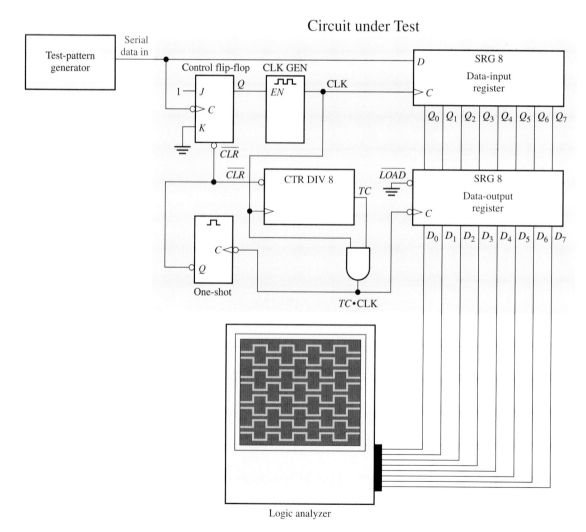

FIGURE 10–40
Basic test setup for the serial-to-parallel data converter of Figure 10–33.

FIGURE 10–41

Proper outputs for the circuit under test in Figure 10–40. The input test pattern is shown.

If one or more outputs of the data-output register are incorrect, then you must back up to the outputs of the data-input register. If these outputs are correct, then the problem is associated with the data-output register. Check the inputs to the data-output register directly on the pins of the IC for an open input line. Check that power and ground are correct (look for the absence of noise on the ground line). Verify that the load line is a solid LOW and that there are clock pulses on the clock input of the correct amplitude. Make sure that the connection to the logic analyzer did not short two output lines together. If all of these checks pass inspection, then it is likely that the output register is defective. If the data-input register outputs are also incorrect, the fault could be associated with the input register itself or with any of the other logic, and additional investigation is necessary to isolate the problem.

When measuring digital signals with an oscilloscope, you should always use dc coupling, rather than ac coupling. The reason that ac coupling is not best for viewing digital signals is that the 0 V level of the signal will appear at the *average* level of the signal, not at true ground or 0 V level. It is much easier to find a "floating" ground or incorrect logic level with dc coupling. If you suspect an open ground in a digital circuit, increase the sensitivity of the scope to the maximum possible. A good ground will never appear to have noise under this condition, but an open will likely show some noise, which appears as a random fluctuation in the 0 V level.

SECTION 10–9 REVIEW

1. What is the purpose of providing a test input to a sequential logic circuit?
2. Generally, when an output waveform is found to be incorrect, what is the next step to be taken?

10–10 ■ LOGIC SYMBOLS WITH DEPENDENCY NOTATION

This section can be omitted without affecting other coverage. Two examples of ANSI/IEEE Standard 91–1984 symbols with dependency notation for shift registers are presented.

After completing this section, you should be able to

■ Understand and interpret the logic symbols with dependency notation for the 74HC164 and the 74HC194 shift registers

The logic symbol for a 74HC164 8-bit parallel output serial shift register is shown in Figure 10–42. The common control inputs are shown on the notched block. The clear ($\overline{CLR}$) input is indicated by an R (for RESET) inside the block. Since there is no dependency prefix to link R with the clock (C1), the clear function is asynchronous. The right arrow symbol after C1 indicates data flow from Q_0 to Q_7. The A and B inputs are ANDed, as indicated by the embedded AND symbol, to provide the synchronous data input, $1D$, to the first stage (Q_0). Note the dependency of D on C, as indicated by the 1 suffix on C and the 1 prefix on D.

FIGURE 10–42

Logic symbol for the 74HC164.

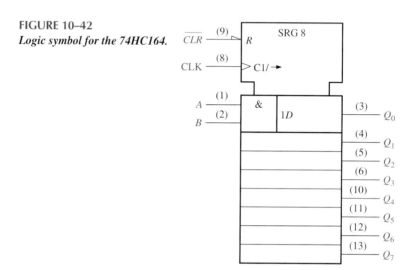

Figure 10–43 is the logic symbol for the 74HC194 4-bit bidirectional universal shift register. Starting at the top left side of the control block, note that the $\overline{CLR}$ input is active-LOW and is asynchronous (no prefix link with C). Inputs S_0 and S_1 are mode inputs that determine the *shift-right, shift-left,* and *parallel load* modes of operation, as indicated by the $\frac{0}{3}$ dependency designation following the M. The $\frac{0}{3}$ represents the binary states of 0, 1, 2, and 3 on the S_0 and S_1 inputs. When one of these digits is used as a prefix for another input, and a dependency is established. The $1{\rightarrow}/2{\leftarrow}$ symbol on the clock input indicates the following: $1{\rightarrow}$ indicates that a right shift (Q_0 toward Q_3) occurs when the mode inputs (S_0, S_1) are in the binary 1 state ($S_0 = 1$, $S_1 = 0$), $2{\leftarrow}$ indicates that a left

FIGURE 10–43

Logic symbol for the 74HC194.

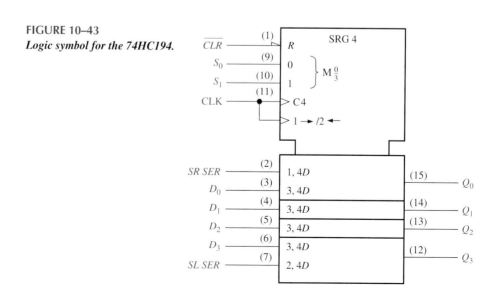

shift (Q_3 toward Q_0) occurs when the mode inputs are in the binary 2 state ($S_0 = 0$, $S_1 = 1$). The shift-right serial input (*SR SER*) is both mode-dependent and clock-dependent, as indicated by 1, 4D. The parallel inputs (D_0, D_1, D_2, and D_3) are all mode-dependent (prefix 3 indicates parallel load mode) and clock-dependent, as indicated by 3, 4D. The shift-left serial input (*SL SER*) is both mode-dependent and clock-dependent, as indicated by 2, 4D.

The four modes for the 74HC194 are summarized as follows:

Do nothing:	$S_0 = 0, S_1 = 0$	(mode 0)
Shift right:	$S_0 = 1, S_1 = 0$	(mode 1, as in 1, 4D)
Shift left:	$S_0 = 0, S_1 = 1$	(mode 2, as in 2, 4D)
Parallel load:	$S_0 = 1, S_1 = 1$	(mode 3, as in 3, 4D)

SECTION 10–10 REVIEW

1. In Figure 10–43, are there any inputs that are dependent on the mode inputs being in the 0 state?
2. Is the parallel load synchronous with the clock?

10–11 ■ DIGITAL SYSTEM APPLICATION

In this system application, you will be working with a security entry system. This system allows you to deactivate (disarm) the alarms so that you can enter the building. Disarming the system is accomplished by entering a specified sequence of four digits on a keypad. The system is rearmed by pressing a switch when you leave the building. The system uses three shift registers as well as other devices that were covered in previous chapters. A memory is also included, but you will not get into specific details until you study memories in Chapter 12.

After completing this section, you should be able to

■ Describe the overall function of the system ■ Explain the purposes of the shift registers as well as other types of devices ■ Determine the specific devices to be used in the system ■ Develop a detailed logic diagram from a general logic diagram ■ Develop a test procedure and troubleshoot the system

Basic Operation of the Security Entry System

A preliminary general logic diagram of the system is shown in Figure 10–44. The system consists of two circuit boards, the panel switches, and the keypad. The focus in this section is on the code entry board and you should be familiar with all of the types of devices used. The memory board and complete system operation is the focus of the system application in Chapter 12.

The security entry system controls the arming and disarming of the sensors and alarms in a given building. Arming or activation of the sensors and alarms is accomplished by pressing the single push-button switch labeled ARM and, while it is pressed, entering any digit on the keypad. Disarming or deactivation is accomplished by entering a predetermined 4-digit number on the keypad.

When the system is armed, shift registers A and B and the memory counter are cleared (reset) to their all-zero state, while shift register C is preset to 11110000. The resulting LOW on the serial output of shift register C enables the alarm system sensors and illuminates the ARMED light.

FIGURE 10–44
Basic logic diagram of the security entry system.

When the first digit of the entry number is entered on the keypad, the corresponding line from the keypad goes to its active level. The decimal-to-BCD encoder produces the 4-bit BCD code corresponding to the decimal digit. At the same time, one-shot A is triggered and produces a clock pulse that causes the 4-bit BCD digit from the encoder to be loaded into register A and the first 4-bit BCD digit from the memory to be loaded into register B. These codes then appear on the A and B inputs of the comparator. The memory stores the BCD codes for the correct 4-digit entry number.

If the correct digit has been entered on the keypad, the four bits on the A inputs of the comparator are the same as the four bits on the B inputs and the comparator produces a HIGH on its $A = B$ output. The pulse from one-shot A also triggers one-shot B. The pulse from one-shot B clocks the state of the comparator output, $A = B$, into the flip-flop, causing its Q output to go HIGH. The pulse from one-shot C clocks, register C and shifts the contents to the right so the register contains 11111000. Since there is still a 0 in the last stage (rightmost) of the register, the alarm system remains armed.

After the second correct digit is entered on the keypad, shift register C contains 11111100. After the third correct digit is entered, the shift register contains 11111110. After the fourth, and final, correct digit is entered, the shift register contains all 1s. As the 1 is shifted into the last stage, the serial output of the register goes HIGH and disarms the system. The ARMED light goes off at this time and you can enter the building without setting off the alarm.

Now let's briefly see what happens if you enter an incorrect digit. If, at any point while entering the 4-digit number an incorrect digit is entered, the BCD code will not agree with the code stored in the memory. As a result, the comparator will produce a LOW on its $A = B$ output. This will result in the parallel loading of shift register C with 11110000 and the clearing of shift registers A and B and the memory counter to all zeros. At this point, you must start over and reenter the 4-digit number.

■ SYSTEM ASSIGNMENTS

The basic logic diagram in Figure 10–44 indicates the general logic for the system and does not represent the detailed logic required to implement the system. In the following three system assignments, you will analyze the basic operation of the code entry board logic to make sure that you understand the functional requirements and the basic system operation. You will then complete the design of the logic by selecting appropriate devices and then developing a detailed logic diagram incorporating the selected devices. Finally, you will compare your logic diagram to an implemented circuit board and then develop a test procedure.

On the circuit board, interconnections on the back side of the board feed through to the pads on the component side. Supply voltage and ground connections are on the back side and are not shown for simplicity.

■ SYSTEM ASSIGNMENT 1: Analysis

■ *Activity A: The Decimal-to-BCD Encoder*

(a) State the purpose of the encoder.

(b) Describe what happens on the encoder inputs when you enter a digit on the keypad.

(c) Describe what happens on the encoder outputs when you enter a digit on the keypad.

■ *Activity 2:   The Timing Circuits*

(a) State the purpose of the block labeled G_1 and the type of gate that it represents.

(b) Explain when and how one-shot A is triggered.

(c) State the purpose of each of the one-shots.

■ *Activity 3:   Shift Registers A and B*

(a) State the purposes of shift register A and shift register B.

(b) Determine the capacity of shift registers A and B (number of bits).

(c) Describe when and how the parallel data are entered into shift register A and into shift register B.

■ *Activity 4:   Comparator*

(a) State the purpose of the comparator.

(b) Define each of the two sets (*A* and *B*) of four input bits.

(c) Describe what happens on the outputs when the two inputs are the same.

(d) Describe what happens on the outputs when the two inputs are different.

■ *Activity 5:   Shift Register C*

(a) State the purpose of shift register C.

(b) Discuss the bit pattern that is initially loaded into the register and state its purpose.

(c) Describe what happens when the serial output is LOW. Describe what happens when the serial output is HIGH.

(d) Describe what happens to shift register C each time a correct digit is entered on the keypad.

■ SYSTEM ASSIGNMENT 2: Design

This system assignment focuses on the design of the code entry circuit board. For information on devices, reference to earlier chapters and/or device data sheets is necessary.

■ *Activity 1:   Decimal-to-BCD Encoder*
The particular keypad used in this system produces active-LOW outputs. When a key is pressed, the corresponding line is LOW and the other lines are HIGH. Specify an IC encoder to properly interface with the keypad.

■ *Activity 2:   The Timing Circuits*
The basic logic diagram shows a block (G_1) for the gate used to trigger one-shot A when a key is pressed. One-shot A is to produce a 1 ms pulse. One-shot B is to produce a 1 ms pulse triggered at the trailing edge of the one-shot A pulse. One-shot C is to produce a 1 ms pulse triggered at the trailing edge of the one-shot B pulse.
(a) Specify the type of gate to be used to implement gate G_1.
(b) Specify the type of IC one-shots and determine the external component values required for each one-shot to produce a 1 ms pulse.

■ *Activity 3:   Shift Registers A and B*
Specify an IC shift register that can accept 4 bits in parallel and provide a parallel output. The shift register must be triggered at the trailing edge of the clock pulse from one-shot A.

■ *Activity 4:   Comparator*
Specify an IC comparator that will compare two 4-bit numbers and produce the three required outputs.

Code Entry Board

■ SYSTEM ASSIGNMENT 2: Design *(continued)*

■ *Activity 5: Shift Register C*

This shift register must store an initial bit pattern of 11110000 which is parallel loaded. It must have a serial input and shift-right capability so that the 1 on the serial input is shifted right on each clock pulse and the register fills with 1s from the serial input. The serial output is used to control the alarm system and ARMED light. An off-board drive circuit is required for this purpose.

Specify the IC device or devices for implementing this shift register.

■ *Activity 6: Other Logic*

Specify the types of ICs required for the AND gate and flip-flop.

■ *Activity 7:* Develop a detailed logic diagram using the devices specified in preceding activities. Show all necessary details including device pin numbers.

■ SYSTEM ASSIGNMENT 3: Verification and Testing

■ *Activity 1:* Draw the logic diagram for the entry code logic board below and compare to your logic diagram from Activity 7 of System Assignment 2. Discuss any differences between your design and the design of this logic board and verify that both work.

■ *Activity 2:* Identify the outputs and inputs on the circuit board connector. If there are more inputs or outputs than shown in the basic diagram in System Assignment 2, they are for additional require-

ments for interfacing the memory board and will be discussed in Chapter 12.

■ *Activity 3:* Develop a test procedure for checking the board shown.

The complete entry code logic board circuit (modified) is in file SA10.EWB on your EWB CD. Use EWB to test this circuit.

**SECTION
10–11 REVIEW**

1. What steps are required to arm the system?
2. How is the system disarmed?
3. What happens if an incorrect digit is entered after two correct digits?

All or portions of Chapter 15, "Integrated Circuit Technologies," may be covered at this point. The coverage in Chapter 15 is not a prerequisite to any other material in this text, and its omission or partial omission will not affect any other material.

■ SUMMARY

- The basic types of data movement in shift registers are illustrated in Figure 10–45.
- Shift register counters are shift registers with feedback that exhibit special sequences. Examples are the Johnson counter and the ring counter.
- The Johnson counter has $2n$ states in its sequence, where n is the number of stages.
- The ring counter has n states in its sequence.

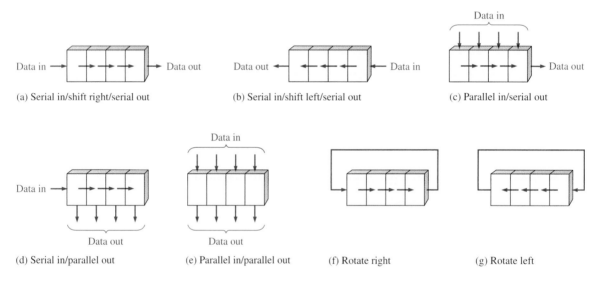

(a) Serial in/shift right/serial out

(b) Serial in/shift left/serial out

(c) Parallel in/serial out

(d) Serial in/parallel out

(e) Parallel in/parallel out

(f) Rotate right

(g) Rotate left

FIGURE 10–45

■ KEY TERMS

These terms are also in the end-of-book glossary.

Bidirectional   Having two directions. In a bidirectional shift register, the stored data can be shifted right or left.

Johnson counter   A type of register in which a specific prestored pattern of 1s and 0s is shifted through the stages, creating a unique sequence of bit patterns.

Load   To enter data into a shift register.

Ring counter   A register in which a certain pattern of 1s and 0s is continuously recirculated.

Shift   To move binary data from state to stage within a shift register or other storage device or to move binary data into or out of the device.

Stage   One storage element in a register.

Universal shift register   A register that has both serial and parallel input and output capability.

■ SELF-TEST

1. A stage in a shift register consists of
   (a) a latch (b) a flip-flop (c) a byte of storage (d) four bits of storage

2. To serially shift a byte of data into a shift register, there must be
   (a) one clock pulse (b) one load pulse
   (c) eight clock pulses (d) one clock pulse for each 1 in the data

3. To parallel load a byte of data into a shift register with a synchronous load, there must be
   (a) one clock pulse (b) one clock pulse for each 1 in the data
   (c) eight clock pulses (d) one clock pulse for each 0 in the data

4. The group of bits 10110101 is serially shifted (right-most bit first) into an 8-bit parallel output shift register with an initial state of 11100100. After two clock pulses, the register contains
   (a) 01011110 (b) 10110101 (c) 01111001 (d) 00101101

5. With a 100 kHz clock frequency, eight bits can be serially entered into a shift register in
   (a) 80 μs (b) 8 μs (c) 80 ms (d) 10 μs

6. With a 1 MHz clock frequency, eight bits can be parallel entered into a shift register in
   (a) 8 μs (b) in the propagation delay time of eight flip-flops
   (c) 1 μs (d) in the propagation delay time of one flip-flop

7. A modulus-10 Johnson counter requires
   (a) ten flip-flops (b) four flip-flops (c) five flip-flops (d) twelve flip-flops

8. A modulus-10 ring counter requires a minimum of
   (a) ten flip-flops (b) five flip-flops (c) four flip-flops (d) twelve flip-flops

9. When an 8-bit serial in/serial out shift register is used for a 24 μs time delay, the clock frequency must be
   (a) 41.67 kHz (b) 333 kHz (c) 125 kHz (d) 8 MHz

10. The purpose of the ring counter in the keyboard encoding circuit of Figure 10–38 is
    (a) to sequentially apply a HIGH to each row for detection of key closure
    (b) to provide trigger pulses for the key code register
    (c) to sequentially apply a LOW to each row for detection of key closure
    (d) to sequentially reverse bias the diodes in each row

■ PROBLEMS

SECTION 10–1 Basic Shift Register Functions

1. Why are shift registers considered basic memory devices?
2. What is the storage capacity of a register that can retain two bytes of data?

SECTION 10–2 Serial In/Serial Out Shift Registers

3. For the data input and clock in Figure 10–46, determine the states of each flip-flop in the shift register of Figure 10–3 and show the Q waveforms. Assume that the register contains all 1s initially.
4. Solve Problem 3 for the waveforms in Figure 10–47.

FIGURE 10–46

FIGURE 10–47

5. What is the state of the register in Figure 10–48 after each clock pulse if it starts in the 101001111000 state?

6. For the serial in/serial out shift register, determine the data-output waveform for the data-input and clock waveforms in Figure 10–49. Assume that the register is initially cleared.

FIGURE 10–48

FIGURE 10–49

7. Solve Problem 6 for the waveforms in Figure 10–50.

8. A leading-edge clocked serial in/serial out shift register has a data-output waveform as shown in Figure 10–51. What binary number is stored in the 8-bit register if the first data bit out (left-most) is the LSB?

FIGURE 10–50

FIGURE 10–51

SECTION 10–3   Serial In/Parallel Out Shift Registers

9. Show a complete timing diagram showing the parallel outputs for the shift register in Figure 10–8. Use the waveforms in Figure 10–49 with the register initially clear.

10. Solve Problem 9 for the input waveforms in Figure 10–50.

11. Develop the Q_0 through Q_7 outputs for a 74HC164 shift register with the input waveforms shown in Figure 10–52.

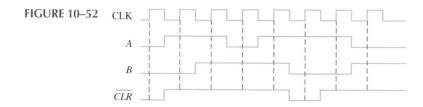

FIGURE 10–52

SECTION 10–4 Parallel In/Serial Out Shift Registers

12. The shift register in Figure 10–53(a) has $SHIFT/\overline{LOAD}$ and CLK inputs as shown in part (b). The serial data input (SER) is a 0. The parallel data inputs are $D_0 = 1$, $D_1 = 0$, $D_2 = 1$, and $D_3 = 0$ as shown. Develop the data-output waveform in relation to the inputs.

13. The waveforms in Figure 10–54 are applied to a 74HC165 shift register. The parallel inputs are all 0. Determine the Q_7 waveform.

FIGURE 10–53

FIGURE 10–54

14. Solve Problem 13 if the parallel inputs are all 1.

15. Solve Problem 13 if the SER input is inverted.

SECTION 10–5 Parallel In/Parallel Out Shift Registers

16. Determine all the Q output waveforms for a 74HC195 4-bit shift register when the inputs are as shown in Figure 10–55.

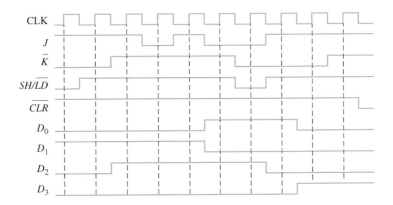

FIGURE 10–55

17. Solve Problem 16 if the $SH/\overline{LD}$ input is inverted and the register is initially clear.

18. Use two 74HC195 shift registers to form an 8-bit shift register. Show the required connections.

SECTION 10–6   Bidirectional Shift Registers

19. For the 8-bit bidirectional register in Figure 10–56, determine the state of the register after each clock pulse for the *RIGHT/LEFT* control waveform given. A HIGH on this input enables a shift to the right, and a LOW enables a shift to the left. Assume that the register is initially storing the decimal number seventy-six in binary, with the right-most position being the LSB. There is a LOW on the data-input line.

20. Solve Problem 19 for the waveforms in Figure 10–57.

FIGURE 10–56

21. Use two 74HC194 4-bit bidirectional shift registers to create an 8-bit bidirectional shift register. Show the connections.

22. Determine the Q outputs of a 74HC194 with the inputs shown in Figure 10–58. Inputs D_0, D_1, D_2, and D_3 are all HIGH.

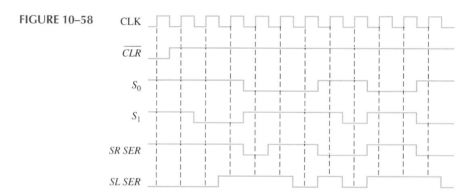

FIGURE 10–58

SECTION 10–7   Shift Register Counters

23. How many flip-flops are required to implement each of the following in a Johnson counter configuration:

   (a) modulus-6   **(b)** modulus-10   **(c)** modulus-14   **(d)** modulus-16

24. Draw the logic diagram for a modulus-18 Johnson counter. Show the timing diagram and write the sequence in tabular form.

25. For the ring counter in Figure 10–59, show the waveforms for each flip-flop output with respect to the clock. Assume that FF0 is initially SET and that the rest are RESET. Show at least ten clock pulses.

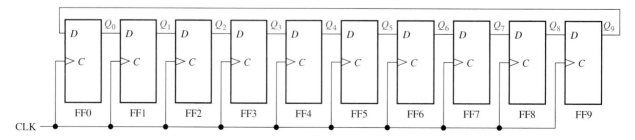

FIGURE 10–59

26. The waveform pattern in Figure 10–60 is required. Devise a ring counter, and indicate how it can be preset to produce this waveform on its Q_9 output. At CLK16 the pattern begins to repeat.

FIGURE 10–60

SECTION 10–8 Shift Register Applications

27. Use 74HC195 4-bit shift registers to implement a 16-bit ring counter. Show the connections.

28. What is the purpose of the power-on $\overline{LOAD}$ input in Figure 10–38?

29. What happens when two keys are pressed simultaneously in Figure 10–38?

SECTION 10–9 Troubleshooting

30. Based on the waveforms in Figure 10–61(a), determine the most likely problem with the register in part (b) of the figure.

31. Refer to the parallel in/serial out shift register in Figure 10–12. The register is in the state where $Q_0Q_1Q_2Q_3 = 1001$, and $D_0D_1D_2D_3 = 1010$ is loaded in. When the $SHIFT/\overline{LOAD}$ input is taken HIGH, the data shown in Figure 10–62 are shifted out. Is this operation correct? If not, what is the most likely problem?

(a)   (b)

FIGURE 10–61

FIGURE 10–62

32. You have found that the bidirectional register in Figure 10–19 will shift data right but not left. What is the most likely fault?

33. For the keyboard encoder in Figure 10–38, list the possible faults for each of the following symptoms:

(a) The state of the key code register does not change for any key closure.

(b) The state of the key code register does not change when any key in the third row is closed. A proper code occurs for all other key closures.

(c) The state of the key code register does not change when any key in the first column is closed. A proper code occurs for all other key closures.

(d) When any key in the second column is closed, the left three bits of the key code ($Q_0Q_1Q_2$) are correct, but the right three bits are all 1s.

34. Develop a test procedure for exercising the keyboard encoder in Figure 10–38. Specify the procedure on a step-by-step basis, indicating the output code from the key code register that should be observed at each step in the test.

35. What symptoms are observed for the following failures in the serial-to-parallel converter in Figure 10–33:

   (a) AND gate output stuck in HIGH state

   (b) clock generator output stuck in LOW state

   (c) third stage of data-input register stuck in SET state

   (d) terminal count output of counter stuck in HIGH state

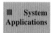

SECTION 10–11   Digital System Application

36. Develop a waveform timing diagram of the three one-shot outputs as specified in System Assignment 2.

37. Assume the entry code is 1939. Determine the states of shift register A and shift register C after the second correct digit has been entered.

38. Assume the entry code is 7646 and the digits 7645 are entered. Determine the states of shift register A and shift register C after each of the digits is entered.

39. A problem has developed with the security entry system such that it cannot be deactivated (the serial output of shift register C stays LOW). Upon checking several points, you find that one-shot A is never triggered. What are the possible faults?

40. As in Problem 39, the security entry system cannot be deactivated. This time, however, you find that the one-shots are all triggering properly. What devices could possibly be faulty in this case?

Special Design Problems

41. Specify the devices that can be used to implement the serial-to-parallel data converter in Figure 10–33. Develop the complete logic diagram, showing any modifications necessary to accommodate the specific devices used.

42. Modify the serial-to-parallel converter in Figure 10–33 to provide 16-bit conversion.

43. Design an 8-bit parallel-to-serial data converter that produces the data format in Figure 10–34. Show a logic diagram and specify the devices.

44. Design a power-on $\overline{LOAD}$ circuit for the keyboard encoder in Figure 10–38. This circuit must generate a short-duration LOW pulse when the power switch is turned on.

45. Implement the test-pattern generator used in Figure 10–40 to troubleshoot the serial-to-parallel converter.

46. Review the tablet-counting and control system that was introduced in Chapter 1 and was the topic of the system applications in Chapters 2, 3, and 4. Utilizing the knowledge gained in this chapter, implement registers A and B in that system using specific IC devices.

Electronics Workbench Troubleshooting

47. Open file PR10_47.EWB and test the 4-bit shift register to determine if there is a fault. Identify the fault if possible.

48. Open file PR10_48.EWB and test the 74164 8-bit serial in/parallel out shift register to determine if there is a fault. If there is a fault, identify it if possible.

49. Open file PR10_49.EWB and test the 74165 8-bit parallel load shift register to determine if there is a fault. If there is a fault, identify it if possible.

50. Open file PR10_50.EWB and test the 74195 4-bit parallel access shift register to determine if there is a fault. If there is a fault, identify it if possible.

51. Open file PR10_51.EWB and test the 10-bit ring counter to determine if there is a fault. If there is a fault, identify it if possible.

■ ANSWERS TO SECTION REVIEWS

SECTION 10–1

1. A counter has a specified sequence of states, but a shift register does not.
2. Storage and data movement are two functions of a shift register.

SECTION 10–2

1. FF0: data input to J_0, $\overline{\text{data input}}$ to K_0: FF1: Q_0 to J_1, $\overline{Q}_0$ to K_1; FF2: Q_1 to J_2, $\overline{Q}_1$ to K_2; FF3: Q_2 to J_3, $\overline{Q}_2$ to K_3
2. Eight clock pulses

SECTION 10–3

1. 0100 after 2 clock pulses
2. Take the serial output from the right-most flip-flop for serial out operation.

SECTION 10–4

1. When *SHIFT/$\overline{LOAD}$* is HIGH, the data are shifted right one bit per clock pulse. When *SHIFT/$\overline{LOAD}$* is LOW, the data on the parallel inputs are loaded into the register.
2. The parallel load operation is asynchronous, so it is not dependent on the clock.

SECTION 10–5

1. The data outputs are 1001.   2. $Q_0 = 1$ after one clock pulse

SECTION 10–6

1. 1111 after the fifth clock pulse

SECTION 10–7

1. Sixteen states are in an 8-bit Johnson counter sequence.
2. For a 3-bit Johnson counter: 000, 100, 110, 111, 011, 001, 000

SECTION 10–8

1. 625 scans/second   2. $Q_5Q_4Q_3Q_2Q_1Q_0 = 011011$
3. The diodes provide unidirectional paths for pulling the ROWs LOW and preventing HIGHs on the ROW lines from being connected to the switch matrix. The resistors pull the COLUMN lines HIGH.

SECTION 10–9

1. A test input is used to sequence the circuit through all of its states.
2. Check the input to that portion of the circuit. If the signal on that input is correct, the fault is isolated to the circuitry between the good input and the bad output.

SECTION 10–10

1. No inputs are dependent on the mode inputs being in the 0 state.
2. Yes, the parallel lead is synchronous with the clock as indicated by the 4D label.

SECTION 10–11

1. Press ARM switch; then press any key to arm system.
2. Enter correct 4-digit number to disarm system.
3. The system is reinitialized.

■ **ANSWERS**
TO RELATED
PROBLEMS
FOR
EXAMPLES

10–1 See Figure 10–63.

10–2 The state of the register after three additional clock pulses is 0000.

10–3 See Figure 10–64.   **10–4** See Figure 10–65.

10–5 See Figure 10–66.   **10–6** $f = 1/3 \ \mu s = 333$ kHz

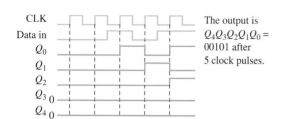

The output is $Q_4 Q_3 Q_2 Q_1 Q_0 = $ 00101 after 5 clock pulses.

FIGURE 10–63

FIGURE 10–64

FIGURE 10–65

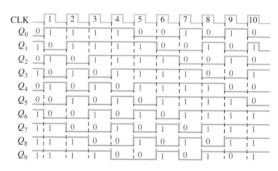

FIGURE 10–66

■ **ANSWERS**
TO SELF-TEST

1. (b)   **2.** (c)   **3.** (a)   **4.** (c)   **5.** (a)   **6.** (d)   **7.** (c)   **8.** (a)

9. (b)   **10.** (c)

11

SEQUENTIAL LOGIC APPLICATIONS OF PLDs

■ CHAPTER OBJECTIVES

❑ Describe the OLMCs in the GAL22V10 and the GAL16V8
❑ Explain the difference between the registered mode and the combinational mode
❑ Develop ABEL input files for the design of shift registers
❑ Use ABEL to implement counters in a PLD
❑ Use ABEL state diagrams to implement logic systems in PLDs
❑ Implement sequential logic in the traffic light control system using a GAL

■ CHAPTER OVERVIEW

The basics of programmable logic devices were covered in Chapter 7 with emphasis on using PLDs to implement combinational logic. The coverage of PLDs is continued in this chapter with an introduction to the registered mode and sequential logic applications. If necessary, review Chapter 7 during or prior to your study of this chapter. Like Chapter 7, this chapter can be omitted without affecting any other material.

As in Chapter 7, the ABEL programming language is used to illustrate the implementation of sequential logic, but this approach can readily be transferred to CUPL or other similar PLD languages.

 SPECIFIC DEVICES

GAL16V8   GAL22V10

■ DIGITAL SYSTEM APPLICATION

This Digital System Application illustrates concepts taught in this chapter. The sequential logic portion of the traffic light control system from Chapter 9 is redesigned using a programmable logic device, and the complete system except for the timing circuits is implemented with a single GAL22V10.

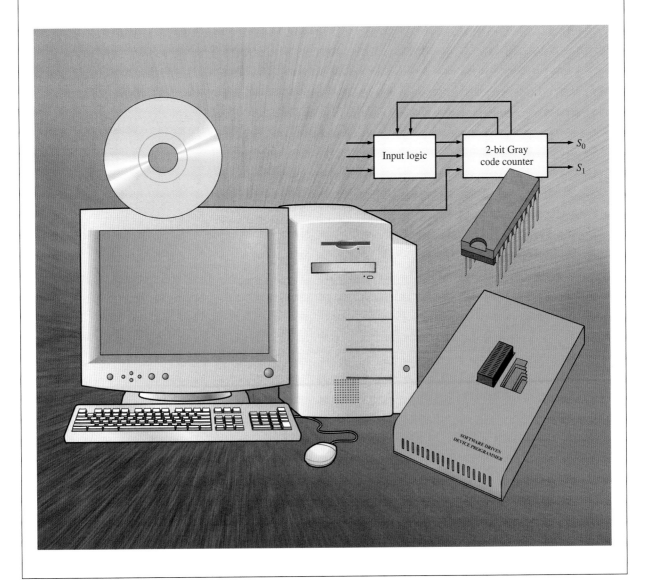

11–1 ■ THE COMPLETE OLMC

In Chapter 7 the output logic macrocells (OLMC) for the GAL22V10 and the GAL16V8 were examined in terms of their combinational modes, and the flip-flop contained in the OLMC was only briefly mentioned. In this section, we will discuss the complete OLMC which includes a flip-flop, again using the GAL22V10 and the GAL16V8 as representative PLD devices.

After completing this section, you should be able to

■ Discuss the basic operation of an OLMC   ■ Discuss local and global cells

The block diagrams of the GAL22V10 and the GAL16V8 that were introduced in Chapter 7 are repeated in Figure 11–1. The GAL22V10 has ten output logic macrocells (OLMCs) and the GAL16V8 has eight.

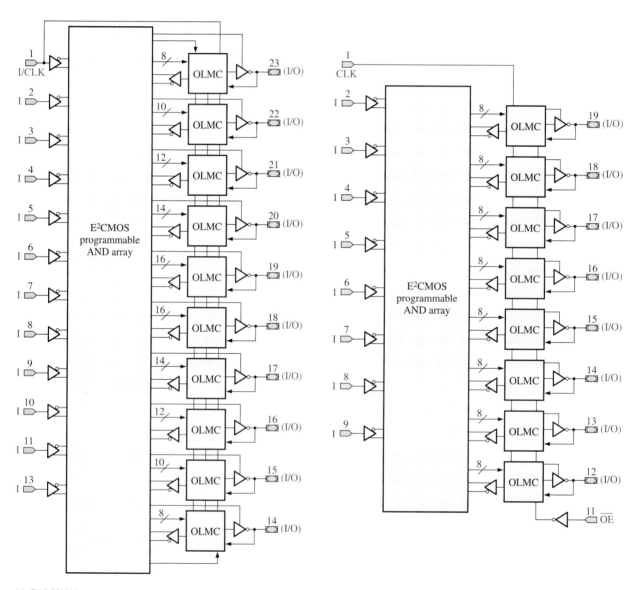

(a) GAL22V10

(b) GAL16V8

FIGURE 11–1
GAL block diagrams.

The Output Logic Macrocell (OLMC) in the GAL22V10

As stated in Chapter 7, an OLMC contains programmable logic circuits that can be configured for either a **combinational** logic output or input or for a **registered** output, which is a sequential logic output. In the registered mode, pins that are defined as 'reg' are obtained from the Q output of a flip-flop. As you know, the OLMCs are internally configured automatically by programming a set of cells that are separate from the logic array cells.

Logic Diagram   Figure 11–2 shows a basic logic diagram for the GAL22V10 OLMC. The inputs from the AND gates in the array to the OR gate in the OLMC vary from ten to sixteen. The other logic consists of a flip-flop and two multiplexers.

FIGURE 11–2
The GAL22V10 OLMC.

The 1-of-4 multiplexer connects one of its four input lines to the tristate output buffer based on the states of two select inputs, S_0 and S_1. The inputs to the 1-of-4 multiplexer are the OR gate output, the complement of the OR gate output, the flip-flop output Q, and the complement of the flip-flop output $\overline{Q}$. This allows the output of the OLMC to be either active-HIGH or active-LOW in each mode. The 1-of-2 multiplexer connects either the output of the tristate buffer or the $\overline{Q}$ output of the flip-flop back through a buffer to the AND array, based on the state of S_1. The select bits, S_0 and S_1, for each OLMC are programmed into a dedicated group of cells in the array by the compiler software, so the user does not have to directly manipulate these bits.

The flip-flop is a positive edge-triggered D type. The asynchronous reset (AR) input resets the flip-flop to the logic 0 state ($Q = 0$) independent of the clock. The synchronous preset (SP) input sets the flip-flop to the logic 1 state ($Q = 1$) on the rising edge of the clock pulse. The GAL22V10 has product terms for the AR and the SP inputs.

The four OLMC configurations are

- Combinational mode with active-LOW output
- Combinational mode with active-HIGH output
- Registered mode with active-LOW output
- Registered mode with active-HIGH output

The combinational modes were covered in Chapter 7. The registered modes are covered in Section 11–2.

The OLMC in the GAL16V8

As you learned in Chapter 7, the OLMC in the GAL16V8 is somewhat different from that in the GAL22V10. A GAL16V8 OLMC can be programmed in one of three available modes to emulate most of the existing PALs; thus, a GAL16V8 can replace the PAL that it is programmed to emulate. There are two combinational modes, *simple* and *complex*, and one registered mode.

Logic Diagram Figure 11–3 shows a basic logic diagram for the GAL16V8 OLMC. Each GAL16V8 OLMC contains two **global cells,** $\overline{SYN}$ and ACO, which affect all of the OLMCs. Also, there are two **local cells** for each OLMC, $AC1(n)$ and XOR, which affect only the associated OLMC. Both global and local cells select various paths in the OLMC and determine the combinational or registered mode.

FIGURE 11–3
The GAL16V8 OLMC.

The $\overline{SYN}$ bit and the ACO bit control the mode configuration. $\overline{SYN} = 0$ and $ACO = 1$ set the registered mode, and $\overline{SYN} = 1$ and $ACO = 0$ set the combinational modes. The $AC1$ bit controls the input/output configuration. The XOR bit selects either active-LOW ($XOR = 0$) or active-HIGH ($XOR = 1$) output polarity.

**SECTION 11–1
REVIEW**

1. What are the two modes for the GAL22V10?
2. How is the mode determined?
3. What are the modes for the GAL16V8?

11–2 ■ OLMC MODE SELECTION

In this section, we will examine the mode configurations using the GAL22V10 for illustration and discuss the difference between the registered mode and the combinational mode. As in Chapter 7, ABEL is used to illustrate the software aspects of PLD applications.

After completing this section, you should be able to

■ Discuss the role of the control bits S_0 and S_1 in mode selection   ■ Show how active-LOW and active-HIGH outputs are achieved

The Combinational Mode

The modes in the GAL22V10 are determined by the S_0 and S_1 bits, which are controlled by programming. For the combinational mode, $S_1S_0 = 10$ or $S_1S_0 = 11$. Figure 11–4 shows the logic paths through the OLMC for active-LOW and active-HIGH combinational outputs along with the resulting effective OLMC logic in each case. The flip-flop is not used in the combinational mode.

In Figure 11–4(a), the inversion of the tristate output buffer produces the active-LOW output. In part (b), the inversions of the 1-of-4 multiplexer and the tristate output buffer cancel, resulting in the active-HIGH output.

The Registered Mode

For the registered mode, $S_1S_0 = 00$ or $S_1S_0 = 01$. Figure 11–5 (p. 575) shows the logic paths through the OLMC for active-LOW and active-HIGH registered outputs along with the resulting effective OLMC logic in each case. In part (a), the inversion of the tristate output buffer produces the active-LOW output. In part (b), the inversions of the flip-flop ($\overline{Q}$ output) and the tristate output buffer cancel, resulting in the active-HIGH output. Notice that the feedback to the AND array through a buffer is from the $\overline{Q}$ output of the flip-flop and not from the output pin as in the combinational mode, so a registered output cannot be used as an input.

Software Mode Specification

The modes of an OLMC are established with software. This is done using statements in the declarations part of the input file and in the way logic descriptions are written. Several basic elements of ABEL declarations and expressions are important in programming a given output as either combinational or registered.

1. The ISTYPE statement is used to declare an output as either combinational or registered with the use of the attributes 'com' or 'reg'. For example, to specify a given output as registered, an ISTYPE statement can be included in a pin declaration as follows:

   ```
   Q0 PIN 23 ISTYPE 'reg';
   ```

   Also, the output can be specified as either not inverted or inverted (active-HIGH or active-LOW) using the attributes 'buffer' or 'invert' in the ISTYPE statement. For example,

   ```
   Q0 PIN 23 ISTYPE 'reg,buffer';
   ```

   indicates that the registered output Q_0 is not inverted. Pin numbers are optional and if not stated, ABEL can determine them for you.

2. The assignment operators : = and : > are used in logic descriptions to indicate a registered output. The registered assignment operator : = is analogous to the operator = used in combinational logic expressions. The registered assignment operator : > is analogous

(a) OLMC in the active-LOW combinational mode and the effective logic diagram

(b) OLMC in the active-HIGH combinational mode and the effective logic diagram

FIGURE 11–4

Combinational mode for active-LOW and active-HIGH outputs. The red lines show the logic paths in each case.

to the operator –> used in combinational logic truth tables. For example, the registered output expression

```
Q0 := D0;
```

indicates that the output Q_0 will assume the value of the D_0 input on the clock pulse and will hold that value until the next clock pulse. As a comparison, the combinational output expression

```
X = A;
```

indicates that the output X is equal to the input A.

(a) OLMC in the active-LOW registered mode and the effective logic diagram

(b) OLMC in the active-HIGH registered mode and the effective logic diagram

FIGURE 11–5

Registered mode for active-LOW and active-HIGH outputs. The red lines show the logic paths in each case.

3. The dot extension .CLK is used to indicate that the register device is a clocked flip-flop. An equation using this dot extension must accompany the output equation. For example, the output equation

   ```
   Q0 := D0;
   ```

   must be followed by a clock equation such as

   ```
   Q0.CLK = Clock;
   ```

   ABEL is an extensive language and the elements just introduced represent only a few of the many that are available. However, these few are essential in programming a PLD for operation as a sequential circuit.

EXAMPLE 11–1   Write the ABEL pin declarations and the equations to specify the GAL22V10 in Figure 11–6 as a parallel input/parallel output clocked register with D_0 through D_7 as inputs and Q_0 through Q_7 as registered outputs that are not inverted.

FIGURE 11–6

Solution   The pin declaration statements can be written as (pin numbers are optional)

```
Clock,D0,D1,D2,D3,D4,D5,D6,D7  PIN 1,2,3,4,5,6,7,8,9;
Q0,Q1,Q2,Q3,Q4,Q5,Q6,Q7  PIN 22,21,20,19,18,17,16,15 ISTYPE 'reg,buffer';
```

The equations can be written using set notation as follows:

```
[Q0, Q1, Q2, Q3, Q4, Q5, Q6, Q7] := [D0, D1, D2, D3, D4, D5, D6, D7];
[Q0, Q1, Q2, Q3, Q4, Q5, Q6, Q7].CLK = Clock;
```

This is equivalent to, but much simpler than, writing individual equations such as

```
Q0 := D0;
Q0.CLK = Clock;
Q1 := D1;
Q1.CLK = Clock;
```

```
                      Q2 := D2;
                      Q2.CLK = Clock;
                      Q3 := D3;
                      Q3.CLK = Clock;
                      Q4 := D4;
                      Q4.CLK = Clock;
                      Q5 := D5;
                      Q5.CLK = Clock;
                      Q6 := D6;
                      Q6.CLK = Clock;
                      Q7 := D7;
                      Q7.CLK = Clock;
```

Related Problem Change the pin declaration statements to make the outputs inverted.

SECTION 11–2 REVIEW

1. What is the basic difference between the registered mode and the combinational mode?
2. How can ISTYPE statements be used?
3. What is the difference between the operator $:=$ and the operator $=$ in an equation?
4. What is the purpose of the .CLK extension?

11–3 ■ IMPLEMENTING SHIFT REGISTERS WITH PLDs

In the last section, you learned about the registered mode and how the OLMC flip-flop is used. In this section, you will see how shift registers can be implemented with a PLD. Again, the GAL22V10 is used for illustration.

After completing this section, you should be able to

■ Use the .CLK extension ■ Use the .AR extension ■ Use the .c. and .x. special constants in test vectors ■ Implement an 8-bit serial in/parallel out shift register ■ Implement a 4-bit parallel in/serial out shift register

An 8-Bit Serial In/Parallel Out Shift Register

Figure 11–7 shows an 8-bit shift register in which the data are entered serially. When the Enable input to the AND gate is HIGH, the data are clocked into the left-most flip-flop and then shifted from left to right. The Q output of each flip-flop is available. The Clear input is active-LOW and resets all of the flip-flops ($Q = 0$) independent of the clock.

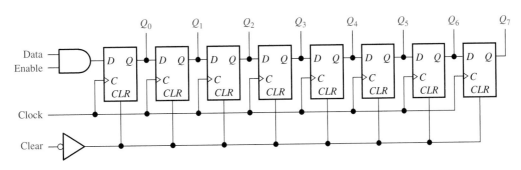

FIGURE 11–7
Eight-bit serial in/parallel out shift register.

Implementing the Shift Register with a PLD

The shift register inputs and outputs are assigned to pins on the GAL22V10 as shown in Figure 11–8. Since the D input of the first flip-flop comes from the AND gate, the output of the first flip-flop, Q_0, is expressed as the following ABEL registered equation:

```
Q0 := Data & Enable;
```

Since the Q output of each flip-flop is connected to the D input of the next flip-flop, the remaining flip-flop outputs are expressed as follows:

```
Q1 := Q0;
Q2 := Q1;
Q3 := Q2;
Q4 := Q3;
Q5 := Q4;
Q6 := Q5;
Q7 := Q6;
```

FIGURE 11–8

GAL22V10 input/output assignment for the 8-bit shift register.

Developing the ABEL Input File

Recall that an ABEL input file module typically contains three parts: declarations, logic descriptions, and test vectors. In the input file for the 8-bit shift register, the declarations section makes the pin assignments for the inputs and outputs.

The equations section expresses the output logic equations, utilizing the set notation that was introduced in Chapter 7. For the Clock equation, the dot extension .CLK specifies the clock input for each of the flip-flops associated with the specified outputs. For the Clear equation, the dot extension .AR specifies an asynchronous reset for each of the flip-flops associated with the specified outputs.

Entries in the test vectors section consist of numeric values, the special constant .c. that is used to indicate the clock pulse, and .x. that indicates a "don't care". Recall that test vectors are used to test the PLD to make sure that it works as expected when programmed. The ABEL input file is as follows:

```
Module  Eight_bit_shift_register

Title  '8-bit shift register in a GAL22V10'

"Device Declaration

    Register    device    'P22V10'

"Pin Declaration

    Clock, Clear              pin 1,2;
    Data, Enable              pin 3,4;
    Q0,Q1,Q2,Q3,Q4,Q5,Q6,Q7   pin 16,17,18,19,20,21,22,23 ISTYPE 'reg,buffer';

Equations

    Q0 := Data & Enable;

    [Q1,Q2,Q3,Q4,Q5,Q6,Q7] := [Q0,Q1,Q2,Q3,Q4,Q5,Q6];

    [Q0,Q1,Q2,Q3,Q4,Q5,Q6,Q7].CLK = Clock;

    [Q0,Q1,Q2,Q3,Q4,Q5,Q6,Q7].AR = !Clear;

Test_Vectors

    ([Clock, Clear, Data, Enable] -> [Q0,Q1,Q2,Q3,Q4,Q5,Q6,Q7])
     [ .x. ,  0  , .x. ,  .x. ] -> [ 0, 0, 0, 0, 0, 0, 0, 0];
     [ .c. ,  1  ,  1  ,   0  ] -> [ 0, 0, 0, 0, 0, 0, 0, 0];
     [ .c. ,  1  ,  0  ,   1  ] -> [ 0, 0, 0, 0, 0, 0, 0, 0];
     [ .c. ,  1  ,  1  ,   1  ] -> [ 1, 0, 0, 0, 0, 0, 0, 0];
     [ .c. ,  1  ,  0  ,   1  ] -> [ 0, 1, 0, 0, 0, 0, 0, 0];
     [ .c. ,  1  ,  1  ,   1  ] -> [ 1, 0, 1, 0, 0, 0, 0, 0];
     [ .c. ,  1  ,  0  ,   1  ] -> [ 0, 1, 0, 1, 0, 0, 0, 0];
     [ .c. ,  1  ,  1  ,   1  ] -> [ 1, 0, 1, 0, 1, 0, 0, 0];
     [ .c. ,  1  ,  0  ,   1  ] -> [ 0, 1, 0, 1, 0, 1, 0, 0];
     [ .c. ,  1  ,  1  ,   1  ] -> [ 1, 0, 1, 0, 1, 0, 1, 0];
     [ .c. ,  1  ,  0  ,   1  ] -> [ 0, 1, 0, 1, 0, 1, 0, 1];
     [ .c. ,  1  ,  1  ,   1  ] -> [ 1, 0, 1, 0, 1, 0, 1, 0];
     [ .c. ,  0  ,  1  ,   1  ] -> [ 0, 0, 0, 0, 0, 0, 0, 0];

END
```

EXAMPLE 11–2 Implement the 4-bit parallel in/serial out shift register in Figure 11–9 by developing an ABEL input file for a GAL22V10. The $SH/\overline{LD}$ input is the Shift/Load. When this input is LOW, the parallel data are loaded into the register, and when it is HIGH, data are shifted from left to right. D_0 through D_3 are the parallel data inputs and Q_3 is the serial data output.

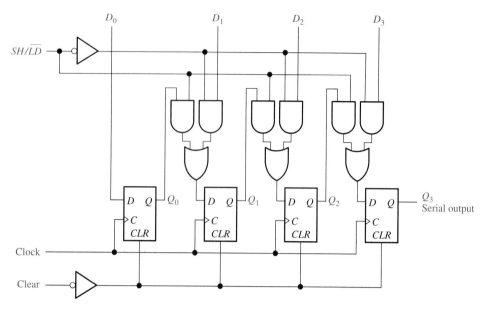

FIGURE 11–9
Four-bit parallel in/serial out shift register.

Solution   The ABEL input file is as follows:

```
Module  Four_bit_shift_register

Title  '4-bit shift register in a GAL22V10'

"Device Declaration

   Register   device   'P22V10'

"Pin Declaration

   Clock, Clear    pin 1,2;
   SHLD            pin 3;
   DO, D1, D2, D3  pin 4,5,6,7 ISTYPE 'reg,buffer';
   Q0, Q1, Q2, Q3  pin 14,15,16,17 ISTYPE 'reg,buffer';

Equations

   Q0 := D0;

   Q1 := Q0 & SHLD # D1 & !SHLD;

   Q2 := Q1 & SHLD # D2 & !SHLD;

   Q3 := Q2 & SHLD # D3 & !SHLD;

   [Q0, Q1, Q2, Q3].CLK = Clock;

   [Q0, Q1, Q2, Q3].AR = !Clear;

Test_Vectors

   ([Clock, Clear, SHLD, D0 , D1 , D2 , D3 ] -> [Q3])
    [ .x. ,   0  , .x. , .x., .x., .x., .x.] -> [ 0];
    [ .c. ,   1  ,  0  ,  0 ,  1 ,  0 ,  1 ] -> [ 1];
    [ .c. ,   1  ,  0  ,  1 ,  0 ,  1 ,  0 ] -> [ 0];
    [ .c. ,   1  ,  1  ,  1 ,  0 ,  1 ,  0 ] -> [ 1];
    [ .c. ,   1  ,  1  ,  1 ,  0 ,  1 ,  0 ] -> [ 0];
    [ .c. ,   1  ,  1  ,  1 ,  0 ,  1 ,  0 ] -> [ 1];
    [ .c. ,   0  ,  0  ,  1 ,  0 ,  1 ,  0 ] -> [ 0];

END
```

Related Problem   Modify the pin declaration and the equation sections of the ABEL
input file to change the shift register in Figure 11–9 from four bits to five bits.

1. What is the meaning of the ABEL equation Q0 := A & !B # !A & B;?
2. What is the purpose of the dot extension .AR?
3. When is the special constant .c. used?

11–4 ■ IMPLEMENTING COUNTERS WITH PLDs

As you have learned, counters consist of flip-flops just as shift registers do; but counters have a prescribed sequence of states, whereas registers do not. Counters are in a category of digital circuits known as state machines as mentioned briefly in Chapter 9. In this section, the general types of state machines are discussed and three approaches to implementing a counter in a PLD are covered.

After completing this section, you should be able to

■ Define the two categories of state machines ■ Implement a counter using equation entry ■ Implement a counter using truth table entry ■ Implement a counter using state diagram entry

Types of State Machines

Sequential logic circuits can be classified into a category of circuits known as **state machines** of which there are two basic types. In the Moore state machine shown in Figure 11–10(a), the outputs depend only on the internal state and any inputs that are synchronized with the circuit. Counters are examples of the Moore type of state machine, where the flip-flops are the memory portion.

In the Mealy state machine, the outputs are determined by both the internal state and by inputs that are not synchronized with the circuit, as indicated in Figure 11–10(b).

(a) Moore state machine

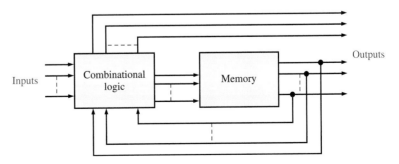

(b) Mealy state machine

FIGURE 11–10
Generalized block diagrams for the two categories of state machines.

A 3-Bit Up/Down Gray Code Counter Design

To illustrate the process of implementing a counter design in a PLD, we will use a 3-bit up/down Gray code counter.

The first step is to design the counter with D flip-flops. D flip-flops are used because most PLDs including the GAL22V10 and the GAL16V8 have this type of device in their OLMCs. A similar counter was designed in Chapter 9 with J-K flip-flops.

The State Diagram Figure 11–11 shows the state diagram for a 3-bit up/down Gray code counter. The $Y = 1$ or $Y = 0$ beside each arrow indicates the state of the up/down control input.

FIGURE 11–11
State diagram for a 3-bit up/down Gray code counter.

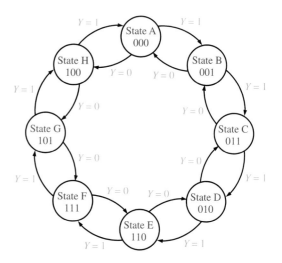

The Next-State Table The next-state table is derived from the state diagram and is shown in Table 11–1. For each present state there are two possible next states, depending on the UP/DOWN control variable, Y.

TABLE 11–1
Next-state table for the 3-bit up/down Gray code counter.

Present State			Next State					
			$Y = 0$ (DOWN)			$Y = 1$ (UP)		
Q_2	Q_1	Q_0	Q_2	Q_1	Q_0	Q_2	Q_1	Q_0
0	0	0	1	0	0	0	0	1
0	0	1	0	0	0	0	1	1
0	1	1	0	0	1	0	1	0
0	1	0	0	1	1	1	1	0
1	1	0	0	1	0	1	1	1
1	1	1	1	1	0	1	0	1
1	0	1	1	1	1	1	0	0
1	0	0	1	0	1	0	0	0

The Flip-Flop Transition Table The transition table for the D flip-flop is shown in Table 11–2. This table indicates the required D input for each of the four possible state transitions. Q_N is the present state and Q_{N+1} is the next state.

TABLE 11–2
Transition table for a D flip-flop.

Output Transitions		Flip-Flop Input
Q_N	Q_{N+1}	D
0 ⟶ 0		0
0 ⟶ 1		1
1 ⟶ 0		0
1 ⟶ 1		1

The Karnaugh Maps D inputs of the three flip-flops in the Gray code counter are plotted on the Karnaugh maps as shown in Figure 11–12. The up/down control input Y is treated as one of the state variables along with Q_0, Q_1, and Q_2. Using the next-state table (Table 11–1), the flip-flop input from Table 11–2 is transferred onto the maps for each present state of the counter.

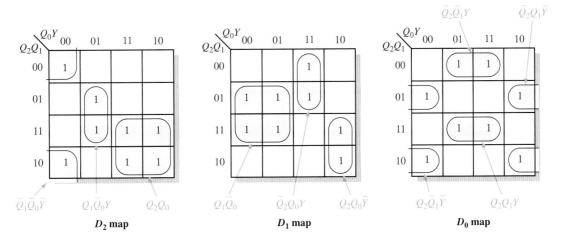

FIGURE 11–12
D input maps for the 3-bit up/down Gray code counter.

The Counter Circuit   Figure 11–13 is the logic diagram for the 3-bit up/down Gray code counter. The combinational logic is used to implement the three equations for the D inputs to the flip-flops.

The Boolean Expressions for the D Inputs

The 1s on the maps are combined in the largest possible groupings as illustrated in Figure 11–12 and the groups are factored, resulting in the following expressions:

$$D_0 = Q_2 Q_1 Y + \overline{Q}_2 \overline{Q}_1 Y + Q_2 \overline{Q}_1 \overline{Y} + \overline{Q}_2 Q_1 \overline{Y}$$
$$D_1 = Q_2 Q_0 \overline{Y} + \overline{Q}_2 Q_0 Y + Q_1 \overline{Q}_0$$
$$D_2 = \overline{Q}_1 \overline{Q}_0 \overline{Y} + Q_1 \overline{Q}_0 Y + Q_2 Q_0$$

Implementing the Counter with ABEL Equations

You are already familiar with using ABEL equations to describe a sequential circuit with registered outputs. The process for counters is basically the same as for shift registers in terms of the structure of the ABEL input file, as you will see in Example 11–3.

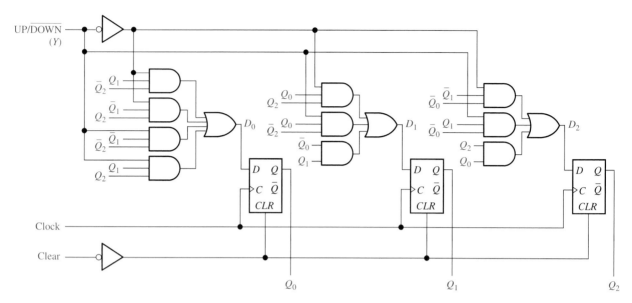

FIGURE 11–13
Logic diagram of the 3-bit up/down Gray code counter.

In addition to the .c. special constant for a clock pulse, another special constant is .x., which indicates a "don't care" for an input or output and can be used in truth tables and test vectors.

EXAMPLE 11–3

Develop an ABEL input file using the logic equation format to implement the 3-bit up/down Gray code counter in Figure 11–13. A GAL22V10 is the target device.

Solution Y represents the UP/$\overline{\text{DOWN}}$ control input. Since the Q output goes to the value of the D input on the next clock pulse, the ABEL equations are expressed in terms of the Q outputs instead of the D inputs. The input file is as follows:

```
Module   Three_bit_U/D_Gray_Code_Counter
Title    '3-bit Gray Code Counter in a GAL22V10'
"Device Declaration
   Counter      device      'P22V10'
"Pin Declaration
   Clock, Clear    pin 1,2;
   Y               pin 3;
   Q0,Q1,Q2        pin 21,22,23 ISTYPE 'reg,buffer';
Equations
   Q0 := Q2 & Q1 & Y # !Q2 & !Q1 & Y # Q2 & !Q1 & !Y # !Q2 & Q1 & !Y;
   Q1 := Q2 & Q0 & !Y # !Q2 & Q0 & Y # Q1 & !Q0;
   Q2 := !Q1 & !Q0 & !Y # Q1 & !Q0 & Y # Q2 & Q0;
   [Q0,Q1,Q2].CLK = Clock;
   [Q0,Q1,Q2].AR = !Clear;
```

```
Test_Vectors ([Clock, Clear,   Y ] -> [Q2,Q1,Q0])
                [ .c. ,   0  , .x.] -> [ 0, 0, 0];
                [ .c. ,   1  ,  0 ] -> [ 1, 0, 0];
                [ .c. ,   1  ,  0 ] -> [ 1, 0, 1];
                [ .c. ,   1  ,  0 ] -> [ 1, 1, 1];
                [ .c. ,   1  ,  0 ] -> [ 1, 1, 0];
                [ .c. ,   1  ,  0 ] -> [ 0, 1, 0];
                [ .c. ,   1  ,  0 ] -> [ 0, 1, 1];
                [ .c. ,   1  ,  0 ] -> [ 0, 0, 1];
                [ .c. ,   1  ,  0 ] -> [ 0, 0, 0];
                [ .c. ,   1  ,  1 ] -> [ 0, 0, 1];
                [ .c. ,   1  ,  1 ] -> [ 0, 1, 1];
                [ .c. ,   1  ,  1 ] -> [ 0, 1, 0];
                [ .c. ,   1  ,  1 ] -> [ 1, 1, 0];
                [ .c. ,   1  ,  1 ] -> [ 1, 1, 1];
                [ .c. ,   1  ,  1 ] -> [ 1, 0, 1];
                [ .c. ,   1  ,  1 ] -> [ 1, 0, 0];
                [ .c. ,   1  ,  1 ] -> [ 0, 0, 0];

END
```

Related Problem   Explain what the test vectors do line by line in this input file.

Implementing the Counter with an ABEL Truth Table

Instead of using equations, the counter can be described using the truth table entry method, which was covered in Section 7–7. The entry method used is basically a matter of personal preference.

EXAMPLE 11–4

Develop an ABEL input file using the truth table format to implement the 3-bit up/down Gray code counter in Figure 11–13. A GAL22V10 is the target device.

Solution   In a truth table, the inputs are on the left of the assignment operator and the outputs are on the right. The assignment operator :> is used in truth tables for circuits with registered outputs and indicates that, for the given input values, the outputs will assume the given state after the next clock pulse. The ABEL input file is as follows. Notice that we still have the equations defining Clock and Clear.

```
Module   Three_bit_U/D_Gray_Code_Counter
Title    '3-bit Gray Code Counter in a GAL22V10'
"Device Declaration
   Counter      device      'P22V10'
"Pin Declaration
   Clock, Clear   pin 1,2;
   Y              pin 3;
   Q0,Q1,Q2       pin 21,22,23 ISTYPE 'reg,buffer';
Equations
   [Q0,Q1,Q2].CLK = Clock;
   [Q0,Q1,Q2].AR = !Clear;
```

```
Truth_Table ([Clear,  Y  ,  Q2  ,  Q1  ,  Q0 ] :> [Q2, Q1, Q0])
           [  0  , .x. , .x. , .x. , .x.] :> [ 0,  0,  0];
           [  1  ,  0  ,  0  ,  0  ,  0 ] :> [ 1,  0,  0];
           [  1  ,  0  ,  1  ,  0  ,  0 ] :> [ 1,  0,  1];
           [  1  ,  0  ,  1  ,  0  ,  1 ] :> [ 1,  1,  1];
           [  1  ,  0  ,  1  ,  1  ,  1 ] :> [ 1,  1,  0];
           [  1  ,  0  ,  1  ,  1  ,  0 ] :> [ 0,  1,  0];
           [  1  ,  0  ,  0  ,  1  ,  0 ] :> [ 0,  1,  1];
           [  1  ,  0  ,  0  ,  1  ,  1 ] :> [ 0,  0,  1];
           [  1  ,  0  ,  0  ,  0  ,  1 ] :> [ 0,  0,  0];
           [  1  ,  1  ,  0  ,  0  ,  0 ] :> [ 0,  0,  1];
           [  1  ,  1  ,  0  ,  0  ,  1 ] :> [ 0,  1,  1];
           [  1  ,  1  ,  0  ,  1  ,  1 ] :> [ 0,  1,  0];
           [  1  ,  1  ,  0  ,  1  ,  0 ] :> [ 1,  1,  0];
           [  1  ,  1  ,  1  ,  1  ,  0 ] :> [ 1,  1,  1];
           [  1  ,  1  ,  1  ,  1  ,  1 ] :> [ 1,  0,  1];
           [  1  ,  1  ,  1  ,  0  ,  1 ] :> [ 1,  0,  0];
           [  1  ,  1  ,  1  ,  0  ,  0 ] :> [ 0,  0,  0];

Test_Vectors ([Clock, Clear,  Y ] -> [Q2,Q1,Q0])
             [ .c. ,   0  , .x.] -> [ 0,  0,  0];
             [ .c. ,   1  ,  0 ] -> [ 1,  0,  0];
             [ .c. ,   1  ,  0 ] -> [ 1,  0,  1];
             [ .c. ,   1  ,  0 ] -> [ 1,  1,  1];
             [ .c. ,   1  ,  0 ] -> [ 1,  1,  0];
             [ .c. ,   1  ,  0 ] -> [ 0,  1,  0];
             [ .c. ,   1  ,  0 ] -> [ 0,  1,  1];
             [ .c. ,   1  ,  0 ] -> [ 0,  0,  1];
             [ .c. ,   1  ,  0 ] -> [ 0,  0,  0];
             [ .c. ,   1  ,  1 ] -> [ 0,  0,  1];
             [ .c. ,   1  ,  1 ] -> [ 0,  1,  1];
             [ .c. ,   1  ,  1 ] -> [ 0,  1,  0];
             [ .c. ,   1  ,  1 ] -> [ 1,  1,  0];
             [ .c. ,   1  ,  1 ] -> [ 1,  1,  1];
             [ .c. ,   1  ,  1 ] -> [ 1,  0,  1];
             [ .c. ,   1  ,  1 ] -> [ 1,  0,  0];
             [ .c. ,   1  ,  1 ] -> [ 0,  0,  0];

    END
```

Related Problem Explain the use of the "don't care" special constant in the first line of the truth table.

Implementing the Counter with an ABEL State Diagram

A third alternative for ABEL design entry is the state diagram method, which is often better for more complex state machines. However, the 3-bit counter will serve to introduce this method.

The state diagram for the 3-bit up/down Gray code counter was shown in Figure 11–11. There are eight states labeled A through H, each representing a 3-bit Gray code. In each state, the counter can go either to the next higher state or to the next lower state, depending on the value of the UP/$\overline{\text{DOWN}}$ control variable Y.

IF-THEN-ELSE Statements ABEL provides a conditional statement to describe to which state the counter will go next based on certain conditions. In this case there are two conditions, $Y = 0$ and $Y = 1$. So, for example, when the counter is in State A (000), it will go to State B (001) if $Y = 1$ else it will go to State H ($Y = 0$). The ABEL IF-THEN-ELSE statement for State A is

```
IF Y THEN B ELSE H;
```

The statements can also be written using lowercase letters.

WITH Statement   This statement can be used in equations for outputs associated with specific transitions or with equations for registered outputs that must become active in a destination state.

Defining the States   The use of a state diagram must include a definition of all the states in the counter. First, the variable *QSTATE* is defined as the set of *Q* outputs, which are the states of the flip-flops:

```
QSTATE = [Q2,Q1,Q0];
```

This is followed by the name and values for each of the eight states.

```
A = [0 , 0 , 0];
B = [0 , 0 , 1];
C = [0 , 1 , 1];
D = [0 , 1 , 0];
E = [1 , 1 , 0];
F = [1 , 1 , 1];
G = [1 , 0 , 1];
H = [1 , 0 , 0];
```

The next example shows how some of these basic ABEL state diagram elements can be used to describe the counter.

EXAMPLE 11–5

Develop an ABEL input file using the state diagram format to implement the 3-bit up/down Gray code counter in Figure 11–13. A GAL22V10 is the target device.

Solution   The ABEL input file is as follows. Notice that we still have the equations defining Clock and Clear.

```
Module   Three_bit_U/D_Gray_Code_Counter
Title    '3-bit Gray Code Counter in a GAL22V10'
"Device Declaration
    Counter      device       'P22V10'
"Pin Declaration
    Clock, Clear   pin 1,2;
    Y              pin 3;
    Q0,Q1,Q2       pin 21,22,23 ISTYPE 'reg,buffer';
"State Definitions
    QSTATE = [Q2,Q1,Q0];
    A      = [ 0, 0, 0];
    B      = [ 0, 0, 1];
    C      = [ 0, 1, 1];
    D      = [ 0, 1, 0];
    E      = [ 1, 1, 0];
    F      = [ 1, 1, 1];
    G      = [ 1, 0, 1];
    H      = [ 1, 0, 0];

Equations
    QSTATE.CLK = Clock;

    QSTATE.AR = !Clear;
```

```
State_diagram QSTATE

    State A:    if Y then B else H;

    State B:    if Y then C else B;

    State C:    if Y then D else B;

    State D:    if Y then E else C;

    State E:    if Y then F else D;

    State F:    if Y then G else E;

    State G:    if Y then H else F;

    State H:    if Y then A else G;

Test_Vectors ([Clock, Clear,  Y ] -> [Q2,Q1,Q0])
              [ .c. ,   0  , .x.] -> [ 0, 0, 0];
              [ .c. ,   1  ,  0 ] -> [ 1, 0, 0];
              [ .c. ,   1  ,  0 ] -> [ 1, 0, 1];
              [ .c. ,   1  ,  0 ] -> [ 1, 1, 1];
              [ .c. ,   1  ,  0 ] -> [ 1, 1, 0];
              [ .c. ,   1  ,  0 ] -> [ 0, 1, 0];
              [ .c. ,   1  ,  0 ] -> [ 0, 1, 1];
              [ .c. ,   1  ,  0 ] -> [ 0, 0, 1];
              [ .c. ,   1  ,  0 ] -> [ 0, 0, 0];
              [ .c. ,   1  ,  1 ] -> [ 0, 0, 1];
              [ .c. ,   1  ,  1 ] -> [ 0, 1, 1];
              [ .c. ,   1  ,  1 ] -> [ 0, 1, 0];
              [ .c. ,   1  ,  1 ] -> [ 1, 1, 0];
              [ .c. ,   1  ,  1 ] -> [ 1, 1, 1];
              [ .c. ,   1  ,  1 ] -> [ 1, 0, 1];
              [ .c. ,   1  ,  1 ] -> [ 1, 0, 0];
              [ .c. ,   1  ,  1 ] -> [ 0, 0, 0];

END
```

Related Problem Write the IF-THEN-ELSE statement for State A using $\overline{Y}$ instead of Y.

GOTO Statements In addition to the IF-THEN-ELSE statement for conditional state transitions, ABEL provides the GOTO statement for unconditional state transitions. If a counter goes to only one next state from a given present state, a GOTO statement can be used. For example, a 3-bit counter that sequences through a straight binary count can be described by the following ABEL state definitions and state diagram:

```
QSTATE = [Q2,Q1,Q0];
A      = [ 0, 0, 0];
B      = [ 0, 0, 1];
C      = [ 0, 1, 0];
D      = [ 0, 1, 1];
E      = [ 1, 0, 0];
F      = [ 1, 0, 1];
G      = [ 1, 1, 0];
H      = [ 1, 1, 1];

State_diagram QSTATE

    State A:    GOTO B;
    State B:    GOTO C;
    State C:    GOTO D;
    State D:    GOTO E;
    State E:    GOTO F;
    State F:    GOTO G;
    State G:    GOTO H;
    State H:    GOTO A;
```

11–5 ■ PLD SYSTEM IMPLEMENTATION

In this section, a specific example is used to illustrate the use of the state diagrams and other entry methods to implement a complete system with PLDs using ABEL. The example system is an elevator control for a two-floor building and uses two GAL16V8 devices. This system is implemented from a description of the system requirements, a block diagram, and a state diagram.

After completing this section, you should be able to

- Describe an elevator control system
- Implement a design from a state diagram
- Use the dot extension .FB

The System

A block diagram of a control system for a two-floor elevator is shown in Figure 11–14. The control system monitors the status of the elevator push-button switches, responds to service calls, and displays the floor number and direction of the elevator. The system uses two GAL16V8s, one for the control logic and one for the display logic, as indicated. The amount of logic required could be implemented with a single GAL. However, one GAL16V8 does not have enough outputs to accommodate the system requirements.

FIGURE 11–14
Block diagram of the elevator control system.

There is a "request" push-button on each floor (REQ1 and REQ2) beside the elevator door. When the elevator arrives at your floor, the doors open for a set period of time to let passengers on or off and then close automatically. Inside the elevator you press either the FLOOR2 or the FLOOR1 push-button switches. If you are on the first floor, you press the FLOOR2 push-button to go to the second floor; and if you are on the second floor, you press the FLOOR1 push-button to go to the first floor. Pressing the Open push-button inside the elevator will cause the door to reopen or stay open longer than the preset time as long as the elevator is not moving. If none of the push-button switches is activated, the elevator rests at the last floor serviced. A very slow clock signal controls both the movement of the elevator and the time that the door remains open when the elevator stops at a floor.

The state control logic consists of clock-independent latches that store a momentary closure of the push-button switches and a state machine that produces three elevator control outputs synchronized with the clock: *DOOR, MOTION,* and *DIR* (direction). Table 11–3 shows the states of these synchronized outputs.

TABLE 11–3

Output Variable	State	
DOOR	0 = Open	1 = Closed
MOTION	0 = Wait	1 = Move
DIR	0 = Up	1 = Down

The display unit is basically a decoder consisting only of combinational logic that produces six outputs to a 7-segment display (segment *f* is not used) to indicate either floor 1 or floor 2 and two outputs to an arrow indicator light to indicate the direction that the elevator is moving.

Inputs to the State Control Logic

- CLK   The clock signal establishes the length of time that the door remains open. The period of the clock is 10 s.
- *REQ1*   Request for the elevator to come to the first floor. The REQ push-button is located outside of the elevator beside the door.
- *REQ2*   Request for the elevator to come to the second floor. The REQ push-button is located outside of the elevator beside the door.
- *FLOOR2*   Instruction for the elevator to go to the second floor. The FLOOR2 push-button is located inside the elevator.
- *FLOOR1*   Instruction for the elevator to go to the first floor. The FLOOR1 push-button is located inside the elevator.
- *OPEN*   Instruction for the door to open when the elevator is not moving. The OPEN push-button is located inside the elevator.
- *ARRIVE*   Indication that the elevator has arrived at a new floor. Sensor switches on each floor which are activated by the elevator provide this input.

Outputs from the State Control Logic

- *L1PB* and *L1PB_*   First floor latch outputs for push-button switches. When one of the push-button switches for the first floor is pressed, the request is immediately stored in an S-R latch, independent of the clock. *L1PB* and *L1PB_* are complements.
- *L2PB* and *L2PB_*   Second floor latch outputs for push-button switches. When one of the push-button switches for the second floor is pressed, the request is immediately stored in an S-R latch, independent of the clock. *L2PB* and *L2PB_* are complements.

- *DOOR* Controls the status of the elevator door. When *DOOR* = 0, the door is open. When *DOOR* = 1, the door is closed.
- *MOTION* Controls the movement of the elevator. When *MOTION* = 0, the elevator waits (does not move). When *MOTION* = 1, the elevator moves either up or down.
- *DIR* Controls the direction of movement. When *DIR* = 0, the elevator moves up on the next call for second floor service. When *DIR* = 1, the elevator moves down on the next call for first floor service.

The State Diagram for the Elevator State Control Logic

The six states of the system are as follows:

- REST1 The elevator is waiting on the first floor with the door open.
- CLOSE1 The elevator is waiting on the first floor with the door closed.
- UP The elevator is moving from first floor to second floor.
- REST2 The elevator is waiting on the second floor with the door open.
- CLOSE2 The elevator is waiting on the second floor with the door closed.
- DOWN The elevator is moving from second floor to first floor.

The state diagram in Figure 11–15 shows the conditions for the system to change from one state to the next.

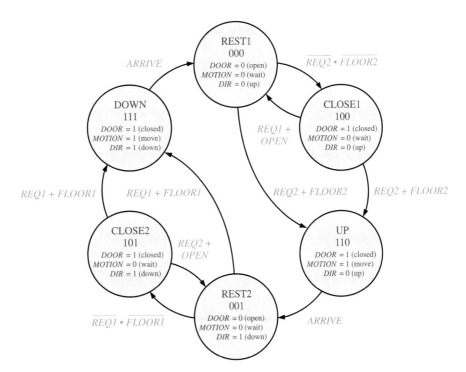

FIGURE 11–15
State diagram for the elevator control system.

Description of Operation

Refer to the state diagram in Figure 11–15. To begin, assume the system is in the REST1 state in which the elevator is waiting on the first floor with the door open. If, after a period of time established by the clock, there is no request from the second floor and the FLOOR2 button is not pressed ($\overline{REQ2} \cdot \overline{FLOOR2}$), the system goes to the CLOSE1 state in which the elevator remains waiting on the first floor with the door closed. When the system is in the REST1 state and there is a request from the second floor or the FLOOR2 button is pressed ($REQ2 + FLOOR2$), the system bypasses the CLOSE1 state and goes to the UP state on the next clock pulse. In the UP state, the door closes and the elevator moves upward.

When the system is in the CLOSE1 state, it remains there until there is a request from the second floor or the FLOOR2 button is pressed ($REQ2 + FLOOR2$), which causes the system to go to the UP state. Alternatively, it remains in the CLOSE1 state until there is a request from the first floor or the OPEN button is pressed ($REQ1 + OPEN$), which causes the system to return to the REST1 state.

The system remains in the UP state until the *ARRIVE* signal is received. An *ARRIVE* signal is produced when the elevator reaches the second floor and activates a switch, putting the system into the REST2 state. In the REST2 state, the elevator is waiting on the second floor with the door open. If, after a period of time established by the clock, there is no request from the first floor and the FLOOR1 button is not pressed ($\overline{REQ1} \cdot \overline{FLOOR1}$), the system goes to the CLOSE2 state in which the elevator remains waiting on the second floor with the door closed. When the system is in the REST2 state and there is a request from the first floor or the FLOOR1 button is pressed ($REQ1 + FLOOR1$), the system bypasses the CLOSE2 state and goes to the DOWN state on the next clock pulse. In the DOWN state, the door closes and the elevator moves downward.

When the system is in the CLOSE2 state, it remains there until there is a request from the first floor or the FLOOR1 button is pressed ($REQ1 + FLOOR1$), which causes the system to go to the DOWN state. Alternatively, it remains in the CLOSE2 state until there is a request from the second floor or the OPEN button is pressed ($REQ2 + OPEN$), which causes the system to return to the REST2 state.

The system remains in the DOWN state until an *ARRIVE* signal is received when the elevator reaches the first floor, which activates a switch and puts the system back into the REST1 state.

Implementing the State Control Logic with a PLD

Refer to the block diagram for the GAL16V8 in Figure 11–1(b). Figure 11–16 shows the pin assignments for the GAL16V8 used in the state control logic portion of the system. Each of the elements required to describe the system with ABEL will be developed and then put together to form a complete input file. Notice that the state control logic con-

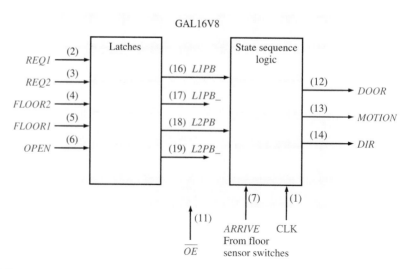

FIGURE 11–16
State control logic block diagram showing GAL16V8 pin assignments for inputs and outputs.

sists of two parts, the latches for the push-button switch inputs and the state sequence logic.

State Control Logic Inputs and Outputs

The statements defining the inputs can be written as follows. $\overline{OE}$ is the Output Enable that is required for the GAL16V8. Specification of pin numbers is optional.

```
CLK, !OE                     pin 1,11;
REQ1, REQ2                   pin 2,3;
FLOOR2, FLOOR1, OPEN, ARRIVE  pin 4,5,6,7;
```

Because of the very long clock period ($T = 10$ s), the latches for the first and second floor push-buttons (PB) are implemented with the logic array gates to achieve an immediate response instead of using the clock-dependent D flip-flops in the OLMCs. The statements defining the outputs can be written as follows:

```
L1PB, L1PB_      pin 16,17 ISTYPE 'com,buffer;
L2PB, L2PB_      pin 18,19 ISTYPE 'com,buffer;
DOOR, MOTION, DIR pin 12,13,14 ISTYPE 'reg,buffer';
```

The variables, *L1PB*, *L1PB_*, *L2PB*, and *L2PB_*, produced by the latches are listed as outputs only because they are used as inputs by the sequential logic portion of the state control logic and must have pin designations in order to be fed back to the array logic. These outputs are not used for any external purpose.

State Definitions

The six states in the state control system must be defined in terms of their binary values. As you have seen, the state of the system is determined by the control output variables *DOOR, MOTION,* and *DIR.* First, these state control variables must be defined with a statement such as

```
CONSTATE = [DOOR, MOTION, DIR];
```

Next, the binary values must be stated and there are two ways to do this. The first way uses set notation. For example, the REST1 state is

```
REST1 = [ 0 , 0 , 0];
```

The second way uses ABEL binary notation where the binary number is prefixed with a $^\wedge$B as follows:

```
REST1  = ^B000;
CLOSE1 = ^B100;
UP     = ^B110;
REST2  = ^B001;
CLOSE2 = ^B101;
DOWN   = ^B111;
```

The Push-Button Latch Logic

The two latches for the push-button switch inputs are implemented with combinational logic. This is necessary because a push-button switch may be pressed and released at any time and, therefore, the clocked flip-flops in the OLMC cannot be used. For example, a button may be pressed and released between clock pulses (they are 10 seconds apart) and completely missed if a flip-flop (which requires a clock pulse) is used to store the switch closure. An S-R latch, however, will immediately store a switch closure and hold it until it is reset.

The latch for the first floor is SET ($L1PB = 1$) if the first floor request button is pressed ($REQ1$) or if the elevator is on the first floor and the OPEN button is pressed ($\overline{DIR}$)($OPEN$) or if the elevator is on the second floor and the FLOOR1 button is pressed (DIR)($FLOOR1$).

The elevator is on the first floor when the system is in the REST1 state or the CLOSE1 state. The elevator is on the second floor when the system is in the REST2 state or the CLOSE2 state. From the state diagram you can see that the DIR variable defines the floor because $DIR = 0$ when the system is in REST1 or CLOSE1 and $DIR = 1$ when the system is in REST2 or CLOSE2. The SET condition for the first-floor latch is expressed by the following Boolean equation:

$$L1PB = REQ1 + (\overline{DIR})(OPEN) + (DIR)(FLOOR1) + \overline{L1PB\_}$$

The latch is RESET if the system is in the REST1 state (000). The RESET condition is expressed by the following Boolean equation:

$$L1PB\_ = (\overline{DOOR})(\overline{MOTION})(\overline{DIR}) + \overline{L1PB}$$

Equations for the second floor latch are similar.

$$L2PB = REQ2 + (DIR)(OPEN) + (\overline{DIR})(FLOOR2) + \overline{L2PB\_}$$

$$L2PB\_ = (\overline{DOOR})(\overline{MOTION})(DIR) + \overline{L2PB}$$

The combinational logic for the first-floor and second-floor latches is shown in Figure 11–17.

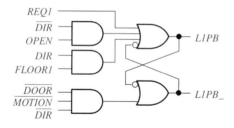

(a) Latch for push-button inputs, first floor

(b) Latch for push-button inputs, second floor

FIGURE 11–17
Latch circuits in the state control logic.

The Boolean equations for the latches are translated into ABEL as follows:

```
L1PB  = REQ1 # !DIR.FB & OPEN # DIR.FB & FLOOR1 # !L1PB_;
L1PB_ = (!DOOR.FB & !MOTION.FB & !DIR.FB) # !L1PB;
L2PB  = REQ2 # DIR.FB & OPEN # !DIR.FB & FLOOR2 # !L2PB_;
L2PB_ = (!DOOR.FB & !MOTION.FB & DIR.FB) # !L2PB;
```

The use of the dot extension .FB with the state variables indicates that the variable is to be fed back to the array logic directly from the associated flip-flop. OLMC flip-flops produce the state variables *DOOR, MOTION,* and *DIR* in the sequential portion of the system.

The State Sequence Logic

The sequential portion of the state control logic produces the state variable outputs *DOOR*, *MOTION*, and *DIR*. This logic can be defined using state diagram entry without having to develop equations as we did for the combinational latch logic. The IF-THEN-ELSE statements for each of the six states are as follows. Input/output variables are enclosed in parentheses.

```
State REST1:     if (L2PB) then UP else CLOSE1;
State CLOSE1:    if (L2PB) then UP else if L1PB then REST1 else CLOSE1;
State UP:        if (ARRIVE) then REST2 else UP;
State REST2:     if (L1PB) then DOWN else CLOSE2;
State CLOSE2:    if (L1PB) then DOWN else if L2PB then REST2 else CLOSE2;
State DOWN:      if (ARRIVE) then REST1 else DOWN;
```

As you have seen, a sequential system can be implemented in a PLD without having to first design the logic because the software automatically implements the design based on a description of the state diagram.

EXAMPLE 11–6

Write a complete ABEL input file for the system control logic using the elements already developed.

Solution

```
Module    Elevator_State_Control

Title    'Control logic for two-floor elevator system'

"Device Declaration

    Elevator_cont        device        'P16V8'

"Pin Declaration
    CLK, !OE                      pin 1,11;
    REQ1, REQ2                    pin 2,3;
    FLOOR2, FLOOR1, OPEN, ARRIVE  pin 4,5,6,7;

    L1PB, L1PB_          pin 16,17 ISTYPE 'com,buffer;
    L2PB, L2PB_          pin 18,19 ISTYPE 'com,buffer;
    DOOR, MOTION, DIR    pin 12,13,14 ISTYPE 'reg,buffer';

"State Definitions
    CONSTATE = [DOOR, MOTION, DIR];
    REST1   = ^B000;
    CLOSE1  = ^B100;
    UP      = ^B110;
    REST2   = ^B001;
    CLOSE2  = ^B101;
    DOWN    = ^B111;

Equations

    CONSTATE.CLK = CLK;
    L1PB = REQ1 # !DIR.FB & OPEN # DIR.FB & FLOOR1 # !L1PB_;
    L1PB_ = (!DOOR.FB & !MOTION.FB & !DIR.FB) # !L1PB;
    L2PB = REQ2 # DIR.FB & OPEN # !DIR.FB & FLOOR2 # !L2PB_;
    L2PB_ = (!DOOR.FB & !MOTION.FB & DIR.FB) # !L2PB;
State_diagram CONSTATE

    State REST1:   if (L2PB) then UP else CLOSE1;
    State CLOSE1:  if (L2PB) then UP else if L1PB then REST1 else CLOSE1;
    State UP:      if (ARRIVE) then REST2 else UP;
    State REST2:   if (L1PB) then DOWN else CLOSE2;
    State CLOSE2:  if (L1PB) then DOWN else if L2PB then REST2 else CLOSE2;
    State DOWN:    if (ARRIVE) then REST1 else DOWN;
```

```
Test_Vectors ([CLK, REQ1, REQ2, FLOOR2, FLOOR1, OPEN, ARRIVE] -> [DOOR,MOTION,DIR])
              [.c.,  0 ,  0 ,   0  ,   0  ,  0 ,   1   ] -> [ .x. ,  .x.  ,.x.];
              [.c.,  1 ,  0 ,   0  ,   0  ,  0 ,   1   ] -> [ .x. ,  .x.  ,.x.];
              [.c.,  1 ,  0 ,   0  ,   0  ,  0 ,   1   ] -> [  0  ,   0   , 0 ];
              [.c.,  0 ,  0 ,   0  ,   0  ,  0 ,   0   ] -> [  1  ,   0   , 0 ];
              [.c.,  1 ,  0 ,   0  ,   0  ,  0 ,   0   ] -> [  0  ,   0   , 0 ];
              [.c.,  0 ,  0 ,   0  ,   0  ,  0 ,   1   ] -> [  1  ,   0   , 0 ];
              [.c.,  0 ,  0 ,   0  ,   0  ,  1 ,   0   ] -> [  0  ,   0   , 0 ];
              [.c.,  0 ,  1 ,   0  ,   0  ,  0 ,   0   ] -> [  1  ,   1   , 0 ];
              [.c.,  0 ,  0 ,   0  ,   0  ,  0 ,   1   ] -> [  0  ,   0   , 1 ];
              [.c.,  0 ,  0 ,   0  ,   0  ,  0 ,   0   ] -> [  1  ,   0   , 1 ];
              [.c.,  0 ,  1 ,   0  ,   0  ,  0 ,   0   ] -> [  0  ,   0   , 1 ];
              [.c.,  0 ,  0 ,   0  ,   0  ,  0 ,   0   ] -> [  1  ,   0   , 1 ];
              [.c.,  0 ,  0 ,   0  ,   0  ,  1 ,   0   ] -> [  0  ,   0   , 1 ];
              [.c.,  1 ,  0 ,   0  ,   0  ,  0 ,   0   ] -> [  1  ,   1   , 1 ];
              [.c.,  0 ,  0 ,   0  ,   0  ,  0 ,   1   ] -> [  0  ,   0   , 0 ];
              [.c.,  0 ,  0 ,   1  ,   0  ,  0 ,   0   ] -> [  1  ,   1   , 0 ];
              [.c.,  0 ,  0 ,   0  ,   0  ,  0 ,   1   ] -> [  0  ,   0   , 1 ];
              [.c.,  0 ,  0 ,   0  ,   1  ,  0 ,   0   ] -> [  1  ,   1   , 1 ];
              [.c.,  0 ,  0 ,   0  ,   0  ,  0 ,   1   ] -> [  0  ,   0   , 0 ];
              [.c.,  0 ,  0 ,   0  ,   0  ,  0 ,   0   ] -> [  1  ,   0   , 0 ];
              [.c.,  0 ,  1 ,   0  ,   0  ,  0 ,   0   ] -> [  1  ,   1   , 0 ];
              [.c.,  0 ,  0 ,   0  ,   0  ,  0 ,   1   ] -> [  0  ,   0   , 1 ];
              [.c.,  0 ,  0 ,   0  ,   0  ,  0 ,   0   ] -> [  1  ,   0   , 1 ];
              [.c.,  1 ,  0 ,   0  ,   0  ,  0 ,   0   ] -> [  1  ,   1   , 1 ];
              [.c.,  0 ,  0 ,   0  ,   0  ,  0 ,   1   ] -> [  0  ,   0   , 0 ];
              [.c.,  0 ,  0 ,   0  ,   0  ,  0 ,   0   ] -> [  1  ,   0   , 0 ];
              [.c.,  0 ,  0 ,   1  ,   0  ,  0 ,   0   ] -> [  1  ,   1   , 0 ];
              [.c.,  0 ,  0 ,   0  ,   0  ,  0 ,   1   ] -> [  0  ,   0   , 1 ];
              [.c.,  0 ,  0 ,   0  ,   0  ,  0 ,   0   ] -> [  1  ,   0   , 1 ];
              [.c.,  0 ,  0 ,   0  ,   1  ,  0 ,   0   ] -> [  1  ,   1   , 1 ];
              [.c.,  0 ,  0 ,   0  ,   0  ,  0 ,   1   ] -> [  0  ,   0   , 0 ];
END
```

Related Problem The test vectors should take the system through all possible paths in the state diagram. Verify that this happens.

Implementing the Display Logic with a PLD

The pin assignments for the GAL16V8 used in the display logic portion of the system are shown in Figure 11–18. The inputs to the display logic come from the state control logic. Only six active-LOW outputs are required for the 7-segment readout because only the digit 1 and the digit 2 are displayed; therefore, segment *f* is not needed. Two other active-LOW outputs activate the up and down arrows. All of the logic in this portion of the system is combinational.

Input and Output Pin Declarations

The ABEL statement designating the input pins can be written as follows. $\overline{\text{OE}}$ is the Output Enable that is required for the GAL16V8.

```
!OE, MOTION, DIR   pin 11,2,3;
```

The ABEL statements designating the output pins can be written as follows:

```
SEGA, SEGB, SEGC   pin 12,13,14   ISTYPE 'com,invert';
SEGD, SEGE, SEGG   pin 15,16,17   ISTYPE 'com,invert';
UPARR, DWNARR      pin 18,19      ISTYPE 'com,invert';
```

FIGURE 11–18
Display logic block diagram showing GAL16V8 pin assignments for inputs and outputs.

The Up Arrow and Down Arrow Logic

The Up arrow must be illuminated when the elevator is moving ($MOTION = 1$) in the upward direction ($DIR = 0$). The Boolean equation is

$$UPARR = MOTION \cdot \overline{DIR} \cdot \overline{OE}$$

The Down arrow must be illuminated when the elevator is moving ($MOTION = 1$) in the downward direction ($DIR = 1$). The Boolean equation is

$$DWNARR = MOTION \cdot DIR \cdot \overline{OE}$$

As you can see, these functions can be implemented with 2-input AND gates.

The Boolean equations for the latches are translated into ABEL as follows:

```
UPARR  = MOTION & !DIR & !OE;
DWNARR = MOTION & DIR & !OE;
```

The 7-Segment Logic

The display must show a "1" when the system is in the REST1 state and in the CLOSE1 state. The display must show a "2" in the REST2 state and in the CLOSE2 state.

Segments *b* and *c* are used for a "1" and segments *a*, *b*, *d*, *e*, and *g* are used for a "2", as shown in Figure 11–18. Segment *b* is common to both digits. Therefore, the ABEL equations for the segments are as follows:

```
SEGA = !MOTION & DIR & !OE;
SEGB = !MOTION & !OE;
SEGC = !MOTION & !DIR & !OE;
SEGD = !MOTION & DIR & !OE;
SEGE = !MOTION & DIR & !OE;
SEGG = !MOTION & DIR & !OE;
```

Notice that the equations for segments *a*, *d*, *e*, and *g* are the same, so one output could be used for all of these. However, this output will have to be buffered externally in order to drive the four segments.

EXAMPLE 11–7

Write a complete ABEL input file for the system display logic using the elements already developed.

Solution

```
Module    Elevator_Display_Logic

Title    'Display logic for two-floor elevator system'

"Device Declaration

    Elevator_disp        device       'P16V8'

"Pin Declaration

    !OE, MOTION, DIR    pin 11,2,3;

    SEGA, SEGB, SEGC    pin 12,13,14  ISTYPE 'com,invert';
    SEGD, SEGE, SEGG    pin 15,16,17  ISTYPE 'com,invert';
    UPARR, DWNARR       pin 18,19     ISTYPE 'com,invert';

Equations

    UPARR = MOTION & !DIR & !OE;
    DWNARR = MOTION & DIR & !OE;
    SEGA = !MOTION & DIR & !OE;
    SEGB = !MOTION & !OE;
    SEGC = !MOTION & !DIR & !OE;
    SEGD = !MOTION & DIR & !OE;
    SEGE = !MOTION & DIR & !OE;
    SEGG = !MOTION & DIR & !OE;

Test_Vectors
([!OE, MOTION, DIR] -> [SEGA,SEGB,SEGC,SEGD,SEGE,SEGG,UPARR,DWNARR])
 [ 0 ,   0  ,  0 ] -> [ 1 , 0 , 0 , 1 , 1 , 1 , 1 , 1 ];
 [ 0 ,   0  ,  1 ] -> [ 0 , 0 , 1 , 0 , 0 , 0 , 1 , 1 ];
 [ 0 ,   1  ,  0 ] -> [ 1 , 1 , 1 , 1 , 1 , 1 , 0 , 1 ];
 [ 0 ,   1  ,  1 ] -> [ 1 , 1 , 1 , 1 , 1 , 1 , 1 , 0 ];
 [ 1 ,  .x. ,  .x.] -> [ 1 , 1 , 1 , 1 , 1 , 1 , 1 , 1 ];

END
```

Related Problem Redo the input file to combine the four equivalent segment outputs into one output called SEGADEG.

SECTION 11–5 REVIEW

1. Explain why the latches are required in the system control logic.
2. In the test vectors in Example 11–6, explain how the system is initialized to the REST1 state. Why are the output variables "don't cares" in the first two rows?
3. Can you use the state diagram entry method instead of equations for the display logic? Explain.
4. Can you use the truth table entry method for all of the logic design discussed in this section?

11–6 ■ DIGITAL SYSTEM APPLICATION

In Chapter 9 the sequential logic portion of the traffic light control system was implemented using SSI and MSI devices. You will now apply your knowledge of PLDs and PLD programming to implement the sequential logic with a GAL22V10. You should review Section 9–10 before proceeding.

After completing this section, you should be able to

■ Program the GAL22V10 from a state diagram ■ Explain how the GAL replaces the logic previously implemented in Chapter 9. ■ Develop an ABLE input file to implement the traffic light control system

Block Diagram of the Sequential Logic

The block diagram of the traffic light control system is shown in Figure 11–19, and the block diagram of the sequential logic is shown in Figure 11–20.

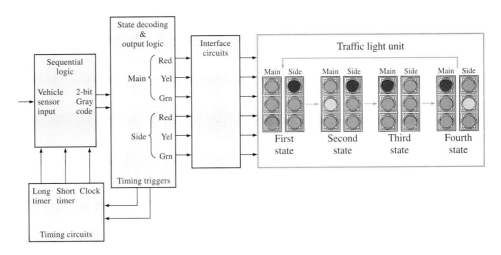

FIGURE 11–19
Traffic light control system block diagram.

FIGURE 11–20
Sequential logic block diagram.

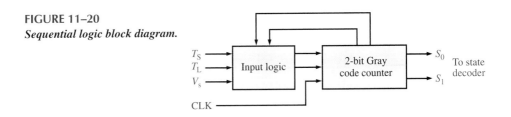

The sequential logic portion of the system uses a 2-bit Gray code counter to produce the sequence of states described by the state diagram in Figure 11–21. A review of the input variables is as follows:

V_S	Vehicle on side street
$\overline{V}_S$	No vehicle on side street
T_L	Long timer on (25 s)
$\overline{T}_L$	Long timer off (25 s)
T_S	Short timer on (4 s)
$\overline{T}_S$	Short timer off (25 s)

FIGURE 11–21
State diagram for the traffic light control system.

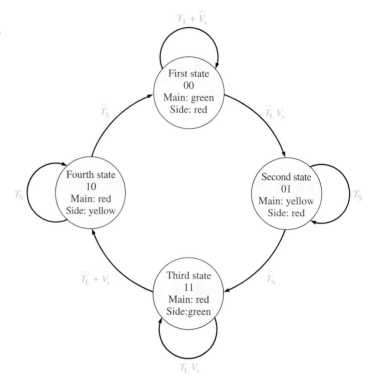

■ **SYSTEM ASSIGNMENT 1: GAL22V10 Programming**

The GAL22V10 that was used for the state decoding and output logic in the Chapter 7 system application can be programmed for the addition of the sequential logic because there are enough inputs and outputs still available.

■ *Activity 1:* From the state diagram in Figure 11–21 and the pin assignment diagram for the GAL22V10 shown here, write an ABEL input file to program the sequential logic into the device. The existing inputs and outputs for the previously programmed state decoding and output logic are shown in red. The inputs and outputs for the sequential logic to be added are shown in blue.

■ *Activity 2:* Write a single input file for the complete system by properly combining the input file for the state decoding and output logic from Chap-

ter 7 and the sequential logic input file that you just completed. The test vectors for this complete input file are developed in System Assignment 2.

■ SYSTEM ASSIGNMENT 2: Verification and Testing

The single GAL22V10 package replaces the 74LS74A, 74LS08, 74LS153 that were originally used to implement the sequential logic in the Chapter 9 system application as well as the 74LS139 and 74LS08 used for the state decoding and output logic in the Chapter 6 system application. The timing circuits are the same.

- **Activity 1:** Develop test vectors to check the programmed GAL and add it to the input file from System Assignment 1.

- **Activity 2:** Identify the function of each of the input and output terminals on the system PC board shown below.

- **Activity 3:** Draw a logic diagram of the system board.

SECTION 11–6 REVIEW

1. After programming the GAL22V10 in this section, how many outputs are still unused and available?
2. How many inputs are unused and available?
3. Compare the GAL implementation to the implementation using SSI and MSI devices for all of the logic.

■ SUMMARY

- The GAL16V8 and the GAL22V10 are examples of reprogrammable logic devices.
- The OLMC in a GAL is the Output Logic Macrocell that can be configured by programming for either a combinational or a registered output.
- In the OLMC registered mode, the output comes from a D flip-flop.
- The ISTYPE statement in ABEL can be used to declare an output either combinational or registered as ISTYPE 'com'; or ISTYPE 'reg';.
- The ABEL assignment operators := and :> are used to indicate a registered output.
- The dot extension .CLK is used to indicate that the register device is a clocked flip-flop.
- The dot extension .AR is used to asynchronously reset a flip-flop.
- Two basic types of state machines are the Moore and the Mealy.
- In a Moore type of state machine, the outputs are determined only on the internal state and any synchronized inputs.
- In a Mealy type of state machine, the outputs are determined by both the internal state and by inputs that are not synchronized with the system.
- A counter is a simple example of a Moore type of state machine.
- State machines can be implemented in a PLD using the state diagram entry method in ABEL. The IF-THEN-ELSE statement is an important element of the state diagram syntax.

■ KEY TERMS

These terms are also in the end-of-book glossary.

Global cell   A programmable cell in a PLD array that affects all of the OLMCs when programmed.

Local cell   A programmable cell in a PLD array that affects individual OLMCs when programmed.

Registered   A PLD output configuration where the output comes from a flip-flop.

■ SELF-TEST

1. The GAL22V10 has
   (a) 10 inputs and 22 outputs   (b) 22 inputs and 10 outputs
   (c) 11 inputs and 10 input/outputs   (d) 11 inputs and 10 outputs
2. The GAL16V8 has
   (a) 8 inputs and 16 outputs   (b) 16 inputs and 8 outputs
   (c) 8 inputs and 8 input/outputs   (d) 8 inputs and 8 outputs
3. An OLMC is an
   (a) Output Latch Main Cell
   (b) Odd Logic Multiplexer Cell
   (c) Optimum Logic Minimization Method
   (d) Output Logic Macrocell
4. A typical OLMC consists of
   (a) gates, multiplexers, and a flip-flop   (b) gates and a shift register
   (c) a Gray code counter   (d) a fixed logic array

5. Which of the following ABEL equations refers to a registered output?

   (a) X = A & B;   (b) !X = !(A & B);

   (c) X := A & B;   (d) X –> A & B;

6. A counter can be implemented in a PLD using ABEL by

   (a) equation entry   (b) truth table entry

   (c) state diagram entry   (d) all of the above

   (e) answers (b) and (c)

7. Combinational logic can be implemented in a PLD using ABEL by

   (a) equation entry   (b) truth table entry

   (c) state diagram entry   (d) all of the above

   (e) answers (a) and (b)

8. A valid IF-THEN-ELSE statement is

   (a) if A then C else D

   (b) if A then C else D;

   (c) IF A THEN C ELSE D

   (d) IF-A-THEN-C-ELSE-D:

   (e) answers (b) and (c)

9. The ABEL expression A = [0, 1, 0]

   (a) means that state A is defined by the binary value 010

   (b) means that variable A has a sequence of values 0,1,0

   (c) must be preceded by a state definition statement

   (d) answers (a) and (c)

10. .CLK and .AR are examples of

   (a) special constants   (b) ABEL variables

   (c) dot extensions   (d) none of the above

■ PROBLEMS

SECTION 11–1   The Complete OLMC

1. List the four GAL22V10 OLMC configurations.

2. Determine the GAL16V8 mode configuration for each combination of "cell" values.

   (a) $\overline{SYN} = 0, AC0 = 1, XOR = 1$

   (b) $\overline{SYN} = 1, AC0 = 0, XOR = 0$

3. Besides the flip-flop and logic gates, what type of device is used in an OLMC?

SECTION 11–2   OLMC Mode Selection

4. Determine the GAL22V10 mode for each of the bit combinations.

   (a) $S_1S_0 = 10$ (b) $S_1S_0 = 00$

   (c) $S_1S_0 = 11$ (d) $S_1S_0 = 01$

5. Explain each of the ABEL output pin declarations.

   (a) X pin 15 ISTYPE 'com';

   (b) Q2 pin 20 ISTYPE 'reg';

   (c) A PIN 18 ISTYPE 'com,invert';

   (d) Q1 pin 21 ISTYPE 'reg,buffer';

6. Identify each of the ABEL expressions as registered or combinational.

   (a) Y = A # B;

   (b) X := A & !B;

   (b) Q1 := D1 # D0;

   (d) Q2 = !C;

7. Write an ABEL clock equation to accompany the following expression: Q1 := (A & B) # C;

SECTION 11–3 Implementing Shift Registers with PLDs

8. Write an ABEL input file, excluding the test vectors, to implement the serial in/serial out shift register from Figure 11–22 in a GAL22V10.

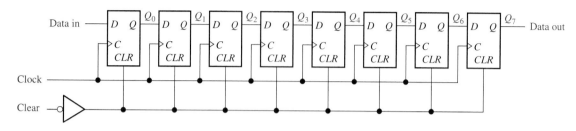

FIGURE 11–22

9. Develop the test vectors for the input file from Problem 8.

10. Repeat Problem 8 using a GAL16V8 as the target device.

11. Repeat Problem 9 using a GAL16V8 as the target device.

12. Write an ABEL input file to expand the 4-bit parallel in/serial out register in Figure 11–23 to an 8-bit register. The target device is to be a GAL22V10. Include test vectors.

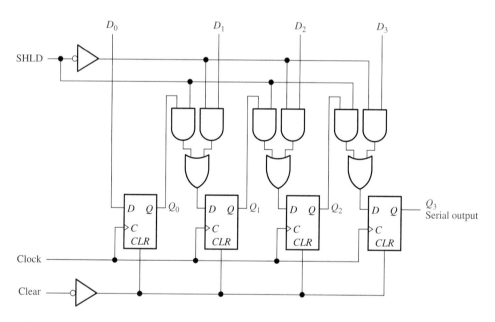

FIGURE 11–23

13. Determine the feasibility of implementing the 8-bit register in Problem 12 with A GAL16V8.

SECTION 11–4 Implementing Counters with PLDs

14. Explain the basic difference between a Moore and a Mealy state machine.

15. What is the largest counter in terms of the number of stages or bits that can be implemented with a single GAL16V8?

16. What is the largest counter in terms of the number of stages or bits that can be implemented with a single GAL22V10?

17. Convert the 3-bit up-down Gray code counter in Figure 11–24 to a 3-bit up only Gray code counter and implement it in a GAL22V10 using logic equation entry.

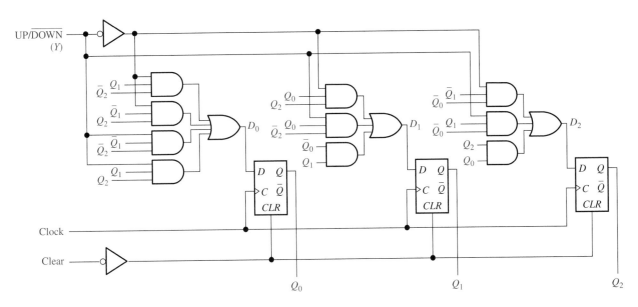

FIGURE 11–24

18. Repeat Problem 17 using truth table entry.

19. Repeat Problem 17 using state diagram entry.

20. Repeat Problem 17 with a GAL16V8 as the target device.

21. Repeat Problem 18 with a GAL16V8 as the target device.

22. Repeat Problem 19 with a GAL16V8 as the target device.

SECTION 11–6   PLD System Implementation

23. Referring to the state diagram in Figure 11–25, if the OPEN push-button is pressed, what happens when the system is in the

(a) CLOSE1 state   (b) REST1 state

(c) UP state   (d) REST2 state

(e) CLOSE2 state   (f) DOWN state

FIGURE 11–25

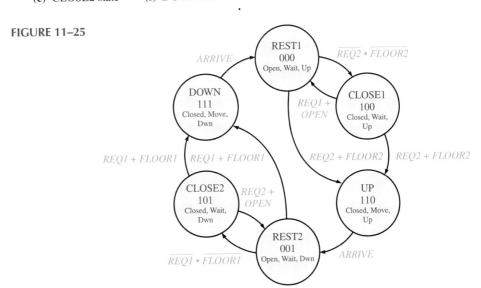

24. Repeat Problem 23 for each of the following cases:
  (a) The FLOOR1 push-button is pressed.
  (b) The FLOOR2 push-button is pressed.
  (c) The REQ1 push-button is pressed.
  (d) The REQ2 push-button is pressed.

25. Write a complete ABEL input file to implement the state control logic portion of the elevator control system that was developed in this section with a GAL22V10 instead of a GAL16V8.

26. Repeat Problem 25 for the display logic in the elevator control system.

SECTION 11–6 Digital System Application

27. Can the traffic light control system be implemented with a single GAL16V8? If not, explain.

28. Assume the short timer input (T_S) to the sequential logic portion of the traffic light control system fails such that it is always HIGH ($T_S = 1$). How is the system operation affected?

29. Assume the long timer input (T_L) to the sequential logic portion of the traffic light control system fails such that it is always LOW ($T_L = 0$). How is the system operation affected?

30. If the vehicle sensor always produces a HIGH, how is the system operation affected?

Special Design Problems

31. Design an 8-bit serial in/serial out shift register and write an ABEL input file to implement it using GAL16V8s.

32. Modify the design in Problem 31 for a 16-bit serial in/serial out shift register.

33. Implement a 4-bit counter that will sequentially produce the excess-3 code using a GAL22V10 as the target device.

34. Modify the design in Problem 33 so that the counter can be preloaded in parallel to any specified 4-bit excess-3 code number.

35. Develop a state diagram for a three-floor elevator control system.

36. Using the state diagram from Problem 35, write an ABEL input file to implement the state control logic for the three-floor elevator system using a GAL22V10.

Electronics Workbench Troubleshooting

37. Open file PR11_37.EWB on your EWB CD. Determine if the shift register is working properly. If there is a fault, try to isolate it.

38. Open file PR11_38.EWB on your EWB CD and test the counter for proper operation. If there is a fault, try to isolate it.

39. Open file PR11_39.EWB on your EWB CD and test the counter for proper operation. If there is a fault, try to isolate it.

40. Open file PR11_40.EWB on your EWB CD and test the counter for proper operation. If there is a fault, try to isolate it.

▪ ANSWERS TO SECTION REVIEWS

SECTION 11–1

1. The GAL22V10 modes are combinational and registered.

2. The mode is set by the select bits S_0 and S_1.

3. The GAL16V8 has two combinational modes (simple and complex) and a registered mode.

SECTION 11–2

1. In the registered mode, the output is from a flip-flop. In the combinational mode, the output is not from a flip-flop.

2. An ISTYPE statement is used to declare an output either combinational or registered.

3. The := operator indicates a registered function while the = operator is for a combinational function.

4. The dot extension .CLK indicates that the register device is a clocked flip-flop.

SECTION 11–3

1. Q0 := A & !B # !A & B; means that output Q_0 is registered and equal to the exclusive-OR of A and B.

2. The dot extension .AR indicates asynchronous reset of a flip-flop.

3. The special constant .c. indicates a clock pulse.

SECTION 11–4

1. The categories of state machines are Moore and Mealy.

2. A counter is a Moore state machine.

3. Entry methods are equations, truth table, and state diagram.

4. The special constant .x. represents "don't care".

SECTION 11–5

1. The latches are required to store a temporary switch closure.

2. A maximum of three clock pulses are required to get the system to the REST1 state regardless of its initial state. Since three clock pulses are used, we don't care what the outputs are for the first two.

3. State diagram entry cannot be used for the display logic because the logic is all combinational.

4. Truth table entry can be used for all logic.

SECTION 11–6

1. No outputs are unused and available.

2. Six inputs are unused and available.

3. One GAL22V10 replaces five ICs in the original design.

■ **ANSWERS TO RELATED PROBLEMS FOR EXAMPLES**

11–1 Q0,Q1,Q2,Q3,Q4,Q5,Q6,Q7
PIN 22,21,20,19,18,17,16,15 ISTYPE 'reg,invert';

11–2 'Pin Declaration

```
Clock, Clear        pin 1,2;
SHLD                pin 3;
D0, D1, D2, D3, D4  pin 4,5,6,7,8 ISTYPE 'reg,buffer';
Q0, Q1, Q2, Q3, Q4  pin 14,15,16,17,18 ISTYPE 'reg,buffer';

Equations
Q0 := D0;
Q1 := Q0 & SHLD # D1 & !SHLD;
Q2 := Q1 & SHLD # D2 & !SHLD;
Q3 := Q2 & SHLD # D3 & !SHLD;
Q4 := Q3 & SHLD # D4 & !SHLD;

[Q0, Q1, Q2, Q3, Q4].CLK = Clock;
[Q0, Q1, Q2, Q3, Q4].AR = !Clear;
```

11–3 Line 1 clears the counter to 000.
Line 2 decrements the counter to 100.
Line 3 decrements the counter to 101.
Line 4 decrements the counter to 111.
Line 5 decrements the counter to 110.
Line 6 decrements the counter to 010.
Line 7 decrements the counter to 011.
Line 8 decrements the counter to 001.
Line 9 decrements the counter to 000.
Line 10 increments the counter to 001.
Line 11 increments the counter to 011.
Line 12 increments the counter to 010.

Line 13 increments the counter to 110.
Line 14 increments the counter to 111.
Line 15 increments the counter to 101.
Line 16 increments the counter to 100.
Line 17 increments the counter to 000.

11–4 The .x. don't-care constant in the first line allows the truth table to show that the outputs are re-set when clear is LOW regardless of the states of Q_0, Q_1, and Q_2.

11–5 State A: if !Y then H else B

11–6 The test vector lines trace the paths as shown in Figure 11–26.

FIGURE 11–26

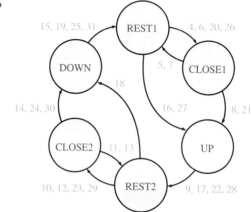

11–7
```
Module    Elevator_display_logic
Title     'Display logic for the two-story elevator system'
"Device declaration
   Elevator_disp   DEVICE   'P16V8'

"Pin declarations
   MOTION, DIR           PIN 2,3;
   SEGADEG, SEGB, SEGC   PIN 12,13,14 ISTYPE 'com,invert';
   UPARR, DWNARR         PIN 18,19 ISTYPE 'com,invert';

Equations
   UPARR   = MOTION & !DIR;
   DWNARR  = MOTION & DIR;
   SEGADEG = !MOTION & DIR;
   SEGB    = !MOTION;
   SEGC    = !MOTION & !DIR;

Test_vectors
   ([MOTION,DIR] -> [SEGADEG, SEGB, SEGC, UPARR, DWNARR])
   [ 0  , 0 ] -> [  1  , 0 , 0 , 1 , 1 ];
   [ 0  , 1 ] -> [  0  , 1 , 1 , 1 , 1 ];
   [ 1  , 0 ] -> [  1  , 1 , 1 , 0 , 1 ];
   [ 1  , 1 ] -> [  1  , 0 , 0 , 1 , 0 ];

END
```

■ **ANSWERS**
TO SELF-TEST

1. (b)	**2.** (b)	**3.** (d)	**4.** (a)	**5.** (c)	**6.** (d)	**7.** (e)	**8.** (b)
9. (d)	**10.** (c)						

12

MEMORY AND STORAGE

■ CHAPTER OBJECTIVES

❑ Define the basic memory characteristics
❑ Explain what a RAM is and how it works
❑ Explain the difference between static RAMs (SRAMs) and dynamic RAMs (DRAMs)
❑ Explain what a ROM is and how it works
❑ Explain how a ROM is programmed
❑ Describe the various types of PROMs
❑ Discuss the characteristics of a flash memory
❑ Describe the expansion of ROMs and RAMs to increase word length and word capacity
❑ Describe the basic organization of magnetic disks and magnetic tapes
❑ Describe the basic operation of magneto-optical disks and optical disks
❑ Describe basic methods for memory testing
❑ Develop flowcharts for memory testing
❑ Apply a memory device in a system application

■ CHAPTER OVERVIEW

Chapter 10 covered shift registers, which are a type of storage device; in fact, a shift register is essentially a small-scale memory. The memory devices covered in this chapter are generally used for longer-term storage of larger amounts of data than registers can provide.

Computers and other types of systems require the permanent or semipermanent storage of large amounts of binary data. Microprocessor-based systems rely on storage devices and memories for their operation be-cause of the necessity for storing programs and for retaining data during processing.

In computer terminology, *memory* usually refers to RAM and ROM and *storage* refers to hard disk, floppy disk, and CD-ROM. In this chapter semiconductor, magnetic, and optical memories are covered.

SPECIFIC DEVICES

27C16B   74F189   μPD43256B

■ DIGITAL SYSTEM APPLICATION

This Digital System Application illustrates concepts taught in this chapter. The system application in Section 12–10 completes the security entry system from Chapter 10. The focus in this chapter is the memory logic portion of the system, which stores the entry code. Once the memory logic is developed, it is interfaced with the security code logic from Chapter 10 to form the complete system.

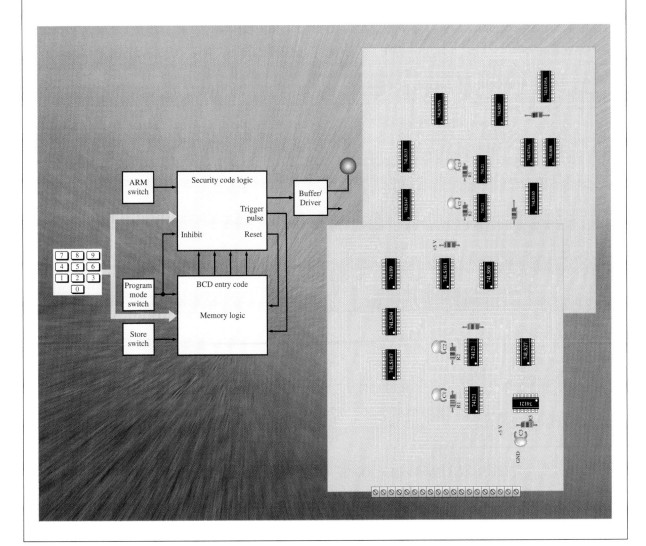

12–1 ■ BASICS OF SEMICONDUCTOR MEMORY

Memory is the portion of a system for storing binary data in large quantities. Semiconductor memories consist of arrays of storage elements that are either latches or capacitors.

After completing this chapter, you should be able to

■ Explain how a memory stores binary data ■ Discuss the basic organization of a memory ■ Describe the write operation ■ Describe the read operation ■ Describe the addressing operation ■ Explain what RAMs and ROMs are

COMPUTER NOTE

The general definition of *word* is a complete unit of information consisting of a unit of binary data. When applied to computer instructions, a word is more specifically defined as two bytes (16 bits). As a very important part of assembly language used in computers, the DW (Define Word) directive means to define data in 16-bit units. This definition is independent of the particular microprocessor or the size of its data bus. Assembly language also allows definitions of bytes (8 bits) with the DB directive, double words (32 bits) with the DD directive, and quadwords (64 bits) with the QD directive.

Units of Binary Data: Bits, Bytes, Nibbles, and Words

As a rule, memories store data in units that have from one to eight bits. The smallest unit of binary data, as you know, is the **bit.** In many applications, data are handled in an 8-bit unit called a **byte** or in multiples of 8-bit units. The byte can be split into two 4-bit units which are called **nibbles.** A complete unit of information is called a **word** and generally consists of one or more bytes. Some memories store data in 9-bit groups; a 9-bit group consists of a byte plus a parity bit.

The Basic Semiconductor Memory Array

Each storage element in a memory can retain either a 1 or a 0 and is called a **cell.** Memories are made up of arrays of cells as illustrated in Figure 12–1 using 64 cells as an example. Each block in the **memory array** represents one storage cell, and its location can be identified by specifying a row and a column.

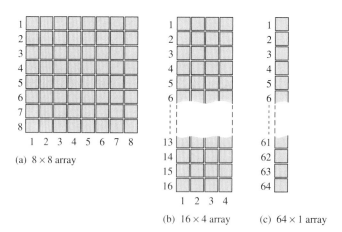

(a) 8 × 8 array

(b) 16 × 4 array (c) 64 × 1 array

FIGURE 12–1

A 64-cell memory array organized in three different ways.

The 64-cell array can be organized in several ways based on units of data. Figure 12–1(a) shows an 8 × 8 array, which can be viewed as either a 64-bit memory or an 8-byte memory. Part (b) shows a 16 × 4 array, which is a 16-nibble memory, and part (c) shows a 64 × 1 array which is a 64-bit memory. A memory is identified by the number of words it can store times the word size. For example, a 16k × 8 memory can store 16,384 words of eight bits each. The inconsistency here is common in memory terminology. The actual number of words is always a power of 2, which, in this case, is $2^{14} = 16,384$. However, it is common practice to state the number to the nearest thousand, in this case, 16k.

Memory Address and Capacity

The *location* of a unit of data in a memory array is called its **address.** For example, in Figure 12–2(a), the address of a bit in the array is specified by the row and column as shown. In Figure 12–2(b), the address of a byte is specified only by the row. So, as you can see, the address depends on how the memory is organized into units of data. Personal computers have memories organized in bytes. This means that the smallest group of bits that can be addressed is eight.

(a) The address of the bit highlighted in blue is row 5, column 4.

(b) The address of the byte highlighted in blue is row 3.

FIGURE 12–2
Examples of memory address.

The **capacity** of a memory is the total number of data units that can be stored. For example, in the bit-organized memory array in Figure 12–2(a), the capacity is 64 bits. In the byte-organized memory array in Figure 12–2(b), the capacity is 8 bytes, which is also 64 bits. Computer memories typically have 32 Mbytes (Mbyte is megabyte) or more of internal memory.

Basic Memory Operations

Since a memory stores binary data, data must be put into the memory and data must be copied from the memory when needed. The **write** operation puts data into a specified address in the memory, and the **read** operation takes data out of a specified address in the memory. The addressing operation, which is part of both the write and the read operations, selects the specified memory address.

Data units go into the memory during a write operation and come out of the memory during a read operation on a set of lines called the *data bus.* As indicated in Figure 12–3, the data bus is bidirectional, which means that data can go in either direction (into the

FIGURE 12–3
Block diagram of a memory showing address bus, address decoder, bidirectional data bus, and read/write inputs.

memory or out of the memory). In this case of a byte-organized memory, the data bus has at least eight lines so that all eight bits in a selected address are transferred in parallel. For a write or a read operation, an address is selected by placing a binary code representing the desired address on a set of lines called the *address bus*. The address code is decoded internally and the appropriate address is selected. The number of lines in the address bus depends on the capacity of the memory. For example, a 15-bit address code can select 32,768 locations (2^{15}) in the memory, a 16-bit address code can select 65,536 locations (2^{16}) in the memory, and so on. In personal computers, a 32-bit address bus can select 4,294,967,296 locations (2^{32}), expressed as 4G.

The Write Operation   A simplified write operation is illustrated in Figure 12–4. To store a byte of data in the memory, a code held in the address register is placed on the address bus. Once the address code is on the bus, the address decoder decodes the address and selects the specified location in the memory. The memory then gets a write command, and the data byte held in the data register is placed on the data bus and stored in the selected memory address, thus completing the write operation. When a new data byte is written into a memory address, the current data byte stored at that address is overwritten and destroyed.

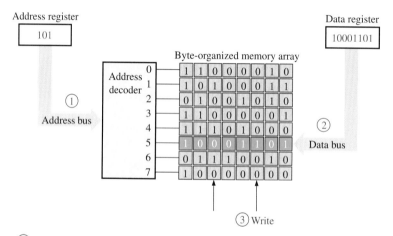

① Address code 101 is placed on the address bus and address 5 is selected.
② Data byte is placed on the data bus.
③ Write command causes the data byte to be stored in address 5, replacing previous data.

FIGURE 12–4
Illustration of the write operation.

The Read Operation   A simplified read operation is illustrated in Figure 12–5. Again, a code held in the address register is placed on the address bus. Once the address code is on the bus, the address decoder decodes the address and selects the specified location in the memory. The memory then gets a read command, and a "copy" of the data byte that is stored in the selected memory address is placed on the data bus and loaded into the data register, thus completing the read operation. When a data byte is read from a memory address, it also remains stored at that address and is not destroyed. This is called *nondestructive read*.

RAMs and ROMs

The two major categories of semiconductor memories are the RAM and the ROM. **RAM** (random-access memory) is a type of memory in which all addresses are accessible in an equal amount of time and can be selected in any order for a read or write operation. All RAMs have both *read* and *write* capability. Because RAMs lose stored data when the power is turned off, they are **volatile** memories.

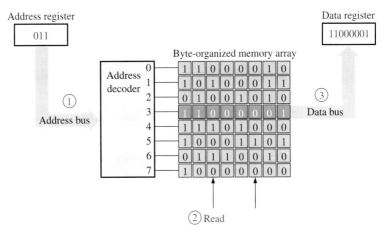

① Address code 011 is placed on the address bus and address 3 is selected.

② Read command is applied.

③ The contents of address 3 is placed on the data bus and shifted into data register. The contents of address 3 is not destroyed by the read operation.

FIGURE 12–5
Illustration of the read operation.

ROM (read-only memory) is a type of memory in which data are stored permanently or semipermanently. Data can be read from a ROM, but there is no write operation as in the RAM. The ROM, like the RAM, is a random-access memory but the term RAM traditionally means a random-access *read/write* memory. Several types of RAMs and ROMs will be covered in this chapter. Because ROMs retain stored data even if power is turned off, they are **nonvolatile** memories.

SECTION 12–1 REVIEW

1. What is the smallest unit of data that can be stored in a memory?
2. What is the bit capacity of a memory that can store 256 bytes of data?
3. What is a write operation?
4. What is a read operation?
5. How is a given unit of data located in a memory?
6. Describe the difference between a RAM and a ROM.

12–2 ■ RANDOM-ACCESS MEMORIES (RAMs)

RAMs are read/write memories in which data can be written into or read from any selected address in any sequence. When a data unit is written into a given address in the RAM, the data unit previously stored at that address is replaced by the new data unit. When a data unit is read from a given address in the RAM, the data unit remains stored and is not destroyed by the read operation. This nondestructive read operation can be viewed as copying the content of an address while leaving the content intact. A RAM is typically used for short-term data storage because it cannot retain stored data when power is turned off.

After completing this section, you should be able to

■ Name the two categories of RAM ■ Explain what a SRAM is ■ Describe the SRAM storage cell ■ Explain the difference between an asynchronous SRAM and a synchronous burst SRAM ■ Explain what a DRAM is ■ Describe the DRAM storage cells ■ Discuss the types of DRAM ■ Compare the SRAM with the DRAM

The RAM Family

The two categories of RAM are the *static RAM* (SRAM) and the *dynamic RAM* (DRAM). Static RAMs use flip-flops as storage elements and can therefore store data indefinitely *as long as dc power is applied.* Dynamic RAMs use capacitors as storage elements and cannot retain data very long without the capacitors being recharged by a process called **refreshing.** Both SRAMs and DRAMs will lose stored data when dc power is removed and, therefore, are classified as volatile memories.

Data can be read much faster from SRAMs than from DRAMs. However, DRAMs can store much more data than SRAMs for a given physical size and cost because the DRAM cell is much simpler, and more cells can be crammed into a given area than in the SRAM.

The basic types of SRAM are the *asynchronous SRAM* and the *synchronous burst SRAM.* The basic types of DRAM are the *Fast Page Mode DRAM* (FPM DRAM), the *Extended Data Out DRAM* (EDO DRAM), the *Burst EDO DRAM* (BEDO DRAM), and the *synchronous DRAM* (SDRAM). These are shown in Figure 12–6.

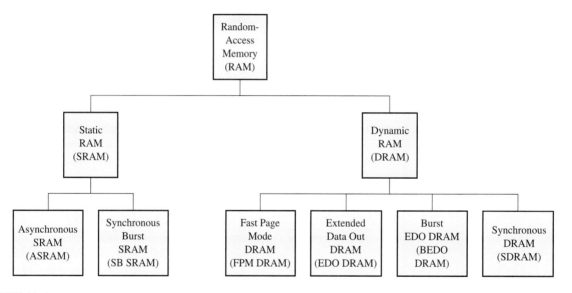

FIGURE 12–6
The RAM family.

Static RAMs (SRAMs)

Storage Cell   All static RAMs are characterized by flip-flop storage cells that are typically implemented in integrated circuits with several MOS transistors **(MOSFETs).** There are also some smaller SRAMs implemented with bipolar transistors. As long as dc power is applied to a **static memory** cell, it can retain a 1 or 0 state indefinitely. If power is removed, the stored data bit is lost.

Figure 12–7 shows a basic **SRAM** flip-flop storage cell. The cell is selected by an active level on the Bit Select line and a data bit (1 or 0) is written into the cell by placing it on the Data and $\overline{\text{Data}}$ lines. A data bit is read by taking it off the Data and $\overline{\text{Data}}$ lines. Input data and output data can share the same lines because write and read operations occur at different times.

Basic Static Memory Cell Array   The storage cells in a SRAM are organized in rows and columns as illustrated in Figure 12–8 for the case of an $n \times 4$ array. All the cells in a row share the same Row Select line. Each set of Data and $\overline{\text{Data}}$ lines go to each cell in a given column and are connected to a single data line that serves as both an input and output (Data I/O) through the data input and data output buffers.

FIGURE 12–7
A typical static RAM storage cell showing simplified transistor symbols.

FIGURE 12–8
Basic SRAM array.

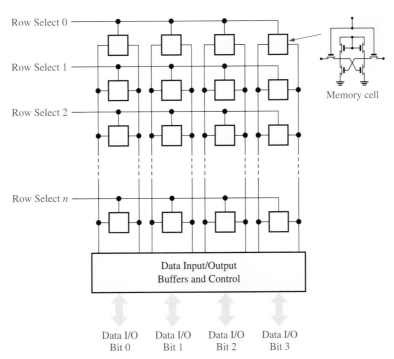

To write a data unit, in this case a nibble, into a given row of cells in the memory array, the Row Select line is taken to its active state and four data bits are placed on the Data Input lines. The Write line is then taken to its active state, which causes each data bit to be stored in a selected cell in the associated column. To read a data unit, the Read line is taken to its active state, which causes the four data bits stored in the selected row to appear on the Data Output lines.

Basic Asynchronous SRAM Organization

An asynchronous SRAM is one in which the operation is not synchronized with a system clock. To illustrate the general organization of a SRAM, a 32k × 8 bit memory such as the μPD43256B is used. A logic symbol for this memory is shown in Figure 12–9.

In the READ mode, the eight data bits that are stored in a selected address appear on the data output lines. In the WRITE mode, the eight data bits that are applied to the data

FIGURE 12–9
Logic diagram for a 32k × 8 SRAM.

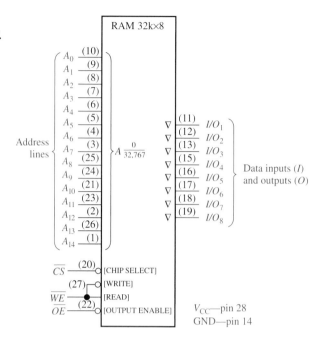

input lines are stored at a selected address. The data input and data output lines (I/O_1 through I/O_8) are the same lines. During READ, they act as output lines (I_1 through I_8) and during WRITE they act as input lines (O_1 through O_8).

Tristate Outputs and Buses   Tristate buffers in a memory allow the data lines to act as either input or output lines and connect the memory to the data bus in a computer. These buffers have three output states: HIGH (1), LOW (0), and HIGH-Z (open). Tristate outputs are indicated on logic symbols by a small inverted triangle (∇), as shown in Figure 12–9, and are used for compatibility with bus structures such as those found in microprocessor-based systems.

Physically, a **bus** is a set of conductive paths that serve to interconnect two or more functional components of a system or several diverse systems. Electrically, a bus is a collection of specified voltage levels and/or current levels and signals that allow the various devices connected to the bus to communicate and work properly together.

For example, a microprocessor is connected to memories and input/output devices by certain bus structures. An address bus allows the microprocessor to address the memories, and the data bus provides for transfer of data between the microprocessor, the memories, and the input/output devices such as monitors, printers, keyboards, and modems. The control bus allows the microprocessor to control data transfers and timing for the various components.

Memory Array   SRAM chips can be organized in single bits, nibbles (4 bits), bytes (8 bits), or multiple bytes (16, 24, 32 bits, etc.).

Figure 12–10 shows the organization of a typical 32k × 8 SRAM. The memory cell array is arranged in 256 rows and 128 × 8 columns (128 columns, each with 8 bits) as shown in part (a). There are actually $2^{15} = 32,768$ addresses and each address contains 8 bits. The capacity of this example memory is 32,768 bytes (typically expressed as 32 kbytes).

The SRAM in Figure 12–10(b) works as follows. First, the chip select, $\overline{CS}$, must be LOW for the memory to operate. Eight of the fifteen address lines are decoded by the row decoder to select one of the 256 rows. Seven of the fifteen address lines are decoded by the column decoder to select one of the 128 8-bit columns.

(a) Memory array configuration

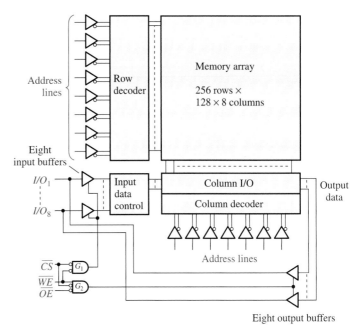

(b) Memory block diagram

FIGURE 12–10
Basic organization of a 32k × 8 SRAM.

READ In the READ mode, the write enable input, $\overline{WE}$, is HIGH and the output enable, $\overline{OE}$, is LOW. The input tristate buffers are disabled by gate G_1, and the column output tristate buffers are enabled by gate G_2. Therefore, the eight data bits from the selected address are routed through the column I/O to the data lines (I/O_1 though I/O_8), which are acting as data output lines.

WRITE In the WRITE mode, $\overline{WE}$ is LOW and $\overline{OE}$ is HIGH. The input buffers are enabled by gate G_1, and the output buffers are disabled by gate G_2. Therefore, the eight input data bits on the data lines are routed through the input data control and the column I/O to the selected address and stored.

Read and Write Cycles Figure 12–11 shows typical timing diagrams for a memory read cycle and a write cycle. For the read cycle shown in part (a), a valid address code is applied to the address lines for a specified time interval called the *read cycle time, t_{RC}*. Next, the chip select ($\overline{CS}$) and the output enable ($\overline{OE}$) inputs go LOW. One time interval after the $\overline{OE}$ input goes LOW, a valid data byte from the selected address appears on the data lines. This time interval is called the *output enable access time, t_{GQ}*.

Two other access times for the read cycle are the *address access time, t_{AQ}*, measured from the beginning of a valid address to the appearance of valid data on the data lines and the *chip enable access time, t_{EQ}*, measured from the HIGH-to-LOW transition of $\overline{CS}$ to the appearance of valid data on the data lines.

During each read cycle, one unit of data, a byte in this case, is read from the memory.

For the write cycle shown in Figure 12–11(b), a valid address code is applied to the address lines for a specified time interval called the *write cycle time, t_{WC}*. Next, the chip select ($\overline{CS}$) and the write enable ($\overline{WE}$) inputs go LOW. The required time interval from the beginning of a valid address until the $\overline{WE}$ input goes LOW is called the *address setup time, $t_{s(A)}$*. The time that the $\overline{WE}$ input must be LOW is the write pulse width. The time that the $\overline{WE}$ input must remain LOW after valid data are applied to the data inputs is

(a) Read cycle ($\overline{WE}$ HIGH)

(b) Write cycle ($\overline{WE}$ LOW)

FIGURE 12–11
Basic read and write cycle timing for the SRAM in Figure 12–10.

designated t_{WD}; the time that the valid input data must remain on the data lines after the $\overline{WE}$ input goes HIGH is the *data hold time,* $t_{h(D)}$.

During each write cycle, one unit of data is written into the memory.

Basic Synchronous Burst SRAM Organization

Unlike the asynchronous SRAM, a synchronous SRAM is synchronized with the system clock. For example, in a computer system, the synchronous SRAM operates with the same clock signal that operates the microprocessor so that the microprocessor and memory are synchronized for faster operation.

The fundamental concept of the synchronous feature of a SRAM can be shown with Figure 12–12, which is a simplified block diagram of a 32k $\times$ 8 memory for purposes of illustration. The synchronous SRAM is very similar to the asynchronous SRAM in terms of the memory array, address decoder, and read/write and enable inputs. The basic difference is that the synchronous SRAM uses clocked registers to synchronize all inputs with the system clock. The address, the read/write input, the chip enable, and the input data are all latched into their respective registers on an active clock pulse edge. Once this information is latched, the memory operation is in sync with the clock.

For the purpose of simplification, a notation for multiple parallel lines or bus lines is introduced in Figure 12–12, as an alternative to drawing each line separately. A set of parallel lines can be indicated by a single heavy line with a slash and the number of

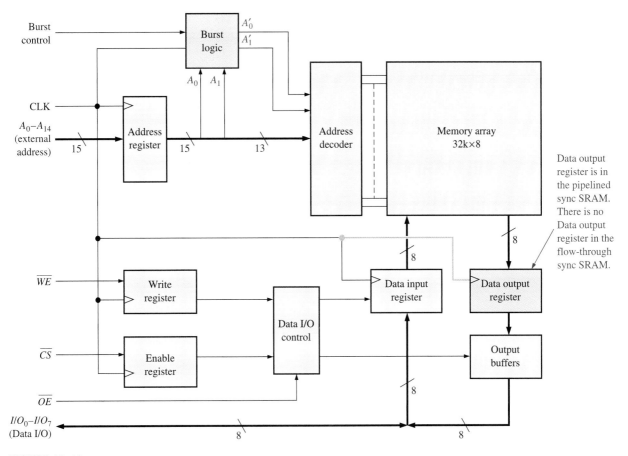

FIGURE 12–12

A basic block diagram of a synchronous burst SRAM.

separate lines in the set. For example, the following notation represents a set of 8 parallel lines:

The address bits A_0 through A_{14} are latched into the Address register on the positive edge of a clock pulse. On the same clock pulse, the state of the write enable ($\overline{WE}$) line and chip select ($\overline{CS}$) are latched into the Write register and the Enable register respectively. These are one-bit registers or simply flip-flops. Also, on the same clock pulse the input data are latched into the Data input register for a Write operation, and data in a selected memory address are latched into the Data output register for a Read operation, as determined by the Data I/O control based on inputs from the Write register, Enable register, and the Output enable ($\overline{OE}$).

Two basic types of synchronous SRAM are the *flow-through* and the *pipelined*. The flow-through synchronous SRAM does not have a Data output register, so the output data flow asynchronously to the data I/O lines through the output buffers. The **pipelined** synchronous SRAM has a Data output register, as shown in Figure 12–12, so the output data are synchronously placed on the data I/O lines.

The Burst Feature

As shown in Figure 12–12, synchronous SRAMs normally have an address burst feature, which allows the memory to read or write at up to four locations using a single address. When an external address is latched in the address register, the two lowest-order address bits, A_0 and A_1, are applied to the burst logic. This produces a sequence of four internal addresses by adding 00, 01, 10, and 11 to the two lowest-order address bits on successive clock pulses. The sequence always begins with the base address, which is the external address held in the address register.

Burst Logic   The address burst logic in a typical synchronous SRAM consists of a binary counter and exclusive-OR gates, as shown in Figure 12–13. For 2-bit burst logic, the internal burst address sequence is formed by the base address bits A_2–A_{14} plus the two burst address bits A_1' and A_0'.

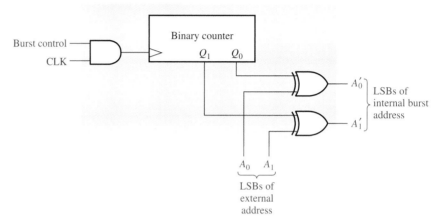

FIGURE 12–13
Address burst logic.

To begin the burst sequence, the counter is in its 00 state and the two lowest-order address bits are applied to the inputs of the XOR gates. Assuming that A_0 and A_1 are both 0, the internal address sequence in terms of its two lowest-order bits is 00, 01, 10, and 11.

Cache Memory

One of the major applications of SRAMs is in cache memories in computers. **Cache memory** is a relatively small, high-speed memory that stores the most recently used instructions or data from the larger but slower main memory. Cache memory uses dynamic RAM (DRAM), which is covered next. Typically, SRAM is several times faster than DRAM. Overall, a cache memory gets stored information to the microprocessor much faster than if only high-capacity DRAM is used. Cache memory is basically a cost-effective method of improving system performance without having to resort to the expense of making all of the memory faster.

    The concept of cache memory is based on the idea that computer programs tend to get instructions or data from one area of main memory before moving to another area. Basically, the cache controller "guesses" which area of the slow dynamic memory the CPU (central-processing unit) will need next and moves it to the cache memory so that it is ready when needed. If the cache controller guesses right, the data are immediately available to the microprocessor. If the cache controller guesses wrong, the CPU must go to the main memory and wait much longer for the correct instructions or data. Fortunately, the cache controller is right most of the time.

Cache Analogy   There are many analogies that can be used in describing a cache memory, but comparing it to a home refrigerator is perhaps the most effective. A home refrigerator can be thought of as a "cache" for certain food items while the supermarket is the main memory where all foods are kept. Each time you want something to eat or drink, you can go to the refrigerator (cache) first to see if the item you want is there. If it is, you save a lot of time. If it is not there, then you have to spend extra time to get it from the supermarket.

L1 and L2 Caches A first-level cache (L1 cache) is usually integrated into the processor chip and has a very limited storage capacity. L1 cache is also known as *primary* cache. A second-level cache (L2 cache) is a separate memory chip or set of chips external to the processor and usually has a larger storage capacity than an L1 cache. L2 cache is also known as *secondary* cache. Some systems may have higher-level caches (L3, L4, etc.), but L1 and L2 are the most common. Also, some systems employ a disk cache, which is located in the main DRAM memory and is used to enhance the performance of the hard disk because DRAM, although much slower than SRAM, is much faster than the hard disk drive. Figure 12–14 illustrates L1 and L2 cache memories in a computer system.

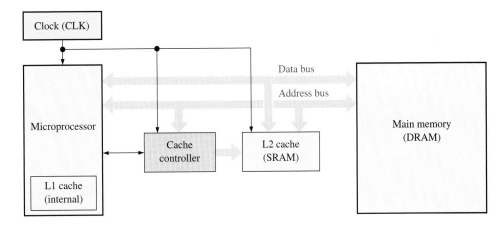

FIGURE 12–14
Block diagram showing L1 and L2 cache memories in a computer system.

Dynamic RAMs (DRAMs) Storage Cells

Dynamic memory cells store a data bit in a small capacitor rather than in a latch. The advantage of this type of cell is that it is very simple, thus allowing very large memory arrays to be constructed on a chip at a lower cost per bit than in static memories. The disadvantage is that the storage capacitor cannot hold its charge over an extended period of time and will lose the stored data bit unless its charge is refreshed periodically. To refresh requires additional memory circuitry and complicates the operation of the **DRAM.** Figure 12–15 shows a typical DRAM cell consisting of a single MOS transistor (MOSFET) and a capacitor.

FIGURE 12–15
A MOS dynamic RAM cell.

In this type of cell, the transistor acts as a switch. The basic simplified operation is illustrated in Figure 12–16 and is as follows. A LOW on the $R/\overline{W}$ line (WRITE mode) enables the tristate input buffer and disables the output buffer. For a 1 to be written into the cell, the D_{IN} line must be HIGH, and the transistor must be turned on by a HIGH on the row line. The transistor acts as a closed switch connecting the capacitor to the bit line. This connection allows the capacitor to charge to a positive voltage, as shown in Figure 12–16(a). When a 0 is to be stored, a LOW is applied to the D_{IN} line. If the capacitor is storing a 0, it remains uncharged, or if it is storing a 1, it discharges as indicated in Figure 12–16(b). When the row line is taken back LOW, the transistor turns off and disconnects the capacitor from the bit line, thus "trapping" the charge (1 or 0) on the capacitor.

To read from the cell, the $R/\overline{W}$ (Read/$\overline{Write}$) line is HIGH, enabling the output buffer and disabling the input buffer. When the row line is taken HIGH, the transistor turns on and connects the capacitor to the bit line and thus to the output buffer (sense amplifier), so the data bit appears on the data-output line (D_{OUT}). This process is illustrated in Figure 12–16(c).

(a) Writing a 1 into the memory cell

(b) Writing a 0 into the memory cell

(c) Reading a 1 from the memory cell

(d) Refreshing a stored 1

FIGURE 12–16

Basic operation of a DRAM cell.

For refreshing the memory cell, the $R/\overline{W}$ line is HIGH, the row line is HIGH, and the refresh line is HIGH. The transistor turns on, connecting the capacitor to the bit line. The output buffer is enabled, and the stored data bit is applied to the input of the refresh buffer, which is enabled by the HIGH on the refresh input. This produces a voltage on the bit line corresponding to the stored bit, thus replenishing the capacitor as illustrated in Figure 12–16(d).

Basic Organization of a Dynamic RAM (DRAM)

The major application of DRAMs is in the main memory of computers. The difference between DRAMs and SRAMs is the type of memory cell. As you have seen, the DRAM memory cell consists of one transistor and a capacitor and is much simpler than the SRAM cell. This allows much greater densities in DRAMs and results in greater bit capacities for a given chip area, although much slower access time.

Again, because charge stored in a capacitor will leak off, the DRAM cell requires a frequent refresh operation to preserve the stored data bit. This requirement results in more complex circuitry than in a SRAM. Several features common to most DRAMs are now discussed using a generic 1 M × 1 bit DRAM as an example.

Address Multiplexing   DRAMs use a technique called *address multiplexing* to reduce the number of address lines. Figure 12–17 shows the block diagram of a 1,048,576-bit (1 Mbit) DRAM with a 1 M × 1 organization. We will focus on the beige blocks to illustrate address multiplexing. The blue blocks represent the refresh logic.

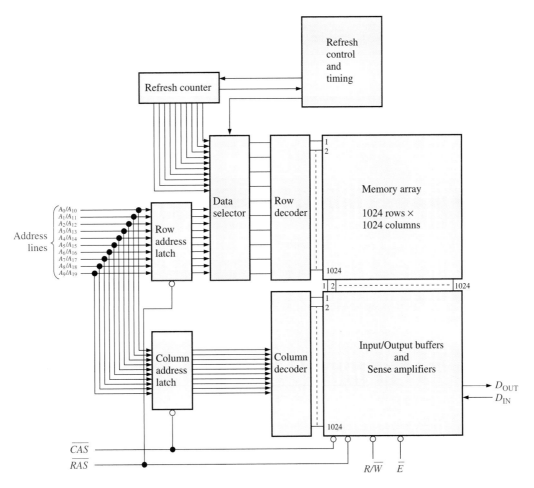

FIGURE 12–17
Simplified block diagram of a 1M × 1 DRAM.

The ten address lines are time multiplexed at the beginning of a memory cycle by the row address strobe ($\overline{RAS}$) and the column address strobe ($\overline{CAS}$) into two separate 10-bit address fields. First, the 10-bit row address is latched into the row address latch. Next, the 10-bit column address is latched into the column address latch. The row address and the column address are decoded to select one of the 1,048,576 addresses ($2^{20} = 1{,}048{,}576$) in the memory array. The basic timing for the address multiplexing operation is shown in Figure 12–18.

FIGURE 12–18
Basic timing for address multiplexing.

Read and Write Cycles At the beginning of each read or write memory cycle, $\overline{RAS}$ and $\overline{CAS}$ go active (LOW) to multiplex the row and column addresses into the latches and decoders. For a read cycle, the $R/\overline{W}$ input is HIGH. For a write cycle, the $R/\overline{W}$ input is LOW. This is illustrated in Figure 12–19.

Fast Page Mode In the normal read or write cycle described previously, the row address for a particular memory location is first loaded by an active-LOW $\overline{RAS}$ and then the column address for that location is loaded by an active-LOW $\overline{CAS}$. The next location is selected by another $\overline{RAS}$ followed by a $\overline{CAS}$, and so on.

A "page" is a section of memory available at a single row address and consists of all the columns in a row. Fast page mode allows fast successive read or write operations at each column address in a selected row. A row address is first loaded by $\overline{RAS}$ going LOW and remaining LOW while $\overline{CAS}$ is toggled between HIGH and LOW. A single row address is selected and remains selected while $\overline{RAS}$ is active. Each successive $\overline{CAS}$ selects another column in the selected row. So, after a fast page mode cycle, all of the addresses in the selected row have been read from or written into, depending on $R/\overline{W}$. For example, a fast page mode cycle for the DRAM in Figure 12–17 requires $\overline{CAS}$ to go active 1024 times for each row selected by $\overline{RAS}$.

Fast page mode operation for read is illustrated by the timing diagram in Figure 12–20. When $\overline{CAS}$ goes to its nonasserted state (HIGH), it disables the data outputs. Therefore, the transition of $\overline{CAS}$ to HIGH must occur only after valid data are latched by the external system.

Refresh Cycles As you know, DRAMs are based on capacitor charge storage for each bit in the memory array. This charge degrades (leaks off) with time and temperature, so each bit must be periodically refreshed (recharged) to maintain the correct bit state. Typically, a DRAM must be refreshed every 8 ms to 16 ms, although for some devices the refresh period can exceed 100 ms.

A read operation automatically refreshes all the addresses in the selected row. However, in typical applications, you cannot always predict how often there will be a read cycle and so you cannot depend on a read cycle to occur frequently enough to prevent data loss. Therefore, special refresh cycles must be implemented in DRAM systems.

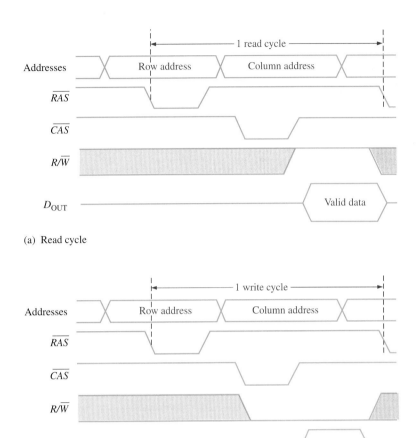

(a) Read cycle

(b) Write cycle

FIGURE 12–19
Normal read and write cycle timing.

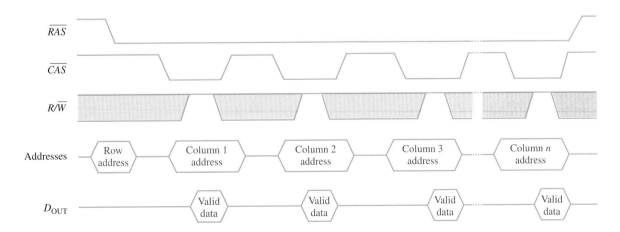

FIGURE 12–20
Fast page mode timing for a read operation.

Burst refresh and *distributed refresh* are the two basic refresh modes for refresh operations. In burst refresh, all rows in the memory array are refreshed consecutively each refresh period. For a memory with a refresh period of 8 ms, a burst refresh of all rows occurs once every 8 ms. The normal read and write operations are suspended during a burst refresh cycle.

In distributed refresh, each row is refreshed at intervals interspersed between normal read or write cycles. For example, the memory in Figure 12–17 has 1024 rows. As an example, for an 8 ms refresh period, each row must be refreshed every 8 ms/1024 = 7.8 μs when distributed refresh is used.

The two types of refresh operations are $\overline{RAS}$-*only refresh* and $\overline{CAS}$ *before* $\overline{RAS}$ *refresh.* $\overline{RAS}$-only refresh consists of a $\overline{RAS}$ transition to the LOW (active) state which latches the address of the row to be refreshed while $\overline{CAS}$ remains HIGH (inactive) throughout the cycle. An external counter is used to provide the row addresses for this type of operation.

The $\overline{CAS}$ before $\overline{RAS}$ refresh is initiated by $\overline{CAS}$ going LOW before $\overline{RAS}$ goes LOW. This sequence activates an internal refresh counter that generates the row address to be refreshed. This address is switched by the data selector into the row decoder.

Types of DRAMs

Now that you have learned the basic concept of a DRAM, let's briefly look at the major types. These are the *Fast Page Mode (FPM) DRAM,* the *Extended Data Output (EDO) DRAM,* the *Burst Extended Data Output (BEDO) DRAM,* and the *Synchronous (S) DRAM.*

FPM DRAM   Fast page mode operation was described earlier. This type of DRAM traditionally has been the most common and has been the type used in computers until the development of the EDO DRAM. Recall that a page in memory is all of the column addresses contained within one row address.

As you have seen, the basic idea of the **FPM DRAM** is based on the probability that the next several memory addresses to be accessed are in the same row (on the same page). Fortunately, this happens a large percentage of the time. FPM saves time over pure random accessing because in FPM the row address is specified only once for access to several successive column addresses whereas for pure random accessing, a row address is specified for each column address.

Recall that in a fast page mode read operation, the $\overline{CAS}$ signal has to wait until the valid data from a given address are accepted (latched) by the external system (CPU) before it can go to its nonasserted state. When $\overline{CAS}$ goes to its nonasserted state, the data outputs are disabled. This means that the next column address cannot occur until after the data from the current column address are transferred to the CPU. This limits the rate at which the columns within a page can be addressed.

EDO DRAM   The Extended Data Output DRAM, sometimes called *hyper page mode DRAM,* is very similar to the FPM DRAM. The key difference is that the $\overline{CAS}$ signal in the **EDO DRAM** does not disable the output data when it goes to its nonasserted state because the valid data from the current address can be held until $\overline{CAS}$ is asserted again. This means that the next column address can be accessed before the external system accepts the current valid data. The idea is to speed up the access time.

BEDO DRAM   The Burst Extended Data Output DRAM is an EDO DRAM with address burst capability. Recall from the discussion of the synchronous burst SRAM that the address burst feature allows up to four addresses to be internally generated from a single external address, which saves some access time. This same concept applies to the **BEDO DRAM.**

SDRAM Faster DRAMs are needed to keep up with the ever-increasing speed of microprocessors. The Synchronous DRAM is one of the most recent efforts to accomplish this. Like the synchronous static RAM discussed earlier, the operation of the **SDRAM** is synchronized with the system clock, which also runs the microprocessor in a computer system. The same basic ideas described in relation to the Synchronous burst SRAM, also apply to the SDRAM.

This synchronized operation makes the SDRAM totally different from the other asynchronous DRAM types previously discussed. With asynchronous memories, the microprocessor must wait for the DRAM to complete its internal operations. However, with synchronous operation, the DRAM latches addresses, data, and control information from the processor under control of the system clock. This allows the processor to handle other tasks while the memory read or write operations are in progress, rather than having to wait for the memory to do its thing as is the case in asynchronous systems.

SECTION 12–2 REVIEW

1. List two types of SRAM.
2. What is a cache?
3. Explain how SRAMs and DRAMs differ.
4. Describe the refresh operation in a DRAM.
5. List four types of DRAM.

12–3 ■ READ-ONLY MEMORIES (ROMs)

A ROM contains permanently or semipermanently stored data, which can be read from the memory but either cannot be changed at all or cannot be changed without specialized equipment. A ROM stores data that are used repeatedly in system applications, such as tables, conversions, or programmed instructions for system initialization and operation. ROMs retain stored data when the power is off and are therefore nonvolatile memories.

After completing this section, you should be able to

■ List the types of ROMs ■ Describe a basic mask ROM storage cell ■ Explain how data are read from a ROM ■ Discuss internal organization of a typical ROM ■ Discuss some ROM applications

The ROM Family

Figure 12–21 shows how semiconductor ROMs are categorized. The mask ROM is the type in which the data are permanently stored in the memory during the manufacturing process. The **PROM,** or programmable ROM, is the type in which the data are electrically stored by the user with the aid of specialized equipment. Both the mask ROM and the PROM can be of either MOS or bipolar technology. The **EPROM,** or erasable PROM, is strictly an MOS device. The **UV EPROM** is electrically programmable by the user, but the stored data must be erased by exposure to ultraviolet light over a period of several minutes. The electrically erasable PROM (**EEPROM** or E^2PROM) can be erased in a few milliseconds.

The Mask ROM

The mask ROM is usually referred to simply as a ROM. It is permanently programmed during the manufacturing process to provide widely used standard functions, such as popular conversions, or to provide user-specified functions. Once the memory is programmed,

FIGURE 12–21
The ROM family.

it cannot be changed. Most IC ROMs utilize the presence or absence of a transistor connection at a row/column junction to represent a 1 or a 0.

Figure 12–22 shows MOS ROM cells. The presence of a connection from a row line to the gate of a transistor represents a 1 at that location because when the row line is taken HIGH, all transistors with a gate connection to that row line turn on and connect the HIGH (1) to the associated column lines. At row/column junctions where there are no gate connections, the column lines remain LOW (0) when the row is addressed.

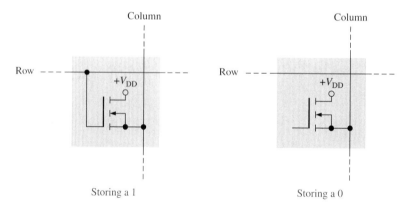

FIGURE 12–22
ROM cells.

A Simple ROM

To illustrate the ROM concept, Figure 12–23 shows a small, simplified ROM array. The blue squares represent stored 1s, and the gray squares represent stored 0s. The basic read operation is as follows: When a binary address code is applied to the address input lines, the corresponding row line goes HIGH. This HIGH is connected to the column lines through the transistors at each junction (cell) where a 1 is stored. At each cell where a 0 is stored, the column line stays LOW because of the terminating resistor. The column lines form the data output. The eight data bits stored in the selected row appear on the output lines.

As you can see, the example ROM in Figure 12–23 is organized into 16 addresses, each of which stores 8 data bits. Thus, it is a 16 × 8 (16-by-8) ROM, and its total capacity is 128 bits or 16 bytes.

FIGURE 12–23
A 16 × 8-bit ROM array.

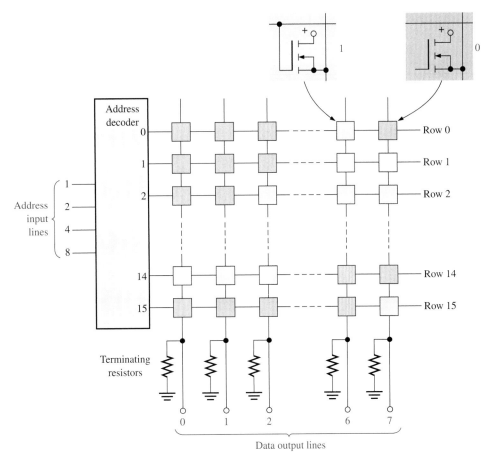

EXAMPLE 12–1

Show a basic ROM, similar to the one in Figure 12–23, programmed for a 4-bit binary-to-Gray conversion.

Solution Review Chapter 2 for the Gray code. Table 12–1 is developed for use in programming the ROM.

TABLE 12–1

Binary				Gray			
B_3	B_2	B_1	B_0	G_3	G_2	G_1	G_0
0	0	0	0	0	0	0	0
0	0	0	1	0	0	0	1
0	0	1	0	0	0	1	1
0	0	1	1	0	0	1	0
0	1	0	0	0	1	1	0
0	1	0	1	0	1	1	1
0	1	1	0	0	1	0	1
0	1	1	1	0	1	0	0
1	0	0	0	1	1	0	0
1	0	0	1	1	1	0	1
1	0	1	0	1	1	1	1
1	0	1	1	1	1	1	0
1	1	0	0	1	0	1	0
1	1	0	1	1	0	1	1
1	1	1	0	1	0	0	1
1	1	1	1	1	0	0	0

The resulting ROM array is shown in Figure 12–24. You can see that a binary code on the address input lines produces the corresponding Gray code on the output lines (columns). For example, when the binary number 0110 is applied to the address input lines, address 6, which stores the Gray code 0101, is selected.

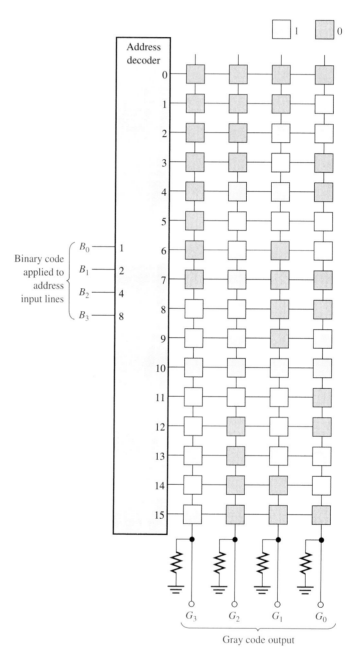

FIGURE 12–24

Representation of a ROM programmed as a binary-to-Gray code converter.

Related Problem Using Figure 12–24, determine the Gray code output when a binary code of 1011 is applied to the address input lines.

Internal ROM Organization

Most IC ROMs have a more complex internal organization than that in the basic simplified example just presented. To illustrate how an IC ROM is structured, let's use a 1024-bit device with a 256 × 4 organization. The logic symbol is shown in Figure 12–25. When any one of 256 binary codes (eight bits) is applied to the address lines, four data bits appear on the outputs if the chip enable inputs are LOW. (There are eight address lines because $2^8 = 256$.)

Although the 256 × 4 organization of this device implies that there are 256 rows and 4 columns in the memory array, this is not actually the case. The memory cell array is actually a 32 × 32 matrix (32 rows and 32 columns), as shown in the block diagram in Figure 12–26.

FIGURE 12–25
A 256 × 4 ROM logic symbol. The A $\frac{0}{255}$ designator means that the 8-bit address code selects addresses 0 through 255.

COMPUTER NOTE
ROM is used in the personal computer to store what is referred to as the BIOS (Basic Input/Output Services). These are programs that are used to perform fundamental supervisory and support functions for the computer. For example, BIOS programs stored in the ROM control certain video monitor functions, provide for disk formatting, scan the keyboard for inputs, and control certain printer functions.

FIGURE 12–26
A typical ROM. This particular example is a 1024-bit ROM with a 256 × 4 organization based on a 32 × 32 array.

The ROM in Figure 12–26 works as follows: Five of the eight address lines (A_0 through A_4) are decoded by the row decoder (often called the Y decoder) to select one of the 32 rows. Three of the eight address lines (A_5 through A_7) are decoded by the column decoder (often called the X decoder) to select four of the 32 columns. Actually, the column decoder consists of four 1-of-8 decoders (data selectors), as shown in Figure 12–26.

The result of this structure is that when an 8-bit address code (A_0 through A_7) is applied, a 4-bit data word appears on the data outputs when the chip enable lines ($\overline{E}_0$ and $\overline{E}_1$) are LOW to enable the output buffers. This type of internal organization (architecture) is typical of IC ROMs of various capacities.

ROM Access Time

A typical timing diagram that illustrates ROM access time is shown in Figure 12–27. The **access time,** t_a, of a ROM is the time from the application of a valid address code on the input lines until the appearance of valid output data. Access time can also be measured from the activation of the chip enable ($\overline{E}$) input to the occurrence of valid output data when a valid address is already on the input lines.

FIGURE 12–27
ROM access time (t_a) from address change to data output with chip enable already active.

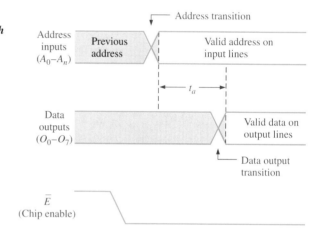

SECTION 12–3 REVIEW

1. What is the bit storage capacity of a ROM with a 512 × 8 organization?
2. List the types of read-only memories.
3. How many address bits are required for a 2048-bit memory organized as a 256 × 8 memory?

12–4 ■ PROGRAMMABLE ROMs (PROMs AND EPROMs)

PROMs are basically the same as mask ROMs, once they have been programmed. The difference is that PROMs come from the manufacturer unprogrammed and are custom programmed in the field to meet the user's needs.

After completing this section, you should be able to

■ Distinguish between a mask ROM and a PROM ■ Describe a basic PROM storage cell ■ Discuss EPROMs including UV EPROMs and EEPROMs ■ Analyze an EPROM programming cycle

PROMs

PROMs use some type of fusing process to store bits, in which a memory *link* is burned open or left intact to represent a 0 or a 1. The fusing process is irreversible; once a PROM is programmed, it cannot be changed.

Figure 12–28 illustrates a MOS PROM array with fusible links. The fusible links are manufactured into the PROM between the source of each cell's transistor and its column line. In the programming process, a sufficient current is injected through the fusible link to burn it open to create a stored 0. The link is left intact for a stored 1.

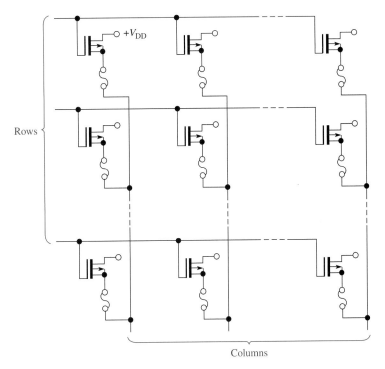

FIGURE 12–28
MOS PROM array with fusible links. (All drains are commonly connected to V_{DD}.)

Three basic fuse technologies used in PROMs are metal links, silicon links, and *pn* junctions. A brief description of each of these follows.

1. Metal links are made of a material such as nichrome. Each bit in the memory array is represented by a separate link. During programming, the link is either "blown" open or left intact. This is done basically by first addressing a given cell and then forcing a sufficient amount of current through the link to cause it to open.
2. Silicon links are formed by narrow, notched strips of polycrystalline silicon. Programing of these fuses requires melting of the links by passing a sufficient amount of current through them. This amount of current causes a high temperature at the fuse location that oxidizes the silicon and forms an insulation around the now-open link.
3. Shorted junction, or avalanche-induced migration, technology consists basically of two *pn* junctions arranged back-to-back. During programming, one of the diode junctions is avalanched, and the resulting voltage and heat cause aluminum ions to migrate and short the junction. The remaining junction is then used as a forward-biased diode to represent a data bit.

PROM Programming

A PROM is normally programmed by inserting it into a special instrument called a *PROM programmer*. Basically, the programming is accomplished as shown by the simplified setup in Figure 12–29. An address is selected by the electronic switches on the address lines, and then a pulse is applied to those output lines corresponding to bit locations where 0s are to be stored (the PROM starts out with all 1s). These pulses blow the fusible links, thus creating the desired bit pattern. The next address is then selected and the process is repeated. This sequence is actually done automatically by a software-driven PROM programmer.

FIGURE 12–29
Simplified concept of PROM programming setup.

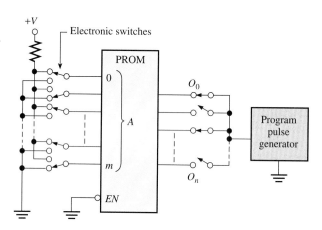

EPROMs

An EPROM is an erasable PROM. Unlike an ordinary PROM, an EPROM can be reprogrammed if an existing program in the memory array is erased first.

An EPROM uses an NMOSFET array with an isolated-gate structure. The isolated transistor gate has no electrical connections and can store an electrical charge for indefinite periods of time. The data bits in this type of array are represented by the presence or absence of a stored gate charge. Erasure of a data bit is a process that removes the gate charge.

Two basic types of erasable PROMs are the ultraviolet light erasable PROM (UV EPROM) and the electrically erasable PROM (EEPROM).

UV EPROMs You can recognize the UV EPROM device by the transparent quartz lid on the package, as shown in Figure 12–30. The isolated gate in the **FET** of an ultraviolet EPROM is "floating" within an oxide insulating material. The programming process causes electrons to be removed from the floating gate. Erasure is done by exposure of the memory array chip to high-intensity ultraviolet radiation through the quartz window on top of the package. The positive charge stored on the gate is neutralized after several minutes to an hour of exposure time.

FIGURE 12–30
Ultraviolet erasable PROM package.

EEPROMs An electrically erasable PROM can be both erased and programmed with electrical pulses. Since it can be both electrically written into and electrically erased, the EEPROM can be rapidly programmed and erased in circuit for reprogramming.

Two types of EEPROMs are the floating-gate MOS and the metal nitride-oxide silicon (MNOS). The application of a voltage on the control gate in the floating-gate structure permits the storage and removal of charge from the floating gate.

An Example EPROM

The 27C16B is an example of an EPROM device. Its operation is representative of that of other typical EPROMs of various sizes. As the logic symbol in Figure 12–31 shows, this device has 2048 addresses ($2^{11} = 2048$), each with eight bits. Notice that the eight outputs are tristate (∇).

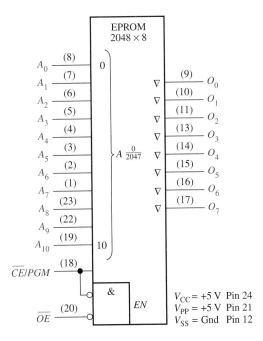

FIGURE 12–31
The logic symbol for a 2048 × 8 EPROM.

To read from the memory, the output enable input ($\overline{OE}$) must be LOW and the power-down/program ($\overline{CE/PGM}$) input LOW. To erase the stored data, the device is exposed to high-intensity ultraviolet light through the transparent lid. A typical 12 mW/cm^2 filterless UV lamp will erase the data in about 20 to 25 minutes. As in most EPROMs after erasure, all bits are 1s. Normal ambient light contains the correct wavelength of UV light for erasure over a period of time. Therefore, the transparent lid must be kept covered.

To program the device, a high dc voltage is applied to V_{PP} (which is normally +5 V or +3 V), and $\overline{OE}$ is HIGH. The eight data bits to be programmed into a given address are applied to the outputs (O_0 through O_7), and the address is selected on inputs A_0 through A_{10}. Next, a 10 ms to 55 ms HIGH level pulse is applied to the $\overline{CE/PGM}$ input. The addresses can be programmed in any order.

A timing diagram for the programming mode is shown in Figure 12–32. These signals are normally produced by an EPROM programmer.

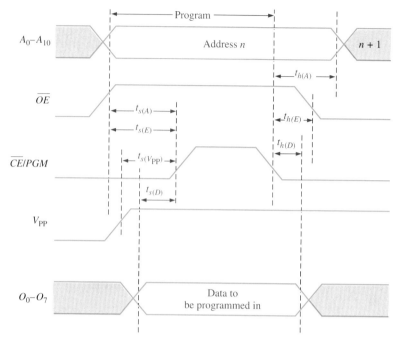

FIGURE 12–32

Timing diagram for a 2048 × 8 EPROM programming cycle, with critical setup times (t_s) and hold times (t_h) indicated.

SECTION 12–4 REVIEW

1. How do PROMs differ from ROMs?
2. After erasure, all bits are (1s, 0s) in a typical EPROM.
3. What is the normal mode of operation for a PROM?

12–5 ■ FLASH MEMORIES

The ideal memory has high storage capacity, nonvolatility, in-system read and write capability, comparatively fast operation, and cost effectiveness. The traditional memory technologies such as ROM, PROM, EPROM, EEPROM, SRAM, and DRAM individually exhibit one or more of these characteristics, but no single technology has all of them except the flash memory.

After completing this section, you should be able to

■ Discuss the basic characteristics of a flash memory ■ Describe the basic operation of a flash memory cell ■ Compare flash memories with other types of memories

Flash memories are high-density read/write memories (high-density translates into large bit storage capacity) that are nonvolatile, which means that data can be stored indefinitely without power. They are frequently used in place of floppy or small-capacity hard disk drives in portable computers.

High-density means that a large number of cells can be packed into a given surface area on the chip; that is, the higher the density, the more bits that can be stored on a given size chip. This high density is achieved in flash memories with a storage cell that consists of a single floating-gate MOS transistor. A data bit is stored as charge or the absence of charge on the floating gate depending if a 0 or a 1 is stored.

Flash Memory Cell

A single-transistor cell in a flash memory is represented in Figure 12–33. The stacked gate MOS transistor consists of a control gate and a floating gate in addition to the drain and source. The floating gate stores electrons (charge) as a result of a sufficient voltage applied to the control gate. A *0 is stored when there is more charge* and a *1 is stored when there is less or no charge.* The amount of charge present on the floating gate determines if the transistor will turn on and conduct current from the drain to the source when a control voltage is applied during a read operation.

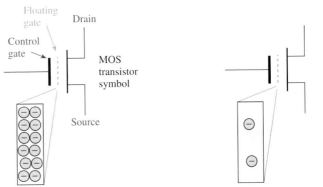

Many electrons = more charge = stored 0.   Few electrons = less charge = stored 1.

FIGURE 12–33
The storage cell in a flash memory.

Basic Flash Memory Operation

There are three major operations in a flash memory: the *programming* operation, the *read* operation, and the *erase* operation.

Programming   Initially, all cells are at the 1 state because charge was removed from each cell in a previous erase operation. The programming operation adds electrons (charge) to the floating gate of those cells that are to store a 0. No charge is added to those cells that are to store a 1. Application of a sufficient positive voltage to the control gate with respect to the source during programming attracts electrons to the floating gate as indicated in Figure 12–34. Once programmed, a cell can retain the charge for up to 100 years without any external power.

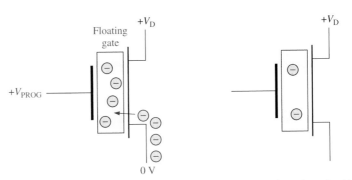

To store a 0, a sufficient positive voltage is applied to the control gate with respect to the source to add charge to the floating gate during programming.

To store a 1, no charge is added and the cell is left in the erased condition.

FIGURE 12–34
Simplified illustration of storing a 0 or a 1 in a flash cell during the programming operation.

Read During a read operation, a positive voltage is applied to the control gate. The amount of charge present on the floating gate of a cell determines whether or not the voltage applied to the control gate will turn on the transistor. If a 1 is stored, the control gate voltage is sufficient to turn the transistor on. If a 0 is stored, the transistor will not turn on because the control gate voltage is not sufficient to overcome the negative charge stored in the floating gate. Think of the charge on the floating gate as a voltage source that opposes the voltage applied to the control gate during read. So the floating gate charge associated with a stored 0 prevents the control gate voltage from reaching the turn-on threshold, whereas the small or zero charge associated with a stored 1 allows the control gate voltage to exceed the turn-on threshold.

When the transistor turns on, there is current from the drain to the source of the cell transistor. The presence of this current is sensed to indicate a 1, and the absence of this current is sensed to indicate a 0. This basic idea is illustrated in Figure 12–35.

When a 0 is read, the transistor remains off because the charge on the floating gate prevents the read voltage from exceeding the turn-on threshold.

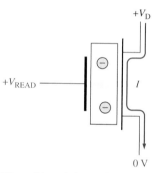

When a 1 is read, the transistor turns on because the absence of charge on the floating gate allows the read voltage to exceed the turn-on threshold.

FIGURE 12–35
The read operation of a flash cell in an array.

Erase During an erase operation, charge is removed from all the memory cells. A sufficient positive voltage is applied to the transistor source with respect to the control gate. This is opposite in polarity to that used in programming. This voltage attracts electrons from the floating gate and depletes it of charge, as illustrated in Figure 12–36. A flash memory is always erased prior to being reprogrammed.

FIGURE 12–36
Simplified illustration of removing charge from a cell during erase.

To erase a cell, a sufficient positive voltage is applied to the source with respect to the control gate to remove charge from the floating gate during the erase operation.

Basic Flash Memory Array

A simplified array of flash memory cells is shown in Figure 12–37. Only one row line is accessed at a time. When a cell in a given bit line turns on (stored 1) during a read operation, there is current through the bit line, which produces a voltage drop across the active load. This voltage drop is compared to a reference voltage with a comparator circuit and an output level indicating a 1 is produced. If a 0 is stored, then there is no current or little current in the bit line and an opposite level is produced on the comparator output.

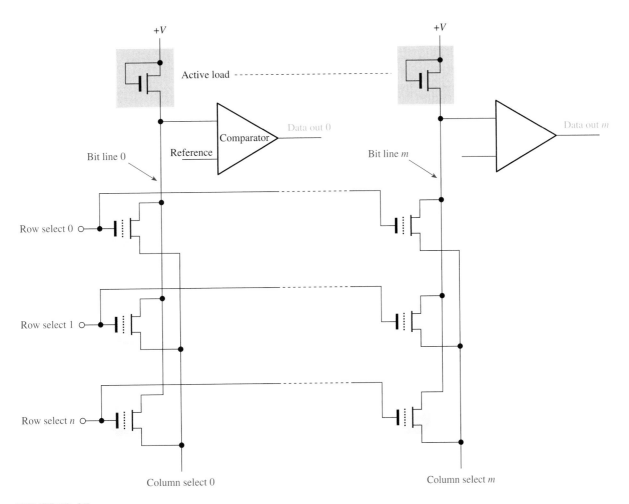

FIGURE 12–37
Basic flash memory array.

Comparison of Flash Memories with Other Memories

Let's compare flash memories with other types of memories with which you are already familiar.

Flash vs. ROM, EPROM, and EEPROM   Read-only memories are high-density, non-volatile devices. However, once programmed the contents of a ROM can never be altered. Also, the initial programming is a time-consuming and costly process.

Although the EPROM is a high-density, nonvolatile memory, it can be erased only by removing it from the system and using ultraviolet light. It can be reprogrammed only with specialized equipment.

The EEPROM has a more complex cell structure than either the ROM or EPROM and so the density is not as high, although it can be reprogrammed without being removed from the system. Because of its lower density, the cost/bit is higher than ROMs or EPROMs.

A flash memory can be reprogrammed easily in the system because it is essentially a READ/WRITE device. The density of a flash memory compares with the ROM and EPROM because both have single transistor cells. A flash memory (like a ROM, EPROM, or EEPROM) is nonvolatile, which allows data to be stored indefinitely with power off.

Flash vs. SRAM As you have learned, static random-access memories are volatile READ/WRITE devices. A SRAM requires constant power to retain the stored data. In many applications, a battery backup is used to prevent data loss if the main power source is turned off. However, since battery failure is always a possibility, indefinite retention of the stored data in a SRAM cannot be guaranteed. Because the memory cell in a SRAM is basically a flip-flop consisting of several transistors, the density is relatively low.

A flash memory is also a READ/WRITE memory, but unlike the SRAM it is nonvolatile. Also, a flash memory has a much higher density than a SRAM.

Flash vs. DRAM Dynamic random-access memories are volatile high-density READ/WRITE devices. DRAMs require not only constant power to retain data, but also the stored data must be refreshed frequently. In many applications, backup storage such as hard disk must be used with a DRAM.

Flash memories exhibit higher densities than DRAMs because a flash memory cell consists of one transistor and does not need refreshing, whereas a DRAM cell is one transistor plus a capacitor that has to be refreshed. Typically, a flash memory consumes much less power than an equivalent DRAM and can be used as a hard disk replacement in many applications.

Table 12–2 provides a summary of the comparison of the memory technologies.

TABLE 12–2

Comparison of types of memories.

Memory Type	Nonvolatile	High-density	One-transistor Cell	In-system Writability
Flash	Yes	Yes	Yes	Yes
SRAM	No	No	No	Yes
DRAM	No	Yes	Yes	Yes
ROM	Yes	Yes	Yes	No
EPROM	Yes	Yes	Yes	No
EEPROM	Yes	No	No	Yes

SECTION 12–5 REVIEW

1. What types of memories are nonvolatile?
2. What is a major advantage of a flash memory over a SRAM or DRAM?
3. List the three modes of operation of a flash memory.

12–6 ■ MEMORY EXPANSION

Available memory can be expanded to increase the word length (number of bits in each address) or the word capacity (number of different addresses) or both. Memory expansion is accomplished by adding an appropriate number of memory chips to the address, data, and control buses as explained in this section.

After completing this section, you should be able to

- Define *word-length expansion* ■ Show how to expand the word length of a memory
- Define *word-capacity expansion* ■ Show how to expand the word capacity of a memory

Word-Length Expansion

To increase the **word length** of a memory, the number of bits in the data bus must be increased. For example, an 8-bit word length can be achieved by using two memories, each with 4-bit words as illustrated in Figure 12–38(a). As you can see in part (b), the 16-bit address bus is commonly connected to both memories so that the combination memory still has the same number of addresses ($2^{16} = 65,536$) as each individual memory. The 4-bit data buses from the two memories are combined to form an 8-bit data bus. Now when an address is selected, eight bits are produced on the data bus—four from each memory.

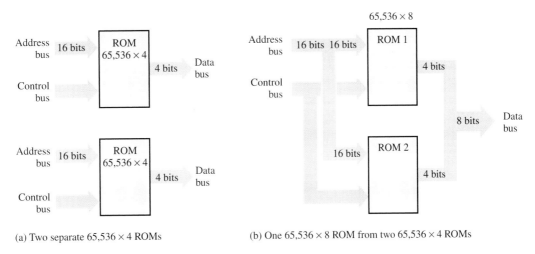

(a) Two separate $65,536 \times 4$ ROMs   (b) One $65,536 \times 8$ ROM from two $65,536 \times 4$ ROMs

FIGURE 12–38
Expansion of two $65,536 \times 4$ ROMs into a $65,536 \times 8$ ROM to illustrate word-length expansion.

The following example shows the details of $65,536 \times 4$ to $65,536 \times 8$ expansion.

EXAMPLE 12–2

Expand the $65,536 \times 4$ ROM ($64k \times 4$) in Figure 12–39 to form a $64k \times 8$ ROM. Note that "64k" is the accepted shorthand for 65,536. Why not "65k"? Maybe it's because 64 is also a power-of-two.

FIGURE 12–39
A $64k \times 4$ ROM.

Solution Two 64k × 4 ROMs are connected as shown in Figure 12–40. Notice that a specific address is accessed in ROM 1 and ROM 2 at the same time. The four bits from a selected address in ROM 1 and the four bits from the corresponding address in ROM 2 go out in parallel to form an 8-bit word on the data bus. Also notice that a LOW on the chip enable line, $\overline{E}$, which forms a simple control bus, enables *both* memories.

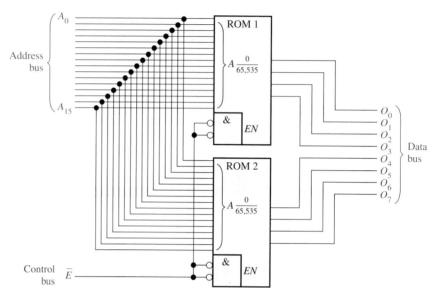

FIGURE 12–40

Related Problem Describe how you would expand a 64k × 1 ROM to a 64k × 8 ROM.

EXAMPLE 12–3

Use the memories in Example 12–2 to form a 64k × 16 ROM.

Solution In this case you need a memory that stores 65,536 16-bit words. Four 64k × 4 ROMs are required to do the job, as shown in Figure 12–41.

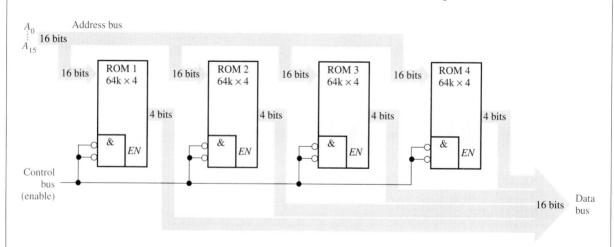

FIGURE 12–41

Related Problem How many 64k × 1 ROMs would be required to implement the memory shown in Figure 12–41?

A ROM has only data outputs, but a RAM has both data inputs and data outputs. For word-length expansion in a RAM (SRAM or DRAM), the data inputs *and* data outputs form the data bus. Because the same lines are used for data input and data output, tristate buffers are required. Most RAMs provide internal tristate circuitry. Figure 12–42 illustrates RAM expansion to increase the data word length.

FIGURE 12–42

Illustration of word-length expansion with two $2^m \times n$ RAMs forming a $2^m \times 2n$ RAM.

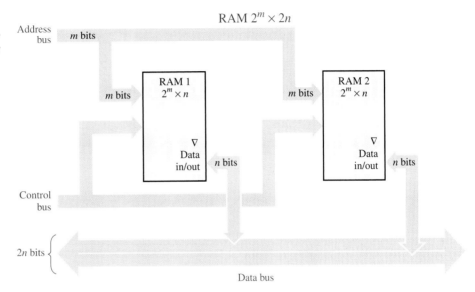

EXAMPLE 12–4 Use $1M \times 4$ SRAMs to create a $1M \times 8$ SRAM.

Solution Two $1M \times 4$ SRAMs are connected as shown in the simplified block diagram of Figure 12–43.

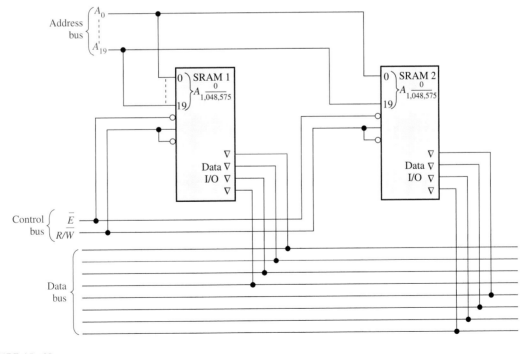

FIGURE 12–43

Related Problem Use $1M \times 8$ SRAMs to create a $1M \times 16$ SRAM.

Word-Capacity Expansion

When memories are expanded to increase the **word capacity,** the *number of addresses is increased.* To achieve this increase, the number of address bits must be increased, as illustrated in Figure 12–44, (where two 1M × 8 RAMs are expanded to form a 2M × 8 memory).

(a) Individual memories each store 1,048,576 8-bit words

(b) Memories expanded to form a 2M × 8 RAM requiring a 21-bit address bus

FIGURE 12–44
Illustration of word-capacity expansion.

Each individual memory has 20 address bits to select its 1,048,576 addresses, as shown in part (a). The expanded memory has 2,097,152 addresses and therefore requires 21 address bits, as shown in part (b). The twenty-first address bit is used to enable the appropriate memory chip. The data bus for the expanded memory remains eight bits wide. Details of this expansion are illustrated in Example 12–5.

EXAMPLE 12–5   Use 512k × 4 RAMs to implement a 1M × 4 memory.

FIGURE 12–45

Solution The expanded addressing is achieved by connecting the chip enable ($\overline{E}_0$) input to the twentieth address bit (A_{19}), as shown in Figure 12–45. Input $\overline{E}_1$ is used as an enable input common to both memories. When the twentieth address bit (A_{19}) is LOW, RAM 1 is selected (RAM 2 is disabled), and the nineteen lower-order address bits (A_0–A_{18}) access each of the addresses in RAM 1. When the twentieth address bit (A_{19}) is HIGH, RAM 2 is enabled by a LOW on the inverter output (RAM 1 is disabled), and the nineteen lower-order address bits (A_0–A_{18}) access each of the RAM 2 addresses.

Related Problem What are the ranges of addresses in RAM 1 and in RAM 2 in Figure 12–45?

Memory components are extremely sensitive to static electricity. Use the following precautions when handling memory chips or modules such as SIMMs and DIMMs:

❏ Before handling, discharge your body's static charge by touching a grounded surface or wear a grounding wrist strap containing a high-value resistor if available. A convenient, reliable ground is the ac outlet ground.
❏ Do not remove components from their antistatic bags until you are ready to install them.
❏ Do not lay parts on the antistatic bags because only the inside is antistatic.
❏ When handling SIMMs or DIMMs, hold by the edges or the metal mounting bracket. Do not touch components on the boards or the edge connector pins.
❏ Never slide any part over any type of surface.
❏ Avoid plastic, vinyl, styrofoam, and nylon in the work area.

SIMMs and DIMMs

Memories are commonly supplied as single in-line memory modules (**SIMMs**) or as dual in-line memory modules (**DIMMs**). SIMMs and the more recent DIMMs are small circuit boards on which memory chips (ICs) are mounted with the inputs and outputs connected to an edge connector on the bottom of the board. DIMMs are generally faster, but they can only be installed in newer machines that are designed for them.

Two classifications of SIMMs are the 30-pin and the 72-pin. These are illustrated in Figure 12–46. Although available memory capacities for SIMMs vary anywhere from 256 kbytes to 32 Mbytes, the key difference in the two pin configurations is the size of the data path. Generally, 30-pin SIMMs are designed for 8-bit data buses, and more SIMMs are required for handling more data bits. The 72-pin SIMMs can accommodate a 32-bit data bus, so for 64-bit data buses a pair of SIMMS is required.

FIGURE 12–46
30-pin and 72-pin SIMMs.

DIMMs look similar to SIMMs but provide an increase in memory density with only a relatively slight increase in physical size. The key difference is that DIMMs distribute the input and output pins on both sides of the PC board, whereas SIMMs use only one side. Common DIMM configurations are 72-pin, 100-pin, 144-pin, and 168-pin that accommodate both 32-bit and 64-bit data paths. Generally, DIMM capacities range from 4 Mbytes to 512 Mbytes.

SIMMs and DIMMs plug into sockets on a system board such as those illustrated in Figure 12–47 where several sockets are generally available for memory expansion. The sockets for SIMMs and DIMMs, of course, are different and not interchangeable.

FIGURE 12–47
A SIMM/DIMM inserted into a socket on a system board.

When installing SIMMs or DIMMs, always observe the precautionary rules in the previous HOTip. Then follow the steps listed.

1. Line up the notches on the SIMM or DIMM board with the notches in the memory socket.
2. Push firmly on the module until it is securely seated in the socket.
3. Generally, the latches on both sides of the socket will snap into place when the module is completely inserted. These latches also release the module, so it can be removed from the socket.

SECTION 12–6 REVIEW

1. How many 16k $\times$ 1 RAMs are required to achieve a memory with a word capacity of 16k and a word length of eight bits?
2. To expand the 16k $\times$ 8 memory in question 1 to a 32k $\times$ 8 organization, how many more 16k $\times$ 1 RAMs are required?
3. What does SIMM stand for?
4. What does DIMM stand for?

12–7 ■ SPECIAL TYPES OF MEMORIES

In this section, the first in–first out (FIFO) memory, the last in–first out (LIFO) memory, the memory stack, and the charge-coupled device memory are covered.

After completing this section, you should be able to

- Describe a FIFO memory   ■ Describe a LIFO memory   ■ Discuss memory stacks
- Explain how to use a portion of RAM as a memory stack   ■ Describe a basic CCD memory

First In–First Out (FIFO) Memories

This type of memory is formed by an arrangement of shift registers. The term **FIFO** refers to the basic operation of this type of memory, in which the first data bit written into the memory is the first to be read out.

One important difference between a conventional shift register and a FIFO register is illustrated in Figure 12–48. In a conventional register, a data bit moves through the register only as new data bits are entered; in a FIFO register, a data bit immediately goes through the register to the right-most bit location that is empty.

Conventional shift register							FIFO shift register					
Input	**X**	**X**	**X**	**X**	**Output**		**Input**	—	—	—	—	**Output**
0	0	X	X	X	⟶		0	—	—	—	0	⟶
1	1	0	X	X	⟶		1	—	—	1	0	⟶
1	1	1	0	X	⟶		1	—	1	1	0	⟶
0	0	1	1	0	⟶		0	0	1	1	0	⟶

X = unknown data bits.

In a conventional shift register, data stay to the left until "forced" through by additional data.

— = empty positions.

In a FIFO shift register, data "fall" through (go right).

FIGURE 12–48
Comparison of conventional and FIFO register operation.

Figure 12–49 is a block diagram of a FIFO serial memory. This particular memory has four serial 64-bit data registers and a 64-bit control register (marker register). When data are entered by a shift-in pulse, they move automatically under control of the marker register to the empty location closest to the output. Data cannot advance into occupied positions. However, when a data bit is shifted out by a shift-out pulse, the data bits remaining in the registers automatically move to the next position toward the output. In an asynchronous FIFO, data are shifted out independent of data entry, with the use of two separate clocks.

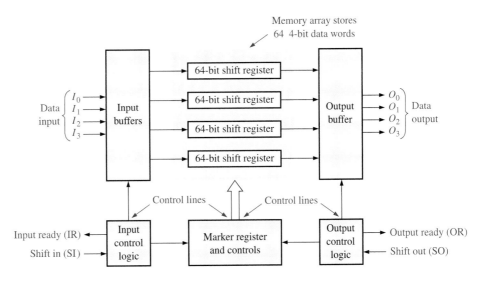

FIGURE 12–49
Block diagram of a typical FIFO serial memory.

FIFO Applications

One important application area for the FIFO register is the case in which two systems of differing data rates must communicate. Data can be entered into a FIFO register at one rate and taken out at another rate. Figure 12–50 illustrates how a FIFO register might be used in these situations.

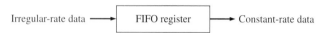

(a) Irregular telemetry data can be stored and retransmitted at a constant rate.

(b) Data input at a slow keyboard rate can be stored and then transferred at a higher rate for processing.

(c) Data input at a constant rate can be stored and then output in bursts.

(d) Data in bursts can be stored and reformatted into a constant-rate output.

FIGURE 12–50
Examples of the FIFO register in data-rate buffering applications.

Last In–First Out (LIFO) Memories

The last in–first out (**LIFO**) memory is found in applications involving microprocessors and other computing systems. It allows data to be stored and then recalled in reverse order; that is, the last data byte to be stored is the first data byte to be retrieved.

Register Stacks   A LIFO memory is commonly referred to as a push-down stack. In some systems, it is implemented with a group of registers as shown in Figure 12–51. A stack can consist of any number of registers, but the register at the top is called the *top-of-stack*.

FIGURE 12–51
Register stack.

To illustrate the principle, a byte of data is loaded in parallel onto the top of the stack. Each successive byte pushes the previous one down into the next register. This process is illustrated in Figure 12–52. Notice that the new data byte is always loaded into the top register and the previously stored bytes are pushed deeper into the stack. The name *push-down stack* comes from this characteristic.

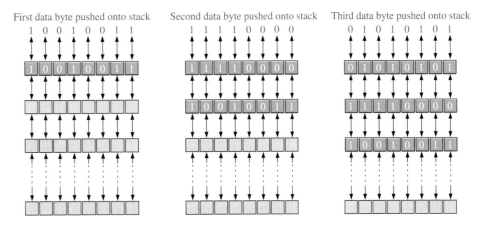

FIGURE 12–52
Simplified illustration of pushing data onto the stack.

Data bytes are retrieved in the reverse order. The last byte entered is always at the top of the stack, so when it is pulled from the stack, the other bytes pop up into the next higher locations. This process is illustrated in Figure 12–53.

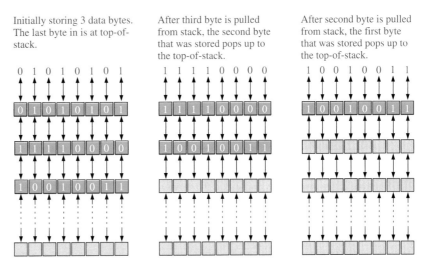

FIGURE 12–53
Simplified illustration of pulling data from the stack.

RAM Stack   Another approach to LIFO memory used in microprocessor-based systems is the allocation of a section of RAM as the stack rather than the use of a dedicated set of registers. As you have seen, for a register stack the data moves up or down from one location to the next. In a RAM stack, the data itself does not move but the top-of-stack moves under control of a register called the stack pointer.

Consider a random-access memory that is byte organized—that is, one in which each address contains eight bits—as illustrated in Figure 12–54. The binary address

FIGURE 12–54
Representation of a 64 kbyte memory with the
16-bit addresses expressed in hexadecimal.

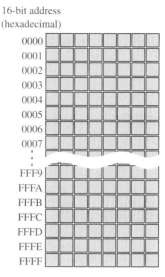

16-bit address
(hexadecimal)

0000000000001111, for example, can be written as 000F in hexadecimal. A 16-bit address can have a *minimum* hexadecimal value of 0000_{16} and a *maximum* value of $FFFF_{16}$. With this notation, a 64 kbyte memory array can be represented as shown in Figure 12–54. The lowest memory address is 0000_{16} and the highest address is $FFFF_{16}$.

Now, consider a section of RAM set aside for use as a stack. A special separate register, the stack pointer, contains the address of the top of the stack, as illustrated in Figure 12–55. A 4-digit hexadecimal representation is used for the binary addresses. In the figure, the addresses are chosen for purposes of illustration.

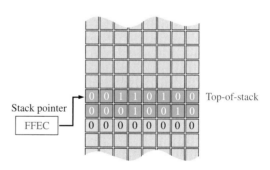

(a) The stack pointer is initially at FFEE before the data word 0001001000110100 (1234) is pushed onto the stack.

(b) The stack pointer is decremented by two and the data word 0001001000110100 is placed in the two locations prior to the original stack pointer location.

FIGURE 12–55
Illustration of the PUSH operation for a RAM stack.

Now let's see how data are pushed onto the stack. The stack pointer is initially at address $FFEE_{16}$, which is the top of the stack as shown in Figure 12–55(a). The stack pointer is then decremented (decreased) by two to $FFEC_{16}$. This moves the top of the stack to a lower memory address, as shown in Figure 12–55(b). Notice that the top of the stack is not stationary as in the fixed register stack but moves downward (to lower addresses) in the RAM as data words are stored. Figure 12–55(b) shows that two bytes (one data word) is then pushed onto the stack. After the data word is stored, the top of the stack is at $FFEC_{16}$.

Figure 12–56 illustrates the POP operation for the RAM stack. The last data word stored in the stack is read first. The stack pointer which is at FFEC is incremented (increased) by two to address $FFEE_{16}$ and a POP operation is performed as shown in part (b).

(a) The stack pointer is at FFEC before the data word is copied (popped) from the stack.

(b) The stack pointer is incremented by two and the last data word stored is copied (popped) from the stack.

FIGURE 12–56
Illustration of the POP operation for the RAM stack.

Keep in mind that RAMs are nondestructive when read, so the data word still remains in the memory after a POP operation. A data word is destroyed only when a new word is written over it.

A Ram stack can be of any depth, depending on the number of continuous memory addresses assigned for that purpose.

CCD Memories

The **CCD** (charge-coupled device) memory stores data as charges on capacitors. Unlike the dynamic RAM, however, the storage cell does not include a transistor. High density is the main advantage of CCDs.

The CCD memory consists of long rows of semiconductor capacitors, called *channels*. Data are entered into a channel serially by depositing a small charge for a 0 and a large charge for a 1 on the capacitors. These charge packets are then shifted along the channel by clock signals as more data are entered.

As with the DRAM, the charges must be refreshed periodically. This process is done by shifting the charge packets serially through a refresh circuit. Figure 12–57 shows the basic concept of a CCD channel. Because data are shifted serially through the channels, the CCD memory has a relatively long access time. CCD arrays are used in some modern cameras to capture video images in the form of light-induced charge.

FIGURE 12–57
A CCD (charge-coupled device) channel.

SECTION 12–7 REVIEW

1. What is a FIFO memory?
2. What is a LIFO memory?
3. Explain the PUSH operation in a memory stack.
4. Explain the POP operation in a memory stack.
5. What does the term *CCD* stand for.

12–8 ■ MAGNETIC AND OPTICAL STORAGE

In this section, you will be introduced to the basics of magnetic disks, magnetic tape, magneto-optical disks, and optical disks. These storage media are very important, particularly in computer applications, where they are used for mass nonvolatile storage of data and programs.

After completing this section, you should be able to

■ Describe a magnetic hard disk ■ Describe a floppy disk ■ Discuss removable hard disks ■ Explain the principle of magneto-optical disks ■ Discuss the CD-ROM, CD-R, and CD-RW disks ■ Describe the WORM ■ Discuss the DVD-ROM

Magnetic Storage

Magnetic Hard Disks Computers use hard disks as the internal mass storage media. **Hard disks** are rigid "platters" made of aluminum alloy or a mixture of glass and ceramic covered with a magnetic coating. Hard disk drives mainly come in two diameter sizes, 5.25 in. and 3.5 in. although 2.5 in. and 1.75 in. are also available. A hard disk drive is hermetically sealed to keep the disks dust-free.

Typically, two or more platters are stacked on top of each other on a common shaft or spindle that turns the assembly at several thousand rpm. There is a separation between each disk to allow for a magnetic read/write head that is mounted on the end of an actuator arm, as shown in Figure 12–58. There is a read/write head for both sides of each disk since data are recorded on both sides of the disk surface. The drive actuator arm synchronizes all the read/write heads to keep them in perfect alignment as they "fly" across the disk surface with a separation of only a fraction of a millimeter from the disk. A small dust particle could cause a head to "crash," causing damage to the disk surface.

FIGURE 12–58
Simplified drawing of a hard disk drive.

Basic Read/Write Head Principles The hard drive is a random-access device because it can retrieve stored data anywhere on the disk in any order. A simplified diagram of the magnetic surface read/write operation is shown in Figure 12–59. The direction or polarization of the magnetic domains on the disk surface is controlled by the direction of the magnetic flux lines (magnetic field) produced by the write head according to the direction of a current pulse in the winding. This magnetic flux magnetizes a small spot on the disk surface in the direction of the magnetic field. A magnetized spot of one polarity represents a

FIGURE 12–59
Simplified read/write head operation.

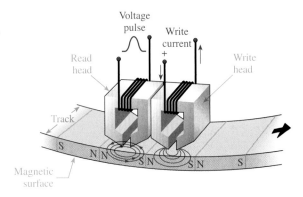

binary 1, and one of the opposite polarity represents a binary 0. Once a spot on the disk surface is magnetized, it remains until written over with an opposite magnetic field.

When the magnetic surface passes a read head, the magnetized spots produce magnetic fields in the read head, which induce voltage pulses in the winding. The polarity of these pulses depends on the direction of the magnetized spot and indicates whether the stored bit is a 1 or a 0. The read and write heads are usually combined in a single unit.

Hard Disk Format A hard disk is organized or formatted into tracks and sectors, as shown in Figure 12–60(a). Each track is divided into a number of sectors, and each track and sector has a physical address that is used by the operating system to locate a particular data record. Hard disks typically have from a few hundred to a few thousand tracks. As you can see in the figure, there is a constant number of tracks/sector, with outer sectors using more surface area than the inner sectors. The arrangement of tracks and sectors on a disk is known as the *format*.

A hard disk stack is illustrated in Figure 12–60(b). Hard disk drives differ in the number of platters in a stack, but there is always a minimum of two. All of the same corresponding tracks on each platter are collectively known as a cylinder, as indicated.

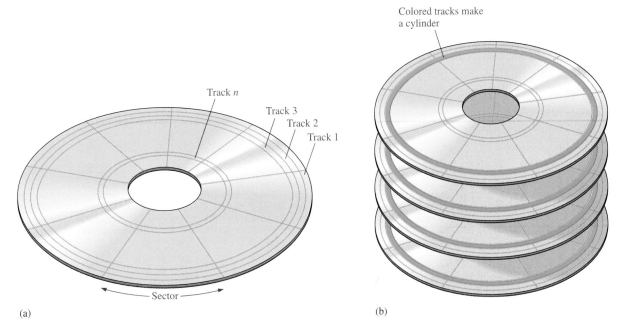

FIGURE 12–60
Hard disk organization and formatting.

COMPUTER NOTE

Data are stored on a hard drive in the form of files. Keeping track of the location of files is the job of the device driver that manages the hard drive (sometimes referred to as hard drive BIOS). The device driver and the computer's operating system can access two tables to keep track of files and file names. The first table is called the FAT (File Allocation Table). The FAT shows what is assigned to specific files and keeps a record of open sectors and bad sectors. The second table is the Root Directory which has file names, type of file, time and date of creation, starting cluster number, and other information about the file.

Hard Disk Performance   Several basic parameters determine the performance of a given hard disk drive. A *seek* operation is the movement of the read/write head to the desired track. The **seek time** is the average time for this operation to be performed. Typically, hard disk drives have an average seek time of from about 7 ms to about 14 ms, depending on the particular drive.

The **latency period** is the time it takes for the desired sector to spin under the head once the head is positioned over the desired track. A worst case is when the desired sector is just past the head position and spinning away from it. The sector must rotate almost a full revolution back to the head position. *Average latency period* assumes that the disk must make half of a revolution. Obviously, the latency period depends on the constant rotational speed of the disk. Disk rotation speeds are different for different disk drives but typically are 3600 rpm, 4500 rpm, 5400 rpm, and 7200 rpm. Some recent disk drives rotate at 10,033 rpm and have an average latency period of less than 3 ms.

The sum of the average seek time and the average latency period is the *access time* for the disk drive.

Floppy Disks   The floppy disk derives its name because it is made of a flexible polyester material with a magnetic coating on both sides. The early floppy disks were 5.25 inches in diameter and were packaged in a semiflexible jacket. Current **floppy disks** or diskettes are 3.5 inches in diameter and are encased in a rigid plastic jacket, as shown in Figure 12–61. A spring-loaded shutter covers the access window and remains closed until the disk is inserted into a disk drive. A metal hub has one hole to center the disk and another for spinning it within the protective jacket. Obviously, floppy disks are removable disks, whereas hard disks are not. Floppy disks are formatted into tracks and sectors similar to hard disks except for the number of tracks and sectors. The high-density 1.44 Mbyte floppies have 80 tracks per side with 18 sectors. With the advent of other types of removable disks such as Zip, the floppy has serious competition, but its low cost may continue to make the floppy disk competitive for smaller storage applications.

FIGURE 12–61
The 3.5 inch floppy disk (diskette).

Zip©™   The Zip drive is one type of removable magnetic storage device that is the apparent replacement for the limited-capacity floppy. Like the floppy disk, the **Zip disk** cartridge is a flexible disk housed in a rigid case about the same size as that of the floppy disk but thicker. The Zip drive is much faster than the floppy drive because it has a 3000 rpm spin rate compared to the floppy's 300 rpm. The Zip drive has a storage capacity of 100 Mbytes, about 69 times more than the 1.44 Mbyte floppy. A typical external Zip drive and cartridge are shown in Figure 12–62.

Jaz™   Another type of removable magnetic storage device is the Jaz drive, which is similar to a hard disk drive except that two platters are housed in a removable cartridge protected by a dust-proof shutter. The **Jaz cartridges** are available with storage capacities of 1 or 2 Gbytes. A Jaz drive and cartridge are shown in Figure 12–63.

FIGURE 12–62
A Zip™ drive and cartridges (courtesy of Iomega Corporation).

FIGURE 12–63
A Jaz™ drive and disk cartridges (courtesy of Iomega Corporation).

Other Removable Drive Technologies

In addition to the popular Zip and Jaz removable drives, there are a few others worthy of mention. Keep in mind that the technology is changing so rapidly that there most likely will be further advancements at the time you are reading this.

The SuperDisk drive uses a 120 Mbyte diskette that can also accept the 3.5 in., 1.44 Mbyte floppy disks. Also the SparQ drive features 1 Gbyte cartridges and the SyJet drive accepts 1.5 Gbyte cartridges. One difference between the SparQ drive and the 1 Gbyte Jaz drive is that the SparQ is an external drive only and interfaces with the computer via a parallel port, while the Jaz drive is available in both internal and external versions.

Magnetic Tape　Tape is used for backup data from mass storage devices and typically is slower than disks because data on tape is accessed serially rather than randomly. There are several types that are available, including QIC, DAT, 8 mm, and DLT.

QIC is an abbreviation for quarter-inch cartridge and looks much like audio tape cassettes with two reels inside. Various QIC standards have from 36 to 72 tracks that can store from 80 Mbytes to 1.2 Gbytes. More recent renovations under the Travan standard have lengthened the tape and increased its width allowing storage capacities up to 4 Gbytes. QIC tape drives use read/write heads that have a single write head with a read head on each side. This allows the tape drive to verify data just written when the tape is running in either direction. In the record mode the tape moves past the read/write heads at approximately 100 inches/second, as indicated in Figure 12–64.

DAT, which is an abbreviation for digital audio tape, uses a technique called helical scan recording. DATs offer storage capacities ranging up to 12 Gbyte but is more expensive than QIC.

A third type of tape format, the 8 mm tape, was originally designed for the video industry but has been adopted by the computer industry as a reliable way to store large amounts of computer data. 8 mm is similar to DAT but offers storage capacities up to 25 Gbytes.

FIGURE 12–64
QIC tape.

DLT is an abbreviation for digital linear tape. DLT is a half-inch wide tape, which is 60% wider than 8 mm and, of course, twice as wide as standard QIC. Basically, DLT differs in the way the tape-drive mechanism works to minimize tape wear compared to other systems. DLT offers the highest storage capacity of all the tape formats with capacities ranging up to 35 Gbytes. A recent advance in DLT technology promises to up the capacity to 500 Gbytes.

Magneto-Optical Storage

As the name implies, magneto-optical (MO) storage devices use a combination of magnetic and optical (laser) technologies. A **magneto-optical disk** is formatted into tracks and sectors similar to magnetic disks.

The basic difference between a purely magnetic disk and an MO disk is that the magnetic coating used on the MO disk requires heat to alter the magnetic polarization. Therefore, the MO is extremely stable at ambient temperature making data unchangeable. To write a data bit, a high-power laser beam is focused on a tiny spot on the disk, and the temperature of that tiny spot is raised above a temperature level called the Curie point (about 200° C). Once heated, the magnetic particles at that spot can easily have their direction (polarization) changed by a magnetic field generated by the write head. Information is read from the disk with a less-powerful laser than used for writing, making use of the Kerr effect where the polarity of the reflected laser light is altered depending on the orientation of the magnetic particles. Spots of one polarity represents 0s and spots of the opposite polarity represent 1s. Basic MO operation is shown in Figure 12–65, which represents a small cross-sectional area of a disk.

Optical Storage

CD-ROM   The basic Compact Disk–Read-Only Memory is a 120 mm diameter disk with a sandwich of three coatings: a polycarbonate plastic on the bottom, a thin aluminum sheet for reflectivity, and a top coating of lacquer for protection. The **CD-ROM** disk is formatted in a single spiral track with sequential 2 kbyte sectors and has a capacity of 680 Mbytes. Data is prerecorded at the factory in the form of minute indentations called *pits* and the flat area surrounding the pits called *lands*. The pits are stamped into the plastic layer and cannot be erased.

A CD player reads data from the spiral track with a low-power infrared laser, as illustrated in Figure 12–66. The data is in the form of pits and lands as shown. Laser light reflected from a pit is 180° out-of-phase with the light reflected from the lands. As the disk rotates, the narrow laser beam strikes the series of pits and lands of varying lengths, and a photodiode detects the difference in the reflected light. The result is a series of 1s and 0s corresponding to the configuration of pits and lands along the track.

(a) Unrecorded disk

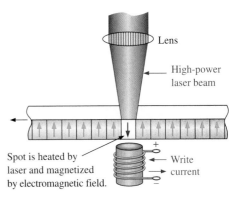

(b) Writing: A high-power laser beam heats the spot, causing the magnetic particles to align with the electromagnetic field.

(c) Reading: A low-power laser beam reflects off of the reversed-polarity magnetic particles and its polarization shifts. If the particles are not reversed, the polarization of the reflected beam is unchanged.

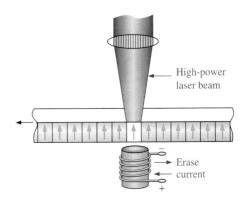

(d) Erasing: The electromagnetic field is reversed as the high-power laser beam heats the spot, causing the magnetic particles to be restored to the original polarity.

FIGURE 12–65
Basic principle of a magneto-optical disk.

FIGURE 12–66
Basic operation of reading data from a CD-ROM.

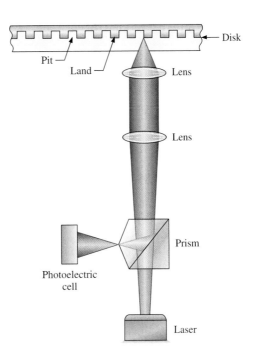

WORM Write Once/Read Many (**WORM**) is a type of optical storage that can be written onto one time after which the data cannot be erased but can be read many times. To write data, a low-power laser is used to burn microscopic pits on the disk surface. 1s and 0s are represented by the burned and nonburned areas.

CD-R This is essentially a type of WORM. The difference is that the CD-Recordable allows multiple write sessions to different areas of the disk. The **CD-R** disk has a spiral track like the CD-ROM, but instead of mechanically pressing indentations on the disk to represent data, the CD-R uses a laser to burn microscopic spots into an organic dye surface. When heated beyond a critical temperature with a laser during read, the burned spots change color and reflect less light than the nonburned areas. Therefore, 1s and 0s are represented on a CD-R by burned and nonburned areas, whereas on a CD-ROM they are represented by pits and lands. Like the CD-ROM, the data cannot be erased once it is written.

CD-RW The CD-Rewritable disk can be used to read and write data. Instead of the dye-based recording layer in the CD-R, the **CD-RW** commonly uses a crystalline compound with a special property. When it is heated to a certain temperature, it becomes crystalline when it cools; but if it is heated to a certain higher temperature, it melts and becomes amorphous when it cools. To write data, the focused laser beam heats the material to the melting temperature resulting in an amorphous state. The resulting amorphous areas reflect less light than the crystalline areas, allowing the read operation to detect 1s and 0s. The data can be erased or overwritten by heating the amorphous areas to a temperature above the crystallization temperature but lower than the melting temperature that causes the amorphous material to revert to a crystalline state.

DVD-ROM Originally DVD was an abbreviation for Digital Video Disk but eventually came to represent *Digital Versatile Disk*. Like the CD-ROM, **DVD-ROM** data is prestored on the disk. However, the pit size is smaller than for the CD-ROM, allowing more data to be stored on a track. The major difference between CD-ROM and DVD-ROM is that the CD is single-sided, while the DVD has data on both sides. Also, in addition to double-sided DVD disks, there are also multiple-layer disks that use semitransparent data layers placed over the main data layers, providing storage capacities up to 17 Gbytes. To access all the layers, the laser beam requires refocusing going from one layer to the other.

SECTION 12–8 REVIEW

1. List the major types of magnetic storage.
2. What is the storage capacity of floppy disks currently in use?
3. Generally, how is a magnetic disk organized?
4. How are data written on and read from a magneto-optical disk?
5. What are the storage capacities of Zip and Jaz disks?
6. List the types of optical storage.

12–9 ■ TESTING AND TROUBLESHOOTING

Because memories can contain large numbers of storage cells, testing each cell can be a very lengthy and frustrating process. Fortunately, memory testing is usually an automated process performed with a programmable test instrument or with the aid of software for in-system testing. Most microprocessor-based systems provide automatic memory testing as part of their system software.

After completing this section, you should be able to

■ Discuss the checksum method of testing ROMs   ■ Discuss the checkerboard pattern method of testing RAMs

ROM Testing

Since ROMs contain known data, they can be checked for the correctness of the stored data by reading each data word from the memory and comparing it with a data word that is known to be correct. One way of doing this is illustrated in Figure 12–67. This process requires a reference ROM that contains the same data as the ROM to be tested. A special test instrument is programmed to read each address in both ROMs simultaneously and to compare the contents. A flowchart in Figure 12–68 illustrates the basic sequence.

FIGURE 12–67
Block diagram for a complete contents check of a ROM.

FIGURE 12–68
Flowchart for a complete contents check of a ROM.

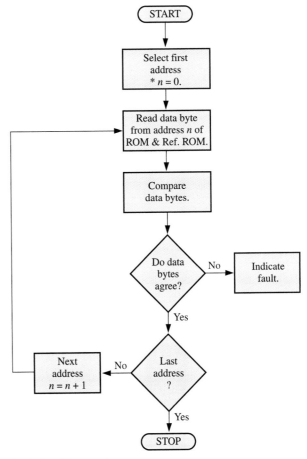

* n is the address number.

Checksum Method Although the previous method checks each ROM address for correct data, it has the disadvantage of requiring a reference ROM for each different ROM to be tested. Also, a failure in the reference ROM can produce a fault indication.

In the checksum method a number, the sum of the contents of all the ROM addresses, is stored in a designated ROM address when the ROM is programmed. To test the ROM, the contents of all the addresses except the checksum are added, and the result is compared with the checksum stored in the ROM. If there is a difference, there is definitely a fault. If the checksums compare, the ROM is most likely good. However, there is a remote possibility that a combination of bad memory cells could cause the checksums to compare.

This process is illustrated in Figure 12–69 with a simple example. The checksum in this case is produced by taking the sum of each column of data bits and discarding the carries. This is actually an XOR sum of each column. The flowchart in Figure 12–70 illustrates the basic checksum test.

The checksum test can be implemented with a special test instrument, or it can be incorporated as a test routine in the built-in (system) software or microprocessor-based systems. In that case, the ROM test routine is automatically run on system start-up.

FIGURE 12–69
Simplified illustration of a programmed ROM with the checksum stored at a designated address.

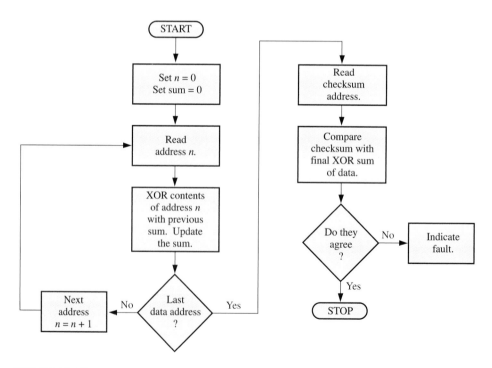

FIGURE 12–70
Flowchart for a basic checksum test.

RAM Testing

To test a RAM for its ability to store both 0s and 1s in each cell, first 0s are written into all the cells in each address and then read out and checked. Next, 1s are written into all the cells in each address and then read out and checked. This basic test will detect a cell that is stuck in either a 1 state or a 0 state.

Some memory faults cannot be detected with the all-0s–all-1s test. For example, if two adjacent memory cells are shorted, they will always be in the same state, both 0s or both 1s. Also, the all-0s–all-1s test is ineffective if there are internal noise problems such that the contents of one or more addresses are altered by a change in the contents of another address.

The Checkerboard Pattern Test One way to more fully test a RAM is by using a checkerboard pattern of 1s and 0s, as illustrated in Figure 12–71. Notice that all adjacent cells have opposite bits. This pattern checks for a short between two adjacent cells; if there is a short, both cells will be in the same state.

FIGURE 12–71
The RAM checkerboard test pattern.

(a)

(b)

After the RAM is checked with the pattern in Figure 12–71(a), the pattern is reversed, as shown in part (b). This reversal checks the ability of all cells to store both 1s and 0s.

A further test is to alternate the pattern one address at a time and check all the other addresses for the proper pattern. This test will catch a problem in which the contents of an address are dynamically altered when the contents of another address change.

A basic procedure for the checkerboard test is illustrated by the flowchart in Figure 12–72. The procedure can be implemented with the system software in microprocessor-based systems so that either the tests are automatic when the system is powered up or they can be initiated from the keyboard.

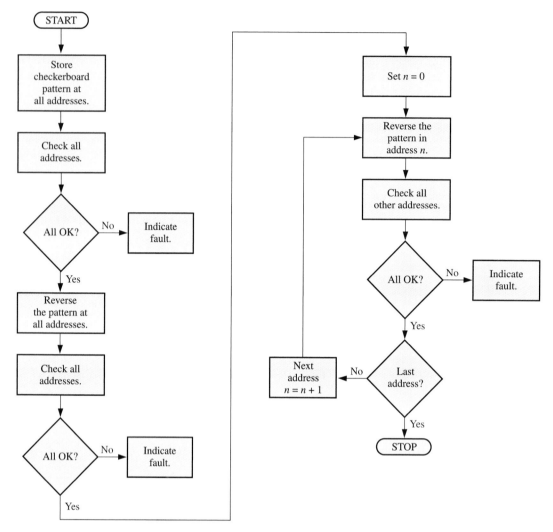

FIGURE 12–72
Flowchart for basic RAM checkerboard test.

SECTION 12–9 REVIEW

1. Describe the checksum method of ROM testing.
2. Why can the checksum method not be applied to RAM testing?
3. List the three basic faults that the checkerboard pattern test can detect in a RAM.

12–10 ■ DIGITAL SYSTEM APPLICATION

In this system application, you will complete the security entry system. In Chapter 10, the security code logic was developed. In this section, the memory logic will be developed and the two parts of the system interfaced.

After completing this section, you should be able to

■ Implement the memory portion of the system and develop a logic diagram
■ Interface the memory logic with the security code logic and with the external switches ■ Explain the overall operation of the system ■ Develop a test procedure for the memory board ■ Develop a test procedure for the complete system

General System Operation

A basic block diagram of the complete security entry system is shown in Figure 12–73. The security code logic was developed in Chapter 10 and you may want to review that material before proceeding.

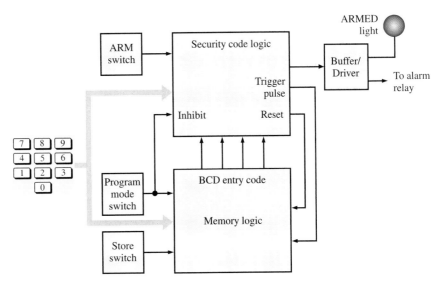

FIGURE 12–73
Basic block diagram of the security entry system.

The memory logic uses a SRAM to store the 4-digit entry code in BCD format. In the switch-selected programming mode, four digits are entered into the memory from the keypad. Once stored in the memory, the four BCD digits become the permanent entry code. If it is necessary to change the entry code, the memory is reprogrammed with a new one. During the programming mode, the security code logic is inhibited.

The memory is programmed by first putting the system into the program mode with the program mode switch and using the store switch and keypad to enter the desired 4-digit code in the memory.

Once the memory is programmed with the desired code, the program mode switch is turned off. The system is then armed by pressing the arm switch while selecting any digit on the keypad.

Basic Operation of the Memory Logic

A simplified diagram of the memory logic is shown in Figure 12–74. To initially store a code in the SRAM, the program mode switch is closed, triggering one-shot B, and it remains closed during programming. The pulse output of the one-shot goes through OR gate G_2 to reset (clear) the counter. In the RESET state, the counter contains all zeros (000) which selects address 0 in the SRAM. Address 0 is not used for storing a code.

To enter the first security code digit, the desired key on the keypad is pressed and held for a few seconds while the store switch is closed momentarily, triggering one-shot A. The pulse output of the one-shot goes through OR gate G_1, to clock the counter to the 001 state, thus selecting address 1 in the SRAM. At the end of the pulse from one-shot A, one-shot C is triggered and generates a write pulse to the memory, which stores the selected digit code in address 1.

FIGURE 12–74
Simplified basic logic diagram of the memory logic.

To enter the second digit, the desired key is pressed and held, again triggering one-shot A and advancing the counter to the 010 state, which selects memory address 2. One-shot C is then triggered and generates a write pulse, which stores the digit code in memory address 2.

Similar steps are repeated to store the third digit in address 3 and the fourth digit in address 4 in the SRAM. Once the four digits are stored, the program mode switch is opened. The SRAM stores the code as long as there is power. The system is assumed to have a battery backup in the event of power failure.

■ APPLICATION ASSIGNMENTS

The basic logic diagram in Figure 12–74 indicates the general logic required for the system but does not represent the detailed logic necessary to implement the system. In the following application assignments, you will analyze the basic operation of the memory logic to make sure that you understand the functional requirements, complete the design of the logic by selecting appropriate devices, and then develop a detailed logic diagram incorporating the selected devices. You will compare your logic diagram to an implemented circuit board and then develop a test procedure for the memory logic board. Finally, you will interface the security code logic board, memory logic board, and the front panel components (switches, keypad, and ARMED light with its buffer/driver board) to form the complete system.

On the circuit board, interconnections on the back side of the board feed through to the pads on the component side. When you trace out the circuit in Application Assignment 2, common pads that are connected on the back side are generally oriented vertically or horizontally with respect to each other and not at an angle.

■ APPLICATION ASSIGNMENT 1: Design

The focus of this assignment is the design of the memory logic. For information on devices, reference to earlier chapters and/or device data sheets is necessary.

■ *Activity 1:   Decimal-to-BCD Encoder*
The particular keypad used in this system produces active-LOW outputs. When a key is pressed, the corresponding line is LOW and the other lines are HIGH. Specify an IC encoder to properly interface with the keypad.

■ *Activity 2:   The SRAM*
The requirement for the memory to store 4-bit BCD codes establishes the word length. The requirement for the memory to store four BCD codes in addresses 1 through 4 and the requirement for address 0 to be reserved for the reset or armed conditions establishes the minimum word capacity. A SRAM device of minimum capacity that meets the storage requirements in terms of its organization is the 74F189. A partial data sheet for this device is given in Figure 12–75.

■ *Activity 3:   Counter*
The counter must be capable of producing a 3-bit binary sequence for addressing the memory, and it must have a reset or clear input. Select an appropriate IC counter.

■ *Activity 4:   One-shots*
One-shot A is to produce an output pulse with a width of approximately 10 ms. One-shots B and C are to each produce an output pulse with a width of approximately 1 ms. Select an appropriate type of IC one-shot and determine the values of the resistor and capacitor for each one.

■ *Activity 5:   Gating Logic*
Based on your selection of the IC for the counter and the inputs from the security code logic, specify the types of logic gates (G_1 and G_2) required to provide the proper active levels.

■ *Activity 6:*   Develop a detailed logic diagram using the devices specified in the preceding activities. Show all necessary details including device pin numbers.

■ *Activity 7:*   Write a detailed technical report describing the operation of the memory logic and specifying a parts list.

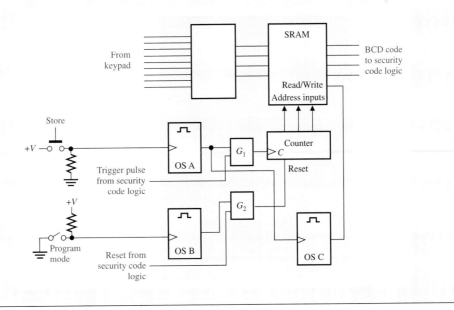

54F/74F189 64-Bit Random Access Memory with Tri-State Outputs

General Description

The 'F189 is a high-speed 64-bit RAM organized as a 16-word by 4-bit array. Address inputs are buffered to minimize loading and are fully decoded on-chip. The outputs are TRI-STATE and are in the high impedance state whenever the Chip Select ($\overline{CS}$) input is HIGH. The outputs are active only in the Read mode and the output data is the complement of the stored data.

Features

- TRI-STATE outputs for data bus applications
- Buffered inputs minimize loading
- Address decoding on-chip
- Diode clamped inputs minimize ringing
- Available in SOIC, (300 mil only)

Commercial	Military	Package Number	Package Description
74F189PC		N16E	16-Lead (0.300" Wide) Molded Dual-In-Line
	54F189DL (Note 2)	J16A	16-Lead Ceramic Dual-In-Line
74F189SC (Note 1)		M16A	16-Lead (0.150" Wide) Molded Small Outline, JEDEC
74F189SJ (Note 1)		M16D	16-Lead (0.300" Wide) Molded Small Outline, EIAJ
	54F189FL (Note 2)	W16A	16-Lead Cerpack
	54F189LL (Note 2)	E20A	20-Lead Ceramic Leadless Chip Carrier, Type C

Note 1: Devices also available in 13" reel. Use suffix = SCX and SJX.
Note 2: Military grade device with environmental and burn-in processing. Use suffix = DLQB, FLQB and LLQB.

Logic Symbols

Connection Diagrams

Pin Assignment for DIP, SOIC and Flatpak

Pin Assignment for LCC

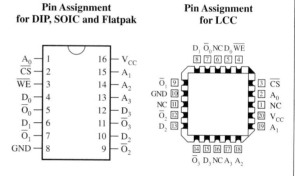

Unit Loading/Fan Out

Pin Names	Description	54F/74F U.L. HIGH/LOW	54F/74F Input I_{IH}/I_{IL} Output I_{OH}/I_{OL}
A_0–A_3	Address inputs	1.0/1.0	20 μA/ – 0.6 mA
$\overline{CS}$	Chip select input (active LOW)	1.0/1.0	20 μA/ – 1.2 mA
$\overline{WE}$	Write enable input (active LOW)	1.0/1.0	20 μA/ – 0.6 mA
D_0–D_3	Data inputs	1.0/1.0	20 μA/ – 0.6 mA
$\overline{O}_0$–$\overline{O}_3$	Inverted data outputs	150/40 (33.3)	–3.0 mA/24 mA (20 mA)

Function Table

Inputs		Operation	Condition of Outputs
$\overline{CS}$	$\overline{WE}$		
L	L	Write	High impedance
L	H	Read	Complement of stored data
H	X	Inhibit	High impedance

H = HIGH voltage level
L = LOW voltage level
X = Immaterial

Block Diagram

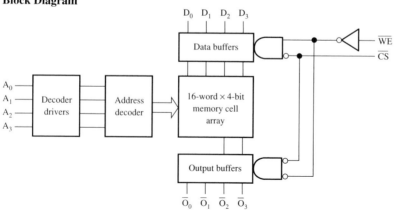

FIGURE 12–75
Partial data sheet for a 74F189 16 × 4 SRAM.

■ APPLICATION ASSIGNMENT 2: Verification and Testing

■ *Activity 1:* Draw the logic diagram for the memory logic board below and compare to your logic diagram from Activity 6 of Application Assignment 1. Discuss any differences between your design and the design of this logic board and verify that both work.

■ *Activity 2:* Identify the outputs and inputs on the circuit board connector.

■ *Activity 3:* Develop a test procedure for checking the board shown.

■ APPLICATION ASSIGNMENT 3:   The Complete System

The components and logic boards for the security entry system are shown. When interconnected, these elements form the complete system for activating and deactivating the alarm system.

■ *Activity 1:*   Create a point-to-point list for properly interconnecting all of the components and boards.

■ *Activity 2:*   In Chapter 10 and in Application Assignment 2 of this chapter, you developed test procedures for checking each board individually. Now, develop a test procedure for the complete system that includes all of the parts shown below.

■ *Activity 3:*   Write a detailed technical report describing the overall system operation. Provide detailed step-by-step instructions for operating the system.

■ *Activity 4:*   List possible reasons for the following system malfunctions:
  (a) ARMED light will not come on when the ARM switch is pressed.
  (b) ARMED light will not turn off when proper code is entered.
  (c) ARMED light turns off after the first code digit is entered.

Toggle switch   Push-button switch A   Push-button switch B   Buffer/Driver board for ARMED LED and alarm relay   LED   Keypad

SECTION 12–10 REVIEW

1. Why does the system have to be reprogrammed with the entry code if there is a power failure and there is no battery backup?
2. How many addresses in the SRAM on the memory logic board are unused?
3. How does the program mode switch inhibit the security code logic during the programming mode?

■ SUMMARY

■ Types of semiconductor memories:

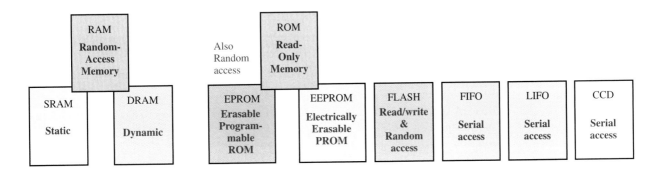

■ Types of SRAMs (Static RAMs) and DRAMs (Dynamic RAMs):

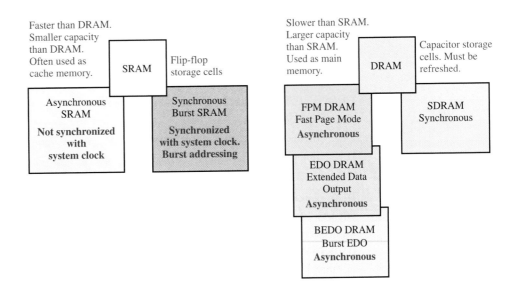

■ Types of magnetic storage:

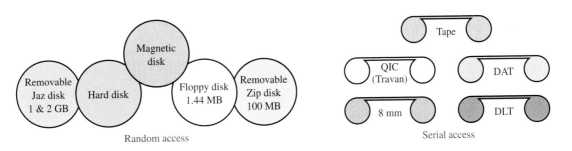

■ Types of optical (laser) storage:

Magneto-Optical Disk

Cross between magnetic and optical

CD-ROM

Prerecorded at factory

CD-R

Recordable

CD-RW

Rewritable

WORM

Write once read many

DVD-ROM

Digital versatile disk

■ KEY TERMS

These terms are also in the end-of-book glossary.

Access time The time from the application of a valid memory address to the appearance of valid output data.

Address The location of a given storage cell or group of cells in a memory.

BEDO DRAM Burst extended data output dynamic random-access memory.

Bus A set of interconnections that interface one or more devices based on a standardized specification.

Byte A group of eight bits.

Cache memory A relatively small, high-speed memory that stores the most recently used instructions or data from the larger but slower main memory.

Capacity The total number of data units (bits, nibbles, bytes, words) that a memory can store.

CCD Charge-coupled device; a type of semiconductor memory that stores data in the form of charge packets and is serially accessed.

CD-R CD-Recordable; an optical disk storage device on which data can be stored once.

CD-ROM An optical disk storage device on which data is prestored.

CD-RW CD-Rewritable; an optical disk storage on which data can be written and over-written many times.

Cell A single storage element in a memory.

DAT Digital audio tape; a type of magnetic tape format.

DIMM Dual in-line memory module.

DLT Digital linear tape; a type of magnetic tape format.

DRAM Dynamic random-access memory; a type of semiconductor memory that uses capacitors as the storage elements and is a volatile, read/write memory.

DVD-ROM Digital versatile disk-ROM; also known as digital video disk-ROM; a type of optical storage device on which data is prestored with a much higher capacity than a CD-ROM.

Dynamic memory A type of semiconductor memory having capacitive storage cells that lose stored data over a period of time and, therefore, must be refreshed.

EDO DRAM Extended data output dynamic random-access memory.

EEPROM Electrically erasable programmable read-only memory; a type of semiconductor memory device.

EPROM Erasable programmable read-only memory; a type of semiconductor memory device that typically uses ultraviolet light to erase data.

FET Field-effect transistor.

FIFO First in–first out memory.

Flash memory A nonvolatile read/write random-access semiconductor memory in which data is stored as charge on the floating gate of a certain type of FET.

Floppy disk A magnetic storage device; a flexible disk with a diameter of 3.5 inches and a storage capacity of 1.44 Mbytes encased in a rigid plastic housing.

FPM DRAM Fast page mode dynamic random-access memory.

Hard disk A magnetic storage device; typically, a stack of two or more rigid disks enclosed in a sealed housing.

Jaz cartridge A magnetic storage device; hard disks encased in a rigid plastic cartridge with storage capacities of 1 Gbyte or 2 Gbytes.

Latency period The time it takes for the desired sector to spin under the head once the head is positioned over the desired track of a magnetic hard disk.

LIFO Last in–first out memory; a memory stack.

Magneto-Optical disk A storage device that uses both electromagnetism and a laser beam to read and write data.

Memory array An array of memory cells arranged in rows and columns.

MOSFET Metal-oxide semiconductor field-effect transistor.

Nibble A group of four bits.

Nonvolatile A term that describes a memory that can retain stored data when the power is removed.

Pipeline As applied to memories, an implementation that allows a read or write operation to be initiated before the previous operation is completed.

PROM Programmable read-only memory; a type of semiconductor memory.

QIC Quarter-inch cassette; a type of magnetic tape.

RAM Random-access memory; a volatile read/write semiconductor memory.

Read The process of retrieving data from a memory.

Refresh To renew the contents of a dynamic memory by recharging the capacitor storage cells.

ROM Read-only memory; a nonvolatile random-access semiconductor memory.

Seek time The time for the read/write head in a hard drive to position itself over the desired track for a read operation.

SDRAM Synchronous dynamic random-access memory.

SIMM Single in-line memory module.

SRAM Static random-access memory; a type of volatile read/write semiconductor memory.

Static memory A volatile semiconductor memory that uses flip-flops as the storage cells and is capable of retaining data without refreshing.

UV PROM Ultraviolet erasable programmable ROM.

Volatile A term that describes a memory that loses its stored data when the power is removed.

Word A complete unit of binary data.

Word capacity The number of words that a memory can store.

Word length The number of bits in a word.

WORM Write once–read many; a type of optical storage device.

Write The process of storing data in a memory.

Zip disk A type of magnetic storage; a flexible disk with a capacity of 100 Mbytes housed in a rigid plastic cartridge about the size of a floppy.

■ **SELF-TEST**

1. The bit capacity of a memory that has 1024 addresses and can store 8 bits at each address is
    (a) 1024   (b) 8192   (c) 8   (d) 4096

2. A 32-bit data word consists of
    (a) 2 bytes   (b) 4 nibbles   (c) 4 bytes   (d) 3 bytes and 1 nibble

3. Data are stored in a random-access memory (RAM) during the
    (a) read operation   (b) enable operation
    (c) write operation   (d) addressing operation

4. Data that are stored at a given address in a random-access memory (RAM) is lost when
    (a) power goes off   (b) the data are read from the address
    (c) new data are written at the address   (d) answers (a) and (c)

5. A ROM is a
    (a) nonvolatile memory   (b) volatile memory
    (c) read/write memory   (d) byte-organized memory

6. A memory with 256 addresses has
    (a) 256 address lines   (b) 6 address lines   (c) 1 address line   (d) 8 address lines

7. A byte-organized memory has
    (a) 1 data output line   (b) 4 data output lines
    (c) 8 data output lines   (d) 16 data output lines

8. The storage cell in a SRAM is
    (a) a flip-flop   (b) a capacitor   (c) a fuse   (d) a magnetic domain

9. A DRAM must be
    (a) replaced periodically   (b) refreshed periodically
    (c) always enabled   (d) programmed before each use

10. A flash memory is
    (a) volatile   (b) a read-only memory
    (c) a read/write memory   (d) nonvolatile
    (e) answers (a) and (c)   (f) answers (c) and (d)

11. Hard disk, floppy disk, Zip disk, and Jaz disk are all
    (a) magneto-optical storage devices   (b) semiconductor storage devices
    (c) magnetic storage devices   (d) optical storage devices

12. Optical storage devices employ
    (a) ultraviolet light   (b) electromagnetic fields
    (c) optical couplers   (d) lasers

■ **PROBLEMS**   SECTION 12–1   **Basics of Semiconductor Memory**

1. Identify the ROM and the RAM in Figure 12–76.

FIGURE 12–76

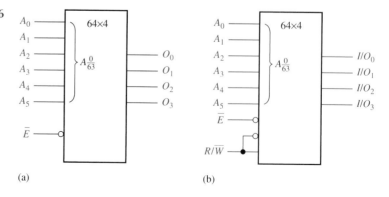

(a)   (b)

2. Explain why RAMs and ROMs are both random-access memories.

3. Explain the purposes of the address bus and the data bus.

4. What memory address (0 through 256) is represented by each of the following hexadecimal numbers:

   (a) $0A_{16}$ (b) $3F_{16}$ (c) CD_{16}

SECTION 12–2   Random-Access Memories (RAMs)

5. A static memory array with four rows similar to the one in Figure 12–8 is initially storing all 0s. What is its content after the following conditions? Assume a 1 selects a row.
   Row 0 = 1, Data in (Bit 0) = 1,
   Row 1 = 0, Data in (Bit 1) = 1,
   Row 2 = 1, Data in (Bit 2) = 1,
   Row 3 = 0, Data in (Bit 3) = 0.

6. Draw a basic logic diagram for a 512 × 8-bit static RAM, showing all the inputs and outputs.

7. Assuming that a 64k × 8 SRAM has a structure similar to that of the SRAM in Figure 12–10, determine the number of rows and 8-bit columns in its memory cell array.

8. Redraw the block diagram in Figure 12–10 for a 64k × 8 memory.

9. Explain the difference between a SRAM and a DRAM.

10. What is the capacity of a DRAM that has twelve address lines?

SECTION 12–3   Read-Only Memories (ROMs)

11. For the ROM array in Figure 12–77, determine the outputs for all possible input combinations, and summarize them in tabular form (Blue cell is a 1, gray cell is a 0).

12. Determine the truth table for the ROM in Figure 12–78.

13. Using a procedure similar to that in Example 12–1, design a ROM for conversion of single-digit BCD to excess-3 code.

14. What is the total *bit* capacity of a ROM that has 14 address lines and 8 data outputs?

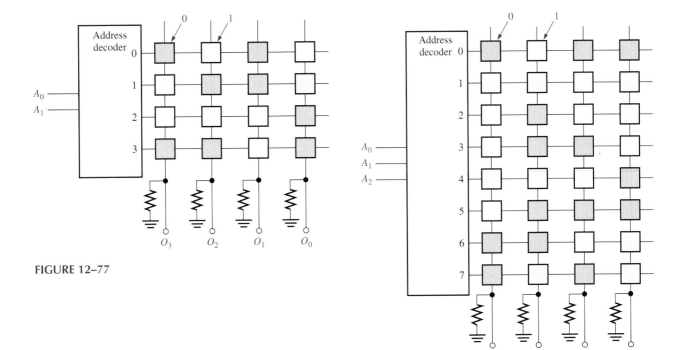

FIGURE 12–77

FIGURE 12–78

SECTION 12–4 Programmable ROMs (PROMs and EPROMs)

15. Assuming that the PROM matrix in Figure 12–79 is programmed by blowing a fuse link to create a 0, indicate the links to be blown to program an X^3 look-up table, where X is a number from 0 through 7.

FIGURE 12–79

16. Determine the addresses that are programmed and the contents of each address after the programming sequence in Figure 12–80 has been applied to a PROM like the one shown in Figure 12–31.

SECTION 12–6 Memory Expansion

17. Use 16k × 4 DRAMs to build a 64k × 8 RAM. Show the logic diagram.

18. Using a block diagram, show how 64k × 1 dynamic RAMs can be expanded to build a 256k × 4 RAM.

19. What is the word length and the word capacity of the memory of Problem 17? Problem 18?

SECTION 12–7 Special Types of Memories

20. Complete the timing diagram in Figure 12–81 by showing the output waveforms that are initially all LOW for a FIFO serial memory like that shown in Figure 12–49.

21. Consider a 4096 × 8 RAM in which the last 64 addresses are used as a LIFO stack. If the first address in the RAM is 000_{16}, designate the 64 addresses used for the stack.

22. In the memory of Problem 21, sixteen bytes are pushed into the stack. At what address is the first byte in located? At what address is the last byte in located?

FIGURE 12–80

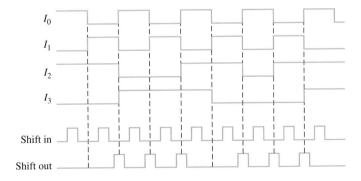

FIGURE 12–81

SECTION 12–8   Magnetic and Optical Storage

23. Describe the general format of a hard disk.

24. Explain seek time and latency period in a hard disk drive.

25. Why does magnetic tape require a much longer access time than does a disk?

26. Explain the differences in a magneto-optical disk, a CD-ROM, and a WORM.

SECTION 12–9   Testing and Troubleshooting

27. Determine if the contents of the ROM in Figure 12–82 are correct.

FIGURE 12–82

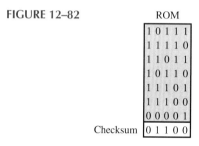

28. A 128 × 8 ROM is implemented as shown in Figure 12–83. The decoder decodes the two most significant address bits to enable the ROMs one at a time, depending on the address selected.

   (a) Express the lowest address and the highest address of each ROM as hexadecimal numbers.

   (b) Assume that a single checksum is used for the entire memory and it is stored at the highest address. Develop a flowchart for testing the complete memory system.

   (c) Assume that each ROM has a checksum stored at its highest address. Modify the flowchart developed in part (b) to accommodate this change.

   (d) What is the disadvantage of using a single checksum for the entire memory rather than a checksum for each individual ROM?

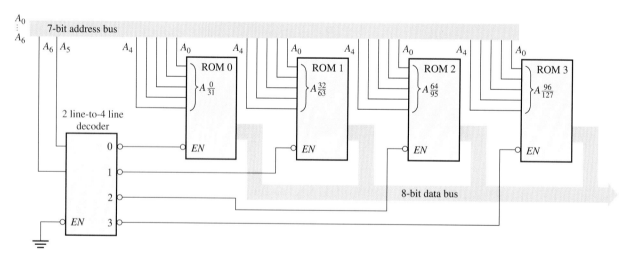

FIGURE 12–83

29. Suppose that a checksum test is run on the memory in Figure 12–83 and each individual ROM has a checksum at its highest address. What IC or ICs will you replace for each of the following error messages that appear on the system's video monitor?

   (a) ADDRESSES 40 − 5F FAULTY

   (b) ADDRESSES 20 − 3F FAULTY

   (c) ADDRESSES 00 − 7F FAULTY

SECTION 12–10 Digital System Application

30. Develop a timing diagram for the basic memory logic in Figure 12–74 to illustrate the entry of the digits 4321 into the SRAM. Include all the inputs and outputs of each device.

31. In the programming mode, what is the state of the counter in Figure 12–74 after two code digits have been entered?

32. What is the purpose of the 74LS27 on the memory logic board in Application Assignment 2?

33. Discuss the advantages and the disadvantages of using a PROM instead of a SRAM in the memory logic.

Special Design Problems

34. Develop a basic block diagram for the ROM tester for the setup in Figure 12–67 to test the ROM in Figure 12–84.

35. Repeat Problem 34 for the ROM in Figure 12–85.

FIGURE 12–84

FIGURE 12–85

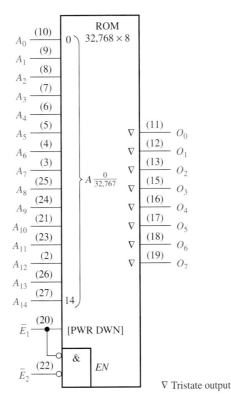

∇ Tristate output

36. Modify, if necessary, the design of the memory logic of the security entry system to accommodate a 5-digit entry code.

37. Make the appropriate modifications to the entry code logic of the security entry system for a 5-digit entry code. Refer to Chapter 10.

■ ANSWERS TO SECTION REVIEWS

SECTION 12–1

1. Bit is the smallest unit of data.   **2.** 256 bytes is 2048 bits.

3. A write operation stores data in memory.

4. A read operation takes a copy of data from memory.

5. A unit of data is located by its address.

6. A RAM is volatile and has read/write capability. A ROM is nonvolatile and has only read capability.

SECTION 12–2

1. Asynchronous and synchronous burst

2. A small fast memory between the CPU and main memory

3. SRAMs have latch storage cells that can retain data indefinitely while power is applied. DRAMs have capacitive storage cells that must be periodically refreshed.

4. The refresh operation prevents data from being lost because of capacitive discharge. A stored bit is restored periodically by recharging the capacitor to its nominal level.

5. FPM, EDO, BEDO, Synchronous

SECTION 12–3

1. 512×8 equals 4096 bits.

2. Mask ROM, PROM, EPROM, UV EPROM, EEPROM

3. Eight bits of address are required for 256 byte locations ($2^8 = 256$).

SECTION 12–4

1. PROMs are field-programmable; ROMs are not.

2. 1s are left after EPROM erasure.

3. Read is the normal mode of operation for a PROM.

SECTION 12–5

1. Flash, ROM, EPROM, and EEPROM are nonvolatile.

2. Flash is nonvolatile; SRAM and DRAM are volatile.

3. Programming, read, erase

SECTION 12–6

1. Eight RAMs   **2.** Eight RAMs   **3.** SIMM; Single in-line memory module

4. DIMM: Dual in-line memory module

SECTION 12–7

1. In a FIFO memory the *first* bit (or word) *in* is the *first* bit (or word) *out*.

2. In a LIFO memory the *last* bit (or word) *in* is the *first* bit (or word) *out*. A stack is a LIFO.

3. The operation or instruction that adds data to the memory stack

4. The operation or instruction that removes data from the memory stack

5. CCD is a charge-coupled device.

SECTION 12–8

1. Magnetic storage: floppy disk, hard disk, tape, and magneto-optical disk.
2. Floppy disk storage capacity: 1.44 Mbytes
3. A magnetic disk is organized in tracks and sectors.
4. A magneto-optical disk uses a laser beam and an electromagnet.
5. Zip: 100 Mbytes; and Jaz: 1 Gbyte and 2 Gbytes
6. Optical storage: CD-ROM, CD-R, CD-RW, DVD-ROM, WORM

SECTION 12–9

1. The contents of the ROM are added and compared with a prestored checksum.
2. Checksum cannot be used because the contents of a RAM are not fixed.
3. (1) a short between adjacent cells; (2) an inability of some cells to store both 1s and 0s; (3) dynamic altering of the contents of one address when the contents of another address change.

SECTION 12–10

1. The system must be reprogrammed because the SRAM is volatile and loses stored data when power is off.
2. Five of the sixteen addresses are used. Eleven addresses are unused.
3. The program mode switch applies a LOW to the negative-OR gate (NAND) that triggers one-shot A in the security code logic.

■ **ANSWERS TO RELATED PROBLEMS FOR EXAMPLES**

12–1 $G_3G_2G_1G_0 = 1110$
12–2 Connect eight 64k × 1 ROMs in parallel to form a 64k × 8 ROM.
12–3 Sixteen 64k × 1 ROMs
12–4 See Figure 12–86.
12–5 ROM 1: 0 to 524,287;   ROM 2: 524,288 to 1,048,575

FIGURE 12–86

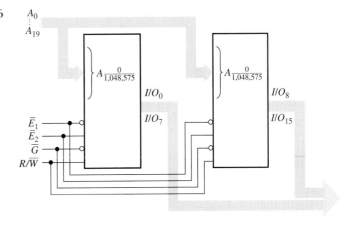

■ **ANSWERS TO SELF-TEST**

1. (b)	2. (c)	3. (c)	4. (d)	5. (a)	6. (d)
7. (c)	8. (a)	9. (b)	10. (f)	11. (c)	12. (d)

13

INTERFACING

■ CHAPTER OBJECTIVES

❑ Describe applications of digital-to-analog and analog-to-digital conversion

❑ Explain D/A conversion by the binary-weighted input method and by the $R/2R$ ladder method

❑ Explain A/D conversion by any of the following processes: flash (simultaneous), stairstep ramp, tracking, single-slope, dual-slope, and successive-approximation

❑ Recognize such conversion errors as nonmonotonicity, differential nonlinearity, offset, improper gain, and missing or incorrect codes in DACs and ADCs

❑ Troubleshoot DAC and ADC systems

❑ Explain how digital devices are interfaced by the use of bus systems

❑ Define the basic characteristics and applications of the PCI and ISA internal bus standards

❑ Define the basic characteristics and applications of the RS-232C, IEEE 1394 (FireWire), USB, IEEE 488 (GPIB), and SCSI external bus standards

■ CHAPTER OVERVIEW

Interfacing is the process of making two or more devices or systems operationally compatible with each other so that they function together as required.

Many quantities are analog in nature; that is, they are continuous quantities. Physical quantities such as temperature, pressure, time, and velocity are examples of analog quantities. In this chapter, methods of converting from digital codes to analog quantities are introduced. This is called digital-to-analog conversion (D/A conversion). Also, methods for converting analog quantities to digital codes are covered. This is called analog-to-digital conversion (A/D conversion).

In addition to D/A and A/D conversion, several types of standard buses are discussed.

 SPECIFIC DEVICES

DAC0808   ADC0804   AD9300
AD673

■ DIGITAL SYSTEM APPLICATION

This Digital System Application illustrates concepts taught in this chapter and focuses on both the analog and the digital portions of an antenna positioning system. A/D conversion is required to convert analog voltages representing the antenna position into digital form and logic circuits are required to process and display the angular positions.

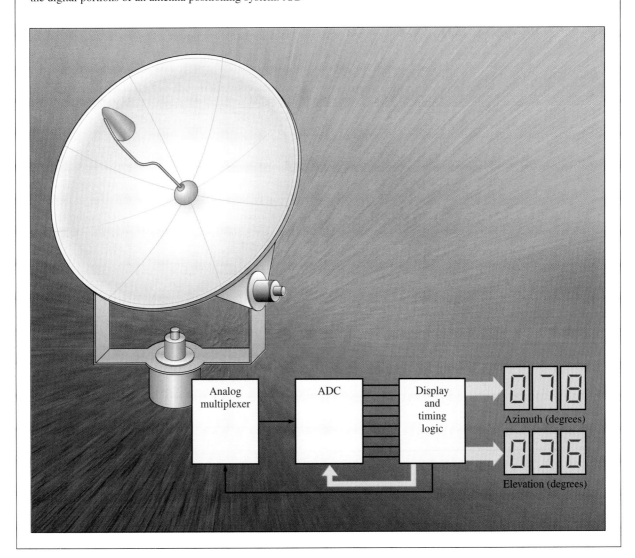

13–1 ■ DIGITAL AND ANALOG INTERFACING

In Chapter 1, we briefly described analog and digital quantities and pointed out the difference between them. Analog quantities are sometimes called real-world quantities because most physical quantities that we are interested in measuring are analog in nature. Many applications of computers and other digital systems require the input of real-world quantities, such as temperature, speed, position, pressure, and force. Real-world quantities can even include graphic images. Also, digital systems often must produce outputs to control real-world quantities.

After completing this section, you should be able to

- Define an analog quantity ■ State the purpose of analog-to-digital conversion
- State the purpose of digital-to-analog conversion ■ Give examples of both analog-to-digital and digital-to-analog conversion

Digital and Analog Signals

An analog quantity is one that has continuous values over a given range, as contrasted with a discrete set of values for the digital case. To illustrate the difference between an analog and a digital representation of a quantity, let's take the case of a voltage that varies over a range from 0 V to +15 V. The analog representation of this quantity takes in *all* values between 0 and +15 V of which there is an infinite number.

In the case of a digital representation of the voltage using a 4-bit binary code, only sixteen values can be defined. More values between 0 and +15 can be represented by using more bits in the digital code. So an analog quantity can be represented to some degree of accuracy with a digital code that specifies discrete values within the range. This concept is illustrated in Figure 13–1, where the analog function shown is a smoothly changing curve that takes on values between 0 V and +15 V. If a 4-bit code is used to represent this curve, each binary number represents a discrete point on the curve.

In Figure 13–1 the voltage on the analog curve is measured, or sampled, at each of thirty-five equal intervals. The voltage at each of these intervals is represented by a 4-bit code as indicated. At this point, we have a series of binary numbers representing various voltage values along the analog curve. This is the basic idea of analog-to-digital (A/D) conversion.

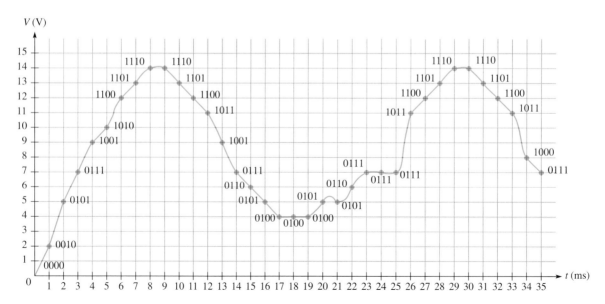

FIGURE 13–1
Discrete (digital) points on an analog curve.

An approximation of the analog function in Figure 13–1 can be reconstructed from the sequence of digital numbers that has been generated. Obviously, there will be some error in the reconstruction because only certain values are represented (thirty-six in this example) and not the continuous set of values. If the digital values at all of the thirty-six points are graphed as shown in Figure 13–2, we have a reconstructed function. As you can see, the graph only approximates the original curve because values between the points are not known. Accuracy can be increased by sampling the curve more often and by increasing the number of bits used to represent each sampled value.

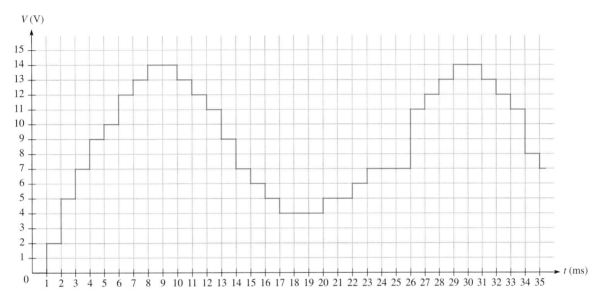

FIGURE 13–2
A rough digital reproduction of the analog curve in Figure 13–1.

Application Examples

To interface between the digital and analog worlds, two basic processes are required: **analog-to-digital (A/D) conversion** and **digital-to-analog (D/A) conversion.** The following three system examples illustrate applications of these conversion processes.

An Electronic Thermostat   A simplified block diagram of an electronic thermostat is shown in Figure 13–3. The room temperature sensor produces an analog voltage that is proportional to the temperature. This voltage is increased by the linear amplifier and applied to the **analog-to-digital converter (ADC),** where it is converted to a digital code and periodically sampled by the control logic. For example, suppose the room temperature is 67°F. A specific voltage value corresponding to this temperature appears on the ADC input and is converted to an 8-bit binary number, 01000011.

The control logic compares this binary number with a binary number representing the desired temperature (say 01001000 for 72°F). This desired value has been previously entered from the keypad and stored in a register. The comparison shows that the actual room temperature is less than the desired temperature. As a result, the control logic instructs the unit control circuit to turn the furnace on. As the furnace runs, the control logic continues to monitor the actual temperature via the ADC. When the actual temperature equals or exceeds the desired temperature, the control logic turns the furnace off. Both the actual temperature and the desired temperature are displayed.

CD Player   The compact-disk (CD) player is an example of a system that uses a **digital-to-analog converter (DAC).** A basic block diagram is shown in Figure 13–4. An audio signal is digitally recorded on the CD in the form of pits that are sensed by

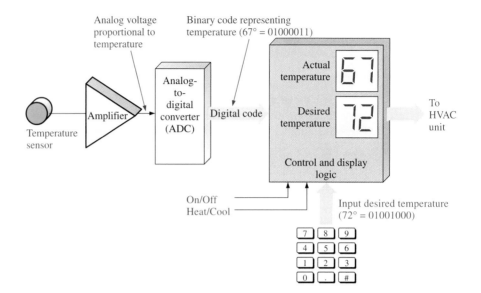

FIGURE 13–3
Basic block diagram of an electronic thermostat that uses an ADC.

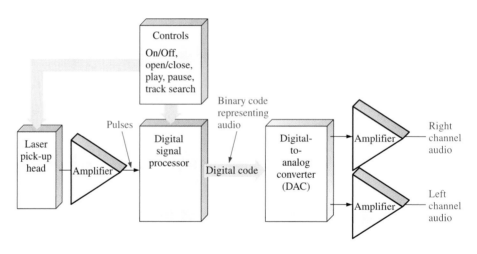

FIGURE 13–4
Basic block diagram of a CD player.

the laser pickup and amplified. The amplified digital signal is converted to a sequence of binary codes that represent the originally recorded audio signal. The binary representation of the audio is converted to analog form by the DAC, amplified, and sent to the speakers.

A Digital Audiotape (DAT) Player/Recorder Another system example that includes both A/D and D/A conversion is the **DAT** player/recorder. A basic block diagram is shown in Figure 13–5.

The audio input signal, of course, is an analog quantity. In the record mode, sound is picked up, amplified, and converted to digital form by the ADC. The digital codes representing the audio signal are processed and recorded on the tape.

In the play mode, the digitized audio signal is read from the tape, processed, and converted back to analog form by the DAC. It is then amplified and sent to the speaker system.

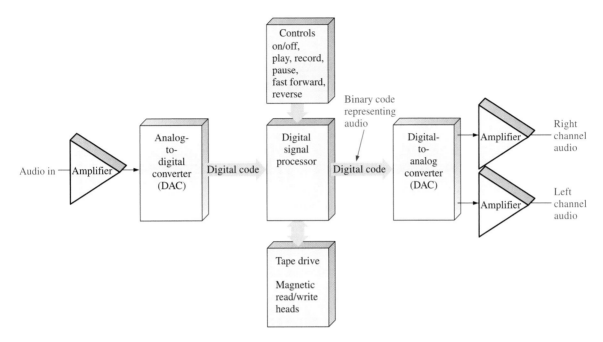

FIGURE 13–5
Basic block diagram of a DAT system.

SECTION 13–1 REVIEW

1. In nature, quantities normally appear in what form?
2. Explain the basic purpose of A/D conversion.
3. Explain the basic purpose of D/A conversion.

13–2 ■ DIGITAL-TO-ANALOG (D/A) CONVERSION

D/A conversion is an important part of many systems. In this section, we will examine the theory of operation of two basic types of digital-to-analog converters (DACs) and learn about their performance characteristics.

After completing this section, you should be able to

■ Explain basically what an operational amplifier is ■ Show how the op-amp can be used as an inverting amplifier or a comparator ■ Explain the operation of a binary-weighted-input DAC ■ Explain the operation of an $R/2R$ ladder DAC ■ Discuss resolution, accuracy, linearity, monotonicity, and settling time in a DAC

The Operational Amplifier

Before getting into digital-to-analog converters (DACs), let's look briefly at an element that is common to most types of DACs and analog-to-digital converters (ADCs). This element is the operational amplifier, or op-amp for short.

An **op-amp** is a linear amplifier that has two inputs (inverting and noninverting) and one output. It has a very high voltage gain and a very high input impedance, as well as a very low output impedance. The op-amp symbol is shown in Figure 13–6(a). When used as an inverting amplifier, the op-amp is configured as shown in part (b). The feedback resistor, R_f, and the input resistor, R_i, control the voltage gain according to the

(a) Op-amp symbol

(b) Op-amp as an inverting amplifier with gain of R_f/R_i

(c) Op-amp as a comparator

FIGURE 13–6
The operational amplifier (op-amp).

formula in Equation (13–1), where V_{out}/V_{in} is the closed-loop voltage gain (closed loop refers to the feedback from output to input provided by R_f). The negative sign indicates inversion.

$$\frac{V_{out}}{V_{in}} = -\frac{R_f}{R_i}$$

(13–1)

In the inverting amplifier configuration, the inverting input of the op-amp is approximately at ground potential (0 V) because feedback and the extremely high open-loop gain make the differential voltage between the two inputs extremely small. Since the noninverting input is grounded, the inverting input is at approximately 0 V, which is called *virtual ground.*

When the op-amp is used as a comparator, as shown in Figure 13–6(c), two voltages are applied to the inputs. When these input voltages differ by a very small amount, the op-amp is driven into one of its two saturated output states, either HIGH or LOW, depending on which input voltage is greater.

Binary-Weighted-Input Digital-to-Analog Converter

One method of D/A conversion uses a resistor network with resistance values that represent the binary weights of the input bits of the digital code. Figure 13–7 shows a 4-bit DAC of this type. Each of the input resistors will either have current or have no current, depending on the input voltage level. If the input voltage is zero (binary 0), the current is also

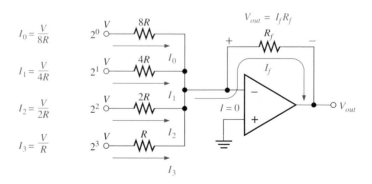

FIGURE 13–7
A 4-bit DAC with binary-weighted inputs.

zero. If the input voltage is HIGH (binary 1), the amount of current depends on the input resistor value and is different for each input resistor, as indicated in the figure.

Since there is practically no current into the op-amp inverting ($-$) input, all of the input currents sum together and go through R_f. Since the inverting input is at 0 V (virtual ground), the drop across R_f is equal to the output voltage, so $V_{out} = I_f R_f$.

The values of the input resistors are chosen to be inversely proportional to the binary weights of the corresponding input bits. The lowest-value resistor (R) corresponds to the highest binary-weighted input (2^3). The other resistors are multiples of R (that is, $2R$, $4R$, and $8R$) and correspond to the binary weights 2^2, 2^1, and 2^0, respectively. The input currents are also proportional to the binary weights. Thus, the output voltage is proportional to the sum of the binary weights because the sum of the input currents is through R_f.

One of the disadvantages of this type of DAC is the number of different resistor values. For example, an 8-bit converter requires eight resistors, ranging from some value of R to $128R$ in binary-weighted steps. This range of resistors requires tolerances of one part in 255 (less than 0.5%) to accurately convert the input, making this type of DAC very difficult to mass-produce.

EXAMPLE 13–1

Determine the output of the DAC in Figure 13–8(a) if the waveforms representing a sequence of 4-bit numbers in Figure 13–8(b) are applied to the inputs. Input D_0 is the least significant bit (LSB).

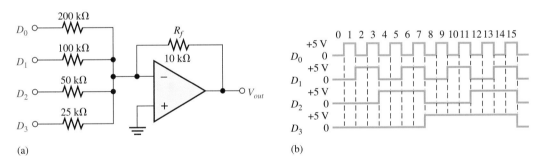

(a)　　　　　(b)

FIGURE 13–8

Solution First, determine the current for each of the weighted inputs. Since the inverting ($-$) input of the op-amp is at 0 V (virtual ground) and a binary 1 corresponds to $+5$ V, the current through any of the input resistors is 5 V divided by the resistance value.

$$I_0 = \frac{5 \text{ V}}{200 \text{ k}\Omega} = 0.025 \text{ mA}$$

$$I_1 = \frac{5 \text{ V}}{100 \text{ k}\Omega} = 0.05 \text{ mA}$$

$$I_2 = \frac{5 \text{ V}}{50 \text{ k}\Omega} = 0.1 \text{ mA}$$

$$I_3 = \frac{5 \text{ V}}{25 \text{ k}\Omega} = 0.2 \text{ mA}$$

Almost no current goes into the inverting op-amp input because of its extremely high impedance. Therefore, assume that all of the current goes through the feedback resistor R_f. Since one end of R_f is at 0 V (virtual ground), the drop across R_f equals the output voltage, which is negative with respect to virtual ground.

$$V_{out(D0)} = (10 \text{ k}\Omega)(-0.025 \text{ mA}) = -0.25 \text{ V}$$
$$V_{out(D1)} = (10 \text{ k}\Omega)(-0.05 \text{ mA}) = -0.5 \text{ V}$$
$$V_{out(D2)} = (10 \text{ k}\Omega)(-0.1 \text{ mA}) = -1 \text{ V}$$
$$V_{out(D3)} = (10 \text{ k}\Omega)(-0.2 \text{ mA}) = -2 \text{ V}$$

From Figure 13–8(b), the first binary input code is 0000, which produces an output voltage of 0 V. The next input code is 0001, which produces an output voltage of −0.25 V. For this, the output voltage is −0.25 V. The next code is 0010, which produces an output voltage of −0.5 V. The next code is 0011, which produces an output voltage of −0.25 V + −0.5 V = −0.75 V. Each successive binary code increases the output voltage by −0.25 V, so for this particular straight binary sequence on the inputs, the output is a stairstep waveform going from 0 V to −3.75 V in −0.25 V steps. This is shown in Figure 13–9.

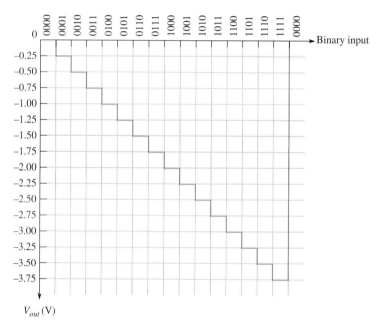

FIGURE 13–9
Output of the DAC in Figure 13–8.

Related Problem Reverse the input waveforms to the DAC in Figure 13–8 (D_3 to D_0, D_2 to D_1, D_1 to D_2, D_0 to D_3) and determine the output.

The *R*/2*R* Ladder Digital-to-Analog Converter

Another method of D/A conversion is the *R*/2*R* ladder, as shown in Figure 13–10 for four bits. It overcomes one of the problems in the binary-weighted-input DAC in that it requires only two resistor values.

Start by assuming that the D_3 input is HIGH (+5 V) and the others are LOW (ground, 0 V). This condition represents the binary number 1000. A circuit analysis will show that this reduces to the equivalent form shown in Figure 13–11(a) on page 692.

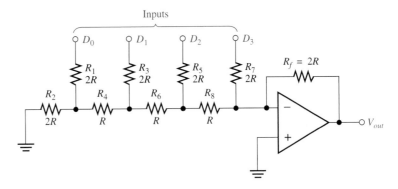

FIGURE 13–10
An R/2R ladder DAC.

Essentially no current goes through the $2R$ equivalent resistance because the inverting input is at virtual ground. Thus, all of the current ($I = 5 \text{ V}/2R$) through R_7 also goes through R_f, and the output voltage is -5 V. The operational amplifier keeps the inverting ($-$) input near zero volts (≈ 0 V) because of negative feedback. Therefore, all current goes through R_f rather than into the inverting input.

Figure 13–11(b) shows the equivalent circuit when the D_2 input is at $+5$ V and the others are at ground. This condition represents 0100. If we thevenize looking from R_8, we get 2.5 V in series with R, as shown. This results in a current through R_f of $I = 2.5 \text{ V}/2R$, which gives an output voltage of -2.5 V. Keep in mind that there is no current into the op-amp inverting input and that there is no current through the equivalent resistance to ground because it has 0 V across it, due to the virtual ground.

Figure 13–11(c) shows the equivalent circuit when the D_1 input is at $+5$ V and the others are at ground. This condition represents 0010. Again thevenizing looking from R_8, you get 1.25 V in series with R as shown. This results in a current through R_f of $I = 1.25 \text{ V}/2R$, which gives an output voltage of -1.25 V.

In part (d) of Figure 13–11, the equivalent circuit representing the case where D_0 is at $+5$ V and the other inputs are at ground is shown. This condition represents 0001. Thevenizing from R_8 gives an equivalent of 0.625 V in series with R as shown. The resulting current through R_f is $I = 0.625 \text{ V}/2R$, which gives an output voltage of -0.625 V.

Notice that each successively lower-weighted input produces an output voltage that is halved, so that the output voltage is proportional to the binary weight of the input bits.

Performance Characteristics of Digital-to-Analog Converters

The performance characteristics of a DAC include resolution, accuracy, linearity, monotonicity, and settling time, each of which is discussed in the following list:

■ *Resolution.* The resolution of a DAC is the reciprocal of the number of discrete steps in the output. This, of course, is dependent on the number of input bits. For example, a 4-bit DAC has a resolution of one part in $2^4 - 1$ (one part in fifteen). Expressed as a percentage, this is $(1/15)100 = 6.67\%$. The total number of discrete steps equals $2^n - 1$, where n is the number of bits. Resolution can also be expressed as the number of bits that are converted.

■ *Accuracy.* Accuracy is a comparison of the actual output of a DAC with the expected output. It is expressed as a percentage of a full-scale, or maximum, output voltage. For example, if a converter has a full-scale output of 10 V and the accuracy is $\pm 0.1\%$, then the maximum error for any output voltage is $(10 \text{ V})(0.001) = 10 \text{ mV}$. Ideally, the accuracy should be no worse than $\pm\frac{1}{2}$ of a least significant bit. For an 8-bit converter, the least significant bit is 0.39% of full scale. The accuracy should be approximately $\pm 0.2\%$.

(a) Equivalent circuit for $D_3 = 1, D_2 = 0, D_1 = 0, D_0 = 0$

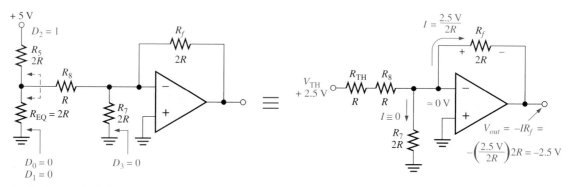

(b) Equivalent circuit for $D_3 = 0, D_2 = 1, D_1 = 0, D_0 = 0$

(c) Equivalent circuit for $D_3 = 0, D_2 = 0, D_1 = 1, D_0 = 0$

(d) Equivalent circuit for $D_3 = 0, D_2 = 0, D_1 = 0, D_0 = 1$

FIGURE 13–11
Analysis of the R/2R ladder DAC.

- *Linearity.* A linear error is a deviation from the ideal straight-line output of a DAC. A special case is an offset error, which is the amount of output voltage when the input bits are all zeros.
- *Monotonicity.* A DAC is **monotonic** if it does not take any reverse steps when it is sequenced over its entire range of input bits.
- *Settling time.* Settling time is normally defined as the time it takes a DAC to settle within $\pm\frac{1}{2}$ LSB of its final value when a change occurs in the input code.

EXAMPLE 13–2

Determine the resolution, expressed as a percentage, of the following:
(a) an 8-bit DAC **(b)** a 12-bit DAC

Solution
(a) For the 8-bit converter,

$$\frac{1}{2^8 - 1} \times 100 = \frac{1}{255} \times 100 = \textbf{0.392\%}$$

(b) For the 12-bit converter,

$$\frac{1}{2^{12} - 1} \times 100 = \frac{1}{4095} \times 100 = \textbf{0.0244\%}$$

Related Problem Calculate the resolution for a 16-bit DAC.

 A Specific Digital-to-Analog Converter

The DAC0808 is an example of an *R/2R* ladder-based DAC. A block diagram is shown in Figure 13–12. This device operates with a dc supply voltage range for V_{CC} from $+4.5$ V to $+5.5$ V and for V_{EE} from -4.5 V to -16.5 V. The resolution is 8 bits with an output settling time of 150 ns. The settling time is the time for the analog output to stabilize after the digital input is applied. The DAC0808 will interface directly with TTL or CMOS and its accuracy is $+0.19\%$.

FIGURE 13–12
The DAC0808 digital-to-analog converter.

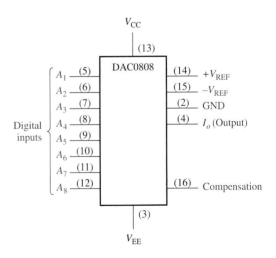

In a typical application, the DAC0808 is connected to an op-amp, as shown in Figure 13–13, and the analog output is taken from the op-amp. The $+V_{REF}$ input establishes the analog output voltage. For example, a 10 V reference results in a maximum output voltage of 10 V as shown in the figure.

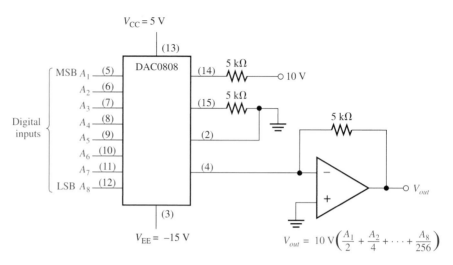

FIGURE 13–13
The DAC0808 in a typical DAC configuration.

SECTION 13–2 REVIEW

1. What is the disadvantage of the DAC with binary-weighted inputs?
2. What is the resolution of a 4-bit DAC?

13–3 ■ ANALOG-TO-DIGITAL (A/D) CONVERSION

As you have seen, analog-to-digital conversion is the process by which an analog quantity is converted to digital form. It is necessary when measured quantities must be in digital form for processing in a computer or for display or storage. Several types of analog-to-digital converters (ADCs) are now examined.

After completing this section, you should be able to

■ Explain how a flash ADC works ■ Describe the operation of a stairstep-ramp ADC
■ Explain the operation of a tracking ADC ■ Discuss single-slope and dual-slope ADCs ■ Describe the operation of a successive-approximation ADC

Flash (Simultaneous) Analog-to-Digital Converter

The flash method utilizes comparators that compare reference voltages with the analog input voltage. When the analog voltage exceeds the reference voltage for a given comparator, a HIGH is generated. Figure 13–14 shows a 3-bit converter that uses seven comparator circuits; a comparator is not needed for the all-0s condition. A 4-bit converter of this type requires fifteen comparators. In general, $2^n - 1$ comparators are required for conversion to an n-bit binary code. The large number of comparators necessary for a reasonable-sized binary number is one of the disadvantages of the **flash ADC.** Its chief advantage is that it provides a fast conversion time.

The reference voltage for each comparator is set by the resistive voltage-divider circuit. The output of each comparator is connected to an input of the priority encoder. The encoder is sampled by a pulse on the enable input, and a 3-bit code representing the value of the analog input appears on the encoder's outputs. The binary code is determined by the highest-order input having a HIGH level.

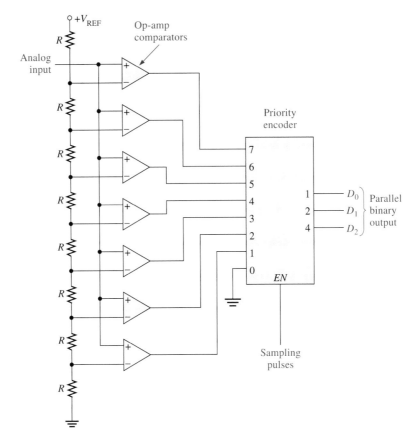

FIGURE 13–14
A 3-bit flash ADC.

The sampling rate determines the accuracy with which the sequence of digital codes represents the analog input of the ADC. The more samples taken in a given unit of time, the more accurately the analog signal is represented in digital form.

Example 13–3 illustrates the basic operation of the flash ADC in Figure 13–14.

EXAMPLE 13–3

Determine the binary code output of the 3-bit flash ADC for the analog input signal in Figure 13–15 and the sampling pulses (encoder enable) shown. For this example, $V_{REF} = +8$ V.

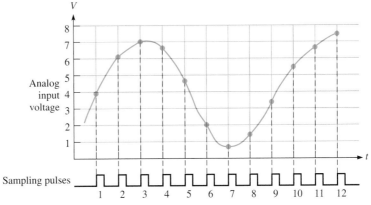

FIGURE 13–15
Sampling of values on an analog waveform for conversion to digital form.

Solution The resulting digital output sequence is listed as follows and shown in the waveform diagram of Figure 13–16 in relation to the sampling pulses:

100, 110, 111, 110, 100, 010, 000, 001, 011, 101, 110, 111

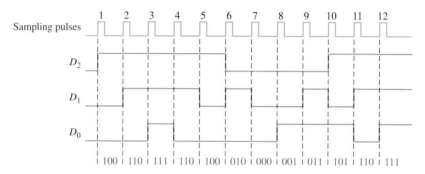

FIGURE 13–16
Resulting digital outputs for sampled values. Output D_0 is the LSB.

Related Problem If the sampling pulse frequency in Figure 13–15 is doubled, determine the binary numbers represented by the resulting digital output sequence for 24 pulses.

Stairstep-Ramp Analog-to-Digital Converter

The stairstep-ramp method of A/D conversion is also known as the *digital-ramp* or the *counter* method. It employs a DAC and a binary counter to generate the digital value of an analog input. Figure 13–17 shows a diagram of this type of converter.

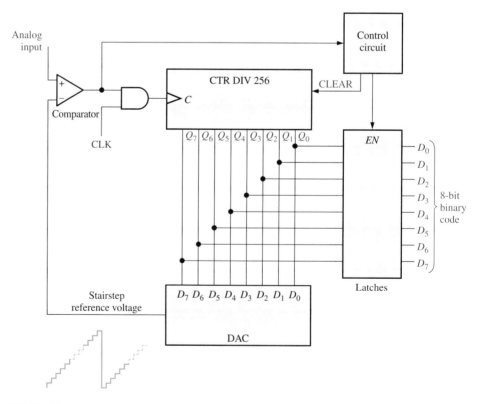

FIGURE 13–17
Stairstep-ramp ADC (8 bits).

COMPUTER NOTE
Sound cards used in computers use an ADC to convert sound from a micro-phone, audio CD player, or other source into a digital signal. The ADC sends the digital sig-nal to a digital signal processor (DSP). Based on instructions from a ROM, one function of the DSP is to compress the digi-tal signal so it uses less storage space. The DSP then sends the compressed data to the computer's processor which, in turn, sends the data to a hard drive or CD ROM for storage. To play a recorded sound, the stored data is retrieved by the processor and sent to the DSP where it is decom-pressed and sent to a DAC. The output of the DAC, which is a reproduction of the original sound signal, is applied to the speakers.

Assume that the counter begins RESET and the output of the DAC is zero. Now assume that an analog voltage is applied to the input. When it exceeds the reference voltage (output of DAC), the comparator switches to a HIGH output state and enables the AND gate. The clock pulses begin advancing the counter through its binary states, producing a stairstep reference voltage from the DAC. The counter continues to advance from one binary state to the next, producing successively higher steps in the reference voltage. When the stairstep reference voltage reaches the analog input voltage, the comparator output will go LOW and disable the AND gate, thus cutting off the clock pulses to stop the counter. The binary state of the counter at this point equals the number of steps in the reference voltage required to make the reference equal to or greater than the analog input. This binary number, of course, represents the value of the analog input. The control logic loads the binary count into the latches and resets the counter, thus beginning another count sequence to sample the input value.

The stairstep-ramp method is slower than the flash method because, in the worst case of maximum input, the counter must sequence through its maximum number of states before a conversion occurs. For an 8-bit conversion, this means a maximum of 256 counter states. Figure 13–18 illustrates a conversion sequence for a 4-bit conversion. Notice that for each sample, the counter must count from zero up to the point at which the stairstep reference voltage reaches the analog input voltage. The conversion time varies, depending on the analog voltage.

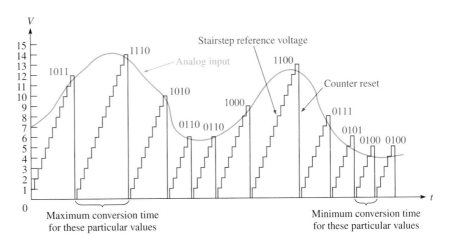

FIGURE 13–18
Example of a 4-bit conversion, showing an analog input and the stairstep reference voltage for a stairstep-ramp ADC.

Tracking Analog-to-Digital Converter

The tracking method uses an up/down counter and is faster than the stairstep-ramp method because the counter is not reset after each sample, but rather tends to track the analog input. Figure 13–19 shows a typical 8-bit tracking ADC.

As long as the output reference voltage is less than the analog input, the comparator output is HIGH, putting the counter in the UP mode, which causes it to produce an up sequence of binary counts. This causes an increasing stairstep reference voltage out of the DAC, which continues until the ramp reaches the value of the input voltage.

When the reference voltage equals the analog input, the comparator's output switches LOW and puts the counter in the DOWN mode, causing it to back down one count. If the analog input is decreasing, the counter will continue to back down in its sequence and effectively track the input. If the input is increasing, the counter will back down one count after the comparison occurs and then will begin counting up again. When the input is constant, the counter backs down one count when a comparison occurs. The

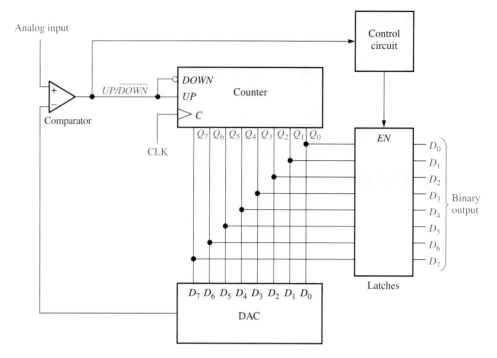

FIGURE 13–19
An 8-bit tracking ADC.

reference output is now less than the analog input, and the comparator output goes HIGH, causing the counter to count up. As soon as the counter increases one state, the reference voltage becomes greater than the input, switching the comparator to its LOW state. This causes the counter to back down one count. This back-and-forth action continues as long as the analog input is a constant value, thus causing an oscillation between two binary states in the output. This is a disadvantage of this type of converter.

Figure 13–20 illustrates the tracking action of this type of ADC for a 4-bit conversion.

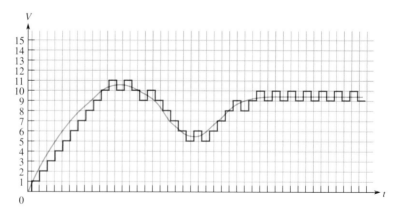

FIGURE 13–20
Tracking action of a tracking ADC.

Single-Slope Analog-to-Digital Converter

Unlike the stairstep-ramp and tracking methods, the single-slope converter does not require a DAC. It uses a linear ramp generator to produce a constant-slope reference voltage. A diagram is shown in Figure 13–21. At the beginning of a conversion cycle, the counter is reset and the ramp generator output is 0 V. The analog input is greater than the reference voltage at this point and therefore produces a HIGH output from

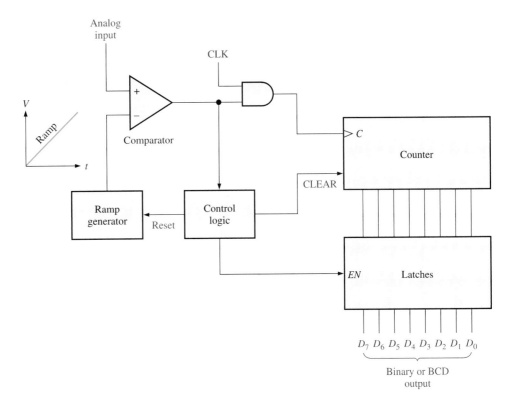

FIGURE 13–21
Single-slope ADC.

the comparator. This HIGH enables the clock to the counter and starts the ramp generator.

Assume that the slope of the ramp is 1 V/ms. The ramp will increase until it equals the analog input; at this point the ramp is reset, and the binary or BCD count is stored in the latches by the control logic. Let's assume that the analog input is 2 V at the point of comparison. This means that the ramp is also 2 V and has been running for 2 ms. Since the comparator output has been HIGH for 2 ms, 200 clock pulses have been allowed to pass through the gate to the counter (assuming a clock frequency of 100 kHz). At the point of comparison, the counter is in the binary state that represents decimal 200. With proper scaling and decoding, this binary number can be displayed as 2.00 V. This basic concept is used in some digital voltmeters.

Dual-Slope Analog-to-Digital Converter

The operation of the dual-slope ADC is similar to that of the single-slope type except that a variable-slope ramp and a fixed-slope ramp are both used. This type of converter is common in digital voltmeters and other types of measurement instruments.

A ramp generator (integrator), A_1, is used to produce the dual-slope characteristic. A block diagram of a dual-slope ADC is shown in Figure 13–22 for reference.

Figure 13–23 (p. 701) illustrates dual-slope conversion. Start by assuming that the counter is reset and the output of the integrator is zero. Now assume that a positive input voltage is applied to the input through the switch (SW) as selected by the control logic. Since the inverting input of A_1 is at virtual ground, and assuming that V_{in} is constant for a period of time, there will be constant current through the input resistor R and therefore through the capacitor C. Capacitor C will charge linearly because the current is constant, and as a result, there will be a negative-going linear voltage ramp on the output of A_1, as illustrated in Figure 13–23(a).

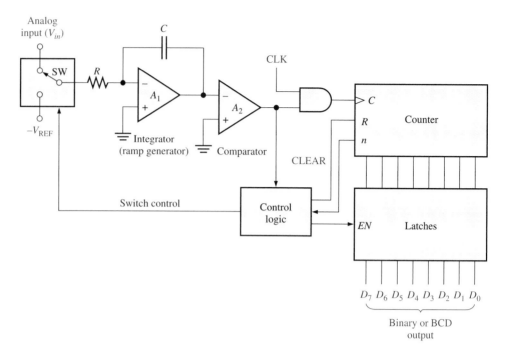

FIGURE 13–22
Dual-slope ADC.

When the counter reaches a specified count, it will be reset, and the control logic will switch the negative reference voltage ($-V_{REF}$) to the input of A_1 as shown in Figure 13–23(b). At this point the capacitor is charged to a negative voltage ($-V$) proportional to the input analog voltage.

Now the capacitor discharges linearly because of the constant current from the $-V_{REF}$ as shown in Figure 13–23(c). This linear discharge produces a positive-going ramp on the A_1 output, starting at $-V$ and having a constant slope that is independent of the charge voltage. As the capacitor discharges, the counter advances from its RESET state. The time it takes the capacitor to discharge to zero depends on the initial voltage $-V$ (proportional to V_{in}) because the discharge rate (slope) is constant. When the integrator (A_1) output voltage reaches zero, the comparator (A_2) switches to the LOW state and disables the clock to the counter. The binary count is latched, thus completing one conversion cycle. The binary count is proportional to V_{in} because the time it takes the capacitor to discharge depends only on $-V$, and the counter records this interval of time.

Successive-Approximation Analog-to-Digital Converter

Perhaps the most widely used method of A/D conversion is successive-approximation. It has a much faster conversion time than the other methods with the exception of the flash method. It also has a fixed conversion time that is the same for any value of the analog input.

Figure 13–24 (p. 702) shows a basic block diagram of a 4-bit successive-approximation ADC. It consists of a DAC, a successive-approximation register (SAR), and a comparator. The basic operation is as follows: The input bits of the DAC are enabled (made equal to a 1) one at a time, starting with the most significant bit (MSB). As each bit is enabled, the comparator produces an output that indicates whether the analog input voltage is greater or less than the output of the DAC. If the DAC output is greater than the analog input, the comparator's output is LOW, causing the bit in the register to reset. If the output is less than the analog input, the 1 bit is retained in the register. The system does this with the MSB first, then the next most significant bit, then the next, and so on. After all the bits of the DAC have been tried, the conversion cycle is complete.

(a) Fixed-interval, negative-going ramp (while the counter counts up to n)

(b) End of fixed-interval when the counter sends a pulse to control logic to switch SW to the $-V_{REF}$ input

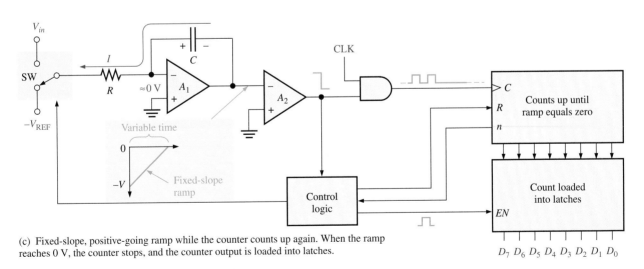

(c) Fixed-slope, positive-going ramp while the counter counts up again. When the ramp reaches 0 V, the counter stops, and the counter output is loaded into latches.

FIGURE 13–23

Illustration of dual-slope conversion.

FIGURE 13–24
Successive-approximation ADC.

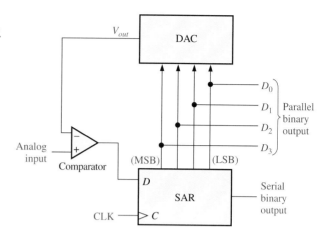

In order to better understand the operation of the successive-approximation ADC, let's take a specific example of a 4-bit conversion. Figure 13–25 illustrates the step-by-step conversion of a constant analog input voltage (5.1 V in this case). Let's assume that the DAC has the following output characteristic: $V_{out} = 8$ V for the 2^3 bit (MSB), $V_{out} = 4$ V for the 2^2 bit, $V_{out} = 2$ V for the 2^1 bit, and $V_{out} = 1$ V for the 2^0 bit (LSB).

Figure 13–25(a) shows the first step in the conversion cycle with the MSB = 1. The output of the DAC is 8 V. Since this is greater than the analog input of 5.1 V, the output of the comparator is LOW, causing the MSB in the SAR to be reset to a 0.

Figure 13–25(b) shows the second step in the conversion cycle with the 2^2 bit equal to a 1. The output of the DAC is 4 V. Since this is less than the analog input of 5.1 V, the output of the comparator switches to a HIGH, causing this bit to be retained in the SAR.

Figure 13–25(c) shows the third step in the conversion cycle with the 2^1 bit equal to a 1. The output of the DAC is 6 V because there is a 1 on the 2^2 bit input and on the 2^1 bit

FIGURE 13–25
Illustration of the successive-approximation conversion process.

input; 4 V + 2 V = 6 V. Since this is greater than the analog input of 5.1 V, the output of the comparator switches to a LOW, causing this bit to be reset to a 0.

Figure 13–25(d) shows the fourth and final step in the conversion cycle with the 2^0 bit equal to a 1. The output of the DAC is 5 V because there is a 1 on the 2^2 bit input and on the 2^0 bit input; 4 V + 1 V = 5 V.

The four bits have all been tried, thus completing the conversion cycle. At this point the binary code in the register is 0101, which is approximately the binary value of the analog input of 5.1 V. Additional bits will produce an even more accurate result. Another conversion cycle now begins, and the basic process is repeated. The SAR is cleared at the beginning of each cycle.

A Specific Analog-to-Digital Converter

The ADC0804 is an example of a successive-approximation ADC. A block diagram is shown in Figure 13–26. This device operates from a +5 V supply and has a resolution of eight bits with a conversion time of 100 μs. Also, it has guaranteed monotonicity and an on-chip clock generator. The data outputs are tristate, so they can be interfaced with a microprocessor bus system.

FIGURE 13–26
The ADC0804 analog-to-digital converter.

The basic operation of the device is as follows: The ADC0804 contains the equivalent of a 256-resistor DAC network. The successive-approximation logic sequences the network to match the analog differential input voltage ($V_{in+} - V_{in-}$) with an output from the resistive network. The MSB is tested first. After eight comparisons (sixty-four clock periods), an 8-bit binary code is transferred to output latches, and the interrupt ($\overline{INTR}$) output goes LOW. The device can be operated in a free-running mode by connecting the $\overline{INTR}$ output to the write ($\overline{WR}$) input and holding the conversion start ($\overline{CS}$) LOW. To ensure startup under all conditions, a LOW $\overline{WR}$ input is required during the power-up cycle. Taking $\overline{CS}$ low anytime after that will interrupt the conversion process.

When the $\overline{WR}$ input goes LOW, the internal successive-approximation register (SAR) and the 8-bit shift register are reset. As long as both $\overline{CS}$ and $\overline{WR}$ remain LOW, the ADC remains in a RESET state. Conversion starts one to eight clock periods after $\overline{CS}$ or $\overline{WR}$ makes a LOW-to-HIGH transition.

When a LOW is at both the $\overline{CS}$ and $\overline{RD}$ inputs, the tristate output latch is enabled and the output code is applied to the D_0 through D_7 lines. When either the $\overline{CS}$ or the $\overline{RD}$ input returns to a HIGH, the D_0 through D_7 outputs are disabled.

SECTION 13–3 REVIEW

1. What is the fastest method of analog-to-digital conversion?
2. Which A/D conversion method uses an up/down counter?
3. Does the successive-approximation converter have a fixed conversion time?

13–4 ■ TROUBLESHOOTING DACs AND ADCs

Basic testing of DACs and ADCs includes checking their performance characteristics—such as monotonicity, offset, linearity, and gain—and checking for missing or incorrect codes. In this section, the fundamentals of testing these analog interfaces are introduced.

After completing this section, you should be able to

■ Discuss the testing of DACs for nonmonotonicity, differential nonlinearity, low or high gain, and offset error   ■ Discuss testing ADCs for a missing code, incorrect code, and offset

Testing Digital-to-Analog Converters

The concept of DAC testing is illustrated in Figure 13–27. In this basic method, a sequence of binary codes is applied to the inputs, and the resulting output is observed. The binary code sequence extends over the full range of values from 0 to $2^n - 1$ in ascending order, where n is the number of bits.

FIGURE 13–27
Basic test setup for a DAC.

The ideal output is a straight-line stairstep as indicated. As the number of bits in the binary code is increased, the resolution is improved. That is, the number of discrete steps increases, and the output approaches a straight-line linear ramp.

D/A Conversion Errors

Several D/A conversion errors to be checked for are shown in Figure 13–28, which uses a 4-bit conversion for illustration purposes. A 4-bit conversion produces fifteen discrete steps. Each graph in the figure includes an ideal stairstep ramp for comparison with the faulty outputs.

(a) Nonmonotonic output (blue)

(b) Differential nonlinearity (blue)

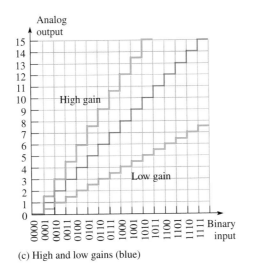

(c) High and low gains (blue)

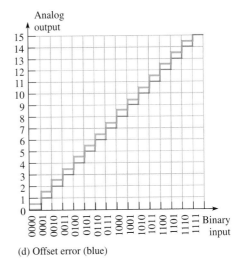

(d) Offset error (blue)

FIGURE 13–28
Illustrations of several D/A conversion errors.

Nonmonotonicity The step reversals in Figure 13–28(a) indicate nonmonotonic performance, which is a form of nonlinearity. In this particular case, the error occurs because the 2^1 bit in the binary code is interpreted as a constant 0. That is, a short is causing the bit input line to be stuck LOW.

Differential Nonlinearity Figure 13–28(b) illustrates differential nonlinearity in which the step amplitude is less than it should be for certain input codes. This particular output could be caused by the 2^2 bit having an insufficient weight, perhaps because of a faulty input resistor. We could also see steps with amplitudes greater than normal, if a particular binary weight were greater than it should be.

Low or High Gain Output errors caused by low or high gain are illustrated in Figure 13–28(c). In the case of low gain, all of the step amplitudes are less than ideal. In the case of high gain, all of the step amplitudes are greater than ideal. This situation may be caused by a faulty feedback resistor in the op-amp circuit.

Offset Error An offset error is illustrated in Figure 13–28(d). Notice that when the binary input is 0000, the output voltage is nonzero and that this amount of offset is the same for all steps in the conversion. A faulty op-amp may be the culprit in this situation.

EXAMPLE 13–4   The DAC output in Figure 13–29 is observed when a straight 4-bit binary sequence is applied to the inputs. Identify the type of error, and suggest an approach to isolate the fault.

FIGURE 13–29

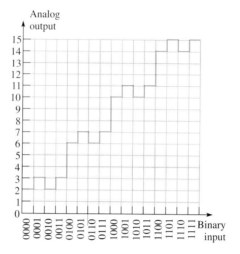

Solution   The DAC in this case is nonmonotonic. Analysis of the output reveals that the device is converting the following sequence, rather than the actual binary sequence applied to the inputs.

> 0010, 0011, 0010, 0011, 0110, 0111, 0110, 0111, 1010, 1011, 1010, 1011,
> 1110, 1111, 1110, 1111

Apparently, the 2^1 bit is stuck in the HIGH (1) state. To find the problem, first monitor the bit input pin to the device. If it is changing states, the fault is internal to the DAC and it should be replaced. If the external pin is not changing states and is always HIGH, check for an external short to $+V$ that may be caused by a solder bridge somewhere on the circuit board.

Related Problem   Determine the output of a DAC when a straight 4-bit binary sequence is applied to the inputs and the 2^0 bit is stuck HIGH.

Testing Analog-to-Digital Converters

One method for testing ADCs is shown in Figure 13–30. A DAC is used as part of the test setup to convert the ADC output back to analog form for comparison with the test input.

FIGURE 13–30
A method for testing ADCs.

A test input in the form of a linear ramp is applied to the input of the ADC. The resulting binary output sequence is then applied to the DAC test unit and converted to a stairstep ramp. The input and output ramps are compared for any deviation.

A/D Conversion Errors

Again, a 4-bit conversion is used to illustrate the principles. Let's assume that the test input is an ideal linear ramp.

Missing Code The stairstep output in Figure 13–31(a) indicates that the binary code 1001 does not appear on the output of the ADC. Notice that the 1000 value stays for two intervals and then the output jumps to the 1010 value.

In a flash ADC, for example, a failure of one of the comparators can cause a missing-code error.

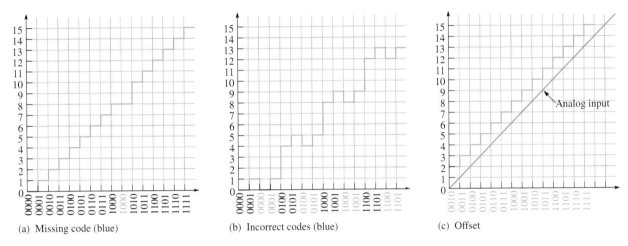

(a) Missing code (blue)   (b) Incorrect codes (blue)   (c) Offset

FIGURE 13–31
Illustrations of A/D conversion errors.

Incorrect Codes The stairstep output in Figure 13–31(b) indicates that several of the binary code words coming out of the ADC are incorrect. Analysis indicates that the 2^1-bit line is stuck in the LOW (0) state in this particular case.

Offset Offset conditions are shown in 13–31(c). In this situation the ADC interprets the analog input voltage as greater than its actual value. This error is probably due to a faulty comparator circuit.

EXAMPLE 13–5

A 4-bit flash ADC is shown in Figure 13–32(a). It is tested with a setup like the one in Figure 13–30. The resulting reconstructed analog output is shown in Figure 13–32(b). Identify the problem and the most probable fault.

Solution The binary code 0011 is missing from the ADC output, as indicated by the missing step. Most likely, the output of comparator 3 is stuck in its inactive state (LOW).

(a)

(b)

FIGURE 13–32

Related Problem Reconstruct the analog output in a test setup like in Figure 13–30 if the ADC in Figure 13–32(a) has comparator 8 stuck in the HIGH output state.

SECTION 13–4 REVIEW

1. How do you detect nonmonotonic behavior in a DAC?
2. What effect does low gain have on a DAC output?
3. Name two types of output errors in an ADC.

13–5 ■ INTERNAL SYSTEM INTERFACING

The concept of the bus was introduced in Chapter 12 in relation to memories. All the components in a given digital system such as a computer are interconnected by buses, which serve as communication paths. Physically, a bus is a set of conductive paths that serves to interconnect two or more functional components of a system or several diverse systems. Electrically, a bus is a collection of specified voltage levels and/or current levels and signals that allow the various devices connected to the bus to work properly together.

After completing this section, you should be able to

■ Discuss the concept of a multiplexed bus ■ Explain the reason for tristate outputs

Basic Multiplexed Buses

Many digital systems such as computers are microprocessor-based. In such systems the microprocessor controls and communicates with other devices, such as memories and input/output (I/O) devices, via the *internal bus structure,* as indicated in Figure 13–33.

FIGURE 13–33
The interconnection of microprocessor-based system components by a bidirectional, multiplexed bus.

A **bus** is multiplexed so that any of the devices connected to it can either send or receive data to or from one of the other devices. A sending device is often called a **source,** and a receiving device is often called an **acceptor.** At any given time, there is only one source active. For example, the RAM may be sending data to the input/output (I/O) interface under control of the microprocessor.

Bus Signals

With synchronous bus control, the microprocessor usually originates all control and timing signals. The other devices then synchronize their operations to those control and timing signals. With asynchronous bus control, the control and timing signals are generated jointly by a source and an acceptor. The process of jointly establishing communication is called **handshaking.** A simple example of a handshaking sequence is given in Figure 13–34.

An important control function is called **bus arbitration.** Arbitration prevents two sources from trying to use the bus at the same time.

FIGURE 13–34
An example of a handshaking sequence.

Connecting Devices to a Bus

Tristate drivers (see Chapter 15) are normally used to interface the outputs of a source device to a bus. Usually more than one source is connected to a bus, but only one can have access at any given time. All the other sources must be disconnected from the bus to prevent **bus contention.**

Tristate circuits are used to connect a source to a bus or disconnect it from a bus, as illustrated in Figure 13–35(a) for the case of two sources. The select input is used to connect either source A or source B but not both at the same time to the bus. When the select input is LOW, source A is connected and source B is disconnected. When the select input is HIGH, source B is connected and source A is disconnected. A switch equivalent of this action is shown in part (b) of the figure.

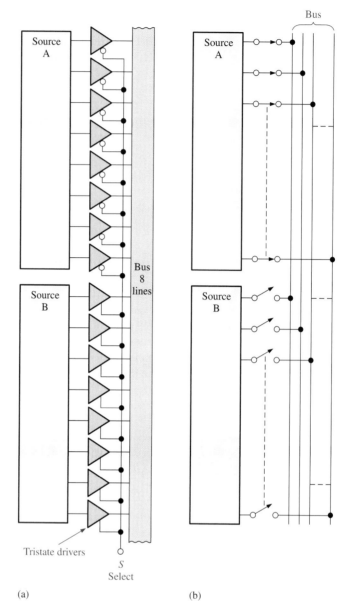

FIGURE 13–35
Tristate driver interface to a bus.

When the enable input of a tristate circuit is not active, the device is in a high-impedance **(high-Z)** state and acts like an open switch. Many digital ICs provide internal tristate drivers for the output lines. A tristate output is indicated by a ∇ symbol as shown in Figure 13–36.

Tristate Buffer Operation Figure 13–37(a) shows the logic symbol for a noninverting tristate buffer with an active-HIGH enable. Part (b) of the figure shows one with an active-LOW enable.

The basic operation of a tristate buffer can be understood in terms of switching action as illustrated in Figure 13–38. When the enable input is active, the gate operates as a normal noninverting circuit. That is, the output is HIGH when the input is HIGH and LOW when the input is LOW, as shown in parts (a) and (b) respectively. The HIGH and LOW levels represent two of the states. The buffer operates in its third state when the enable input is not active. In this state, the circuit acts as an open switch, and the output is com-

FIGURE 13–36
Method of indicating tristate outputs on an IC device.

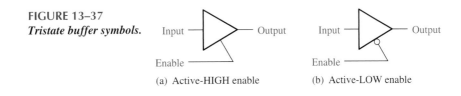

FIGURE 13–37
Tristate buffer symbols.

Input — Output   Input — Output

Enable   Enable

(a) Active-HIGH enable   (b) Active-LOW enable

FIGURE 13–38
Tristate buffer operation.

HIGH — HIGH

HIGH

(a)

LOW — LOW

HIGH

(b)

LOW or HIGH — Disconnected (high-Z)

LOW

(c)

pletely disconnected from the input, as shown in part (c). This is sometimes called the *high-impedance* or *high-Z* state. A more detailed coverage of tristate circuitry is found in Chapter 15.

Many microprocessors, memories, and other integrated circuit functions have tristate buffers that serve to interface with the buses. Such buffers are necessary when two or more devices are connected to a common bus. To prevent the devices from interfering with each other, the tristate buffers are used to disconnect all devices except the ones that are communicating at any given time.

Bus Contention

Bus contention occurs when two or more devices try to output opposite logic levels on the same common bus line. The most common form of bus contention is when one device has not completely turned off before another device connected to the bus line is turned on. This generally occurs in memory systems when switching from the READ mode to the WRITE mode or vice versa and is the result of a timing problem.

Multiplexed I/Os

Some devices that send and receive data have combined input and output lines, called I/O **ports,** that must be multiplexed onto the data bus. Bidirectional tristate drivers interface this type of device with the bus as illustrated in Figure 13–39(a).

Each I/O port has a pair of tristate drivers. When the $\overline{SND}/RCV$ (Send/Receive) line is LOW, the upper tristate driver in each pair is enabled and the lower one disabled. In this state, the device is acting as a source and sending data to the bus. When the $\overline{SND}/RCV$ line is HIGH, the lower tristate driver in each pair is enabled so that the device is acting as an acceptor and receiving data from the bus. This operation is illustrated in Figure 13–39(b). Some devices provide for multiplexed I/O operation with internal circuitry.

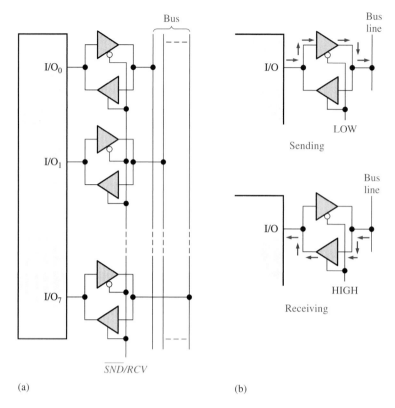

FIGURE 13–39
Multiplexed I/O operation.

SECTION 13–5 REVIEW

1. Why are tristate drivers required to interface digital devices to a bus?
2. What is the purpose of a bus system?

13–6 ■ STANDARD BUSES

A bus is a "highway" for digital signals; it consists of a set of physical connections (printed circuit traces or wires) over which the data and other information are moved from one place to another. A bus also consists of a standard set of specifications that designate the characteristics and types of signals that can travel along its pathway. Internal buses interconnect the various components within a computer system: the processor, memory, disk drive, controller, and interface cards. External or I/O buses provide for transfer of digital signals between a computer and the "outside world" and interface the computer with peripheral equipment (a video monitor, keyboard, mouse, and printer) or with other equipment that is to be controlled by a computer, such as test and measurement instruments.

After completing this section, you should be able to

■ Discuss various internal and external serial and parallel buses ■ Define local bus
■ Describe PCI and ISA internal bus standards ■ Describe the RS-232C bus ■ Describe the FireWire ■ Discuss the USB ■ Explain the GPIB ■ Discuss SCSI

Internal Buses

Internal buses in a computer carry addresses, data, and control signals between the microprocessor, cache memory, SRAM, DRAM, disk drives, expansion slots, and other internal devices. Most personal computers today consist of three types of internal buses: the *local bus,* the *PCI bus,* and the *ISA bus.* Figure 13–40 shows the basic arrangement of a bus system.

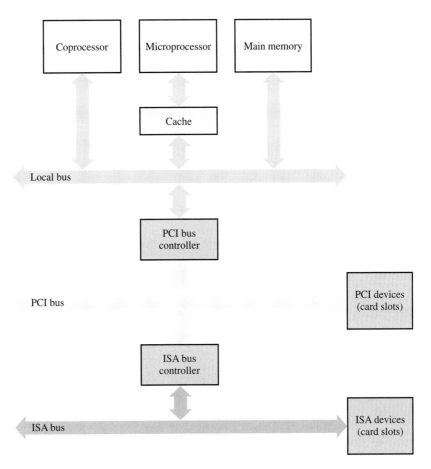

FIGURE 13–40
Simplified illustration of the basic bus system in a typical personal computer.

Local Bus   This bus directly connects the microprocessor to the cache memory, the main memory, the coprocessor, and the PCI bus controller. The **local bus** is the only internal bus that connects directly to the microprocessor. Generally, this bus includes the data bus, the address bus, and the control bus that allows the microprocessor to communicate with the other devices. The local bus can be considered as the *primary* bus in a computer system. For example, the Pentium local bus consists of 32 memory address lines, 64 data lines, and numerous control lines.

PCI (Peripheral Control Interconnect) Bus   This bus is for interfacing the microprocessor with external devices via expansion slots (connectors). The **PCI** was developed by Intel; and since it was first introduced in 1993, it has become the standard personal computer interface bus, replacing several earlier bus standards. PCI is a 64-bit bus, although it is often implemented as a 32-bit bus in which the address and data buses are multiplexed. It can operate at clock speeds of 33 MHz or 66 MHz.

The PCI bus is isolated from the local bus by a bus controller unit that acts as a "bridge" between the two buses. PCI is considered a *secondary* bus and is clocked independently of the microprocessor. The PCI can connect the microprocessor to **peripheral** devices, such as a hard drive, via expansion slots with adapter cards. The PCI card slot pin diagram is shown in Figure 13–41 for the 5 V specification. There is also a 3.3 V version for newer 3.3 V CMOS expansion cards and a universal version.

PCI supports "plug-and-play," the ability of a computer to automatically configure expansion boards and other devices. This allows a device to be connected to a computer without concern about setting switches, changing jumper wires, or dealing with any other configuration elements. This is accomplished with a 256-byte memory that allows the computer to interrogate the PCI interface.

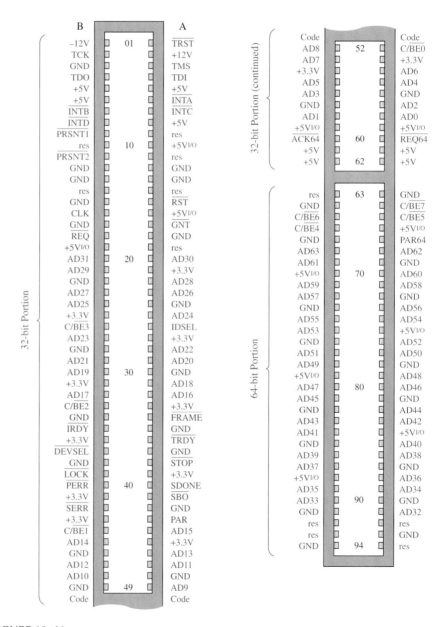

FIGURE 13–41
Pin diagram for the PCI 5 V connector slot for interface cards. The multiplexed address bus and data bus pins are prefixed by AD.

ISA (Industry Standard Architecture) Bus   This expansion bus was developed by IBM for its AT personal computer and is the standard bus into which virtually all printed circuit cards made before 1993 plug. The **ISA** is currently incorporated into most modern personal computers, as a companion to the PCI bus, for purposes of backward compatibility. Eventually, PCI will probably replace ISA altogether.

The ISA has either an 8-bit or a 16-bit data bus and can operate at 8.33 MHz. An expanded version called the EISA provides a 32-bit data bus but has largely been discontinued due to the slow speed and replaced with the PCI bus. The ISA slot is actually a two-part connector, as shown in Figure 13–42. The upper part is for 8-bit cards; both parts combined provide for 16-bit data transfer.

FIGURE 13–42
Pin diagram for the ISA connector slot.

External Buses

External devices are connected to a computer via an input/output (I/O) interface called a *port.* There are two basic types of computer ports, the *serial port* and the *parallel port,* and most computers have a parallel port and at least one serial port for connecting modems, printers, mice, and other peripheral devices.

A serial port is used for serial data communication, where only 1 bit is transferred at a time. Modems and mice are examples of typical serial devices. Also, serial ports are sometimes used for interfacing test and measurement equipment with a computer. A parallel port is used for parallel data communication, where at least 1 byte (8 bits) is transferred at a time. There are several bus standards currently in use for both serial and parallel ports. The most prominent ones are described next.

Serial I/O Interface Buses

RS-232C   This is one of the oldest and most common standards for serial interface approved by the Electronic Industries Association (EIA). The RS-232C is also referred to as the EIA-232. Most all **modems** conform to the EIA-232 standard, and most personal computers have an RS-232C port. The mouse, some display screens, and serial printers—in addition to modems—are designed to connect to the RS-232C port. The RS-232C is commonly used for interfacing data terminal equipment (**DTE**) with data communications equipment (**DCE**). For example, a computer is classified as DTE and a modem is classified as DCE.

The EIA-232 standard specifies twenty-five lines between DTE and DCE requiring a twenty-five pin plug (DB-25), as shown in Figure 13–43. In personal computer applications, not all of the RS-232C signals are required. A minimum of three and a maximum of eleven are typically used. For this reason, a 9-pin connection (DB-9) was defined by IBM for its serial interface.

FIGURE 13–43
The RS-232C 25-pin connector plug.

Figure 13–44(a) lists the signals and pin assignments for a 25-pin RS-232C connector, and part (b) lists the signals and pin assignments for a 9-pin connector. The eleven pins and signals shown in blue in part (a) indicate the signals typically used for personal computer applications, and the three minimum signals are indicated by an asterisk (pins 2, 3, 7).

The specified maximum cable length for the RS-232C is 50 feet with a rate of data transfer of 20 kbaud. If a shorter cable is used, the baud rate can be higher. The specification of data transfer rate in baud and bits per second (bps) are not necessarily always equal. This is because the baud rate is modem terminology and is defined as the number of signal changes per second, which is called the modulation rate. In modems, one signal change sometimes transfers several bits of data. At lower rates, the baud is equal to bits per second but at higher rates the baud may be less than the bits per second.

To overcome the limitations of the RS-232C, two other standards, the RS-422 and RS-423 were developed. These newer standards specify much longer cable lengths and higher data transfer rates under certain conditions. For example, both the RS-422 and RS-423 specify a maximum cable length of 4000 feet. The maximum RS-422 data transfer rate is 10 Mbaud for 40 feet of cable and 100 kbaud for 4000 feet. For the RS-423, the data

(a) Full RS-232C 25-pin interface with typical personal computer
    configuration indicated by blue and three minimum signals
    marked by an asterisk (pins 2, 3, 7).

(b) 9-pin RS-232C interface

FIGURE 13–44
The RS-232C pin assignments and signals for both connector versions.

transfer rate is 100 kbaud for 30 feet and 1 kbaud for 4000 feet. The RS-232C still remains the most common.

IEEE 1394   This is a relatively new, very fast external serial bus standard that supports data transfer rates of up to 400 Mbps and is typically used for, although not limited to, interfacing with graphics and video peripherals, such as digital cameras. The **IEEE 1394** standard is often called *FireWire,* a name trademarked by Apple computer who first developed it. Other companies use other names to describe their IEEE 1394 products. IEEE stands for *Institute of Electrical and Electronics Engineers.*

Up to 63 devices can be connected to FireWire based on a daisy-chain arrangement. The FireWire cable consists of six wires, two twisted pair for data and two for power. Also this standard allows "hot plugging," the ability to add and remove devices connected to a computer while the computer is running.

USB (Universal Serial Bus)   This relatively new external serial bus standard may eventually replace most other existing low- to medium-speed serial and parallel bus standards with, perhaps the exception of IEEE 1394, which is projected to fill a different niche in high-speed applications and play a companion role with USB. The **USB** supports two data transfer rates, a high-speed rate of 12 Mbps and a low-speed rate of 1.5 Mbps. A USB port can be used to

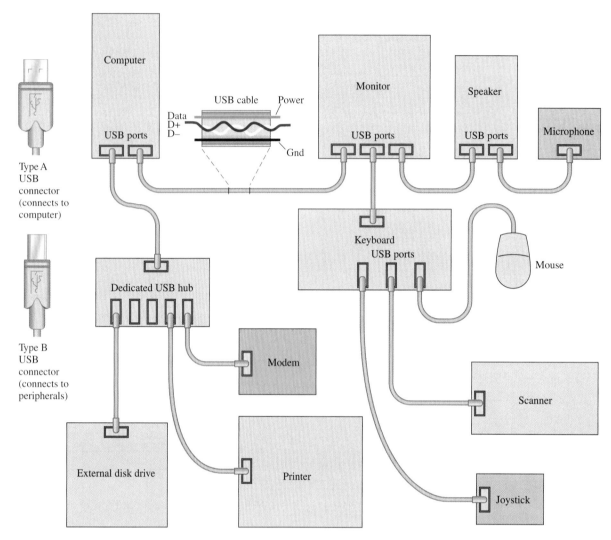

FIGURE 13–45
Example of a computer system with USB interfacing.

connect up to 127 peripheral devices and permits both plug-and-play and hot plugging. The USB cable has four wires, two for data and two for power, and connects the computer to USB peripheral devices, any of which can also act as "hubs" for connecting to other USB peripheral devices. Figure 13–45 illustrates a computer system with USB interfacing.

Parallel I/O Interface Buses

IEEE 488 This bus standard has been around a long time and is also known as the General-Purpose Interface Bus **(GPIB).** Widely used in test and measurement applications, it was developed by Hewlett-Packard in the 1960s. The **IEEE 488** specifies 24 lines that are used to transfer eight parallel data bits at a time and provide eight control signals that include three handshake lines and five bus-management lines. Also included are eight ground lines used for shielding and ground returns. Currently, the maximum data transfer rate is 1 Mbyte/s with a proposal to increase the data rate to 8 Mbytes/s (8 Mbps).

To connect test equipment to a computer using the IEEE 488 bus, an interface card is installed in the computer, which turns the computer into a system **controller.** In a typical GPIB setup, up to 14 controlled devices (test and measurement instruments) can be connected to the system controller. When the system controller issues a command for a controlled device to perform a specified operation, such as a frequency measurement, it is said that the controller "talks" and the controlled device "listens."

A **listener** is an instrument capable of receiving data over the GPIB when it is addressed by the system controller (computer). Examples of listeners are printers, monitors, programmable power supplies, and programmable signal generators. A **talker** is an instrument capable of sending data over the GPIB. Examples are DMMs and frequency counters that can output bus-compatible data. Some instruments can send and receive data and are called talker/listeners; examples are computers, modems, and certain measurement instruments. The system controller can specify each of the other instruments on the bus as either a talker or a listener for the purpose of data transfer. The controller is usually a talker/listener.

A typical GPIB arrangement is shown in Figure 13–46 as an example. The three basic bus signal groupings are shown as the *data bus, data transfer control bus,* and *interface management bus.*

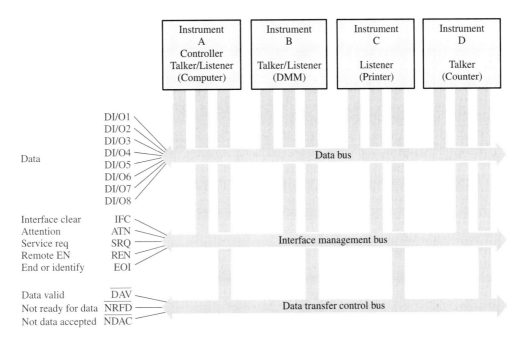

FIGURE 13–46
A typical IEEE 488 (GPIB) connection.

The parallel data lines are designated DI/O1 through DI/O8 (data input/output). One byte of data is transferred on this bidirectional part of the bus. Every byte that is transferred undergoes a handshaking operation via the data transfer control. The three active-LOW handshaking lines indicate if data are valid ($\overline{\text{DAV}}$), if the addressed instrument is not ready for data ($\overline{\text{NRFD}}$), or if the data are not accepted ($\overline{\text{NDAC}}$). More than one instrument can accept data at the same time, and the slowest instrument sets the rate of transfer. Figure 13–47 shows the timing diagram for the GPIB handshaking sequence, and Table 13–1 describes the handshaking signals.

FIGURE 13–47
Timing diagram for the GPIB handshaking sequence.

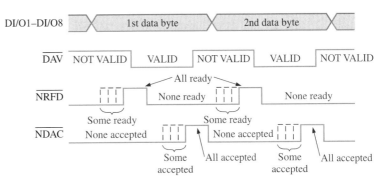

TABLE 13–1
The GPIB handshaking signals.

Name	Description
$\overline{\text{DAV}}$	**Data Valid:** After the talker detects a HIGH on the $\overline{\text{NRFD}}$ line, a LOW is placed on this line by the talker when the data on its I/O are settled and valid.
$\overline{\text{NRFD}}$	**Not Ready for Data:** The listener places a LOW on this line to indicate that it is not ready for data. A HIGH indicates that it is ready. The $\overline{\text{NRFD}}$ line will not go HIGH until all addressed listeners are ready to accept data.
$\overline{\text{NDAC}}$	**Not Data Accepted:** The listener places a LOW on this line to indicate that it has not accepted data. When it accepts data from its I/O, it releases its $\overline{\text{NDAC}}$ line. The $\overline{\text{NDAC}}$ line to the talker does not go HIGH until the last listener has accepted data.

The five signals of the interface management bus control the orderly flow of data. The ATN (attention) line is monitored by all instruments on the bus. When ATN is active, the system controller selects the specific interface operation, designates the talkers and the listeners, and provides specific addressing for the listeners. Each GPIB instrument has a specific identifying address that is used by the system controller. Table 13–2 describes the GPIB interface management lines and their functions.

TABLE 13–2
The GPIB management lines.

Name	Description
ATN	**Attention:** Causes all the devices on the bus to interpret data, as a controller command or address and activates the handshaking function.
IFC	**Interface Clear:** Initializes the bus.
SRQ	**Service Request:** Alerts the controller that a device needs to communicate.
REN	**Remote Enable:** Enables devices to respond to remote program control.
EOI	**End or Identify:** Indicates the last byte of data to be transferred.

The GPIB is limited to a maximum cable length of 15 meters, and there can be no more than one instrument per meter with a maximum capacitive loading of 50 pF each. The cable length limitation can be overcome by the use of bus extenders and modems. A bus extender provides for cable-interfacing of instruments that are separated by a distance greater than allowed by the GPIB specifications or for communicating over greater distances via modem-interfaced telephone lines. The use of bus extenders and/or modems is illustrated in Figure 13–48.

SCSI (Small Computer System Interface) Pronounced *scuzzy,* this is a widely used standard for interfacing personal computers and peripherals. Although **SCSI** is an ANSI (American National Standards Institute) standard, there are several variations and connector types originating from a variety of manufacturers. One type of SCSI may not be compatible with another type. SCSI-1 is the 25-pin connector version that provides an 8-bit

FIGURE 13–48
A bus extender and modem can be used for interfacing remote GPIB systems.

data bus and supports data transfer rates of 4 Mbytes/s. Some other versions of the SCSI bus standard are listed as follows:

- *SCSI-2*   This version is the same as SCSI-1, but uses a 50-pin connector and supports multiple devices.
- *Wide SCSI*   This uses a wider connector than SCSI-2 to support 16-bit data transfers.
- *Fast SCSI*   This provides for 8-bit data transfer but supports data transfer rates of 10 Mbytes/s.
- *Fast Wide SCSI*   This version allows 16-bit data transfer at 20 Mbytes/s.
- *Ultra SCSI*   This version transfers 8 bits of data at 20 Mbytes/s.
- *SCSI-3*   This version has 16 data lines and runs at 40 Mbytes/s.
- *Ultra SCSI-2*   This version transfers 8 bits at 40 Mbytes/s.
- *Wide Ultra SCSI-2*   This version provides for 16-bite data transfer and operates at 80 Mbytes/s.

The signal descriptions for a SCSI 25-pin connector are given in Table 13–3, and the pin configuration is shown in Figure 13–49.

TABLE 13–3
SCSI signals.

Pin Number	Signal Name	Signal Description	Pin Number	Signal Name	Signal Description
1	REQ/	Request	14	GND	Signal ground
2	MSG/	Message	15	C/D/	Command/Data
3	I/O/	Input/Output	16	GND	Signal ground
4	RST/	SCSI bus reset	17	ATN/	Attention
5	ACK/	Acknowledge	18	GND	Signal ground
6	BSY/	Busy	19	SEL/	Select
7	GND	Signal ground	20	DBP/	Data parity
8	DB0/	Data bit 0	21	DB1/	Data bit 1
9	GND	Signal ground	22	DB2/	Data bit 2
10	DB3/	Data bit 3	23	DB4/	Data bit 4
11	DB5/	Data bit 5	24	GND	Signal ground
12	DB6/	Data bit 6	25	TPWR	Terminator power
13	DB7/	Data bit 7			

FIGURE 13–49
SCSI 25-pin connector.

13 12 11 10 9 8 7 6 5 4 3 2 1

25 24 23 22 21 20 19 18 17 16 15 14

**SECTION 13–6
REVIEW**

1. Name the two major bus categories in terms of the way data are transferred.
2. Classify each of the following buses as serial or parallel:
   (a) SCSI   (b) RS-232C   (c) USB   (d) GPIB
3. Explain the basic difference between a serial bus and a parallel bus.
4. How many devices can be connected to the USB?
5. Is the FireWire a faster bus than the USB in terms of data transfer?

13–7 ■ DIGITAL SYSTEM APPLICATION

In this system application, an ADC is applied in a satellite antenna positioning system. Analog inputs indicating the position of the antenna in terms of elevation and azimuth are fed into an ADC that converts the positions into digital code for display. The antenna can be adjusted remotely until it is in a desired position.

After completing this section, you should be able to

■ Explain the basic operation of an analog multiplexer   ■ Explain the overall operation of the system   ■ Discuss the use of the ADC in this particular application
■ Develop a test procedure

General System Operation

Two angular positions are required to properly aim the antenna at a desired satellite. The azimuth position is along an arc from east to west (or north to south) parallel with the horizon. The elevation position is from horizon to horizon in an overhead arc. These angular movements of the antenna are illustrated in Figure 13–50. Two motors drive the antenna to the proper position, one for azimuth and one for elevation. Also, two position transducers (potentiometers are used as angle sensors in this application) produce voltages proportional to the current directions of the antenna as determined by the angular positions of the motor shafts.

A basic block diagram of the antenna positioning system is shown in Figure 13–51. The analog multiplexer has one input from the azimuth potentiometer and one from the elevation potentiometer. The select input alternately switches the two analog voltages to the ADC for conversion to an 8-bit digital code. During one interval, the analog voltage representing the current azimuth position is converted to digital and displayed. During the next interval, the analog voltage for the current elevation position is converted to digital and displayed. The antenna position is controlled by motor controls that can be located away from the antenna and connected by cable to the motors. By observing the azimuth and elevation position readouts, the antenna can be brought to the desired position.

The Analog Multiplexer

A general description of analog multiplexers is provided to familiarize you with this type of device before we apply it in the system.

In data acquisition systems where inputs from several different sources must be independently converted to digital form for processing, a technique called *multiplexing* is

FIGURE 13–50
Antenna dish with position controls and sensors for azimuth and elevation.

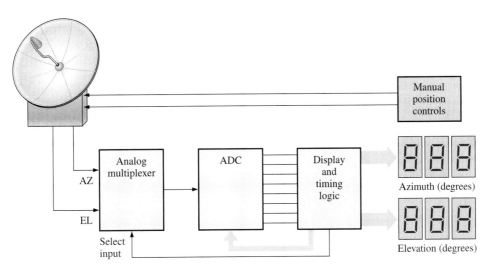

FIGURE 13–51
Simplified block diagram of the antenna positioning system.

used. A separate analog switch is used for each analog source as illustrated in Figure 13–52 for a 4-channel system. In this type of application, all of the outputs of the analog switches are connected together to form a common output and only one switch can be closed at a given time. The common switch outputs are connected to the input of an op-amp voltage follower as indicated.

A good example of an IC analog multiplexer is the AD9300 shown in Figure 13–53. This device contains four analog switches that are controlled by a channel decoder. The inputs A_0 and A_1 determine which one of the four switches is on. If A_0 and A_1 are both LOW, input In_1 is selected. If A_0 is HIGH and A_1 is LOW, input In_2 is selected. If A_0 is LOW and A_1 is HIGH, input In_3 is selected. If A_0 and A_1 are both HIGH, input In_4 is selected. The enable input controls the switch that connects or disconnects the output.

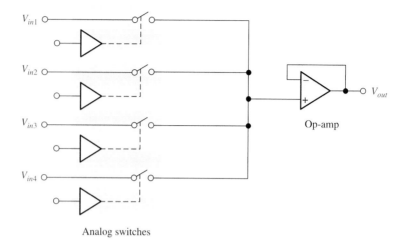

FIGURE 13–52
A basic 4-channel analog multiplexer.

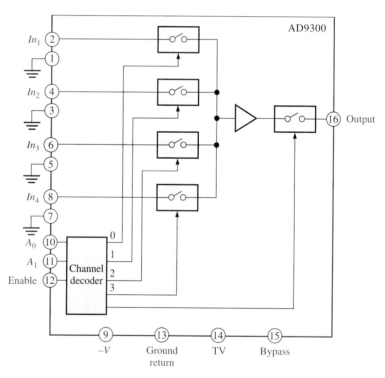

FIGURE 13–53
The AD9300 analog multiplexer.

The Potentiometer

In this application, the potentiometer is used as a position transducer to convert the angular shaft position of the motor to a proportional dc voltage. The potentiometer is mechanically linked to the associated motor so as the motor shaft turns, the wiper of the potentiometer slides along the resistive element. It is calibrated so that the smallest angle produces the smallest voltage. In Figure 13–54, a simplified diagram shows the basic construction and the schematic indicates minimum and maximum angles corresponding to the wiper position.

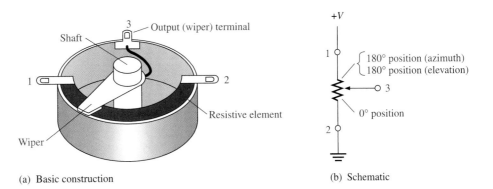

(a) Basic construction   (b) Schematic

FIGURE 13–54
The potentiometer as an angular transducer.

Basic Operation of the Analog-to-Digital Circuits

Figure 13–55 shows the specific analog-to-digital circuits used in the system. The analog voltages from the azimuth and elevation potentiometers on the antenna are fed into two of the inputs to the AD9300 analog multiplexer. The other two analog inputs are not used and are grounded. The digital timing logic produces a square wave that is used as the select input, A_0. When the square wave is LOW, Input 1 is selected and the azimuth voltage is switched to the input of the ADC. When the square wave is HIGH, Input 2 is selected and the elevation voltage is switched to the input of the ADC. Although the AD673 is used here, other ADCs such as the ADC0804 can be used.

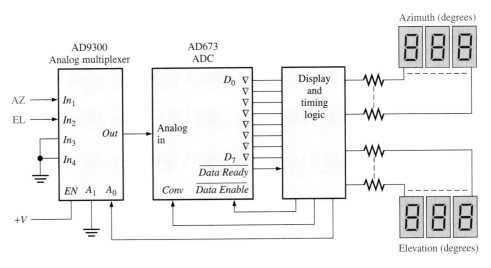

FIGURE 13–55
Conversion circuits and display/timing logic for the antenna positioning system.

As each analog voltage appears on the ADC input, the timing logic issues a convert pulse to initiate a conversion by the ADC. After the conversion is complete, the ADC asserts a LOW on the data ready line to the timing logic, which responds by asserting a LOW on the data enable line. Then the ADC places the 8-bit binary code on its outputs. The timing diagram is illustrated in Figure 13–56.

As the azimuth and elevation analog voltages are alternately converted to a binary code, the display logic stores each code in its associated register. The binary code is then converted to BCD and then to 7-segment format to illuminate the 3-digit display. The displays are updated many times a second by the multiplexing action. The updating rate depends on the frequency of the square wave used as the multiplexer select input.

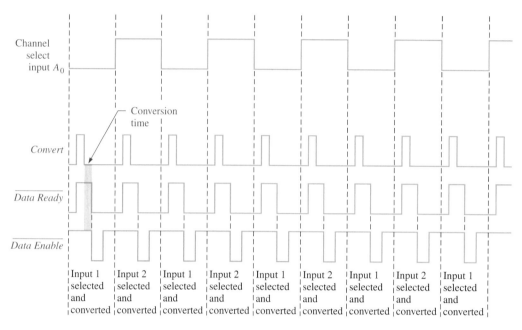

FIGURE 13–56
Basic timing diagram for the system in Figure 13–55.

■ SYSTEM ASSIGNMENTS

The basic logic diagram in Figure 13–55 and the timing diagram in Figure 13–56 indicate the general display and timing logic requirements but do not represent the detailed logic necessary to implement the system. In the following system assignments, you will develop, from given specifications, the detailed circuitry required to implement the display and timing logic functions of the system.

■ SYSTEM ASSIGNMENT 1: Design of Display Logic

This system assignment focuses on the design of the display logic for which the basic block diagram is shown below. Reference to earlier chapters and/or device data sheets is necessary.

■ *Activity 1:* **Binary-to-BCD Code Converter**
The 8-bit binary code from the ADC is to be converted to a 12-bit BCD code (three digits). Specify a circuit or combination of circuits that will accomplish this function.

■ *Activity 2:* **Azimuth and Elevation Registers**
The BCD for the azimuth position is stored in one register and the BCD for the elevation position is stored in another. The registers are alternately updated. Notice that a parallel-in/parallel-out operation is required. Specify IC registers that will meet the requirements.

■ *Activity 3:* **BCD-to-7-Segment Decoder**
A BCD-to-7-segment decoder is required for each of the three azimuth position digits and each of the three elevation position digits. Specify an appropriate device. Also specify a 7-segment display and determine the appropriate resistor values for interfacing the decoders to the displays.

■ *Activity 4:* Develop a detailed logic diagram using the devices specified in the preceding activities. Show all necessary details including device pin numbers.

■ SYSTEM ASSIGNMENT 2: Design of Timing Logic

This system assignment focuses on the design of the timing logic for which the basic timing diagram for two conversions is shown below. Reference to earlier chapters and/or device data sheets is necessary.

■ *Activity 1: Channel Select Waveform*
This waveform alternately selects channel 1 (In_1) and channel 2 (In_2) of the ADC for conversion to binary.
Specifications: Square wave with a frequency that will provide a minimum of 100 conversions of each analog input per second. Develop a circuit for generating this square wave.

■ *Activity 2: Convert Waveform*
Each positive-going convert pulse initiates a conversion in the ADC on its trailing edge. Conversion time for the AD673 is a maximum of 30 μs. There must be one pulse for each conversion. Specify a circuit to generate this waveform.
Specifications: Positive-going pulses with a minimum pulse width of 500 ns delayed from each edge of the channel select waveform by a minimum of 100 ns.

■ *Activity 3: Data Enable Waveform*
When the AD673 completes a conversion, it places a LOW on the data ready line which re-

mains LOW until the leading edge of the next convert pulse. The data enable line must remain HIGH during a conversion. At the end of a conversion, the data enable goes LOW to place the binary code on the ADC output lines. Specify a circuit to produce this waveform.
Specifications: Negative-going pulses triggered from the negative-going transition of the data ready line and with a minimum pulse width of 500 ns.

■ *Activity 4: Azimuth and Elevation Register Clock*
Determine the timing of the signal required to properly clock the parallel data into the registers. Specify how it is to be generated.

■ *Activity 5:* Develop a detailed logic diagram of the timing logic and combine with the diagram of the display logic developed in System Assignment 1.

■ *Activity 6:* Write a technical report to provide a detailed description of the operation of the system and a parts list.

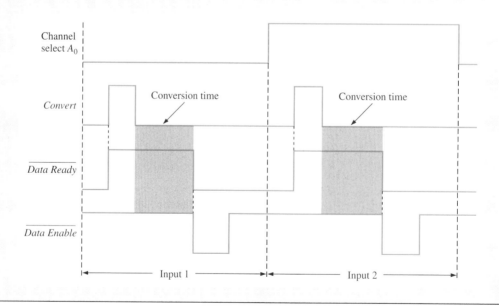

SECTION 13–7 REVIEW

1. Why are three digits required for the azimuth and elevation displays?
2. Discuss how the analog voltages representing the azimuth and elevation positions are generated.
3. Why are two registers required in the display logic?

■ SUMMARY

- ■ Types of digital-to-analog converters (DACs):
  - Binary-weighted input
  - *R*/2*R* ladder (most common)
- ■ Types of analog-to-digital converters (ADCs):
  - Flash (fastest)
  - Stairstep-ramp
  - Tracking
  - Single-slope
  - Dual-slope
  - Successive-approximation (most common)
- ■ Table 13–4 compares standard buses.

TABLE 13–4

	Internal Buses		External Buses				
	PCI	**ISA**	**RS-232C**	**IEEE 1394**	**USB**	**IEEE 488**	**SCSI**
Type	Parallel	Parallel	Serial	Serial	Serial	Parallel	Parallel
Data lines	32/64	8/16	—	—	—	8	8/16
Data rate	33/66 MHz	8.33 MHz	20 kbaud	400 Mbps	1.5/12 Mbps	1 Mbps	4 Mbps (1)
							10 Mbps (Fast)
							20 Mbps (Ultra)
							40 Mbps (3)
Number of devices	—	—	1	63	127	14	16

■ KEY TERMS

These terms are also in the end-of-book glossary.

Acceptor A receiving device on a bus.

Analog-to-digital converter (ADC) A device used to convert an analog signal to digital form.

Bus A set of interconnections that interface one or more devices based on a standardized specification.

Bus arbitration The process that prevents two sources from using a bus at the same time.

Bus contention An adverse condition that could occur if two or more devices try to communicate at the same time on a bus.

Controller An instrument that can specify each of the other instruments on the bus as either a talker or a listener for the purpose of data transfer.

DCE Data communications equipment.

Digital-to-analog converter (DAC) A device used to convert a digital input to an analog signal.

DTE Data terminal equipment.

Flash ADC A simultaneous analog-to-digital converter.

GPIB General-purpose interface bus based on the IEEE 488 standard.

Handshaking The process of signal interchange by which two digital devices or systems jointly establish communication.

High-Z The high-impedance state of a tristate circuit in which the output is effectively disconnected from the rest of the circuit.

IEEE 488 bus Same as the GPIB. A standard parallel bus used widely for test and measurement interfacing.

IEEE 1394 bus A serial bus for high-speed data transfer; also known as FireWire.

Interfacing The process of making two or more electronic devices or systems operationally compatible with each other so that they can function properly together.

ISA bus Industry standard architecture bus; an internal parallel bus standard.

Listener An instrument capable of receiving data on a GPIB.

Local bus An internal bus that connects the microprocessor to the cache memory, the main memory, the coprocessor, and the PCI bus controller.

Modem A modulator/demodulator for interfacing digital devices to analog transmission systems such as telephone lines.

Monotonicity The characteristic of a DAC defined by the absence of any incorrect step reversals; one type of digital-to-analog linearity.

PCI bus Peripheral control interconnect bus; an internal parallel bus standard.

Peripheral A device or instrument that provides communication with a computer or provides auxiliary services or functions for a computer.

Port The physical interface between a computer and a peripheral.

SCSI Small computer system interface; an external parallel bus standard.

Source A sending device on a bus.

Talker An instrument capable of transmitting data on a GPIB.

Tristate A type of output in logic circuits that exhibits three states: HIGH, LOW, and High-Z.

USB Universal serial bus; an external serial bus standard.

■ SELF-TEST

1. In a binary-weighted digital-to-analog converter (DAC), the resistors on the inputs
   (a) determine the amplitude of the analog signal
   (b) determine the weights of the digital inputs
   (c) limit the power consumption
   (d) prevent loading on the source

2. Each resistor value in a binary-weighted DAC is
   (a) the same as the other values
   (b) half of the feedback resistor
   (c) twice the value of the next lowest value resistor
   (d) half the value of the next lowest value resistor

3. In an $R/2R$ DAC, there are
   (a) four values of resistors   (b) one resistor value
   (c) two resistor values   (d) a number of resistor values equal to the number of inputs

4. An 8-bit DAC has a resolution of
   (a) 0.1%   (b) 0.392%   (c) 1%   (d) 3.92%

5. Ideally, the accuracy of a 4-bit DAC is
   (a) approximately 3.1%   (b) approximately 0.31%
   (c) approximately 6.2%   (d) approximately 0.62%

6. The type of analog-to-digital converter (ADC) with the fastest conversion time is
   (a) flash   (b) stairstep-ramp
   (c) tracking   (d) simultaneous-conversion

7. Most devices are interfaced to a bus with
   (a) totem-pole outputs   (b) tristate drivers
   (c) *pnp* transistors   (d) resistors

8. The PCI bus consists of
   (a) 8 or 16 data lines   (b) 32 or 64 data lines   (c) 1 serial data line

9. The devices operating on a GPIB are called
   (a) source and load   (b) talker and listener
   (c) transmitter and receiver (d) donor and acceptor

10. The RS-232C is
   (a) a standard interface for parallel data   (b) a standard interface for serial data
   (c) an enhancement of the IEEE 488 interface (d) the same as SCSI

11. The FireWire bus is the same as the
   (a) IEEE 488 bus   (b) USB   (c) IEEE 1394
   (d) RS-422   (e) RS-423

12. The USB can support up to
   (a) 63 devices   (b) 14 devices
   (c) 100 devices   (d) 127 devices

■ PROBLEMS

SECTION 13–1   Digital and Analog Interfacing

1. The analog curve in Figure 13–57 is sampled at 1 ms intervals. Represent the total curve by a series of 4-bit binary numbers.

FIGURE 13–57

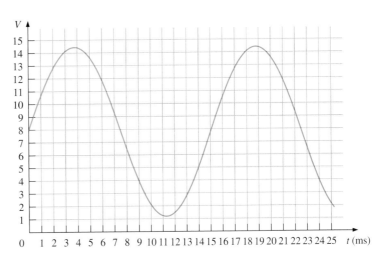

2. Sketch the digital reproduction of the curve in Problem 1.

3. Graph the analog function represented by the following sequence of binary numbers:

   1111, 1110, 1101, 1100, 1010, 1001, 1000, 0111, 0110, 0101, 0100, 0101, 0110, 0111, 1000, 1001, 1010, 1011, 1100, 1100, 1100, 1011, 1010, 1001.

SECTION 13–2   Digital-to-Analog (D/A) Conversion

4. The input voltage to a certain op-amp inverting amplifier is 10 mV, and the output is 2 V. What is the closed-loop voltage gain?

5. To achieve a closed-loop voltage gain of 330 with an inverting amplifier, what value of feedback resistor do you use if $R_i = 1.0 \text{ k}\Omega$?

6. In the 4-bit DAC in Figure 13–7, the lowest-weighted resistor has a value of 10 kΩ. What should the values of the other input resistors be?

7. Determine the output of the DAC in Figure 13–58(a) if the sequence of 4-bit numbers in part (b) is applied to the inputs. The data inputs have a low value of 0 V and a high value of +5 V.

8. Repeat Problem 7 for the inputs in Figure 13–59.

9. Determine the resolution expressed as a percentage, for each of the following DACs:

    (a) 3-bit   **(b)** 10-bit   **(c)** 18-bit

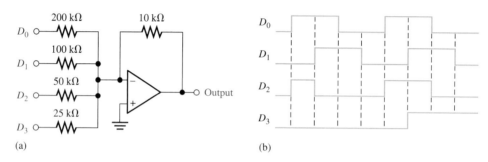

(a)

(b)

FIGURE 13–58

FIGURE 13–59

SECTION 13–3   Analog-to-Digital (A/D) Conversion

10. Determine the binary output code of a 3-bit flash ADC for the analog input signal in Figure 13–60. The sampling rate is 100 kHz.

FIGURE 13–60

FIGURE 13–61

11. Repeat Problem 10 for the analog waveform in Figure 13–61.

12. For a 3-bit stairstep-ramp ADC, the reference voltage advances one step every microsecond. Determine the encoded binary sequence for the analog signal in Figure 13–62.

13. For a 4-bit stairstep-ramp ADC, assume that the clock period is 1 μs. Determine the binary sequence on the output for the analog signal in Figure 13–63.

FIGURE 13–62

FIGURE 13–63

14. Repeat Problem 12 for a tracking ADC.

15. For a certain 4-bit successive-approximation ADC, the maximum ladder output is +8 V. If a constant +6 V is applied to the analog input, determine the sequence of binary states for the SAR.

SECTION 13–4   Troubleshooting DACs and ADCs

16. Develop a circuit for generating an 8-bit binary test sequence for the test setup in Figure 13–27.

17. A 4-bit DAC has failed in such a way that the MSB is stuck in the 0 state. Draw the analog output when a straight binary sequence is applied to the inputs.

FIGURE 13–64

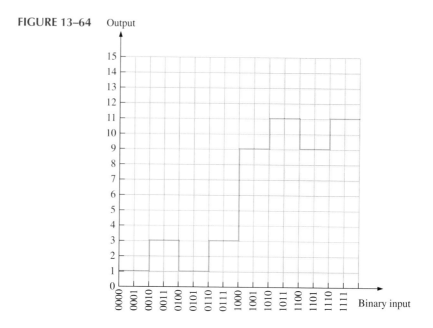

18. A straight binary sequence is applied to a 4-bit DAC, and the output in Figure 13–64 is observed. What is the problem?

19. An ADC produces the following sequence of binary numbers when an analog signal is applied to its input: 0000, 0001, 0010, 0011, 0100, 0101, 0110, 0111, 0110, 0101, 0100, 0011, 0010, 0001, 0000.

   (a) Reconstruct the input digitally.

   (b) If the ADC failed so that the code 0111 were missing, what would the reconstructed output look like?

SECTION 13–5 Internal System Interfacing

20. In a simple serial transfer of eight data bits from a source device to an acceptor device, the handshaking sequence in Figure 13–65 is observed on the four generic bus lines. By analyzing the time relationships, identify the function of each signal, and indicate if it originates at the source or at the acceptor.

21. Determine the signal on the bus line in Figure 13–66 for the data-input and enable waveforms shown.

FIGURE 13–65

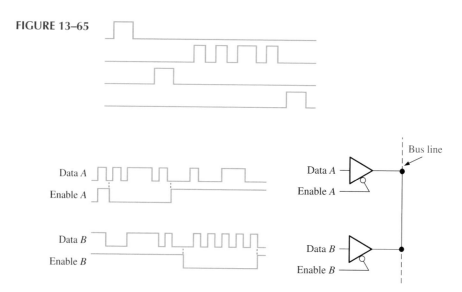

FIGURE 13–66

22. In Figure 13–67(a), data from the two sources are being placed on the data bus under control of the select line. The select waveform is shown in Figure 13–67(b). Determine the data-bus waveforms for the device output codes indicated.

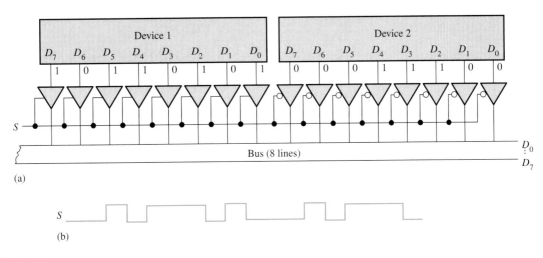

(a)

(b)

FIGURE 13–67

SECTION 13–6 Standard Buses

23. Explain the basic difference between a local bus and the PCI bus.

24. Define "plug-and-play."

25. How does the PCI bus differ from the ISA bus?

26. If a shorter RS-232C is used, can data be transmitted at a faster rate?

27. DCE and DTE are part of which bus specification? Explain the acronyms, DCE and DTE.

28. List the wires in a USB cable.

29. Eight GPIB-compatible instruments are connected to the bus. How many more can be added without exceeding the specifications?

30. Consider the GPIB interface between a talker and a listener as shown in Figure 13–68(a). From the handshaking timing diagram in part (b), determine how many data bytes are actually transferred to the listening device.

(a)

(b)

FIGURE 13–68

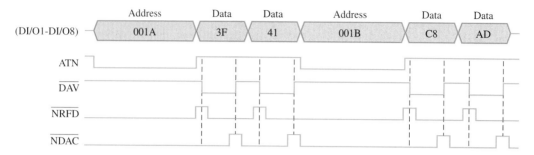

FIGURE 13–69

31. Describe the operations depicted in the GPIB timing diagram of Figure 13–69. Develop a basic block diagram of the system involved in this operation.

32. A talker sends a data byte to a listener in a GPIB system. Simultaneously, a DTE sends a data byte to a DCE on an RS-232C interface. Which system will receive the complete data byte first? Why?

SECTION 13–7   Digital System Application

33. Determine the maximum frequency for the select input waveform with the limitations specified in system assignment 2.

34. Determine the resolution of the ADC used in the antenna positioning system.

35. Determine the smallest incremental change in the azimuth and elevation angles that can be measured assuming ranges from 0° to 180°.

36. The antenna positioning system can be simplified by limiting it to positioning in one dimension at a time; that is, you must first position it in the azimuth and then in the elevation. Develop a basic block diagram for this approach.

EWB Troubleshooting

These problems require your EWB compact disk.

37. Open file PR13_37.EWB and determine the fault, if any, with the circuit.
38. Open file PR13_38.EWB and determine the fault, if any, with the circuit.
39. Open file PR13_39.EWB and determine the fault, if any, with the circuit.
40. Open file PR13_40.EWB and determine the fault, if any, with the circuit.

■ ANSWERS TO SECTION REVIEWS

SECTION 13–1

1. In nature, quantities are analog.
2. To convert an analog quantity to digital form is A/D conversion.
3. To convert a digital quantity to analog form is D/A conversion.

SECTION 13–2

1. In a binary-weighted DAC, each resistor has a different value.
2. $[1/(2^4 - 1)]100 = 6.67\%$

SECTION 13–3

1. Simultaneous (flash) method is faster.
2. Tracking ADC uses an up/down counter.
3. Yes, successive approximation has fixed conversion time.

SECTION 13–4

1. A step reversal indicates nonmonotonic behavior in a DAC.
2. Step amplitudes in a DAC are less than ideal with low gain.
3. Missing code and incorrect code are types of ADC output errors.

SECTION 13–5

1. Tristate drivers allow devices to be completely disconnected from the bus when not in use, thus preventing interference with other devices.
2. A bus interconnects all the devices in a system and makes communication between devices possible.

SECTION 13–6

1. Serial and parallel data transfer
2. **(a)** parallel　**(b)** serial　**(c)** serial　**(d)** parallel
3. Serial—one bit at a time; Parallel—8 or more bits at a time.
4. 127 USB devices
5. FireWire is faster than USB.

SECTION 13–7

1. Three digits are required to represent angles from 0° to 180°.
2. The position of the potentiometer wiper is controlled by the motor shaft position which corresponds to the antenna position. The voltage output is proportional to the wiper position.
3. Two registers are required to store the current azimuth and elevation position codes.

■ ANSWERS TO RELATED PROBLEMS FOR EXAMPLES

13–1 See Figure 13–70.

FIGURE 13–70

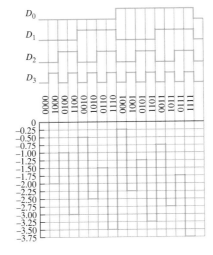

13–2 $[1/(2^{16} - 1]100 = 0.00153\%$

13–3 010, 100, 101, 110, 110, 111, 110, 110, 101, 100, 011, 010, 000, 000, 001, 001, 010, 011, 100, 101, 110, 110, 111, 111

13–4 See Figure 13–71.

13–5 See Figure 13–72.

FIGURE 13–71

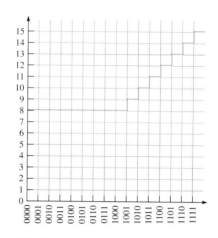

FIGURE 13–72

14

INTRODUCTION TO MICRO-PROCESSORS AND COMPUTERS

■ CHAPTER OBJECTIVES

❏ Name the basic elements of a microprocessor
❏ Name the basic units of a computer
❏ Discuss the characteristics and development of the Intel microprocessor family
❏ Discuss the characteristics and development of the Motorola microprocessor family
❏ Distinguish between assembly language and machine language
❏ Explain the basic operation of an 8086/8088 CPU
❏ Explain the basic architecture of the 8086/8088 microprocessors
❏ Explain the multiplexed bus operation of the 8086/8088 microprocessor
❏ Discuss the software model of the Pentium processors
❏ Describe a simple assembly language program
❏ Describe the seven instruction groups for the 80X86 and Pentium processors
❏ Describe the function of the bus controller in an 8086/8088 CPU
❏ Analyze a timing diagram for a memory-read cycle and a memory-write cycle
❏ Distinguish between a dedicated I/O port and a memory-mapped I/O port
❏ Compare polled I/O, interrupt-driven I/O, and software interrupts
❏ Describe the functions of PIC and PPI devices
❏ Define and explain the advantage of DMA

■ CHAPTER OVERVIEW

This chapter provides a brief introduction to micro-processors and computers. Naturally, a single chapter must be limited because one or more chapters could easily be devoted to each of the section topics. Keep in mind, however, that the purpose here is to give you a basic introduction.

Both the Intel and Motorola microprocessor families are briefly discussed. The Intel 8088 is used as a "model" to illustrate basic microprocessor concepts with recent enhancements described in further detail. This is a valid approach because the 8086/8088 is the first generation of the Intel 80X86 family. Although the newer generations—the 80286, 80386, 80486, and Pentium family—are more powerful and contain advanced features, they are related in architecture and basic functions such as the register structure.

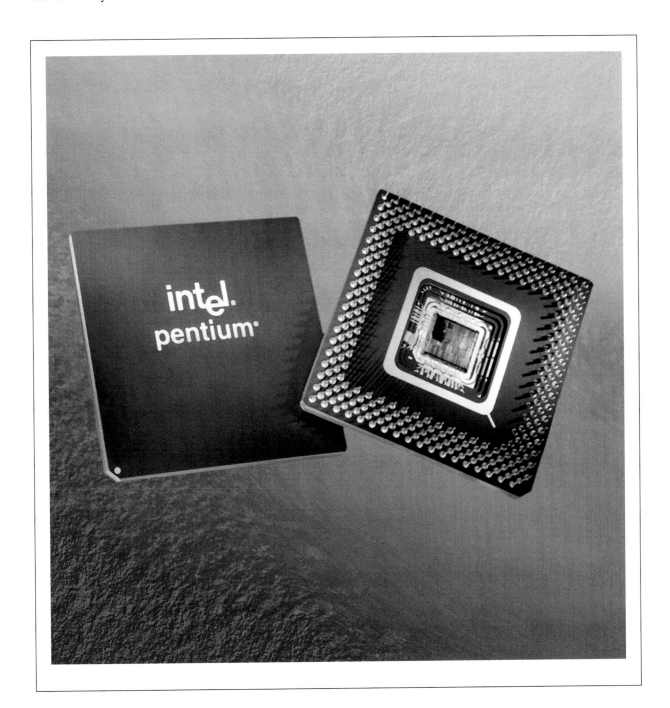

14–1 ■ THE MICROPROCESSOR AND THE COMPUTER

The microprocessor is a digital integrated circuit device that can be programmed with a series of instructions to perform specified functions on data. When a microprocessor is connected to memory and provided with input and output devices, it becomes a computer.

After completing this section, you should be able to

■ Define *microprocessor* and discuss its basic elements ■ Explain the function of the ALU, the register array, and the control unit ■ Describe the address bus, the data bus, and the control bus ■ Determine the amount of memory that a microprocessor can access based on the size of its address bus ■ Define *assembly language* ■ Describe the basic elements in a computer ■ Discuss what each part of a computer does ■ Explain what a peripheral device is

The Computer Revolution

Over the last 25 years, the microprocessor has revolutionized the computer field. Prior to microprocessors, computers were universally very large, very expensive, and almost none were owned by individuals. That all changed with the introduction of the **microprocessor,** a large-scale integrated circuit that contained the entire central processing unit of a computer on a single integrated circuit. The first computers that used microprocessors were not much more than hobby computers and were called "home computers" or microcomputers. In 1981, IBM entered the small computer market with its personal computer (commonly referred to as a PC), which was based on the Intel 8088 microprocessor. Computers that are compatible with the original Intel architecture are still referred to as PCs, but they are much more than "personal" computers. Today, microprocessor-based computers have become the workhorse computers for business and industry as well as individuals.

Basic Elements of a Computer System

All computer systems consist of basic functional blocks that include a *central processing unit* (CPU), *memory,* and *input/output ports.* These functional blocks are connected together with three buses, as shown in the block diagram of Figure 14–1. The three buses are the data bus, the address bus, and the control bus. Input and output devices are connected through the input/output ports. A **port** is a physical interface on a computer through which data are passed to and from peripherals.

Instructions and data are stored in memory in specific locations determined by the **program,** a group of instructions designed to solve a specific problem. Each location has a unique **address** associated with it. Instructions are obtained by the CPU by placing an address on the address bus. Instructions are transferred via the data bus as they are requested by the CPU. The CPU executes the instructions sequentially; frequently, the instructions modify data stored in memory or obtained from an input device. Processed data may be stored back in memory or sent to an output device via the data bus. Signals on the control bus are generated by the CPU to coordinate all of these operations, which are described in more detail in this chapter.

Microprocessors

As previously defined, a microprocessor is a large-scale integrated circuit that contained the entire CPU of a computer on a single integrated circuit. As microprocessor-based computers evolved, they have become much more powerful and have widely replaced traditional mainframe computers in business and industry.

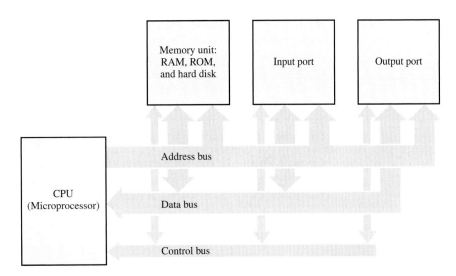

FIGURE 14–1
Basic computer block diagram.

Microprocessors are used in powerful workstations and servers as well as in peripheral devices such as printers and disk drives. In addition, microprocessors are used in many applications involving control and monitoring functions—including industrial and machine control, engine controls, and data collection to name a few. For many of these control applications, a cousin of the microprocessor known as a **microcontroller** has found widespread application.

Within a given microprocessor are several units, each designed for a specific job. The specific units, their design and organization, are called the microprocessor's **architecture.** The architecture determines the instruction set and the process for executing those instructions. Three of the most basic units, common to all microprocessors, are the *arithmetic logic unit* (ALU), the *register array,* and a *control unit,* as shown in Figure 14–2. Other units work in conjunction with these three basic units to form a specific microprocessor.

FIGURE 14–2
Three basic elements of a microprocessor.

One interesting form of microprocessor is called a **coprocessor.** A coprocessor is actually a microprocessor designed with a limited instruction set optimized to perform arithmetic operations very quickly. As microprocessors changed, one of the changes has been to incorporate the coprocessor inside what was originally just the microprocessor. Thus, the coprocessor for the Pentium is actually included inside the Pentium.

Arithmetic Logic Unit The **ALU** is the key processing element of the microprocessor. It is directed by the control unit to perform arithmetic operations such as addition and subtraction and logic operations such as NOT, AND, OR, and exclusive-OR. Data for the ALU are obtained from the register array.

Register Array The **register array** is the collection of registers that are contained within the microprocessor. During the execution of a program, data and addresses are temporarily stored in registers that make up this array. The ALU can access the registers very quickly, making the program run more efficiently. Some registers are classed as general-purpose registers, meaning they can be used for any purpose dictated by the program. Other registers have specific capabilities and functions and cannot be used as general-purpose registers. Still others are called *program invisible registers,* used only by the processor and not available to the programmer.

Control Unit The **control unit** is "in charge" of the processing of instructions. It provides the timing and control signals for getting data into and out of the microprocessor and for synchronizing the execution of instructions.

Microprocessor Buses

The three buses mentioned earlier are part of the internal and external connections for microprocessors to allow data, addresses, and control signals to be moved.

The Address Bus The **address bus** is a "one-way street" over which the microprocessor sends an address code to a memory or other external device. The size or width of the address bus is specified by the number of conductive paths. The first microprocessors used in computers had 16 address lines, which could address 65,536 (2^{16}) unique locations (this is called 64k). The more bits there are in the address bus, the more memory locations a given microprocessor can access. The number of address lines increased to 20, 24, and 32 bits as microprocessor technology advanced. The Pentium, with 32-bit address lines, can address over 4,295,000,000 (4G) memory locations!

The Data Bus The **data bus** is a "two-way street" on which data or instruction codes are transferred into the microprocessor or on which the result of an operation or computation is sent out from the microprocessor. The original microprocessors had data buses that were eight bits wide. Depending on the particular microprocessor, the data bus size is 8 bits, 16 bits, 32 bits, or 64 bits wide.

The Control Bus The **control bus** is used by the microprocessor to coordinate its operations and to communicate with external devices. The control bus has signals that enable either the memory or an input/output port at the proper time to read or write data. Control lines are also used to insert special wait states for slower devices and prevent bus contention, a condition that could occur if two or more devices try to communicate at the same time.

Microprocessor Programming

All microprocessors work with a basic instruction set, formulated by the designers of the processor. The Pentium, for example, has hundreds of variations of its instructions divided into seven basic groups:

■ Data transfer
■ Arithmetic
■ Bit manipulation
■ Loops and jumps
■ Strings
■ Subroutines and interrupts
■ Processor control

Each instruction consists of a binary string (1s and 0s) that is decoded by the microprocessor before being executed. These instruction groups are described in Section 14–4. The binary code instructions are called *machine language* and are all that the microprocessor recognizes. The first computers were programmed by actually writing the instructions in binary code, which was a tedious job and prone to error.

To simplify the task of writing instructions for a computer, assembly language was devised. **Assembly language** is classified as a low-level language because the English-like instructions contained in a given assembly language can be directly translated into binary instructions or data patterns. An **assembler,** which is a program, converts the English-like instructions, called **mnemonics,** in the assembly language into the machine language (binary patterns) used by the microprocessor, as illustrated in Figure 14–3. Other instructions used in assembly language describe data structures and other information needed by the assembler. These instructions are called **pseudo-operations** because they are not translated into actual instructions for the microprocessor.

FIGURE 14–3
Block diagram of microprocessor programming.

Assembly language and the corresponding machine language is specific to the type of microprocessor or microprocessor family. For example, assembly instructions written for the Motorola 600 series processors will not work on the Intel family of processors. The advantage of assembly language is that the programmer has direct control of the process, and usually codes written in it execute faster. In addition, the programmer can manage machine processes, such as interrupts or I/O devices, better with assembly language than most other computer languages.

On the other hand, high-level programming languages such as BASIC, Pascal, C, or FORTRAN are independent of the type of microprocessor in a computer system. A program called a **compiler** or interpreter translates the high-level program statements into machine language. The advantage is that a general-purpose language can be designed for specific types of problems, such as business applications. For general problem solving, high-level languages are superior to assembly language. Programming is discussed further in Section 14–4.

A Computer System

For a computer to accomplish a given task, it must communicate with the "outside world" by interfacing with people, sensing devices, or devices to be controlled. A typical computer system is shown in Figure 14–4. Usually, there is a keyboard for data entry, a video monitor, and a printer. A mouse is also used on most computers. Other types of peripherals (such as an external disk drive, modem, scanner, graphics tablet, or voice input) are often part of a system.

Central Processing Unit The part of the computer in control of running a program is the **CPU** which, for microprocessor-based computers, is the microprocessor itself. Basically, the CPU addresses a memory location, obtains (fetches) a program instruction that is stored

FIGURE 14–4
A typical computer system.

there, and carries out (executes) the instruction. After completing one instruction, the CPU moves on to the next one (although newer microprocessors can operate on more than one instruction simultaneously). This fetch and execute process is repeated until all of the instructions in a specific program have been executed. A simple example of an application program is a set of instructions stored in the memory as binary codes that directs the CPU to fetch a series of numbers also stored in the memory, add them, and store the sum back in the memory.

Memory Unit   The memory unit typically consists of RAM, ROM, and hard disk memories. The RAM stores data and programs temporarily during processing. Data are numbers or other information in binary form, and programs are lists of instructions. Because the RAM is generally a volatile memory, everything must be backed up in nonvolatile disk storage.

The ROM stores system programs such as the **BIOS** (basic input/output system). The **BIOS** contains certain instructions telling the CPU what to do when power is first turned on and directs the power on self-test or POST and reports any errors. It also contains routines for handling video display graphics, the keyboard, asynchronous communications, as well as other peripherals. These programs can be accessed by other programs though a series of software interrupts.

Input Ports   Generally, the computer receives information from an input device via an input port. (Some microprocessors do not use ports; they communicate with I/O devices through locations in memory specifically reserved for the I/O device as will be discussed in Section 14–7). The keyboard, mouse, and hard disk are examples of input devices that are accessed by the CPU typically via individual input ports.

Output Ports   The computer sends information to output devices via an output port (or through reserved memory locations). An example of a peripheral with which a computer communicates through an output port is the video monitor. Some peripherals function as both input and output devices; examples are external disk drives, and modems. For devices that require two-way communication, a separate input and output port is set up for the device.

SECTION 14–1 REVIEW

1. Name the basic elements of a computer.
2. Describe each of the three buses used by a microprocessor.
3. What are two types of information stored in memory?
4. What is assembly language? What two types of operations are in assembly language?

14–2 ■ MICROPROCESSOR FAMILIES

The Intel Corporation is a pioneer in the development and marketing of microprocessors. Today, Intel microprocessor products have a large share of the world market and are used in most PC-compatible computers. This section contains a brief history and description of the evolution of the 80X86 microprocessor family from the 8086/8088 to the Pentium I, II, and III. Motorola also has enjoyed significant success with its line of microprocessors. The first Apple computer was based on the Motorola 6800, and all Apple Macintosh computers today use Motorola microprocessors. As in the 80X86 Intel family, each Motorola microprocessor builds on the basic design established by the first device with added features and improved performance. Our emphasis is on main features and improvement from one generation of devices to the next. Keep in mind that microprocessor technology advances at a fast pace, and performance parameters change as current devices are improved or replaced with new versions.

After completing this section, you should be able to

■ Discuss the Intel 80X86 family of microprocessors ■ Discuss the Motorola 680X0 family of microprocessors ■ Describe a cache memory and state its purpose ■ Describe pipelining and state its purpose

The Early Intel Microprocessors

Intel's first microprocessor was the 4004, introduced in 1971. It was designed for the Busicom calculator and had only a 4-bit data bus. This device was followed by the 8008, which had an 8-bit data bus and was used by Don Lancaster, a computer hobbyist, to create an early computer. Two more 8-bit microprocessors, the 8080 and the 8085 were introduced in the mid-1970s. These two devices could address only $2^{16} = 65,536$ memory locations (64 kbytes of memory) with their 16-bit address buses. The 8080 microprocessor was used in the first computer available to the public, the Altair.

The 80X86 Family of Microprocessors

Since its introduction in 1978, the so-called 80X86 architecture has undergone a number of evolutionary stages. The first generation of the 80X86 family includes the 8086, the 8088, and the 80186. Next came the 80286, followed by the 80386, and then the 80486. The Pentium is the fifth Intel microprocessor; and now it is followed by the Pentium Pro, Pentium II, and Pentium III microprocessors, frequently referred to as simply processors. Another version of the Pentium microprocessor, dubbed the Celeron, was introduced in 1998. Each new microprocessor built upon the basic concepts of the 8086/8088 with additional features and improved performance.

There are three basic methods for improving the ability of a microprocessor to process data. Chip designers are always trying to increase the clock speed, but a major problem with just increasing the speed is that the circuit dissipates more power (increased heat) as the clock speed is increased. A second method is to increase the size of the data bus, to bring more information to the processor. Finally, the internal architecture can be changed to execute instructions more efficiently. An internal architecture change of particular importance was the introduction of pipelining and duplicate execution units for parallel processing. As you review the history of microprocessors, you will see that each of the three methods mentioned has been instrumental in improving the performance as new microprocessors were introduced and are key parts of microprocessor specifications.

The 8086/8088 and 80186   Developed in 1978, the 8086 was the first of the 80X86 family and is the basis for all Intel microprocessors that followed. The 8086 was a 16-bit microprocessor (16-bit data bus) and represented a significant departure from the earlier 8-bit devices. The number of address lines in the 8086 increased to twenty, making it capable of addressing $2^{20} = 1,048,576$ memory locations (1 Mbyte). The 8086 was a factor of ten improvement over the earlier 8080 processor. The various versions of the 8086 operated at clock frequencies of 5 MHz, 8 MHz, or 10 MHz.

The 8088 is essentially an 8086 with the 16-bit internal data bus multiplexed down to an 8-bit external data bus. It was intended to meet the demand for applications in simpler 8-bit systems and was used in the original IBM computer (PC). The success of the IBM computer led Fortune magazine to name the company "one of the business triumphs of the seventies."

The 80186 is an 8086 with several support functions such as clock generator, system controller, interrupt controller, and direct memory access (DMA) controller integrated on the chip. An increased clock frequency of 12.5 MHz was added, and the 5 MHz rate available in the 8086/8088 was dropped, resulting in a selection of 8 MHz, 10 MHz, or 12.5 MHz. The 80186 represented a relatively small increase in performance over the 8086; it was never used as the CPU of a commercial computer.

The 80286 (286)   The 80286 was introduced in 1982, and the memory addressing capability was increased to 24 address lines. This provides for addressing $2^{24} = 16,777,216$ memory locations (16 Mbytes). Even more importantly, an advanced mode of operation called *protected mode,* was first incorporated in the 80286 and used in all of the following Intel processors. This mode allows access to additional memory locations and advanced programming features such as **multitasking,** an operating system environment in which the computer seems to run multiple programs or tasks simultaneously. In addition, the 286 added a new functional addressing unit devoted to virtual addressing, a concept that allows a large program to execute in a smaller physical memory. Although the 80286 operates at the same clock frequencies as the 80186, it represented a major change in architecture and added a number of new instructions devoted to multitasking. Even with all of the improvements, the 286 could execute all of the code written for the earlier 8086.

The 80386 (386)   In 1985, a major step was taken when the first Intel 32-bit microprocessor was introduced. The 80386 has both a 32-bit data bus and a 32-bit address bus. The DX version has a 32-bit external data bus while the SX version has a 16-bit external data bus. With 32 address bits, $2^{32} = 4,294,967,296$ memory locations (4 Gbytes) can be accessed.

A new paging mechanism was added to the 80386 that enabled more efficient use of memory yet retained the upward compatibility for earlier processors that was necessary. The process of forming the physical address was broken into two layers of lookup tables where all of the information about the address was stored. This process enabled any program written for the 8086 to have its own 1 Mbyte memory area, as if it were operating alone in the original 8086 memory. Operating systems (like OS/2) could run several tasks on a rotating basis so that several users had concurrent access to programs.

The 386 is compatible with all earlier models, including the register array. Although the general-purpose registers are 16 bits wide on all earlier machines, they were extended to 32 bits on the 386 and given important new capabilities. In addition to some other features, the 80386 was the first Intel microprocessor to enhance processing speed with the use of instruction pipelining. This technique allows the microprocessor to start working on a new instruction before it has completed the current one. Versions that operate at clock frequencies of 16 MHz, 20 MHz, 25 MHz, or 33 MHz are available in the 80386.

Other more economical versions of the 80386 include the 80386SX and the 80386SL in which the data bus is multiplexed down to 16 bits and the address bus down to 24 bits. All of the 80386 microprocessors can also operate with a math (numeric) coprocessor that can be used to enhance or support the functions of the microprocessor by performing floating-point operations.

The 80486 (486)   The next step in the evolutionary process was the introduction of the 80486 with an internal 8-kbyte cache memory in 1989. As you learned in Chapter 12, cache memory reduces memory access time by storing recently used instructions or data in a very fast SRAM rather than having the information in the relatively slow main memory (DRAM). Also, an internal math coprocessor was incorporated, and the internal architecture was redesigned to speed certain instructions. Although there were relatively few changes to the programming model between the 386 and the 486, the introduction of cache memory and the other improvements resulted in a significantly better processor. The 80486 was Intel's first processor that had over 1 million transistors. Certain versions of the 80486 operate at clock frequencies up to 66 MHz.

The Pentium   The Pentium, introduced in 1993, retains the 32-bit address bus of the 80486 but doubles the data bus to 64 bits. The Pentium retained the 32-bit registers of the 486 but made it possible to deal with the 64-bit data bus by adding a second execution unit. The Pentium also has two 8-kbyte cache memories, one for instructions and one for data. A dual pipeline method, known as *superscalar architecture,* increases the speed at which instructions can be processed above that of the single pipelining method introduced in the 80386. The Pentium has internal circuitry to recognize when both pipelines may be used; in those instances, the Pentium can execute two instructions at once. Another improvement was a much improved math coprocessor. The original Pentium operated at a clock frequency of 60 MHz or 66 MHz.

Pentium Pro   The Pentium Pro, code-named the P6, was introduced in 1995 and followed the basic architecture of the original Pentium. Designed for high-end computers, work stations, and servers, the Pentium Pro contained three integer units, a floating-point unit, and both level one and level two cache memory. The Pentium Pro is packaged with a second, speed-enhancing cache memory chip. It operates at clock speeds up to 200 MHz.

Pentium II   The Pentium II was introduced in 1997 and incorporates Intel MMX technology designed specifically to process video, audio, and graphics data efficiently. The MMX technology improves video compression/decompression, image manipulation, encryption, and I/O processing. The Pentium II has 57 new instructions designed to speed the repetitive sequences often found in multimedia operations. The Pentium II is compatible with all of the earlier processors from the 8086 on. Additional members of the Pentium II family were introduced in 1998, including the Celeron processor, designed for basic PC market. By 1999, the clock speed of the Pentium II was up to 450 MHz, and the clock speed of the Celeron processor was up to 400 MHz. Further speed enhancements are expected in the future.

Pentium III   The Pentium III was introduced in 1999. It was designed for the consumer and business market desktop computers. The Pentium III included 70 new instructions designed to enhance 3D, imaging, and video applications. Initial computers with the Pentium III operated at 500 MHz, but the Pentium III has been tested (by Intel) at speeds of over 1 GHz using special cooling techniques.

The Motorola 6800 and Its Variations

The first microprocessors from Motorola were all 8-bit devices (8-bit data bus). The 6800 appeared in 1975 with a clock frequency of 2 MHz and was capable of addressing 64 kbytes of memory with a 16-bit address bus. The internal architecture included two accumulators, a program counter, index register, and a stack pointer but no general-purpose registers. (An **accumulator** is a register with special capability for arithmetic and logic operations.) For use as a "scratch-pad," the 6800 used RAM memory instead of general-purpose registers.

In the 6802, a 128-byte RAM was added for use in the place of some registers. The clock frequency was increased in the 6803 to 3.58 MHz and a UART (universal asynchronous receiver/transmitter) was added for serial communications. The last of the 8-bit microprocessors was the 6809, which offered an enhanced instruction set that included a multiply instruction. All of the 8-bit devices retained the 16-bit address bus.

The 680X0 Family of Microprocessors

The 68000   The 68000 was the first of Motorola's 16-bit microprocessors and was introduced in 1979. Although it is classified as a 16-bit processor, internally it had 32-bit registers. It also had 24 address lines that could access 16 Mbytes of memory, compared to the 1 Mbyte capability of the 8086/8088. Unlike the 8086, it also had independent address and data lines. With a clock frequency of 16 MHz, the 68000 could run faster than any of the Intel devices at that time. Also, the RAM-based register concept used in the 6800 was abandoned in favor of an internal register array that included eight 32-bit general-purpose registers, seven 32-bit address registers, two 32-bit stack pointers, as well as a 32-bit program counter and a 16-bit status register. The processors that followed were all upward compatible; that is, programs written for the 68000 can run on all of the 680X0 family.

The 68020   Motorola officially entered the world of 32-bit microprocessors in 1985 with the 68020, which had an increase in performance over 16-bit processors. It could address 4 Gbytes of memory (using a 32-bit address bus) and featured a 256-byte cache memory for instructions, an innovative feature at that time. The clock frequency was increased to 33 MHz, and a math coprocessor (the 68882) and other support chips were available.

The 68030   Introduced in 1991, the 68030 added another 256-byte cache for data, and the clock speed was increased to up to 50 MHz. The address bus was increased to 32 bits. With these and other improvements that produced increased efficiency, the 68030 could execute instructions at more than double the speed of its immediate predecessor.

The 68040   Introduced in 1995, the 68040 increased the data and instruction caches to 4 kbytes each, and an on-chip math coprocessor featuring floating-point support was added. The processor also had multiple independent pipelines to enable concurrent execution of instructions and an internal floating-point unit and on-chip memory management.

The 68060   The 68060 was also introduced in 1995 and featured a superscalar architecture characterized by multiple pipelining of instructions and two 8-kbyte caches (for data and for instructions). The power consumption was reduced by lowering the operating voltage to 3.3 V. The top clock frequency is 75 MHz.

The Power PC   The MPC601 was introduced in 1992 as the first of a series 64-bit microprocessors with superscalar architecture that can effectively execute up to three instructions per clock cycle. It has 32 kbytes of cache memory and an internal math coprocessor. The Power PC is a RISC type of microprocessor. RISC means reduced instruction set computer, an approach that reduces the number of instructions available but optimizes the reduced instruction set and results in overall improved performance. The latest addition to the Power PC family is the 750, announced in 1999. It features a 32-kbyte instruction cache and a 32-kbyte data cache. The speed of the 750 is up to 400 MHz.

SECTION 14–2
REVIEW

1. What is meant by the term *32-bit microprocessor?*
2. What was the first 16-bit Intel microprocessor?
3. What is meant by RISC?
4. What is the purpose of a cache memory?
5. What is superscalar architecture?

14–3 ■ THE 8086/8088 MICROPROCESSOR AND SOFTWARE MODEL FOR THE PENTIUM PROCESSOR

As you saw in the last section, the 80X86 microprocessor family has undergone a tremendous change from the 8086/8088 to the Pentium family, both in speed and in complexity. However, the basic register set and other features of the 8086/8088 have been retained (and expanded) throughout the evolutionary process so that all of the newer Intel processors respond to the same instructions (as well as a number of new instructions) as the original 8086/8088. This section starts with a limited introduction of the 8086/8088 to introduce basic concepts of microprocessor architecture, operation, and programming. The section ends with a brief overview of the principal changes to the register structure that forms the software model of the newer processors. The approach is to show a basic processor (the 8086/8088) and discuss the enhancements to the Intel line as it has evolved.

After completing this section, you should be able to

- Discuss the basic microprocessor operation ■ Describe the bus interface unit
- State the purpose of the segment registers ■ State the purpose of the instruction pointer ■ Describe the execution unit ■ Describe the general set of registers
- State the purpose of the flag register

Basic Operation

A microprocessor executes a program (list of instructions) by repeatedly cycling through the following three steps:

1. Fetch an instruction from memory and place it in the CPU.
2. Decode the instruction; if other information is required by the instruction, fetch the other information. In the decode step, the program counter is updated to point to the next instruction.
3. Execute the instruction (do what the instruction says). Results are returned to registers and memory during this step.

Prior to the 8086/8088 microprocessors, these steps were performed sequentially; the 8085 processor could not fetch an instruction at the same time it was executing it. The architecture of the 8086/8088 was fundamentally different. It provided for two separate internal units: the execution unit (**EU**), which executes instructions, and the bus interface unit (**BIU**), which interfaces with the system buses and fetches instructions, reads operands, and writes results. These units are shown in Figure 14–5.

The BIU performs all the bus operations for the EU, such as data transfers from memory or I/O. While the EU is executing instructions, the BIU "looks ahead" and fetches more instructions from memory. This action is called **prefetching** or *pipelining*. The concept of prefetching is to allow the processor to execute instructions at the same time as the next instruction was being fetched, eliminating idle time. The prefetched instructions are stored in an internal high-speed memory called the instruction **queue** (pronounced "Q"). The queue allows the BIU to keep the EU supplied with instructions. The EU does not

FIGURE 14–5
The 8086/8088 has two separate internal units, the EU and the BIU.

have to wait for the next instruction to be fetched from memory; but instead it retrieves the next instruction directly from the queue in much less time. In the Pentium, this process is taken a step further. Two complete execution units enable two instructions to execute at the same time provided they are independent. Certain compilers are designed to take advantage of the two execution units by a process known as **instruction pairing** to remove dependencies.

Basic 8086/8088 Architecture

Figure 14–6 is a block diagram of the architecture (internal organization) of an 8088 microprocessor. Externally, the 8088 has 20 address bits that can address 1 Mbyte (1,048,576 bytes) of memory and uses an 8-bit data bus. Internally, the 8088 has a 16-bit data bus and a 4-byte queue. The 8086 is identical except that it has an external 16-bit data bus and a 6-byte instruction queue.

The Bus Interface Unit (BIU)

The major parts of the BIU are the 4-byte instruction queue, the segment registers (CS, DS, SS, and ES), the instruction pointer (IP), and the address summing block (Σ). The 16-bit internal data buses and the Q bus interconnect the BIU and the EU.

Instruction Queue   The instruction queue increases the average speed with which a program is executed (called the **throughput**) by storing up to four bytes (six in the 8086). As described earlier, this technique allows the 8088 essentially to do two things, fetch and execute, at one time. This feature has been expanded in subsequent processors to include much larger and faster queues.

Segment Registers   The 8086/8088 processors have four segment registers (CS, DS, SS, and ES) that are all 16-bit registers used in the process of forming a 20-bit address. A **segment** is a 64-kbyte block of memory and can begin at any point in the 1 Mbyte (1,048,576 bytes) of memory space, provided it begins on a 16-byte boundary (evenly divisible by 16).

In designing the 8086/8088 and subsequent processors, Intel chose a unique method of generating the required 20-bit physical address using two 16-bit registers. One of the registers that forms the physical address (or actual) is always a segment register; the other register is a 16-bit general register containing address information. The principal advantage of the method selected was to allow codes to be easily relocatable. A **relocatable code** can be moved anywhere within the memory space without changing the basic code.

Each of the four segments identify the starting address of a 64-kbyte (65,536-byte) block representing a "window" in the entire 1-Mbyte (20-bit) memory space. The starting

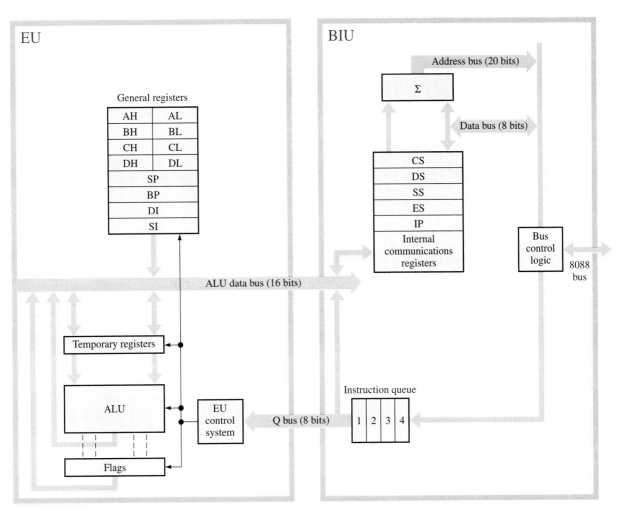

FIGURE 14–6
The internal organization of the 8088 microprocessor.

address of a segment is represented by the 16-bit number in the segment register plus an implied 4 bits appended to the right that are always assumed to be zero. In other words, the segment registers contain the most significant 16 bits that represent the physical starting address of the segment.

The four segment registers (CS, DS, SS, and ES) can be changed by the program to point to other 64-kbyte blocks if necessary. (For small codes, it is normally not necessary to change the segments.) The four segments can be separate locations within the memory space or can overlap, depending on the size and requirements of the particular code. They can even be defined as the same 64-kbyte block. In the 8086/8088, currently addressable memory segments are those defined by the segment address contained in the CS (code segment) register, the DS (data segment) register, the SS (stack segment) register, and the ES (extra segment) register. In later processors, other segment registers were added.

As mentioned, within each segment are 64 kbytes of memory. To find a given memory location, a segment address is combined with an offset address. The segment address represents the most significant sixteen bits (four hex digits) of the physical address which represent the beginning address of a segment. The offset address is sixteen additional bits that represent the distance from the start of the segment to the physical address within the segment. Figure 14–7 illustrates how the memory is be divided into segments and offsets and shows examples of nonoverlapping and overlapping segments.

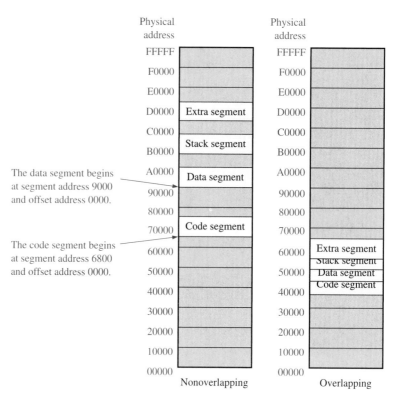

FIGURE 14–7
Nonoverlapping and overlapping segments in the first 1 Mbyte of memory. Each segment represents 64 kbytes.

Instruction Pointer (IP) and Address Summing Block The 16-bit **IP** (instruction pointer) points to the offset of the next instruction to be executed in memory. The IP always references the CS (code segment) register; thus, the physical address of the next instruction is formed by combining the code segment and the instruction pointer. The IP always contains the offset address of the next instruction, and the CS register always contains the segment address. This address is shown in assembly language as CS:IP.

To form the 20-bit physical address of the next instruction, the 16-bit offset address in the IP is added to the segment address contained in the CS register, which has been shifted four bits to the left, as indicated in Figure 14–8. As mentioned earlier, an assumed binary 0000 is the least significant position. The addition is then done by the address summing block.

Figure 14–9 illustrates the addressing of a location in memory by the segment: offset method. In this figure, $A000_{16}$ is in the segment register and $A0B0_{16}$ is in the IP. When the CS register is shifted and added to the IP, we get $A0000_{16} + A0B0_{16} = AA0B0_{16}$ for the physical address.

FIGURE 14–8
Formation of the 20-bit physical address from the segment base address and the offset address.

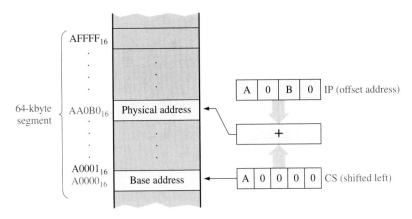

FIGURE 14–9
Illustration of the segmented addressing method.

EXAMPLE 14–1

The hexadecimal contents of the CS register and the IP are shown in Figure 14–10. Determine the physical address in memory of the next instruction.

FIGURE 14–10

| $A034_{16}$ | CS |

| $0FF2_{16}$ | IP |

Solution Shifting the CS base address left four bits (one hex digit) effectively places a 0_{16} in the LSD position, as shown in Figure 14–11. The shifted base address and the offset address are added to produce the 20-bit physical address.

FIGURE 14–11

| A 0 3 4 0 | + | 0 F F 2 | = | A 1 3 3 2 |

Base address   Offset address   Physical address
(shifted left 4 bits)   of next instruction

Related Problem Determine the physical address if the CS register contains $6B4D_{16}$.

It is important to understand how the segment:offset method is used to form the physical address; however, in programming work, it isn't usually necessary for the programmer to specify actual physical addresses. This job is done by the assembler program using labels supplied by the programmer. When a physical address is required, the programmer generally specifies it with the segment:offset method. Thus, the address for Example 14–1 would be given as simply A034:0FF2.

The Execution Unit (EU)

The EU decodes instructions fetched by the BIU, generates appropriate control signals, and executes the instructions. The main parts of the EU are the arithmetic logic unit (ALU), the general registers, and the flags.

The ALU This unit does all the arithmetic and logic operations, working with either 8-bit or 16-bit operands.

The General Registers This set of 16-bit registers is divided into two sets of four registers each, as shown in Figure 14–12. One set consists of the data registers, and the other set consists of the pointer and index registers. The **pointer** and index registers are generally used to keep offset addresses (as used here, a pointer refers to a specific memory location). In the case of the stack pointer (SP) and the base pointer (BP), the default reference to form a physical address is the stack segment (SS). The index pointers (SI and DI) and the base register (BX) generally default to the data segment (DS) register (an exception is made for certain instructions to this general rule).

FIGURE 14–12
The 8086/8088 general register set.

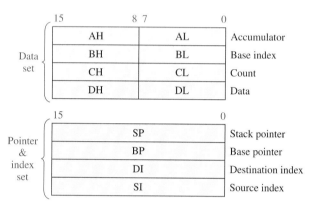

Each of the 16-bit data registers (AX, BX, CX, DX) has two separately accessible 8-bit sections. Depending on the program, they can be used either as a 16-bit register or as two 8-bit registers. The low-order bytes of the data registers are designated as *AL, BL, CL,* and *DL*. The high-order bytes are designated as *AH, BH, CH,* and *DH.* These registers can be used in most arithmetic and logic operations in any manner specified by the programmer for storing data prior to and after processing. Also, some of these registers are used specifically by certain program instructions.

The pointer and index registers are the stack pointer (SP), the base pointer (BP), the source index (SI), and the destination index (DI). These registers are used in various forms of memory addressing under control of the EU.

The Flags The flag register contains nine independent status and control bits **(flags),** as shown in Figure 14–13. A status flag is a one-bit indicator used to reflect a certain condition after an arithmetic or logic operation by the ALU, such as a carry (CF), a zero result (ZF), or the sign of a result (SF), among others. The control flags are used to alter processor operations in certain situations.

FIGURE 14–13
The 8088 status and control flags.

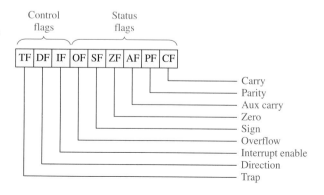

Software Model of the Pentium Family of Processors

As Intel introduced newer microprocessors, capabilities and speed increased dramatically. With the Pentium processor, the earlier pipeline concept introduced in the 8086/8088 was increased to two integer pipelines. The external coprocessor was incorporated within the microprocessor, and address and data buses were greatly expanded. Other improvements (such as clock speed, reduced instruction clock cycles, branch prediction capability, and an integral floating-point unit) made the Pentium a significantly better processor than its predecessors. In addition to processor improvements, many improvements to other parts of computers occurred (such as bus protocols and size, speed, memory size, and cost). Despite all of these changes, the designers of the newer processors maintained upward compatibility; that is, the newest Pentium could still run the software for any of the processors that preceded it. This was done by maintaining the basic software model (register structure) of the original 8086/8088.

The registers described previously for the 8086/8088 are a subset of the registers in the Pentium family of processors. Beginning with the 80386 processor, the register set was expanded to include 32-bit registers. The 32-bit registers kept the original names but an E (for Extended) was added as a prefix to the register names; thus, the 32-bit designation for the AX register is the EAX. In addition, two new segment registers were added. The extended registers are shown in Figure 14–14. The shaded areas represent the registers only available on the 386 and above.

FIGURE 14–14
Registers for the Intel processors from 8086/8088 through Pentium.

In addition to the extended registers, the addressable space in memory was increased dramatically with the introduction of newer processors. To keep the upward compatibility, Intel reserved the first 1 Mbyte of memory for codes running in real mode. **Real mode** is any operation that allows the processor to only access the first 1 Mbyte of memory to simulate the 8086/8088. Code written for an earlier processor can run in real mode on a newer processor (although the reverse is not strictly true). Code written in real mode is generally compatible (with some exceptions) with all of the Intel processors from the 8086/8088 upward.

SECTION 14–3
REVIEW

1. Name the general-purpose registers in the 8086/8088 microprocessor.
2. What is the purpose of the BIU?
3. Does the EU interface with the system buses?
4. What is the function of the instruction queue?
5. What is the advantage of the segment:offset method of forming addresses?
6. What is instruction pairing?

14–4 ■ MICROPROCESSOR PROGRAMMING

All computers must be programmed to perform even the most elementary tasks. This section focuses on the basic concept of programming in assembly language, a low-level language that is a symbolic representation of the basic machine language used by microprocessors.

After completing this section, you should be able to

■ Explain the general purpose of a computer program ■ Describe a simple assembly language program ■ Discuss how a microprocessor executes a program ■ Describe the seven instruction groups for the 80X86 and Pentium processors

The circuitry and physical components (**hardware**) of a computer are useless without programs (**software**). As mentioned earlier, a program is a list of instructions arranged to achieve a specific result. To give you an idea of how a computer carries out a task under program control, we'll look at two simple assembly language programs and use them to illustrate the process that occurs in any computer. A complete description of assembly language is beyond the scope of this text.

The programs in this section should give you some feel for computer software operation. Although only a few instructions will be introduced, the 80X86 family of microprocessors has a very large number of instructions available; this forms the basis of their *Complex Instruction Set Computer* (CISC). These processors, with their complex instruction sets, allow enormous flexibility and programming power. Because of the huge installed base of software that depends on this instruction set, it is not likely to be significantly changed in the future.

Another design philosophy emerged in the 1980s that concentrated on fewer instructions, but optimizing their execution. Processors following this architecture philosophy were called *Reduced Instruction Set Computers* (RISC) and have formed the basis of computers such as Apple's Power PC line. To keep the discussion within reasonable bounds, we will concentrate only on the Intel processors based on the original 8086/8088.

Any program, or series of instructions, must eventually be given to the microprocessor as machine language or **machine code,** a series of binary codes that the processor understands. While it is possible to program directly in machine language, the process is extremely difficult, prone to error, and time consuming. Assembly language is the next level above machine code. In assembly language, machine instructions are replaced with a mnemonic, a short alphabetic code that is easier for humans to remember. A mnemonic is frequently referred to as an **op-code.** The op-code usually is accompanied by one or two operands. An **operand** is a variable, a register, a memory location, or a value. The translation from mnemonics and operands written by the programmer to machine code understood by the microprocessor is accomplished by the assembler program, discussed earlier. There is a one-to-one correspondence between the assembly language instruction and the machine code produced by the assembler.

As an example of the programming process, let's assume we want the computer to add a list of unsigned numbers from memory and place the sum back in memory. The last number in the list is a zero, which is used as a last data indicator. The steps in our example task are as follows:

1. Clear a register for the total.
2. Point to first number in memory.
3. Check to see if number is zero. If zero, end problem.
4. If number is not zero, add the number in memory to the total.
5. Point to next number.
6. Repeat steps 3, 4, and 5.

These steps are shown on the flowchart in Figure 14–15.

FIGURE 14–15
Flowchart for adding a list of unsigned numbers.

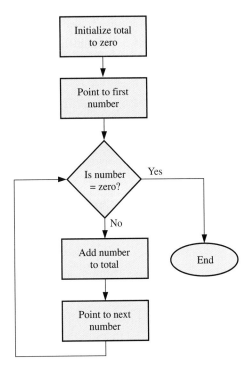

The following assembly language program implements the addition problem represented by the flowchart in Figure 14–15. Notice that the square brackets are used to indicate that a register is "pointing" to a memory location. Also, the program contains an assembler directive (can you find it?). An **assembler directive** (also called a pseudo-op) is an instruction to the assembler, not to the microprocessor. Two labels, "next" and "done" are shown in the program. Labels represent an address. Comments (not used by the processor) are preceded with a semicolon.

```
        mov ax,0             ; clear ax register for total
        mov bx,50H           ; point to a location in memory (50H)
                             ; where the data starts
next:   cmp word ptr [bx],0  ; check if the number pointed to is
                             ; zero
        jz done              ; if zero, go to "done" as this is the
                             ; last data point
        add ax,[bx]          ; add memory location pointed to by bx
                             ; to the total in ax
        add bx,02            ; point to the next number (two bytes
                             ; per word)
        jmp next             ; repeat the process
done:   mov [bx],ax          ; replace the zero in memory with the
                             ; total in ax
        nop                  ; no operation—this is here to signal
                             ; the end
```

Depending on the assembler, most programs in assembly language will have a number of assembler directives that are used by the assembler to do a variety of tasks. These tasks include setting up segments, using the appropriate instruction set, describing data sizes, and performing many other "housekeeping" functions. To simplify the explanation, only one directive (required) was shown in the preceding program. The directive was **word ptr,** which is used to indicate the size of the data pointed to by the BX register.

The Debug Assembler

With a few small changes, you can run the preceding program, if you choose, by using a built-in assembler, present in DOS-based PCs. You will be able to observe it execute step by step. All DOS-based PCs have a program called **Debug** that includes a primitive assembler. To use Debug, go into DOS and type **Debug<cr>** at the DOS prompt. (**<cr>** stands for "carriage return," which means to press the [ENTER] key.) You should see a minus sign, which is the Debug prompt. Debug has a number of commands to observe or enter data or programs. The complete list of Debug commands will be shown if you type **?** at the Debug prompt. Before writing and executing an assembly program, you can enter some data by typing the information shown in bold:

–a 50

This tells Debug to start assembling instructions at the current data segment at an offset of 50H. Although these are data that are being entered, it is simpler to enter a 16-bit word this way. (Keep in mind that all data in Debug is entered in HEX.) Debug responds with a segment address, which will undoubtedly be different than that shown (20D8), but it doesn't matter. The offset address (50H) will be the same. Type the information shown in bold (remember each **<cr>** means press the [ENTER] key).

```
20D8:0050 dw 30 <cr>
20D8:0052 dw 15 <cr>
20D8:0054 dw a0 <cr>
20D8:0056 dw 0c <cr>
20D8:0058 dw 00 <cr>
<cr>
```

The **dw** is an assembler directive. It is not stored itself; it merely informs the assembler that each data point is two bytes long. In the data shown, only one byte is used, but the program will save each point in two locations with a zero in the high-order position.

You can now enter the program at location 100 as follows:

```
          –a 100 <cr>
20D8:0100 mov ax,0 <cr>
20D8:0103 mov bx,50 <cr>
20D8:0106 cmp word ptr [bx],0 <cr>
20D8:0109 jz 112 <cr>
20D8:010B add ax,[bx] <cr>
20D8:010D add bx,2 <cr>
20D8:0110 jmp 106 <cr>
20D8:0112 mov [bx],ax <cr>
20D8:0114 nop <cr>
20D8:0115 <cr>
```

To confirm that the program has been entered correctly, you can type **u 100 114** at the Debug prompt, and the code you typed will be shown on the screen. (It will be shown in capital letters). Now type **r** after the Debug prompt, and a list of the 16-bit registers and condition of the flags will appear. Notice that the IP should have 100, the starting address of the code. Underneath the list of registers, the first instruction (MOV AX, 0000) will be shown.

You can cause Debug to execute this instruction with the **t** (trace) command. This will bring onto the screen the latest condition of all of the registers and show the next instruction (MOV BX, 0050). Executing this with a **t** command will show that the number

0050 has been moved into the BX register. The steps up to this point are shown in Figure 14–16. Notice that the first data point is shown in the lower right.

Continuing in this way, you can execute the entire code and observe the changes to the registers as the microprocessor follows the instructions, as shown in Figure 14–17. This program has a common programming structure called a *loop*. A loop is a repetitive

```
-u 100 114
20D8:0100 B80000          MOV      AX,0000
20D8:0103 BB5000          MOV      BX,0050
20D8:0106 833F00          CMP      WORD PTR [BX],+00
20D8:0109 7407            JZ       0112
20D8:010B 0307            ADD      AX,[BX]
20D8:010D 83C302          ADD      BX,+02
20D8:0110 EBF4            JMP      0106
20D8:0112 8907            MOV      [BX],AX
20D8:0114 90              NOP
-r
AX=0000   BX=0000   CX=0000   DX=0000   SP=FFEE   BP=0000   SI=0000   DI=0000
DS=20D8   ES=20D8   SS=20D8   CS=20D8   IP=0100      NV UP EI PL ZR NA PE NC
20D8:0100 B8000           MOV      AX,0000
-t

AX=0000   BX=0000   CX=0000   DX=0000   SP=FFEE   BP=0000   SI=0000   DI=0000
DS=20D8   ES=20D8   SS=20D8   CS=20D8   IP=0103      NV UP EI PL ZR NA PE NC
20D8:0103 BB5000          MOV      BX,0050
-t

AX=0000   BX=0050   CX=0000   DX=0000   SP=FFEE   BP=0000   SI=0000   DI=0000
DS=20D8   ES=20D8   SS=20D8   CS=20D8   IP=0106      NV UP EI PL ZR NA PE NC
20D8:0106 833F00          CMP      WORD PTR [BX],+00                   DS:0050=0030
-
```

FIGURE 14–16
Steps in beginning to execute the addition program with Debug.

```
DS=20D8   ES=20D8   SS=20D8   CS=20D8   IP=0110      NV UP EI PL NZ NA PO NC
20D8:0110 EBF4            JMP      0106
-t

AX=00F1   BX=0058   CX=0000   DX=0000   SP=FFEE   BP=0000   SI=0000   DI=0000
DS=20D8   ES=20D8   SS=20D8   CS=20D8   IP=0106      NV UP EI PL NZ NA PO NC
20D8:0106 833F00          CMP      WORD PTR [BX],+00                   DS:0058=0000
-t

AX=00F1   BX=0058   CX=0000   DX=0000   SP=FFEE   BP=0000   SI=0000   DI=0000
DS=20D8   ES=20D8   SS=20D8   CS=20D8   IP=0109      NV UP EI PL ZR NA PE NC
20D8:0109 7407            JZ       0112
-t

AX=00F1   BX=0058   CX=0000   DX=0000   SP=FFEE   BP=0000   SI=0000   DI=0000
DS=20D8   ES=20D8   SS=20D8   CS=20D8   IP=0112      NV UP EI PL ZR NA PE NC
20D8:0112 8907            MOV      [BX],AX                             DS:0058=0000
-t

AX=00F1   BX=0058   CX=0000   DX=0000   SP=FFEE   BP=0000   SI=0000   DI=0000
DS=20D8   ES=20D8   SS=20D8   CS=20D8   IP=0114      NV UP EI PL ZR NA PE NC
20D8:0114 90              NOP
-d 0050 005f
20D8:0050    30 00 15 00 A0 00 0C 00-F1 00 00 00 00 20 20 20    0...........
-
```

Original data   Sum

FIGURE 14–17
Last portion of tracing the addition program. The sum 00F1 is shown in blue with the low part (F1) given first.

group of instructions that are executed until some condition is met; in this case, the condition is finding a zero in the data. After the zero has been found, the last instruction will be executed and the sum will be stored (in this case, 00F1 is the hex sum) in place of the zero that indicated the last data point. You can observe this by pressing **d 0050 005F** (display between addresses 0050 and 005F) at the Debug prompt when you reach the last instruction (NOP), as shown in Figure 14–17. The result appears as the 9th and 10th bytes (the 5th word) on the line following the display instruction. Notice that the least significant part of the answer is shown first. When executed in "real time" by the microprocessor, this program actually uses only about 1 μs to do this entire process. If you choose to repeat the process, you will need to reload the zero at location 0058 because it has been replaced with the sum. (Recall that the program uses the zero as a "last data point" sensor.)

EXAMPLE 14–2   Write the instructions for an assembly language program that will find the largest unsigned number in the data and place it in the last position. Assume the last data point is signaled with a zero.

Solution   The flowchart is shown in Figure 14–18.

FIGURE 14–18
Flowchart. The variable BIG represents the largest value.

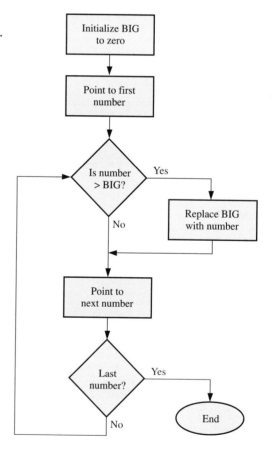

The data is assumed to be the same as before. The program listing (with comments) is as follows:

```
        mov ax,0000         ; initial value of BIG is in the ax
                            ; register
        mov bx,0050         ; point to a location in memory
                            ; (50H) where the data starts
repeat: cmp [bx],ax         ; is data point larger than BIG?
        jbe check           ; if the data point is smaller, go
                            ; to "check"
```

```
        mov ax,[bx]              ; otherwise, put new largest data
                                 ; point in ax
check:  add bx,02                ; point to the next number in
                                 ; memory (two bytes per word)
        cmp word ptr [bx],0      ; test for last data point
        jnz repeat               ; continue if the data point is not
                                 ; a zero
        mov [bx],ax              ; save BIG in memory
        nop                      ; no operation
```

The Debug listing of this program is shown in Figure 14–19. Data is entered in the same way as before starting at location 0050. In this case the same data is used, but you may choose new data if you prefer. It is important that a zero be entered as the last data point because the program continues until it finds this point. The zero is replaced each time the program is run with the largest data point. The program is entered by starting the assembly at location 100 by entering the program after the **a 100** command is issued.

FIGURE 14–19
Listing of Debug portion of program.

```
-a 100
20D8:0100 mov ax,0
20D8:0103 mov bx,50
20D8:0106 cmp [bx],ax
20D8:0108 jbe 10c
20D8:010A mov ax, [bx]
20D8:010C add bx,2
20D8:010F cmp word ptr [bx],0
20D8:0112 jnz 106
20D8:0114 mov [bx],ax
20D8:0116 nop
20D8:0117
```

The program can be traced to watch the execution, one step at a time. Alternatively, you can enter **g= 100 116** to "go" between address 100 and 116. Figure 14–20 shows the data before and after execution. Note that each data point is stored in two

```
-a 100
20D8:0100 mov ax,0
20D8:0103 mov bx,50
20D8:0106 cmp [bx],ax
20D8:0108 jbe 10c
20D8:010A mov ax, [bx]
20D8:010C add bx,2
20D8:010F cmp word ptr [bx],0
20D8:0112 jnz 106
20D8:0114 mov [bx],ax
20D8:0116 nop
20D8:0117                          End of data signal
-d 50 5f                         ⌢‿‿‿⌢
20D8:0050   30 00 15 00 A0 00 0C 00-00 00 00 00 00 20 20 20   0............
-g= 100 116 ‿‿‿‿‿‿‿‿‿‿‿‿
             Data before executing program

AX=00A0  BX=0058  CX=0000  DX=0000  SP=FFEE  BP=0000  SI=0000  DI=0000
DS=20D8  ES=20D8  SS=20D8  CS=20D8  IP=0116  NV UP EI PL ZR NA PE NC
20D8:0116 90              NOP
-d 50 5f
20D8:0050   30 00 15 00 A0 00 0C 00-A0 00 00 00 00 20 20 20   0...........
-                        ‿‿‿‿‿‿‿‿‿‿  ‿‿
-
-            Data are unchanged.
-
                              End of data now has
                              largest data point.
```

FIGURE 14–20
Data before and after a run.

bytes (although the data is only one byte long) because the data was defined as words. Also, note that the low byte preceded the high byte in memory. After the program is run, the last data point (formerly zero) is seen to be equal to the largest value (A0 in this example). This value is seen to be in both the AX register and in memory.

Related Problem   Explain how you could change the flowchart to find the smallest number in the list instead of the largest.

Types of Instructions

The programs in this section only show a few of the hundreds of variations of instructions available to programmers. To simplify learning the Intel instruction set, instructions are divided into seven categories. These categories are described here.

Data Transfer   The most basic data transfer instruction MOV was introduced in the example programs. The MOV instruction, for example, can be used in several ways to copy a byte, a word (16 bits), or a double word (32 bits) between various sources and destinations such as registers, memory, and I/O ports. (A better mnemonic for MOV might have been "COPY" because this is what the instruction actually does.) Other data transfer instructions include IN (get data from a port), OUT (send data to a port), PUSH (copy data onto the stack, a separate area of memory), POP (copy data from the stack), and XCHG (exchange).

Arithmetic   There are a number of instructions and variations of these instructions for addition, subtraction, multiplication, and division. The ADD instruction was used in both example programs. Other arithmetic instructions include INC (increment), DEC (decrement), CMP (compare), SUB (subtract), MUL (multiply), and DIV (divide). Variations of these instructions allow for carry operations and for signed or unsigned arithmetic. These instructions allow for specification of operands located in memory, registers, and I/O ports.

Bit Manipulation   This group of instructions includes those used for three classes of operations: logical (Boolean) operations, shifts, and rotations. The logical instructions are NOT, AND, OR, XOR, and TEST. An example of a shift instruction is SAR (shift arithmetic right). An example of a rotate instruction is ROL (rotate left). When bits are shifted out of an operand, they are lost; but when bits are rotated out of an operand, they are looped back into the other end. These logical, shift, and rotate instructions can operate on bytes or words in registers or memory.

Loops and Jumps   These instructions are designed to alter the normal (one after the other) sequence of instructions. Most of these instructions test the processor's flags to determine which instruction should be processed next. In Example 14–2, the instruction JBE and JNZ were used to alter the path. Other instructions in this group include JMP (unconditional jump), JA (jump above), JO (jump overflow), LOOP (decrement the CX register and repeat if not zero) and many others.

Strings   A **string** is a **contiguous** (one after the other) sequence of bytes or words. Strings are common in computer programs. A simple example is a sentence that the programmer wishes to display on the screen. There are five basic string instructions that are designed to copy, load, store, compare, or scan a string—either as a byte at a time or a word at a time. Examples of string instructions are MOVSB (copy a string, one byte at a time) and MOVSW (copy a string, one word at a time).

Subroutine and Interrupts   A **subroutine** is a miniprogram that can be used repeatedly but programmed only once. For example, if a programmer needs to convert ASCII numbers from a keyboard to a BCD format, a simple programming structure is to make the required instructions a separate process and "call" the process whenever necessary. Instructions in this group include CALL (begin the subroutine) and RET (return to the main program).

Processor Control This is a small group of instructions that allow direct control of some of the processor's flags and other miscellaneous tasks. An example is the STC (set carry flag) instruction.

SECTION 14–4 REVIEW

1. Define *program.*
2. What is a complex instruction set?
3. What is an op-code?
4. What is an operand?
5. What is a string?

14–5 ■ THE CENTRAL PROCESSING UNIT (CPU)

A typical CPU (or MPU) consists of a microprocessor and various associated ICs called support circuits. First, we will look at the general operation of a CPU, and then we will examine a specific implementation using the 8088 microprocessor as a basic model and discuss some differences with it and newer processors.

After completing this section, you should be able to

■ Explain the purpose of the CPU in a computer ■ Discuss system buses ■ Describe a basic CPU using an 8088 microprocessor ■ Explain the purpose of a bus controller ■ Discuss bus interfacing

As you saw in the last section, the instructions for the 80X86 family needed to be upward compatible to avoid the problem of making older programs immediately obsolete when a new processor was introduced. (Motorola did the same with its 68000 family.) Instructions for the 8086/8088 could run on the new processors (but not the reverse). Although the original 8086/8088 has long been obsolete for PCs, many controllers still use it and the related microcontroller ICs. In addition, many of the significant improvements that occurred impacted the speed of the processor (such as clock speed and pipelining), but these improvements had a relatively small effect on the basic programming structure. (An exception is in the addition of protected mode which is complicated.) For these reasons, a study of the 8086/8088 CPU and supporting ICs is a useful starting point for understanding the basic architecture of computers. We will introduce the 8086/8088 processors in this discussion, but highlight some key enhancements that have occurred with newer processors to give you a sense of important changes in technology.

An 8088-Based CPU

A typical CPU (or MPU) is implemented with a microprocessor and some additional support circuits. Figure 14–21 shows an 8088-based CPU.

The 8086/8088 Microprocessor As you know, the 8088 is basically the same as the 8086 microprocessor except that the 8088 has an 8-bit external data bus rather than a 16-bit data bus. The 8088 with its 8-bit data bus makes it compatible with many 8-bit peripheral devices. Internally, however, the 8088 can handle data in 16-bit words or in bytes, depending on how it is programmed. Also, the 8088, like the 8086, has a 20-bit address bus, which allows up to 1 Mbyte of memory to be addressed.

FIGURE 14–21
A simplified 8088-based CPU.

In the 8086/8088 processors, the data bus is combined with the lower address lines. There are a total of twenty lines used for addressing; in the 8088 these lines are AD_0–AD_7 and A_8–A_{19}. The eight lines designated AD_0–AD_7 serve as both data and address lines, using a multiplexed operation; A_8–A_{19} serve strictly for addressing. In the 8086, the lower 16 lines (AD_0–AD_{15}) are multiplexed; A_{16}–A_{19} serve strictly for addressing. The 80186/80188 also use multiplexed data and address lines, but these are the only processors in the 80X86 series that use multiplexed data and address lines. The principal advantage of this method was to eliminate pins, but these processors require extra circuitry to demultiplex these lines.

When a 20-bit address is sent out by the 8088 microprocessor, all twenty lines are used as the address bus, as illustrated in Figure 14–22(a). When the address is sent, a pulse called *ALE* (address latch enable) is also sent to signal external circuitry that a stable address is on the *AD* lines and can be latched. When data are sent or received by the microprocessor, the lower lines, (AD_0–AD_7 on the 8088) are used as a bidirectional data bus, as shown in Figure 14–22(b). The bus multiplexing is controlled by the bus controller. The bus controller is one of the support circuits. Other microprocessor input and output lines will be described in association with the appropriate support circuits.

The Clock Generator A clock generator circuit, such as the 8284, provides the basic timing signals to the 8088 and to other parts of the system including peripherals. It also provides a RESET signal to the 8088 to initialize the internal circuitry and a READY signal to synchronize the microprocessor with the rest of the system. An external resonant crystal is used to establish the frequency of oscillation (14.31818 MHz in this case), as shown in Figure 14–23. This frequency is available on the OSC output. A divide-by-three counter inside the clock generator produces a 4.772727 MHz clock with a duty cycle of 33% to run the microprocessor. This signal is available on the CLK output. Also, a 2.386363 MHz signal is derived from the CLK frequency and is available on the PCLK output of the clock generator. The 8086/8088 was the only processor that used the 8284 clock generator; however, the 80286 used a similar IC named the 82284. Later processors had many of the clock generator functions located internally to reduce the components required for the system. The Pentium requires an external oscillator running at the same frequency (currently several hundred MHz) as the system to control the basic system speed.

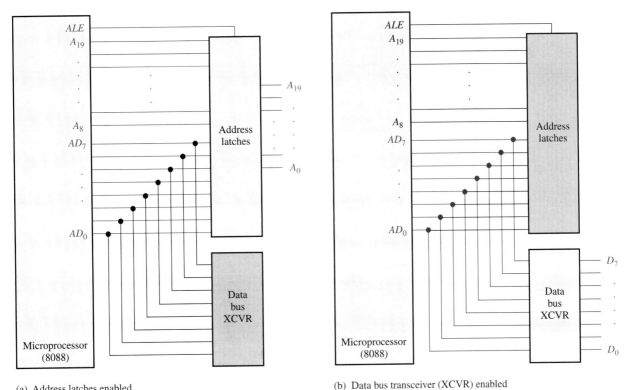

(a) Address latches enabled

(b) Data bus transceiver (XCVR) enabled

FIGURE 14–22
Multiplexed bus operation of the 8088.

FIGURE 14–23
The 8284 clock generator.

The Bus Controller The 8288 bus controller is used to relieve the 8088 microprocessor of most bus control functions in order to increase the processing efficiency. One function of the bus controller in the CPU is to provide the signals to multiplex the AD_0–AD_7 bus lines coming from the microprocessor. The bus controller generates these control signals based on the codes on the status lines ($\overline{S}_0$, $\overline{S}_1$, $\overline{S}_2$) from the microprocessor.

When the microprocessor needs to communicate with memory or I/O, it sends a code on the status lines to the bus controller. The bus controller then generates an *ALE* signal, which latches the address placed on AD_0–AD_7 and A_8–A_{19} by the microprocessor into the

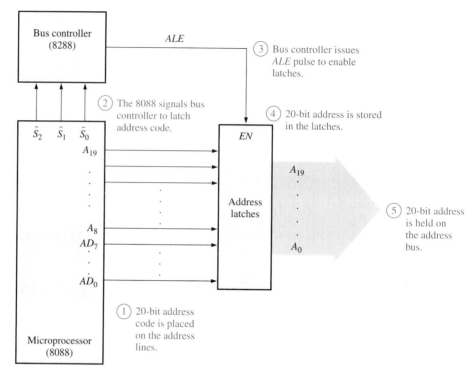

FIGURE 14–24
Steps in placing a valid address on the bus.

address latches. There are twenty latches (A_0–A_{19}), whose tristate outputs form the system address bus. Figure 14–24 illustrates the steps in this operation. During a memory read/write cycle, a valid address is available on the address bus; therefore, the lower eight bits (AD_0–AD_7) of the microprocessor's combined bus are freed up to send or receive data.

After the address has been latched, the microprocessor signals the bus controller via the status lines ($\overline{S}_0$, $\overline{S}_1$, and $\overline{S}_2$) that it wants to read data or write data to the memory location or I/O that is specified by the address that is held on the address bus. The status line codes for the four possible read/write operations are given in Table 14–1.

TABLE 14–1
The 8088 read/write status codes.

| Status Code | | | | Active Bus Controller |
$\overline{S}_2$	$\overline{S}_1$	$\overline{S}_0$	Operation	Output Command
0	0	1	Read I/O port	$\overline{IORC}$
0	1	0	Write I/O port	$\overline{IOWC}$
1	0	1	Read memory	$\overline{MRDC}$
1	1	0	Write to memory	$\overline{MWTC}$

When the bus controller receives one of the status codes, it sends a read command ($\overline{IORC}$ or $\overline{MRDC}$) or a write command ($\overline{IOWC}$ or $\overline{MWTC}$) to the appropriate device, and it also sends two signals to the tristate data bus transceiver (*XCVR*). These two signals are $DT/\overline{R}$ (data transmit/receive) and *DEN* (data enable). The *DEN* signal enables the data bus transceiver, and the $DT/\overline{R}$ signal selects the direction of data on the data bus (D_0–D_7). Figure 14–25 illustrates an 8088 memory-read operation, and Figure 14–26 on page 770 illustrates an 8088 memory-write operation. Timing diagrams for the memory read and write cycles are given in Section 14–6.

FIGURE 14–25
Example of an 8088 memory-read operation.

While many signals have been added to newer microprocessors for various tasks (cache control for example), other interface signals have remained the same. The type of bus cycle and the interrupt signals (with one addition) remain essentially the same as in the 8086/8088.

Processors starting with the 80286 did not multiplex the address or data and therefore did not require address latch circuits, but the number of pins increased and required new packaging. The 8086/8088 was originally packaged in a 40-pin dual-in-line package. Newer processors may have several hundred pins; the Pentium Pro, for example, has 387 pins. Package technology has improved dramatically in the last decade, including increased numbers of pins, increased thermal performance, and smaller sizes with significantly greater density of parts.

CPU Enhancements

A major change in the CPU since the 8086/8088 is the use of "pipelined" architecture. Pipelining speeds up the processing of instructions by fetching and executing instructions in many stages, each working in parallel. Some instructions are able to be executed in a single clock cycle in the Pentium that formerly took as much as 16 or more cycles on the 8086/8088.

A second important enhancement is the addition and expansion of cache memory. Cache memory can be either internal or external to the processor. An internal cache memory is designed to be very fast. It can store instructions or data and increases the processor speed by eliminating delays required to access external memory. The original Pentium was the first CPU to feature separate data and instruction caches.

FIGURE 14–26
Example of an 8088 memory-write operation.

A third enhancement is the inclusion of the internal math coprocessor. The internal coprocessor performs math functions much faster and more efficiently than the coprocessors from the 80486 and earlier. In addition, internal dual integer processors and branch prediction logic are added in the Pentium. These and a number of other enhancements changed the requirements for the bus control logic which is now implemented with PLDs.

SECTION 14–5 REVIEW

1. What is the purpose of the clock generator in the CPU?
2. List two functions of the bus controller.
3. Explain why bus multiplexing is necessary in an 8088-based CPU.
4. What are three enhancements to the Pentium that were unavailable in the 8086/8088?

14–6 ■ THE MEMORY

This section covers the memory portion of a typical computer and discusses how it is used by the CPU.

After completing this section, you should be able to

■ Explain address allocation in computer memories ■ Discuss the read and write operations ■ Describe a memory-read cycle ■ Describe a memory-write cycle

Address Allocation

As you have seen, the 8086/8088 CPU can address up to 1,048,576 locations with a 20-bit address bus (A_0–A_{19}). In a typical computer, a portion of the total address space is allocated to RAM and ROM and a separate portion to I/O. Input and output ports have their own unique (dedicated) memory address space in the Intel processors but use memory mapping in other processors. This is discussed further in Section 14–7.

The designers of the original IBM PC allocated the 1 Mbyte of RAM and ROM memory into three general areas: a user area, a video area, and a system area. Within these three areas, certain reserved locations were set aside for specific purposes. The lower RAM area (called conventional memory) was a 640-kbyte block that started at the bottom of memory (location 0) and was principally available for users (a small portion is set aside for use by the system). This seemed like a very large amount of user memory at the time (since it was ten times more than the 8-bit processors had available), but it is very small by today's standards. (Recall that the Pentium can address 4 Gbytes!).

Immediately above the 640-kbyte block of RAM was a reserved 128-kbyte block of video RAM area between address A0000 and BFFFF. This area was set aside for the information that is displayed on the video screen.

At the top of the available 1 Mbyte of memory, a 256-kbyte block starting at address C0000 and going to FFFFF was assigned for system programs and implemented with ROM. The last part of this ROM included the BIOS (Basic Input Output System). Upon reset, the system addressed memory locations within the BIOS to "boot" the system. BIOS also reserved a small section of RAM (between address 400 and 4FF) as a work area, and the system used other parts of RAM near the bottom of memory to hold certain required information. Most of this structure has been retained on all of the Intel microprocessors.

Figure 14–27 shows a basic memory map for a typical 8086/8088 computer. The RAM is allocated addresses 00000_{16} through $BFFFF_{16}$ (a total of 786,431 locations) which includes the RAM and video areas mentioned. The ROM is allocated addresses $C0000_{16}$ through $FFFFF_{16}$ (a total of 262,143 locations).

FIGURE 14–27
Example of a typical memory address allocation within an 8086/8088 computer.

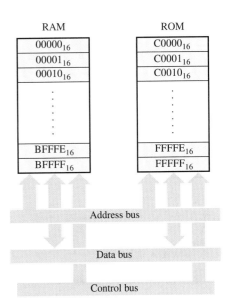

As the number of address lines increased in succeeding processors, much of the earlier memory organization was retained; however, certain significant changes were introduced. With the 80386, a new *paging* mechanism was added to the processor for accessing memory. This unit was part of a memory management unit (MMU) that had its roots in the 80286 microprocessor. Paging allowed programs to be located anywhere in the available space and still operate properly by translating the addresses. This means that a program that required access to a certain location could function even if the location was occupied by another program. In order to keep compatibility with earlier processors, the memory above FFFFF is referred to as extended memory. From 00000 to FFFFF, the memory is referred to as the real memory area. Programs operating in real mode operate in this area.

CPU-Memory Operation

The CPU handles two types of memory transfers, *read* and *write*. Basically, during the read operation, the CPU fetches either a program instruction, another address (that appears as data), or data itself. During the write operation, the CPU sends data resulting from a computation or some other operation back to RAM. A diagram of the 8088 CPU and memory portions of a computer is shown in Figure 14–28.

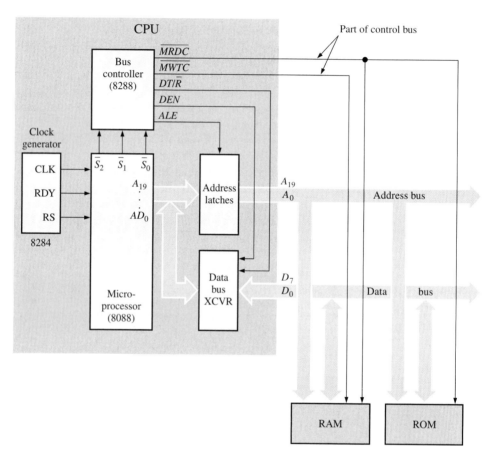

FIGURE 14–28
8088 CPU memory organization.

Memory-Read Cycle   The 8086/8088 microprocessor initiates a memory-read cycle by sending a status code of $\overline{S_2}\,\overline{S_1}\,\overline{S_0} = 101$ to the bus controller and placing a binary address on the AD_0–AD_7 and A_8–A_{19} bus lines. The bus controller then issues an *ALE* signal, which latches the address onto the address bus (A_0–A_{19}), freeing up the AD_0–AD_7 lines for data transfer. Also, the bus controller issues a low $DT/\overline{R}$ signal to the bus transceiver to set it up for receiving data from the memory and passing it on to the microprocessor.

Next, the bus controller issues a memory read command ($\overline{MRDC}$) to the memory. This command instructs the memory to place the contents of the selected address on the data bus. The *DEN* signal from the bus controller then enables the bus transceiver, and a

FIGURE 14–29
Basic timing diagram for an 8088 memory-read cycle.

byte of data flows into the microprocessor. Figure 14–29 shows the basic timing diagram for an 8088 memory-read cycle.

Memory-Write Cycle   The 8088 microprocessor initiates a memory-write cycle by sending status code $\overline{S}_2\,\overline{S}_1\,\overline{S}_0 = 110$ to the bus controller and placing a binary address on the bus lines. The bus controller then issues an *ALE* signal to latch the address onto the address bus (A_0–A_{19}) and a high $DT/\overline{R}$ signal to set up the bus transceiver for sending data to the memory. The microprocessor then places the data byte on the AD_0–AD_7 lines. The *DEN* signal from the bus controller then enables the bus transceiver to place the data byte on the data bus (D_0–D_7).

Next, the bus controller issues a memory-write command ($\overline{MWTC}$) to the memory. This command instructs the RAM to store the data byte in the selected address. Figure 14–30 shows a basic timing diagram for an 8088 memory-write cycle.

FIGURE 14–30
Basic timing diagram for an 8088-based memory-write cycle.

SECTION 14–6 REVIEW	
	1. Name the two cycles used by the 8088 CPU to transfer data to and from the memory.
	2. How many clock cycles does it take to read data from the memory in an 8088-based system?
	3. When does a valid address appear on the address bus (A_0–A_{19}) during a write cycle?

14–7 ■ THE INPUT/OUTPUT (I/O) PORT

In this section, you will see generally how the CPU communicates with input and output devices via I/O ports. An I/O port can be thought of as a "window" between the computer and the outside world through which information is passed.

After completing this section, you should be able to

■ Explain what an I/O port is ■ Discuss the basic types of ports ■ Explain the difference between dedicated and memory-mapped I/O ports ■ State the purpose of a programmable peripheral interface

A block diagram of a basic computer system, including I/O ports and typical peripherals, is shown in Figure 14–31. Notice that a port can be strictly for input, strictly for output, or bidirectional for both input and output. Although only four I/O ports are shown in the figure, the CPU is capable of handling many more than that. All of the Intel 80X86 and Pentium processors can address 64,000 ports!

As mentioned, an **I/O port** provides an interface between the computer and the "outside world" of peripheral equipment. Depending on the system design, the I/O ports can be

FIGURE 14–31
Basic 8088 computer system with I/O ports and peripherals.

either *dedicated* or *memory mapped.* These terms describe the ways in which the ports are accessed by the CPU.

Dedicated I/O Ports

A dedicated I/O port is assigned a unique address within the I/O address space of the computer, and dedicated I/O commands are used for communication. This address space is completely separate from memory addresses. Dedicated I/O ports are accessed by the I/O read and write commands, $\overline{IORC}$ and $\overline{IOWC}$ in an 8086/8088-based system.

Output Port The basic operation of the output port is as follows: The port occupies one location within the I/O address space of the system; that is, it has a unique address different from the addresses of other ports. The lower 16 bits of the address bus are used for port addresses because the I/O space is mapped to 64k ($2^{16} = 64$k).

During a CPU write cycle, the port address is placed on the address bus, and the CPU issues a low $\overline{IOWC}$ (I/O write command) signal to enable the address decoder. The decoder then produces a signal to latch the data byte onto the port output lines. Figure 14–32 shows how a dedicated output port might be implemented.

FIGURE 14–32
A basic dedicated output port.

Input Port A dedicated input port is not quite as simple as an output port because the port output lines are connected to the system data bus. This arrangement requires that the port output lines be disabled when the port is not in use to prevent interference with other activity on the data bus. Figure 14–33 shows an example of a basic input port implementation.

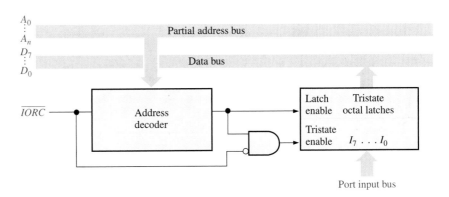

FIGURE 14–33
A basic dedicated input port.

Bidirectional Port   The third type of dedicated I/O port is the bidirectional port, which is basically a combination of the input and output ports with additional enabling controls. A bidirectional port allows data to flow to or from a peripheral device such as a disk drive.

Memory-Mapped I/O Ports

The second approach to accessing I/O ports is the **memory-mapped port.** The port implementation is essentially the same as for the dedicated ports in Figure 14–32 and 14–33 except for two things:

1. The memory-mapped ports are assigned addresses within the computer memory address space and are actually viewed as memory locations by the CPU. All twenty address lines are used for memory-mapped I/O in the 8086/8088.
2. The CPU accesses a memory-mapped I/O port with the $\overline{MRDC}$ and $\overline{MWTC}$ commands rather than the $\overline{IORC}$ and $\overline{IOWC}$ commands. In other words, the memory-mapped I/O ports are treated exactly as memory locations. Instead of using the IN and OUT instruction in assembly language, memory mapped I/O uses the MOV instruction.

Although memory-mapped I/O design is simple, it has the disadvantage of using memory space, with the potential for fragmentation of memory. In addition, decoding the address requires the entire address bus rather than only part of the address bus.

Programmable Peripheral Interface

In many systems, special ICs are used to implement the I/O ports. One such device is the 8255A programmable peripheral interface **(PPI),** shown in Figure 14–34. This PPI chip continues to be very popular and has been used in many I/O applications and on add-on cards that plug into the computer's bus. The address decoder is a separate device for enabling the PPI when a port address is decoded. This particular PPI provides three I/O ports. Ports A and B have 8 lines each, and port C is split into two 4-line groups. The ports can be programmed for various combinations of input ports, output ports, and bidirectional ports and for control and handshaking lines by writing a word to a special control register. This means that the port can be dynamically configured; that is, software can change the size and type of port and can add or delete handshaking. The PPI can be used for either dedicated or memory-mapped ports.

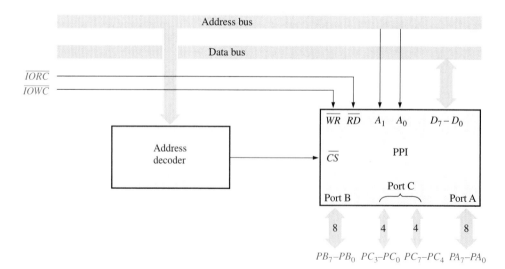

FIGURE 14–34
The 8255A programmable peripheral interface (PPI).

1. What is an I/O port?
2. Explain the difference between a dedicated I/O port and a memory-mapped I/O port.
3. What is the purpose of a PPI?

14–8 ■ INTERRUPTS

The concept of an I/O port was presented in Section 14–7, and both dedicated and memory-mapped ports were introduced. In this section, the establishment of communications between a peripheral and the CPU is presented. Three methods are discussed: polled I/O, interrupt-driven I/O, and software interrupts.

After completing this section, you should be able to

■ Discuss the need for interrupts in a computer system ■ Describe the basic concept of a polled I/O ■ Describe the basic concept of an interrupt-driven I/O ■ Discuss a software interrupt

In microprocessor-based systems such as the personal computer, peripheral devices require periodic service from the CPU. The term *service* generally means sending data to or taking data from the device or performing some updating process. There are three ways that a service routine can be started: *polled I/O, interrupt driven I/O,* or *software interrupts.* An **interrupt** is a signal or instruction that causes the current process to be temporarily stopped while a service routine is run. The three methods for service are described further in this section.

In general, peripheral devices are very slow compared with the CPU. For example, the 8086/8088 CPU can perform about 1,000,000 read or write operations in one second (based on a 4.77 MHz clock frequency). A printer may average only a few characters per second (one character is represented by eight bits), depending on the type of material being printed and the type of printer. A keyboard input rate may be one or two characters per second, depending on the speed of the operator. So, in between the times that the CPU is required to service a peripheral, it can do a lot of processing. In most systems, this processing time must be maximized by using an efficient method of servicing the peripherals.

Polled I/O

One method of servicing the peripherals is called **polling.** In this method, the CPU must test each peripheral device in sequence at certain intervals to see if it needs or is ready for servicing. Figure 14–35 illustrates the basic polled I/O method.

The CPU sequentially selects each peripheral device via the multiplexer to see if it needs service by checking the state of its ready line. Certain peripherals may need service at irregular and unpredictable intervals, that is, more frequently on some occasions than on others. Nevertheless, the CPU must poll the device at the highest rate. For example, let's say that a certain peripheral occasionally needs service every 1000 μs but most of the time requires service only once every 100 ms. As you can see, precious processing time is wasted if the CPU polls the device, as it must, at its maximum rate (every 1000 μs) because most of the time the device will not need service when it is polled.

Each time the CPU polls a device, it must stop the program that it is currently processing, go through the polling sequence, provide service if needed, and then return to the point where it left off in its current program.

Another problem with the sequentially polled I/O approach is that if two or more devices need service at the same time, the first one polled will be serviced first; the other devices will have to wait although they may need servicing much more urgently than the first device polled. As you can see, polling is suitable only for devices that can be serviced at

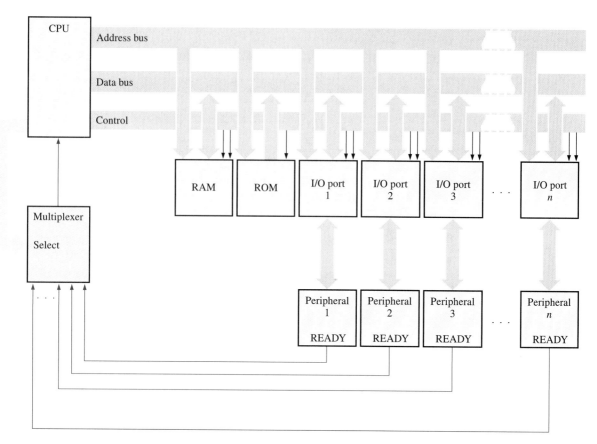

FIGURE 14–35
The basic polled I/O configuration.

regular and predictable intervals and only in situations in which there are no priority considerations.

Interrupt-Driven I/O

This approach overcomes the disadvantages of the polling method. In the interrupt-driven method, the CPU responds to a need for service only when service is requested by a peripheral device. Thus, the CPU can concentrate on running the current program without having to break away unnecessarily to see if a device needs service.

When the CPU receives an I/O interrupt signal, it temporarily stops its current program, acknowledges the interrupt, and fetches a special program (service routine) from memory for the particular device that has issued the interrupt. When the service routine is complete, the CPU returns to where it left off.

A device called a programmable interrupt controller (**PIC**) handles the interrupts on a priority basis. It accepts service requests from the peripherals. If two or more devices request service at the same time, the one assigned the highest priority is serviced first, then one with the next highest priority, and so on. After issuing an interrupt (*INTR*) signal to the CPU, the PIC provides the CPU with information that "points" the CPU to the beginning memory address of the appropriate service routine. This process is called *vectoring*. Figure 14–36 shows a basic interrupt-driven I/O configuration.

Software Interrupts

Another type of interrupt is called a **software interrupt.** Software interrupts are program instructions that can invoke the same service routines described previously. The difference is they are invoked from software rather than from external hardware. When invoked, the

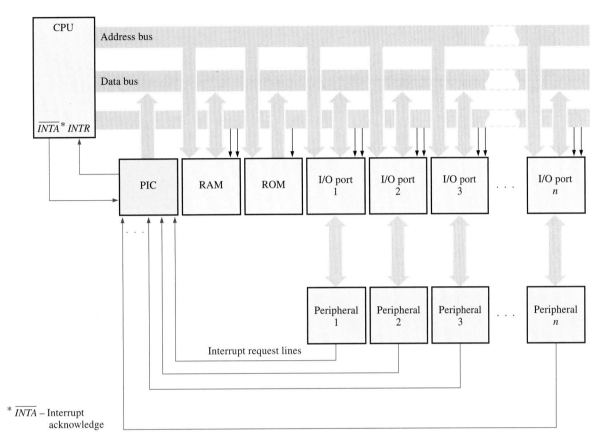

FIGURE 14–36
A basic interrupt-driven I/O configuration.

interrupt service routine executes exactly as if a hardware interrupt had occurred. The first five interrupts are defined by Intel. Others are defined by the BIOS and by DOS to perform many of the I/O operations, such as reading and writing data to the disk, writing data to the display, and reading data from the keyboard.

SECTION 14–8 REVIEW

1. How does an interrupt-driven I/O differ from a polled I/O?
2. What is the main advantage of an interrupt-driven I/O?
3. What is a software interrupt?

14–9 ■ DIRECT MEMORY ACCESS (DMA)

In this short section, we define the technique of data transfer called direct memory access (DMA). A comparison of a CPU-handled transfer and a DMA transfer is presented.

After completing this section, you should be able to

■ Define the term *DMA*   ■ Compare a memory I/O data transfer handled by the CPU to a DMA transfer

All I/O data transfers discussed so far have passed through the CPU. For example, when data are to be transferred from RAM to a peripheral device, the CPU reads the first data byte from the memory and loads it into an internal register within the microprocessor.

FIGURE 14–37
A memory I/O transfer handled by the CPU.

Then the CPU writes the data byte to the appropriate I/O port. This read/write operation is repeated for each byte in the group of data to be transferred. Figure 14–37 illustrates this process.

For large blocks of data, intermediate stops by the microprocessor consume a lot of time. For this reason, many systems use a technique called direct memory access (**DMA**) to speed up data transfers between RAM and certain peripheral devices. Basically, DMA bypasses the CPU for certain types of data transfers, thus eliminating the time consumed by the normal fetch and execute cycles required for each read or write operation.

For direct memory transfers, a device called the DMA controller takes control of the system buses and allows data to flow directly between RAM and the peripheral device, as indicated in Figure 14–38. Transfers between the disk drive and RAM are particularly suited for DMA because of the large amounts of data involved and the serial nature of the transfers. The DMA controller can handle data transfers several times faster than the CPU.

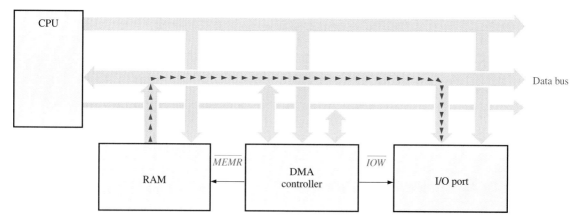

FIGURE 14–38
A DMA transfer.

**SECTION 14–9
REVIEW**

1. What does DMA stand for?
2. Discuss the advantage of DMA, and give an example of a type of transfer for which it is often used.

■ SUMMARY

■ Basic units of a computer are shown in Figure 14–39.

FIGURE 14–39

■ The three basic computer buses are the *address bus, data bus,* and the *control bus.* The size of any bus is specified by the number of separate conductive paths.
■ Typical peripheral devices include the keyboard, external disk drives, mouse, printer, modem, and scanner.
■ The number of address lines increased from 20 for the 8086/8088 to 32 for the Pentium family. The data bus was originally 16 for the 8086 and is 64 for the Pentium family.
■ General registers for the 8086/8088 are a subset of those in all of the Intel 80X86 and Pentium processors. They include

> Accumulator (AX, which includes AH and AL)
> Base (BX, which includes BH and BL)
> Count (CX, which includes CH and CL)
> Data (DX, which includes DH and DL)
> Stack pointer (SP)
> Base pointer (BP)
> Source index (SI)
> Destination index (DI)

Beginning with the 80386, this basic set was expanded to the extended register set.
■ The basic segment registers for the 8086/8088 are a subset of those in all of the Intel 80X86 and Pentium processors. The segment registers are

> Code segment (CS)
> Data segment (DS)
> Stack segment (SS)
> Extra segment (ES)

Beginning with the 80386, two new segment registers were added.
■ The flag registers for the 8086/8088 are a subset of those in all of the Intel processors. They include

> Trap (TF)
> Direction (DF)
> Interrupt enable (IF)
> Overflow (OF)
> Carry (CF)
> Sign (SF)
> Zero (ZF)
> Auxiliary carry (AF)
> Parity (PF)

■ The basic "language" of a computer is called machine code in which instructions are given as a series of binary codes.
■ In assembly language, machine instructions are replaced with a short alphabetic English mnemonic that has a one-to-one correspondence to machine code. Assembly language also uses directives to allow the programmer to specify other parameters that are not translated directly into machine code.
■ Ports are an interface to external devices. They can be set up as input, output, or a combination of both. They can be accessed as dedicated or as memory-mapped and can be serviced by polling, interrupt-driven, or software.

■ KEY TERMS

These terms are also in the end-of-book glossary.

Accumulator A general-purpose register used for arithmetic and logic operations.

Address A unique memory location containing one byte.

Address bus Generally, a one-way group of conductors from the microprocessor to memory, containing the address information.

ALU Arithmetic logic unit; the key processing element of a microprocessor that performs arithmetic and logic operations.

Architecture The internal functional arrangement of the elements that give a device its particular operating characteristics.

Assembler A program that converts English-like mnemonics into machine code.

Assembler directive An instruction to the assembler; also called a pseudo-op.

Assembly language A form of computer program language in which statements expressed in mnemonics represent the instructions.

BIOS Basic input/output system; a set of programs in ROM that interfaces the I/O devices in a computer system.

BIU Bus interface unit; the portion of the CPU that interfaces with the system buses and fetches instructions, reads operands, and writes results.

Compiler A program that translates high-level program statements, such as BASIC, Pascal, or Fortran, into machine language.

Contiguous Joined together.

Control bus A one-way set of conductors that connects the CPU to other parts of the computer to coordinate its operations and to communicate with external devices.

Control unit The portion within the microprocessor that provides the timing and control signals for getting data into and out of the microprocessor and for synchronizing the execution of instructions.

Coprocessor A microprocessor designed with a limited instruction set optimized to perform arithmetic operations very quickly; it generally works in conjunction with a general-purpose microprocessor.

CPU Central processing unit (also called the MPU); the main part of a computer responsible for control and processing of data.

Data bus A bidirectional set of conductive paths on which data or instruction codes are transferred into the microprocessor or on which the result of an operation or computation is sent out from the microprocessor.

Debug A code within DOS that allows various operations on files. It includes a primitive assembler.

DMA Direct memory access; a method to directly interface a peripheral device to memory without using the CPU for control.

EU Execution unit; the portion of a CPU that executes instructions; it contains the arithmetic logic unit (ALU), the general registers, and the flags.

Flag A bit that indicates the result of an arithmetic or logic operation or is used to alter an operation.

Hardware The circuitry and physical components of a computer system (as opposed to the directions called software).

Instruction pairing The process of combining certain independent instructions so that they can be executed simultaneously by two separate execution units.

Interrupt A signal or instruction that causes the current process to be temporarily stopped while a service routine is run.

I/O port   Input/output port; the interface between an internal bus and a peripheral.

IP   Instruction pointer; a special register within the CPU that holds the offset address of the next instruction to be executed.

Machine code   The basic binary instructions understood by the processor.

Memory-mapped I/O   A method of addressing ports by assigning addresses within the computer memory address space to the port. The CPU views the I/O ports as memory locations.

Microcontroller   A specialized microprocessor designed for control functions.

Microprocessor   A digital integrated circuit device that can be programmed with a series of instructions to perform specified functions on data.

Mnemonic   An English-like instruction that is converted by an assembler into a machine code for use by a processor.

Multitasking   An operating system environment in which the computer seems to run multiple programs or tasks simultaneously.

Op-code   The code for an instruction; a mnemonic.

Operand   A variable, a register, a memory location, or a value used in an assembly language program as part of the instruction.

PIC   Programmable interrupt controller; handles the interrupts on a priority basis.

Pointer   The contents of a register (or registers) that contain an address.

Polling   The process of checking a series of peripheral devices to determine if any require service from the CPU.

Port   A physical interface on a computer through which data are passed to or from peripherals.

PPI   Programmable peripheral interface; a special IC used to implement I/O ports.

Prefetching   The process of executing instructions at the same time as other instructions are "fetched," eliminating idle time; also called *pipelining*.

Program   A group of instructions designed to solve a specific problem.

Pseudo-operation   An instruction to the assembler (as opposed to a processor).

Queue   A high-speed memory that stores instructions or data.

Real mode   Operation of an Intel processor in a manner to emulate the 8086's 1 Mbyte of memory.

Register array   A set of temporary storage locations within the microprocessor for keeping data and addresses that need to be accessed quickly by the program.

Relocatable code   A program that can be moved anywhere within the memory space without changing the basic code.

Segment   A 64k block of memory.

Software   Programs and instructions.

Software interrupt   An instruction that invokes an interrupt service routine.

String   A contiguous sequence of bytes or words.

Subroutine   A series of instructions that can be assembled together and used repeatedly by a program but programmed only once.

Throughput   The average speed with which a program is executed.

■ **SELF-TEST**

1. A basic computer does not include
   (a) an arithmetic logic unit   (b) a control unit
   (c) peripheral units   (d) a memory unit

2. A 20-bit address bus supports

(a) 100,000 memory addresses   (b) 1,048,576 memory addresses

(c) 2,097,152 memory addresses   (d) 20,000 memory addresses

3. The size of the data bus on the Pentium processors is

(a) 16   (b) 24   (c) 32   (d) 64

4. A bus that is used to transfer information both to and from the microprocessor is the

(a) address bus   (b) data bus

(c) both of the above   (d) none of the above

5. An example of a peripheral unit is

(a) the address register   (b) the MPU

(c) the video monitor   (d) the interface adapter

6. Two types of memory transfers handled by the CPU are

(a) direct and interrupt   (b) read and write

(c) bussed and multiplexed   (d) input and output

7. Two types of I/O ports are

(a) dedicated and memory-mapped   (b) DMA and PPI

(c) tristate and octal   (d) read and write

8. In the Intel 80X86 family, the maximum number of 8-bit I/O devices is

(a) 64   (b) 1000   (c) 64,000   (d) 1 million   (e) unlimited

9. Polling is a method used for

(a) determining the state of the microprocessor

(b) establishing communication between the CPU and a peripheral

(c) establishing a priority for communication with several peripherals

(d) determining the next instruction

10. Of the following, which is a 8-bit register?

(a) AH   (b) BX   (c) SS   (d) IP

11. Essentially, a mnemonic is a(n)

(a) flowchart   (b) operand

(c) machine code   (d) instruction

12. DMA stands for

(a) digital microprocessor address   (b) direct memory access

(c) data multiplexed access   (d) direct memory addressing

13. A computer program is a list of

(a) memory addresses that contain data to be used in an operation

(b) addresses that contain instructions to be used in an operation

(c) instructions arranged to achieve a specific result

14. A type of assembly language that alters the course of the program is called a

(a) loop   (b) jump

(c) both of the above   (d) none of the above

15. A type of interrupt that is invoked from within a program is called a

(a) software interrupt   (b) polled interrupt

(c) direct interrupt   (d) I/O interrupt

■ QUESTIONS   SECTION 14–1   The Microprocessor and the Computer

1. Name the three basic elements of a microprocessor.

2. (a) What is a coprocessor?

   (b) Where is the coprocessor for the Pentium?

3. Name the three basic buses in a computer and explain what each is used for.

4. What is a port?

SECTION 14–2   Microprocessor Families

5. **(a)** Name the first Intel microprocessor that used a 32-bit address bus.

   (b) How many locations could it address?

6. How many data lines are used in the Pentium?

7. Name the first Motorola microprocessor that used a 32-bit address bus.

8. What was the major difference between the Motorola MPC601 and the 68000 series of processors?

SECTION 14–3   The 8086/8088 Microprocessor and Software Model for the Pentium Processor

9. What are the three basic steps a processor repeatedly cycles through?

10. What is meant by "pipelining"?

11. Name the six segment registers of the 80386 and above.

12. Assume the code segment register contains the hex number 0F05 and the instruction pointer register contains the number 0100. What is the physical address of the next instruction to be executed?

13. Explain the difference between the AH, the AL, the AX, and the EAX registers.

14. **(a)** What is a flag?

    (b) What two purposes are flags used for?

15. **(a)** Explain the advantage of instruction pairing in the Pentium processor.

    (b) Why can't the 80486 use instruction pairing?

SECTION 14–4   Microprocessor Programming

16. What is an assembler?

17. Draw a flowchart for a program that adds the numbers from one to 10 and saves the result in a memory location named TOTAL.

18. Draw a flowchart showing how you could count the number of bytes in a string and place the count in a location in memory called COUNT. Assume the string starts at a location named START and has a 20H (ASCII for a space) to signal the end. You should not count the space character.

19. What does the assembler directive **word ptr** do?

20. Explain what happens when the instruction **mov ax,[bx]** is executed.

21. Describe the seven basic types of instructions for the Intel processors.

SECTION 14–5   The Central Processing Unit (CPU)

22. What is the difference between the $\overline{IORC}$ and $\overline{MRDC}$ signals?

23. Which address lines are used for I/O read and write operations in the 8086/8088?

24. What is the purpose of the *ALE* signal from the bus controller of the 8086/8088.

25. What is the function of the Status bits from the 8086/8088 CPU?

26. Name at least three enhancements to the Pentium processor that were not present in the original 8086/8088.

SECTION 14–6   The Memory

27. What locations in memory are reserved for the BIOS work area?

28. What is meant by the term *extended memory?*

29. When the bus controller sends the status code $\overline{S}_2 \overline{S}_1 \overline{S}_0 = 110$, what type of operation is indicated?

SECTION 14–7   The Input/Output (I/O) Port

30. Explain the difference between dedicated and memory-mapped I/O ports. Which type cannot use the IN and OUT instruction?

31. What is the maximum number of dedicated ports that can be connected to a Pentium processor?

32. Explain the function of an 8255A integrated circuit.

SECTION 14–8   Interrupts

33. Compare polled I/O to interrupt-driven I/O.

34. What is meant by the term *vectoring?*

35. What is meant by a software interrupt?

SECTION 14–9   Direct Memory Access (DMA)

36. Explain what happens in a DMA operation.

■ ANSWERS TO SECTION REVIEWS

SECTION 14–1

1. A computer includes a central processing unit (CPU), memory, and input/output ports.

2. The three buses are (1) the data bus which transfers data to and from the CPU, (2) the address bus which carries either a memory or port address, and (3) the control bus which carries timing and control signals.

3. Data and instructions

4. Assembly language is a low-level language that can be directly translated into binary instructions or data patterns. Two types of operations are instructions and pseudo-operations (directives).

SECTION 14–2

1. An 32-bit microprocessor has an 32-bit data bus.

2. The 8086 was the first 16-bit Intel microprocessor.

3. RISC means reduced instruction set computer.

4. Cache memory stores recently used instructions or data to speed up operations.

5. Superscalar refers to a type of architecture that allows more than one instruction to be executed during one clock cycle.

SECTION 14–3

1. The general-purpose registers in the 8086/8088 are

Accumulator (AX: AH, AL)	Stack pointer (SP)
Base index (BX: BH, BL)	Base pointer (BP)
Count (CX: CH, CL)	Destination index (DI)
Data (DX: DH, DL)	Source index (SI)

2. The BIU provides addressing and data interface.

3. No, the EU does not interface with the buses.

4. The instruction queue stores prefetched instructions for the EU to increase throughput.

5. Codes can be easily relocated within memory.

6. Instruction pairing is the process of combining independent instructions so that they can be executed simultaneously by the two execution units in the Pentium.

SECTION 14–4

1. A program is a list of computer instructions arranged to achieve a specific result.

2. A complex instruction set has a full repertoire of instructions available, giving the programmer flexibility.

3. An op-code is the code for an instruction.

4. An operand is the quantity or element operated on by an instruction.

5. A string is a contiguous sequence of bytes or words.

SECTION 14–5

1. The clock generator provides basic timing signals for the microcomputer.

2. The bus controller issues memory read and write commands, issues I/O read and write commands, and enables the address latches and the data bus transceiver.

3. The twenty bus lines available must be used for both the 20-bit address bus and the 8-bit data bus in the 8088.

4. Pipelining which enabled some instructions to execute in a single clock cycle, addition and expansion of cache memory, inclusion of the internal math coprocessor, internal dual integer processors and branch prediction logic

SECTION 14–6

1. The two CPU cycles are write and read.

2. Four clock cycles

3. A valid address appears on the bus after the leading edge of the *ALE* pulse.

SECTION 14–7

1. An I/O port is the interface between the internal data bus and a peripheral.

2. A dedicated I/O port has an address within the I/O address space; a memory-mapped I/O port has an address within the memory address space.

3. A PPI provides I/O ports.

SECTION 14–8

1. For an interrupt-driven I/O, the CPU provides service to a peripheral only when requested to do so by the peripheral; for a polled I/O, the CPU periodically checks a peripheral to see if it needs service.

2. Interrupt-driven I/Os save CPU time.

3. A software interrupt is an instruction that invokes an interrupt service routine.

SECTION 14–9

1. DMA is direct memory access.

2. A DMA transfer of data from memory to I/O or vice versa saves CPU time. Direct memory access is often used in transferring data between RAM and a disk drive.

■ **ANSWERS TO RELATED PROBLEMS FOR EXAMPLES**

14–1 $6C4C2_{16}$

14–2 Change first block (initialization block) to "BIG = FFFF"; this is the largest possible unsigned number. Change first question to "Is number < BIG?"

■ **ANSWERS TO SELF-TEST**

1. (c)   **2.** (b)   **3.** (d)   **4.** (b)   **5.** (c)   **6.** (b)   **7.** (a)   **8.** (c)

9. (b)   **10.** (a)   **11.** (d)   **12.** (b)   **13.** (c)   **14.** (c)   **15.** (a)

15

INTEGRATED CIRCUIT TECHNOLOGIES

Before beginning this chapter, Section 3–7 should be covered.

■ CHAPTER OBJECTIVES

❏ Determine the noise margin of a device from data sheet parameters

❏ Calculate the power dissipation of a device

❏ Explain how propagation delay affects the frequency of operation or speed of a circuit

❏ Interpret the speed-power product as a measure of performance

❏ Use data sheets to obtain information about a specific device

❏ Explain what the fan-out of a gate means

❏ Describe how basic TTL and CMOS gates operate at the component level

❏ Recognize the difference between TTL totem-pole outputs and TTL open-collector outputs and understand the limitations and uses of each

❏ Connect circuits in a wired-AND configuration

❏ Describe the operation of tristate circuits

❏ Properly terminate unused gate inputs

❏ Compare the performance of TTL and CMOS families

❏ Handle CMOS devices without risk of damage due to electrostatic discharge

❏ State the advantages of ECL

❏ Describe PMOS and NMOS circuits

❏ Describe an E^2CMOS cell

■ CHAPTER OVERVIEW

This chapter is intended to be used as a "floating" chapter. It can be covered at any one of several appropriate points throughout the book or completely omitted, depending on course objectives and personal preference. Two suggested points that seem most appropriate are following Chapters 3 or 6, but it can be placed after other chapters, as indicated in those chapters, if desired.

Also, selected topics from this chapter can be introduced at various points. In any case, Section 3–7 should be covered before beginning this chapter.

Other approaches are to use this chapter as an optional reading assignment or to use it only for reference. Notice that a tab-edge design on the right-hand pages provides for easy reference to this chapter.

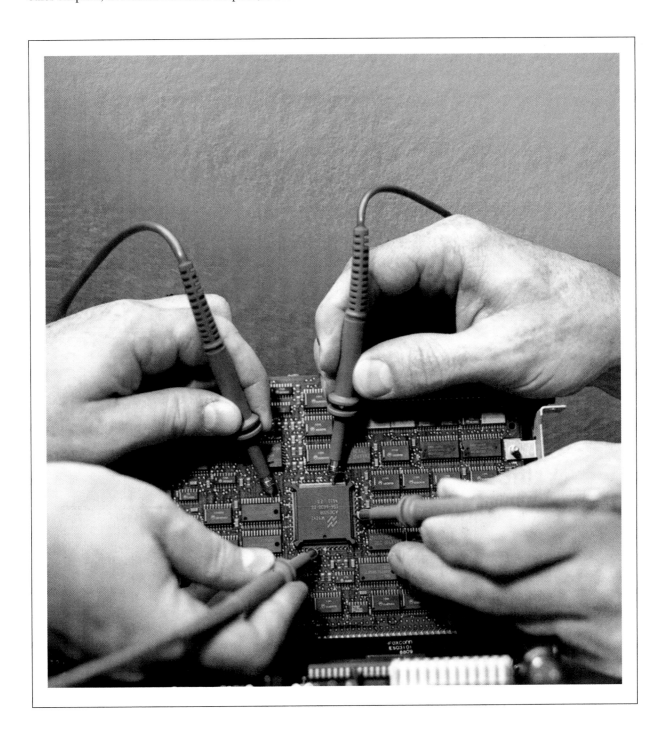

15–1 ■ BASIC OPERATIONAL CHARACTERISTICS AND PARAMETERS

When you work with digital ICs, you should be familiar, not only with their logical operation, but also with such operational properties as voltage levels, noise immunity, power dissipation, fan-out, and propagation delays. In this section, the practical aspects of these properties are discussed.

After completing this section, you should be able to

■ Determine the power and ground connections ■ Describe the logic levels for CMOS and TTL ■ Discuss noise immunity ■ Determine the power dissipation of a logic circuit ■ Define the propagation delays of a logic gate ■ Discuss speed-power product and explain its significance ■ Discuss loading and fan-out of TTL and CMOS

DC Supply Voltage

The nominal value of the dc supply voltage for **TTL** (transistor-transistor logic) devices is +5 V. TTL is also designated T^{2}L. **CMOS** (complementary metal-oxide semiconductor) devices are available in two different supply voltage categories, +5 V and +3.3 V, the latter is known as low-voltage CMOS. Although omitted from logic diagrams for simplicity, the dc supply voltage is connected to the V_{CC} pin of an IC package, and ground is connected to the GND pin. Both voltage and ground are distributed internally to all elements within the package, as illustrated in Figure 15–1 for a 14-pin package.

FIGURE 15–1
Example of V_{CC} and ground connection and distribution in an IC package. Other pin connections are omitted for simplicity.

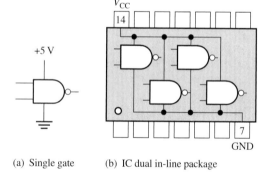

(a) Single gate   (b) IC dual in-line package

CMOS Logic Levels

Logic levels were discussed briefly in Chapter 1. There are four different logic-level specifications: V_{IL}, V_{IH}, V_{OL}, and V_{OH}. For CMOS circuits, the ranges of input voltages (V_{IL}) that can represent a valid LOW (logic 0) are from 0 V to 1.5 V for the +5 V logic and 0 V to 0.8 V for the 3.3 V logic. The ranges of input voltages (V_{IH}) that can represent a valid HIGH (logic 1) are from 3.5 V to 5 V for the 5 V logic and 2 V to 3.3 V for the 3.3 V logic, as indicated in Figure 15–2. The ranges of values from 1.5 V to 3.5 V for 5 V logic and 0.8 V to 2 V for 3.3 V logic are regions of unpredictable performance, and values in these ranges are unallowed. When an input voltage is in one of these ranges, it can be interpreted as either a HIGH or a LOW by the logic circuit. Therefore, CMOS **gates** cannot be operated reliably when the input voltages are in these unallowed ranges.

The ranges of CMOS output voltages (V_{OL} and V_{OH}) for both 5 V and 3.3 V logic are also shown in Figure 15–2. Notice that the minimum HIGH output voltage, $V_{OH(min)}$,

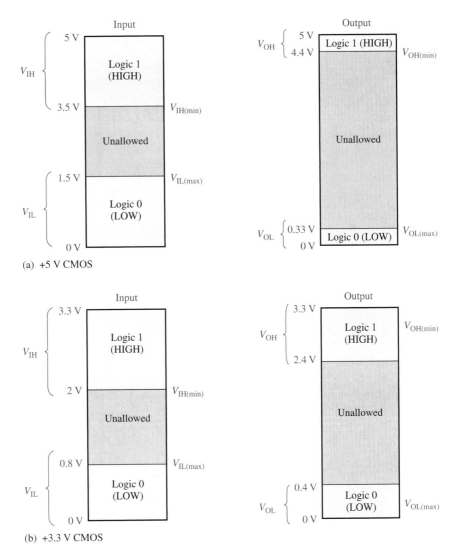

FIGURE 15–2
Input and output logic levels for CMOS.

is greater than the minimum HIGH input voltage, $V_{IH(min)}$. Also, notice that the maximum LOW output voltage, $V_{OL(max)}$, is less than the maximum LOW input voltage, $V_{IL(max)}$.

TTL Logic Levels

The input and output logic levels for TTL are given in Figure 15–3. Just as for CMOS, there are four different logic level specifications: V_{IL}, V_{IH}, V_{OL}, and V_{OH}.

Noise Immunity

Noise is unwanted voltage that is induced in electrical circuits and can present a threat to the proper operation of the circuit. Wires and other conductors within a system can pick up stray high-frequency electromagnetic radiation from adjacent conductors in which currents are changing rapidly or from many other sources external to the system. Also, power-line voltage fluctuation is a form of low-frequency noise.

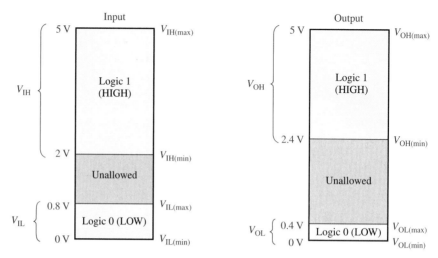

FIGURE 15–3
Input and output logic levels for TTL.

In order not to be adversely affected by noise, a logic circuit must have a certain amount of **noise immunity.** This is the ability to tolerate a certain amount of unwanted voltage fluctuation on its inputs without changing its output state. For example, if noise voltage causes the input of a 5 V CMOS gate to drop below 3.5 V in the HIGH state, the input is in the unallowed region and operation is unpredictable (see Figure 15–2). Thus, the gate may interpret the fluctuation below 3.5 V as a LOW level, as illustrated in Figure 15–4(a). Similarly, if noise causes a gate input to go above 1.5 V in the LOW state, an uncertain condition is created, as illustrated in part (b).

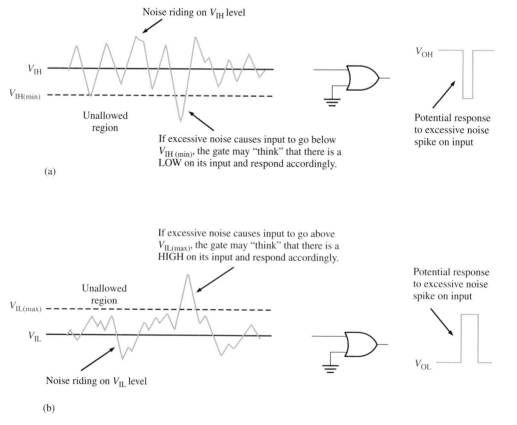

FIGURE 15–4
Illustration of the effects of input noise on gate operation.

Noise Margin

A measure of a circuit's noise immunity is called the **noise margin,** which is expressed in volts. There are two values of noise margin specified for a given logic circuit: the HIGH-level noise margin (V_{NH}) and the LOW-level noise margin (V_{NL}). These parameters are defined by the following equations:

$$V_{NH} = V_{OH(min)} - V_{IH(min)} \qquad (15\text{–}1)$$

$$V_{NL} = V_{IL(max)} - V_{OL(max)} \qquad (15\text{–}2)$$

Sometimes you will see the noise margin expressed as a percentage of V_{CC}. From the equations, V_{NH} is the difference between the lowest possible HIGH output from a driving gate ($V_{OH(min)}$) and the lowest possible HIGH input that the load gate can tolerate ($V_{IH(min)}$). Noise margin V_{NL} is the difference between the maximum possible LOW input that a gate can tolerate ($V_{IL(max)}$) and the maximum possible LOW output of the driving gate ($V_{OL(max)}$). Noise margins are illustrated in Figure 15–5.

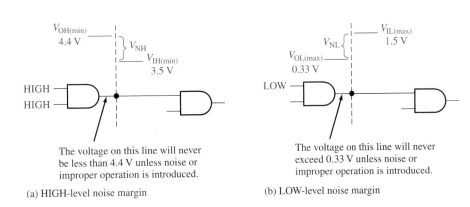

(a) HIGH-level noise margin

(b) LOW-level noise margin

FIGURE 15–5
Illustration of noise margins. Values are for 5 V CMOS, but the principle applies to any logic family.

EXAMPLE 15–1

Determine the HIGH-level and LOW-level noise margins for CMOS and for TTL by using the information in Figures 15–2 and 15–3.

Solution   For 5 V CMOS,

$$V_{IH(min)} = 3.5 \text{ V}$$
$$V_{IL(max)} = 1.5 \text{ V}$$
$$V_{OH(min)} = 4.4 \text{ V}$$
$$V_{OL(max)} = 0.33 \text{ V}$$
$$V_{NH} = V_{OH(min)} - V_{IH(min)} = 4.4 \text{ V} - 3.5 \text{ V} = \textbf{0.9 V}$$
$$V_{NL} = V_{IL(max)} - V_{OL(max)} = 1.5 \text{ V} - 0.33 \text{ V} = \textbf{1.17 V}$$

For TTL,

$$V_{IH(min)} = 2 \text{ V}$$
$$V_{IL(max)} = 0.8 \text{ V}$$
$$V_{OH(min)} = 2.4 \text{ V}$$
$$V_{OL(max)} = 0.4 \text{ V}$$

$$V_{NH} = V_{OH(min)} - V_{IH(min)} = 2.4 \text{ V} - 2 \text{ V} = \textbf{0.4 V}$$
$$V_{NL} = V_{IL(max)} - V_{OL(max)} = 0.8 \text{ V} - 0.4 \text{ V} = \textbf{0.4 V}$$

A TTL gate is immune to up to 0.4 V of noise for both the HIGH and LOW input states.

Related Problem Based on the preceding noise margin calculations, which family of devices, 5 V CMOS or TTL, should be used in a high-noise environment?

Power Dissipation

A logic gate draws current from the dc supply voltage source as indicated in Figure 15–6. When the gate is in the HIGH output state, an amount of current designated by I_{CCH} is drawn, and in the LOW output state, a different amount of current, I_{CCL}, is drawn.

FIGURE 15–6
Currents from the dc supply.

(a)   (b)

As an example, if I_{CCH} is specified as 1.5 mA when V_{CC} is 5 V and if the gate is in a static (nonchanging) HIGH output state, the **power dissipation** (P_D) of the gate is

$$P_D = V_{CC}I_{CCH} = (5 \text{ V})(1.5 \text{ mA}) = 7.5 \text{ mW}$$

When a gate is pulsed, its output switches back and forth between HIGH and LOW, and the amount of supply current varies between I_{CCH} and I_{CCL}. The average power dissipation depends on the duty cycle and is usually specified for a duty cycle of 50%. When the duty cycle is 50%, the output is HIGH half the time and LOW the other half. The average supply current is therefore

$$I_{CC} = \frac{I_{CCH} + I_{CCL}}{2} \qquad \textbf{(15–3)}$$

The average power dissipation is

$$P_D = V_{CC}I_{CC} \qquad \textbf{(15–4)}$$

EXAMPLE 15–2

A certain gate draws 2 μA when its output is HIGH and 3.6 μA when its output is LOW. What is its average power dissipation if V_{CC} is 5 V and the gate is operated on a 50% duty cycle?

Solution The average I_{CC} is

$$I_{CC} = \frac{I_{CCH} + I_{CCL}}{2} = \frac{2.0 \ \mu A + 3.6 \ \mu A}{2} = 2.8 \ \mu A$$

The average power dissipation is

$$P_D = V_{CC}I_{CC} = (5 \text{ V})(2.8 \ \mu A) = \textbf{14 } \mu\textbf{W}$$

Related Problem A certain IC gate has an $I_{CCH} = 1.5 \ \mu A$ and $I_{CCL} = 2.8 \ \mu A$. Determine the average power dissipation for 50% duty cycle operation if V_{CC} is 5 V.

Power dissipation in a TTL circuit is essentially constant over its range of operating frequencies. Power dissipation in CMOS, however, is frequency dependent. It is extremely low under static (dc) conditions and increases as the frequency increases. These characteristics are shown in the general curves of Figure 15–7. For example, the power dissipation of a low-power Schottky (LS) TTL gate is a constant 2.2 mW. The power dissipation of an HCMOS gate is 2.75 μW under static conditions and 170 μW at 100 kHz.

FIGURE 15–7
Power-versus-frequency curves for TTL and CMOS.

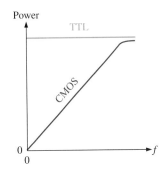

Propagation Delay

When a signal passes (propagates) through a logic circuit, it always experiences a time delay, as illustrated in Figure 15–8. A change in the output level always occurs a short time, called the **propagation delay time,** later than the change in the input level that caused it.

FIGURE 15–8
A basic illustration of propagation delay.

As mentioned in Chapter 3, there are two propagation delay times specified for logic gates:

■ t_{PHL}: The time between a designated point on the input pulse and the corresponding point on the output pulse when the output is changing from HIGH to LOW.
■ t_{PLH}: The time between a designated point on the input pulse and the corresponding point on the output pulse when the output is changing from LOW to HIGH.

These propagation delay times are illustrated in Figure 15–9, with the 50% points on the pulse edges used as references.

FIGURE 15–9
Propagation delay times.

The propagation delay of a gate limits the frequency at which it can be operated. The greater the propagation delay, the lower the maximum frequency. Thus, a higher-speed circuit is one that has a smaller propagation delay. For example, a gate with a delay of 3 ns is faster than one with a 10 ns delay.

Speed-Power Product

The speed-power product provides a basis for the comparison of logic circuits when *both* propagation delay and power dissipation are important considerations in the selection of the type of logic to be used in a certain application. The lower the speed-power product, the better. The unit of speed-power product is the picojoule (pJ). For example, HCMOS has a speed-power product of 1.2 pJ at 100 kHz while LS TTL has a value of 22 pJ.

Loading and Fan-Out

When the output of a logic gate is connected to one or more inputs of other gates, a load on the driving gate is created, as shown in Figure 15–10. There is a limit to the number of load gate inputs that a given gate can drive. This limit is called the **fan-out** of the gate.

FIGURE 15–10
Loading a gate output with gate inputs.

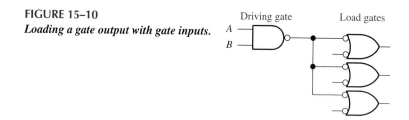

CMOS Loading Loading in CMOS differs from that in TTL because the type of transistors used in CMOS logic present a predominantly capacitive load to the driving gate, as illustrated in Figure 15–11. In this case, the limitations are the charging and discharging times associated with the output resistance of the driving gate and the input capacitance of the load gates. When the output of the driving gate is HIGH, the input capacitance of the load gate is charging through the output resistance of the driving gate. When the output of the driving gate is LOW, the capacitance is discharging, as indicated in Figure 15–11.

When more load gate inputs are added to the driving gate output, the total capacitance increases because the input capacitances effectively appear in parallel. This increase in capacitance increases the charging and discharging times, thus reducing the maximum frequency at which the gate can be operated. Therefore, the fan-out of a CMOS gate

(a) Charging (b) Discharging

FIGURE 15–11
Capacitive loading of a CMOS gate.

depends on the frequency of operation. The fewer the load gate inputs, the greater the maximum frequency.

TTL Loading A TTL driving gate sources current to a load gate input in the HIGH state (I_{IH}) and sinks current from the load gate in the LOW state (I_{IL}). **Current sourcing** and **current sinking** are illustrated in simplified form in Figure 15–12, where the resistors represent the internal input and output resistance of the gate for the two conditions.

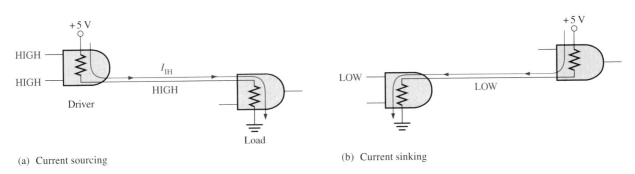

(a) Current sourcing

(b) Current sinking

FIGURE 15–12
Basic illustration of current sourcing and current sinking in logic gates.

As more load gates are connected to the driving gate, the loading on the driving gate increases. The total source current increases with each load gate input that is added, as illustrated in Figure 15–13. As this current increases, the internal voltage drop of the driving gate increases, causing the output, V_{OH}, to decrease. If an excessive number of load gate inputs are connected, V_{OH} drops below $V_{OH(min)}$, and the HIGH-level noise margin is reduced, thus compromising the circuit operation. Also, as the total source current increases, the power dissipation of the driving gate increases.

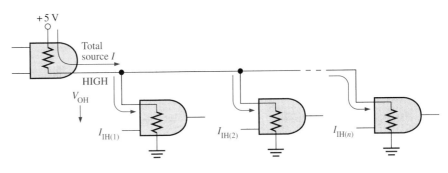

FIGURE 15–13
HIGH-state TTL loading.

The fan-out is the maximum number of load gate inputs that can be connected without adversely affecting the specified operational characteristics of the gate. For example, low-power Schottky (LS) TTL has a fan-out of 20 unit loads. One input of the same logic family as the driving gate is called a **unit load.**

The total sink current also increases with each load gate input that is added, as shown in Figure 15–14. As this current increases, the internal voltage drop of the driving gate increases, causing V_{OL} to increase. If an excessive number of loads are added, V_{OL} exceeds $V_{OL(max)}$, and the LOW-level noise margin is reduced.

In TTL, the current-sinking capability (LOW output state) is the limiting factor in determining the fan-out.

FIGURE 15–14
LOW-state TTL loading.

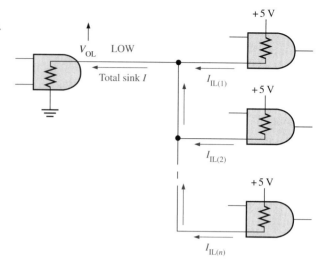

**SECTION 15–1
REVIEW**

1. Define V_{IH}, V_{IL}, V_{OH}, and V_{OL}.
2. Is it better to have a lower value of noise margin or a higher value?
3. Gate A has a greater propagation delay time than gate B. Which gate can operate at a higher frequency?
4. How does excessive loading affect the noise margin of a gate?

15–2 ■ CMOS CIRCUITS

Basic internal CMOS circuitry and its operation are discussed in this section. The abbreviation CMOS stands for complementary metal-oxide semiconductor. The term complementary refers to the use of two types of transistors in the output circuit. An n-channel MOSFET (MOS field-effect transistor) and a p-channel MOSFET are used.

After completing this section, you should be able to

- Identify a MOSFET by its symbol ■ Discuss the switching action of a MOSFET
- Describe the basic operation of a CMOS inverter circuit ■ Describe the basic operation of CMOS NAND and NOR gates ■ Explain the operation of a CMOS gate with an open-drain output ■ Discuss the operation of tristate CMOS gates ■ List the precautions required when handling CMOS devices

The MOSFET

Metal-oxide semiconductor field-effect transistors (**MOSFETs**) are the active switching elements in CMOS circuits. These devices differ greatly in construction and internal operation from bipolar junction transistors used in TTL circuits, but the switching action is basically the same: they function ideally as open or closed switches, depending on the input.

Figure 15–15(a) shows the symbols for both n-channel and p-channel MOSFETs. As indicated, the three terminals of a MOSFET are **gate, drain,** and **source.** When the gate voltage of an n-channel MOSFET is more positive than the source, the MOSFET is on *(saturation),* and there is, ideally, a closed switch between the drain and the source. When the gate-to-source voltage is zero, the MOSFET is off *(cutoff),* and there is, ideally, an open switch between the drain and the source. This operation is illustrated in Figure 15–15(b). The p-channel MOSFET operates with opposite voltage polarities, as shown in part (c).

Sometimes a simplified MOSFET symbol as shown in Figure 15–16 is used.

(a) MOSFET symbols

(b) *n*-channel switch

(c) *p*-channel switch

FIGURE 15–15
Basic symbols and switching action of MOSFETs.

FIGURE 15–16
Simplified MOSFET symbol.

CMOS Inverter

Complementary MOS (CMOS) logic uses the MOSFET in complementary pairs as its basic element. A complementary pair uses both *p*-channel and *n*-channel enhancement MOSFETs, as shown in the inverter circuit in Figure 15–17.

When a HIGH is applied to the input, as shown in Figure 15–18(a), the *p*-channel MOSFET Q_1 is off and the *n*-channel MOSFET Q_2 is on. This condition connects the output to ground through the *on* resistance of Q_2, resulting in a LOW output. When a LOW is

FIGURE 15–17
A CMOS inverter circuit.

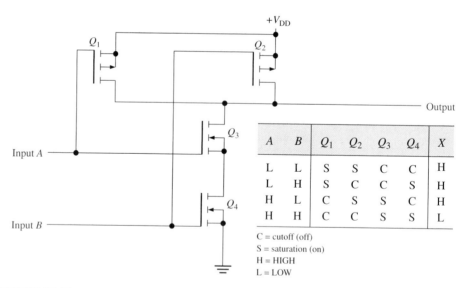

FIGURE 15–18
Operation of a CMOS inverter.

applied to the input, as shown in Figure 15–18(b), Q_1 is on and Q_2 is off. This condition connects the output to $+V_{DD}$ (dc supply voltage) through the *on* resistance of Q_1, resulting in a HIGH output.

CMOS NAND Gate

Figure 15–19 shows a CMOS NAND gate with two inputs. Notice the arrangement of the complementary pairs (*n*-channel and *p*-channel MOSFETs).

A	B	Q_1	Q_2	Q_3	Q_4	X
L	L	S	S	C	C	H
L	H	S	C	C	S	H
H	L	C	S	S	C	H
H	H	C	C	S	S	L

C = cutoff (off)
S = saturation (on)
H = HIGH
L = LOW

FIGURE 15–19
A CMOS NAND gate circuit.

The operation of a CMOS NAND gate is as follows:

■ When both inputs are LOW, Q_1 and Q_2 are on, and Q_3 and Q_4 are off. The output is pulled HIGH through the *on* resistance of Q_1 and Q_2 in parallel.
■ When input *A* is LOW and input *B* is HIGH, Q_1 and Q_4 are on, and Q_2 and Q_3 are off. The output is pulled HIGH through the low *on* resistance of Q_1.
■ When input *A* is HIGH and input *B* is LOW, Q_1 and Q_4 are off, and Q_2 and Q_3 are on. The output is pulled HIGH through the low *on* resistance of Q_2.

■ Finally, when both inputs are HIGH, Q_1 and Q_2 are off, and Q_3 and Q_4 are on. In this case, the output is pulled LOW through the *on* resistance of Q_3 and Q_4 in series to ground.

CMOS NOR Gate

Figure 15–20 shows a CMOS NOR gate with two inputs. Notice the arrangement of the complementary pairs.

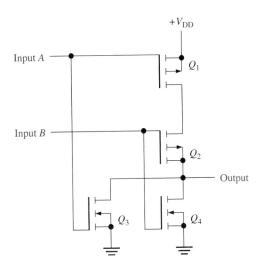

A	B	Q_1	Q_2	Q_3	Q_4	X
L	L	S	S	C	C	H
L	H	S	C	C	S	L
H	L	C	S	S	C	L
H	H	C	C	S	S	L

C = cutoff (off)
S = saturation (on)
H = HIGH
L = LOW

FIGURE 15–20
A CMOS NOR gate circuit.

The operation of a CMOS NOR gate is as follows:

■ When both inputs are LOW, Q_1 and Q_2 are on, and Q_3 and Q_4 are off. As a result, the output is pulled HIGH through the *on* resistance of Q_1 and Q_2 in series.
■ When input A is LOW and input B is HIGH, Q_1 and Q_4 are on, and Q_2 and Q_3 are off. The output is pulled LOW through the low *on* resistance of Q_4 to ground.
■ When input A is HIGH and input B is LOW, Q_1 and Q_4 are off, and Q_2 and Q_3 are on. The output is pulled LOW through the *on* resistance of Q_3 to ground.
■ When both inputs are HIGH, Q_1 and Q_2 are off, and Q_3 and Q_4 are on. The output is pulled LOW through the *on* resistance of Q_3 and Q_4 in parallel to ground.

Open-Drain Gates

The term *open-drain* means that the drain terminal of the output transistor is unconnected and must be connected externally to V_{CC} through a load. An open-drain gate is the CMOS counterpart of an open-collector TTL gate that you will see in the next section. An open-drain output circuit is a single *n*-channel MOSFET as shown in Figure 15–21(a). An external **pull-up resistor** must be used, as shown in part (b), to produce a HIGH output state. Also, open-drain outputs can be connected in a wired-AND configuration, a concept that is discussed in the next section in relation to TTL.

Tristate CMOS Gates

Tristate outputs are available in both CMOS and TTL logic. The **tristate** output combines the advantages of the totem-pole and open-collector circuits. As you recall, the three output states are HIGH, LOW, and high-impedance **(high-Z).** When selected for normal

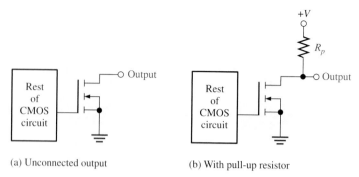

FIGURE 15–21
Open-drain CMOS gates.

HIGH — LOW
LOW (enable)

LOW — HIGH
LOW (enable)

don't care
X — OPEN
HIGH (disable)

(a) Enabled for normal logic operation

(b) High-Z state

FIGURE 15–22
The three states of a tristate circuit.

logic-level operation, as determined by the state of the enable input, a tristate circuit operates in the same way as a regular gate. When a tristate circuit is selected for high-Z operation, the output is effectively disconnected from the rest of the circuit by the internal circuitry. Figure 15–22 illustrates the operation of a tristate circuit. The inverted triangle (∇) designates a tristate output.

The circuitry in a tristate CMOS gate, as shown in Figure 15–23, allows each of the output transistors Q_1 and Q_2 to be turned off at the same time, thus disconnecting the output from the rest of the circuit.

When the enable input is LOW, the device is enabled for normal logic operation. When the enable is HIGH, both Q_1 and Q_2 are off, and the circuit is in the high-Z state.

FIGURE 15–23
A tristate CMOS inverter.

Precautions for Handling CMOS

As you have learned, all CMOS devices are subject to damage from electrostatic discharge (ESD). Therefore, they must be handled with special care. Review the following precautions:

1. All CMOS devices are shipped in conductive foam to prevent electrostatic charge buildup. When they are removed from the foam, the pins should not be touched.
2. The devices should be placed with pins down on a grounded surface, such as a metal plate, when removed from protective material. Do not place CMOS devices in polystyrene foam or plastic trays.
3. All tools, test equipment, and metal workbenches should be earth-grounded. A person working with CMOS devices should, in certain environments, have his or her wrist grounded with a length of cable and a large-value series resistor. The resistor prevents severe shock should the person come in contact with a voltage source.
4. Do not insert CMOS devices (or any other ICs) into sockets or PC boards with the power on.
5. All unused inputs should be connected to the supply voltage or ground as indicated in Figure 15–24. If left open, an input can acquire electrostatic charge and "float" to unpredicted levels.

FIGURE 15–24
Handling unused CMOS inputs.

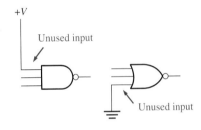

6. After assembly on PC boards, protection should be provided by storing or shipping boards with their connectors in conductive foam. The CMOS input and output pins may also be protected with large-value resistors connected to ground.

SECTION 15–2 REVIEW

1. What type of transistor is used in CMOS logic?
2. What is meant by the term *complementary MOS?*
3. Why must CMOS devices be handled with care?

15–3 ■ TTL CIRCUITS

The internal circuit operation of TTL logic gates with totem-pole outputs is covered in this section. Also, the operation of TTL gates with open-collector outputs and the operation of tristate gates are covered.

After completing this section, you should be able to

■ Identify a bipolar junction transistor (BJT) by its symbol ■ Describe the switching action of a BJT ■ Describe the basic operation of a TTL inverter circuit ■ Describe the basic operation of TTL AND, NAND, OR, and NOR gate circuits ■ Explain what a totem-pole output is ■ Explain the operation and use a TTL gate with an open-collector output ■ Explain the operation of a gate with a tristate output

The Bipolar Junction Transistor

The **bipolar** junction transistor (**BJT**) is the active switching element used in all TTL circuits. Figure 15–25 shows the symbol for an *npn* BJT with its three terminals: **base, emitter,** and **collector.** A BJT has two **junctions,** the base-emitter junction and the base-collector junction.

FIGURE 15–25
The symbol for a BJT.

Collector (C)

Base (B)

Emitter (E)

The basic switching operation is as follows: When the base is approximately 0.7 V more positive than the emitter and when sufficient current is provided into the base, the **transistor** turns on and goes into saturation. In saturation, the transistor ideally acts like a closed switch between the collector and the emitter, as illustrated in Figure 15–26(a). When the base is less than 0.7 V more positive than the emitter, the transistor turns off and becomes an open switch between the collector and the emitter, as shown in part (b). To summarize in general terms, a HIGH on the base turns the transistor on and makes it a closed switch. A LOW on the base turns the transistor off and makes it an open switch. In TTL, some BJTs have multiple emitters.

FIGURE 15–26
The ideal switching action of the BJT.

(a) Saturated (ON) transistor
    and ideal switch equivalent

(b) OFF transistor and
    ideal switch equivalent

TTL Inverter

The logic function of an inverter or any type of gate is always the same, regardless of the type of circuit technology that is used. Figure 15–27 shows a standard TTL circuit for an inverter. In this figure Q_1 is the input coupling transistor, and D_1 is the input clamp **diode.** Transistor Q_2 is called a *phase splitter,* and the combination of Q_3 and Q_4 forms the output circuit often referred to as a **totem-pole** arrangement.

When the input is a HIGH, the base-emitter junction of Q_1 is **reverse biased,** and the base-collector junction is **forward biased.** This condition permits current through R_1 and the base-collector junction of Q_1 into the base of Q_2, thus driving Q_2 into saturation. As a result, Q_3 is turned on by Q_2, and its collector voltage, which is the output, is near ground potential. We therefore have a LOW output for a HIGH input. At the same time, the collector of Q_2 is at a sufficiently low voltage level to keep Q_4 off.

When the input is LOW, the base-emitter junction of Q_1 is forward biased, and the base-collector junction is reverse biased. There is current through R_1 and the base-emitter junction of Q_1 to the LOW input. A LOW provides a path to ground for the current. There is no current into the base of Q_2, so it is off. The collector of Q_2 is HIGH, thus turning Q_4 on. A saturated Q_4 provides a low-resistance path from V_{CC} to the output; we therefore

FIGURE 15–27
A standard TTL inverter circuit.

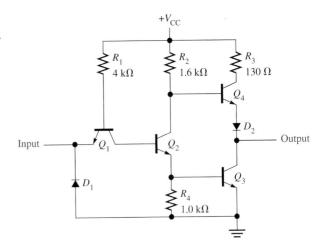

have a HIGH on the output for a LOW on the input. At the same time, the emitter of Q_2 is at ground potential, keeping Q_3 off.

Diode D_1 in the TTL circuit prevents negative spikes of voltage on the input from damaging Q_1. Diode D_2 ensures that Q_4 will turn off when Q_2 is on (HIGH input). In this condition the collector voltage of Q_2 is equal to the base-to-emitter voltage, V_{BE}, of Q_3 plus the collector-to-emitter voltage, V_{CE}, of Q_2. Diode D_2 provides an additional V_{BE} equivalent drop in series with the base-emitter junction of Q_4 to ensure its turn-off when Q_2 is on.

The operation of the TTL inverter for the two input states is illustrated in Figure 15–28. In the circuit in part (a), the base of Q_1 is 2.1 V above ground, so Q_2 and Q_3 are on. In the circuit in part (b), the base of Q_1 is about 0.7 V above ground—not enough to turn Q_2 and Q_3 on.

(a)   (b)

FIGURE 15–28
Operation of a TTL inverter.

TTL NAND Gate

A 2-input TTL NAND gate is shown in Figure 15–29. Basically, it is the same as the inverter circuit except for the additional input emitter of Q_1. In TTL technology multiple-emitter transistors are used for the input devices. These multiple-emitter transistors can be compared to the diode arrangement, as shown in Figure 15–30.

Perhaps you can understand the operation of this circuit better by visualizing Q_1 in Figure 15–29 replaced by the diode arrangement in Figure 15–30. A LOW on either

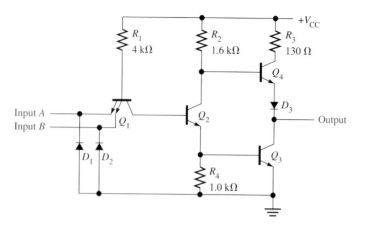

FIGURE 15–29
A TTL NAND gate circuit.

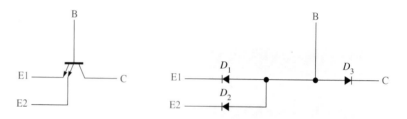

FIGURE 15–30
Diode equivalent of a TTL multiple-emitter transistor.

input A or input B forward-biases the respective diode and reverse-biases D_3 (Q_1 base-collector junction). This action keeps Q_2 off and results in a HIGH output in the same way as described for the TTL inverter. Of course, a LOW on both inputs will do the same thing.

A HIGH on both inputs reverse-biases both input diodes and forward-biases D_3 (Q_1 base-collector junction). This action turns Q_2 on and results in a LOW output in the same way as described for the TTL inverter.

You should recognize this operation as that of the NAND function: The output is LOW only if all inputs are HIGH.

Open-Collector Gates

The TTL gates described in the previous sections all had the totem-pole output circuit. Another type of output available in TTL integrated circuits is the **open-collector** output. This is comparable to the open-drain output of CMOS. A standard TTL inverter with an open-collector is shown in Figure 15–31(a). The other types of gates are also available with open-collector outputs.

Notice that the output is the collector of transistor Q_3 with nothing connected to it, hence the name *open collector.* In order to get the proper HIGH and LOW logic levels out of the circuit, an external pull-up resistor must be connected to V_{CC} from the collector of Q_3, as shown in Figure 15–31(b). When Q_3 is off, the output is pulled up to V_{CC} through the external resistor. When Q_3 is on, the output is connected to near-ground through the saturated transistor.

The ANSI/IEEE standard symbol that designates an open-collector output is shown in Figure 15–32 for an inverter and is the same for an open-drain output.

(a) Open-collector inverter circuit

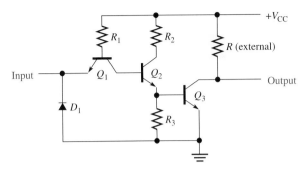

(b) With external pull-up resistor

FIGURE 15–31
TTL inverter with open-collector output.

FIGURE 15–32
Open-collector symbol in an inverter.

Tristate TTL Gates

Figure 15–33 shows the basic circuit for a TTL tristate inverter. When the enable input is LOW, Q_2 is off, and the output circuit operates as a normal totem-pole configuration, in which the output state depends on the input state. When the enable input is HIGH, Q_2 is on. There is thus a LOW on the second emitter of Q_1, causing Q_3 and Q_5 to turn off, and diode D_1 is forward biased, causing Q_4 also to turn off. When both totem-pole transistors are off, they are effectively open, and the output is completely disconnected from the internal circuitry, as illustrated in Figure 15–34.

FIGURE 15–33
Basic tristate inverter circuit.

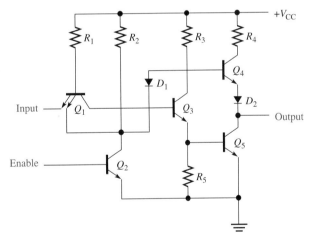

FIGURE 15–34
An equivalent circuit for the tristate output in the high-Z state.

Schottky TTL

The basic or standard TTL NAND gate circuit was discussed earlier. It is a current-sinking type of logic that draws current from the load when in the LOW output state and sources negligible current to the load when in the HIGH output state. Most TTL logic used today is some form of Schottky TTL, which provides a faster switching time by incorporating *Schottky* diodes to prevent the transistors from going into saturation, thereby decreasing the time for a transistor to turn on or off. Figure 15–35 shows a Schottky gate circuit. Notice the symbols for the Schottky transistor and Schottky diodes. Schottky devices are designated by an *S* in their part number, such as 74S00. Other types of Schottky TTL are low-power Schottky designated by LS, advanced Schottky designated by AS, advanced low-power Schottky designated by ALS, and fast designated by F.

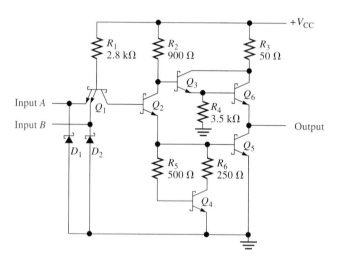

FIGURE 15–35
Schottky TTL NAND gate.

SECTION 15–3 REVIEW

1. An *npn* BJT is on when the base is more negative than the emitter.   (T or F)
2. In terms of switching action, what do the on and off states of a BJT represent?
3. What are the two major types of output circuits in TTL?
4. Explain how tristate logic differs from normal, two-state logic.

15–4 ■ PRACTICAL CONSIDERATIONS IN THE USE OF TTL

Although CMOS is the more predominant IC technology in industry and commercial applications, TTL is still alive and well. In educational applications, TTL is usually preferred because it does not have the handling restrictions that CMOS does due to ESD. Because of this, several practical considerations in the use and application of TTL circuits will be covered using standard TTL for illustration.

After completing this section, you should be able to

■ Describe current sinking and current sourcing   ■ Use an open-collector circuit for wired-AND operation   ■ Describe the effects of connecting two or more totem-pole outputs   ■ Use open-collector gates to drive LEDs and lamps   ■ Explain what to do with unused TTL inputs

Current Sinking and Current Sourcing

The concepts of current sinking and current sourcing were introduced in Section 15–1. Now that you are familiar with the totem-pole-output circuit configuration used in TTL, let's look closer at the sinking and sourcing action.

Figure 15–36 shows a standard TTL inverter with a totem-pole output connected to the input of another TTL inverter. When the driving gate is in the HIGH output state, the driver is sourcing current to the load, as shown in Figure 15–36(a). The input to the load gate is like a reverse-biased diode, so there is practically no current required by the load. Actually, since the input is nonideal, there is a maximum of 40 μA from the totem-pole output of the driver into the load gate input.

When the driving gate is in the LOW output state, the driver is sinking current from the load, as shown in Figure 15–36(b). This current is 1.6 mA maximum for standard TTL and is indicated on the **data sheet** with a negative value because it is *out* of the input.

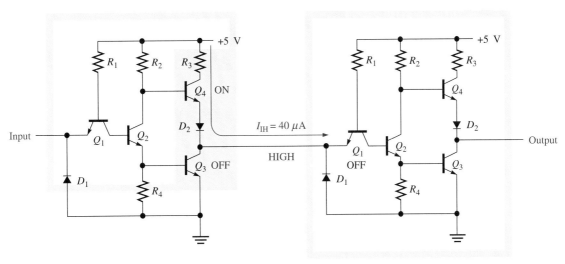

(a) Current sourcing (I_{IH} value is maximum)

(b) Current sinking (I_{IL} value is maximum)

FIGURE 15–36
Current sinking and sourcing action in TTL.

EXAMPLE 15–3

When a TTL NAND gate drives five TTL inputs, how much current does the driver output source, and how much does it sink?

Solution   Total source current (in HIGH output state):

$$I_{IH(max)} = 40 \ \mu A \text{ per input}$$

$$I_{T(source)} = (5 \text{ inputs})(40 \ \mu A/\text{input}) = 5(40 \ \mu A) = \textbf{200} \ \boldsymbol{\mu A}$$

Total sink current (in LOW output state):

$$I_{IL(max)} = -1.6 \text{ mA per input}$$

$$I_{T(sink)} = (5 \text{ inputs})(-1.6 \text{ mA/input}) = 5(-1.6 \text{ mA}) = \textbf{−8.0 mA}$$

Related Problem   Repeat the calculations for an LS TTL gate. Refer to a data sheet.

EXAMPLE 15–4

Refer to the data sheet in Appendix A, and determine the fan-out of the 7400 NAND gate.

Solution   According to the data sheet, the current parameters are as follows:

$$I_{IH(max)} = 40 \ \mu A \qquad I_{OH(max)} = -400 \ \mu A$$

$$I_{IL(max)} = 1.6 \text{ mA} \qquad I_{OL(max)} = 16 \text{ mA}$$

Fan-out for the HIGH output state is figured as follows: Current $I_{OH(max)}$ is the maximum current that the gate can source to a load. Each load input requires an $I_{IH(max)}$ of 40 μA. The HIGH-state fan-out is

$$\left| \frac{I_{OH(max)}}{I_{IH(max)}} \right| = \frac{400 \ \mu A}{40 \ \mu A} = \textbf{10}$$

For the LOW output state, fan-out is calculated as follows: $I_{OL(max)}$ is the maximum current that the gate can sink. Each load input produces an $I_{IL(max)}$ of 1.6 mA. The LOW-state fan-out is

$$\left| \frac{I_{OL(max)}}{I_{IL(max)}} \right| = \frac{16 \text{ mA}}{1.6 \text{ mA}} = \textbf{10}$$

In this case both the HIGH-state fan-out and the LOW-state fan-out are the same.

Related Problem   Determine the fan-out for a 74LS00 NAND gate.

Using Open-Collector Gates for Wired-AND Operation

The outputs of open-collector gates can be wired together to form what is called a *wired-AND* configuration. Figure 15–37 illustrates how four inverters are connected to produce a 4-input negative-AND gate. A single external pull-up resistor, R_p, is required in all wired-AND circuits.

When one (or more) of the inverter inputs is HIGH, the output X is pulled LOW because an output transistor is on and acts as a closed switch to ground, as illustrated in Figure 15–38(a). In this case only one inverter has a HIGH input, but this is sufficient to pull the output LOW through the saturated output transistor Q_1 as indicated.

For the output X to be HIGH, *all* inverter inputs must be LOW so that all the open-collector output transistors are off as indicated in Figure 15–38(b). When this condition exists, the output X is pulled HIGH through the pull-up resistor. Thus, the output X is HIGH only when *all* the inputs are LOW. Therefore, we have a negative-AND function, as expressed in the following equation:

$$X = \overline{A}\,\overline{B}\,\overline{C}\,\overline{D}$$

FIGURE 15–37
A wired-AND configuration of four inverters.

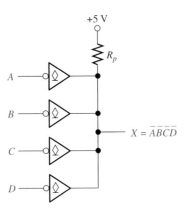

$X = \overline{A}\,\overline{B}\,\overline{C}\,\overline{D}$

FIGURE 15–38
Open-collector wired neg-ative-AND operation with inverters.

(a) When one or more output transistors are on, the output is LOW.

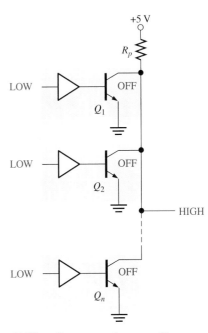

(b) When all output transistors are off, the output is HIGH.

EXAMPLE 15–5

Write the output expression for the wired-AND configuration of open-collector AND gates in Figure 15–39.

FIGURE 15–39

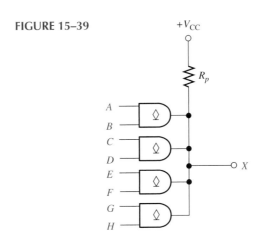

Solution   The output expression is

$$X = ABCDEFGH$$

The wired-AND connection of the four 2-input AND gates creates an 8-input AND gate.

Related Problem   Determine the output expression if NAND gates are used in Figure 15–39.

EXAMPLE 15–6

Three open-collector AND gates are connected in a wired-AND configuration as shown in Figure 15–40. Assume that the wired-AND circuit is driving four standard TTL inputs (-1.6 mA each).
(a) Write the logic expression for X.
(b) Determine the minimum value of R_p if $I_{OL(max)}$ for each gate is 30 mA and $V_{OL(max)}$ is 0.4 V.

FIGURE 15–40

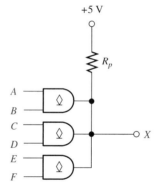

Solution
(a) $X = ABCDEF$
(b) $4(1.6$ mA$) = 6.4$ mA

$$I_{Rp} = I_{OL(max)} - 6.4 \text{ mA} = 30 \text{ mA} - 6.4 \text{ mA} = 23.6 \text{ mA}$$
$$R_p = \frac{V_{CC} - V_{OL(max)}}{I_{Rp}} = \frac{5 \text{ V} - 0.4 \text{ V}}{23.6 \text{ mA}} = \textbf{195 } \boldsymbol{\Omega}$$

Related Problem   Show the wired-AND circuit for a 10-input AND function using 74LS09 quad 2-input AND gates.

Connection of Totem-Pole Outputs

Totem-pole outputs cannot be connected together because such a connection might produce excessive current and result in damage to the devices. For example, in Figure 15–41, when Q_1 in device A and Q_2 in device B are both on, the output of device A is effectively shorted to ground through Q_2 of device B.

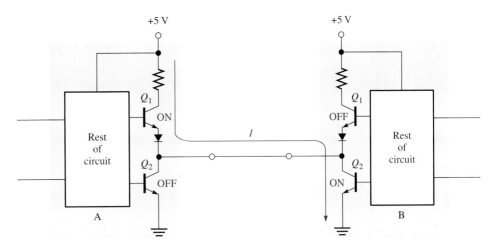

FIGURE 15–41
Totem-pole outputs wired together. Such a connection may cause excessive current through Q_1 of device A and Q_2 of device B and should never be used.

Open-Collector Buffer/Drivers

A TTL circuit with a totem-pole output is limited in the amount of current that it can sink in the LOW state ($I_{OL(max)}$) to 16 mA for standard TTL and 8 mA for LS TTL. In many special applications, a gate must drive external devices, such as LEDs, lamps, or relays, that may require more current than that.

Because of their higher voltage and current-handling capability, circuits with open-collector outputs are generally used for driving LEDs, lamps, or relays. However, totem-pole outputs can be used, as long as the output current required by the external device does not exceed the amount that the TTL driver can sink.

With an open-collector TTL gate, the collector of the output transistor is connected to an **LED** or incandescent lamp as illustrated in Figure 15–42. In part (a) the limiting re-

(a) Driving an LED

(b) Driving a low-current lamp

FIGURE 15–42
Some applications of open-collector drivers.

sistor, R_L, is used to keep the current below maximum LED current. When the output of the gate is LOW, the output transistor is sinking current, and the LED is on. The LED is off when the output transistor is off and the output is HIGH. A typical open-collector buffer gate can sink up to 40 mA. In part (b) of the figure, the lamp requires no limiting resistor because the filament is resistive. Typically, up to +30 V can be used on the open collector, depending on the particular logic family.

EXAMPLE 15–7

Determine the value of the limiting resistor, R_L, in the open-collector circuit of Figure 15–43 if the LED current is to be 20 mA. Assume a 1.5 V drop across the LED when it is forward biased and a LOW-state output voltage of 0.1 V at the output of the gate.

FIGURE 15–43

Solution

$$V_{R_L} = 5 \text{ V} - 1.5 \text{ V} - 0.1 \text{ V} = 3.4 \text{ V}$$

$$R_L = \frac{V_{R_L}}{I} = \frac{3.4 \text{ V}}{20 \text{ mA}} = \textbf{170 } \boldsymbol{\Omega}$$

Related Problem Determine the value of the limiting resistor, R_L, if the LED requires 35 mA.

Unused TTL Inputs

An unconnected input on a TTL gate acts as a HIGH because an open input results in a reverse-biased emitter junction on the input transistor, just as a HIGH level does. This effect is illustrated in Figure 15–44. However, because of noise sensitivity, it is best not to leave unused TTL inputs unconnected (open). There are several alternative ways to handle unused inputs.

Tied-Together Inputs The most common method for handling unused gate inputs is to connect them to a used input of the same gate. For AND gates and NAND gates, all tied-together inputs count as one unit load in the LOW state; but for OR gates and NOR gates, each input tied to another input counts as a separate unit load in the LOW state. In the HIGH state, each tied-together input counts as a separate load for all types of TTL gates. In Figure 15–45(a) are two examples of the connection of two unused inputs to a used input.

The AND and NAND gates present only a single unit load no matter how many inputs are tied together whereas OR and NOR gates present a unit load for each tied-together input. This is because the NAND gate uses a multiple-emitter input transistor; so no matter how many inputs are LOW, the total LOW-state current is limited to a fixed value. The NOR gate uses a separate transistor for each input; therefore, the LOW-state current is the sum of the currents from all the tied-together inputs.

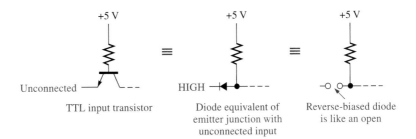

FIGURE 15–44
Comparison of an open TTL input and a HIGH-level input.

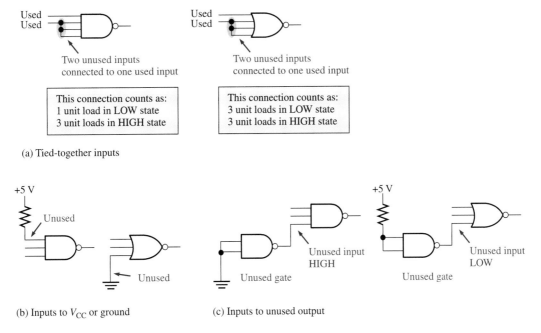

FIGURE 15–45
Methods for handling unused TTL inputs.

Inputs to V_{CC} or Ground   Unused inputs of AND and NAND gates can be connected to V_{CC} through a 1.0 kΩ resistor. This connection pulls the unused inputs to a HIGH level. Unused inputs of OR and NOR gates can be connected to ground. These methods are illustrated in Figure 15–45(b).

Inputs to Unused Output   A third method of terminating unused inputs may be appropriate in some cases when an unused gate or inverter is available. The unused gate output must be a constant HIGH for unused AND and NAND inputs and a constant LOW for unused OR and NOR inputs, as illustrated in Figure 15–45(c).

**SECTION 15–4
REVIEW**

1. In what output state does a TTL circuit sink current from a load?
2. Why does a TTL circuit source less current into a TTL load than it sinks?
3. Why can TTL circuits with totem-pole outputs not be connected together?
4. What type of TTL circuit must be used for a wired-AND configuration?
5. Why type of TTL circuit would you use to drive a lamp?
6. An unconnected TTL input acts as a LOW.   (T or F)

15–5 ■ COMPARISON OF CMOS AND TTL PERFORMANCE

In this section, the main operational and performance characteristics of selected CMOS series are compared with those of the major TTL series and with BiCMOS.

After completing this section, you should be able to

■ Compare TTL (bipolar), BiMOS, and CMOS devices in terms of propagation delay, maximum clock frequency, power dissipation, and drive capability

In the past, the superior characteristic of TTL (bipolar) compared to CMOS was its relatively high speed and output current capability. Now, these advantages of TTL have diminished to the point where CMOS is often equal or superior in many areas and has become the dominant IC technology, although TTL is still widely available and in use, as you know. One family of IC logic devices, BiCMOS, combines CMOS logic with TTL output circuitry in an effort to combine the advantages of both.

Table 15–1 provides a comparison of the performance of several IC logic families.

TABLE 15–1

Comparison of selected performance parameters of several 74XX IC families.

	Bipolar (TTL)			BiCMOS	CMOS					
					5 V			3.3 V		
	F	LS	ALS	ABT	HC	AC	AHC	LV	LVC	ALVC
Speed										
Gate propagation delay, t_p (ns)	3.3	10	7	3.2	7	5	3.7	9	4.3	3
FF maximum clock freq. (MHz)	145	33	45	150	50	160	170	90	100	150
Power Dissipation/Gate										
Bipolar: 50% dc (mW)	6	2.2	1.4							
CMOS: quiescent (μW)				17	2.75	0.55	2.75	1.6	0.8	0.8
Output Drive I_{OL} (mA)	20	8	8	64	4	24	8	12	24	24

SECTION 15–5 REVIEW

1. What is a BiCMOS circuit?
2. In general, what is the main advantage of CMOS over bipolar (TTL)?

15–6 ■ EMITTER-COUPLED LOGIC (ECL) CIRCUITS

Emitter-coupled logic, like TTL, is a bipolar technology. The typical ECL circuit consists of a different amplifier input circuit, a bias circuit, and emitter-follower outputs. ECL is much faster than TTL because the transistors do not operate in saturation and is used in more specialized high-speed applications.

After completing this section, you should be able to

- Describe how ECL differs from TTL and CMOS
- Explain the advantages and disadvantages of ECL

An **ECL** OR/NOR gate is shown in Figure 15–46(a). The emitter-follower outputs provide the OR logic function and its NOR complement, as indicated by Figure 15–46(b).

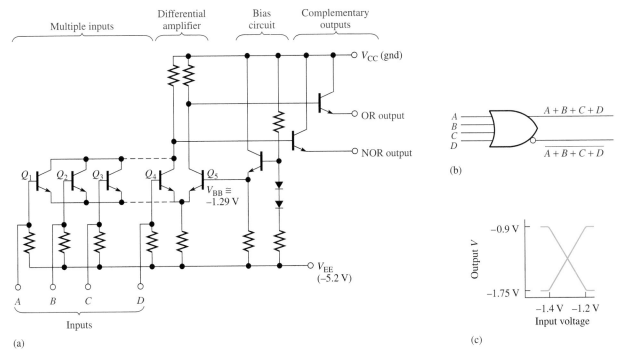

FIGURE 15–46
An ECL OR/NOR gate circuit.

Because of the low output impedance of the emitter-follower and the high input impedance of the differential amplifier input, high fan-out operation is possible. In this type of circuit, saturation is not possible. The lack of saturation results in higher power consumption and limited voltage swing (less than 1 V), but it permits high-frequency switching.

The V_{CC} pin is normally connected to ground, and the V_{EE} pin is connected to -5.2 V from the power supply for best operation. Notice that in Figure 15–46(c) the output varies from a LOW level of -1.75 V to a HIGH level of -0.9 V with respect to ground. In positive logic a 1 is the HIGH level (less negative), and a 0 is the LOW level (more negative).

Noise Margin

As you have learned, the noise margin of a gate is the measure of its immunity to undesired voltage fluctuations (noise). Typical ECL circuits have noise margins from about 0.2 V to 0.25 V. These are less than for TTL and make ECL less suitable in high-noise environments.

Comparison of ECL with TTL and CMOS

Table 15–2 shows a comparison of key performance parameters for F and AHC with those for two ECL series, 10H and E-lite.

TABLE 15–2

Comparison of two ECL series performance parameters with F and AHC.

	Bipolar (TTL)	CMOS	Bipolar (ECL)	
	F	**AHC**	**10H**	**E-lite**
Speed Gate propagation delay, t_p (ns)	3.3	3.7	1	0.22
FF maximum clock freq. (MHz)	145	170	330	2800
Power Dissipation/Gate Bipolar: 50% dc	8.9 mW		25 mW	73 mW
CMOS: quiescent		2.5 μW		

SECTION 15–6 REVIEW

1. What is the primary advantage of ECL over TTL?
2. Name two disadvantages of ECL compared with TTL.

15–7 ■ PMOS, NMOS, AND E²CMOS

The PMOS and NMOS circuits are used largely in LSI functions, such as long shift registers, large memories, and microprocessor products. Such use is a result of the low power consumption and very small chip area required for MOS transistors. E²CMOS is used in reprogrammable PLDs.

After completing this section, you should be able to

■ Describe a basic PMOS gate ■ Describe a basic NMOS gate ■ Describe a basic E²CMOS cell

PMOS

One of the first high-density **MOS** circuit technologies to be produced was **PMOS.** It utilizes enhancement-mode *p*-channel MOS transistors to form the basic gate building blocks. Figure 15–47 shows a basic PMOS gate that produces the NOR function in positive logic.

The operation of the PMOS gate is as follows: The supply voltage V_{GG} is a negative voltage, and V_{CC} is a positive voltage or ground (0 V). Transistor Q_3 is permanently biased to create a constant drain-to-source resistance. Its sole purpose is to function as a current-limiting resistor. If a HIGH (V_{CC}) is applied to input *A* or *B*, then Q_1 or Q_2 is off, and the output is pulled down to a voltage near V_{GG}, which represents a LOW. When a LOW voltage (V_{GG}) is applied to both input *A* and input *B*, both Q_1 and Q_2 are turned on. This causes the output to go to a HIGH level (near V_{CC}). Since a LOW output occurs when either or both inputs are HIGH, and a HIGH output occurs only when all inputs are LOW, we have a NOR gate.

FIGURE 15–47
Basic PMOS gate.

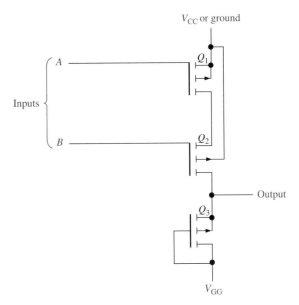

NMOS

The **NMOS** devices were developed as processing technology improved. The *n*-channel MOS transistor is used in NMOS circuits, as shown in Figure 15–48 for a NAND gate and a NOR gate.

In Figure 15–48(a), Q_3 acts as a resistor to limit current. When a LOW (V_{GG} or ground) is applied to one or both inputs, then at least one of the transistors (Q_1 or Q_2) is off, and the output is pulled up to a HIGH level near V_{CC}. When HIGHs (V_{CC}) are applied to both A and B, both Q_1 and Q_2 conduct, and the output is LOW. This action, of course, identifies this circuit as a NAND gate.

In Figure 15–48(b), Q_3 again acts as a resistor. A HIGH on either input turns Q_1 or Q_2 on, pulling the output LOW. When both inputs are LOW, both transistors are off, and the output is pulled up to a HIGH level.

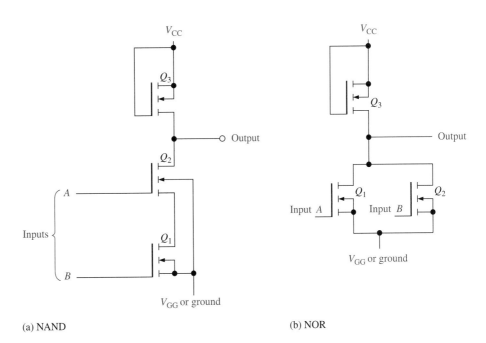

(a) NAND

(b) NOR

FIGURE 15–48
Two NMOS gates.

E²CMOS

E²CMOS technology is based on a combination of CMOS and NMOS technologies and is used in generic array logic devices (GALs) such as those discussed in Chapters 7 and 11 and other PLDs. An E²CMOS cell is built around a MOS transistor with a floating gate that is externally charged or discharged by a small programming current. A schematic of this type of cell is shown in Figure 15–49.

FIGURE 15–49
An E²CMOS cell.

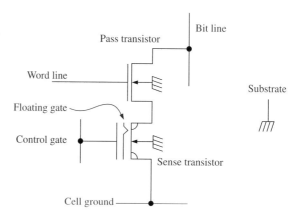

When the floating gate is charged to a positive potential by removing electrons, the sense transistor is turned on, storing a binary zero. When the floating gate is charged to a negative potential by placing electrons on it, the sense transistor is turned off, storing a binary 1. The control gate controls the potential of the floating gate. The pass transistor isolates the sense transistor from the array during read and write operations that use the word and bit lines.

The cell is programmed by applying a programming pulse to either the control gate or the bit line of a cell that has been selected by a voltage on the word line. During the programming cycle, the cell is first erased by applying a voltage to the control gate to make the floating gate negative. This leaves the sense transistor in the *off* state (storing a 1). A write pulse is applied to the bit line of a cell in which a 0 is to be stored. This will charge the floating gate to a point where the sense transistor is on (storing a 0). The bit stored in the cell is read by sensing presence or absence of a small cell current in the bit line. When a 1 is stored, there is no cell current because the sense transistor is off. When a 0 is stored, there is a small cell current because the sense transistor is on. Once a bit is stored in a cell, it will remain indefinitely unless the cell is erased or a new bit is written into the cell.

SECTION 15–7 REVIEW

1. What is the main feature of NMOS and PMOS technology in integrated circuits?
2. What is the mechanism for charge storage in an E²CMOS cell?

■ SUMMARY

■ *Formulas*

(15–1) $V_{NH} = V_{OH(min)} - V_{IH(min)}$ High-level noise margin

(15–2) $V_{NL} = V_{IL(max)} - V_{OL(max)}$ Low-level noise margin

(15–3) $I_{CC} = \dfrac{I_{CCH} + I_{CCL}}{2}$ Average dc supply current

(15–4) $P_D = V_{CC}I_{CC}$ Power dissipation

- Totem-pole outputs of TTL cannot be connected together.
- Open-collector and open-drain outputs can be connected for wired-AND.
- CMOS devices offer lower power dissipation than any of the TTL series.
- A TTL device is not as vulnerable to electrostatic discharge (ESD) as is a CMOS device.
- Because of ESD, CMOS devices must be handled with great care.
- ECL is the fastest type of logic circuit.
- E^2CMOS is used in GALs and other PLDs.

■ KEY TERMS

These terms are also in the end-of-book glossary.

Bipolar A type of transistor technology in which a transistor is constructed with two *pn* junctions.

CMOS Complementary metal-oxide semiconductor; a type of integrated logic circuit that uses *n*- and *p*-channel MOSFETs (metal-oxide semiconductor field-effect transistors).

Current sinking The action of a logic circuit in which it accepts current into its output from a load.

Current sourcing The action of a logic circuit in which it sends current from its output to a load.

Drain One of the three terminals of a MOSFET.

E^2CMOS Electrically erasable CMOS; the IC technology used in programmable logic devices (PLDs).

Gate A logic circuit that performs a specified logic operation such as AND or OR; one of the three terminals of a MOSFET.

Noise immunity The ability of a logic circuit to reject unwanted signals (noise).

Noise margin The difference between the maximum LOW output of a gate and the maximum acceptable LOW input of an equivalent gate; also, the difference between the minimum HIGH output of a gate and the minimum HIGH input of an equivalent gate. Noise margin is sometimes expressed as a percentage of the dc supply voltage.

Open-collector A type of output for a TTL circuit in which the collector of the output transistor is left internally disconnected and is available for connection to an external load that requires relatively high current or voltage.

Propagation delay time The time interval between the occurrence of an input transition and the occurrence of the corresponding output transition in a logic circuit.

Pull-up resistor A resistor with one end connected to the dc supply voltage used to keep a given point in a logic circuit HIGH when in the inactive state.

Source One of the three terminals of a MOSFET.

TTL Transistor-transistor logic. A type of integrated circuit that uses bipolar junction transistors.

Unit load A measure of fan-out. One gate input represents a unit load to a driving gate.

■ SELF-TEST

1. When the frequency of the input signal to a CMOS gate is increased, the average power dissipation
   - **(a)** decreases
   - **(b)** increases
   - **(c)** does not change
   - **(d)** decreases exponentially

2. CMOS operates more reliably than TTL in a high-noise environment because of its
   - **(a)** lower noise margin
   - **(b)** input capacitance
   - **(c)** higher noise margin
   - **(d)** smaller power dissipation

3. Proper handling of a CMOS device is necessary because of its

(a) fragile construction (b) high-noise immunity

(c) susceptibility to electrostatic discharge (d) low power dissipation

4. Which of the following is not a TTL circuit?

(a) 74F00 (b) 74AS00 (c) 74HC00 (d) 74ALS00

5. An open TTL NOR gate input

(a) acts as a LOW (b) acts as a HIGH

(c) should be grounded (d) should be connected to V_{CC} through a resistor

(e) answers (b) and (c) (f) answers (a) and (c)

6. An LS TTL gate can drive a maximum of

(a) 20 unit loads (b) ten unit loads

(c) forty unit loads (d) unlimited unit loads

7. If two unused inputs of a LS TTL gate are connected to an input being driven by another LS TTL gate, the total number of remaining unit loads that can be driven by this gate is

(a) seven (b) eight (c) seventeen (d) unlimited

8. The main advantage of ECL over TTL or CMOS is

(a) ECL is less expensive (b) ECL consumes less power

(c) ECL is available in a greater variety of circuit types (d) ECL is faster

9. ECL cannot be used in

(a) high-noise environments (b) damp environments (c) high-frequency applications

10. The basic mechanism for storing a data bit in an E^2CMOS cell is

(a) control gate (b) floating drain (c) floating gate (d) cell current

■ PROBLEMS

SECTION 15–1 Basic Operational Characteristics and Parameters

1. A certain logic gate has a $V_{OH(min)} = 2.2$ V, and it is driving a gate with a $V_{IH(min)} = 2.5$ V. Are these gates compatible for HIGH-state operation? Why?

2. A certain logic gate has a $V_{OL(max)} = 0.45$ V, and it is driving a gate with a $V_{IL(max)} = 0.75$ V. Are these gates compatible for LOW-state operation? Why?

3. A TTL gate has the following actual voltage level values: $V_{IH(min)} = 2.25$ V, $V_{IL(max)} = 0.65$ V. Assuming it is being driven by a gate with $V_{OH(min)} = 2.4$ V and $V_{OL(max)} = 0.4$ V, what are the HIGH- and LOW-level noise margins?

4. What is the maximum amplitude of noise spikes that can be tolerated on the inputs in both the HIGH state and the LOW state for the gate in Problem 3?

5. Voltage specifications for three types of logic gates are given in Table 15–3. Select the gate that you would use in a high-noise industrial environment.

TABLE 15–3

	$V_{OH(min)}$	$V_{OL(max)}$	$V_{IH(min)}$	$V_{IL(max)}$
Gate A	2.4 V	0.4 V	2 V	0.8 V
Gate B	3.5 V	0.2 V	2.5 V	0.6 V
Gate C	4.2 V	0.2 V	3.2 V	0.8 V

6. A certain gate draws a dc supply current from a $+5$ V source of 2 mA in the LOW state and 3.5 mA in the HIGH state. What is the power dissipation in the LOW state? What is the power dissipation in the HIGH state? Assuming a 50% duty cycle, what is the average power dissipation?

7. Each gate in the circuit of Figure 15–50 has a t_{PLH} and a t_{PHL} of 4 ns. If a positive-going pulse is applied to the input as indicated, how long will it take the output pulse to appear?

8. For a certain gate, $t_{PLH} = 3$ ns and $t_{PHL} = 2$ ns. What is the average propagation delay time?

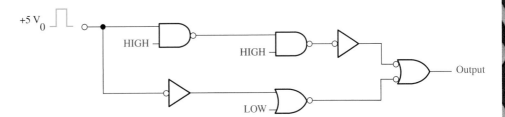

FIGURE 15–50

TABLE 15–4

	t_{PLH}	t_{PHL}	P_D
Gate A	1 ns	1.2 ns	15 mW
Gate B	5 ns	4 ns	8 mW
Gate C	10 ns	10 ns	0.5 mW

9. Table 15–4 lists parameters for three types of gates. Basing your decision on the speed-power product, which one would you select for best performance?

10. Which gate in Table 15–4 would you select if you wanted the gate to operate at the highest possible frequency?

11. A standard TTL gate has a fan-out of 10. Are any of the gates in Figure 15–51 overloaded? If so, which ones?

12. Which CMOS gate network in Figure 15–52 can operate at the highest frequency?

FIGURE 15–51

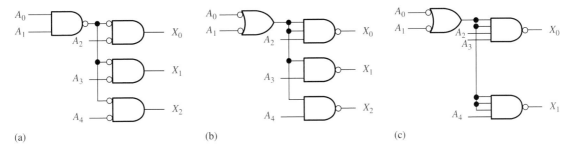

(a)　　　　　　　(b)　　　　　　　(c)

FIGURE 15–52

SECTION 15–2 CMOS Circuits

13. Determine the state (on or off) of each MOSFET in Figure 15–53.

14. The CMOS gate network in Figure 15–54 is incomplete. Indicate the changes that should be made.

15. Devise a circuit, using appropriate CMOS logic gates and/or inverters, with which signals from four different sources can be connected to a common line at different times without interfering with each other.

(a)   (b)   (c)   (d)

FIGURE 15–53

FIGURE 15–54
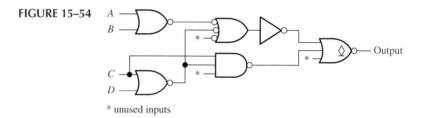

* unused inputs

SECTION 15–3 TTL Circuits

16. Determine which BJTs in Figure 15–55 are off and which are on.

17. Determine the output state of each TTL gate in Figure 15–56.

18. The TTL gate network in Figure 15–57 is incomplete. Indicate the changes that should be made.

(a)   (b)   (c)   (d)

FIGURE 15–55

(a)   (b)   (c)   (d)

FIGURE 15–56

FIGURE 15–57

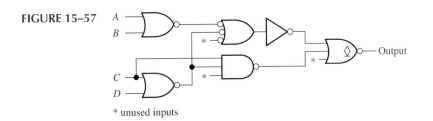

* unused inputs

SECTION 15–4   Practical Considerations in the Use of TTL

19. Determine the output level of each TTL gate in Figure 15–58.

20. For each part of Figure 15–59, tell whether each driving gate is sourcing or sinking current. Specify the maximum current out of or into the output of the driving gate or gates in each case. All gates are standard TTL.

FIGURE 15–58

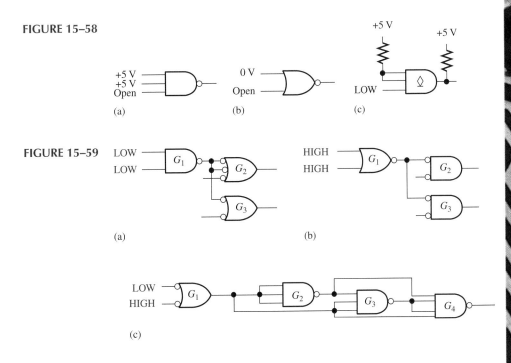

21. Use open-collector inverters to implement the following logic expressions:
    (a) $X = \overline{A}\,\overline{B}\,\overline{C}$ (b) $X = A\overline{B}C\overline{D}$ (c) $X = ABC\overline{D}\,\overline{E}F$

22. Write the logic expression for each of the circuits in Figure 15–60.

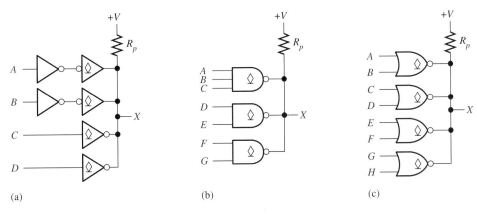

FIGURE 15–60

23. Determine the minimum value for the pull-up resistor in each circuit in Figure 15–60 if $I_{OL(max)} = 40$ mA and $V_{OL(max)} = 0.25$ V for each gate. Assume that 10 standard TTL unit loads are being driven from output X and the supply voltage is 5 V.

24. A certain relay requires 60 mA. Devise a way to use open-collector NAND gates with $I_{OL(max)} = 40$ mA to drive the relay.

SECTION 15–5 Comparison of CMOS and TTL Performance

25. Select the IC family with the best speed-power product in Table 15–1.

26. Determine from Table 15–1 the logic family that is most appropriate for each of the following requirements:

(a) shortest propagation delay time

(b) fastest flip-flop toggle rate

(c) lowest power dissipation

(d) best compromise between speed and power for a logic gate

27. Determine the total propagation delay from each input to each output for each circuit in Figure 15–61.

28. One of the flip-flops in Figure 15–62 may have an erratic output. Which one is it if any and why?

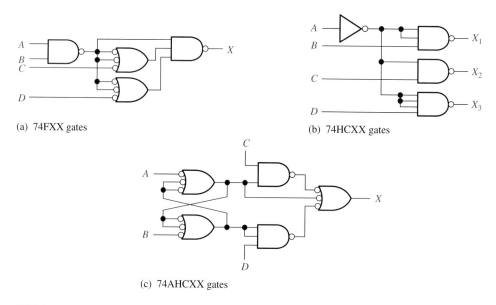

(a) 74FXX gates

(b) 74HCXX gates

(c) 74AHCXX gates

FIGURE 15–61

SECTION 15–6 Emitter-Coupled Logic (ECL) Circuits

29. What is the basic difference between ECL circuitry and TTL circuitry?

30. Select ECL, HCMOS, or the appropriate TTL series for each of the following requirements:

(a) highest speed

(b) lowest power

(c) best compromise between high speed and low power (speed-power product)

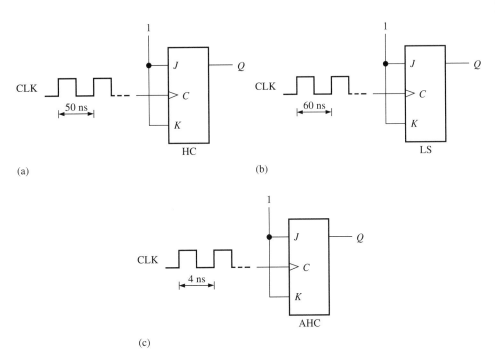

FIGURE 15–62

■ **ANSWERS TO SECTION REVIEWS**

SECTION 15–1

1. V_{IH}: HIGH level input voltage; V_{IL}: LOW level input voltage; V_{OH}: HIGH level output voltage; V_{OL}: LOW level output voltage
2. A higher value of noise margin is better.
3. Gate B can operate at a higher frequency.
4. Excessive loading reduces the noise margin of a gate.

SECTION 15–2

1. MOSFETs are used in CMOS logic.
2. A complementary output circuit consists of an n-channel and a p-channel MOSFET.
3. Because electrostatic discharge can damage CMOS devices

SECTION 15–3

1. False, the npn BJT is off.
2. The on state of a BJT is a closed switch; the off state is an open switch.
3. Totem-pole and open-collector are types of TTL outputs.
4. Tristate logic provides a high-impedance state, in which the output is disconnected from the rest of the circuit.

SECTION 15–4

1. Sink current occurs in a LOW output state.
2. Source current is less than sink current because a TTL load looks like a reverse-biased diode in the HIGH state.
3. The totem-pole transistors cannot handle the current when one output tries to go HIGH and the other is LOW.
4. Wired-AND must use open-collector.   5. Lamp driver must be open-collector.
5. False, an unconnected TTL input generally acts as a HIGH.

SECTION 15–5

1. BiCMOS uses bipolar transistors for input and output circuitry and CMOS in between.

2. CMOS has lower power dissipation than bipolar.

SECTION 15–6

1. ECL is faster than TTL.　　**2.** ECL has more power and less noise margin than TTL.

SECTION 15–7

1. NMOS and PMOS are high density.

2. The floating gate is the mechanism for storing charge in an E^2CMOS cell.

■ **ANSWERS TO RELATED PROBLEMS FOR EXAMPLES**

15–1 CMOS　　**15–2** 10.75 μW

15–3 $I_{T(source)} = 5(20 \ \mu A) = 100 \ \mu A$

$I_{T(sink)} = 5(-0.4 \ mA) = -2.0 \ mA$

15–4 Fan-out = 20　　**15–5** $X = (\overline{AB})(\overline{CD})(\overline{EF})(\overline{GH}) = (\overline{A} + \overline{B})(\overline{C} + \overline{D})(\overline{E} + \overline{F})(\overline{G} + \overline{H})$

15–6 See Figure 15–63.　　**15–7** $R_L = 97 \ \Omega$

FIGURE 15–63

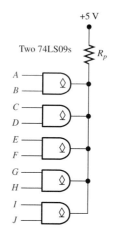

■ **ANSWERS TO SELF-TEST**

1. (b)　　**2.** (c)　　**3.** (c)　　**4.** (c)　　**5.** (e)

6. (a)　　**7.** (c)　　**8.** (d)　　**9.** (a)　　**10.** (c)

A DATA SHEETS

SN5400, SN54LS00, SN54S00,
SN7400, SN74LS00, SN74S00
QUADRUPLE 2-INPUT POSITIVE-NAND GATES

DECEMBER 1983 – REVISED MARCH 1988

- Package Options Include Plastic ''Small Outline'' Packages, Ceramic Chip Carriers and Flat Packages, and Plastic and Ceramic DIPs

- Dependable Texas Instruments Quality and Reliability

description

These devices contain four independent 2-input-NAND gates.

The SN5400, SN54LS00, and SN54S00 are characterized for operation over the full military temperature range of −55°C to 125°C. The SN7400, SN74LS00, and SN74S00 are characterized for operation from 0°C to 70°C.

FUNCTION TABLE (each gate)

INPUTS		OUTPUT
A	B	Y
H	H	L
L	X	H
X	L	H

logic symbol[†]

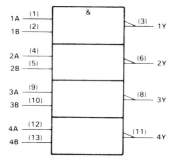

[†]This symbol is in accordance with ANSI-IEEE Std. 91-1984 and IEC Publication 617-12.
Pin numbers shown are for D, J, and N packages.

```
SN5400 . . . J PACKAGE
SN54LS00, SN54S00 . . . J OR W PACKAGE
SN7400    . N PACKAGE
SN74LS00, SN74S00 . . . D OR N PACKAGE
             (TOP VIEW)
```

```
1A [ 1      14 ] VCC
1B [ 2      13 ] 4B
1Y [ 3      12 ] 4A
2A [ 4      11 ] 4Y
2B [ 5      10 ] 3B
2Y [ 6       9 ] 3A
GND [ 7      8 ] 3Y
```

```
SN5400 . . . W PACKAGE
        (TOP VIEW)
```

```
1A  [ 1     14 ] 4Y
1B  [ 2     13 ] 4B
1Y  [ 3     12 ] 4A
VCC [ 4     11 ] GND
2Y  [ 5     10 ] 3B
2A  [ 6      9 ] 3A
2B  [ 7      8 ] 3Y
```

```
SN54LS00, SN54S00 . . . FK PACKAGE
             (TOP VIEW)
```

NC   No internal connection

logic diagram (positive logic)

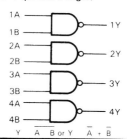

Y   A̅ B̅ or Y   A̅ + B̅

SN5400, SN54LS00, SN54S00,
SN7400, SN74LS00, SN74S00
QUADRUPLE 2-INPUT POSITIVE-NAND GATES

schematics (each gate)

'00

Resistor values shown are nominal.

absolute maximum ratings over operating free-air temperature range (unless otherwise noted)

Supply voltage, V_CC (see Note 1) . . . . 7 V
Input voltage: '00, 'S00 . . . . 5.5 V
         'LS00 . . . . 7 V
Operating free-air temperature range: SN54' . . . . -55°C to 125°C
                           SN74' . . . . 0°C to 70°C
Storage temperature range . . . . -65°C to 150°C

NOTE 1: Voltage values are with respect to network ground terminal.

SN5400, SN7400
QUADRUPLE 2-INPUT POSITIVE-NAND GATES

recommended operating conditions

		SN5400			SN7400			UNIT
		MIN	NOM	MAX	MIN	NOM	MAX	
V_{CC}	Supply voltage	4.5	5	5.5	4.75	5	5.25	V
V_{IH}	High-level input voltage	2			2			V
V_{IL}	Low-level input voltage			0.8			0.8	V
I_{OH}	High-level output current			−0.4			−0.4	mA
I_{OL}	Low-level output current			16			16	mA
T_A	Operating free-air temperature	−55		125	0		70	°C

electrical characteristics over recommended operating free-air temperature range (unless otherwise noted)

PARAMETER	TEST CONDITIONS†	SN5400			SN7400			UNIT
		MIN	TYP‡	MAX	MIN	TYP‡	MAX	
V_{IK}	V_{CC} = MIN, I_I = −12 mA			−1.5			−1.5	V
V_{OH}	V_{CC} = MIN, V_{IL} = 0.8 V, I_{OH} = −0.4 mA	2.4	3.4		2.4	3.4		V
V_{OL}	V_{CC} = MIN, V_{IH} = 2 V, I_{OL} = 16 mA		0.2	0.4		0.2	0.4	V
I_I	V_{CC} = MAX, V_I = 5.5 V			1			1	mA
I_{IH}	V_{CC} = MAX, V_I = 2.4 V			40			40	µA
I_{IL}	V_{CC} = MAX, V_I = 0.4 V			−1.6			−1.6	mA
I_{OS}§	V_{CC} = MAX	−20		−55	−18		−55	mA
I_{CCH}	V_{CC} = MAX, V_I = 0 V		4	8		4	8	mA
I_{CCL}	V_{CC} = MAX, V_I = 4.5 V		12	22		12	22	mA

† For conditions shown as MIN or MAX, use the appropriate value specified under recommended operating conditions.
‡ All typical values are at V_{CC} = 5 V, T_A = 25°C.
§ Not more than one output should be shorted at a time.

switching characteristics, V_{CC} = 5 V, T_A = 25°C

PARAMETER	FROM (INPUT)	TO (OUTPUT)	TEST CONDITIONS	MIN	TYP	MAX	UNIT
t_{PLH}	A or B	Y	R_L = 400 Ω, C_L = 15 pF		11	22	ns
t_{PHL}					7	15	ns

SN54LS00, SN74LS00
QUADRUPLE 2-INPUT POSITIVE-NAND GATES

recommended operating conditions

		SN54LS00 MIN	NOM	MAX	SN74LS00 MIN	NOM	MAX	UNIT
V_{CC}	Supply voltage	4.5	5	5.5	4.75	5	5.25	V
V_{IH}	High-level input voltage	2			2			V
V_{IL}	Low-level input voltage			0.7			0.8	V
I_{OH}	High-level output current			−0.4			−0.4	mA
I_{OL}	Low-level output current			4			8	mA
T_A	Operating free-air temperature	−55		125	0		70	°C

electrical characteristics over recommended operating free-air temperature range (unless otherwise noted)

PARAMETER	TEST CONDITIONS †	SN54LS00 MIN	TYP‡	MAX	SN74LS00 MIN	TYP‡	MAX	UNIT
V_{IK}	$V_{CC} = $ MIN, $I_I = -18$ mA			−1.5			−1.5	V
V_{OH}	$V_{CC} = $ MIN, $V_{IL} = $ MAX, $I_{OH} = -0.4$ mA	2.5	3.4		2.7	3.4		V
V_{OL}	$V_{CC} = $ MIN, $V_{IH} = 2$ V, $I_{OL} = 4$ mA		0.25	0.4		0.25	0.4	V
	$V_{CC} = $ MIN, $V_{IH} = 2$ V, $I_{OL} = 8$ mA					0.35	0.5	
I_I	$V_{CC} = $ MAX, $V_I = 7$ V			0.1			0.1	mA
I_{IH}	$V_{CC} = $ MAX, $V_I = 2.7$ V			20			20	μA
I_{IL}	$V_{CC} = $ MAX, $V_I = 0.4$ V			−0.4			−0.4	mA
I_{OS} §	$V_{CC} = $ MAX	−20		−100	−20		−100	mA
I_{CCH}	$V_{CC} = $ MAX, $V_I = 0$ V		0.8	1.6		0.8	1.6	mA
I_{CCL}	$V_{CC} = $ MAX, $V_I = 4.5$ V		2.4	4.4		2.4	4.4	mA

† For conditions shown as MIN or MAX, use the appropriate value specified under recommended operating conditions.
‡ All typical values are at $V_{CC} = 5$ V, $T_A = 25$°C
§ Not more than one output should be shorted at a time, and the duration of the short-circuit should not exceed one second.

switching characteristics, $V_{CC} = 5$ V, $T_A = 25$°C

PARAMETER	FROM (INPUT)	TO (OUTPUT)	TEST CONDITIONS	MIN	TYP	MAX	UNIT
t_{PLH}	A or B	Y	$R_L = 2$ kΩ, $C_L = 15$ pF		9	15	ns
t_{PHL}					10	15	ns

SN54F00, SN74F00
QUADRUPLE 2-INPUT POSITIVE-NAND GATES

SDFS035A – MARCH 1987 – REVISED OCTOBER 1993

● **Package Options Include Plastic Small-Outline Packages, Ceramic Chip Carriers, and Standard Plastic and Ceramic 300-mil DIPs**

description

These devices contain four independent 2-input NAND gates. They perform the Boolean functions $Y = \overline{A \cdot B}$ or $Y = \overline{A} + \overline{B}$ in positive logic.

The SN54F00 is characterized for operation over the full military temperature range of $-55°C$ to $125°C$. The SN74F00 is characterized for operation from $0°C$ to $70°C$.

FUNCTION TABLE
(each gate)

INPUTS		OUTPUT
A	B	Y
H	H	L
L	X	H
X	L	H

SN54F00 . . . J PACKAGE
SN74F00 . . . D OR N PACKAGE
(TOP VIEW)

1A	1	14	V_{CC}
1B	2	13	4B
1Y	3	12	4A
2A	4	11	4Y
2B	5	10	3B
2Y	6	9	3A
GND	7	8	3Y

SN54F00 . . . FK PACKAGE
(TOP VIEW)

NC – No internal connection

logic symbol†

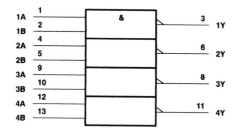

1A	1
1B	2
2A	4
2B	5
3A	9
3B	10
4A	12
4B	13

& → 3 1Y
6 2Y
8 3Y
11 4Y

† This symbol is in accordance with ANSI/IEEE Std 91-1984 and IEC Publication 617-12.

logic diagram (positive logic)

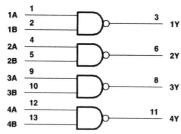

1A	1
1B	2
2A	4
2B	5
3A	9
3B	10
4A	12
4B	13

3 1Y
6 2Y
8 3Y
11 4Y

Pin numbers shown are for the D, J, and N packages.

SN54F00, SN74F00
QUADRUPLE 2-INPUT POSITIVE-NAND GATES

SDFS035A – MARCH 1987 – REVISED OCTOBER 1993

absolute maximum ratings over operating free-air temperature range (unless otherwise noted)†

Supply voltage range, V_{CC} .... −0.5 V to 7 V
Input voltage range, V_I (see Note 1) .... −1.2 V to 7 V
Input current range .... −30 mA to 5 mA
Voltage range applied to any output in the high state .... −0.5 V to V_{CC}
Current into any output in the low state .... 40 mA
Operating free-air temperature range: SN54F00 .... −55°C to 125°C
                                 SN74F00 .... 0°C to 70°C
Storage temperature range .... −65°C to 150°C

† Stresses beyond those listed under "absolute maximum ratings" may cause permanent damage to the device. These are stress ratings only and functional operation of the device at these or any other conditions beyond those indicated under "recommended operating conditions" is not implied. Exposure to absolute-maximum-rated conditions for extended periods may affect device reliability.
NOTE 1: The input voltage ratings may be exceeded provided the input current ratings are observed.

recommended operating conditions

		SN54F00			SN74F00			UNIT
		MIN	NOM	MAX	MIN	NOM	MAX	
V_{CC}	Supply voltage	4.5	5	5.5	4.5	5	5.5	V
V_{IH}	High-level input voltage	2			2			V
V_{IL}	Low-level input voltage			0.8			0.8	V
I_{IK}	Input clamp current			−18			−18	mA
I_{OH}	High-level output current			−1			−1	mA
I_{OL}	Low-level output current			20			20	mA
T_A	Operating free-air temperature	−55		125	0		70	°C

electrical characteristics over recommended operating free-air temperature range (unless otherwise noted)

PARAMETER	TEST CONDITIONS		SN54F00			SN74F00			UNIT
			MIN	TYP‡	MAX	MIN	TYP‡	MAX	
V_{IK}	$V_{CC} = 4.5$ V,	$I_I = -18$ mA			−1.2			−1.2	V
V_{OH}	$V_{CC} = 4.5$ V,	$I_{OH} = -1$ mA	2.5	3.4		2.5	3.4		V
	$V_{CC} = 4.75$ V,	$I_{OH} = -1$ mA				2.7			
V_{OL}	$V_{CC} = 4.5$ V,	$I_{OL} = 20$ mA		0.3	0.5		0.3	0.5	V
I_I	$V_{CC} = 5.5$ V,	$V_I = 7$ V			0.1			0.1	mA
I_{IH}	$V_{CC} = 5.5$ V,	$V_I = 2.7$ V			20			20	μA
I_{IL}	$V_{CC} = 5.5$ V,	$V_I = 0.5$ V			−0.6			−0.6	mA
I_{OS}§	$V_{CC} = 5.5$ V,	$V_O = 0$	−60		−150	−60		−150	mA
I_{CCH}	$V_{CC} = 5.5$ V,	$V_I = 0$		1.9	2.8		1.9	2.8	mA
I_{CCL}	$V_{CC} = 5.5$ V,	$V_I = 4.5$ V		6.8	10.2		6.8	10.2	mA

‡ All typical values are at $V_{CC} = 5$ V, $T_A = 25$°C.
§ Not more than one output should be shorted at a time, and the duration of the short circuit should not exceed one second.

SN54F00, SN74F00
QUADRUPLE 2-INPUT POSITIVE-NAND GATES

SDFS035A – MARCH 1987 – REVISED OCTOBER 1993

switching characteristics (see Note 2)

PARAMETER	FROM (INPUT)	TO (OUTPUT)	$V_{CC} = 5$ V, $C_L = 50$ pF, $R_L = 500$ Ω, $T_A = 25°C$			$V_{CC} = 4.5$ V to 5.5 V, $C_L = 50$ pF, $R_L = 500$ Ω, $T_A = $ MIN to MAX†				UNIT
			'F00			SN54F00		SN74F00		
			MIN	TYP	MAX	MIN	MAX	MIN	MAX	
t_{PLH}	A or B	Y	1.6	3.3	5	2	7	1.6	6	ns
t_{PHL}			1	2.8	4.3	1.5	6.5	1	5.3	

† For conditions shown as MIN or MAX, use the appropriate value specified under recommended operating conditions.
NOTE 2: Load circuits and waveforms are shown in Section 1.

TEXAS INSTRUMENTS

Data sheet acquired from Harris Semiconductor
SCHS117

August 1997

CD54HC04, CD54HCT04, CD74HC04, CD74HCT04

High Speed CMOS Logic Hex Inverter

Features

- Buffered Inputs
- Typical Propagation Delay: 6ns at V_{CC} = 5V, C_L = 15pF, T_A = 25°C
- Fanout (Over Temperature Range)
  - Standard Outputs . . . . 10 LSTTL Loads
  - Bus Driver Outputs . . . . 15 LSTTL Loads
- Wide Operating Temperature Range . . . -55°C to 125°C
- Balanced Propagation Delay and Transition Times
- Significant Power Reduction Compared to LSTTL Logic ICs
- HC Types
  - 2V to 6V Operation
  - High Noise Immunity: N_{IL} = 30%, N_{IH} = 30% of V_{CC} at V_{CC} = 5V
- HCT Types
  - 4.5V to 5.5V Operation
  - Direct LSTTL Input Logic Compatibility, V_{IL}= 0.8V (Max), V_{IH} = 2V (Min)
  - CMOS Input Compatibility, $I_I \leq 1\mu A$ at V_{OL}, V_{OH}

Description

The Harris CD54HC04, CD54HCT04, CD74HC04 and CD74HCT04 logic gates utilize silicon gate CMOS technology to achieve operating speeds similar to LSTTL gates with the low power consumption of standard CMOS integrated circuits. All devices have the ability to drive 10 LSTTL loads. The 74HCT logic family is functionally pin compatible with the standard 74LS logic family.

Ordering Information

PART NUMBER	TEMP. RANGE (°C)	PACKAGE	PKG. NO.
CD74HC04E	-55 to 125	14 Ld PDIP	E14.3
CD74HCT04E	-55 to 125	14 Ld PDIP	E14.3
CD74HC04M	-55 to 125	14 Ld SOIC	M14.15
CD74HCT04M	-55 to 125	14 Ld SOIC	M14.15
CD54HC04F	-55 to 125	14 Ld CERDIP	F14.3
CD54HCT04F	-55 to 125	14 Ld CERDIP	F14.3
CD54HC04W	-55 to 125	Wafer	
CD54HCT04W	-55 to 125	Wafer	
CD54HC04H	-55 to 125	Die	
CD54HCT04H	-55 to 125	Die	

NOTE:

1. When ordering, use the entire part number. Add the suffix 96 to obtain the variant in the tape and reel.

Pinout

CD54HC04, CD54HCT04, CD74HC04, CD74HCT04
(PDIP, CERDIP, SOIC)
TOP VIEW

1A	1	14 V_{CC}
$\overline{1Y}$	2	13 6A
2A	3	12 $\overline{6Y}$
$\overline{2Y}$	4	11 5A
3A	5	10 $\overline{5Y}$
$\overline{3Y}$	6	9 4A
GND	7	8 $\overline{4Y}$

CD54HC04, CD54HCT04, CD74HC04, CD74HCT04

Functional Diagram

TRUTH TABLE

INPUTS	
nA	nY
L	H
H	L

NOTE: H = High Voltage Level, L = Low Voltage Level

Logic Symbol

CD54HC04, CD54HCT04, CD74HC04, CD74HCT04

Absolute Maximum Ratings

DC Supply Voltage, V_{CC} . . . . -0.5V to 7V
DC Input Diode Current, I_{IK}
  For V_I < -0.5V or V_I > V_{CC} + 0.5V . . . . ±20mA
DC Output Diode Current, I_{OK}
  For V_O < -0.5V or V_O > V_{CC} + 0.5V . . . . ±20mA
DC Output Source or Sink Current per Output Pin, I_O
  For V_O > -0.5V or V_O < V_{CC} + 0.5V . . . . ±25mA
DC V_{CC} or Ground Current, I_{CC} or I_{GND} . . . . ±50mA

Operating Conditions

Temperature Range (T_A) . . . . -55°C to 125°C
Supply Voltage Range, V_{CC}
  HC Types . . . . .2V to 6V
  HCT Types . . . . .4.5V to 5.5V
DC Input or Output Voltage, V_I, V_O . . . . 0V to V_{CC}
Input Rise and Fall Time
  2V . . . . 1000ns (Max)
  4.5V . . . . 500ns (Max)
  6V . . . . 400ns (Max)

Thermal Information

Thermal Resistance (Typical, Note 2)	θ_{JA} (°C/W)	θ_{JC} (°C/W)
PDIP Package	100	N/A
CERDIP Package	130	55
SOIC Package.	180	N/A

Maximum Junction Temperature (Hermetic Package or Die) . . . 175°C
Maximum Junction Temperature (Plastic Package) . . . . 150°C
Maximum Storage Temperature Range . . . .-65°C to 150°C
Maximum Lead Temperature (Soldering 10s) . . . . 300°C
  (SOIC - Lead Tips Only)

CAUTION: Stresses above those listed in "Absolute Maximum Ratings" may cause permanent damage to the device. This is a stress only rating and operation of the device at these or any other conditions above those indicated in the operational sections of this specification is not implied.

NOTE:

2. θ_{JA} is measured with the component mounted on an evaluation PC board in free air.

DC Electrical Specifications

PARAMETER	SYMBOL	TEST CONDITIONS V_I (V)	TEST CONDITIONS I_O (mA)	V_{CC} (V)	25°C MIN	25°C TYP	25°C MAX	-40°C TO +85°C MIN	-40°C TO +85°C MAX	-55°C TO 125°C MIN	-55°C TO 125°C MAX	UNITS
HC TYPES												
High Level Input Voltage	V_{IH}	-	-	2	1.5	-	-	1.5	-	1.5	-	V
				4.5	3.15	-	-	3.15	-	3.15	-	V
				6	4.2	-	-	4.2	-	4.2	-	V
Low Level Input Voltage	V_{IL}	-	-	2	-	-	0.5	-	0.5	-	0.5	V
				4.5	-	-	1.35	-	1.35	-	1.35	V
				6	-	-	1.8	-	1.8	-	1.8	V
High Level Output Voltage CMOS Loads	V_{OH}	V_{IH} or V_{IL}	-0.02	2	1.9	-	-	1.9	-	1.9	-	V
			-0.02	4.5	4.4	-	-	4.4	-	4.4	-	V
			-0.02	6	5.9	-	-	5.9	-	5.9	-	V
High Level Output Voltage TTL Loads		-	-	-	-	-	-	-	-	-	-	V
			-4	4.5	3.98	-	-	3.84	-	3.7	-	V
			-5.2	6	5.48	-	-	5.34	-	5.2	-	V
Low Level Output Voltage CMOS Loads	V_{OL}	V_{IH} or V_{IL}	0.02	2	-	-	0.1	-	0.1	-	0.1	V
			0.02	4.5	-	-	0.1	-	0.1	-	0.1	V
			0.02	6	-	-	0.1	-	0.1	-	0.1	V
Low Level Output Voltage TTL Loads		-	-	-	-	-	-	-	-	-	-	V
			4	4.5	-	-	0.26	-	0.33	-	0.4	V
			5.2	6	-	-	0.26	-	0.33	-	0.4	V
Input Leakage Current	I_I	V_{CC} or GND	-	6	-	-	±0.1	-	±1	-	±1	µA

CD54HC04, CD54HCT04, CD74HC04, CD74HCT04

DC Electrical Specifications (Continued)

PARAMETER	SYMBOL	TEST CONDITIONS V$_I$ (V)	I$_O$ (mA)	V$_{CC}$ (V)	25°C MIN	TYP	MAX	-40°C TO +85°C MIN	MAX	-55°C TO 125°C MIN	MAX	UNITS
Quiescent Device Current	I$_{CC}$	V$_{CC}$ or GND	0	6	-	-	2	-	20	-	40	µA
HCT TYPES												
High Level Input Voltage	V$_{IH}$	-	-	4.5 to 5.5	2	-	-	2	-	2	-	V
Low Level Input Voltage	V$_{IL}$	-	-	4.5 to 5.5	-	-	0.8	-	0.8	-	0.8	V
High Level Output Voltage CMOS Loads	V$_{OH}$	V$_{IH}$ or V$_{IL}$	-0.02	4.5	4.4	-	-	4.4	-	4.4	-	V
High Level Output Voltage TTL Loads			-4	4.5	3.98	-	-	3.84	-	3.7	-	V
Low Level Output Voltage CMOS Loads	V$_{OL}$	V$_{IH}$ or V$_{IL}$	0.02	4.5	-	-	0.1	-	0.1	-	0.1	V
Low Level Output Voltage TTL Loads			4	4.5	-	-	0.26	-	0.33	-	0.4	V
Input Leakage Current	I$_I$	V$_{CC}$ and GND	0	5.5	-		±0.1	-	±1	-	±1	µA
Quiescent Device Current	I$_{CC}$	V$_{CC}$ or GND	0	5.5	-	-	2	-	20	-	40	µA
Additional Quiescent Device Current Per Input Pin: 1 Unit Load (Note)	ΔI$_{CC}$	V$_{CC}$ - 2.1	-	4.5 to 5.5	-	100	360	-	450	-	490	µA

NOTE: For dual-supply systems theorectical worst case (V$_I$ = 2.4V, V$_{CC}$ = 5.5V) specification is 1.8mA.

HCT Input Loading Table

INPUT	UNIT LOADS
nB	1.2

NOTE: Unit Load is ΔI$_{CC}$ limit specified in DC Electrical Specifications table, e.g. 360µA max at 25°C.

Switching Specifications Input t$_r$, t$_f$ = 6ns

PARAMETER	SYMBOL	TEST CONDITIONS	V$_{CC}$ (V)	25°C MIN	TYP	MAX	-40°C TO 85°C MIN	MAX	-55°C TO 125°C MIN	MAX	UNITS
HC TYPES											
Propagation Delay, Input to Output (Figure 1)	t$_{PLH}$, t$_{PHL}$	C$_L$ = 50pF	2	-	-	85	-	105	-	130	ns
			4.5	-	-	7	-	21	-	67	ns
			6	-	-	14	-	18	-	22	ns
Propagation Delay, Data Input to Output Y	t$_{PLH}$, t$_{PHL}$	C$_L$ = 15pF	5	-	6	-	-	-	-	-	ns

CD54HC04, CD54HCT04, CD74HC04, CD74HCT04

Switching Specifications Input t_r, t_f = 6ns (Continued)

PARAMETER	SYMBOL	TEST CONDITIONS	V_{CC} (V)	25°C			-40°C TO 85°C		-55°C TO 125°C		UNITS
				MIN	TYP	MAX	MIN	MAX	MIN	MAX	
Transition Times (Figure 1)	t_{TLH}, t_{THL}	C_L = 50pF	2	-	-	75	-	95	18	110	ns
			4.5	-	-	15	-	19	-	22	ns
			6	-	-	13	-	16	-	19	ns
Input Capacitance	C_I	-	-	-	-	10	-	10	-	10	pF
Power Dissipation Capacitance (Notes 3, 4)	C_{PD}	-	5	-	21	-	-	-	-	-	pF
HCT TYPES											
Propagation Delay, Input to Output (Figure 2)	t_{PLH}, t_{PHL}	C_L = 50pF	4.5	-	-	19	-	24	-	29	ns
Propagation Delay, Data Input to Output Y	t_{PLH}, t_{PHL}	C_L = 15pF	5	-	7	-	-	-	-	-	ns
Transition Times (Figure 2)	t_{TLH}, t_{THL}	C_L = 50pF	4.5	-	-	15	-	19	-	22	ns
Input Capacitance	C_I	-	-	-	-	10	-	10	-	10	pF
Power Dissipation Capacitance (Notes 3, 4)	C_{PD}	-	5	-	24	-	-	-	-	-	pF

NOTES:

3. C_{PD} is used to determine the dynamic power consumption, per gate.

4. $P_D = V_{CC}^2 f_i (C_{PD} + C_L)$ where f_i = input frequency, C_L = output load capacitance, V_{CC} = supply voltage.

Test Circuits and Waveforms

FIGURE 1. HC TRANSITION TIMES AND PROPAGATION DELAY TIMES, COMBINATION LOGIC

FIGURE 2. HCT TRANSITION TIMES AND PROPAGATION DELAY TIMES, COMBINATION LOGIC

SN54AHC74, SN74AHC74
DUAL POSITIVE-EDGE-TRIGGERED D-TYPE FLIP-FLOPS
WITH CLEAR AND PRESET
SCLS255E – DECEMBER 1995 – REVISED NOVEMBER 1998

- Operating Range 2-V to 5.5-V V$_{CC}$
- *EPIC*™ (Enhanced-Performance Implanted CMOS) Process
- Latch-Up Performance Exceeds 250 mA Per JESD 17
- ESD Protection Exceeds 2000 V Per MIL-STD-883, Method 3015; Exceeds 200 V Using Machine Model (C = 200 pF, R = 0)
- Package Options Include Plastic Small-Outline (D), Shrink Small-Outline (DB), Thin Very Small-Outline (DGV), Thin Shrink Small-Outline (PW), and Ceramic Flat (W) Packages, Ceramic Chip Carriers (FK), and Standard Plastic (N) and Ceramic (J) DIPs

description

The 'AHC74 devices are dual positive-edge-triggered D-type flip-flops.

A low level at the preset ($\overline{PRE}$) or clear ($\overline{CLR}$) inputs sets or resets the outputs, regardless of the levels of the other inputs. When $\overline{PRE}$ and $\overline{CLR}$ are inactive (high), data at the data (D) input meeting the setup time requirements is transferred to the outputs on the positive-going edge of the clock pulse. Clock triggering occurs at a voltage level and is not directly related to the rise time of the clock pulse. Following the hold-time interval, data at the D input can be changed without affecting the levels at the outputs.

SN54AHC74 . . . J OR W PACKAGE
SN74AHC74 . . . D, DB, DGV, N, OR PW PACKAGE
(TOP VIEW)

SN54AHC74 . . . FK PACKAGE
(TOP VIEW)

NC – No internal connection

The SN54AHC74 is characterized for operation over the full military temperature range of –55°C to 125°C. The SN74AHC74 is characterized for operation from –40°C to 85°C.

FUNCTION TABLE

INPUTS				OUTPUTS	
$\overline{PRE}$	$\overline{CLR}$	CLK	D	Q	$\overline{Q}$
L	H	X	X	H	L
H	L	X	X	L	H
L	L	X	X	H†	H†
H	H	↑	H	H	L
H	H	↑	L	L	H
H	H	L	X	Q$_0$	$\overline{Q}_0$

† This configuration is nonstable; that is, it does not persist when $\overline{PRE}$ or $\overline{CLR}$ returns to its inactive (high) level.

SN54AHC74, SN74AHC74
DUAL POSITIVE-EDGE-TRIGGERED D-TYPE FLIP-FLOPS
WITH CLEAR AND PRESET
SCLS255E – DECEMBER 1995 – REVISED NOVEMBER 1998

logic symbol†

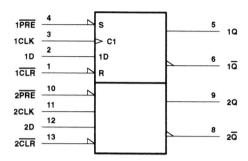

† This symbol is in accordance with ANSI/IEEE Std 91-1984 and IEC Publication 617-12.
Pin numbers shown are for the D, DB, DGV, J, N, PW, and W packages.

logic diagram, each flip-flop (positive logic)

SN54AHC74, SN74AHC74
DUAL POSITIVE-EDGE-TRIGGERED D-TYPE FLIP-FLOPS
WITH CLEAR AND PRESET
SCLS255E – DECEMBER 1995 – REVISED NOVEMBER 1998

absolute maximum ratings over operating free-air temperature range (unless otherwise noted)[†]

Supply voltage range, V_{CC} .... −0.5 V to 7 V
Input voltage range, V_I (see Note 1) .... −0.5 V to 7 V
Output voltage range, V_O (see Note 1) .... −0.5 V to V_{CC} + 0.5 V
Input clamp current, I_{IK} ($V_I < 0$) .... −20 mA
Output clamp current, I_{OK} ($V_O < 0$ or $V_O > V_{CC}$) .... ±20 mA
Continuous output current, I_O ($V_O = 0$ to V_{CC}) .... ±25 mA
Continuous current through V_{CC} or GND .... ±50 mA
Package thermal impedance, θ_{JA} (see Note 2): D package .... 127°C/W
                DB package .... 158°C/W
                DGV package .... 182°C/W
                N package .... 78°C/W
                PW package .... 170°C/W
Storage temperature range, T_{stg} .... −65°C to 150°C

[†] Stresses beyond those listed under "absolute maximum ratings" may cause permanent damage to the device. These are stress ratings only, and functional operation of the device at these or any other conditions beyond those indicated under "recommended operating conditions" is not implied. Exposure to absolute-maximum-rated conditions for extended periods may affect device reliability.

NOTES: 1. The input and output voltage ratings may be exceeded if the input and output current ratings are observed.
        2. The package thermal impedance is calculated in accordance with JESD 51, except for through-hole packages, which use a trace length of zero.

recommended operating conditions (see Note 3)

			SN54AHC74 MIN	SN54AHC74 MAX	SN74AHC74 MIN	SN74AHC74 MAX	UNIT
V_{CC}	Supply voltage		2	5.5	2	5.5	V
V_{IH}	High-level input voltage	$V_{CC} = 2$ V	1.5		1.5		V
		$V_{CC} = 3$ V	2.1		2.1		
		$V_{CC} = 5.5$ V	3.85		3.85		
V_{IL}	Low-level input voltage	$V_{CC} = 2$ V		0.5		0.5	V
		$V_{CC} = 3$ V		0.9		0.9	
		$V_{CC} = 5.5$ V		1.65		1.65	
V_I	Input voltage		0	5.5	0	5.5	V
V_O	Output voltage		0	V_{CC}	0	V_{CC}	V
I_{OH}	High-level output current	$V_{CC} = 2$ V		−50		−50	µA
		$V_{CC} = 3.3$ V ± 0.3 V		−4		−4	mA
		$V_{CC} = 5$ V ± 0.5 V		−8		−8	
I_{OL}	Low-level output current	$V_{CC} = 2$ V		50		50	µA
		$V_{CC} = 3.3$ V ± 0.3 V		4		4	mA
		$V_{CC} = 5$ V ± 0.5 V		8		8	
$\Delta t/\Delta v$	Input transition rise or fall rate	$V_{CC} = 3.3$ V ± 0.3 V		100		100	ns/V
		$V_{CC} = 5$ V ± 0.5 V		20		20	
T_A	Operating free-air temperature		−55	125	−40	85	°C

NOTE 3: All unused inputs of the device must be held at V_{CC} or GND to ensure proper device operation. Refer to the TI application report, *Implications of Slow or Floating CMOS Inputs*, literature number SCBA004.

SN54AHC74, SN74AHC74
DUAL POSITIVE-EDGE-TRIGGERED D-TYPE FLIP-FLOPS
WITH CLEAR AND PRESET
SCLS255E – DECEMBER 1995 – REVISED NOVEMBER 1998

electrical characteristics over recommended operating free-air temperature range (unless otherwise noted)

PARAMETER		TEST CONDITIONS	V_{CC}	$T_A = 25°C$			SN54AHC74		SN74AHC74		UNIT
				MIN	TYP	MAX	MIN	MAX	MIN	MAX	
V_{OH}		$I_{OH} = -50\ \mu A$	2 V	1.9	2		1.9		1.9		V
			3 V	2.9	3		2.9		2.9		
			4.5 V	4.4	4.5		4.4		4.4		
		$I_{OH} = -4\ mA$	3 V	2.58			2.48		2.48		
		$I_{OH} = -8\ mA$	4.5 V	3.94			3.8		3.8		
V_{OL}		$I_{OL} = 50\ \mu A$	2 V			0.1		0.1		0.1	V
			3 V			0.1		0.1		0.1	
			4.5 V			0.1		0.1		0.1	
		$I_{OL} = 4\ mA$	3 V			0.36		0.5		0.44	
		$I_{OL} = 8\ mA$	4.5 V			0.36		0.5		0.44	
I_I	Data inputs	$V_I = V_{CC}$ or GND	5.5 V			±0.1		±1		±1	μA
	Control inputs					±0.1		±1		±1	
I_{CC}		$V_I = V_{CC}$ or GND, $I_O = 0$	5.5 V			2		20		20	μA
C_i		$V_I = V_{CC}$ or GND	5 V		2	10				10	pF

timing requirements over recommended operating free-air temperature range, $V_{CC} = 3.3\ V \pm 0.3\ V$ (unless otherwise noted) (see Figure 1)

			$T_A = 25°C$		SN54AHC74		SN74AHC74		UNIT
			MIN	MAX	MIN	MAX	MIN	MAX	
t_w	Pulse duration	PRE or CLR low	6		7		7		ns
		CLK	6		7		7		
t_{su}	Setup time before CLK↑	Data	6		7		7		ns
		PRE or CLR inactive	5		5		5		
t_h	Hold time, data after CLK↑		0.5		0.5		0.5		ns

timing requirements over recommended operating free-air temperature range, $V_{CC} = 5\ V \pm 0.5\ V$ (unless otherwise noted) (see Figure 1)

			$T_A = 25°C$		SN54AHC74		SN74AHC74		UNIT
			MIN	MAX	MIN	MAX	MIN	MAX	
t_w	Pulse duration	PRE or CLR low	5		5		5		ns
		CLK	5		5		5		
t_{su}	Setup time before CLK↑	Data	5		5		5		ns
		PRE or CLR inactive	3		3		3		
t_h	Hold time, data after CLK↑		0.5		0.5		0.5		ns

SN54AHC74, SN74AHC74
DUAL POSITIVE-EDGE-TRIGGERED D-TYPE FLIP-FLOPS
WITH CLEAR AND PRESET
SCLS255E – DECEMBER 1995 – REVISED NOVEMBER 1998

switching characteristics over recommended operating free-air temperature range,
V_{CC} = 3.3 V ± 0.3 V (unless otherwise noted) (see Figure 1)

PARAMETER	FROM (INPUT)	TO (OUTPUT)	LOAD CAPACITANCE	SN54AHC74					UNIT
				T_A = 25°C			MIN	MAX	
				MIN	TYP	MAX			
f_{max}			C_L = 15 pF*	80	125		70		MHz
			C_L = 50 pF	50	75		45		
t_{PLH}*	$\overline{PRE}$ or $\overline{CLR}$	Q or $\overline{Q}$	C_L = 15 pF		7.6	12.3	1	14.5	ns
t_{PHL}*					7.6	12.3	1	14.5	
t_{PLH}*	CLK	Q or $\overline{Q}$	C_L = 15 pF		6.7	11.9	1	14	ns
t_{PHL}*					6.7	11.9	1	14	
t_{PLH}	$\overline{PRE}$ or $\overline{CLR}$	Q or $\overline{Q}$	C_L = 50 pF		10.1	15.8	1	18	ns
t_{PHL}					10.1	15.8	1	18	
t_{PLH}	CLK	Q or $\overline{Q}$	C_L = 50 pF		9.2	15.4	1	17.5	ns
t_{PHL}					9.2	15.4	1	17.5	

* On products compliant to MIL-PRF-38535, this parameter is not production tested.

switching characteristics over recommended operating free-air temperature range,
V_{CC} = 3.3 V ± 0.3 V (unless otherwise noted) (see Figure 1)

PARAMETER	FROM (INPUT)	TO (OUTPUT)	LOAD CAPACITANCE	SN74AHC74					UNIT
				T_A = 25°C			MIN	MAX	
				MIN	TYP	MAX			
f_{max}			C_L = 15 pF	80	125		70		MHz
			C_L = 50 pF	50	75		45		
t_{PLH}	$\overline{PRE}$ or $\overline{CLR}$	Q or $\overline{Q}$	C_L = 15 pF		7.6	12.3	1	14.5	ns
t_{PHL}					7.6	12.3	1	14.5	
t_{PLH}	CLK	Q or $\overline{Q}$	C_L = 15 pF		6.7	11.9	1	14	ns
t_{PHL}					6.7	11.9	1	14	
t_{PLH}	$\overline{PRE}$ or $\overline{CLR}$	Q or $\overline{Q}$	C_L = 50 pF		10.1	15.8	1	18	ns
t_{PHL}					10.1	15.8	1	18	
t_{PLH}	CLK	Q or $\overline{Q}$	C_L = 50 pF		9.2	15.4	1	17.5	ns
t_{PHL}					9.2	15.4	1	17.5	

SN54AHC74, SN74AHC74
DUAL POSITIVE-EDGE-TRIGGERED D-TYPE FLIP-FLOPS
WITH CLEAR AND PRESET
SCLS255E – DECEMBER 1995 – REVISED NOVEMBER 1998

switching characteristics over recommended operating free-air temperature range, $V_{CC} = 5\ V \pm 0.5\ V$ (unless otherwise noted) (see Figure 1)

PARAMETER	FROM (INPUT)	TO (OUTPUT)	LOAD CAPACITANCE	SN54AHC74 $T_A = 25°C$					UNIT
				MIN	TYP	MAX	MIN	MAX	
f_{max}			$C_L = 15\ pF^*$	130	170		110		MHz
			$C_L = 50\ pF$	90	115		75		
t_{PLH}^*	$\overline{PRE}$ or $\overline{CLR}$	Q or $\overline{Q}$	$C_L = 15\ pF$		4.8	7.7	1	9	ns
t_{PHL}^*					4.8	7.7	1	9	
t_{PLH}^*	CLK	Q or $\overline{Q}$	$C_L = 15\ pF$		4.6	7.3	1	8.5	ns
t_{PHL}^*					4.6	7.3	1	8.5	
t_{PLH}	$\overline{PRE}$ or $\overline{CLR}$	Q or $\overline{Q}$	$C_L = 50\ pF$		6.3	9.7	1	11	ns
t_{PHL}					6.3	9.7	1	11	
t_{PLH}	CLK	Q or Q	$C_L = 50\ pF$		6.1	9.3	1	10.5	ns
t_{PHL}					6.1	9.3	1	10.5	

* On products compliant to MIL-PRF-38535, this parameter is not production tested.

switching characteristics over recommended operating free-air temperature range, $V_{CC} = 5\ V \pm 0.5\ V$ (unless otherwise noted) (see Figure 1)

PARAMETER	FROM (INPUT)	TO (OUTPUT)	LOAD CAPACITANCE	SN74AHC74 $T_A = 25°C$					UNIT
				MIN	TYP	MAX	MIN	MAX	
f_{max}			$C_L = 15\ pF$	130	170		110		MHz
			$C_L = 50\ pF$	90	115		75		
t_{PLH}	$\overline{PRE}$ or $\overline{CLR}$	Q or $\overline{Q}$	$C_L = 15\ pF$		4.8	7.7	1	9	ns
t_{PHL}					4.8	7.7	1	9	
t_{PLH}	CLK	Q or $\overline{Q}$	$C_L = 15\ pF$		4.6	7.3	1	8.5	ns
t_{PHL}					4.6	7.3	1	8.5	
t_{PLH}	$\overline{PRE}$ or $\overline{CLR}$	Q or $\overline{Q}$	$C_L = 50\ pF$		6.3	9.7	1	11	ns
t_{PHL}					6.3	9.7	1	11	
t_{PLH}	CLK	Q or Q	$C_L = 50\ pF$		6.1	9.3	1	10.5	ns
t_{PHL}					6.1	9.3	1	10.5	

noise characteristics, $V_{CC} = 5\ V$, $C_L = 50\ pF$, $T_A = 25°C$ (see Note 4)

PARAMETER		SN74AHC74		UNIT
		MIN	MAX	
$V_{OL(P)}$	Quiet output, maximum dynamic V_{OL}		0.8	V
$V_{OL(V)}$	Quiet output, minimum dynamic V_{OL}		−0.8	V
$V_{OH(V)}$	Quiet output, minimum dynamic V_{OH}	4.7		V
$V_{IH(D)}$	High-level dynamic input voltage	3.5		V
$V_{IL(D)}$	Low-level dynamic input voltage		1.5	V

NOTE 4:   Characteristics are for surface-mount packages only.

operating characteristics, $V_{CC} = 5\ V$, $T_A = 25°C$

PARAMETER		TEST CONDITIONS		TYP	UNIT
C_{pd}	Power dissipation capacitance	No load,	f = 1 MHz	32	pF

SN54AHC74, SN74AHC74
DUAL POSITIVE-EDGE-TRIGGERED D-TYPE FLIP-FLOPS
WITH CLEAR AND PRESET

SCLS255E – DECEMBER 1995 – REVISED NOVEMBER 1998

PARAMETER MEASUREMENT INFORMATION

TEST	S1
t_{PLH}/t_{PHL}	Open
t_{PLZ}/t_{PZL}	V_{CC}
t_{PHZ}/t_{PZH}	GND
Open Drain	V_{CC}

NOTES: A. C_L includes probe and jig capacitance.
   B. Waveform 1 is for an output with internal conditions such that the output is low except when disabled by the output control.
      Waveform 2 is for an output with internal conditions such that the output is high except when disabled by the output control.
   C. All input pulses are supplied by generators having the following characteristics: PRR ≤ 1 MHz, Z_O = 50 Ω, t_r ≤ 3 ns, t_f ≤ 3 ns.
   D. The outputs are measured one at a time with one input transition per measurement.

Figure 1. Load Circuit and Voltage Waveforms

SN54HC138, SN74HC138
3-LINE TO 8-LINE DECODERS/DEMULTIPLEXERS

SCLS107C – DECEMBER 1982 – REVISED MAY 1997

- **Designed Specifically for High-Speed Memory Decoders and Data Transmission Systems**
- **Incorporate Three Enable Inputs to Simplify Cascading and/or Data Reception**
- **Package Options Include Plastic Small-Outline (D), Thin Shrink Small-Outline (PW), and Ceramic Flat (W) Packages, Ceramic Chip Carriers (FK), and Standard Plastic (N) and Ceramic (J) 300-mil DIPs**

SN54HC138 . . . J OR W PACKAGE
SN74HC138 . . . D, N, OR PW PACKAGE
(TOP VIEW)

SN54HC138 . . . FK PACKAGE
(TOP VIEW)

NC – No internal connection

description

The 'HC138 are designed to be used in high-performance memory-decoding or data-routing applications requiring very short propagation delay times. In high-performance memory systems, these decoders can be used to minimize the effects of system decoding. When employed with high-speed memories utilizing a fast enable circuit, the delay times of these decoders and the enable time of the memory are usually less than the typical access time of the memory. This means that the effective system delay introduced by the decoders is negligible.

The conditions at the binary-select inputs at the three enable inputs select one of eight output lines. Two active-low and one active-high enable inputs reduce the need for external gates or inverters when expanding. A 24-line decoder can be implemented without external inverters and a 32-line decoder requires only one inverter. An enable input can be used as a data input for demultiplexing applications.

The SN54HC138 is characterized for operation over the full military temperature range of –55°C to 125°C. The SN74HC138 is characterized for operation from –40°C to 85°C.

Please be aware that an important notice concerning availability, standard warranty, and use in critical applications of Texas Instruments semiconductor products and disclaimers thereto appears at the end of this data sheet.

SN54HC138, SN74HC138
3-LINE TO 8-LINE DECODERS/DEMULTIPLEXERS

SCLS107C – DECEMBER 1982 – REVISED MAY 1997

FUNCTION TABLE

INPUTS						OUTPUTS							
ENABLE			SELECT										
G1	G2A	G2B	C	B	A	Y0	Y1	Y2	Y3	Y4	Y5	Y6	Y7
X	H	X	X	X	X	H	H	H	H	H	H	H	H
X	X	H	X	X	X	H	H	H	H	H	H	H	H
L	X	X	X	X	X	H	H	H	H	H	H	H	H
H	L	L	L	L	L	L	H	H	H	H	H	H	H
H	L	L	L	L	H	H	L	H	H	H	H	H	H
H	L	L	L	H	L	H	H	L	H	H	H	H	H
H	L	L	L	H	H	H	H	H	L	H	H	H	H
H	L	L	H	L	L	H	H	H	H	L	H	H	H
H	L	L	H	L	H	H	H	H	H	H	L	H	H
H	L	L	H	H	L	H	H	H	H	H	H	L	H
H	L	L	H	H	H	H	H	H	H	H	H	H	L

logic symbols (alternatives)†

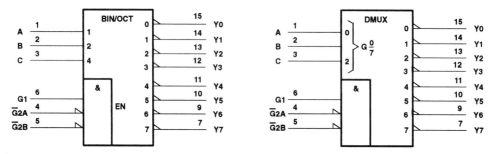

† These symbols are in accordance with ANSI/IEEE Std 91-1984 and IEC Publication 617-12.
Pin numbers shown are for the D, J, N, PW, and W packages.

SN54HC138, SN74HC138
3-LINE TO 8-LINE DECODERS/DEMULTIPLEXERS

SCLS107C – DECEMBER 1982 – REVISED MAY 1997

logic diagram (positive logic)

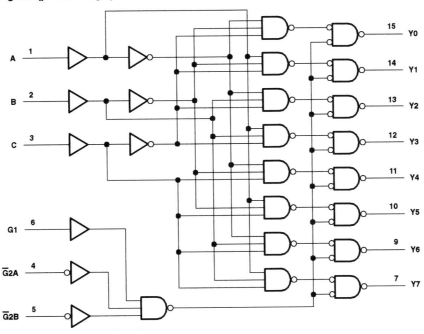

Pin numbers shown are for the D, J, N, PW, and W packages.

absolute maximum ratings over operating free-air temperature range†

Supply voltage range, V_{CC} .... −0.5 V to 7 V
Input clamp current, I_{IK} ($V_I < 0$ or $V_I > V_{CC}$) (see Note 1) .... ±20 mA
Output clamp current, I_{OK} ($V_O < 0$ or $V_O > V_{CC}$) (see Note 1) .... ±20 mA
Continuous output current, I_O (V_O = 0 to V_{CC}) .... ±25 mA
Continuous current through V_{CC} or GND .... ±50 mA
Package thermal impedance, θ_{JA} (see Note 2): D package .... 113°C/W
                             N package .... 78°C/W
                             PW package .... 149°C/W
Storage temperature range, T_{stg} .... −65°C to 150°C

† Stresses beyond those listed under "absolute maximum ratings" may cause permanent damage to the device. These are stress ratings only, and functional operation of the device at these or any other conditions beyond those indicated under "recommended operating conditions" is not implied. Exposure to absolute-maximum-rated conditions for extended periods may affect device reliability.
NOTES: 1. The input and output voltage ratings may be exceeded if the input and output current ratings are observed.
      2. The package thermal impedance is calculated in accordance with JESD 51, except for through-hole packages, which use a trace length of zero.

SN54HC138, SN74HC138
3-LINE TO 8-LINE DECODERS/DEMULTIPLEXERS

SCLS107C – DECEMBER 1982 – REVISED MAY 1997

recommended operating conditions

			SN54HC138			SN74HC138			UNIT
			MIN	NOM	MAX	MIN	NOM	MAX	
V_{CC}	Supply voltage		2	5	6	2	5	6	V
V_{IH}	High-level input voltage	$V_{CC} = 2$ V	1.5			1.5			V
		$V_{CC} = 4.5$ V	3.15			3.15			
		$V_{CC} = 6$ V	4.2			4.2			
V_{IL}	Low-level input voltage	$V_{CC} = 2$ V	0		0.5	0		0.5	V
		$V_{CC} = 4.5$ V	0		1.35	0		1.35	
		$V_{CC} = 6$ V	0		1.8	0		1.8	
V_I	Input voltage		0		V_{CC}	0		V_{CC}	V
V_O	Output voltage		0		V_{CC}	0		V_{CC}	V
t_t	Input transition (rise and fall) time	$V_{CC} = 2$ V	0		1000	0		1000	ns
		$V_{CC} = 4.5$ V	0		500	0		500	
		$V_{CC} = 6$ V	0		400	0		400	
T_A	Operating free-air temperature		−55		125	−40		85	°C

electrical characteristics over recommended operating free-air temperature range (unless otherwise noted)

PARAMETER	TEST CONDITIONS		V_{CC}	$T_A = 25°C$			SN54HC138		SN74HC138		UNIT
				MIN	TYP	MAX	MIN	MAX	MIN	MAX	
V_{OH}	$V_I = V_{IH}$ or V_{IL}	$I_{OH} = -20$ μA	2 V	1.9	1.998		1.9		1.9		V
			4.5 V	4.4	4.499		4.4		4.4		
			6 V	5.9	5.999		5.9		5.9		
		$I_{OH} = -4$ mA	4.5 V	3.98	4.3		3.7		3.84		
		$I_{OH} = -5.2$ mA	6 V	5.48	5.8		5.2		5.34		
V_{OL}	$V_I = V_{IH}$ or V_{IL}	$I_{OL} = 20$ μA	2 V		0.002	0.1		0.1		0.1	V
			4.5 V		0.001	0.1		0.1		0.1	
			6 V		0.001	0.1		0.1		0.1	
		$I_{OL} = 4$ mA	4.5 V		0.17	0.26		0.4		0.33	
		$I_{OL} = 5.2$ mA	6 V		0.15	0.26		0.4		0.33	
I_I	$V_I = V_{CC}$ or 0		6 V		±0.1	±100		±1000		±1000	nA
I_{CC}	$V_I = V_{CC}$ or 0, $I_O = 0$		6 V			8		160		80	μA
C_i			2 V to 6 V		3	10		10		10	pF

SN54HC138, SN74HC138
3-LINE TO 8-LINE DECODERS/DEMULTIPLEXERS

SCLS107C – DECEMBER 1982 – REVISED MAY 1997

switching characteristics over recommended operating free-air temperature range, C_L = 50 pF (unless otherwise noted) (see Figure 1)

PARAMETER	FROM (INPUT)	TO (OUTPUT)	V_{CC}	T_A = 25°C			SN54HC138		SN74HC138		UNIT
				MIN	TYP	MAX	MIN	MAX	MIN	MAX	
t_{pd}	A, B, or C	Any Y	2 V		67	180		270		225	ns
			4.5 V		18	36		54		45	
			6 V		15	31		46		38	
	Enable	Any Y	2 V		66	155		235		195	
			4.5 V		18	31		47		39	
			6 V		15	26		40		33	
t_t		Any	2 V		38	75		110		95	ns
			4.5 V		8	15		22		19	
			6 V		6	13		19		16	

operating characteristics, T_A = 25°C

PARAMETER		TEST CONDITIONS	TYP	UNIT
C_{pd}	Power dissipation capacitance	No load	85	pF

PARAMETER MEASUREMENT INFORMATION

LOAD CIRCUIT

VOLTAGE WAVEFORM
INPUT RISE AND FALL TIMES

VOLTAGE WAVEFORMS
PROPAGATION DELAY AND OUTPUT TRANSITION TIMES

NOTES: A. C_L includes probe and test-fixture capacitance.
B. Phase relationships between waveforms were chosen arbitrarily. All input pulses are supplied by generators having the following characteristics: PRR ≤ 1 MHz, Z_O = 50 Ω, t_r = 6 ns, t_f = 6 ns.
C. The outputs are measured one at a time with one input transition per measurement.
D. t_{PLH} and t_{PHL} are the same as t_{pd}.

Figure 1. Load Circuit and Voltage Waveforms

TEXAS INSTRUMENTS

Data sheet acquired from Harris Semiconductor
SCHS150

September 1997

CD74HC151, CD74HCT151

High Speed CMOS Logic
8-Input Multiplexer

Features

$V_{CC} = 5V$

- Complementary Data Outputs
- Buffered Inputs and Outputs
- Fanout (Over Temperature Range)
  - Standard Outputs . . . . 10 LSTTL Loads
  - Bus Driver Outputs . . . . 15 LSTTL Loads
- Wide Operating Temperature Range . . . $-55^{o}C$ to $125^{o}C$
- Balanced Propagation Delay and Transition Times
- Significant Power Reduction Compared to LSTTL Logic ICs
- Alternate Source is Philips/Signetics
- HC Types
  - 2V to 6V Operation
  - High Noise Immunity: $N_{IL} = 30\%$, $N_{IH} = 30\%$ of V_{CC} at

- HCT Types
  - 4.5V to 5.5V Operation
  - Direct LSTTL Input Logic Compatibility, $V_{IL} = 0.8V$ (Max), $V_{IH} = 2V$ (Min)
  - CMOS Input Compatibility, $I_I \leq 1\mu A$ at V_{OL}, V_{OH}

Description

The Harris CD74HC151 and CD74HCT151 are single 8-channel digital multiplexers having three binary control inputs, S0, S1 and S2 and an active low enable (E) input. The three binary signals select 1 of 8 channels. Outputs are both inverting ($\overline{Y}$) and non-inverting (Y).

Ordering Information

PART NUMBER	TEMP. RANGE (°C)	PACKAGE	PKG. NO.

Pinout

CD74HC151, CD74HCT151
(PDIP, SOIC)
TOP VIEW

CD74HC151, CD74HCT151

Functional Diagram

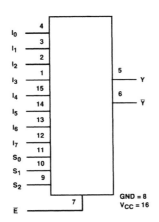

I_0	4
I_1	3
I_2	2
I_3	1
I_4	15
I_5	14
I_6	13
I_7	12
S_0	11
S_1	10
S_2	9

5 — Y
6 — $\bar{Y}$

$\bar{E}$ — 7

GND = 8
V_{CC} = 16

TRUTH TABLE

SELECT INPUTS			DATA INPUTS								ENABLE	OUTPUT	
S2	S1	S0	I0	$\overline{I1}$	I2	I3	I4	I5	I6	I7	$\bar{E}$	$\bar{Y}$	Y
X	X	X	X	X	X	X	X	X	X	X	H	H	L
L	L	L	L	X	X	X	X	X	X	X	L	H	L
L	L	L	H	X	X	X	X	X	X	X	L	L	H
L	L	H	X	L	X	X	X	X	X	X	L	H	L
L	L	H	X	H	X	X	X	X	X	X	L	L	H
L	H	L	X	X	L	X	X	X	X	X	L	H	L
L	H	L	X	X	H	X	X	X	X	X	L	L	H
L	H	H	X	X	X	L	X	X	X	X	L	H	L
L	H	H	X	X	X	H	X	X	X	X	L	L	H
H	L	L	X	X	X	X	L	X	X	X	L	H	L
H	L	L	X	X	X	X	H	X	X	X	L	L	H
H	L	H	X	X	X	X	X	L	X	X	L	H	L
H	L	H	X	X	X	X	X	H	X	X	L	L	H
H	H	L	X	X	X	X	X	X	L	X	L	H	L
H	H	L	X	X	X	X	X	X	H	X	L	L	H
H	H	H	X	X	X	X	X	X	X	L	L	H	L
H	H	H	X	X	X	X	X	X	X	H	L	L	H

NOTE: H = High Voltage Level, L = Low Voltage Level, X = Don't Care

CD74HC151, CD74HCT151

Absolute Maximum Ratings

DC Supply Voltage, V_{CC} . . . . -0.5V to 7V
DC Input Diode Current, I_{IK}
  For V_I < -0.5V or V_I > V_{CC} + 0.5V . . . . ±20mA
DC Output Diode Current, I_{OK}
  For V_O < -0.5V or V_O > V_{CC} + 0.5V . . . . ±20mA
DC Output Source or Sink Current per Output Pin, I_O
  For V_O > -0.5V or V_O < V_{CC} + 0.5V . . . . ±25mA
DC V_{CC} or Ground Current, I_{CC} or I_{GND} . . . . ±50mA

Thermal Information

Thermal Resistance (Typical, Note 3) θ_{JA} (°C/W)
  PDIP Package 90
  SOIC Package 115
Maximum Junction Temperature . . . . 150°C
Maximum Storage Temperature Range . . . . -65°C to 150°C
Maximum Lead Temperature (Soldering 10s) . . . . 300°C
  (SOIC - Lead Tips Only)

Operating Conditions

Temperature Range (T_A) . . . . -55°C to 125°C
Supply Voltage Range, V_{CC}
  HC Types . . . . .2V to 6V
  HCT Types . . . . .4.5V to 5.5V
DC Input or Output Voltage, V_I, V_O . . . . 0V to V_{CC}
Input Rise and Fall Time
  2V . . . . 1000ns (Max)
  4.5V . . . . 500ns (Max)
  6V . . . . 400ns (Max)

CAUTION: Stresses above those listed in "Absolute Maximum Ratings" may cause permanent damage to the device. This is a stress only rating and operation of the device at these or any other conditions above those indicated in the operational sections of this specification is not implied.

NOTE:

3. θ_{JA} is measured with the component mounted on an evaluation PC board in free air.

DC Electrical Specifications

PARAMETER	SYMBOL	TEST CONDITIONS V_I (V)	TEST CONDITIONS I_O (mA)	V_{CC} (V)	25°C MIN	25°C TYP	25°C MAX	-40°C TO 85°C MIN	-40°C TO 85°C MAX	-55°C TO 125°C MIN	-55°C TO 125°C MAX	UNITS
HC TYPES												
High Level Input Voltage	V_{IH}	-	-	2	1.5	-	-	1.5	-	1.5	-	V
				4.5	3.15	-	-	3.15	-	3.15	-	V
				6	4.2	-	-	4.2	-	4.2	-	V
Low Level Input Voltage	V_{IL}	-	-	2	-	-	0.5	-	0.5	-	0.5	V
				4.5	-	-	1.35	-	1.35	-	1.35	V
				6	-	-	1.8	-	1.8	-	1.8	V
High Level Output Voltage CMOS Loads	V_{OH}	V_{IH} or V_{IL}	-0.02	2	1.9	-	-	1.9	-	1.9	-	V
			-0.02	4.5	4.4	-	-	4.4	-	4.4	-	V
			-0.02	6	5.9	-	-	5.9	-	5.9	-	V
High Level Output Voltage TTL Loads			-	-	-	-	-	-	-	-	-	V
			-4	4.5	3.98	-	-	3.84	-	3.7	-	V
			-5.2	6	5.48	-	-	5.34	-	5.2	-	V
Low Level Output Voltage CMOS Loads	V_{OL}	V_{IH} or V_{IL}	0.02	2	-	-	0.1	-	0.1	-	0.1	V
			0.02	4.5	-	-	0.1	-	0.1	-	0.1	V
			0.02	6	-	-	0.1	-	0.1	-	0.1	V
Low Level Output Voltage TTL Loads			-	-	-	-	-	-	-	-	-	V
			4	4.5	-	-	0.26	-	0.33	-	0.4	V
			5.2	6	-	-	0.26	-	0.33	-	0.4	V
Input Leakage Current	I_I	V_{CC} or GND	-	6	-	-	±0.1	-	±1	-	±1	µA
Quiescent Device Current	I_{CC}	V_{CC} or GND	0	6	-	-	8	-	80	-	160	µA

CD74HC151, CD74HCT151

DC Electrical Specifications (Continued)

PARAMETER	SYMBOL	TEST CONDITIONS V_I (V)	I_O (mA)	V_CC (V)	25°C MIN	25°C TYP	25°C MAX	-40°C TO 85°C MIN	-40°C TO 85°C MAX	-55°C TO 125°C MIN	-55°C TO 125°C MAX	UNITS
HCT TYPES												
High Level Input Voltage	V_{IH}	-	-	4.5 to 5.5	2	-	-	2	-	2	-	V
Low Level Input Voltage	V_{IL}	-	-	4.5 to 5.5	-	-	0.8	-	0.8	-	0.8	V
High Level Output Voltage CMOS Loads	V_{OH}	V_{IH} or V_{IL}	-0.02	4.5	4.4	-	-	4.4	-	4.4	-	V
High Level Output Voltage TTL Loads			-4	4.5	3.98	-	-	3.84	-	3.7	-	V
Low Level Output Voltage CMOS Loads	V_{OL}	V_{IH} or V_{IL}	0.02	4.5	-	-	0.1	-	0.1	-	0.1	V
Low Level Output Voltage TTL Loads			4	4.5	-	-	0.26	-	0.33	-	0.4	V
Input Leakage Current	I_I	V_{CC} and GND	0	5.5	-	-	±0.1	-	±1	-	±1	µA
Quiescent Device Current	I_{CC}	V_{CC} or GND	0	5.5	-	-	8	-	80	-	160	µA
Additional Quiescent Device Current Per Input Pin: 1 Unit Load	ΔI_{CC}	V_{CC} -2.1	-	4.5 to 5.5	-	100	360	-	450	-	490	µA

NOTE: For dual-supply systems theoretical worst case (V_I = 2.4V, V_{CC} = 5.5V) specification is 1.8mA.

HCT Input Loading Table

INPUT	UNIT LOADS
Select	1.5
Data	0.45
Enable	0.3

NOTE: Unit Load is ΔI_{CC} limit specified in DC Electrical Table, e.g., 360µA max at 25°C.

Switching Specifications Input t_r, t_f = 6ns

PARAMETER	SYMBOL	TEST CONDITIONS	V_CC (V)	25°C MIN	25°C TYP	25°C MAX	-40°C TO 85°C MIN	-40°C TO 85°C MAX	-55°C TO 125°C MIN	-55°C TO 125°C MAX	UNITS
HC TYPES											
Propagation Delay (Figure 1)	t_{PLH}, t_{PHL}	C_L = 50pF	2	-	-	170	-	215	-	255	ns
Any Data Input to Y			4.5	-	-	34	-	43	-	51	ns
		C_L =15pF	5	-	14	-	-	-	-	-	ns
		C_L = 50pF	6	-	-	29	-	37	-	43	ns

CD74HC151, CD74HCT151

Switching Specifications Input t_r, t_f = 6ns **(Continued)**

PARAMETER	SYMBOL	TEST CONDITIONS	V$_{CC}$ (V)	25°C MIN	25°C TYP	25°C MAX	-40°C TO 85°C MIN	-40°C TO 85°C MAX	-55°C TO 125°C MIN	-55°C TO 125°C MAX	UNITS
Any Data Input to Ȳ	t$_{PLH}$, t$_{PHL}$	C$_L$ = 50pF	2	-	-	185	-	230	-	280	ns
			4.5	-	-	37	-	46	-	56	ns
		C$_L$ =15pF	5	-	15	-	-	-	-	-	ns
		C$_L$ = 50pF	6	-	-	31	-	39	-	48	ns
Any Select to Y	t$_{PLH}$, t$_{PHL}$	C$_L$ = 50pF	2	-	-	185	-	230	-	280	ns
			4.5	-	-	37	-	46	-	56	ns
		C$_L$ =15pF	5	-	15	-	-	-	-	-	ns
		C$_L$ = 50pF	6	-	-	31	-	39	-	48	ns
Any Select to Ȳ	t$_{PLH}$, t$_{PHL}$	C$_L$ = 50pF	2	-	-	205	-	255	-	310	ns
			4.5	-	-	41	-	51	-	62	ns
		C$_L$ =15pF	5	-	17	-	-	-	-	-	ns
		C$_L$ = 50pF	6	-	-	35	-	43	-	53	ns
Enable to Y	t$_{PLH}$, t$_{PHL}$	C$_L$ = 50pF	2	-	-	140	-	175	-	210	ns
			4.5	-	-	28	-	35	-	42	ns
		C$_L$ =15pF	5	-	11	-	-	-	-	-	ns
		C$_L$ = 50pF	6	-	-	24	-	30	-	36	ns
Enable to Ȳ	t$_{PLH}$, t$_{PHL}$	C$_L$ = 50pF	2	-	-	145	-	180	-	220	ns
			4.5	-	-	29	-	36	-	44	ns
		C$_L$ =15pF	5	-	12	-	-	-	-	-	ns
		C$_L$ = 50pF	6	-	-	25	-	31	-	38	ns
Output Transition Time (Figure 1)	t$_{TLH}$, t$_{THL}$	C$_L$ = 50pF	2	-	-	75	-	95	-	110	ns
			4.5	-	-	15	-	19	-	22	ns
			6	-	-	13	-	16	-	19	ns
Input Capacitance	C$_{IN}$	-	-	-	-	10	-	10	-	10	pF
Power Dissipation Capacitance (Notes 4, 5)	C$_{PD}$	-	5	-	59	-	-	-	-	-	pF
HCT TYPES											
Propagation Delay (Figure 2)	t$_{PLH}$, t$_{PHL}$										
Any Data Input to Y		C$_L$ = 50pF	4.5	-	-	38	-	48	-	57	ns
		C$_L$ =15pF	5	-	16	-	-		-	-	ns
Any Data Input to Ȳ	t$_{PLH}$, t$_{PHL}$	C$_L$ = 50pF	4.5	-	-	36	-	45	-	54	ns
		C$_L$ =15pF	5	-	15	-	-	-	-	-	ns
Any Select to Y	t$_{PLH}$, t$_{PHL}$	C$_L$ = 50pF	4.5	-		41	-	51	-	62	ns
		C$_L$ =15pF	5	-	17	-	-	-	-	-	ns
Any Select to Ȳ	t$_{PLH}$, t$_{PHL}$	C$_L$ = 50pF	4.5	-	-	43	-	54	-	65	ns
		C$_L$ =15pF	5	-	18	-	-	-	-	-	ns
Enable to Y	t$_{PLH}$, t$_{PHL}$	C$_L$ = 50pF	4.5	-	-	29	-	36	-	44	ns
		C$_L$ =15pF	5	-	12	-	-	-	-	-	ns

CD74HC151, CD74HCT151

Switching Specifications Input t_r, t_f = 6ns (Continued)

PARAMETER	SYMBOL	TEST CONDITIONS	V_{CC} (V)	25°C			-40°C TO 85°C		-55°C TO 125°C		UNITS
				MIN	TYP	MAX	MIN	MAX	MIN	MAX	
Enable to $\overline{Y}$	C_L = 50pF	C_L = 50pF	4.5	-	-	36	-	46	-	54	ns
	C_L =15pF	C_L =15pF	5	15	-	-	-	-	-	-	ns
Output Transition Time	t_{TLH}, t_{THL}	C_L = 50pF	4.5	-	-	15	-	19	-	22	ns
Input Capacitance	C_{IN}	-	-	-	-	10	-	10	-	10	pF
Power Dissipation Capacitance (Notes 4, 5)	C_{PD}	-	5		58	-	-	-	-	-	pF

NOTES:

4. C_{PD} is used to determine the dynamic power consumption, per gate.

5. $P_D = V_{CC}^2 f_i (C_{PD} + C_L)$ where f_i = input frequency, C_L = output load capacitance, V_{CC} = supply voltage.

Test Circuit and Waveform

FIGURE 1.

TEXAS INSTRUMENTS

Data sheet acquired from Harris Semiconductor
SCHS154

February 1998

CD74HC161, CD74HCT161, CD74HC163, CD74HCT163

High Speed CMOS Logic Presettable Counters

Features

- CD74HC161, CD74HCT161 4-Bit Binary Counter, Asynchronous Reset
- CD74HC163, CD74HCT163 4-Bit Binary Counter, Synchronous Reset
- Synchronous Counting and Loading
- Two Count Enable Inputs for n-Bit Cascading
- Look-Ahead Carry for High-Speed Counting
- Fanout (Over Temperature Range)
  - Standard Outputs . . . . 10 LSTTL Loads
  - Bus Driver Outputs . . . . 15 LSTTL Loads
- Wide Operating Temperature Range . . . -55°C to 125°C
- Balanced Propagation Delay and Transition Times
- Significant Power Reduction Compared to LSTTL Logic ICs
- HC Types
  - 2V to 6V Operation
  - High Noise Immunity: N_{IL} = 30%, N_{IH} = 30% of V_{CC} at V_{CC} = 5V
- HCT Types
  - 4.5V to 5.5V Operation
  - Direct LSTTL Input Logic Compatibility, V_{IL}= 0.8V (Max), V_{IH} = 2V (Min)
  - CMOS Input Compatibility, $I_I \le 1\mu A$ at V_{OL}, V_{OH}

Ordering Information

PART NUMBER	TEMP. RANGE (°C)	PACKAGE	PKG. NO.
CD74HC161E	-55 to 125	16 Ld PDIP	E16.3
CD74HC161M	-55 to 125	16 Ld SOIC	M16.15
CD74HC163E	-55 to 125	16 Ld PDIP	E16.3
CD74HC163M	-55 to 125	16 Ld SOIC	M16.15
CD74HCT161E	-55 to 125	16 Ld PDIP	E16.3
CD74HCT161M	-55 to 125	16 Ld SOIC	M16.15
CD74HCT163E	-55 to 125	16 Ld PDIP	E16.3
CD74HCT163M	-55 to 125	16 Ld SOIC	M16.15

NOTES:

1. When ordering, use the entire part number. Add the suffix 96 to obtain the variant in the tape and reel.

2. Wafer and die for this part number is available which meets all electrical specifications. Please contact your local sales office or Harris customer service for ordering information.

Description

The Harris CD74HC161, CD74HCT161, CD74HC163 and CD74HCT163 are presettable synchronous counters that feature look-ahead carry logic for use in high-speed counting applications. The CD74HC161 and CD74HCT161 are asynchronous reset decade and binary counters, respectively; the CD74HC163 and CD74HCT163 devices decade and binary counters, respectively and are reset synchronously with the clock. Counting and parallel presetting are both accomplished synchronously with the negative-to-positive transition of the clock.

A low level on the synchronous parallel enable input, SPE, disables counting operation and allows data at the P0 to P3 inputs to be loaded into the counter (provided that the setup and hold requirements for SPE are met).

All counters are reset with a low level on the Master Reset input, MR. In the CD74HC163 and CD74HCT163 counters (synchronous reset types), the requirements for setup and hold time with respect to the clock must be met.

Two count enables, PE and TE, in each counter are provided for n-bit cascading. In all counters reset action occurs regardless of the level of the $\overline{SPE}$, PE and TE inputs (and the clock input, CP, in the CD74HC161 and CD74HCT161 types).

If a decade counter is preset to an illegal state or assumes an illegal state when power is applied, it will return to the normal sequence in one count as shown in state diagram.

The look-ahead carry feature simplifies serial cascading of the counters. Both count enable inputs (PE and TE) must be high to count. The TE input is gated with the Q outputs of all four stages so that at the maximum count the terminal count (TC) output goes high for one clock period. This TC pulse is used to enable the next cascaded stage.

Pinout

CD74HC161, CD74HCT161, CD74HC163, CD74HCT163 (PDIP, SOIC) TOP VIEW

$\overline{MR}$	1	16 V_{CC}
CP	2	15 TC
P0	3	14 Q0
P1	4	13 Q1
P2	5	12 Q2
P3	6	11 Q3
PE	7	10 TE
GND	8	9 $\overline{SPE}$

CD74HC161, CD74HCT161, CD74HC163, CD74HCT163

Absolute Maximum Ratings

DC Supply Voltage, V_{CC} . . . . -0.5V to 7V
DC Input Diode Current, I_{IK}
  For V_I < -0.5V or V_I > V_{CC} + 0.5V . . . . ±20mA
DC Output Diode Current, I_{OK}
  For V_O < -0.5V or V_O > V_{CC} + 0.5V . . . . ±20mA
DC Drain Current, per Output, I_O
  For -0.5V < V_O < V_{CC} + 0.5V . . . . ±25mA
DC Output Source or Sink Current per Output Pin, I_O
  For V_O > -0.5V or V_O < V_{CC} + 0.5V . . . . ±25mA
DC V_{CC} or Ground Current, I_{CC} . . . . ±50mA

Operating Conditions

Temperature Range, T_A . . . . -55°C to 125°C
Supply Voltage Range, V_{CC}
  HC Types . . . . .2V to 6V
  HCT Types . . . . .4.5V to 5.5V
DC Input or Output Voltage, V_I, V_O . . . . 0V to V_{CC}
Input Rise and Fall Time
  2V . . . . 1000ns (Max)
  4.5V . . . . 500ns (Max)
  6V . . . . 400ns (Max)

Thermal Information

Thermal Resistance (Typical, Note 6)	θ_{JA} (°C/W)
PDIP Package	90
SOIC Package	160

Maximum Junction Temperature . . . . 150°C
Maximum Storage Temperature Range . . . . -65°C to 150°C
Maximum Lead Temperature (Soldering 10s) . . . . 300°C
  (SOIC - Lead Tips Only)

CAUTION: Stresses above those listed in "Absolute Maximum Ratings" may cause permanent damage to the device. This is a stress only rating and operation of the device at these or any other conditions above those indicated in the operational sections of this specification is not implied.

NOTE:

6. θ_{JA} is measured with the component mounted on an evaluation PC board in free air.

DC Electrical Specifications

PARAMETER	SYMBOL	TEST CONDITIONS V_I (V)	TEST CONDITIONS I_O (mA)	V_{CC} (V)	25°C MIN	25°C TYP	25°C MAX	-40°C TO 85°C MIN	-40°C TO 85°C MAX	-55°C TO 125°C MIN	-55°C TO 125°C MAX	UNITS
HC TYPES												
High Level Input Voltage	V_{IH}	-	-	2	1.5	-	-	1.5	-	1.5	-	V
				4.5	3.15	-	-	3.15	-	3.15	-	V
				6	4.2	-	-	4.2	-	4.2	-	V
Low Level Input Voltage	V_{IL}	-	-	2	-	-	0.5	-	0.5	-	0.5	V
				4.5	-	-	1.35	-	1.35	-	1.35	V
				6	-	-	1.8	-	1.8	-	1.8	V
High Level Output Voltage CMOS Loads	V_{OH}	V_{IH} or V_{IL}	-0.02	2	1.9	-	-	1.9	-	1.9	-	V
			-0.02	4.5	4.4	-	-	4.4	-	4.4	-	V
			-0.02	6	5.9	-	-	5.9	-	5.9	-	V
High Level Output Voltage TTL Loads			-	-	-	-	-	-	-	-	-	V
			-4	4.5	3.98	-	-	3.84	-	3.7	-	V
			-5.2	6	5.48	-	-	5.34	-	5.2	-	V
Low Level Output Voltage CMOS Loads	V_{OL}	V_{IH} or V_{IL}	0.02	2	-	-	0.1	-	0.1	-	0.1	V
			0.02	4.5	-	-	0.1	-	0.1	-	0.1	V
			0.02	6	-	-	0.1	-	0.1	-	0.1	V
Low Level Output Voltage TTL Loads			-	-	-	-	-	-	-	-	-	V
			4	4.5	-	-	0.26	-	0.33	-	0.4	V
			5.2	6	-	-	0.26	-	0.33	-	0.4	V
Input Leakage Current	I_I	V_{CC} or GND	-	6	-	-	±0.1	-	±1	-	±1	µA

CD74HC161, CD74HCT161, CD74HC163, CD74HCT163

Functional Diagram

MODE SELECT - FUNCTION TABLE FOR CD74HC/HCT161

OPERATING MODE	INPUTS						OUTPUTS	
	$\overline{MR}$	CP	PE	TE	$\overline{SPE}$	P_n	Q_n	TC
Reset (Clear)	L	X	X	X	X	X	L	L
Parallel Load	H	↑	X	X	l	l	L	L
	H	↑	X	X	l	h	H	(Note 3)
Count	H	↑	h	h	h (Note 5)	X	Count	(Note 3)
Inhibit	H	X	l (Note 4)	X	h (Note 5)	X	q_n	(Note 3)
	H	X	X	l (Note 4)	h (Note 5)	X	q_n	L

MODE SELECT - FUNCTION TABLE FOR CD74HC/HCT163

OPERATING MODE	INPUTS						OUTPUTS	
	$\overline{MR}$	CP	PE	TE	$\overline{SPE}$	P_n	Q_n	TC
Reset (Clear)	l	↑	X	X	X	X	L	L
Parallel Load	h (Note 5)	↑	X	X	l	l	L	L
	h (Note 5)	↑	X	X	l	h	H	(Note 3)
Count	h (Note 5)	↑	h	h	h (Note 5)	X	Count	(Note 3)
Inhibit	h (Note 5)	X	l (Note 4)	X	h (Note 5)	X	q_n	(Note 3)
	h (Note 5)	X	X	l (Note 4)	h (Note 5)	X	q_n	L

NOTE: H = High voltage level steady state; L = Low voltage level steady state; h = High voltage level one setup time prior to the Low-to-High clock transition; l = Low voltage level one setup time prior to the Low-to-High clock transition; X = Don't Care; q = Lower case letters indicate the state of the referenced output prior to the Low-to-High clock transition; ↑ = Low-to-High clock transition.

3. The TC output is High when TE is High and the counter is at Terminal Count (HHHH for CD74HC/HCT161 and CD74HC/HCT163).

4. The High-to-Low transition of PE or TE on the CD74HC/HCT161 and the CD74HC/HCT163 should only occur while CP is HIGH for conventional operation.

5. The Low-to-High transition of $\overline{SPE}$ on the CD74HC/HCT161 and $\overline{SPE}$ or $\overline{MR}$ on the CD74HC/HCT163 should only occur while CP is HIGH for conventional operation.

CD74HC161, CD74HCT161, CD74HC163, CD74HCT163

DC Electrical Specifications (Continued)

PARAMETER	SYMBOL	TEST CONDITIONS V_I (V)	TEST CONDITIONS I_O (mA)	V_{CC} (V)	25°C MIN	25°C TYP	25°C MAX	-40°C TO 85°C MIN	-40°C TO 85°C MAX	-55°C TO 125°C MIN	-55°C TO 125°C MAX	UNITS
Quiescent Device Current	I_{CC}	V_{CC} or GND	0	6	-	-	8	-	80	-	160	μA
HCT TYPES												
High Level Input Voltage	V_{IH}	-	-	4.5 to 5.5	2	-	-	2	-	2	-	V
Low Level Input Voltage	V_{IL}	-	-	4.5 to 5.5	-	-	0.8	-	0.8	-	0.8	V
High Level Output Voltage CMOS Loads	V_{OH}	V_{IH} or V_{IL}	-0.02	4.5	4.4	-	-	4.4	-	4.4	-	V
High Level Output Voltage TTL Loads			-4	4.5	3.98	-	-	3.84	-	3.7	-	V
Low Level Output Voltage CMOS Loads	V_{OL}	V_{IH} or V_{IL}	0.02	4.5	-	-	0.1	-	0.1	-	0.1	V
Low Level Output Voltage TTL Loads			4	4.5	-	-	0.26	-	0.33	-	0.4	V
Input Leakage Current	I_I	V_{CC} and GND	0	5.5	-	-	±0.1	-	±1	-	±1	μA
Quiescent Device Current	I_{CC}	V_{CC} or GND	0	5.5	-	-	8	-	80	-	160	μA
Additional Quiescent Device Current Per Input Pin: 1 Unit Load	ΔI_{CC} (Note)	V_{CC} -2.1	-	4.5 to 5.5	-	100	360	-	450	-	490	μA

NOTE: For dual-supply systems theoretical worst case (V_I = 2.4V, V_{CC} = 5.5V) specification is 1.8mA.

HCT Input Loading Table

INPUT	UNIT LOADS
P0 - P3	0.25
PE	0.65
CP	1.05
$\overline{MR}$	0.8
$\overline{SPE}$	0.5
TE	1.05

NOTE: Unit Load is ΔI_{CC} limit specified in DC Electrical Table, e.g., 360μA max at 25°C.

Prerequisite For Switching Specifications

PARAMETER	SYMBOL	TEST CONDITIONS	V_{CC} (V)	25°C MIN	25°C TYP	25°C MAX	-40°C TO 85°C MIN	-40°C TO 85°C MAX	-55°C TO 125°C MIN	-55°C TO 125°C MAX	UNITS
HC TYPES											
Maximum CP Frequency (Note7)	f_{MAX}	-	2	6	-	-	5	-	4	-	MHz
			4.5	30	-	-	24	-	20	-	MHz
			6	35	-	-	28	-	24	-	MHz

CD74HC161, CD74HCT161, CD74HC163, CD74HCT163

Prerequisite For Switching Specifications (Continued)

PARAMETER	SYMBOL	TEST CONDITIONS	V_{CC} (V)	25°C			-40°C TO 85°C		-55°C TO 125°C		UNITS
				MIN	TYP	MAX	MIN	MAX	MIN	MAX	
CP Width (Low)	$t_{W(L)}$	-	2	80	-	-	100	-	120	-	ns
			4.5	16	-	-	20	-	24	-	ns
			6	14	-	-	17	-	20	-	ns
$\overline{MR}$ Pulse Width (161)	t_W	-	2	100	-	-	125	-	150	-	ns
			4.5	20	-	-	25	-	30	-	ns
			6	17	-	-	21	-	26	-	ns
Setup Time, Pn to CP	t_{SU}	-	2	60	-	-	75	-	90	-	ns
			4.5	12	-	-	15	-	18	-	ns
			6	10	-	-	13	-	15	-	ns
Setup Time, PE or TE to CP	t_{SU}	-	2	50	-	-	65	-	75	-	ns
			4.5	10	-	-	13	-	15	-	ns
			6	9	-	-	11	-	13	-	ns
Setup Time, $\overline{SPE}$ to CP	t_{SU}	-	2	60	-	-	75	-	90	-	ns
			4.5	12	-	-	15	-	18	-	ns
			6	10	-	-	13	-	15	-	ns
Setup Time, $\overline{MR}$ to CP (163)	t_{SU}	-	2	65	-	-	80	-	100	-	ns
			4.5	13	-	-	16	-	20	-	ns
			6	11	-	-	14	-	17	-	ns
Hold Time, PN to CP	t_H	-	2	3	-	-	3	-	3	-	ns
			4.5	3	-	-	3	-	3	-	ns
			6	3	-	-	3	-	3	-	ns
Hold Time, TE or PE to CP	t_H	-	2	0	-	-	0	-	0	-	ns
			4.5	0	-	-	0	-	0	-	ns
			6	0	-	-	0	-	0	-	ns
Hold Time, $\overline{SPE}$ to CP	t_H	-	2	0	-	-	0	-	0	-	ns
			4.5	0	-	-	0	-	0	-	ns
			6	0	-	-	0	-	0	-	ns
Recovery Time, $\overline{MR}$ to CP (161)	t_{REC}	-	2	75	-	-	95	-	110	-	ns
			4.5	15	-	-	19	-	22	-	ns
			6	13	-	-	16	-	19	-	ns
HCT TYPES											
Maximum CP Frequency	f_{MAX}	-	4.5	30	-	-	24	-	20	-	MHz
CP Width (Low) (Note 7)	$t_{W(L)}$	-	4.5	16	-	-	20	-	24	-	ns
$\overline{MR}$ Pulse Width (161)	t_W	-	4.5	20	-	-	25	-	30	-	ns
Setup Time, Pn to CP	t_{SU}	-	4.5	10	-	-	13	-	15	-	ns
Setup Time, PE or TE to CP	t_{SU}	-	4.5	13	-	-	16	-	20	-	ns
Setup Time, $\overline{SPE}$ to CP	t_{SU}	-	4.5	12	-	-	15	-	18	-	ns
Setup Time, $\overline{MR}$ to CP (163)	t_{SU}	-	4.5	13	-	-	16	-	20	-	ns
Hold Time, PN to CP	t_H	-	4.5	5	-	-	5	-	5	-	ns
Hold Time, TE or PE to CP	t_H	-	4.5	3	-	-	3	-	3	-	ns

CD74HC161, CD74HCT161, CD74HC163, CD74HCT163

Prerequisite For Switching Specifications   (Continued)

PARAMETER	SYMBOL	TEST CONDITIONS	V_{CC} (V)	25°C			-40°C TO 85°C		-55°C TO 125°C		UNITS
				MIN	TYP	MAX	MIN	MAX	MIN	MAX	
Hold Time, $\overline{SPE}$ to CP	t_H	-	4.5	3	-	-	3	-	3	-	ns
Recovery Time, $\overline{MR}$ to CP (161)	t_{REC}	-	4.5	15	-	-	19	-	22	-	ns

NOTE:

7. Applies to non-cascaded operation only. With cascaded counters clock to terminal count propagation delays, count enables (PE or TE)-to-clock setup times, and count enables (PE or TE)-to-clock hold times determine maximum clock frequency. For example with these HC devices:

$$f_{MAX} (CP) = \frac{1}{\text{CP-to-TC prop. delay} + \overline{\text{TE-to-CP setup}} + \overline{\text{TE-to-CP Hold}}} = \frac{1}{37 + 10 + 0} \approx 21\,\text{MHz(min)}$$

Switching Specifications C_L = 50pF, Input t_r, t_f = 6ns

PARAMETER	SYMBOL	TEST CONDITIONS	V_{CC} (V)	25°C			-40°C TO 85°C		-55°C TO 125°C		UNITS
				MIN	TYP	MAX	MIN	MAX	MIN	MAX	
HC TYPES											
Propagation Delay CP to TC	t_{PHL}, t_{PLH}	C_L = 50pF	2	-	-	185	-	230	-	280	ns
			4.5	-	-	37	-	46	-	56	ns
		C_L = 15pF	5	-	15	-	-	-	-	-	ns
		C_L = 50pF	6	-	-	31	-	39	-	48	ns
CP to Qn	t_{PHL}, t_{PLH}	C_L = 50pF	2	-	-	185	-	230	-	280	ns
			4.5	-	-	37	-	46	-	56	ns
		C_L = 15pF	5	-	15	-	-	-	-	-	ns
		C_L = 50pF	6	-	-	31	-	39	-	48	ns
TE to TC	t_{PHL}, t_{PLH}	C_L = 50pF	2	-	-	120	-	150	-	180	ns
			4.5	-	-	24	-	30	-	36	ns
		C_L = 15pF	5	-	9	-	-	-	-	-	ns
		C_L = 50pF	6	-	-	20	-	26	-	31	ns
$\overline{MR}$ to Qn (161)	t_{PHL}	C_L = 50pF	2	-	-	210	-	265	-	315	ns
			4.5	-	-	42	-	53	-	63	ns
		C_L = 15pF	5	-	18	-	-	-	-	-	ns
		C_L = 50pF	6	-	-	36	-	45	-	54	ns
$\overline{MR}$ to TC (161)	t_{PHL}	C_L = 50pF	2	-	-	210	-	265	-	315	ns
			4.5	-	-	42	-	53	-	63	ns
		C_L = 50pF	6	-	-	36	-	45	-	54	ns
Output Transition Time	t_{THL}, t_{TLH}	C_L = 50pF	2	-	-	75	-	95	-	110	ns
			4.5	-	-	15	-	19	-	22	ns
			6	-	-	13	-	16	-	19	ns
Power Dissipation Capacitance (Notes 8, 9)	C_{PD}	-	5	-	60	-	-	-	-	-	pF

CD74HC161, CD74HCT161, CD74HC163, CD74HCT163

Switching Specifications $C_L = 50pF$, Input t_r, $t_f = 6ns$ (Continued)

PARAMETER	SYMBOL	TEST CONDITIONS	V_{CC} (V)	25°C			-40°C TO 85°C		-55°C TO 125°C		UNITS
				MIN	TYP	MAX	MIN	MAX	MIN	MAX	
Input Capacitance	C_{IN}	$C_L = 50pF$	-	10	-	10	-	10	-	10	pF
HCT TYPES											
Propagation Delay											
CP to TC	t_{PHL}, t_{PLH}	$C_L = 50pF$	4.5	-	-	42	-	53	-	63	ns
		$C_L = 15pF$	5	-	18	-	-	-	-	-	ns
CP to Qn	t_{PHL}, t_{PLH}	$C_L = 50pF$	4.5	-	-	39	-	49	-	59	ns
		$C_L = 15pF$	5	-	16	-	-	-	-	-	ns
TE to TC	t_{PHL}, t_{PLH}	$C_L = 50pF$	4.5	-	-	32	-	40	-	48	ns
		$C_L = 15pF$	5	-	13	-	-	-	-	-	ns
$\overline{MR}$ to Qn (161)	t_{PHL}	$C_L = 50pF$	4.5	-	-	50	-	63	-	75	ns
		$C_L = 15pF$	5	-	21	-	-	-	-	-	ns
$\overline{MR}$ to TC (161)	t_{PHL}	$C_L = 50pF$	4.5	-	-	50	-	63	-	75	ns
Output Transition Time	t_{THL}, t_{TLH}	$C_L = 50pF$	4.5	-	-	15	-	19	-	22	ns
Power Dissipation Capacitance (Notes 6, 7)	C_{PD}	-	5	-	63	-	-	-	-	-	pF
Input Capacitance	C_{IN}	$C_L = 50pF$	-	10	-	10	-	10	-	10	pF

NOTES:

8. C_{PD} is used to determine the dynamic power consumption, per package.

9. $P_D = C_{PD} V_{CC}^2 f_i + \Sigma(C_L V_{CC}^2 f_O)$ where f_i = Input Frequency, f_O = Output Frequency, C_L = Output Load Capacitance, V_{CC} = Supply Voltage.

CD74HC161, CD74HCT161, CD74HC163, CD74HCT163

Timing Diagram

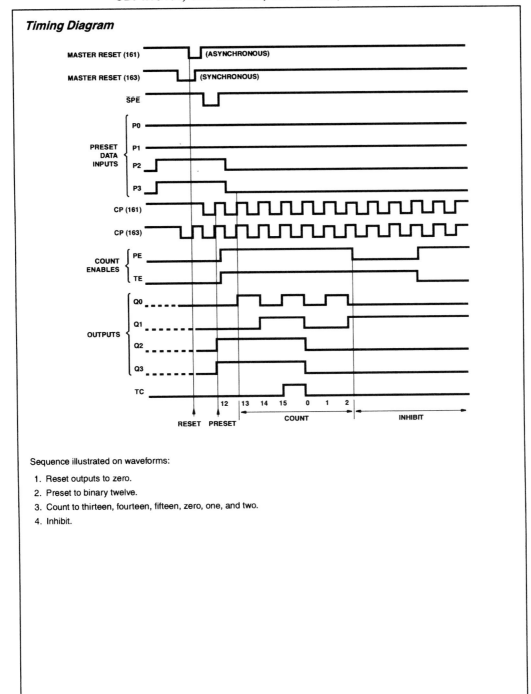

Sequence illustrated on waveforms:

1. Reset outputs to zero.
2. Preset to binary twelve.
3. Count to thirteen, fourteen, fifteen, zero, one, and two.
4. Inhibit.

TEXAS INSTRUMENTS

Data sheet acquired from Harris Semiconductor
SCHS156

February 1998

CD74HC165, CD74HCT165

High Speed CMOS Logic
8-Bit Parallel-In/Serial-Out Shift Register

Features

- Buffered Inputs
- Asynchronous Parallel Load
- Complementary Outputs
- Typical f_{MAX} = 60MHz at V_{CC} = 5V, C_L = 15pF, T_A = 25°C
- Fanout (Over Temperature Range)
  - Standard Outputs . . . . 10 LSTTL Loads
  - Bus Driver Outputs . . . . 15 LSTTL Loads
- Wide Operating Temperature Range . . . -55°C to 125°C
- Balanced Propagation Delay and Transition Times
- Significant Power Reduction Compared to LSTTL Logic ICs
- HC Types
  - 2V to 6V Operation
  - High Noise Immunity: N_{IL} = 30%, N_{IH} = 30% of V_{CC} at V_{CC} = 5V
- HCT Types
  - 4.5V to 5.5V Operation
  - Direct LSTTL Input Logic Compatibility, V_{IL} = 0.8V (Max), V_{IH} = 2V (Min)
  - CMOS Input Compatibility, $I_I \leq 1\mu A$ at V_{OL}, V_{OH}

Pinout

CD74HC165, CD74HCT165
(PDIP, SOIC)
TOP VIEW

Pin			Pin
$\overline{PL}$	1	16	V_{CC}
CP	2	15	$\overline{CE}$
D4	3	14	D3
D5	4	13	D2
D6	5	12	D1
D7	6	11	D0
$\overline{Q_7}$	7	10	DS
GND	8	9	Q_7

CD74HC165, CD74HCT165

Description

The Harris CD74HC165 and CD74HCT165 are 8-bit parallel or serial-in shift registers with complementary serial outputs (Q_7 and $\overline{Q_7}$) available from the last stage. When the parallel load ($\overline{PL}$) input is LOW, parallel data from the D0 to D7 inputs are loaded into the register asynchronously. When the $\overline{PL}$ is HIGH, data enters the register serially at the DS input and shifts one place to the right ($Q_0 \rightarrow Q_1 \rightarrow Q_2$, etc.) with each positive-going clock transition. This feature allow parallel-to-serial converter expansion by typing the Q_7 output to the DS input of the succeeding device.

For predictable operation the LOW-to-HIGH transition of $\overline{CE}$ should only take place while CP is HIGH. Also, CP an d $\overline{CE}$ should be LOW before the LOW-to-HIGH transition of PL to prevent shifting the data when $\overline{PL}$ goes HIGH.

Ordering Information

PART NUMBER	TEMP. RANGE (°C)	PACKAGE	PKG. NO.
CD74HC165E	-55 to 125	16 Ld PDIP	E16.3
CD74HCT165E	-55 to 125	16 Ld PDIP	E16.3
CD74HC165M	-55 to 125	16 Ld SOIC	M16.15
CD74HCT165M	-55 to 125	16 Ld SOIC	M16.15

NOTES:

1. When ordering, use the entire part number. Add the suffix 96 to obtain the variant in the tape and reel.

2. Wafer and die is available which meets all electrical specifications. Please contact your local sales office or Harris customer service for ordering information.

Functional Diagram

TRUTH TABLE

OPERATING MODE	INPUTS					Q_n REGISTER		OUTPUTS	
	$\overline{PL}$	$\overline{CE}$	CP	DS	D0 - D7	Q_0	Q_1 - Q_6	Q_7	$\overline{Q_7}$
Parallel Load	L	X	X	X	L	L	L-L	L	H
	L	X	X	X	H	H	H-H	H	L
Serial Shift	H	L	↑	l	X	L	$q_0 - q_5$	q_6	$\overline{q_6}$
	H	L	↑	h	X	H	$q_0 - q_5$	q_6	$\overline{q_6}$
Hold Do Nothing	H	H	X	X	X	q_0	$q_1 - q_6$	q_7	$\overline{q_7}$

CD74HC165, CD74HCT165

	TRUTH TABLE								
	INPUTS					Q_n REGISTER		OUTPUTS	
OPERATING MODE	$\overline{PL}$	$\overline{CE}$	CP	DS	D0 - D7	Q_0	Q_1 - Q_6	Q_7	$\overline{Q_7}$

NOTE:
H = High Voltage Level
h = High Voltage Level One Set-up Time Prior To The Low-to-high Clock Transition
I = Low Voltage Level One Set-up Time Prior To The Low-to-high Clock Transition
L = Low Voltage Level
X = Don't Care
↑ = Transition from Low to High Level
q_n = Lower Case Letters Indicate The State Of the Reference Output Clock Transition

CD74HC165, CD74HCT165

Absolute Maximum Ratings

DC Supply Voltage, V_{CC} . . . . -0.5V to 7V
DC Input Diode Current, I_{IK}
  For V_I < -0.5V or V_I > V_{CC} + 0.5V . . . .±20mA
DC Output Diode Current, I_{OK}
  For V_O < -0.5V or V_O > V_{CC} + 0.5V . . . .±20mA
DC Drain Current per Output, IO
  For V_O < -0.5V V_O > V_{CC} + 0.5V . . . .±25mA
DC Output Source or Sink Current per Output Pin, I_O
  For V_O > -0.5V or V_O < V_{CC} + 0.5V . . . .±25mA
DC V_{CC} or Ground Current, I_{CC} or I_{GND} . . . .±50mA

Operating Conditions

Temperature Range (T_A) . . . . -55°C to 125°C
Supply Voltage Range, V_{CC}
  HC Types . . . . .2V to 6V
  HCT Types . . . . .4.5V to 5.5V
DC Input or Output Voltage, V_I, V_O . . . . 0V to V_{CC}
Input Rise and Fall Time
  2V . . . . 1000ns (Max)
  4.5V . . . . 500ns (Max)
  6V . . . . 400ns (Max)

Thermal Information

Thermal Resistance (Typical, Note 3)	θ_{JA} (°C/W)
PDIP Package	90
SOIC Package	115

Maximum Junction Temperature . . . . 150°C
Maximum Storage Temperature Range . . . . -65°C to 150°C
Maximum Lead Temperature (Soldering 10s) . . . . 300°C
  (SOIC - Lead Tips Only)

CAUTION: Stresses above those listed in "Absolute Maximum Ratings" may cause permanent damage to the device. This is a stress only rating and operation of the device at these or any other conditions above those indicated in the operational sections of this specification is not implied.

NOTE:

3. θ_{JA} is measured with the component mounted on an evaluation PC board in free air.

DC Electrical Specifications

PARAMETER	SYMBOL	TEST CONDITIONS V_I (V)	TEST CONDITIONS I_O (mA)	V_{CC} (V)	25°C MIN	25°C TYP	25°C MAX	-40°C TO 85°C MIN	-40°C TO 85°C MAX	-55°C TO 125°C MIN	-55°C TO 125°C MAX	UNITS
HC TYPES												
High Level Input Voltage	V_{IH}	-	-	2	1.5	-	-	1.5	-	1.5	-	V
				4.5	3.15	-	-	3.15	-	3.15	-	V
				6	4.2	-	-	4.2	-	4.2	-	V
Low Level Input Voltage	V_{IL}	-	-	2	-	-	0.5	-	0.5	-	0.5	V
				4.5	-	-	1.35	-	1.35	-	1.35	V
				6	-	-	1.8	-	1.8	-	1.8	V
High Level Output Voltage CMOS Loads	V_{OH}	V_{IH} or V_{IL}	-0.02	2	1.9	-	-	1.9	-	1.9	-	V
			-0.02	4.5	4.4	-	-	4.4	-	4.4	-	V
			-0.02	6	5.9	-	-	5.9	-	5.9	-	V
High Level Output Voltage TTL Loads			-4	4.5	3.98	-	-	3.84	-	3.7	-	V
			-5.2	6	5.48	-	-	5.34	-	5.2	-	V
Low Level Output Voltage CMOS Loads	V_{OL}	V_{IH} or V_{IL}	0.02	2	-	-	0.1	-	0.1	-	0.1	V
			0.02	4.5	-	-	0.1	-	0.1	-	0.1	V
			0.02	6	-	-	0.1	-	0.1	-	0.1	V
Low Level Output Voltage TTL Loads			4	4.5	-	-	0.26	-	0.33	-	0.4	V
			5.2	6	-	-	0.26	-	0.33	-	0.4	V
Input Leakage Current	I_I	V_{CC} or GND	-	6	-	-	±0.1	-	±1	-	±1	µA

CD74HC165, CD74HCT165

DC Electrical Specifications (Continued)

PARAMETER	SYMBOL	TEST CONDITIONS V_I (V)	I_O (mA)	V_{CC} (V)	25°C MIN	25°C TYP	25°C MAX	-40°C TO 85°C MIN	-40°C TO 85°C MAX	-55°C TO 125°C MIN	-55°C TO 125°C MAX	UNITS
Quiescent Device Current	I_{CC}	V_{CC} or GND	0	6	-	-	8	-	80	-	160	μA
HCT TYPES												
High Level Input Voltage	V_{IH}	-	-	4.5 to 5.5	2	-	-	2	-	2	-	V
Low Level Input Voltage	V_{IL}	-	-	4.5 to 5.5	-	-	0.8	-	0.8	-	0.8	V
High Level Output Voltage CMOS Loads	V_{OH}	V_{IH} or V_{IL}	-0.02	4.5	4.4	-	-	4.4	-	4.4	-	V
High Level Output Voltage TTL Loads			-4	4.5	3.98	-	-	3.84	-	3.7	-	V
Low Level Output Voltage CMOS Loads	V_{OL}	V_{IH} or V_{IL}	0.02	4.5	-	-	0.1	-	0.1	-	0.1	V
Low Level Output Voltage TTL Loads			4	4.5	-	-	0.26	-	0.33	-	0.4	V
Input Leakage Current	I_I	V_{CC} to GND	0	5.5	-	-	±0.1	-	±1	-	±1	μA
Quiescent Device Current	I_{CC}	V_{CC} or GND	0	5.5	-	-	8	-	80	-	160	μA
Additional Quiescent Device Current Per Input Pin: 1 Unit Load (Note 4)	ΔI_{CC} (Note 4)	V_{CC} -2.1	-	4.5 to 5.5	-	100	360	-	450	-	490	μA

NOTE:

4. For dual-supply systems theoretical worst case (V_I = 2.4V, V_{CC} = 5.5V) specification is 1.8mA.

HCT Input Loading Table

INPUT	UNIT LOADS
DS, D0 to D7	0.35
CP, $\overline{PL}$	0.65

NOTE: Unit Load is ΔI_{CC} limit specified in DC Electrical Specifications table, e.g. 360μA max at 25°C.

Prerequisite For Switching Specifications

PARAMETER	SYMBOL	V_{CC} (V)	25°C MIN	25°C MAX	-40°C TO 85°C MIN	-40°C TO 85°C MAX	-55°C TO 125°C MIN	-55°C TO 125°C MAX	UNITS
HC TYPES									
CP Pulse Width	t_{WL}, t_{WH}	2	80	-	100	-	120	-	ns
		4.5	16	-	20	-	24	-	ns
		6	14	-	17	-	20	-	ns

CD74HC165, CD74HCT165

Prerequisite For Switching Specifications (Continued)

PARAMETER	SYMBOL	V_{CC} (V)	25°C MIN	25°C MAX	-40°C TO 85°C MIN	-40°C TO 85°C MAX	-55°C TO 125°C MIN	-55°C TO 125°C MAX	UNITS
$\overline{PL}$ Pulse Width	t_{WL}	2	80	-	100	-	120	-	ns
		4.5	16	-	20	-	24	-	ns
		6	14	-	17	-	20	-	ns
Set-up Time	t_{SU}	2	80	-	100	-	120	-	ns
DS to CP		4.5	16	-	20	-	24	-	ns
		6	14	-	17	-	20	-	ns
$\overline{CE}$ to CP	$t_{SU(L)}$	2	80	-	100	-	120	-	ns
		4.5	16	-	20	-	24	-	ns
		6	14	-	17	-	20	-	ns
D0-D7 to $\overline{PL}$	t_{SU}	2	80	-	100	-	120	-	ns
		4.5	16	-	20	-	24	-	ns
		6	14	-	17	-	20	-	ns
Hold Time	t_H	2	35	-	45	-	55	-	ns
DS to CP or $\overline{CE}$		4.5	7	-	9	-	11	-	ns
		6	6	-	8	-	9	-	ns
$\overline{CE}$ to CP	t_H	2	0	-	0	-	0	-	ns
		4.5	0	-	0	-	0	-	ns
		6	0	-	0	-	0	-	ns
Recovery Time	t_{REC}	2	100	-	125	-	150	-	ns
$\overline{PL}$ to CP		4.5	20	-	25	-	30	-	ns
		6	17	-	21	-	26	-	ns
Maximum Clock Pulse	f_{MAX}	2	6	-	5	-	4	-	MHz
Frequency		4.5	30	-	24	-	20	-	MHz
		6	35	-	28	-	24	-	MHz
HCT TYPES									
CP Pulse Width	t_{WL}, t_{WH}	4.5	18	-	23	-	27	-	ns
$\overline{PL}$ Pulse Width	t_{WL}	4.5	20	-	25	-	30	-	ns
Set-up Time DS to CP	t_{SU}	4.5	20	-	25	-	30	-	ns
$\overline{CE}$ to CP	$t_{SU(L)}$	4.5	20	-	25	-	30	-	ns
D0-D7 to $\overline{PL}$	t_{SU}	6	20	-	25	-	30	-	ns
Hold Time DS to CP or $\overline{CE}$	t_H	4.5	7	-	9	-	11	-	ns
$\overline{CE}$ to CP	t_S, t_H	4.5	0	-	0	-	0	-	ns
Recovery Time $\overline{PL}$ to CP	t_{REC}	4.5	20	-	25	-	30	-	ns
Maximum Clock Pulse Frequency	f_{MAX}	4.5	27	-	22	-	18	-	MHz

CD74HC165, CD74HCT165

Switching Specifications Input t_r, t_f = 6ns

PARAMETER	SYMBOL	TEST CONDITIONS	V_{CC} (V)	25°C		-40°C TO 85°C	-55°C TO 125°C	UNITS
				TYP	MAX	MAX	MAX	
HC TYPES								
Propagation Delay	t_{PLH}, t_{PHL}	C_L = 50pF	2	-	165	205	250	ns
CP or $\overline{CE}$ to Q_7 or $\overline{Q_7}$			4.5	-	33	41	50	ns
		C_L = 15pF	5	13	-	-	-	ns
		C_L = 50pF	6	-	28	35	43	ns
$\overline{PL}$ to Q_7 or $\overline{Q_7}$	t_{PLH}, t_{PHL}	C_L = 50pF	2	-	175	220	265	ns
			4.5	-	35	44	53	ns
		C_L = 15pF	5	14	-	-	-	ns
		C_L = 50pF	6	-	30	37	45	ns
D7 to Q_7 or $\overline{Q_7}$	t_{PLH}, t_{PHL}	C_L = 50pF	2	-	150	190	225	ns
			4.5	-	30	38	45	ns
		C_L = 15pF	5	12	-	-	-	ns
		C_L = 50pF	6	-	26	33	38	ns
Output Transition Times	t_{TLH}, t_{THL}	C_L = 50pF	2	-	75	95	110	ns
			4.5	-	15	19	22	ns
			6	-	13	16	19	ns
Input Capacitance	C_{IN}	-	-	-	10	10	10	pF
Power Dissipation Capacitance (Notes 5, 6)	C_{PD}	-	5	17	-	-	-	pF
HCT TYPES								
Propagation Delay	t_{PLH}, t_{PHL}	C_L = 50pF	4.5	-	40	50	60	ns
CP or $\overline{CE}$ to Q_7 or $\overline{Q_7}$		C_L = 15pF	5	17	-	-	-	ns
$\overline{PL}$ to Q_7 or $\overline{Q_7}$	t_{PLH}, t_{PHL}	C_L = 50pF	4.5	-	40	50	60	ns
		C_L = 15pF	5	17	-	-	-	ns
D7 to Q_7 or $\overline{Q_7}$	t_{PLH}, t_{PHL}	C_L = 50pF	4.5	-	35	44	53	ns
		C_L = 15pF	5	14	-	-	-	ns
Output Transition Times	t_{TLH}, t_{THL}	C_L = 50pF	4.5	-	15	19	22	ns
Input Capacitance	C_{IN}	C_L = 50pF	-	-	10	10	10	pF
Power Dissipation Capacitance (Notes 5, 6)	C_{PD}	-	5	24	-	-	-	pF

NOTES:

5. C_{PD} is used to determine the dynamic power consumption, per package.

6. $P_D = V_{CC}^2 f_i + \Sigma (C_L V_{CC}^2 + f_O)$ where f_i = Input Frequency, f_O = Output Frequency, C_L = Output Load Capacitance, V_{CC} = Supply Voltage.

Test Circuits and Waveforms

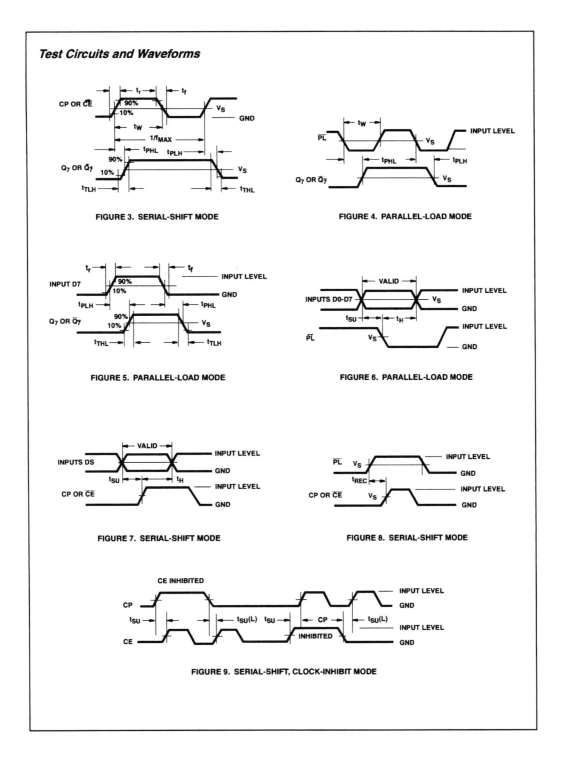

FIGURE 3. SERIAL-SHIFT MODE

FIGURE 4. PARALLEL-LOAD MODE

FIGURE 5. PARALLEL-LOAD MODE

FIGURE 6. PARALLEL-LOAD MODE

FIGURE 7. SERIAL-SHIFT MODE

FIGURE 8. SERIAL-SHIFT MODE

FIGURE 9. SERIAL-SHIFT, CLOCK-INHIBIT MODE

B CONVERSIONS

Decimal	BCD(8421)	Octal	Binary	Decimal	BCD(8421)	Octal	Binary	Decimal	BCD(8421)	Octal	Binary
0	0000	0	0	34	00110100	42	100010	68	01101000	104	1000100
1	0001	1	1	35	00110101	43	100011	69	01101001	105	1000101
2	0010	2	10	36	00110110	44	100100	70	01110000	106	1000110
3	0011	3	11	37	00110111	45	100101	71	01110001	107	1000111
4	0100	4	100	38	00111000	46	100110	72	01110010	110	1001000
5	0101	5	101	39	00111001	47	100111	73	01110011	111	1001001
6	0110	6	110	40	01000000	50	101000	74	01110100	112	1001010
7	0111	7	111	41	01000001	51	101001	75	01110101	113	1001011
8	1000	10	1000	42	01000010	52	101010	76	01110110	114	1001100
9	1001	11	1001	43	01000011	53	101011	77	01110111	115	1001101
10	00010000	12	1010	44	01000100	54	101100	78	01111000	116	1001110
11	00010001	13	1011	45	01000101	55	101101	79	01111001	117	1001111
12	00010010	14	1100	46	01000110	56	101110	80	10000000	120	1010000
13	00010011	15	1101	47	01000111	57	101111	81	10000001	121	1010001
14	00010100	16	1110	48	01001000	60	110000	82	10000010	122	1010010
15	00010101	17	1111	49	01001001	61	110001	83	10000011	123	1010011
16	00010110	20	10000	50	01010000	62	110010	84	10000100	124	1010100
17	00010111	21	10001	51	01010001	63	110011	85	10000101	125	1010101
18	00011000	22	10010	52	01010010	64	110100	86	10000110	126	1010110
19	00011001	23	10011	53	01010011	65	110101	87	10000111	127	1010111
20	00100000	24	10100	54	01010100	66	110110	88	10001000	130	1011000
21	00100001	25	10101	55	01010101	67	110111	89	10001001	131	1011001
22	00100010	26	10110	56	01010110	70	111000	90	10010000	132	1011010
23	00100011	27	10111	57	01010111	71	111001	91	10010001	133	1011011
24	00100100	30	11000	58	01011000	72	111010	92	10010010	134	1011100
25	00100101	31	11001	59	01011001	73	111011	93	10010011	135	1011101
26	00100110	32	11010	60	01100000	74	111100	94	10010100	136	1011110
27	00100111	33	11011	61	01100001	75	111101	95	10010101	137	1011111
28	00101000	34	11100	62	01100010	76	111110	96	10010110	140	1100000
29	00101001	35	11101	63	01100011	77	111111	97	10010111	141	1100001
30	00110000	36	11110	64	01100100	100	1000000	98	10011000	142	1100010
31	00110001	37	11111	65	01100101	101	1000001	99	10011001	143	1100011
32	00110010	40	100000	66	01100110	102	1000010				
33	00110011	41	100001	67	01100111	103	1000011				

Power of Two

2^n	n	2^{-n}
1	0	1.0
2	1	0.5
4	2	0.25
8	3	0.125
16	4	0.062 5
32	5	0.031 25
64	6	0.015 625
128	7	0.007 812 5
256	8	0.003 906 25
512	9	0.001 953 125
1 024	10	0.000 976 562 5
2 048	11	0.000 488 281 25
4 096	12	0.000 244 140 625
8 192	13	0.000 122 070 312 5
16 384	14	0.000 061 035 156 25
32 768	15	0.000 030 517 578 125
65 536	16	0.000 015 258 789 062 5
131 072	17	0.000 007 629 394 531 25
262 144	18	0.000 003 814 697 265 625
524 288	19	0.000 001 907 348 632 812 5
1 048 576	20	0.000 000 953 674 316 406 25
2 097 152	21	0.000 000 476 837 158 203 125
4 194 304	22	0.000 000 238 418 579 101 562 5
8 388 608	23	0.000 000 119 209 289 550 781 25
16 777 216	24	0.000 000 059 604 644 775 390 625
33 554 432	25	0.000 000 029 802 322 387 695 312 5
67 108 864	26	0.000 000 014 901 161 193 847 656 25
134 217 728	27	0.000 000 007 450 580 596 923 828 125
268 435 456	28	0.000 000 003 725 290 298 461 914 062 5
536 870 912	29	0.000 000 001 862 645 149 230 957 031 25
1 073 741 824	30	0.000 000 000 931 322 574 615 478 515 625
2 147 483 648	31	0.000 000 000 465 661 287 307 739 257 812 5
4 294 967 296	32	0.000 000 000 232 830 643 653 869 628 906 25
8 589 934 592	33	0.000 000 000 116 415 321 826 934 814 453 125
17 179 869 184	34	0.000 000 000 058 207 660 913 467 407 226 562 5
34 359 738 368	35	0.000 000 000 029 103 830 456 733 703 613 281 25
68 719 476 736	36	0.000 000 000 014 551 915 228 366 851 806 640 625
137 438 953 472	37	0.000 000 000 007 275 957 614 183 425 903 320 312 5
274 877 906 944	38	0.000 000 000 003 637 978 807 091 712 951 660 156 25
549 755 813 888	39	0.000 000 000 001 818 989 403 545 856 475 830 078 125
1 099 511 627 776	40	0.000 000 000 000 909 494 701 772 928 237 915 039 062 5
2 199 023 255 552	41	0.000 000 000 000 454 747 350 886 464 118 957 519 531 25
4 398 046 511 104	42	0.000 000 000 000 227 373 675 443 232 059 478 759 765 625
8 796 093 022 208	43	0.000 000 000 000 113 686 837 721 616 029 739 379 882 812 5
17 592 186 044 416	44	0.000 000 000 000 056 843 418 860 808 014 869 689 941 406 25
35 184 372 088 832	45	0.000 000 000 000 028 421 709 430 404 007 434 844 970 703 125
70 368 744 177 664	46	0.000 000 000 000 014 210 854 715 202 003 717 422 485 351 562 5
140 737 488 355 328	47	0.000 000 000 000 007 105 427 357 601 001 858 711 242 675 781 25
281 474 976 710 656	48	0.000 000 000 000 003 552 713 678 800 500 929 355 621 337 890 625
562 949 953 421 312	49	0.000 000 000 000 001 776 356 839 400 250 464 677 810 668 945 312 5
1 125 899 906 842 624	50	0.000 000 000 000 000 888 178 419 700 125 232 338 905 334 472 656 25
2 251 799 813 685 248	51	0.000 000 000 000 000 444 089 209 850 062 616 169 452 667 236 328 125
4 503 599 627 370 496	52	0.000 000 000 000 000 222 044 604 925 031 308 084 726 333 618 164 062 5
9 007 199 254 740 992	53	0.000 000 000 000 000 111 022 302 462 515 654 042 363 166 809 082 031 25
18 014 398 509 481 984	54	0.000 000 000 000 000 055 511 151 231 257 827 021 181 583 404 541 015 625
36 028 797 018 963 968	55	0.000 000 000 000 000 027 755 575 615 628 913 510 590 791 702 270 507 812 5
72 057 594 037 927 936	56	0.000 000 000 000 000 013 877 787 807 814 456 755 295 395 851 135 253 906 25
144 115 188 075 855 872	57	0.000 000 000 000 000 006 938 893 903 907 228 377 647 697 925 567 626 953 125
288 230 376 151 711 744	58	0.000 000 000 000 000 003 469 446 951 953 614 188 823 848 962 783 813 476 562 5
576 460 752 303 423 488	59	0.000 000 000 000 000 001 734 723 475 976 807 094 411 924 481 391 906 738 281 25
1 152 921 504 606 846 976	60	0.000 000 000 000 000 000 867 361 737 988 403 547 205 962 240 695 953 369 140 625
2 305 843 009 213 693 952	61	0.000 000 000 000 000 000 433 680 868 994 201 773 602 981 120 347 976 684 570 312 5
4 611 686 018 427 387 904	62	0.000 000 000 000 000 000 216 840 434 497 100 886 801 490 560 173 988 342 285 156 25
9 223 372 036 854 775 808	63	0.000 000 000 000 000 000 108 420 217 248 550 443 400 745 280 086 994 171 142 578 125
18 446 744 073 709 551 616	64	0.000 000 000 000 000 000 054 210 108 624 275 221 700 372 640 043 497 085 571 289 062 5
36 893 488 147 419 103 232	65	0.000 000 000 000 000 000 027 105 054 312 137 610 850 186 320 021 748 542 785 644 531 25
73 786 976 294 838 206 464	66	0.000 000 000 000 000 000 013 552 527 156 088 805 425 093 160 010 874 271 392 822 265 625
147 573 952 589 676 412 928	67	0.000 000 000 000 000 000 006 776 263 578 034 402 712 546 580 005 437 135 696 411 132 812 5
295 147 905 179 352 825 856	68	0.000 000 000 000 000 000 003 388 131 789 017 201 356 273 290 002 718 567 848 205 566 406 25
590 295 810 358 705 651 712	69	0.000 000 000 000 000 000 001 694 065 894 508 600 678 136 645 001 359 283 924 102 783 203 125
1 180 591 620 717 411 303 424	70	0.000 000 000 000 000 000 000 847 032 947 254 300 339 068 322 500 679 641 962 051 391 601 562 5
2 361 183 241 434 822 606 848	71	0.000 000 000 000 000 000 000 423 516 473 627 150 169 534 161 250 339 820 981 025 695 800 781 25
4 722 366 482 869 645 213 696	72	0.000 000 000 000 000 000 000 211 758 236 813 575 084 767 080 625 169 910 490 512 847 900 390 625

ANSWERS TO SELECTED ODD-NUMBERED PROBLEMS

CHAPTER 1

1. Digital can be transmitted and stored more efficiently and reliably.

3. (a) 11010001   (b) 000101010

5. (a) 550 ns   (b) 600 ns   (c) 2.7 μs   (d) 10 V

7. 250 Hz   **9.** 50%   **11.** 8 μs; 1 μs   **13.** AND gate

15. (a) adder   (b) multiplier   (c) multiplexer   (d) comparator

17. 01010000

19. DIP pins go through holes in a circuit board. SMT pins connect to surface pads.

21. 7 V   **23.** A collection of circuits interconnected to perform a specified function

25. Enter a new value on the keypad.

CHAPTER 2

1. (a) 1   (b) 100   (c) 100,000

3. (a) 400; 70; 1   (b) 9000; 300; 50; 6   (c) 100,000; 20,000; 5000; 0; 0; 0

5. (a) 3   (b) 4   (c) 7   (d) 8   (e) 9   (f) 12   (g) 11   (h) 15

7. (a) 51.75   (b) 42.25   (c) 65.875   (d) 120.625
(e) 92.65625   (f) 113.0625   (g) 90.625   (h) 127.96875

9. (a) 5 bits   (b) 6 bits   (c) 6 bits   (d) 7 bits
(e) 7 bits   (f) 7 bits   (g) 8 bits   (h) 8 bits

11. (a) 1010   (b) 10001   (c) 11000   (d) 110000
(e) 111101   (f) 1011101   (g) 1111101   (h) 10111010

13. (a) 1111   (b) 10101   (c) 11100   (d) 100010
(e) 101000   (f) 111011   (g) 1000001   (h) 1001001

15. (a) 100   (b) 100   (c) 1000   (d) 1101   (e) 1110   (f) 11000

17. (a) 1001   (b) 1000   (c) 100011
(d) 110110   (e) 10101001   (f) 10110110

19. (a) 010   (b) 001   (c) 0101   (d) 00101000   (e) 0001010   (f) 11110

21. (a) 00011101   (b) 11010101   (c) 01100100   (d) 11111011

23. (a) 00001100   (b) 10111100   (c) 01100101   (d) 10000011

25. (a) -102 (b) $+116$ (c) -64

27. (a) 0 10001101 11110000101011000000000
(b) 1 10001010 10000011000000000000000

29. (a) 00110000   (b) 00011101   (c) 11101011   (d) 100111110

31. (a) 11000101   (b) 11000000

33. 100111001010

35. (a) 00111000   (b) 01011001   (c) 101000010100
(d) 010111001000   (e) 0100000100000000   (f) 1111101100010111
(g) 1000101010011101

37. (a) 35   (b) 146   (c) 26   (d) 141
(e) 243   (f) 235   (g) 1474   (h) 1792

39. (a) 60_{16} (b) $10B_{16}$ (c) $1BA_{16}$

41. (a) 10   (b) 23   (c) 46   (d) 52   (e) 67
    (f) 367   (g) 115   (h) 532   (i) 4085

43. (a) 001011   (b) 101111   (c) 001000001
    (d) 011010001   (e) 101100000   (f) 100110101011
    (g) 001011010111001   (h) 100101110000000   (i) 00100000010001011

45. (a) 00010000   (b) 00010011   (c) 00011000   (d) 00100001
    (e) 00100101   (f) 00110110   (g) 01000100   (h) 01010111
    (i) 01101001   (j) 10011000   (k) 000100100101   (l) 000101010110

47. (a) 000100000100   (b) 000100101000   (c) 000100110010
    (d) 000101010000   (e) 000110000110   (f) 001000010000
    (g) 001101011001   (h) 010101000111   (i) 0001000001010001

49. (a) 80   (b) 237   (c) 346   (d) 421   (e) 754
    (f) 800   (g) 978   (h) 1683   (i) 9018   (j) 6667

51. (a) 00010100   (b) 00010010   (c) 00010111   (d) 00010110
    (e) 01010010   (f) 000100001001   (g) 000110010101   (h) 0001001001101001

53. The Gray code makes only one bit change at a time when going from one number in the sequence to the next.

55. (a) 1100   (b) 00011   (c) 10000011110

57. (a) CAN   (b) J   (c) =   (d) #   (e) >   (f) B

59. 48 65 6C 6C 6F 2E 20 48 6F 77 20 61 72 65 20 79 6F 75 3F

61. (b) is incorrect.

63. (a) 110100100   (b) 000001001   (c) 111111110

65. (a) 00100101 (25 BCD)   (b) 00011001 (25 binary)
    (c) 00001111 (15)   (d) 0110000110010010 (24,975)
    (e) 0110000110100001 (24,990)

CHAPTER 3

1. See Figure P–1.   3. See Figure P–2.   5. See Figure P–3.
7. See Figure P–4.   9. See Figure P–5.   11. See Figure P–6.

FIGURE P–1   FIGURE P–2   FIGURE P–3   FIGURE P–4

FIGURE P–5   FIGURE P–6

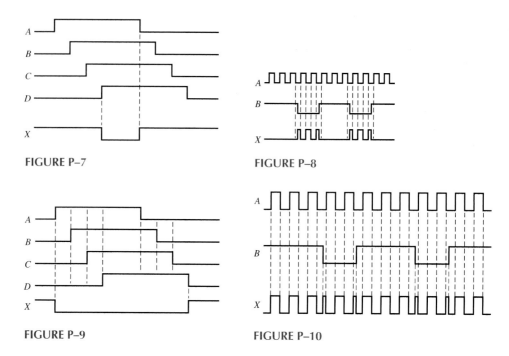

FIGURE P–7

FIGURE P–8

FIGURE P–9

FIGURE P–10

13. See Figure P–7.   **15.** See Figure P–8.   **17.** See Figure P–9.

19. XOR $= A\overline{B} + \overline{A}B$; OR $= A + B$ **21.** See Figure P–10.

23. CMOS   **25.** $t_{PLH} = 4.3$ ns; $t_{PHL} = 10.5$ ns   **27.** 20 mW

29. The gates in parts (b), (c), (e) are faulty.

31. (a) defective output (stuck LOW or open)   **(b)** Pin 4 input or pin 6 output internally open.

33. The seat belt input to the AND gate is open.

35. See Figure P–11.

FIGURE P–11

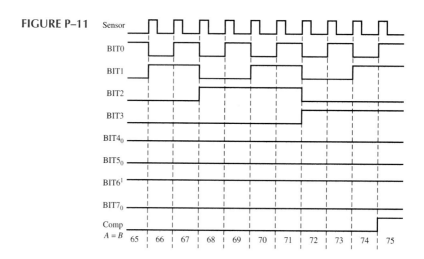

37. BIT5 input stuck at a 1 or open which could be caused by an open input to the corresponding XOR gate or BIT6 input stuck at a 0 due to short.

39. See Figure P–12.

41. Add an inverter to the enable input line of the AND gate.

43. See Figure P–13.

45. The valve and conveyor line must be taken from the NAND gate output. The register line still comes from inverter.

47. Gate inputs shorted together

49. Gate output open

FIGURE P–12   **FIGURE P–13**

CHAPTER 4

1. $X = A + B + C + D$ **3.** $X = \overline{A} + \overline{B} + \overline{C}$

5. **(a)** $AB = 1$ when $A = 1, B = 1$ **(b)** $A\overline{B}C = 1$ when $A = 1, B = 0, C = 1$

   (c) $A + B = 0$ when $A = 0, B = 0$ **(d)** $\overline{A} + B + \overline{C} = 0$ when $A = 1, B = 0, C = 1$

   (e) $\overline{A} + \overline{B} + C = 0$ when $A = 1, B = 1, C = 0$ **(f)** $\overline{A} + B = 0$ when $A = 1, B = 0$

   (g) $A\overline{B}\,\overline{C} = 1$ when $A = 1, B = 0, C = 0$

7. **(a)** Commutative   **(b)** Commutative   **(c)** Distributive

9. **(a)** $\overline{A}B$ **(b)** $A + \overline{B}$ **(c)** $\overline{A}\,\overline{B}\,\overline{C}$ **(d)** $\overline{A} + \overline{B} + \overline{C}$

   (e) $\overline{A} + \overline{B}\,\overline{C}$ **(f)** $\overline{A} + \overline{B} + \overline{C} + \overline{D}$ **(g)** $(\overline{A} + \overline{B})(\overline{C} + \overline{D})$ **(h)** $\overline{A}B + C\overline{D}$

11. **(a)** $(\overline{A} + \overline{B} + \overline{C})(\overline{E} + \overline{F} + \overline{G})(\overline{H} + \overline{I} + \overline{J})(\overline{K} + \overline{L} + \overline{M})$

   (b) $\overline{A}B\overline{C}(\overline{C} + D) + BC = \overline{A}B\overline{C} + BC$ **(c)** $\overline{A}\,\overline{B}\,\overline{C}\overline{D}\overline{E}\,\overline{F}\,\overline{G}\overline{H}$

13. **(a)** $X = ABCD$ **(b)** $X = AB + C$ **(c)** $X = \overline{\overline{A}B}$ **(d)** $X = (A + B)C$

15. See Figure P–14.

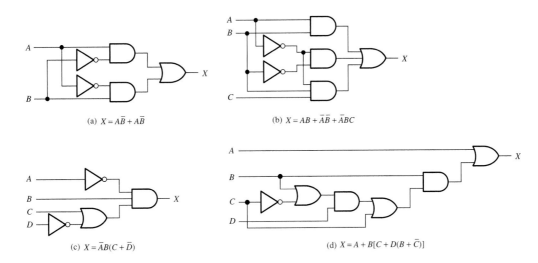

(a) $X = A\overline{B} + A\overline{B}$

(b) $X = AB + \overline{A}\overline{B} + \overline{A}BC$

(c) $X = \overline{A}B(C + \overline{D})$

(d) $X = A + B[C + D(B + \overline{C})]$

FIGURE P–14

17. **(a)** A **(b)** AB **(c)** C **(d)** A **(e)** $\overline{A}C + \overline{B}C$

19. **(a)** $BD + BE + \overline{D}F$ **(b)** $\overline{A}\,\overline{B}C + \overline{A}\,\overline{B}D$ **(c)** B **(d)** $AB + CD$ **(e)** ABC

21. **(a)** $A\overline{B} + AC + BC$ **(b)** $AC + \overline{B}C$ **(c)** $AB + AC$

23. **(a)** Domain: A, B, C
     Standard SOP: $A\overline{B}C + A\overline{B}\,\overline{C} + ABC + \overline{A}BC$

   (b) Domain: A, B, C
     Standard SOP: $ABC + A\overline{B}C + \overline{A}\,\overline{B}C$

   (c) Domain: A, B, C
     Standard SOP: $ABC + AB\overline{C} + A\overline{B}C$

25. (a) $101 + 100 + 111 + 011$ **(b)** $111 + 101 + 001$ **(c)** $111 + 110 + 101$

27. (a) $(A + B + C)(A + B + \overline{C})(A + \overline{B} + C)(\overline{A} + \overline{B} + C)$

   (b) $(A + B + C)(A + \overline{B} + C)(A + \overline{B} + \overline{C})(\overline{A} + B + C)(\overline{A} + \overline{B} + C)$

   (c) $(A + B + C)(A + B + \overline{C})(A + \overline{B} + C)(A + \overline{B} + \overline{C})(\overline{A} + B + C)$

29. (a) See Table P–1.   **(b)** See Table P–2.

31. (a) See Table P–3.   **(b)** See Table P–4.

33. (a) See Table P–5.   **(b)** See Table P–6.

35. See Figure P–15.   **37.** See Figure P–16.

TABLE P–1

A	B	C	X
0	0	0	0
0	0	1	0
0	1	0	1
0	1	1	0
1	0	0	0
1	0	1	1
1	1	0	0
1	1	1	1

TABLE P–2

X	Y	Z	Q
0	0	0	1
0	0	1	1
0	1	0	0
0	1	1	1
1	0	0	0
1	0	1	1
1	1	0	1
1	1	1	0

TABLE P–3

A	B	C	X
0	0	0	1
0	0	1	0
0	1	0	1
0	1	1	1
1	0	0	0
1	0	1	1
1	1	0	1
1	1	1	0

TABLE P–4

W	X	Y	Z	Q
0	0	0	0	1
0	0	0	1	1
0	0	1	0	1
0	0	1	1	1
0	1	0	0	0
0	1	0	1	1
0	1	1	0	1
0	1	1	1	0
1	0	0	0	1
1	0	0	1	1
1	0	1	0	1
1	0	1	1	1
1	1	0	0	0
1	1	0	1	1
1	1	1	0	1
1	1	1	1	1

TABLE P–5

A	B	C	X
0	0	0	0
0	0	1	0
0	1	0	0
0	1	1	1
1	0	0	1
1	0	1	1
1	1	0	1
1	1	1	1

TABLE P–6

A	B	C	D	X
0	0	0	0	1
0	0	0	1	0
0	0	1	0	1
0	0	1	1	1
0	1	0	0	0
0	1	0	1	0
0	1	1	0	0
0	1	1	1	0
1	0	0	0	1
1	0	0	1	0
1	0	1	0	0
1	0	1	1	1
1	1	0	0	1
1	1	0	1	1
1	1	1	0	1
1	1	1	1	1

FIGURE P–15

FIGURE P–16

39. (a) No simplification   **(b)** AC **(c)** $\overline{D}\,\overline{F} + E\overline{F}$

41. (a) $AB + AC$

    (b) $A + BC$

    (c) $B\overline{C}D + A\overline{C}D + BC\overline{D} + AC\overline{D}$

    (d) $A\overline{B} + CD$

43. $\overline{B} + C$ **45.** $\overline{A}\,\overline{B}\,\overline{C}D + C\overline{D} + BC + A\overline{D}$

47. (a) $(A + \overline{B} + C + \overline{D})(\overline{A} + B + \overline{C} + D)(\overline{A} + \overline{B} + \overline{C} + D)$

    (b) $(W + \overline{Z})(W + X)(\overline{Y} + \overline{Z})(X + \overline{Y})$

49. $(A + C + D)(A + \overline{B} + C)(\overline{A} + B + \overline{D})(B + \overline{C} + \overline{D})(\overline{A} + \overline{B} + \overline{C} + D)$

51. $X = \overline{A}\,\overline{B}\,\overline{C}\,\overline{D}\,\overline{E} + \overline{A}BC\overline{D}\,\overline{E} + AB\overline{C}\,\overline{D}\,\overline{E} + \overline{A}\,\overline{B}DE + \overline{A}BD\overline{E} + \overline{A}\,\overline{C}DE + AB\overline{C}D$

53. LED. LEDs emit light, LCDs do not.   **55.** One less inverter and six fewer gates.

57. Add an inverter to the output of the OR gate in each of the segment logic circuits.

59. See Figure P–17.

61. An inverter input or output open

63. b-segment OR gate output open

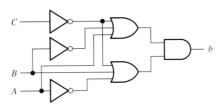

segment $b = (\overline{C} + B + \overline{A})(\overline{C} + \overline{B} + A)$

segment $c = C + \overline{B} + A$

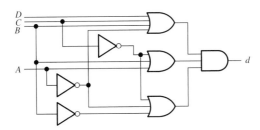

segment $d = (D + C + B + \overline{A})(\overline{C} + B + A)(\overline{C} + \overline{B} + \overline{A})$

segment $e = \overline{A}(\overline{C} + B)$

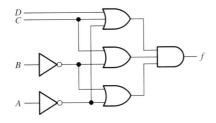

segment $f = (D + C + \overline{A})(C + \overline{B})(\overline{B} + \overline{A})$

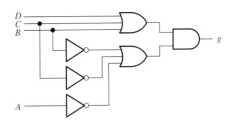

segment $g = (D + C + B)(\overline{C} + \overline{B} + \overline{A})$

FIGURE P–17

CHAPTER 5

  1. See Figure P–18.

  3. (a) $X = ABB$ **(b)** $X = AB + B$ **(c)** $X = \overline{A} + B$

    (d) $X = (A + B) + AB$ **(e)** $X = \overline{\overline{\overline{A}\,\overline{B}}\,\overline{B}C}$ **(f)** $X = A + B)(\overline{B} + C)$

FIGURE P–18

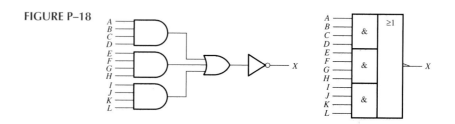

5. (a)

A	B	X
0	0	0
0	1	0
1	0	0
1	1	1

(b)

A	B	X
0	0	0
0	1	1
1	0	0
1	1	1

(c)

A	B	X
0	0	1
0	1	1
1	0	0
1	1	1

(d)

A	B	X
0	0	0
0	1	1
1	0	1
1	1	1

(e)

A	B	C	X
0	0	0	0
0	0	1	0
0	1	0	0
0	1	1	0
1	0	0	0
1	0	1	0
1	1	0	0
1	1	1	0

(f)

A	B	C	X
0	0	0	0
0	0	1	0
0	1	0	0
0	1	1	1
1	0	0	1
1	0	1	1
1	1	0	0
1	1	1	1

7. $X = \overline{A\overline{B} + \overline{A}B} = (\overline{A} + B)(A + \overline{B})$ **9.** See Figure P–19.

11. See Figure P–20. **13.** $X = AB$

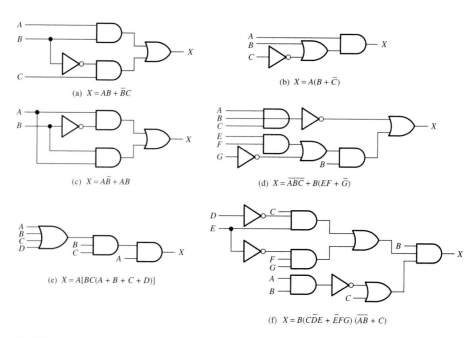

(a) $X = AB + \overline{B}C$

(b) $X = A(B + \overline{C})$

(c) $X = A\overline{B} + AB$

(d) $X = \overline{ABC} + B(EF + \overline{G})$

(e) $X = A[BC(A + B + C + D)]$

(f) $X = B(C\overline{D}E + \overline{E}FG)\,(\overline{AB} + C)$

FIGURE P–19

FIGURE P–20

$X = AB + \overline{C}$

15. (a) No simplification   **(b)** No simplification

   (c) $X = A$ **(d)** $X = \overline{A} + \overline{B} + \overline{C} + EF + \overline{G}$

   (e) $X = ABC$ **(f)** $X = BC\overline{D}E + \overline{A}B\overline{E}FG + BC\overline{E}FG$

17. (a) $X = AC + AD + BC + BD$ **(b)** $X = \overline{A}CD + \overline{B}CD$

   (c) $X = ABD + CD + E$ **(d)** $X = \overline{A} + B + D$ **(e)** $X = ABD + \overline{C}D + \overline{E}$

   (f) $X = \overline{A}\,\overline{C} + \overline{A}\,\overline{D} + \overline{B}\,\overline{C} + \overline{B}\,\overline{D} + \overline{E}\,\overline{G} + \overline{E}\,\overline{H} + \overline{F}\,\overline{G} + \overline{F}\,\overline{H}$

19. See Figure P–21.   **21.** See Figure P–22.   **23.** See Figure P–23.

FIGURE P–21

FIGURE P–22

FIGURE P–23

FIGURE P–24

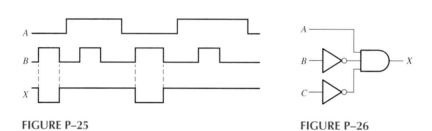

FIGURE P–25   **FIGURE P–26**

25. See Figure P–24.   **27.** $X = A + \overline{B}$; see Figure P–25.   **29.** $X = A\overline{B}\,\overline{C}$; see Figure P–26.

31. The output pulse width is greater than the specified minimum.

33. $X = ABC + D\overline{E}$. Since X is the same as the G_3 output, either G_1 or G_2 has failed, with its output stuck LOW.

35. Pin 10 of the right 74AHC00 is shorted to ground.

37. **(a)** $X = E$. See Figure P–27.   **(b)** $X = E$ **(c)** $X = E$ **39.** See Figure P–28.

41. See Figure P–29 for modified M_4 logic. Other motor logic remains the same.

43. $X =$ lamp on, $A =$ front door switch on, $B =$ back door switch on. See Figure P–30.

45. See Figure P–31. Inverters (not shown) are used to convert each HIGH key closure to LOW.

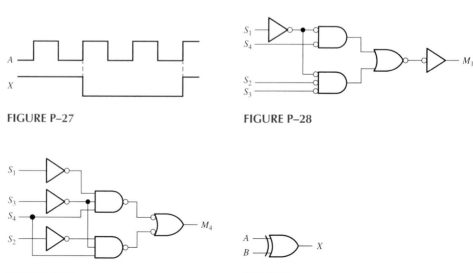

FIGURE P–27   **FIGURE P–28**

FIGURE P–29   **FIGURE P–30**

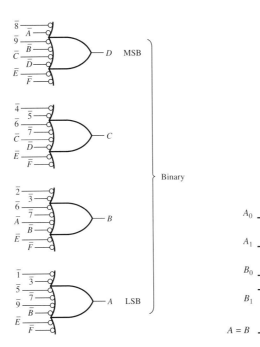

FIGURE P–31

FIGURE P–32

CHAPTER 6

1. **(a)** $A \oplus B = 0, \Sigma = 1, (A \oplus B)C_{in} = 0, AB = 1, C_{out} = 1$
   (b) $A \oplus B = 1, \Sigma = 0, (A \oplus B)C_{in} = 1, AB = 0, C_{out} = 1$
   (c) $A \oplus B = 1, \Sigma = 1, (A \oplus B)(C_{in} = 0, AB = 0, C_{out} = 0$

3. **(a)** $\Sigma = 1, C_{out} = 0;$ **(b)** $\Sigma = 1, C_{out} = 0;$ **(c)** $\Sigma = 0, C_{out} = 1;$ **(d)** $\Sigma = 1, C_{out} = 1$

5. 11100

7. $\Sigma_1 = 0110; \Sigma_2 = 1011; \Sigma_3 = 0110; \Sigma_4 = 0001; \Sigma_5 = 1010$

9. $A = B$ is HIGH when $A_0 = B_0$ and $A_1 = B_1$; see Figure P–32.

11. **(a)** $A > B = 1; A = B = 0; A < B = 0$ **(b)** $A < B = 1; A = B = 0; A > B = 0$
    (c) $A = B = 1; A < B = 0; A > B = 0$

13. See Figure P–33.

FIGURE P–33

FIGURE P–34

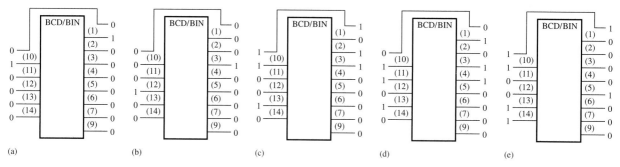

FIGURE P–35

15. $X = A_3A_2\overline{A}_1\overline{A}_0 + \overline{A}_3\overline{A}_2\overline{A}_1A_0 + A_3\overline{A}_2A_1$

17. See Figure P–34.

19. $A_3A_2A_1A_0 = 1011$, invalid BCD

21. See Figure P–35.

23. **(a)** 1010000000 Gray
        1100000000 binary
   (b) 0011001100 Gray
         0010001000 binary

    (c) 1111000111 Gray
        1010000101 binary
   (d) 0000000001 Gray
         0000000001 binary

    See Figure P–36.

25. See Figure P–37.   **27.** See Figure P–38.   **29.** See Figure P–39.

FIGURE P–36

FIGURE P–37

FIGURE P–38

FIGURE P–39

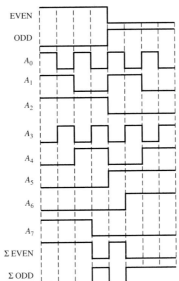

31. (a) OK   **(b)** segment g burned out; output G open   **(c)** Segment b output stuck LOW

33. (a) The A_1 input of the top adder is open:
All binary values corresponding to a BCD number having a value of 0, 1, 4, 5, 8, or 9 will be off by 2. This will first be seen for a BCD value of 0000 0000.

(b) The carry out of the top adder is open:
All values not normally involving an output carry will be off by 32. This will first be seen for a BCD value of 0000 0000.

(c) The Σ_4 output of the top adder is shorted to ground:
Same binary values above 1s will be short by 16. The first BCD value to indicate this will be 0001 1000.

(d) The Σ_3 output of the bottom adder is shorted to ground:
Every other set of 16 values starting with 16 will be short 16. The first BCD value to indicate this will be 0001 0110.

TABLE P–7

S_2	S_1	S_0	Y	$\overline{Y}$
0	0	0	1	0
0	0	1	0	1
0	1	0	0	1
0	1	1	0	1
1	0	0	0	1
1	0	1	0	1
1	1	0	0	1
1	1	1	0	1

35. 1. Place a LOW on pin 7 (Enable).
2. Apply a HIGH to D_0 and a LOW to D_1 through D_7.
3. Go through the binary sequence on the select inputs and check Y and $\overline{Y}$ according to Table P–7.
4. Repeat the binary sequence of select inputs for each set of data inputs listed in Table P–8. A HIGH on the Y output should occur only for the corresponding combinations of select inputs shown.

TABLE P–8

D_0	D_1	D_2	D_3	D_4	D_5	D_6	D_7	Y	$\overline{Y}$	S_2	S_1	S_0
L	H	L	L	L	L	L	L	1	0	0	0	1
L	L	H	L	L	L	L	L	1	0	0	1	0
L	L	L	H	L	L	L	L	1	0	0	1	1
L	L	L	L	H	L	L	L	1	0	1	0	0
L	L	L	L	L	H	L	L	1	0	1	0	1
L	L	L	L	L	L	H	L	1	0	1	1	0
L	L	L	L	L	L	L	H	1	0	1	1	1

37. Apply a HIGH in turn to each Data input, D_0 through D_7 with LOWs on all the other inputs. For each HIGH applied to a data input, sequence through all eight binary combinations of select inputs ($S_2 S_1 S_0$) and check for HIGH on the corresponding data output and LOWs on all the other data outputs.

39. Replace each 74LS00 with a 74LS08 AND gate.

41. $\Sigma = \overline{A}\,\overline{B}C_{in} + \overline{A}B\overline{C}_{in} + A\overline{B}\,\overline{C}_{in} + ABC_{in}$
$C_{out} = \overline{A}BC_{in} + A\overline{B}C_{in} + AB\overline{C}_{in} + ABC_{in}$
See Figure P–40.

43. See the block diagram in Figure P–41.   **45.** See Figure P–42.   **47.** See Figure P–43.

CHAPTER 7

1. 1. Programmable read-only memory (PROM): fixed AND array and programmable OR array
2. Programmable logic array (PLA): programmable AND and OR arrays
3. Programmable array logic (PAL): programmable AND and fixed OR arrays
4. Generic array logic (GAL): reprogrammable AND and fixed OR arrays

3. For $X_1 = \overline{A}BC$, blow: Row 1, Columns 1, 3, 4, 5, 6
Row 2, Columns 1, 2, 4, 5, 6
Row 3, Columns 1, 2, 3, 4, 6
For $X_2 = AB\overline{C}$, blow: Row 4, Columns 2, 3, 4, 5, 6
Row 5, Columns 1, 2, 4, 5, 6
Row 6, Columns 1, 2, 3, 4, 5
For $X_3 = \overline{A}B\overline{C}$, blow: Row 7, Columns 1, 3, 4, 5, 6
Row 8, Columns 1, 2, 4, 5, 6
Row 9, Columns 1, 2, 3, 4, 5

Σ = No simplification $C_{out} = BC_{in} + AB + AC_{in}$

FIGURE P–40

FIGURE P–41

FIGURE P–42

FIGURE P–43

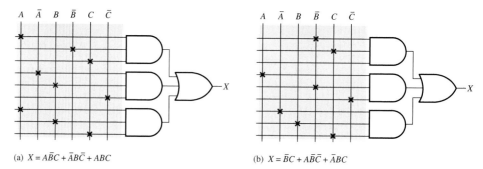

(a) $X = A\bar{B}C + \bar{A}B\bar{C} + ABC$

(b) $X = \bar{B}C + A\bar{B}\bar{C} + \bar{A}BC$

FIGURE P–44

5. See Figure P–44.

7. The programmed polarity output in a PAL works by selectively opening or leaving intact a cell on one input of an XOR gate, causing the gate to invert its other input or not.

9. See Figure P–45.   11. 7 available outputs

13. The select inputs determine combinational or registered mode, input or output, inverted or non-inverted output.

15. **(a)** 8   **(b)** 8   **(c)** 16

17. 8 product terms, 15 variables

19. Computer, programmer, compiler

21. The ispGAL22V10 can be programmed in the circuit, whereas the GAL22V10 must be inserted in a programmer.

23. **(a)** !A & !B & !C;   **(b)** A # B # !C;   **(c)** A & !(B # C);

25. **(a)** X = A & !B & C & !D & E # A & !B & C # !A & !B;

   (b) Y = A & B & !C # D & E & F # !G & H & !I # J & !K & L;

   (c) Z = (!A # !B # C # D) & (E # !F) # G;

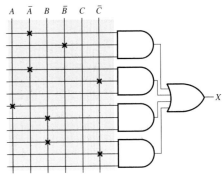

(a) $X = A\bar{B}C + \bar{A}B\bar{C} + A\bar{B} + BC$

(b) $X = \bar{A}\bar{B} + \bar{A}\bar{C} + AB + B\bar{C}$

FIGURE P–45

27. TRUTH_TABLE ([A, B] -> [X])
        [0, 0] -> [1];
        [0, 1] -> [0];
        [1, 0] -> [0];
        [1, 1] -> [1];

29. Test_Vectors (BCD -> [a, b, c, d, e, f, g, h, i, j])
        0 -> [1, 0, 0, 0, 0, 0, 0, 0, 0, 0];
        1 -> [0, 1, 0, 0, 0, 0, 0, 0, 0, 0];
        2 -> [0, 0, 1, 0, 0, 0, 0, 0, 0, 0];
        3 -> [0, 0, 0, 1, 0, 0, 0, 0, 0, 0];
        4 -> [0, 0, 0, 0, 1, 0, 0, 0, 0, 0];
        5 -> [0, 0, 0, 0, 0, 1, 0, 0, 0, 0];
        6 -> [0, 0, 0, 0, 0, 0, 1, 0, 0, 0];
        7 -> [0, 0, 0, 0, 0, 0, 0, 1, 0, 0];
        8 -> [0, 0, 0, 0, 0, 0, 0, 0, 1, 0];
        9 -> [0, 0, 0, 0, 0, 0, 0, 0, 0, 1];

31. Yes, the design requires only 2 inputs and 8 outputs.

33. Module   Decimal_to_BCD_Priority_Encoder
    Title   'Decimal-to-BCD Priority Encoder in a GAL16V8'

    Encoder_1   device   'P16V8';

    I0, I1, I2, I3, I4   pin 1,2,3,4,5;
    I5, I6, I7, I8, I9   pin 6,7,8,9,12;
    Q0, Q1, Q2, Q3   pin 16,17,18,19;

    Equations
        Q0 = I1 & !I2 & !I4 & !I6 & !I8 # I3 & !I4 & !I6 & !I8 # I5 & !I6 &
            !I8 # I7 & !I8 # I9;
        Q1 = I2 & !I4 & !I5 & !I8 & !I9 # I3 & !I4 & !I5 & !I8 & !I9 #
            I6 & !I8 & !I9 # I7 & !I8 & !I9;
        Q2 = I4 & !I8 & !I9 # I5 & !I8 & !I9 # I6 & !I8 & !I9 # I7 & !I8 &
            !I9;
        Q3 = I8 # I9;

CHAPTER 8

1. See Figure P–46.   **3.** See Figure P–47.

FIGURE P–46

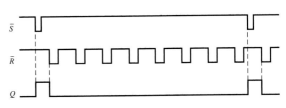

FIGURE P–47

5. See Figure P–48.   **7.** See Figure P–49.   **9.** See Figure P–50.

11. See Figure P–51.   **13.** See Figure P–52.   **15.** See Figure P–53.

17. See Figure P–54.   **19.** See Figure P–55.   **21.** See Figure P–56.

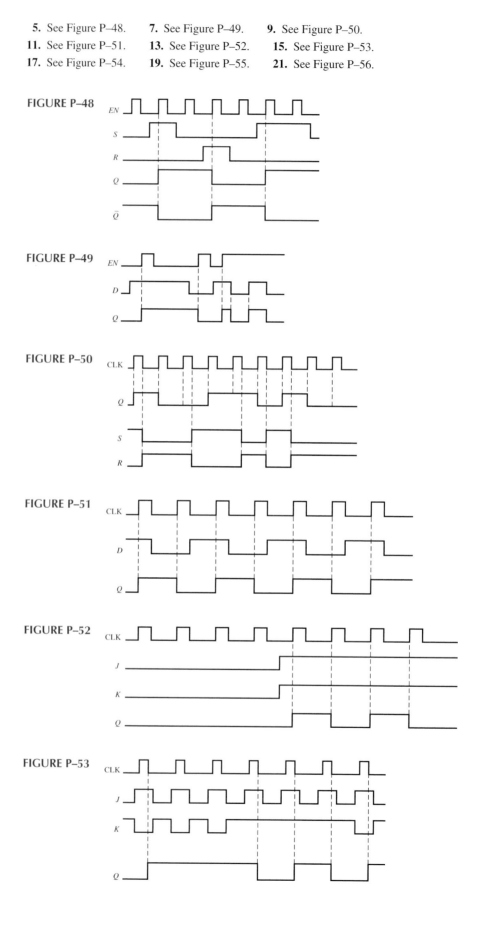

FIGURE P–48

FIGURE P–49

FIGURE P–50

FIGURE P–51

FIGURE P–52

FIGURE P–53

23. See Figure P–57.   **25.** 14.9 MHz   **27.** 150 mA, 750 mW

29. divide-by-2; see Figure P–58.   **31.** 4.62 μs

33. $C_1 = 1\ \mu$F, $R_1 = 227\ k\Omega$ (use 220 kΩ). See Figure P–59.

35. $R_1 = 18\ k\Omega$, $R_2 = 9.1\ k\Omega$.

37. The wire from pin 6 to pin 10 and ground wire are reversed on the protoboard.

FIGURE P–54

FIGURE P–55

FIGURE P–56

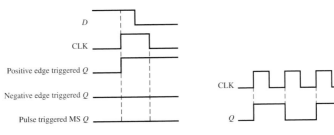

FIGURE P–57

FIGURE P–58

FIGURE P–59

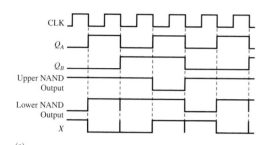

(a)

(b) Same as (a)

(e)

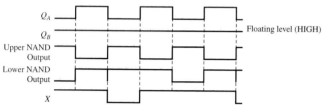

(c)

(d) $X = $ LOW if $Q_B = 1$; $X = \bar{Q}_A$ if $Q_B = 0$

FIGURE P–60

FIGURE P–61

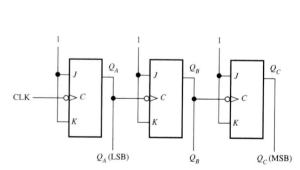

FIGURE P–62

39. $\overline{CLR}$ shorted to ground. **41.** See Figure P–60. Delays not shown.

43. See Figure P–61.
   4 s: $C_1 = 1 \, \mu F$, $R_1 = 3.63 \, M\Omega$ (use 3.9 MΩ)
   25 s: $C_1 = 2.2 \, \mu F$, $R_1 = 10.3 \, M\Omega$ (use 10 MΩ)

45. See Figure P–62.

47. Changes include (1) a 15 s timer (2) decoding to trigger the 15 s timer (3) a decoder to incorporate the turn signal state.

CHAPTER 9

1. See Figure P–63.

3. Worst-case delay is 24 ns; it occurs when all flip-flops change state from 011 to 100 or from 111 to 000.

FIGURE P–63

5. 8 ns

7. Initially, each flip-flop is reset.

At CLK1: $J_0 = K_0 = 1$ Therefore Q_0 goes to a 1.

$J_1 = K_1 = 0$ Therefore Q_1 remains a 0.

$J_2 = K_2 = 0$ Therefore Q_2 remains a 0.

$J_3 = K_3 = 0$ Therefore Q_3 remains a 0.

At CLK2: $J_0 = K_0 = 1$ Therefore Q_0 goes to a 0.

$J_1 = K_1 = 1$ Therefore Q_1 goes to a 1.

$J_2 = K_2 = 0$ Therefore Q_2 remains a 0.

$J_3 = K_3 = 0$ Therefore Q_3 remains a 0.

At CLK3: $J_0 = K_0 = 1$ Therefore Q_0 goes to a 1.

$J_1 = K_1 = 0$ Therefore Q_1 remains a 1.

$J_2 = K_2 = 0$ Therefore Q_2 remains a 0.

$J_3 = K_3 = 0$ Therefore Q_3 remains a 0.

A continuation of this procedure for the next seven clock pulses will show that the counter progresses through the BCD sequence.

9. See Figure P–64.

11. See Figure P–65.

13. See Figure P–66.

15. The sequence is 0000, 1111, 1110, 1101, 1010, 0101. The counter "locks up" in the 1010 and 0101 states and alternates between them.

FIGURE P–64

FIGURE P–65

FIGURE P–66

17. See Figure P–67.   **19.** See Figure P–68.

21. See Figure P–69 for divide-by-10,000. Add one more DIV10 counter to create a divide-by-100,000.

23. See Figure P–70.

25. CLK2, output 0; CLK4, outputs 2, 0; CLK6, output 4; CLK8, outputs 6, 4, 0; CLK10, output 8; CLK12, outputs 10, 8; CLK14, output 12; CLK16, outputs 14, 12, 8

27. A glitch of the AND gate output occurs on the 111 to 000 transition. Eliminate by ANDing $\overline{CLK}$ with counter outputs (strobe) or use Gray code.

FIGURE P–67

FIGURE P–68

FIGURE P–69

FIGURE P–70

29. Hours tens: 0001
Hours units: 0010
Minutes tens: 0000
Minutes units: 0001
Seconds tens: 0000
Seconds units: 0010

31. 64

33. **(a)** Q_0 and Q_1 will not change from their initial state.

(b) normal operation except Q_0 floating

(c) Q_0 waveform is normal; Q_1 remains in initial state.

(d) normal operation

(e) The counter will not change from its initial state.

35. The K input of FF1 must be connected to ground rather than to the J input. Check for a wiring error.

37. Q_0 input to AND gate open and acting as a HIGH

39. See Table P–9.

TABLE P–9

Stage	Open	Loaded Count	f_{out}
1	0	63C1	250.006 Hz
1	1	63C2	250.012 Hz
1	2	63C4	250.025 Hz
1	3	63C8	250.050 Hz
2	0	63D0	250.100 Hz
2	1	63E0	250.200 Hz
2	2	63C0	250 Hz
2	3	63C0	250 Hz
3	0	63C0	250 Hz
3	1	63C0	250 Hz
3	2	67C0	256.568 Hz
3	3	6BC0	263.491 Hz
4	0	73C0	278.520 Hz
4	1	63C0	250 Hz
4	2	63C0	250 Hz
4	3	E3C0	1.383 kHz

41. The decode 6 gate interprets count 4 as a 6 (0110) and clears the counter back to 0 (actually 0010 since Q_1 is open). The apparent sequence of the tens portion of the counter is 0010, 0011, 0010, 0011, 0110.

43. See Figure P–71.

FIGURE P–71

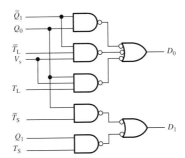

45. Increase the $R_{EXT}C_{EXT}$ time constant of the 25 s one-shot by 2.4 times.

47. See Figure P–72.

49. See Figure P–73.

51. See Figure P–74.

53. See Figure P–75.

FIGURE P–72

FIGURE P–73

FIGURE P–74

FIGURE P–75

FIGURE P–76

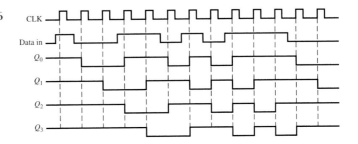

CHAPTER 10

1. Shift registers store binary data.   **3.** See Figure P–76.

5. Initially: 101001111000
    CLK1:   010100111100
    CLK2:   001010011110
    CLK3:   000101001111
    CLK4:   000010100111
    CLK5:   100001010011
    CLK6:   110000101001
    CLK7:   111000010100
    CLK8:   011100001010
    CLK9:   001110000101
    CLK10:   000111000010
    CLK11:   100011100001
    CLK12:   110001110000

7. See Figure P–77.   **9.** See Figure P–78.   **11.** See Figure P–79.

FIGURE P–77

FIGURE P–78

FIGURE P–79

Q_3 through Q_7 remain LOW.

FIGURE P–80

FIGURE P–81

13. See Figure P–80. **15.** See Figure P–81. **17.** See Figure P–82.

19. Initially (76): 01001100

CLK1:	10011000	left
CLK2:	01001100	right
CLK3:	00100110	right
CLK4:	00010011	right
CLK5:	00100110	left
CLK6:	01001100	left
CLK7:	00100110	right
CLK8:	01001100	left
CLK9:	00100110	right
CLK10:	01001100	left
CLK11:	10011000	left

21. See Figure P–83.

23. (a) 3   (b) 5   (c) 7   (d) 8

25. See Figure P–84.

27. See Figure P–85. **29.** An incorrect code may be produced. **31.** D_3 input open

33. (a) No clock at switch closure because of faulty NAND (negative-OR) gate or one-shot; open clock (C) input to key code register; open $SH/\overline{LD}$ input to key code register

(b) Diode in third row open; Q_2 output of ring counter open

(c) The NAND (negative-OR) gate input connected to the first column is open or shorted.

(d) The "2" input to the column encoder is open.

FIGURE P–82

FIGURE P–83

FIGURE P–84

FIGURE P–85

35. (a) Contents of data output register remain constant.

(b) Contents of both registers do not change.

(c) Third stage output of data output register remains HIGH.

(d) Clock generator is disabled after each pulse by the flip-flop being continuously SET and then RESET.

37. shift register A: 1001
shift register B: 11100

39. Gate G_1 has an open or shorted output or an open input. One-shot A is faulty.

41. Control flip-flop: 7476
Clock generator: 555
Counter: 74LS163
Data input register: 74LS164
Data output register: 74LS199
One-shot: 74121

43. See Figure P–86.   **45.** See Figure P–87.

FIGURE P–86

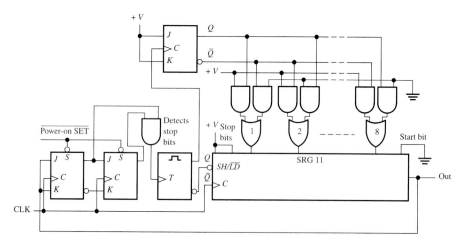

FIGURE P–87

CHAPTER 11

1. Combinational with active-LOW outputs, combinational with active-HIGH outputs, registered with active-LOW outputs, and registered with active-HIGH outputs

3. Multiplexer

5. (a) Pin 15 is a combinational output associated with the variable X.

(b) Pin 20 is a registered output associated with the variable Q_2.

(c) Pin 18 is an active-LOW combinational output associated with the variable A.

(d) Pin 21 is an active-HIGH registered output associated with the variable Q_1.

7. Q1.CLK = Clock;

9.
```
Test_vectors ([Clock,Clear,Din]->[Q0,Q1,Q2,Q3,Q4,Q5,Q6,Qout])
             [ .x. ,  0  ,.x.]->[ 0 , 0 , 0 , 0 , 0 , 0 , 0 ,  0 ];
             [ .c. ,  1  , 1 ]->[ 1 , 0 , 0 , 0 , 0 , 0 , 0 ,  0 ];
             [ .c. ,  1  , 0 ]->[ 0 , 1 , 0 , 0 , 0 , 0 , 0 ,  0 ];
             [ .c. ,  1  , 1 ]->[ 1 , 0 , 1 , 0 , 0 , 0 , 0 ,  0 ];
             [ .c. ,  1  , 0 ]->[ 0 , 1 , 0 , 1 , 0 , 0 , 0 ,  0 ];
             [ .c. ,  1  , 1 ]->[ 1 , 0 , 1 , 0 , 1 , 0 , 0 ,  0 ];
             [ .c. ,  1  , 0 ]->[ 0 , 1 , 0 , 1 , 0 , 1 , 0 ,  0 ];
             [ .c. ,  1  , 1 ]->[ 1 , 0 , 1 , 0 , 1 , 0 , 1 ,  0 ];
             [ .c. ,  1  , 0 ]->[ 0 , 1 , 0 , 1 , 0 , 1 , 0 ,  1 ];
```

11. The test vectors are identical to those for the GAL22V10 in Problem 9.

13. The 8-bit register in Problem 12 cannot be implemented in a GAL16V8 because of insufficient input pins (9 are required; the GAL16V8 has 8).

15. 8-bit counter

17. For an up-only Gray code counter the logic equations are (using ABEL notation)

Q0 =　!Q2 & !Q1 & !Q0 # !Q2 & !Q1 & Q0 # Q2 & Q1 & !Q0 # Q2 & Q1 & Q0

=> !Q2 & !Q1 # Q2 & Q1

Q1 =　!Q2 & !Q1 & Q0 # !Q2 & Q1 & Q0 # !Q2 & Q1 & !Q0 # Q2 & Q1 & !Q0

=> !Q2 & Q0 # Q1 & !Q0

Q2 =　!Q2 & Q1 & !Q0 # Q2 & Q1 & !Q0 # Q2 & Q1 & Q0 # Q2 & !Q1 & Q0

=> Q1 & !Q0 # Q2 & Q0

An ABEL input file for a GAL22V10 circuit is then

```
Module    3-bit_Gray_code_up_counter

Title    '3-bit Gray code up counter in a GAL22V10'

"Device declaration
   Counter   DEVICE    'P22V10'

"Pin Declarations
   Clock,Clear    PIN 1,2;
   Q0,Q1,Q2       Pin 21,22,23 ISTYPE 'reg,buffer';

Equations
   [Q0,Q1,Q2].CLK = Clock;
   [Q0,Q1,Q2].AR = !Clear;
   Q0 := !Q2.FB&!Q1.FB # Q2.FB&Q1.FB;
   Q1 := !Q2.FB&Q0.FB # Q1.FB&!Q0.FB;
   Q2 := Q1.FB&Q0.FB # Q2.FB&Q0.FB;

Test_vectors ([Clock,Clear]->[Q2,Q1,Q0])
             [ .x. ,  0  ]->[ 0 , 0 , 0 ];
             [ .c. ,  1  ]->[ 0 , 0 , 1 ];
             [ .c. ,  1  ]->[ 0 , 1 , 1 ];
             [ .c. ,  1  ]->[ 0 , 1 , 0 ];
             [ .c. ,  1  ]->[ 1 , 1 , 0 ];
             [ .c. ,  1  ]->[ 1 , 1 , 1 ];
             [ .c. ,  1  ]->[ 1 , 0 , 1 ];
             [ .c. ,  1  ]->[ 1 , 0 , 0 ];
             [ .c. ,  1  ]->[ 0 , 0 , 0 ];

   END
```

19.
```
Module    3-bit_Gray_code_up_counter

Title    '3-bit Gray code up counter in a GAL22V10'

"Device declaration
   Counter   DEVICE    'P22V10'
```

```
"Pin Declarations
    Clock,Clear   PIN 1,2;
    Q0,Q1,Q2       Pin 21,22,23 ISTYPE 'reg,buffer';
    COUNT = [Q2,Q1,Q0];
    A     = ^B000;
    B     = ^B001;
    C     = ^B011;
    D     = ^B010;
    E     = ^B110;
    F     = ^B111;
    G     = ^B101;
    H     = ^B100;

Equations
    COUNT.CLK = Clock;
    COUNT.AR = !Clear;

State_diagram COUNT
    State A : GOTO B;
    State B : GOTO C;
    State C : GOTO D;
    State D : GOTO E;
    State E : GOTO F;
    State F : GOTO G;
    State G : GOTO H;
    State H : GOTO A;

Test_vectors ([Clock,Clear]->[Q2,Q1,Q0])
            [ .x. ,  0  ]->[ 0 , 0 , 0 ];
            [ .c. ,  1  ]->[ 0 , 0 , 1 ];
            [ .c. ,  1  ]->[ 0 , 1 , 1 ];
            [ .c. ,  1  ]->[ 0 , 1 , 0 ];
            [ .c. ,  1  ]->[ 1 , 1 , 0 ];
            [ .c. ,  1  ]->[ 1 , 1 , 1 ];
            [ .c. ,  1  ]->[ 1 , 0 , 1 ];
            [ .c. ,  1  ]->[ 1 , 0 , 0 ];
            [ .c. ,  1  ]->[ 0 , 0 , 0 ];

END
```

21.
```
Module   3-bit_Gray_code_up_counter

Title    '3-bit Gray code up counter in a GAL16V8'

"Device declaration
    Counter   DEVICE   'P16V8'

"Pin Declarations
    Clock,Clear  PIN 1,11;
    Q0,Q1,Q2     Pin 17,18,19 ISTYPE 'reg,buffer';

Equations
    [Q0,Q1,Q2].CLK = Clock;
    [Q0,Q1,Q2].AR  = !Clear;

Truth_table ([Clock,Q2,Q1,Q0]:>[Q2,Q1,Q0])
            [ .c. ,0 ,0 ,0] :>[ 0 , 0 , 1 ];
            [ .c. ,0 ,0 ,1] :>[ 0 , 1 , 1 ];
            [ .c. ,0 ,1 ,1] :>[ 0 , 1 , 0 ];
            [ .c. ,0 ,1 ,0] :>[ 1 , 1 , 0 ];
            [ .c. ,1 ,1 ,0] :>[ 1 , 1 , 1 ];
            [ .c. ,1 ,1 ,1] :>[ 1 , 0 , 1 ];
            [ .c. ,1 ,0 ,1] :>[ 1 , 0 , 0 ];
            [ .c. ,1 ,0 ,0] :>[ 0 , 0 , 0 ];
```

```
Test_vectors ([Clock,Clear]->[Q2,Q1,Q0])
            [ .x. ,  0  ]->[ 0 , 0 , 0 ];
            [ .c. ,  1  ]->[ 0 , 0 , 1 ];
            [ .c. ,  1  ]->[ 0 , 1 , 1 ];
            [ .c. ,  1  ]->[ 0 , 1 , 0 ];
            [ .c. ,  1  ]->[ 1 , 1 , 0 ];
            [ .c. ,  1  ]->[ 1 , 1 , 1 ];
            [ .c. ,  1  ]->[ 1 , 0 , 1 ];
            [ .c. ,  1  ]->[ 1 , 0 , 0 ];
            [ .c. ,  1  ]->[ 0 , 0 , 0 ];

END
```

23. **(a)** CLOSE1 state: System goes to REST1 state.

  (b) REST1 state: System stays in REST1 state.

  (c) UP state: System stays in UP state until "ARRIVE" signal.

  (d) REST2 state: System stays in REST2 state.

  (e) CLOSE2 state: System goes to REST2 state.

  (f) DOWN state: System stays in DOWN state until "ARRIVE" signal.

25. Module   Elevator_state_control

```
    Title    'Control logic for a two-story elevator system'

    "Device declaration
        Elevator_cont    DEVICE    'P22V10'

    "Pin declarations
        CLK, !OE                   PIN 1,13;
        REQ1,REQ2                  PIN 2,3;
        FLOOR2,FLOOR1,OPEN,ARRIVE  PIN 4,5,6,7;
        L1PB,L1PB_                 PIN 20,21 ISTYPE 'com,buffer';
        L2PB,L2PB_                 PIN 22,23 ISTYPE 'com,buffer';
        DOOR,MOTION,DIR            PIN 14,15,16 ISTYPE 'reg,buffer';

    "State definitions
        CONSTATE = [DOOR,MOTION,DIR];
        REST1    = ^B000;
        OPEN1    = ^B100;
        UP       = ^B110;
        REST2    = ^B001;
        OPEN2    = ^B101;
        DOWN     = ^B111;

    Equations
        CONSTATE.CLK = CLK;
        CONSTATE.AR  = OE;
        L1PB         = REQ1 # !DIR.FB&OPEN # DIR.FB&FLOOR1 # !L1PB_;
        L1PB_        = !DOOR.FB&!MOTION.FB&!DIR.FB # !L1PB;
        L2PB         = REQ2 # DIR.FB&OPEN # !DIR.FB&FLOOR2 # !L2PB_;
        L2PB_        = !DOOR.FB&!MOTION.FB&DIR.FB # !L2PB;

    State_diagram CONSTATE
        State REST1  : if (L2PB) then UP else CLOSE1;
        State CLOSE1 : if (L2PB) then UP else if (L1PB) then REST1 else CLOSE1;
        State UP     : if (ARRIVE) then REST2 else UP;
        State REST2  : if (L1PB) then DOWN else CLOSE2;
        State CLOSE2 : if (L1PB) then DOWN else if (L2PB) then REST2 else CLOSE2;
        State DOWN   : if (ARRIVE) then REST1 else DOWN;
```

```
Test_vectors
    ([CLK,REQ1,REQ2,FLOOR2,FLOOR1,OPEN,ARRIVE]->[DOOR,MOTION,DIR])
    [.c., 0 , 0 , 0 , 0 , 0 , 1 ]->[ .x. , .x. , .x.];
    [.c., 1 , 0 , 0 , 0 , 0 , 1 ]->[ .x. , .x. , .x.];
    [.c., 1 , 0 , 0 , 0 , 0 , 1 ]->[ 0 , 0 , 0 ];
    [.c., 0 , 0 , 0 , 0 , 0 , 0 ]->[ 1 , 0 , 0 ];
    [.c., 1 , 0 , 0 , 0 , 0 , 0 ]->[ 0 , 0 , 0 ];
    [.c., 0 , 0 , 0 , 0 , 0 , 1 ]->[ 1 , 0 , 0 ];
    [.c., 0 , 0 , 0 , 0 , 1 , 0 ]->[ 0 , 0 , 0 ];
    [.c., 0 , 1 , 0 , 0 , 0 , 0 ]->[ 1 , 1 , 0 ];
    [.c., 0 , 0 , 0 , 0 , 0 , 1 ]->[ 0 , 0 , 1 ];
    [.c., 0 , 0 , 0 , 0 , 0 , 0 ]->[ 1 , 0 , 1 ];
    [.c., 0 , 1 , 0 , 0 , 0 , 0 ]->[ 0 , 0 , 1 ];
    [.c., 0 , 0 , 0 , 0 , 0 , 0 ]->[ 1 , 0 , 1 ];
    [.c., 0 , 0 , 0 , 0 , 1 , 0 ]->[ 0 , 0 , 1 ];
    [.c., 1 , 0 , 0 , 0 , 0 , 0 ]->[ 1 , 1 , 1 ];
    [.c., 0 , 0 , 0 , 0 , 0 , 1 ]->[ 0 , 0 , 0 ];
    [.c., 0 , 0 , 1 , 0 , 0 , 0 ]->[ 1 , 1 , 0 ];
    [.c., 0 , 0 , 0 , 0 , 0 , 1 ]->[ 0 , 0 , 1 ];
    [.c., 0 , 0 , 0 , 1 , 0 , 0 ]->[ 1 , 1 , 1 ];
    [.c., 0 , 0 , 0 , 0 , 0 , 1 ]->[ 0 , 0 , 0 ];
    [.c., 0 , 0 , 0 , 0 , 0 , 0 ]->[ 1 , 0 , 0 ];
    [.c., 0 , 1 , 0 , 0 , 0 , 0 ]->[ 1 , 1 , 0 ];
    [.c., 0 , 0 , 0 , 0 , 0 , 1 ]->[ 0 , 0 , 1 ];
    [.c., 0 , 0 , 0 , 0 , 0 , 0 ]->[ 1 , 0 , 1 ];
    [.c., 1 , 0 , 0 , 0 , 0 , 0 ]->[ 1 , 1 , 1 ];
    [.c., 0 , 0 , 0 , 0 , 0 , 1 ]->[ 0 , 0 , 0 ];
    [.c., 0 , 0 , 0 , 0 , 0 , 0 ]->[ 1 , 0 , 0 ];
    [.c., 0 , 0 , 1 , 0 , 0 , 0 ]->[ 1 , 1 , 0 ];
    [.c., 0 , 0 , 0 , 0 , 0 , 1 ]->[ 0 , 0 , 1 ];
    [.c., 0 , 0 , 0 , 0 , 0 , 0 ]->[ 1 , 0 , 1 ];
    [.c., 0 , 0 , 0 , 1 , 0 , 0 ]->[ 1 , 1 , 1 ];
    [.c., 0 , 0 , 0 , 0 , 0 , 1 ]->[ 0 , 0 , 0 ];

END
```

27. No, Ten inputs are required and the GAL16V8 has only eight.

29. The first state will be skipped if there is a vehicle on the side street. The third state will always be skipped.

31.
```
Module    8-bit_SISO_shift_register

Title    '8-bit SISO shift register in a GAL16V8'

"Device declaration
    Register    DEVICE    'P16V8'

"Pin declarations
    Clock,Clear      PIN 1,11;
    Din              PIN 2;
    Q0,Q1,Q2,Q3      PIN 12,13,14,15 ISTYPE 'reg,buffer';
    Q4,Q5,Q6,Qout    PIN 16,17,18,19 ISTYPE 'reg,buffer';

Equations
    [Q0,Q1,Q2,Q3,Q4,Q5,Q6,Qout].CLK = Clock;
    [Q0,Q1,Q2,Q3,Q4,Q5,Q6,Qout].AR  = !Clear;
    Q0 := Din;
    Q1 := Q0.FB;
    Q2 := Q1.FB;
    Q3 := Q2.FB;
    Q4 := Q3.FB;
    Q5 := Q4.FB;
    Q6 := Q5.FB;
    Qout := Q6.FB;
```

```
Test_vectors ([Clock,Clear,Din]->[Q0,Q1,Q2,Q3,Q4,Q5,Q6,Qout])
                [ .x. ,  0  ,.x.]->[ 0, 0, 0, 0, 0, 0, 0,  0 ];
                [ .c. ,  1  , 1 ]->[ 1, 0, 0, 0, 0, 0, 0,  0 ];
                [ .c. ,  1  , 0 ]->[ 0, 1, 0, 0, 0, 0, 0,  0 ];
                [ .c. ,  1  , 1 ]->[ 1, 0, 1, 0, 0, 0, 0,  0 ];
                [ .c. ,  1  , 0 ]->[ 0, 1, 0, 1, 0, 0, 0,  0 ];
                [ .c. ,  1  , 1 ]->[ 1, 0, 1, 0, 1, 0, 0,  0 ];
                [ .c. ,  1  , 0 ]->[ 0, 1, 0, 1, 0, 1, 0,  0 ];
                [ .c. ,  1  , 1 ]->[ 1, 0, 1, 0, 1, 0, 1,  0 ];
                [ .c. ,  1  , 0 ]->[ 0, 1, 0, 1, 0, 1, 0,  1 ];

     END
```

33.
```
Module    4-bit_Excess-3_code_counter

Title    '4-bit Excess-3 code counter in a GAL22V10'

"Device declaration
     Counter   DEVICE    'P22V10'

"Pin declarations
     Clock,Reset    PIN 1,2;
     Q0,Q1,Q2,Q3    PIN 12,13,14,15 ISTYPE 'reg,buffer';
     COUNT = [Q3,Q2,Q1,Q0];

Equations
     COUNT.CLK = Clock;

Truth_table([Clock,Reset, COUNT  ]:>[ COUNT ])
              [ .C. ,  0  ,  .X.   ]:>[ ^B0011];
              [ .C. ,  1  , ^B0000 ]:>[ ^B0011];
              [ .C. ,  1  , ^B0001 ]:>[ ^B0011];
              [ .C. ,  1  , ^B0010 ]:>[ ^B0011];
              [ .C. ,  1  , ^B0011 ]:>[ ^B0100];
              [ .C. ,  1  , ^B0100 ]:>[ ^B0101];
              [ .C. ,  1  , ^B0101 ]:>[ ^B0110];
              [ .C. ,  1  , ^B0110 ]:>[ ^B0111];
              [ .C. ,  1  , ^B0111 ]:>[ ^B1000];
              [ .C. ,  1  , ^B1000 ]:>[ ^B1001];
              [ .C. ,  1  , ^B1001 ]:>[ ^B1010];
              [ .C. ,  1  , ^B1010 ]:>[ ^B1011];
              [ .C. ,  1  , ^B1011 ]:>[ ^B1100];
              [ .C. ,  1  , ^B1100 ]:>[ ^B0000];
              [ .C. ,  1  , ^B1101 ]:>[ ^B0011];
              [ .C. ,  1  , ^B1110 ]:>[ ^B0011];
              [ .C. ,  1  , ^B1111 ]:>[ ^B0011];

Test_vectors([Clock,Reset ]->[ COUNT])
              [ .C. ,  0   ]->[^B0011];
              [ .C. ,  1   ]->[^B0100];
              [ .C. ,  1   ]->[^B0101];
              [ .C. ,  1   ]->[^B0110];
              [ .C. ,  1   ]->[^B0111];
              [ .C. ,  1   ]->[^B1000];
              [ .C. ,  1   ]->[^B1001];
              [ .C. ,  1   ]->[^B1010];
              [ .C. ,  1   ]->[^B1011];
              [ .C. ,  1   ]->[^B1100];
              [ .C. ,  1   ]->[^B0011];

     END
```

CHAPTER 12

1. (a) ROM   **(b)** RAM

3. *Address bus* provides for transfer of address code to memory for accessing any memory location in any order for a read or write operation. *Data bus* provides for transfer of data between the microprocessor and the memory or I/O.

5.

	Bit 0	Bit 1	Bit 2	Bit 3
Row 0	1	0	0	0
Row 1	0	0	0	0
Row 2	0	0	1	0
Row 3	0	0	0	0

7. 512 rows $\times$ 128 8-bit columns

9. A SRAM stores bits in flip-flops indefinitely as long as power is applied. A DRAM stores bits in capacitors which must be refreshed periodically to retain the data.

11. See Table P–10.　**13.** See Figure P–88.

TABLE P–10

Inputs		Outputs			
A_1	A_0	O_3	O_2	O_1	O_0
0	0	0	1	0	1
0	1	1	0	0	1
1	0	1	1	1	0
1	1	0	0	1	0

FIGURE P–88

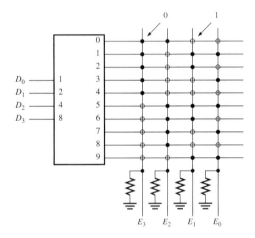

15. Blown links: 1–17, 19–23, 25–31, 34, 37, 38, 40–47, 53, 55, 58, 61, 62, 63, 65, 67, 69

17. Use eight 16k $\times$ 4 RAMs with sixteen address lines. Two of the address lines are decoded to enable the selected memory chips. Four data lines go to each chip.

19. 8 bits, 64k words; 4 bits, 256k words

21. lowest address: $FC0_{16}$
highest address: FFF_{16}

23. A hard disk is formatted into tracks and sectors. Each track is divided into a number of sectors with each sector of a track having a physical address. Hard disks typically have from a few hundred to a few thousand tracks.

25. Magnetic tape has a longer access time than disk because data must be accessed sequentially rather than randomly.

27. Checksum content is in error.

29. (a) ROM 2　**(b)** ROM 1　**(c)** All ROMs

31. 010

33. PROM will retain code when power is off. The code in PROM cannot be readily changed.

35. See Figure P–89.

37. To accommodate a 5-bit entry code, shift register C must be loaded with five 0s instead of four. Connect pin 14 of the 74LS165 to ground rather than to the HIGH.

FIGURE P–89

CHAPTER 13

1. 1000, 1010, 1100, 1110, 1110, 1101, 1011, 1001, 0110, 0011, 0010, 0001, 0001, 0010, 0101, 0111, 1011, 1100, 1110, 1110, 1101, 1011, 1001, 0110, 0011, 0001.

3. See Figure P–90.   **5.** 330 kΩ   **7.** See Figure P–91.

FIGURE P–90

FIGURE P–91

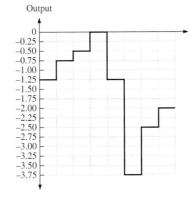

9. (a) 14.29%   **(b)** 0.098%   **(c)** 0.00038%

11. 000, 001, 100, 110, 101, 100, 011, 010, 001, 001, 011, 110, 111, 111, 111, 111, 111, 111, 111, 100

13. 1111, 1100, 1000, 1000, 0011, 0010, 0100, 1110

15. See Table P–11.   **17.** See Figure P–92.

19. See Figure P–93.   **21.** See Figure P–94.

23. The local bus is the collection of buses interfacing directly with the processor. The PCI bus is used for expansion devices and is connected to the local bus through a bus controller.

25. The PCI bus is a 33- or 66 MHz, 32- or 64-bit expansion bus. The ISA bus is an 8.33 MHz, 8- or 16-bit expansion bus.

27. DCE stands for data communications equipment, such as a modem. DTE stands for data terminal equipment, such as a computer. Both acronyms are associated with the RS-232/EIA-232 standard.

29. seven

31. A controller is sending data to two listeners. The first two bytes of data (3F and 41) go to the listener with address 001A. The second two bytes go to the listener with address 001B. The handshaking signals ($\overline{DAV}$, $\overline{NRFD}$, and $\overline{NDAC}$) indicate the data transfer is successful. See Figure P–95.

33. 32.2 kHz   **35.** 0.71°

TABLE P–11

SAR	Comment
1000	Greater than V_{in}, reset MSB
0100	Less than V_{in}, keep the 1
0110	Equal to V_{in}, keep the (final state)

FIGURE P–92

FIGURE P–93

FIGURE P–94

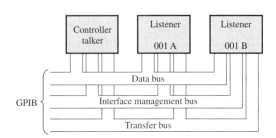

FIGURE P–95

CHAPTER 14

1. Arithmetic logic unit, register array, control unit

3. Address: a one-way bus that carries address information to the memory or to the I/O.

   Data: a two-way bus on which data are transferred into the microprocessor or on which the result of an operation is sent out from the microprocessor.

   Control: a bus used to coordinate the microprocessor's operations and to communicate with external devices.

5. **(a)** 80386   **(b)** over 4 Gbytes

7. 68020

9. Fetch, decode, execute

11. CS, DS, SS, ED, FS, GS

13. AH and AL are 8-bit registers and represent the high and low part of the 16-bit AX register. The EAX is a 32-bit register which includes the AX register as the lower 16 bits.

15. **(a)** Pairing allows two instructions to execute at the same time.

    (b) Instruction pairing requires two execution units; the 80486 has only one.

17. See Figure P–96.

FIGURE P–96

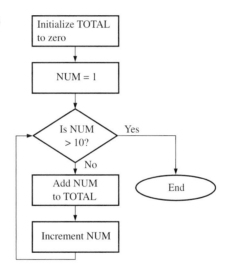

19. The directive tells the assembler that the location pointed to is 16 bits.

21. *Data transfer:* These instructions are used to copy a byte, a word (16 bits), or a double word (32 bits) between various sources and destinations. An example is the MOV instruction.

    Arithmetic: These are instructions for addition, subtraction, multiplication, and division with signed or unsigned numbers. An example is the ADD instruction.

    Bit manipulation: These instructions include logical operations, shifts, and rotations. An example is the OR instruction.

    Loops and jumps: These instructions are designed to alter the normal sequence of instructions. An example is the JNZ instruction.

    Strings: A string instruction can copy, load, store, compare, or scan a string—either as a byte at a time or a word at a time. An example is the MOVSW instruction.

    Subroutine and interrupts: A subroutine is a miniprogram that can be used repeatedly but programmed only once. An example is the CALL instruction.

    Processor control: These instructions allow direct control of some of the processor's flags and other miscellaneous tasks. An example is the STC instruction.

23. A_0–A_{15}

25. Status bits signal the bus controller for an I/O or read operation.

27. 400H–4FFH

29. Memory write

31. 64k

33. In a polled I/O, the CPU polls each device in turn to see if it needs service; in an interrupt-driven system, the peripheral device signals the CPU when it requires service.

35. A program instruction that invokes an interrupt service routine.

CHAPTER 15

1. No; $V_{OH(min)} < V_{IH(min)}$

3. 0.15 V in HIGH state; 0.25 V in LOW state.

5. Gate C **7.** 16 ns   **9.** Gate C **11.** Yes, G_2

13. (a) on   (b) off   (c) off   (d) on

15. See Figure P–97 for one possible circuit.

17. (a) HIGH   (b) Floating   (c) HIGH   (d) High-Z

19. (a) LOW   (b) LOW   (c) LOW   **21.** See Figure P–98.

23. (a) $R_p = 198\ \Omega$ (b) $R_p = 198\ \Omega$ (c) $R_p = 198\ \Omega$

25. ALVC

27. (a) A, B to X: 9.9 ns   (b) A to X_1, X_2, X_3: 14 ns   (c) A to X: 11.1 ns
  C, D to X: 6.6 ns B to X_1: 7 ns B to X: 11.1 ns
                                   C to X_2: 7 ns C to X: 7.4 ns
                                   D to X_3; 7 ns D to X: 7.4 ns

29. ECL operates with nonsaturated BJTs

FIGURE P–97   74HC125 (Tristate)

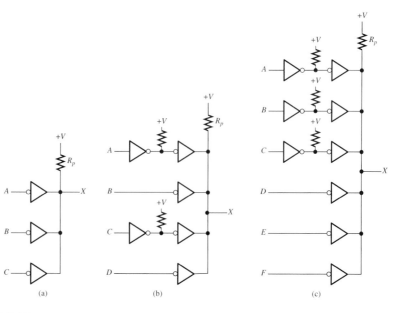

FIGURE P–98

GLOSSARY

ABEL Advanced Boolean Expression Language. A software compiler language for PLD programming; a type of hardware description language (HDL).

Acceptor A receiving device on a bus.

Access time The time from the application of a valid memory address to the appearance of valid output data.

Accumulator A general-purpose register used for arithmetic and logic operations.

Addend In addition, the number that is added to another number called the augend.

Adder A logic circuit used to add two binary numbers.

Address The location of a given storage cell or group of cells in a memory; a unique memory location containing one byte.

Address allocation The arrangement of memory space (addresses) in a computer.

Address bus Generally, a one-way group of conductors from the microprocessor to memory, containing the address information.

Adjacency Characteristic of cells in a Karnaugh map in which there is a single-variable change from one cell to another cell next to it on any of its four sides.

Alphanumeric Consisting of numerals, letters, and other characters.

ALU Arithmetic logic unit; the key processing element of a microprocessor that performs arithmetic and logic operations.

Amplitude In a pulse waveform, the height or maximum value of the pulse as measured from its low level.

Analog Being continuous or having continuous values, as opposed to having a set of discrete values.

Analog-to-digital (A/D) conversion The process of converting an analog signal to digital form.

Analog-to-digital converter (ADC) A device used to convert an analog signal to a sequence of digital codes.

AND A basic logic operation in which a true (HIGH) output occurs only if all the input conditions are true (HIGH).

AND gate A logic gate that produces a HIGH output only when all of the inputs are HIGH.

ANSI American National Standards Institute.

Architecture The internal functional arrangement of the elements that give a device its particular operating characteristics.

Array In a PLD, a matrix formed by rows of product-term lines and columns of input lines with a programmable cell at each junction.

ASCII American Standard Code for Information Interchange; the most widely used alphanumeric code.

Assembler A program that converts English-like mnemonics into machine code.

Assembler directive An instruction to the assembler; also called a pseudo-op.

Assembly language A form of computer program language in which statements expressed in mnemonics represent the instructions.

Associative law In addition (ORing) and multiplication (ANDing) of three or more variables, the order in which the variables are grouped makes no difference.

Astable Having no stable state. An astable multivibrator oscillates between two quasi-stable states.

Asynchronous Having no fixed time relationship.

Asynchronous counter A type of counter in which each stage is clocked from the output of the preceding stage.

Augend In addition, the number to which the addend is added.

Base One of the three regions in a bipolar junction transistor.

Base address The beginning address of a segment of memory.

BCD Binary coded decimal; a digital code in which each of the decimal digits, 0 through 9, is represented by a group of four bits.

BEDO DRAM Burst extended data output dynamic random-access memory.

Bidirectional Having two directions. In a bidirectional shift register, the stored data can be shifted right or left.

Binary Having two values or states; describes a number system that has a base of two and utilizes 1 and 0 as its digits.

BIOS Basic input/output system; a set of programs in ROM that interfaces the I/O devices in a computer system.

Bipolar Having two opposite charge carriers within the transistor structure.

Bistable Having two stable states. Flip-flops and latches are bistable multivibrators.

Bit A binary digit, which can be either a 1 or 0.

Bit time The interval of time occupied by a single bit in a sequence of bits; the period of the clock.

BIU Bus interface unit; the portion of the CPU that interfaces with the system buses and fetches instructions, reads operands, and writes results.

BJT Bipolar junction transistor; a semiconductor device used for switching or amplification. A BJT has two junctions, the base-emitter junction and the base-collector junction.

Boolean addition In Boolean algebra, the OR operation.

Boolean algebra The mathematics of logic circuits.

Boolean expression An expression of variables and operators used to express the operation of a logic circuit.

Boolean multiplication In Boolean algebra, the AND operation.

Buffer A circuit that prevents loading of an input or output.

Bus A set of interconnections that interface one or more devices based on a standardized specification.

Bus arbitration The process that prevents two sources from using a bus at the same time.

Bus contention An adverse condition that could occur if two or more devices try to communicate at the same time on a bus.

Byte A group of eight bits.

Cache memory A relatively small, high-speed memory that stores the most recently used instructions or data from the larger but slower main memory.

Capacity The total number of data units (bits, nibbles, bytes, words) that a memory can store.

Carry The digit generated when the sum of two binary digits exceeds 1.

Carry generation The process of producing an output carry in a full-adder when both input bits are 1s.

Carry propagation The process of rippling an input carry to become the output carry in a full-adder when either or both of the input bits are 1s and the input carry is a 1.

Cascade To connect "end-to-end" as when several counters are connected from the terminal count output of one counter to the enable input of the next counter.

Cascading Connecting the output of one device to the input of a similar device, allowing one device to drive another in order to expand the operational capability.

CCD Charge-coupled device; a type of semiconductor memory that stores data in the form of charge packets and is serially accessed.

CD-R CD-Recordable; an optical disk storage device on which data can be stored once.

CD-ROM An optical disk storage device on which data are prestored and can only be read.

CD-RW CD-Rewritable; an optical disk storage on which data can be written and overwritten many times.

Cell An area on a Karnaugh map that represents a unique combination of variables in product form; a single storage element in a memory; a fused cross point of a row and column in a PLD; a single storage element in a memory.

Character A symbol, letter, or numeral.

Circuit An arrangement of electrical and/or electronic components interconnected in such a way as to perform a specified function.

Clear An asynchronous input used to reset a flip-flop (make the Q output 0); to place a register or counter in the state in which it contains all 0s.

Clock The basic timing signal in a digital system.

CMOS Complementary metal oxide semiconductor; a class of integrated logic circuits that is implemented with a type of field-effect transistor.

Code A set of bits arranged in a unique pattern and used to represent such information as numbers, letters, and other symbols.

Code converter A digital device that converts one type of coded information into another coded form.

Collector One of the three regions in a bipolar transistor.

Combinational logic A combination of logic gates interconnected to produce a specified Boolean function with no storage or memory capability; sometimes called *combinatorial logic*.

Commutative law In addition (ORing) and multiplication (ANDing) of two variables, the order in which the variables are ORed or ANDed makes no difference.

Comparator A digital circuit that compares the magnitudes of two quantities and produces an output indicating the relationship of the quantities.

Compiler A program that translates high-level program statements, such as BASIC, Pascal, or Fortran, into machine language.

Complement The inverse or opposite of a number; in Boolean algebra, the inverse function, expressed with a bar over the variable. The complement of a 1 is a 0, and vice versa.

Complementation Inversion. LOW is the complement of HIGH, and 0 is the complement of 1.

Contiguous Joined together.

Control bus A one-way set of conductors that connects the CPU to other parts of the computer to coordinate its operations and to communicate with external devices.

Controller An instrument that can specify each of the other instruments on the bus as either a talker or a listener for the purpose of data transfer.

Control unit The portion within the microprocessor that provides the timing and control signals for getting data into and out of the microprocessor and for synchronizing the execution of instructions.

Coprocessor A microprocessor designed with a limited instruction set optimized to perform arithmetic operations very quickly; it generally works in conjunction with a general-purpose microprocessor.

Counter A digital circuit capable of counting electronic events, such as pulses, by progressing through a sequence of binary states.

CPU Central processing unit (also called the MPU); the main part of a computer responsible for control and processing of data.

CUPL Compiler for Universal Programmable Logic. A software compiler language for PLD programming; a type of hardware description language.

Current sinking The action of a circuit in which it accepts current into its output from a load.

Current sourcing The action of a circuit in which it sends current out of its output and into a load.

DAT Digital audio tape; a type of magnetic tape format.

Data Information in numeric, alphabetic, or other form.

Data bus A bidirectional set of conductive paths on which data or instruction codes are transferred into the microprocessor or on which the result of an operation or computation is sent out from the microprocessor.

Data selector A circuit that selects data from several inputs one at a time in a sequence and places them on the output; also called a multiplexer.

Data sheet A document that specifies parameter values and operating conditions for an integrated circuit or other device.

DCE Data communications equipment.

Debug A code within DOS that allows various operations on files. It includes a primitive assembler.

Decade Characterized by ten states or values.

Decade counter A digital counter having ten states.

Decimal Describes a number system with a base of ten.

Decoder A digital circuit (device) that converts coded information into another (familiar) or noncoded form.

Decrement To decrease the binary state of a counter by one.

D flip-flop A type of bistable multivibrator in which the output assumes the state of the D input on the triggering edge of a clock pulse.

Demultiplexer (DEMUX) A circuit (digital device) that switches digital data from one input line to several output lines in a specified time sequence.

Dependency notation A notational system for logic symbols that specifies input and output relationships, thus fully defining a given function; an integral part of ANSI/IEEE Std. 91–1984.

Difference The result of a subtraction.

Digit A symbol used to express a quantity.

Digital Related to digits or discrete quantities; having a set of discrete values as opposed to continuous values.

Digital-to-analog (D/A) conversion The process of converting a sequence of digital codes to an analog form.

Digital-to-analog converter (DAC) A device in which information in digital form is converted to analog form.

DIMM Dual in-line memory module.

Diode A semiconductor device that conducts current in only one direction.

DIP Dual in-line package. A type of IC package whose leads must pass through holes to the other side of a PC board.

Distributive law The law that states that ORing several variables and then ANDing the result with a single variable is equivalent to ANDing the single variable with each of the several variables and then ORing the product.

Dividend In a division operation, the quantity that is being divided.

Divisor In a division operation, the quantity that is divided into the dividend.

DLT Digital linear tape; a type of magnetic tape format.

DMA Direct memory access; a method to directly interface a peripheral device to memory without using the CPU for control.

Documentation file The information from a computer that documents the final design after the input file has been processed.

Domain All of the variables in a Boolean expression.

"Don't care" A combination of input literals that cannot occur and are used as 1s or 0s on a Karnaugh map.

Drain One of the terminals of a field-effect transistor.

DRAM Dynamic random-access memory; a type of semiconductor memory that uses capacitors as the storage elements and is a volatile, read/write memory.

DTE Data terminal equipment.

Dual symbols The negative-AND is the dual symbol for the NOR gate, and the negative-OR is the dual symbol for the NAND gate.

Duty cycle The ratio of pulse width to period expressed as a percentage.

DVD-ROM Digital versatile disk-ROM; also known as digital video disk-ROM; a type of optical storage device on which data is prestored with a much higher capacity than a CD-ROM.

Dynamic memory A type of semiconductor memory having capacitive storage cells that lose stored data over a period of time and, therefore, must be refreshed.

ECL Emitter-coupled logic; a class of integrated logic circuit that is implemented with nonsaturating bipolar junction transistors.

E^{2}CMOS Electrically erasable CMOS (EECMOS). The circuit technology used for the reprogrammable cells in a GAL.

Edge-triggered flip-flop A type of flip-flop in which the data are entered and appear on the output on the same clock edge.

EDO DRAM Extended data output dynamic random-access memory.

EEPROM Electrically erasable programmable read-only memory; a type of semiconductor memory device.

Emitter One of the three regions in a bipolar junction transistor.

Enable To activate or put into an operational mode; an input on a logic circuit that enables its operation.

Encoder A digital circuit (device) that converts information to a coded form.

EPROM Erasable programmable read-only memory; a type of semiconductor memory device that typically uses ultraviolet light to erase data.

Error detection The process of detecting bit errors in a digital code.

EU Execution unit; the portion of a CPU that executes instructions; it contains the arithmetic logic unit (ALU), the general registers, and the flags.

Even parity The condition of having an even number of 1s in every group of bits.

Exclusive-NOR (XNOR) gate A logic gate that produces a LOW only when the two inputs are at opposite levels.

Exclusive-OR (XOR) A basic logic operation in which a HIGH occurs when the two inputs are at opposite levels.

Exclusive-OR (XOR) gate   A logic gate that produces a HIGH only when the two inputs are at opposite levels.

Execute   A CPU process in which an instruction is carried out.

Exponent   The part of a floating-point number that represents the number of places that the decimal point (or binary point) is to be moved.

Fall time   The time interval between the 90% point and the 10% point on the negative-going edge of a pulse.

Fan-out   The number of equivalent gate inputs of the same family series that a logic gate can drive.

Feedback   The output voltage or a portion of it that is connected back to the input of a circuit.

FET   Field-effect transistor.

Fetch   A CPU process in which an instruction is obtained from the memory.

FIFO   First in—first out memory.

Flag   A bit that indicates the result of an arithmetic or logic operation or is used to alter an operation.

Flash ADC   A simultaneous analog-to-digital converter.

Flash memory   A nonvolatile read/write random-access semiconductor memory in which data is stored as charge on the floating gate of a certain FET.

Flat pack   An SMT package whose leads extend straight out from the body.

Flip-flop   A basic storage circuit that can store only one bit at a time; a synchronous bistable device.

Floating-point number   A number representation based on scientific notation in which the number consists of an exponent and a mantissa.

Floppy disk   A magnetic storage device; a flexible disk with a diameter of 3.5 inches and a storage capacity of 1.44 Mbytes encased in a rigid plastic housing.

Forward bias   A voltage polarity condition that allows a semiconductor *pn* junction in a transistor or diode to conduct current.

FPM DRAM   Fast page mode dynamic random-access memory.

Frequency   The number of pulses in one second for a periodic waveform. The unit of frequency is the hertz.

Full-adder   A digital circuit that adds two bits and an input carry to produce a sum and an output carry.

Fuse   The programmable element in certain types of PLDs; also called a fusible link.

GAL   Generic array logic. A PLD with a reprogrammable AND array, a fixed OR array, and programmable output logic macrocells.

Gate   A logic circuit that performs a specified logic operation, such as AND or OR; one of the three terminals of a field-effect transistor.

Glitch   A voltage or current spike of short duration, usually unintentionally produced and unwanted.

Global cell   A programmable cell in a PLD array that affects all of the OLMCs when programmed.

GPIB   General-purpose interface bus based on the IEEE 488 standard.

Gray code   An unweighted digital code characterized by a single bit change between adjacent code numbers in a sequence.

Half-adder   A digital circuit that adds two bits and produces a sum and an output carry. It cannot handle input carries.

Handshaking   The process of signal interchange by which two digital devices or systems jointly establish communication.

Hard disk   A magnetic disk storage device; typically, a stack of two or more rigid disks enclosed in a sealed housing.

Hardware   The circuitry and physical components of a computer system (as opposed to the directions called software).

HDL   Hardware Description Language. ABEL and CUPL are examples.

Hexadecimal   Describes a number system with a base of 16.

High-Z   The high-impedance state of a tristate circuit in which the output is effectively disconnected from the rest of the circuit.

Hold time   The time interval required for the control levels to remain on the inputs to a flip-flop after the triggering edge of the clock in order to reliably activate the device.

HPIB   Hewlett-Packard interface bus; same as GPIB (general-purpose interface bus).

Hysteresis   A characteristic of a threshold-triggered circuit, such as the Schmitt trigger, where the device turns on and off at different input levels.

IEEE   Institute of Electrical and Electronic Engineers.

IEEE 488 bus   Same as GPIB (general-purpose interface bus); a standard parallel bus used widely for test and measurement interfacing.

IEEE 1394   A serial bus for high-speed data transfer; also known as FireWire.

I^2L   Integrated injection logic; an IC technology.

Increment   To increase the binary state of a counter by one.

Input   The signal or line going into a circuit; a signal that controls the operation of a circuit.

Input file   The information entered in a computer that describes a logic design using a PLD programming language such as ABEL or CUPL.

Input/Output (I/O)   A terminal of a device that can be used as either an input or as an output.

Instruction   One step in a computer program; a unit of information that tells the CPU what to do.

Instruction pairing   The process of combining certain independent instructions so that they can be executed simultaneously by two separate execution units.

Integer   A whole number.

Integrated circuit (IC)   A type of circuit in which all of the components are integrated on a single chip of semiconductive material of very small size.

Interfacing   The process of making two or more electronic devices or systems operationally compatible with each other so that they function properly together.

Interrupt   A signal or instruction that causes the current process to be temporarily stopped while a service routine is run.

Inversion   The conversion of a HIGH level to a LOW level or vice versa.

Inverter   A NOT circuit; a circuit that changes a HIGH to a LOW or vice versa.

I/O port   Input/output port; the interface between an internal bus and a peripheral.

IP   Instruction pointer; a special register within the CPU that holds the offset address of the next instruction to be executed.

ISA bus   Industry standard architecture bus; an internal parallel bus standard.

Jaz cartridge   A magnetic storage device; hard disks encased in a rigid plastic cartridge with storage capacities of 1 Gbyte or 2 Gbytes.

JEDEC file   A Joint Electronic Device Engineering Council software file generated from the compiler software that is used by a programming device to implement a logic design in a PLD; also called a fuse map or cell map.

J-K flip-flop   A type of flip-flop that can operate in the SET, RESET, no-change, and toggle modes.

Johnson counter   A type of register in which a specific pre-stored pattern of 1s and 0s is shifted through the stages, creating a unique sequence of bit patterns.

Junction   The boundary between an *n* region and a *p* region in a BJT.

Karnaugh map   An arrangement of cells representing the combinations of literals in a Boolean expression and used for a systematic simplification of the expression.

Latch   A bistable digital circuit used for storing a bit.

Latency period   The time it takes for the desired sector to spin under the head once the head is positioned over the desired track of a magnetic hard disk.

LCCC   Leadless ceramic chip carrier; an SMT package that has metallic contacts molded into its body.

LCD   Liquid crystal display.

Leading edge   The first transition of a pulse.

Least significant bit (LSB)   Generally, the right-most bit in a binary whole number or code.

LED   Light-emitting diode.

LIFO   Last in—first out memory, memory stack.

Listener   An instrument capable of receiving data on a GPIB (general-purpose interface bus).

Literal   A variable or the complement of a variable.

Load   To enter data into a shift register.

Local bus   An internal bus that connects the microprocessor to the cache memory, the main memory, the coprocessor, and the PCI bus controller.

Local cell   A programmable cell in a PLD array that affects individual OLMCs when programmed.

Logic   In digital electronics, the decision-making capability of gate circuits, in which a HIGH represents a true statement and a LOW represents a false one.

Look-ahead carry   A method of binary addition whereby carries from preceding adder stages are anticipated, thus eliminating carry propagation delays.

LSI   Large-scale integration; a level of IC complexity in which there are 100 to 9999 equivalent gates per chip.

Machine code   The basic binary instructions understood by the processor.

Machine language   The representation of computer program instructions by a binary code.

Magneto-optical disk   A storage device that uses electromagnetism and a laser beam to read and write data.

Magnitude   The size or value of a quantity.

Mantissa   The magnitude of a floating-point number.

Master-slave flip-flop   A type of flip-flop in which the input data are entered into the device on the leading edge of the clock pulses and appear at the output on the trailing edge. Master-slave flip-flops have, for the most part, been replaced by edge-triggered types.

Memory array   An array of memory cells arranged in rows and columns.

Memory-mapped I/O   A method of addressing ports by assigning addresses within the computer memory address space to the port. The CPU views the I/O ports as memory locations.

Microcomputer   A computer in which the CPU is built around a microprocessor.

Microcontroller   A specialized microprocessor designed for control functions.

Microprocessor   A digital integrated circuit device that can be programmed with a series of instructions to perform specified functions on data.

Minimization   The process that results in an SOP or POS Boolean expression containing the fewest possible terms with the fewest possible literals per term.

Minuend   The number from which another number is subtracted.

Mnemonic   An English-like instruction that is converted by an assembler into a machine code for use by a processor.

Modem   A modulator/demodulator for interfacing digital devices to analog transmission systems such as telephone lines.

Modulus   The number of unique states that a counter will sequence through.

Monostable   Having only one stable state. A monostable multivibrator, commonly called a one-shot, produces a single pulse in response to a triggering input.

Monotonicity   The characteristic of a DAC defined by the absence of any incorrect step reversals; one type of digital-to-analog linearity.

MOS   Metal-oxide semiconductor; a type of transistor technology.

MOSFET   Metal-oxide semiconductor field-effect transistor.

Most significant bit (MSB)   The left-most bit in a binary whole number or code.

MSI   Medium-scale integration; a level of IC complexity in which there are 12 to 99 equivalent gates per chip.

Multiplexer (MUX)   A circuit (digital device) that switches digital data from several input lines onto a single output line in a specified time sequence.

Multiplicand   The number that is being multiplied by another number.

Multiplier The number that multiplies the multiplicand.

Multitasking An operating system environment in which the computer seems to run multiple programs or tasks simultaneously.

Multivibrator A class of digital circuits in which the output is connected back to the input (an arrangement called feedback) to produce either two stable states, one stable state, or no stable states, depending on the configuration.

NAND gate A logic circuit in which a LOW output occurs only if all the inputs are HIGH.

Negative-AND An equivalent NOR gate operation in which the HIGH is the active input when all inputs are LOW.

Negative-OR An equivalent NAND gate operation in which the HIGH is the active input when one or more of the inputs are LOW.

Nibble A group of four bits.

NMOS An *n*-channel metal-oxide semiconductor.

Node A common connection point in a circuit in which a gate output is connected to one or more gate inputs.

Noise immunity The ability of a circuit to reject unwanted signals.

Noise margin The difference between the maximum LOW output of a gate and the maximum acceptable LOW input of an equivalent gate; also, the difference between the minimum HIGH output of a gate and the minimum HIGH input of an equivalent gate.

Nonvolatile A term that describes a memory that can retain stored data when the power is removed.

NOR gate A logic gate in which the output is LOW when any or all of the inputs are HIGH.

NOT A basic logic operation that performs inversions.

Numeric Related to numbers.

Octal Describes a number system with a base of eight.

Odd parity The condition of having an odd number of 1s in every group of bits.

Offset address The distance in number of bytes of a physical address from the base address.

OLMC Output logic macrocell. The programmable output logic in a GAL.

One-shot A monostable multivibrator.

Op code Operation code; the code representing a particular microprocessor instruction; a mnemonic.

Open-collector A type of output in a logic circuit in which the collector of the output transistor is left disconnected from any internal circuitry and is available for external connection; normally used for driving higher-current or higher-voltage loads.

Operand A variable, a register, a memory location, or a value used in an assembly language program as part of the instruction.

Operational amplifier (op-amp) A device with two differential inputs that has very high gain, very high input impedance, and very low output impedance.

OR A basic logic operation in which a true (HIGH) output occurs when one or more of the input conditions are true (HIGH).

OR gate A logic gate that produces a HIGH output when any of the inputs is HIGH.

Oscillator An electronic circuit that is based on the principle of regenerative feedback and produces a repetitive output waveform; a signal source.

Output The signal or line coming out of a circuit.

Overflow The condition that occurs when the number of bits in a sum exceeds the number of bits in each of the numbers added.

PAL Programmable array logic. A PLD with a programmable AND array and a fixed OR array.

Parallel In digital systems, data occurring simultaneously on several lines; the transfer or processing of several bits simultaneously.

Parity In relation to binary codes, the condition of evenness or oddness of the number of 1s in a code group.

Parity bit A bit attached to each group of information bits to make the total number of 1s odd or even for every group of bits.

PCI bus Peripheral control interconnect bus; an internal parallel bus standard.

Period The time required for a periodic waveform to repeat itself.

Periodic Describes a waveform that repeats itself at a fixed interval.

Peripheral A device or instrument that provides communication with a computer or provides auxiliary services or functions for the computer.

Physical address The actual location of a data unit in memory.

PIC Programable interrupt controller; handles the interrupts on a priority basis.

Pipeline As applied to memories, an implementation that allows a read or write operation to be initiated before the previous operation is completed.

PLA Programmable logic array. A PLD with programmable AND and OR arrays.

PLCC Plastic leaded chip carrier; an SMT package whose leads are turned up under its body in a J-type shape.

PLD Programmable logic device.

PMOS A *p*-channel metal-oxide semiconductor.

Pointer The contents of a register (or registers) that contain an address.

Polling The process of checking a series of peripheral devices to determine if any require service from the CPU.

Port A physical interface on a computer through which data are passed to or from peripherals.

Positive logic The system of representing a binary 1 with a HIGH and a binary 0 with a LOW.

Power dissipation The product of the dc supply voltage and the dc supply current in an electronic circuit; the amount of power required by a circuit.

PPI Programmable peripheral interface; a special IC used to implement I/O ports.

Prefetching The process of executing instructions at the same time as other instructions are "fetched," eliminating idle time; also called pipelining.

Preset   An asynchronous input used to set a flip-flop (make the Q output 1).

Priority encoder   An encoder in which only the highest value input digit is encoded and any other active input is ignored.

Product   The result of a multiplication.

Product-of-sums (POS)   A form of Boolean expression that is basically the ANDing of ORed terms.

Product term   The Boolean product of two or more literals equivalent to an AND operation.

Program   A list of computer instructions arranged to achieve a specific result; software.

Programmer   An instrument that programs a PLD using a JEDEC file downloaded from a computer running hardware description language (HDL) software.

PROM   Programmable read-only semiconductor memory. A PLD with a fixed AND array and programmable OR array; used as a memory device and normally not as a logic circuit device.

Propagation delay time   The time interval between the occurrence of an input transition and the occurrence of the corresponding output transition in a logic circuit.

Pseudo-operation   An instruction to the assembler (as opposed to a processor).

Pull-up resistor   A resistor with one end connected to the dc supply voltage used to keep a given point in a circuit HIGH when in the inactive state.

Pulse   A sudden change from one level to another, followed after a time, called the pulse width, by a sudden change back to the original level.

Pulse width   The time interval between the 50% points of the leading and trailing edges of the pulse; the duration of the pulse.

QIC   Quarter-inch cassette; a type of magnetic tape.

Queue   A high-speed memory that stores instructions or data.

Quotient   The result of a division.

Race   A condition in a logic network in which the difference in propagation times through two or more signal paths in the network can produce an erroneous output.

RAM   Random-access memory; a volatile read/write semiconductor memory.

Read   The process of retrieving data from a memory.

Real mode   Operation of an Intel processor in a manner to emulate the 8086's 1 Mbyte of memory.

Recycle   To undergo transition (as in a counter) from the final or terminal state back to the initial state.

Refresh   To renew the contents of a dynamic memory by recharging the capacitor storage cells.

Register   A digital circuit capable of storing and shifting binary information; typically used as a temporary storage device.

Register array   A set of temporary storage locations within the microprocessor for keeping data and addresses that need to be accessed quickly by the program.

Registered   A PLD output configuration where the output comes from a flip-flop.

Relocatable code   A program that can be moved anywhere within the memory space without changing the basic code.

Remainder   The amount left over after a division.

RESET   The state of a flip-flop or latch when the output is 0; the action of producing a RESET state.

Reverse bias   A voltage polarity condition that prevents a *pn* junction of a transistor or diode from conducting current.

Ring counter   A register in which a certain pattern of 1s and 0s is continuously recirculated.

Ripple carry   A method binary addition in which the output carry from each adder becomes the input carry of the next higher-order adder.

Ripple counter   An asynchronous counter.

RISC   Reduced instruction set computing.

Rise time   The time required for the positive-going edge of a pulse to go from 10% of its full value to 90% of its full value.

ROM   Read-only semiconductor memory, accessed randomly; also referred to as mask-ROM.

Schottky   A specific type of transistor-transistor logic circuit technology.

SCSI   Small computer system interface; an external parallel bus standard.

SDRAM   Synchronous dynamic random-access memory.

Seek time   The time for the read/write head in a hard drive to position itself over the desired track for a read operation.

Segment   A 64k block of memory.

Sequence   The order in which several things occur in a specified time relationship.

Sequential circuit   A digital circuit whose logic states follow a specified time sequence.

Serial   Having one element following another, as in a serial transfer of bits; occurring, as pulses, in sequence rather than simultaneously.

SET   The state of a flip-flop or latch when the output is 1; the action of producing a SET state.

Set-up time   The time interval required for the control levels to be on the inputs to a digital circuit, such as a flip-flop, prior to the triggering edge of clock pulse.

Shift   To move binary data from stage to stage within a shift register or other storage device or to move binary data into or out of the device.

Signal tracing   A troubleshooting technique in which waveforms are observed in a step-by-step manner beginning at the input and working toward the output or vice versa. At each point the observed waveform is compared with the correct signal for that point.

Sign bit   The left-most bit of a binary number that designates whether the number is positive (0) or negative (1).

SIMM   Single-in-line memory module.

SMT   Surface-mount technology; An IC package technique in which the packages are smaller than DIPs and are mounted on the printed surface of the PC board.

Software   Computer programs; programs that instruct a computer what to do in order to carry out a given set of tasks.

Software interrupt   An instruction that invokes an interrupt service routine.

SOIC Small-outline integrated circuit; an SMT package that resembles a small DIP but has its leads bent out in a "gull-wing" shape.

Source A sending device of a bus; one of the terminals of a field-effect transistor.

Speed-power product A performance parameter that is the product of the propagation delay time and the power dissipation in a digital circuit.

SRAM Static random-access memory; a type of volatile read/write semiconductor memory.

S-R flip-flop A SET-RESET flip-flop.

SSI Small-scale integration; a level of IC complexity in which there are twelve fewer equivalent gates per chip.

SSOP Shrink small-outline package.

Stage One storage element (flip-flop) in a register.

State diagram A graphic depiction of a sequence of states or values.

State machine A logic system exhibiting a sequence of states conditioned by internal logic and external inputs; any sequential circuit exhibiting a specified sequence of states.

Static memory A volatile semiconductor memory that uses flip-flops as the storage cells and is capable of retaining data without refreshing.

Storage The capability of a digital device to retain bits; the process of retaining digital data for later use.

String A contiguous sequence of bytes or words.

Strobing A process of using a pulse to sample the occurrence of an event at a specified time in relation to the event.

Subroutine A series of instructions that can be assembled together and used repeatedly by a program but programmed only once.

Subtracter A logic circuit used to subtract two binary numbers.

Subtrahend The number that is being subtracted from the minuend.

Sum The result when two or more numbers are added together.

Sum-of-products (SOP) A form of Boolean expression that is basically the ORing of ANDed terms.

Sum term The Boolean sum of two or more literals equivalent to an OR operation.

Synchronous Having a fixed time relationship.

Synchronous counter A type of counter in which each stage is clocked by the same pulse.

Syntax The prescribed format and symbology used in describing a category of functions.

Synthesis The software process of converting a circuit description to a standard JEDEC file for PLD programming.

Talker An instrument capable of transmitting data on a GPIB (general-purpose interface bus).

Terminal count The final state in a counter's sequence.

Test vectors A section of a PLD program that consists of a set of input levels and the resulting output levels and used to test a given logic design prior to committing it to hardware.

Throughput The average speed with which a program is executed.

Timer A circuit that can be used as a one-shot or as an oscillator.

Timing diagram A graph of digital waveforms showing the proper time relationship of all the waveforms and how each waveform changes in relation to the others.

Toggle The action of a flip-flop when it changes state on each clock pulse.

Totem-pole A type of output in TTL circuits.

Trailing edge The second transition of a pulse.

Transistor A semiconductor device exhibiting current and/or voltage gain. When used as a switching device, it approximates an open or closed switch.

Trigger A pulse used to initiate a change in the state of a logic circuit.

Tristate A type of output in logic circuits that exhibits three states: HIGH, LOW, and high-Z; also known as 3-state.

Tristate output buffer A logic circuit having three output states: HIGH, LOW, and high-impedance (open).

Troubleshooting The technique of systematically identifying, isolating, and correcting a fault in a circuit or system.

Truncated Shortened.

Truncated sequence A sequence that does not include all of the possible states of a counter.

Truth table A table showing the inputs and corresponding output level of a logic circuit.

TSSOP Thin shrink small-outline package.

TTL Transistor-transistor logic; a class of integrated logic circuit that uses bipolar junction transistors.

TVSOP Thin very small-outline package.

ULSI Ultra large-scale integration; a level of IC complexity in which there are more than 100,000 equivalent gates per chip.

Unit load A measure of fan-out. One gate input represents a unit load to the output of a gate within the same IC family.

Universal gate Either a NAND gate or a NOR gate. The term *universal* refers to the property of a gate that permits any logic function to be implemented by that gate or by a combination of gates of that kind.

Universal shift register A register that has both serial and parallel input and output capability.

Up/Down counter A counter that can progress in either direction through a certain sequence.

UV EPROM Ultraviolet erasable programmable ROM.

UV PROM Ultraviolet erasable programmable ROM.

Variable A symbol used to represent a logical quantity that can have a value of 1 or 0, usually designated by an italic letter.

VLSI Very large-scale integration; a level of IC complexity in which there are 10,000 to 99,999 equivalent gates per chip.

Volatile A term that describes a memory that loses its stored data when the power is removed.

Weight The value of a digit in a number based on its position in the number.

Word   A complete unit of binary data.

Word capacity   The number of words that a memory can store.

Word length   The number of bits in a word.

WORM   Write once–read many; a type of optical storage device.

Write   The process of storing data in a memory.

Zero suppression   The process of blanking out leading or trailing zeros in a digital display.

ZIF socket   Zero insertion force socket. A type of socket used in most programmers that accepts a PLD package.

Zip disk   A type of magnetic storage; a flexible disk with a capacity of 100 Mbytes housed in a rigid plastic cartridge about the size of a floppy.

INDEX